Studies in Surface Science and Catalysis 117

MESOPOROUS MOLECULAR SIEVES 1998

Studies in Surface Science and Catalysis

Advisory Editors: B. Delmon and J.T. Yates

Vol. 117

MESOPOROUS MOLECULAR SIEVES 1998

Proceedings of the 1st International Symposium,
Baltimore, MD, U.S.A., July 10-12, 1998

Editors

L. Bonneviot, F. Béland, C. Danumah, S. Giasson, S. Kaliaguine
Département de Génie Chimique, Faculté des Sciences et de Génie,
Université Laval, Ste-Foy, Québec, Canada, G1K7P4

1998
ELSEVIER
Amsterdam — Lausanne — New York — Oxford — Shannon — Singapore — Tokyo

ELSEVIER SCIENCE B.V.
Sara Burgerhartstraat 25
P.O. Box 211, 1000 AE Amsterdam, The Netherlands

Library of Congress Cataloging-in-Publication Data

Mesoporous molecular sieves 1998 : proceedings of the first international symposium, Baltimore, MD, U.S.A., July 10-12, 1998 / editors, L. Bonneviot ... [et al.].
p. cm. -- (Studies in surface science and catalysis ; 117)
Includes bibliographical references and index.
ISBN 0-444-82826-5 (alk. paper)
1. Molecular sieves--Congresses. I. Bonneviot, Laurent. II. Series.
TP159.M6M45 1998
660'.2842--dc21 98-22444
CIP

ISBN 0-444-82826-5

Transferred to digital printing 2005

FOREWORD

The idea of using surfactant liquid crystals as templates to generate mesoporous adsorbents and catalysts may be compared to another Mobil generated breakthrough, namely the synthesis in the 1970's of the highly shape selective zeolite ZSM-5. In both cases a rigorous control of pore size (and shape) was achieved in order to yield solids with molecular sieving properties. In a zeolite micropore the reactants are however limited to small molecules with kinetic diameters lower than 1.2 nm. Moreover the kind of functionalizing entities which may be introduced in the channels to generate catalytic active phases is also restricted by spatial constraints. Thus the silicate and aluminosilicate mesoporous molecular sieves (MMS) introduced in the 1990's, having pore diameters tunable from 1.5 up to 30 nm, provide new opportunities in that respect. The spectacular specific surface areas and pore volumes reached by mesoporous molecular sieves constitute another remarkable advantage. In addition since the walls need not be made of crystalline material, the restrictions on the chemical nature of the constitutive atoms are much less than in zeolites. Potentially, this opens unexpectedly large perspectives, because there is no *a priori* reason why any oxides, mixed oxides, or even metals or other inorganic compounds could not be templated into mesoporous molecular sieves. In fact several of the papers in this meeting report recent examples of such materials.

In these proceedings the reader will indeed find feature articles and regular papers from many groups worldwide, covering all aspects of synthesis, physical characterization and catalytic reactivity of MMS and their chemically modified forms. It is actually remarkable that this recent development brought together a broader ever spectrum of scientists from traditionally unrelated fields such as those of liquid crystals, surfactants, sol-gels, amorphous oxides and mixed oxides, solid state, adsorbents and heterogeneous catalysts. Obviously this is a fast growing research area which triggers the imagination and creativity at the cross-road between material design, molecular surface tailoring and catalytic applications.

As every new scientific development at its early stage, this one meets currently the problem of definitions. For instance it is not clear at the present time what distinguishes imperfectly ordered mesoporous materials such as HMS, KIT or MSU from well ordered FSM, MCM or SBA types of materials, in terms of local environment. Materials such as HMS and MSU are claimed to be more active in diffusion limited reactions due to faster effective diffusion compared to MCM or SBA type of materials. It is however not really known if the wall material is of same local structure in both types of solids. In other words when the template is changed and the surfactant (S) - inorganic (I) interaction is modified (S^+I^-, $S^+X^-I^+$, S^oI^o, S^-I^+...), how is the pore wall chemistry affected? How is the surface silanol (hydroxyl) concentration affected? How is the thermal and hydrothermal stability of the resulting solid affected? The difficulty in rationalizing these effects is a result of the large number of parameters involved in the inorganic-surfactant interface. The combination of electrostatic and dispersive forces is highly dependent on such factors as pH, temperature, precursors, surfactants, solvent swelling additives nature and concentration and ionic strength. For example it is possible to change from the synthesis of an hexagonal (MCM-41) to a cubic (MCM-48) structure either by switching from a standard CTMA surfactant to a gemini $C_{16\text{-}12\text{-}16}$ analog, or by merely adding ethanol to the standard recipe. Similar effects may also be obtained by changing the hydrothermal treatment time or even by changing the time evolution of the agitation intensity during sol-gel preparation. Thus up to now MMS synthesis procedures appear as a series of recipes rather than the result of a general understanding of the complex interactions at stake.

Some papers in this symposium deal with the genesis of these new materials and the nucleation-growth mechanism as related to the transformation of the micelles to the composite bicontinuous mesophase. Not much was done yet and the current works in this area deal essentially with the synthesis of hexagonal mesophases in basic media.

Not withstanding the above considerations much is known about the final structure of the calcined forms due to the combined use of such characterization methods as XRD, TEM and N_2 physisorption techniques. A brand new kind of materials having undisputably original new properties was discovered. For the first time was it possible to create regularity in pore size and wall geometry in an amorphous phase. It may be a lucky circumstance that most of the mesoporous molecular sieves synthesized have their pore diameters in the range 1.5 - 4.0 nm, where the Kelvin equation is actually not valid with liquid nitrogen. In spite of this, clear vertical hysteresis loops were observed in nitrogen adsorption plots. This raised the theoretical problem of understanding the particularities of the interaction of physisorbed nitrogen with surfaces of these curvatures. In this meeting several groups discuss various approaches to this problem.

The original properties of MMS are so unique that the design of most existing catalysts could be reconsidered. It might indeed be of interest to introduce MMS either as a support or as the active phase, merely on the basis of their high surface areas, narrow pore size distribution and flexibility in composition. The recent literature provides examples of MMS based catalysts of many types such as acid-base solids, supported metals and supported oxides, mixed oxides, anchored complexes and clusters, grafted organic functional groups and others. Examples of all these developments are documented in the present proceedings including some spectacular new proposals. The new metallic (Pt) mesophases are specially worth mentioning because they represent a new approach to producing non-supported highly dispersed metals.

The matter of thermal and hydrothermal stability used to be a real concern as it is crucial to most potential application. Several of the contributions to this symposium propose new solutions to this problem.

The enthusiastic response of the international research community was indeed most rewarding for the organizers of ISMMS98. It showed that mesoporous molecular sieves have now attracted the attention of many due to the novelty of both the fundamental questions raised and the perspectives of application. The future of this field of activity may prove to be as florishing as the one of zeolites. We have therefore reasons to hope that this first international symposium will be followed by many others.

Finally we wish to acknowledge the friendly cooperation of David Olson and Ray Gorte respectively chairman and program chairman of the 12th IZC.

We would also like to thank Mrs. Hélène Michel who has a large part of responsibility in the success of both the symposium and the proceedings of ISMMS98.

L. B. F. B. C. D. - S. G. S. K.

PAPER SELECTION COMMITTEE

CONTENTS

MESOPOROUS MOLECULAR SIEVES 1998
Studies in Surface Science and Catalysis, Vol. 117
L. Bonneviot, F. Béland, C. Danumah, S. Giasson and S. Kaliaguine (Editors)

Using the organic-inorganic interface to define pore and macroscale structure

G. D. Stucky[a], D. Zhao[a], P. Yang[a], W. Lukens[a], N. Melosh[b], B. F. Chmelka[b],

[a] Department of Chemistry and Materials,
[b] Department of Chemical Engineering,
University of California, Santa Barbara, California 93106 U.S.A.

1. INTRODUCTION

One of the most fascinating underlying aspects of the biogenesis of materials is the space/time definition of structure [1], function, and morphology at multiple length scales from complex mixtures of reactants and accessible processes. Paradoxically, spatial and kinetic incompatibilities during assembly result in instabilities that make it easy to temporally modify product composition, assembly, and macroscale form. In the same manner, in the synthesis of ordered mesoporous materials, competing spatial sequences in the solvent media as well as the reactants are useful in defining domain separation and ultimately pore structure and morphology in mesoporous phases. In practice this means hydrophobic/hydrophilic, polar/nonpolar, and organic/inorganic interfaces, but it should be emphasized that the differences in property (e.g. hydrophobicity or polarity) need not be large to generate highly structured composite phases with well-defined periodic domain definition [2]. Competing interactions and assembly processes generally have different entropic and kinetic handles that can be used to orchestrate hierarchical structure and functionality in the composite product. Entropic control of the system processes can be implemented, for example, by the solvent media used in the reaction or by temperature. When weak assembly forces are used, the structural evolution can be directed by external fields or the co-assembly of a competing process.

All this synthesis freedom presents unprecedented opportunities in mesopore structure design but also some formidable synthetic challenges in reproducibly creating processible materials with designed pore and macroscale morphology. The following discussion focuses on two approaches to achieving this goal: (1) the use of competing interactions, which includes the thermodynamics and kinetics of the assembly processes in the microphases and their interfaces; and (2) the use of designed competing spatial regions in the molecular species that have inherent chemical incompatibilities to direct structural organization. The latter can be used both for structure templating and for creating long-range order.

2. COMPETING INTERACTIONS: THE SILICA-ORGANIC INTERFACE

The mesostructured materials of interest to this Conference have spatially distinct inorganic and organic regions which at the very least have dimensionally well-defined cross sections, which result in uniform pore sizes. In order for this to be achieved, there must be some collective organization of the organic and inorganic components, and an integral part of that process lies at the interface between the spatially distinct organic and inorganic regions. The extent to which the organic and inorganic domains have properties and structure that are characteristic of the corresponding bulk phases, and more importantly their processing into different shapes and forms, depends on the strength of the interface inorganic-organic (IO) interactions relative to entropic and the self-assembly energetics within the organic and inorganic domains. The chemist's challenge is to balance the organic-inorganic and organic-organic interactions, solvation, and inorganic-inorganic interactions at two or more interfaces in such a way as to control structure and form on the desired multiple length scales.

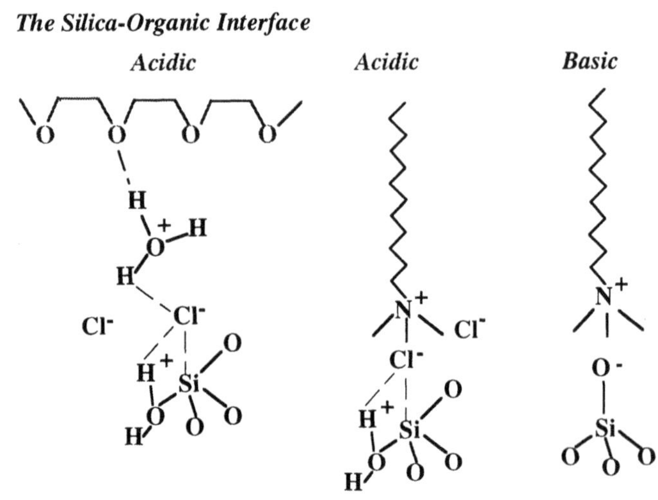

Figure 1

One example of how this can be done is the kinetic limitation of silica polymerization under basic conditions by using low temperatures and careful pH control. Low temperatures are also favorable to organic assembly: optimum organic organization can be more readily achieved with molecular organic/inorganic species at relatively low temperatures. This combination has been used to show that in the absence of inorganic polymerization, these mesophases have liquid crystalline properties similar to those of conventional aqueous lyotropic liquid crystal systems [3],[4]. In order to maintain these liquid crystal-like properties and optimize long-range composite ordering during polymerization of the inorganic species, the inorganic and organic domains must be able to reorganize on the same kinetic time scale into mutually compatible configurations.

The use of weak interface interactions such as those shown in Figure 1 (Acidic) permits one to use the competing kinetics and rate of silica polymerization (through changes in pH, biphase mixing, hydrophobicity of the silicon alkoxide precursors, etc.) and composite assembly to define different shapes and form on longer length scales, such as fibers [5],[6], spheres [7],[8], films [9],[10],[11],[12],[13],[14],[15],[16], and other designed morphologies [17]. This approach has been reviewed in some detail recently [18], but in view of the increasing application of acid silica synthesis chemistry [19],[20] for topological structure control as well as the creation of both new mesostructured phases and ultra-large mesopore structures, a few comments which hopefully reflect a more recent perspective seem appropriate.

2.1. Strong Interface Interactions

At one end of the scale, the use of strong covalent [21],[22],[23] organometallic or direct ionic organic-inorganic interactions [24],[3],[4], (Figure 1, Basic) in surfactant-mesopore synthesis works well to generate molecular inorganic-organic units that form their own liquid crystal arrays independent of the organization that one would get from the organic component alone. The structural phase is defined by the liquid crystal behavior associated with the molecular inorganic/organic species and the assembly of mesophases with different structural symmetry can be carried out starting with these species.

A good deal of the controversy concerning the mechanism of the assembly of mesoporous silica structures, and for that matter pure silica molecular sieve structures in basic aqueous media with anionic silica species has arisen because of disagreements on the role of charges and Coulombic interactions. Classical liquid crystal scientists treat amphiphiles, such as the quaternary ammonium surfactants, as being neutral species with the head group charge completely shielded by the anions surrounding it in solution or in a liquid crystal array. The driving force for the assembly and the structural organization that takes place is totally attributed to van der Waals interactions involving the organic tail and the shape/size of the organic groups that are present.

Inorganic chemists, brought up on the relatively good modelling and predictions that can be made using simple Coulombic interactions for ionic lattices, including ones like tetramethyl ammonium chloride in which the cations and anions are also well screened by their first coordination spheres, know that the only way the energetics of such a array can be fit is through the use of Ewald summations and Madelung potentials that take into account long-range Coulombic interactions. Furthermore it is well established that the strong short-range Coulombic forces, attractive and repulsive, generate a highly organized, crystalline lattice. The thought that adding an organic tail to one of the four tetrahedral positions of an ammonium ion in an ammonium salt somehow completely eliminates the role of Coulombic forces in the assembly of and order of the two-dimensional array of charges that results seems at best provincial. Attempts to relate "cloud formation" during the synthesis of mesoporous structures in basic media to the critical micelle concentrations (cmc) [25] may be misleading in that phase separation may depend as much or more on the Coulombic lattice energy component associated with the charged head group as on the van der Waals forces. Solubility products of salts have a Coulombic lattice energy component. In support of this perspective, the association of cationic surfactant micelles and their cmc1 and cmc2 values are very sensitive to the anion that is present [20]. Both Coulombic and van der Waals components are important, and neither should be neglected as is frequently the case. It should also be noted that during the past six years of research in the area of mesoporous silica materials, in aqueous media the most highly organized and structurally well-defined materials are formed only by introducing a significant amount of charge into the assembly process, either in basic media with anionic silica species as is done in the synthesis of the M41S family [26],[27] and FSM-16 [28] or in acid media using cationic silica species as we have done for the synthesis of the UCSB and SBA family of mesoporous phases.

2.2. Weak Interface Interactions

At the other extreme from the strong interaction scenario in basic media, if the organic-inorganic interface interaction is relatively weak (e.g., hydrogen bonding as in the acidic intermediate case shown in Figure 1), the initial organization of the organic phase may only be slightly modified, but still capable of directing the silica assembly at long length scales. In this way amphiphilic or non-ionic surfactants and polymers can be used to interface through hydrogen bonding between an organized organic array (e.g., oil-in-water emulsion) and charged inorganic species so that the identity and structure-directing ability and dynamics of the organic organization are more fully utilized in the creation of processed macrostructures with ordered mesoporous arrays such as oriented thin films, fibers, or other designed morphological shapes.

The synthesis of hexagonal and cubic mesoporous silica phases at $pH < 2$ (~ the aqueous isoelectric point of silica) [19,20] was an important demonstration that cationic silica species could be used to create periodic porous silica phases that were stable to calcination; and, that hydrogen bonding interactions between inorganic species and the organic structure-directing agents could be used to create 3-d periodic mesoporous silica structures. This chemistry has led to highly organized mesoporous silica structures with large pore sizes (> 300 Å) [29] and numerous new phases including those with $P6_3/mmc$ [21],[30],[31], Pm3m [16,31], Im3m [31] and Pm3n [19,20] symmetries. In addition, almost all the advances that have been made in controlling the macrostructure of ordered silica mesostructured materials use acid media syntheses with cationic pH values below the silica aqueous isoelectric point [19] or acid catalysis in a more non-aqueous media [15]. The latter case may simply reflect a shift of the isoelectric point for forming the cationic silica species in a non-aqueous environment. The different entropic and interaction energies in the nanoscale organic, inorganic, and interface regions during the assembly result in structure frustration with incompatible local packing constraints that forbid an optimal geometry where the free energy is everywhere minimized. During synthesis and processing, before there is extensive silica polymerization, the inorganic/organic structures may readily undergo structural changes or transformations [32] to relieve stress through rotational displacements of the surfaces (disclination defects) (Figure 3) [33],[34]. The use of weak hydrogen-bonding interactions at the silica/organic interface enhances precursor fluidity and synthesis processing by hydrodynamics, shear fluid flow, cosolvents, emulsions, two-phase oil/water interfaces, or by the application of external fields [35]. Thus, the acid synthesis route conveniently provides control over the topological construction process of mesoporous materials at both the molecular and macroscale levels.

The successful application of the silica anion chemistry that is the basis for MCM-41 [26,27] or FSM-16 [28] syntheses [3,4] for morphogenesis purposes has only been realized in a few instances [36],[37],[38],[39]. The direct Coulombic interaction of the cationic surfactant with the anionic silica species gives a salt-like character to the interface that limits easy processing into the smooth continuous surfaces that are accessible with the acid synthesis approach. The consequences of this are also reflected in other properties, including composition. As-synthesized MCM-41 [26,27] does not have the same composition as that of the corresponding hexagonal phase made using the same surfactant under acid synthesis conditions, which has one halide ion/surfactant molecule in the mesostructured product, has generally thinner walls, smaller surface area, and surfactant species that require acid wash (ion exchange) for their removal as opposed to a simple water/ethanol wash for the acid synthesized mesostructured material [19,20]. The acid-synthesized hexagonal honeycomb phase, SBA-3, is not the same as the base-synthesized MCM-41 [26,27] or FSM-16 [28] mesoporous materials although they appear to have the same crystallographic symmetry.

3. INTERFACE CHARGE DENSITY, SPATIALLY COMPETING SEQUENCES AND DOMAIN DEFINITION

3.1. Organic/Inorganic Framework Charge Matching

Composite materials, including nanostructured biomaterials, have by definition spatially distinct domains. The relative sizes and patterning of these domains can be determined by using templates with spatially competing hydrophobic and hydrophilic sequences and by charge-matching the organic phase charge density with that of the framework as well as through space/time control of the kinetics of the assembly. Interface charge matching of amphiphilic (ionic and non-ionic) surfactants with the inorganic species and domains (Figure 1) not only is an important controlling factor in the assembly of mesoporous materials, but also, as noted above, is a primary reason for the differences in composition and surfactant removal of base-synthesized, compared to silica acid-synthesized, mesostructures.

Along with the use of competing hydrophobic-hydrophilic or polar-nonpolar molecular sequences, charge matching during synthesis is a powerful tool in the creation of porous materials on all length scales. Until recently there were only two zeolite cage structures, faujasite and hexagonal faujasite, that had intersecting twelve-ring micropores. This use of charge-matching, multi-process assembly using organic templates that have spatially competing hydrophobic/hydrophilic regions has resulted in three new large-micropore zeolite structures [40]. One of these falls in the mesostructure classification with a cage that is 20 Å in two mutually perpendicular directions. Transition-metal-based zeolite analogs of analcime, chabazite, gismondine, sodalite, faujasite, and rho are examples of some of the zeolite structure types that have been synthesized and characterized by *single-crystal* X-ray diffraction. The concentration of transition metal atoms in the framework can be controlled by varying the charge and geometry of the organic templates. Mn, Zn, Co, Mg, and Al phosphates can be used, confirming the applicability of the approach to different inorganic compositions. In this synthesis, divalent and trivalent ions are both introduced and the siting of the transition metal atoms within the cage is determined by their proximity to the charge on the organic cation. When the hydrophobic sequences of the organic molecule are small, the symmetry of the cage is determined by the symmetry of the organic species. Thus the symmetry of the inorganic framework *and* selective metal-atom siting in the inorganic framework are determined by *templating*, not just 'structure direction' [41],[42],[43].

This is a new generalized synthesis method which has been applied for the first time to zeolite-type structures in which the majority of metal atom sites are occupied by transition metal atoms. It should also apply to other non-oxide compositions and structures. The methodology has already made it possible to synthesize as single crystals not only zeolite analogs that have never been synthesized in any composition or in a non-aluminosilicate system (merlinoite, phillipsite, and thomsonite), but also zeolite-like structures that were theoretically predicted but never found as minerals or previously synthesized.

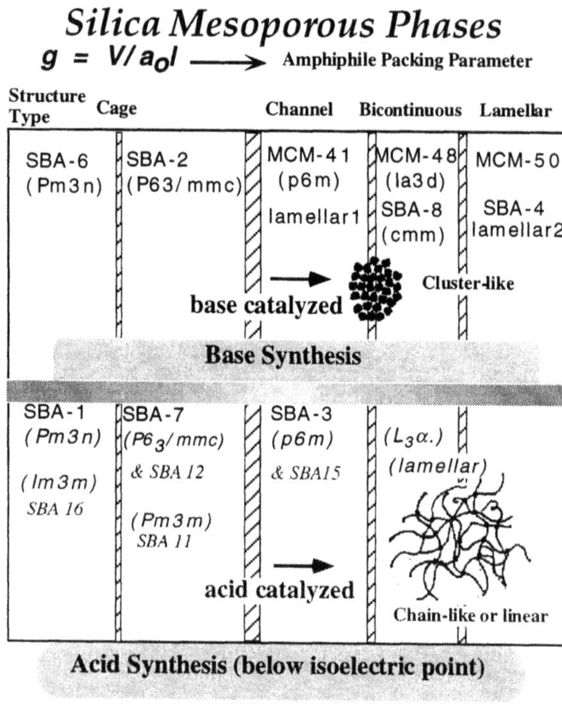

Figure 2

3.2. Spatially Competing Sequences and Domain Definition

From a molecular perspective the structural function of surfactants in mesophase formation can to a first approximation be related to that of classical surfactants in water or other solvents with parallel roles for organic additives [21]. The effective surfactant ion pair packing parameter, $g = V/a_o l$, as proscribed by the inorganic species or solid inorganic framework charge density remains a useful molecular index to predict the geometry of the mesophase products, and phase transitions may be viewed as a variation of g in the liquid crystal-like solid phase. A majority of the known liquid crystal morphologies have been observed in our laboratory for the mesosilicates, including previously unknown liquid crystal phase structures. A schematic showing the phases that have been identified is given in Figure 2. The phases shown in italics in Figure 2 are highly ordered, periodic structures made using non-ionic block copolymers.

The combination of molecular design of surfactant molecules and the metastable organization that results from the collective interactions of the organic/inorganic assembly has led to numerous new mesoporous structural phases, including a 3-d hexagonal cage structure [30], several new cubic cage structures (Pm3n [20,19], Im3m [31] and Pm3m [16,31]), and at least one orthorhombic channel mesoporous structure [44]. The structures of several other new phases are currently being investigated in our laboratory.

The role of solvent, including water and organic solvents or cosolvents, is particularly important [30,21]. Organic co-solvent species, such as EtOH, penetrate into the hydrophobic region of surfactant arrays, and thereby induce a structural rearrangement of the surfactant phase to re-optimize the interface charge-density matching and the surfactant packing. Solvent and co-solvent structure direction can be very effectively used by varying polarity and hydrophobic/hydrophilic properties and by functionalizing the surfactant molecule, for example with hydroxy group or variable charge.

The ultimate periodic symmetry for composite assembly in which the inorganic phase is disordered on a short-range scale is determined by organic molecular packing requirements (and charge matching with the inorganic interface) for both the strong (S^+I^-) and weak ($S^+X^-I^+$) or (S^0H^+)(X^-I^+) organic-inorganic interface interactions. This means for example that composite phases that have similar space group and lattice symmetries can be made with the acid and base synthesis procedures (Figure 1) even though the compositions, diffraction intensities, and other properties are different.

Phase transformations are easily accessible and very useful in mesoporous synthesis [3,21] and will not be discussed here. However, in relation to the differences in properties of acid- versus base-organized silica assemblies, a general pattern of phase transformation behavior is often observed using different surfactants with different *g* values under different reaction conditions. For acidic (pH < 0) synthesis media the phase sequence is: cubic (Pm3n, SBA-1), 3-d hexagonal ($P6_3$/mmc, SBA-2) > 2-d hexagonal (p6m, SBA-3) > lamellar. For basic syntheses media the phase sequence changes to: 3-d hexagonal (P63/mmc, SBA-2) > 2-d hexagonal (p6m, MCM-41) > cubic (Ia3d, MCM-48) > lamellar (MCM-50).

3.3 Spatially Competing Sequences and Domain Definition with Block Copolymers

The organization that takes place in the assembly of amphiphilic surfactants, ionic or neutral arises because of chemical incompatibilities and differences in the chemical potentials of the different species that are present. By attaching a hydrophilic head group to an organic tail a homogeneous composite material can be formed, since the charge or polarity of the organic head group is compatible with the charged silica species present in solution. However, the organic parts of the assembly will still organize and define the hydrophobic domains separate from the hydrophilic head group region.

The differences in polarity or hydrophobicity does not need to be large for organization to take place. An alkylene oxide block copolymer which is segmented by regions that differ only by a single carbon atom as in the $EO_xPO_yEO_x$ block copolymers (EO = ethylene oxide, PO = propylene oxide) will wish to segregate the EO segments from the PO segments simply because of the differing degrees of solvation by water molecules. However the segregation is limited due to the polymer connectivity. A resulting compromise for a collection of these polymers is the formation of periodic microstructures which organize to minimize the interface area between unlike chemical species.

The potential usefulness of these polymer phases is manifold. Length scales of a resulting mesostructured silica phase will be comparable to the size of the block copolymer molecules, typically tens of nanometers. The microstructure is directly coupled to the geometry and chemical and physical properties of the molecules within the polymer segments so that it can be modified in a variety of ways. The domain interfaces are also readily adjusted by modifying the polymer segment incompatibility. The consequences are that even for these non-ionic systems, a large variety of 3-d ordered phases are accessible.

Work is progressing rapidly in this area in several laboratories, with several groups investigating the formation of hybrid materials using polymeric surfactants. Pinnavaia and co-workers [45],[46] first used nonionic surfactants to synthesize worm-like disordered mesoporous silica and alumina in neutral media assembled by hydrogen-bonding interactions [45],[46],[47]. Attard *et al.*[48] have synthesized hexagonal mesoporous silica phases using concentrated (~50 wt.%) $C_{12}EO_8$ solutions and suggested that the formation of mesoporous silica under these conditions occurs by a "real" liquid crystal template route. Göltner and Antonietti have recently followed up on this approach using concentrated non-ionic surfactant and polymer arrays [49],[50] and have shown that if amphiphilic block copolymers are used, the block copolymer properties enhance the ductility and elasticity of the resulting composite so that it can be molded into monolithic blocks [49].

Templin *et al.*[51] have used high concentrations of poly(isoprene-b-ethyleneoxide) di-block copolymers (PI-b-PEO) to make lamellar and hexagonal aluminosilicate-polymer mesostructures that are highly ordered on length scales to ~40 nm. The syntheses were carried out in an acidic and non-aqueous solution (a mixture of $CHCl_3$ and tetrahydrofuran). The thermal stability of these materials and removal of the organic phase to create mesoporous structures has not yet been described.

In our laboratory, a family of mesoporous (20 ~ 300 Å) silica structures have been synthesized by the use of commercially available nonionic alkyl polyethylene oxide oligomeric surfactants and polyoxyalkylene block copolymers in acid media [29,31].The pore structures are highly ordered, and the phases identified in this family include cubic Im $\bar{}$ m, cubic Pm $\bar{}$ m (or others), 3-d hexagonal ($P6_3/mmc$), 2-d hexagonal (p6mm), continuous sponge L_3 phase, and lamellar (L_α) periodic symmetries. The hexagonal, p6mm, SBA-15 mesoporous structures have BET surface areas of 690 ~ 1040 m^2/g, pore sizes of 46 ~ 300 Å, silica wall thicknesses of 31 ~

64 Å, and pore volumes as large as 2.5 cm^3/g. Four or more diffraction peaks are observed even for periodicities over 200 Å. Rather remarkably, the calcined ordered mesoporous silica are thermally stable in boiling water for at least 48 hours. Several phases have been made into fibers, oriented thin films, and other morphologies with excellent pore fidelity, and their applications in several areas are currently under investigation. The assembly of the inorganic and organic periodic composite materials appears to take place by a hydrogen bonding $(S^0H^+)(X^-I^+)$ pathway and the kinetics of this assembly have been determined.

4. SUMMARY

The intent of this brief overview is to present a selected review of some of the current research in the area of mesoporous materials. Not included are porous materials in the range of 50 nm to several microns which also have been described recently and are currently being studied. There is now a continuum of nearly monodispersed porosities that can be created for almost any mean pore size for a variety of compositions and structural phases through the mesoscale regime. The evolution of the development of these materials and their applications will be an exciting part of the future.

REFERENCES

1. See for example, H. Meinhardt, *The Algorithmic Beauty of Sea Shells*, Springer-Verlag, Berlin (1995)

2. M. W. Matsen and F. S. Bates, *Macromolecules 29*, 1091 (1996)

3. A. Firouzi, D. Kumar, L. M. Bull, T. Besier, P. Sieger, Q. Huo, S. A. Walker, J. A. Zasadzinski, C. Glinka, J. Nicol, D. Margolese, G. D. Stucky, and B. F. Chmelka, *Science 267,* 1138 (1995)

4. A. Firouzi, F. Atef, A. G. Oertli, G. D. Stucky, and B. F. Chmelka, *J. Amer. Chem. Soc. 119,* 3596 (1997)

5. Q. Huo, D. Zhao, J. Feng, K. Weston, S. K. Buratto, G. D. Stucky, S. Schacht, and F. Schüth, *Adv. Mater. 9*, 974 (1997)

6. P. J. Bruinsma, A. Y. Kim, J. Liu, and S. Baskaran, *Chem. Mater. 9*, 2507 (1997)

7. S. Schacht, Q. Huo, I. G. Voigt-Martin, G. D. Stucky, and F. Schüth, *Science 273*, 768 (1997)

8. F. Schüth, U. Ciesla, S. Schacht, M. Thieme, Q. Huo, and G. D. Stucky, *Materials Research Society Bulletin*, to appear (1998)

9. H. Yang, A. Kuperman, N. Coombs, S. Mamiche-Afara, and G. A. Ozin, *Nature 379*, 703 (1996)

10. H. Yang, N. Coombs, I. Sokolov, and G. A. Ozin, *Nature 381*, 589 (1996)

11. I. A. Aksay, M. Trau, I. Honma, N. Yao, L. Zhou, P. Fenter, P. M. Eisenberger, and S. M. Gruner, *Science 273* , 892 (1996)

12. M. Ogawa, *J. Chem. Soc., Chem. Commun.*, 1149 (1996)

13. S. H. Tolbert, T. E. Schäffer, J. Feng, P. K. Hansma, and G. D. Stucky, *Chem. Mater. 9*, 1962 (1997)

14. J. E. Martin, M. T. Anderson, J. G. Odinek, and P. P. Newcomer, *Langmuir 13*, 4133 (1997)

15. R. Ryoo, C. H. Ko, S. J. Cho, and J. M. Kim, *J. Phys. Chem. B 101*, 10610 (1997)

16. Y. Lu, R. Gangli, C. A. Drewien, M. T. Anderson, C. J. Brinker, W. Gong, Y. Guo, H. Soyez, B. Dunn, M. H. Huang, and J. I. Zink, *Nature 389*, 364 (1997)

17. H-P. Lin and C-Y. Mou, *Science 273*, 765 (1996)

18. G. D. Stucky, Q. Huo, A. Firouzi, B. F. Chmelka, S. Schacht, I. G. Voigt-Martin, and F. Schüth, in *Progress in Zeolite and Microporous Materials*, H. Chon, S-K. Ihm and Y. S. Uh, eds., Studies in Surface Science and Catalysis, Volume 105, Elsevier, Amsterdam, 1997, pp. 3-28

19. Q. Huo, D. I. Margolese, U. Ciesla, P. Feng, T. E. Gier, P. Sieger, R. Leon, P. M. Petroff, F. Schüth, and G. D. Stucky, *Nature 368*, 317 (1994)

20. Q. Huo, D. I. Margolese, U. Ciesla, D. G. Demuth, P. Feng, T. E. Gier, P. Sieger, A. Firouzi, B. F. Chmelka, F. Schüth, and G. D. Stucky, *Chem. Mater. 6*, 1176 (1994)

21. Q. Huo, D. I. Margolese, and G. D. Stucky, *Chem. Mater. 8*, 1147 (1996)

22. S. L. Burkett, S. D. Sims, and S. Mann, *J. Chem. Soc., Chem. Commun.*, 1367 (1996)

23. C. E. Fowler, S. L. Burkett, and S. Mann, *J. Chem. Soc., Chem. Commun.*, 1769 (1997)

24. A. Monnier, F. Schüth, Q. Huo, D. Kumar, D. Margolese, R. S. Maxwell, G. D. Stucky, M. Krishnamurty, P. M. Petroff, A. Firouzi, M. Janicke, and B. F. Chmelka,

Science 261, 1299 (1993)

25. M. T. Anderson, J. E. Martin, J. G. Odinek, and P. P. Newcomer, *Chem. Mater. 10*, 311 (1998)

26. C. T. Kresge, M. E. Leonowicz, W. J. Roth, and J. S. Beck, *Nature 359*, 710 (1992)

27. J. S. Beck, J. C. Vartuli, W. J. Roth, M. E. Leonowicz, C. T. Kresge, K. D. Schmitt, CT.-W. Chu, D. H. Olson, E. W. Sheppard, S. B. McCullen, J. B. Higgins, and J. L. Schlenker, *J. Amer. Chem. Soc. 114*, 10834 (1992)

28. T. Yanagisawa, T. Shimizu, K. Kuroda, and C. Kato, *Bull. Chem. Soc. Jpn. 63* , 988 (1990)

29. D. Zhao, J. Feng, Q. Huo, N. Melosh, G. H. Fredrickson, B. F. Chmelka, and G. D. Stucky, *Science 279*, 548 (1998)

30. Q. Huo, R. Leon, P. M. Petroff, and G. D. Stucky, *Science 268*, 1324 (1995)

31. D. Zhao, Q. Huo, J. Feng, B. F. Chmelka, and G. D. Stucky, *J. Amer. Chem. Soc.*, to appear (1998)

32. K. W. Gallis and C. C. Landry, *Chem. Mater. 9*, 2035 (1997)

33. J. Feng, Q. Huo, P. M. Petroff, and G. D. Stucky, *Appl. Phys. Lett. 71*, 620 (1997)

34. H. Yang, N. Coombs, O. Dag, I. Sokolov, and G. A. Ozin, *J. Mater. Chem. 7*, 1755 (1997)

35. M. Trau, N. Yao, E. Kim, Y. Xia, G. M. Whitesides, and I. A. Aksay, *Nature 390*, 674 (1997)

36. Q. Huo, J. Feng, F. Schüth, and G. D. Stucky, *Chem. Mater. 9*, 14 (1997)

37. M. Grün, I. Lauer, and K. K. Unger, *Adv. Mater. 9*, 254 (1997)

38. A. Firouzi, D. J. Schaefer, S. H. Tolbert, G. D. Stucky, and B. F. Chmelka, *J. Amer. Chem. Soc. 119*, 9466 (1997)

39. S. H. Tolbert, A. Firouzi, G. D. Stucky, and B. F. Chmelka, *Science 278*, 264 (1997)

40. X. Bu, P. Feng, and G. D. Stucky, *Science 278*, 2080 (1997)

41. P. Feng, X. Bu, and G. D. Stucky, *Angew. Chem. Int. Ed. Engl. 34*, 1745 (1996)

42. X. Bu, T. E. Gier, P. Feng, and G. D. Stucky, *Microporous Materials*, to appear (1998)

43. P. Feng, X. Bu, and G. D. Stucky, *Nature 388,* 735 (1997)

44. Q. Huo, D. Zhao, J. L. Feng, and G. D. Stucky, submitted for publication

45. S. A. Bagshaw, E. Prouzet, and T. J. Pinnavaia, *Science 269*, 1242 (1995)

46. S. A. Bagshaw and T. J. Pinnavaia, *Angew. Chem. Int. Ed. Engl. 35*, 1102 (1996)

47. E. Prouzet and T. J. Pinnavaia, *Angew. Chem. Int. Ed. Engl. 36*, 516 (1997)

48. G. S. Attard, J. C. Glyde, and C. G. Göltner, *Nature 378*, 366 (1995)

49. C. G. Göltner and M. Antonietti, *Adv. Mater. 9*, 431 (1997)

50. M. Antonietti and C. G. Göltner, *Angew. Chem. Int. Ed. Engl. 36*, 910 (1997)

51. M. Templin, A. Franck, A. D. Chesne, H. Leist, Y. Zgang, R. Ulrich, U. Schädler, and U. Wiesner, *Science 278*, 1795 (1997)

MESOPOROUS MOLECULAR SIEVES 1998
Studies in Surface Science and Catalysis, Vol. 117
L. Bonneviot, F. Béland, C. Danumah, S. Giasson and S. Kaliaguine (Editors)

Potential Applications for M41S Type Mesoporous Molecular Sieves

J. C. Vartuli, S. S. Shih, C. T. Kresge, and J. S. Beck
Mobil Technology Company, Paulsboro Technical Center, Paulsboro, New Jersey 08066

The past several years have seen explosive growth in research on ordered mesoporous molecular sieves. Major advances in understanding and exploiting the synthesis protocols and mechanism of formation of these materials have allowed designed tailoring of composition, pore size, structure, texture and active site location and density through functionalization. The concept of 'supramolecular templating' with molecular aggregates of surfactants, proposed as a key mechanistic step in the formation of these materials, has expanded our idea of classical single molecular interaction in the formation of molecular sieves and give us new synthetic tools to engineer porous, inorganic solids. Simplified synthetic preparations of these materials have allowed study of their potential applications in: shape-selective catalysis of large organic molecules, sorption and chromatographic separations, encapsulation media for controlled deposition of polymers and atomic arrays of metal atoms, thin film applications and as nano-repositories for electronic materials. Herein we review the methodology in designing a mesoporous material for a particular application, and illustrate, with examples, the utility of the designed materials. In addition to reviewing highlights of applications of M41S in the open literature, we will focus on some recent applications emanating from work at Mobil. These applications include the use of functionalized M41S for use as chromatographic media as well as catalytic applications of metal loaded M41S in hydrodesulfurization (HDS) and hydrodenitrogenation reactions (HDN).

1. INTRODUCTION

Control of molecular sieve pore size, especially in the mesopore range, had been long sought after from an applications standpoint, yet presented an elusive synthesis challenge. The discovery of the first ordered, mesoporous molecular sieves (designated as M41S) has sparked interest throughout the scientific community [1,2]. These materials, which can be synthesized with pore sizes from 15 to over 100 Å, have a wide range of potential applications including: shape-selective catalysis and sorption of large organic molecules, chromatographic separations, and uses as hosts to confine guest molecules and atomic arrays. Several reviews on the general classification and properties of mesoporous materials [3,4], synthesis [5,6], and potential applications [3,5,7-9] have already appeared. The ability to tailor these novel molecular sieve materials with engineered structure, pore diameter, and composition has provided an excellent avenue for further advances in this area. From these intensive investigations of the physical properties of these new materials, applications ranging from the more traditional fields such as shape-selective catalysis and sorption/separation to more novel areas such as membrane technology and quantum confinement have emerged. Herein we review the recent work in these areas.

2. SYNTHESIS

The synthesis of the various mesoporous structures is well documented [1,2,10-13]. Thermally stable silicate based forms of these molecular sieves are readily prepared by the variation of synthesis conditions, including choice of surfactant and inorganic reagents. Other non-siliceous forms of these materials have also been prepared. A list of some of the possible variations in composition is shown in Table I [14-26]. Control of pore size by variation of synthesis conditions and choice of surfactant system has also been well documented. Narrow pore size distribution (+/- 5%) can be obtained for pore diameters of about 15Å to over 100Å. Recently this range of pore sizes was extended to 300Å using triblock polymers [27]. Finally, post functionalization within the pore walls allows for additional material modification for the desired application. These functionalization methods range from changing the sorptive properties of the material to anchoring active sites or molecules within the pore structure. The examples of potential applications reviewed herein are of MCM-41, the hexagonal form, because it has been the structure most studied and easiest to prepare. Further work is needed to determine if the flexibility of synthesizing various forms of MCM-41 for a specific application could be translated to the other mesoporous structures.

3. CATALYTIC APPLICATIONS

Reactions of large molecules with the mesoporous materials, either in acid form (via aluminum incorporation) or via impregnation with catalytically active metals/complexes have received much attention. Hydrocracking of vacuum gasoil on nickel and molybdenum containing MCM-41 has been demonstrated [28]. It was found that these materials were quite effective in hydrodesulfurization (HDS) and hydrodenitrogenation (HDN) reactions. In this comparative study, a NiMo-MCM-41 catalyst was found to be more effective for the removal of nitrogen and sulfur impurities, and more efficient at hydrocracking of gasoil than NiMo versions of USY and an amorphous silica-alumina with the same metal loadings. These effects were attributed to the higher surface acidity, the presence of large, uniform mesopores and the presence of mild acid activity. In a related study, we have examined the demetalization and asphaltene removal of residuum over NiMo-MCM-41 of various pore sizes ranging from 30 to 80Å. Asphaltene removal increased with increasing pore diameter of the MCM-41 catalysts. The 80Å pore MCM-41 catalyst was more

Table I
Varied Composition of the M41S Mesoporous Molecular Sieves

Structure	Composition	Reference
MCM-41	Si	Mobil (1992), Kuroda (1993)
(Hexagonal)	M, Si	
	M = Ni, Cr, Fe, Co, Ti, Al, Al-P	Mobil (1992)
	M = V	Reddy (1994)
	M = Mn	Goldfarb (1995)
	W	Mallouk (1993), Stucky (1994)
	Fe	Stucky (1994)
	Pb, Sb	Stucky (1994)
	V	Luca (1995)
	V, P	Abe (1995)
	Ti	Ying (1995)
	Zr	Schuth (1995), Kim (1997)
	Al	Pinnavaia (1996), Davis (1996)
MCM-48	Si	Mobil (1992), Stucky (1993), Davis (1994)
(Cubic)	Si, Al	Mobil (1992)
	Si, Ti	Mobil (1992), Koyano (1996)
	Sb	Stucky (1994)
	Si, Mn	Goldfarb (1995)
	Si, V	Stucky (1996)
MCM-50	Si	Mobil (1994)
(Lamellar)	Si Ti	Mobil (1994)
	Al, W, Sb, Fe, Pb, Zn – P Mg, MN, Co, Ni, Zn, Ga, Ni, SnS	Stucky (1994)
	Si, Mn	Goldfarb (1995)

active than a conventional demetalation catalyst. The 40Å MCM-41 material was comparable to the conventional demetalation catalyst in asphaltene removal over a range of reactor temperatures of 320 to 400°C. Interestingly, the average pore diameter of the conventional demetalation catalyst was 150Å, much larger that the pore diameter of any one of the MCM-41 containing catalysts. We also observed that the metal deposition capacity and tolerance improved with increasing pore diameter. These results were attributed to the large uniform mesopores that allowed access of the bulky feedstock molecules.

MCM-41 materials have also demonstrated the ability to convert bulky molecules that are too large to react within the classical microporous molecular sieve pore structure. For example, it was reported that aluminosilicate MCM-41 was a suitable catalyst for the alkylation of the bulky 2,4-di-tert-butylphenol with cinnamyl alcohol [29]. It was shown that while this alkylation does not occur in the restricted environment of an HY zeolite (pore size 7.4 Å), the primary alkylation product 6,8-di-tert-2-phenyl-2,3-dihydro[4H]benzopyran is formed using the large pore diameter mesoporous molecular sieve.

Other work on metals incorporation included the introduction of heteropolyacids such as $H_3PW_{12}O_{40}$ [30]. Here it was shown that the acid incorporation allowed higher selectivity (to less bulky alkylation products) in the alkylation of tert-butyl-phenol with styrene. It was shown that the less bulky products were more comparable to the internal pore dimensions than the larger products normally formed over completely open amorphous substrates containing the same acid species.

Selective partial oxidation of molecules over MCM-41 materials has also received much attention. For example, titanium-containing mesoporous materials (incorrectly referred to as a Ti-zeolite), isomorphous with hexagonal MCM-41, and synthesized with titanium-tetraethoxide as the titanium source, were found to selectively oxidize 1-hexene (H_2O_2, oxidant) to the resulting epoxide [15]. Large molecule oxidation was also studied. It was found that selective epoxidation of norbornene could be achieved with THP as oxidant. The same reaction could not be achieved over Ti-ZSM-5 because of the bulky nature of the reactants. Unfortunately, the work gives no comparative data for an amorphous Ti/silica analog which would contribute to the shape-selectivity claims ascribed to the mesoporous sieve. Further work shows that alpha-terpineol can be epoxidized with these catalysts [31]. In other papers, a Ti-containing mesoporous molecular sieve was capable of the shape-selective conversion of 2,6-di-tert-butyl phenol to the corresponding quinone [14,32]. The sum of these works shows that the oxidations of aromatics and olefins can be effected with these Ti-mesoporous materials, but the more elusive shape-selective oxidations of alkanes and alcohols remain the exclusive domain of the microporous Ti-containing molecular sieves.

Finally mesoporous silicates has been used for the controlled polymerization of styrene, methylmethacrylate and vinyl acetate [33]. In the study, 25 and 40 Å MCM-41 materials were used as hosts. It was observed that the properties of the polymerized guests varied with the degree of confinement. Thus, an increase in the average molecular mass of polymethylmethacrylate was observed with an increase in confinement on polymerization within 40 and 25 Å pores, respectively. The authors suggest that the increase in confinement leads to a smaller quantity of growing polymer chains, which are unable to interact, resulting in the formation of longer chains. A decrease of 20 Kelvin in the glass transition temperature of polyvinyl acetate formed in 25 Å pore diameter material was observed. It was suggested that the decrease leads to a suppression of the cooperative motion of the polymer.

4. SEPARATIONS APPLICATIONS

Another area that the mesoporous structures have received a great deal of attention is in the area of separation and adsorption. The uniform pore structure within the mesopore range and the resultant high pore volume offer materials for separations that vary from the removal of organic and inorganic contaminants in waste streams to chromatographic media. Early work demonstrated the extraordinary high sorption capacity of the M41S materials for hydrocarbon molecules such as benzene [1,2]. It was also shown that modification of the pore walls by post functionalization could substantially alter the sorption capacity and behavior of these materials [2]. Replacing the pore wall silanol groups with trimethylsilyl groups created a more hydrophobic environment within the pore structure. The resulting material demonstrated a substantial reduction in sorption capacity of polar molecules such as water.

A recent article illustrated the use of functionalized MCM 41 for the removal of heavy metals from waste streams. These researchers functionalized the pore walls of a silica MCM-41 with a mercaptopropylsilane [34]. The resultant material demonstrated a high affinity to extract mercury and other heavy metals from both aqueous and nonaqueous waste streams. Furthermore these researchers also demonstrated that this functionalized MCM-41 material could be regenerated and reused. Another recent article indicates that enzymes could be attached to the pore walls of MCM-41 [35]. The authors of this paper were able to attach globular enzymes, cytochrome C and trypsin to siliceous MCM-41.

Modification of the pore walls by coating the mesoporous structure can also alter the adsorption behavior of the material. In collaboration with the University of

Delaware, we were able to produce an anionic-exchange chromatographic material by using siliceous MCM-41 as the support media and coating it with a polyethyleimine. The coated material demonstrated excellent selectivity and high static adsorption capacity in the separation of acidic nucleotides. This static capacity demonstrated by the MCM-41 material was higher than a comparable coated material prepared with amorphous silica. However this static capacity advantage of MCM-41 was offset to some extent with a lower dynamic capacity relative to the same amorphous silica media.

Ultimately, uniform mesostructures synthesized as membranes would have great utility for separation applications ranging from chemicals to pharmaceuticals. Several research groups are actively pursuing the synthesis protocol for this type of material. A recent article detailed the use of magnetic fields to align the pore structure of the resultant mesoporous silica illustrating the techniques used to tackle this difficult synthesis problem [36].

5. NOVEL APPLICATIONS

Turning to more novel applications, it has been speculated that MCM-41 materials could show utility in a variety of electronic, optical and other advanced applications. It has been demonstrated that 2,4,6-triphenylpyrylium (TP+) ion incorporated in MCM-41 can be used as a highly efficient electron-transfer material [37]. In this paper, it was shown that the TP ion, when so incorporated, could be used as a sensitizer (in the presence of long wave radiation) to convert cis-stilbene to trans-stilbene. In a comparative study, TP+ in MCM-41 was much more effective in this conversion than $TP^+BF_4^-$ or TP+ incorporated into the smaller Y zeolite or TP+ on amorphous silica-alumina. In summary, it was shown that the mesoporous hosts are promising for photoinduced electron transfer reactions.

MCM-41 material used as a mesoporous host can be used to encapsulate and anchor an organometallic host, which may subsequently be decomposed to metal clusters [38]. In this paper, a trimethyl tin-molybdenum carbonyl organometallic complex is sorbed in the pores of MCM-41 where it is subsequently anchored via ligand loss (methyl groups + silanol protons). Upon thermal treatment the complex begins to decompose at 200°C and by 300°C converts to nanometer sized Sn-Mo metal clusters. The Sn-Mo clusters were described as having high activity for olefin hydrogenation, although no data were reported.

Finally the mesoporous materials were used as nanosized reactors to form polyaniline within the pore system of copper or iron exchanged MCM-41 by adsorption of aniline vapor followed by oxidative polymerization [39,40]. The resulting encapsulated polyaniline was examined for electrical conductivity. Results showed that the conductivity of the encapsulated polyaniline is greatly reduced compared to polyaniline prepared in the bulk, supporting the supposition of the materials location inside the channel system and not on the external silicate surface. When the polyaniline is extracted, the conductivity increases significantly as compared with bulk samples.

6. CONCLUSIONS

Mesoporous molecular sieves with well-defined pore systems are a recent advance in the molecular sieve science. Potential applications using these novel materials span fields of sorption and separations, catalysis, polymer chemistry, quantum confinement, etc. Detailed mechanistic studies have allowed the precise control of structure and pore diameter in these materials, so that the systems are virtually tunable for a desired application, an ability unparalleled in the synthesis of microporous materials. With these new structures, pore sizes and compositions at hand, researchers in many fields are armed with a new arsenal of materials with which to attack the nanoscopic world.

7. ACKNOWLEDGMENTS

The authors are grateful to the staff at the Paulsboro Technical Center for their invaluable discussions and effort. We also thank Mobil Technology Company for its support.

REFERENCES

1. C.T. Kresge, M.E. Leonowicz, W.J. Roth, J.C. Vartuli, and J.S. Beck, Nature, 359 (1992) 710-712.
2. J.S. Beck, J.C. Vartuli, W.J. Roth, M.E. Leonowicz, C.T. Kresge, K.D. Schmitt, C.T.W. Chu, D.H. Olson, E.W. Sheppard, S.B. McCullen, J.B. Higgins, and J.L. Schlenker, J Am Chem Soc, 114 (1992) 10835.
3. P. Behrens, Adv Mater, 5(2) (1993) 127-132.
4. P. Behrens and G.D. Stucky, Angew Chem Int Ed Engl, 32(5) (1993) 696-699.
5. J.L. Casci, Advanced Zeolite Science and Applications, Stud in Surf Sci and Catal 1994, Edited by J.C. Jansen, M. Stöcker, H.G. Karge, and J. Weitkamp: Elsevier Science, 85, 329-356.
6. D.M. Antonelli, and J.Y. Ying, Curr. Opin. Colloid. Interface Sci., 1(4) (1996) 523-29,
7. G.A Ozin, Adv Mater, 4(10) (1992) 612-649.
8. X.S. Zhao, G.Q. Lu, and G.J. Millar, Ind. Eng. Chem. Res, 35 (1996) 2075-90.
9. A. Corma, Chem. Rev, 97 (1997) 2373-2419.

10. C.Y. Chen, H.Y Li, and M.E. Davis, Microporous Mater, 2 (1993) 17-26.
11. C.Y. Chen, S.L. Burkett, H.X. Li, and M.E. Davis, Microporous Mater, 2 (1993) 27-34.
12. Q. Huo, D.I. Margolese, U. Ciesla, P. Feng, T.E. Gier, P. Sieger, R. Leon, P.M. Petroff, F. Schüth, and G.D. Stucky, Nature, 368 (1994) 317-323.
13. Q. Huo, D.L. Margolese, U. Ciesla, D.G. Demuth, P. Feng, T.E. Gier, P. Sieger, A. Firouzi, B.F. Chmelka, F. Schüth, and G.D. Stucky, Chem Mater, 6(8) (1994) 1176-1191.
14. P.T. Tanev, M. Chibwe, and T.J. Pinnavaia, Nature, 368 (1994) 321.
15. A. Corma, M.T. Navarro, and J.P. Pariente, J. Chem. Soc. Chem. Commun, 147 (1994)
16. A. Corma, V. Fornés, M.T. Navarro, and J. Pérez-Pariente, J. Catalysis, 148 (1994) 569.
17. A. Stein, M. Fendorf, T.P. Jarvie, K.T. Mueller, A.J. Benesi, and T.E. Mallouk, Chem. Mater, 7 (1995) 304.
18. A. Sayari, C. Danumah, and I.L. Moudrakovski, Chem. Mater, 7 (1995) 813.
19. D. Zhao and D. Goldfarb, J. Chem. Soc., Chem. Commun, 875 (1995).
20. V. Luca, D.J. MacLauchlan, J.M. Hook, and R. Withers, Chem. Mater, 7 (1995) 2220.
21. T. Abe, A. Taguchi, and M. Iwamoto, Chem. Mater, 7 (1995) 1429.
22. D.M. Antonelli, and J.Y. Ying, Angew. Chem. Int. Ed. Engl, 34 (1995) 204.
23. F. Schüth, Ber. Bunsenges Phys. Chem, 99 (1995) 1306.
24. F. Vaudry, S. Khodabandeh, and M.E. Davis, Chem. Mater, 8 (1996) 1451-64.
25. A. Kim, P. Bruinsma, Y. Chen, L.-Q. Wang, and J. Liu, J. Chem. Soc., Chem. Commun, 161-2 (1997).
26. S. A. Bagshaw and T.J. Pinnavaia, Angew. Chem. Int. Ed. Engl, 35(10) (1996) 1102-05.
27. D. Zhao, J. Feng, Q. Huo, N. Melosh, G.H. Fredrickson, B.F. Chmelka, and G.D. Stucky, Science, 279 (1998) 548-52.
28. A. Corma, A. Martínez, V. Martínez-Soria, and J.B. Montón, J Catal, 153 (1995) 25-31.
29. E. Armengol, M.L. Cano, A. Corma, H. García, and M.T. Navarro, J Chem Soc Chem Commun, 519-520 (1995).
30. I.V. Kozhevnikov, A. Sinnema, R.J.J. Jansen, K. Pamin, and H. van Bekkum, Catal Lett, 30 (1995) 241-252.
31. A. Corma, M.T. Navarro, J. Pérez-Pariente, and F. Sánchez, In Zeolites and Related Microporous Materials: State of the Art 1994., Stud in Surf Sci and Catal 1994, Proc 10th Intl Zeolite Conf Garmisch-Partenkirchen Germany 1994, Edited by J. Weitkamp, H.G. Karge, H. Pfeifer, and W. Hölderich: Elsevier Science, 84, 69.
32. A. Sayari, R.J. Sudhakar, and A. Dicko, Symp on Synthesis of Zeolites. Layered Compounds and other Microporous Solids, Div. of Petrol Chem 209th National Am Chem Soc Meeting, Anaheim, CA, 1995.
33. P.L. Llewellyn, U. Ciesla, H. Decher, R. Stadler, F. Schüth, and K. Unger, In Zeolites and Related Microporous Materials: State of the Art 1994, Stud in Surf Sci and Catal 1994. Proc 10th Intl Zeolite Conf Garmisch-Partenkirchen Germany 1994, Edited by J. Weitkamp, H.G. Karge, H. Pfeifer, and W. Hölderich, Elsevier Science, 84, 2013.
34. X. Feng, G.E. Fryxell, L.-Q. Wang, A.Y. Kim, J. Liu, and K.M. Kemner, Science, 276 (1997) 923-26.

35. J.F. Diaz and K.J. Balkus, J. Mol. Catal. B: Enzyme, 2(2-3) (1996) 115-26.
36. S.H. Tolbert, A. Firouzi, G.D. Stucky, and B.F. Chmelka, Science, 278 (1997) 264-69.
37. A. Corma, V. Fornés, H. García, M.A. Miranda, and M.J. Sabater, J Am Chem Soc, 116(21) (1994) 9767-9768.
38. C. Huber, K. Moller, and T. Bein, J Chem Soc Chem Commun, 2619-2620 (1994).
39. C.G. Wu and T. Bein, Chem Mater, 6(8) (1994) 1109-1112.
40. C.B. Wu and T. Bein, Science, 264 (1994)1757-1759.

MESOPOROUS MOLECULAR SIEVES 1998
Studies in Surface Science and Catalysis, Vol. 117
L. Bonneviot, F. Béland, C. Danumah, S. Giasson and S. Kaliaguine (Editors)

Catalytic properties of mesoporous molecular sieves prepared by neutral surfactant assembly

Thomas J. Pinnavaia* and Wenzhong Zhang

Department of Chemistry, Center for Fundamental Materials Research, Michigan State University, East Lansing, MI 48824

Chemically modified derivatives of mesoporous HMS molecular sieves silicas, prepared by neutral surfactant assembly pathways, have been used as catalysts for a variety of chemical conversions, including peroxide oxidations of olefins and phenols, ring opening polymerization of lactide dimers, selective reduction of NO_x, and cumene cracking. These HMS catalysts are often more active than analogous catalysts prepared through electrostatic assembly mechanisms. The wormhole-like framework mesopores, along with the presence of complementary textural mesopores, most likely facilitate access to the catalytic active centers in the framework.

1. INTRODUCTION

A major advance in the design of mesoporous molecular sieve catalysts was provided by Mobil researchers in reporting [1] the supramolecular surfactant assembly of a broad family of M41S materials with uniform pore sizes in the mesopore range 2.0-10.0 nm. Also, Kuroda and his co-workers [2] described a structurally related mesoporous silica of potential catalytic utility by quaternary ammonium surfactant rearrangement of a layered precursor (kanemite). The more versatile Mobil approach was based on the supramolecular assembly of cationic surfactants (S^+) and anionic inorganic precursors (I^-). Stucky and his co-workers [3] greatly extended the S^+I^- electrostatic pathway of Mobil to include a charge-reversed S^-I^+ assembly mechanism, as well as counterion-mediated $S^+X^-I^+$ and $S^-M^+I^-$ pathways, where $X^- = Cl^-, Br^-$ and $M^+ = Na^+, K^+$.

We have developed yet another pathway for the preparation of mesoporous molecular sieves based on the hydrolysis of an electrically neutral inorganic

*The support of the National Science Foundation through CRG grant CHE-9633798 is gratefully acknowledged.

precursor (I°) in the presence of a neutral amine (S°) surfactant as the predominate structure directing agent. This S°I° pathway was first used to prepare a mesoporous sieve silica and a Ti-substituted analog. A small amount of protonated amine was used as a co-surfactant in the original synthesis [4], but subsequent studies [5] showed that the protonated co-surfactant component was not needed to achieve framework assembly. Electrostatic forces do not play an important role in S°I° assembly. Instead, the assembly forces at the surfactant-inorganic precursor interfaces are based on hydrogen bonding. An equivalent H-bonding pathway, denoted N°I° has also been demonstrated for nonionic polyethylene oxide surfactants and I° precursors [6]. The S°I° and N°I° pathways are complementary to the above electrostatic templating pathways. These pathways also are distinguishable from the S-I pathway developed by Ying and her co-workers for the synthesis of mesostructured transition metal oxides [7]. In the S-I pathway a coordinate covalent bond is formed between the surfactant (usually, an amine) and the metal center.

The present paper provides an overview of the catalytic properties of mesostructured molecular sieve silicas prepared under reaction conditions where the S°I° pathway predominates. These materials are denoted "HMS silicas", in keeping with their original designation [4], to distinguish them from the anionic M41S framework silicas prepared by electrostatic assembly pathways.

2. PHYSICAL PROPERTIES OF HMS SILICAS

Having defined HMS silicas as mesoporous molecular sieves formed by a supramolecular assembly pathway that is *predominately* S°I° in character, we next describe some of the physical properties of these materials which distinguishes them from the hexagonal, cubic and lamellar structures formed by electrostatic assembly pathways. The first example of an HMS silica was prepared at ambient temperature in the presence of a 13.5:1 molar mixture of dodecylamine and dodecylammonium ion as the structure directing co-surfactants [5]. The product formed under these reaction conditions exhibited only one resolved XRD reflection, which precluded the assignment of a long range structure. Selected area electron diffraction studies provided evidence for the occasional occurrence of very small domains of hexagonal symmetry, but the vast majority of the sample was highly disordered and lacking in a long range regular structure.

Subsequent studies revealed that equivalent HMS silicas could be prepared by omitting the onium ion form of the reaction mixture and using only the neutral amine as the structure director [5]. This S°I° pathway afforded silicas with N_2 adsorption properties, pore sizes, and XRD patterns virtually identical to the original HMS products formed using a mixture of S° and S^+ surfactants. Also, the sparsely occurring small domains of hexagonal order were absent. In fact, hexagonal regions are very rarely formed even when protonated surfactant is present. Instead, the wormhole channel motif shown in Figure 1 is formed almost exclusively [8] even when up to 15% of the amine is protonated. The onium ion can be introduced by adding a protonic acid. Alternatively, the introduction of certain Lewis acid centers, as in the replacement of some Si^{4+} sites by Al^{3+}, Fe^{3+} or B^{3+}, will result in the formation of some protonated amine surfactant during the assembly process in order to balance the resulting framework. However, this small electrostatic participation of the surfactant is structurally inconsequential, and it does not alter the wormhole channel motif.

Figure 2 compares the N_2 adsorption properties of pure HMS silica and Ti-substituted (1.0 mol % Ti) HMS silica prepared by ambient temperature through a predominately S°I° pathway in 5.5:1 (mol/mol) water:ethanol as a solvent. The co-surfactants were dodecylamine (0.25 mol) and dodecylammonium chloride (0.02 mol) and the silica and titania precursors were $Si(OC_2H_5)_4$ and $Ti(\textit{iso}\text{-}OC_3H_7)_4$, respectively. Included for comparison is the isotherms for Ti-MCM-41 (1.0 mol % Ti) prepared by $S^+X^-I^+$ assembly under the same reaction conditions, except that the surfactant was cetyltrimethyl-ammonium bromide, $[C_{16}H_{33}N(CH_3)_3]Br$. The insert to the figure compares the Horvath-Kawasoe pore size distribution curves for the three mesoporous materials.

The curves in Figure 2 illustrate two important differences between HMS and MCM-41 when the mesostructures are assembled at ambient temperature in a solvent with a high water content (>50 vol %). Firstly, the predominantly neutral S°I° pathway affords HMS mesostructures with pore sizes that are comparable to Ti-MCM-41, even though the S° surfactant used to assemble HMS is smaller than the S^+ surfactant used for MCM-41 assembly. This distinction in the relationship between framework pore size and surfactant size reflects the fundamental difference in H bonding and electrostatic bonding mechanisms at the surfactant-silica interface during the assembly of the mesostructures. Secondly, the large hysteresis loop for the HMS derivatives at relative pressures above 0.70 reflect the high textural porosity associated with these mesostructures (cf. Figure 1).

Table 1 compares the surface areas and pore volumes for Ti-HMS and two Ti-MCM-41 catalysts prepared by the electrostatic S^+I^- and $S^+X^-I^+$ pathways at ambient temperature [9]. Each mesostructure contains ~2 mole % Ti and exhibits a HK pore size near 28Å. The values in parenthesis in the Table are for the corresponding pure silicas. Note the very high ratio of textural to framework mesoporosity for the HMS molecular sieves ($V_{tx}/V_{fr} \cong 1.06$) compared to MCM-41 molecular sieves ($V_{tx}/V_{fr} \cong 0.03$). As will be shown below, the textural porosity of HMS catalysts can improve catalytic activity by facilitating substrate transport to the active sites in the mesostructure framework.

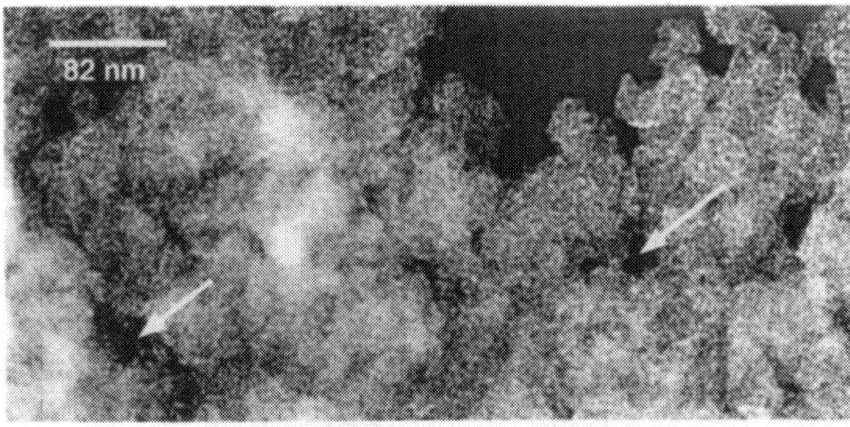

Figure 1. TEM image of HMS showing wormhole framework and textural mesopores (arrows)

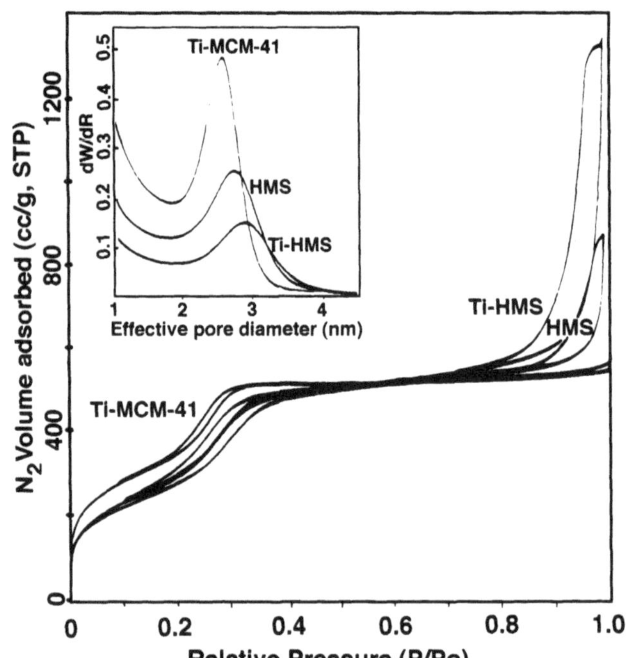

Figure 2. N_2 isotherms and pore size distributions for HMS materials

3. PEROXIDE OXIDATIONS

Metal-substituted HMS silicas have received considerable attention as catalysts for the peroxide oxidation of aromatics to phenols and quinones and of alkenes to epoxides and diols. One of the first reactions to be investigated was the Ti-HMS catalyzed oxidation of 2,6-di-*tert*-butylphenol (2,6-DTBP) to the corresponding mono-and dinuclear quinones using H_2O_2 as the oxidant [4]. This substrate was of particular interest to us, because it was too large to access the framework Ti sites of TS-1, an industrial microporous molecular sieve catalyst. Other oxidation reactions that have been investigated using

Table 1
Textural Properties of Mesoporous Ti-Substituted Molecular Sieve Silicas

parameter[a]	MCM-41 (S^+I^-)	MCM-41 ($S^+X^-I^+$)	HMS (S°I°)
surfactant	$C_{16}H_{33}N(CH_3)_3^+$	$C_{16}H_{33}N(CH_3)_3^+$	$C_{12}H_{25}NH_2$
Mole % Ti			
initial gel	2.0	10	2.2
calcined	2.2	2.5	2.4
product			
d (Å)	38.1 (36.0)	36.5 (33.0)	40.2 (36.0)
S_{BET}(m^2/g)	859 (923)	1354 (1345)	1075 (1108)
V_{total} (cm^3/g)	0.70 (0.72)	0.92 (0.95)	1.40 (1.42)
V_{fr} (cm^3/g)	0.68 (0.70)	0.90 (0.92)	0.68 (0.70)
V_{tx} (cm^3/g)	0.02 (0.02)	0.02 (0.03)	0.72 (0.72)
V_{tx}/V_{fr}	0.03 (0.03)	0.02 (0.03)	1.06 (1.03)

[a]The total liquid pore volume, V_{total}, was estimated at a relative pressure of 0.95 assuming full surface saturation. The volume of framework-confined mesopores, V_{fr}, was determined from the upper inflection point of the corresponding adsorption step. The volume of textural mesopores, V_{tx}, was obtained from the equation $V_{tx} = V_{total} - V_{fr}$. The data in parentheses are for the pure silica analogs.

Ti-HMS as a catalyst includes the conversion of benzene to phenol, methyl metharcylate to methyl pyruvate and styrene to the corresponding epoxide, diol and benzaldehyde [9].

The catalytic properties of mesoporous Ti-HMS and of Ti-MCM-41 for the liquid phase oxidations of methylmethacrylate, styrene and 2,6-di-*tert*-butylphenol are described in Table 2. Included in the Table for comparison are the conversions and selectivities obtained with microporous TS-1 as the catalyst. As expected based on pore size considerations, the conversions observed for all three substrates are substantially larger for the mesoporous catalysts than for the microporous catalyst. The S°I°-assembled Ti-HMS catalyst exhibited consistently greater reactivity than the two Ti-MCM-41 catalysts prepared by electrostatic assembly. The superior performance of Ti-HMS is especially pronounced in the case of the large 2,6-di-*tert*-butylphenol substrate.

The differences in catalytic reactivity between Ti-HMS and Ti-MCM-41 cannot be attributed to differences in Ti siting. XANES and EXAFS studies showed that the titanium center adopt primarily tetrahedral coordination in all three catalysts [9]. Also, the coordination environment is very similar for

Table 2
Catalytic Activity of Ti-substituted (2 mol%) Mesoporous Molecular Sieve Silica

Catalyst		TS-1	Ti-MCM-41 (S^+I^-)	Ti-MCM-41 ($S^+X^-I^+$)	Ti-HMS (S^0I^0)
MMA oxidation	conv. (mol%)	2.5	4.0	6.2	6.8
	MPV[a] select. (mol%)	78	93	93	93
Styrene oxidation	conv. (mol%)	8.4	10	23	28
	PhCHO select. (mol%)	71	82	78	77
	Epoxide select. (mol%)	14	6.2	4.1	4.7
	Diol select. (mol%)	4.5	3.8	8.2	9.6
2,6-DTBP oxidation	conv. (mol%)	5.0	39	22	55
	Quinone select. (mol%)		91	90	91

[a]MPV is methyl pyruvate; b. Quinone selectivity is expressed as the cumulative selectivity of monomer and dimer quinone.

the three catalysts, as judged from the similarities in the EXAFS features. Also, UV-VIS adsorption spectra showed no phase segregation of titania, the spectral features being consistent with site-isolated titanium centers. Because the framework walls of HMS tend to be thicker than MCM-41, the superior reactivity of Ti-HMS cannot be due to an enhancement in the fraction of Ti available for reaction on the pore walls. Thicker walls should bury more titanium at inaccessible sites within the walls. The most distinguishing feature between Ti-HMS and Ti-MCM-41 is the greater textural (interparticle) mesoporosity for Ti-HMS (cf. Figure 2, Table 1). This complementary textural mesoporosity facilitates substrate transport and access to the active sites in the framework-confined mesopores, thus enhancing catalytic efficiency compared to MCM-41 mesostructures with little or no textural porosity.

Ti-, V-, Cr-, Mo- and Mn- substituted HMS silicas have been compared with analogous metal-substituted MCM-41 catalysts for the peroxide oxidation of benzene to phenol [10]. Dodecylamine and cetyltrimethylammonium bromide were used as the surfactants for the preparation of the M-HMS and M-MCM-41 mesostructures via the $S^{\circ}I^{\circ}$ and $S^{+}X^{-}I^{+}$ assembly pathways at ambient temperature. Aqueous solutions of Na_3VO_4, CrO_3, K_2MoO_4, and $MnCl_2$, as well as Ti(*iso*-$OC_3H_7)_{44}$ in isopropyl alcohol, served as the transition metal ion precursors. Each mesostructure was calcined at 650°C for 4h to remove the surfactant prior to use as a catalyst.

All of the above metal-substituted HMS and MCM-41 derivatives are active for benzene hydroxylation to phenol in the absence of a solvent at 72°C, whereas the pure silica analogs are completely inactive. This result signifies that site isolated transition metal centers can be imbedded in the frameworks of the HMS and MCM-41 mesostructures. However, the catalytic properties are very different from those of TS-1, because the hydroxylation of benzene to phenol is accompanied by over-oxidation to very complex mixtures, especially at conversions >10%. We should note that hydroxylation of benzene over Ti-HMS and Ti-MCM-41 also has been investigated in acetone [4], the solvent normally used for benzene hydroxylation with TS-1 and related silicalite catalysts[11]. However, Ti-HMS and Ti-MCM-41, unlike TS-1, catalyze the oxidation of acetone at rates that are competitive with benzene oxidation. One of the oxidation products, namely 3,3,6,6-tetramethyl-1,2,4,5 tetraoxacyclohexane, has a GC retention time very near that of phenol. The coincidence in retention times led to an initial overestimation of the phenol yields obtained in this solvent system [4].

Sayari and his co-workers [12-14] have investigated Ti- and V-substituted HMS silicas as liquid phase oxidation catalysts for large organic molecules. Their results verify that the activity of HMS derivatives is typically higher than the corresponding MCM-41 analogs. He also has emphasized the importance of the accessibility of Ti-active sites in determining reactivity toward large molecules. He further noted that a number of oxidation reactions which occur readily over small pore TS-1 catalysts do not take place in the presence of larger pore Ti-β or ultra large pore Ti-MCM-41 and Ti-HMS, suggesting that differences in surface hydrophilicity and Ti redox potential also plays a role in determining the reactivity of tetrahedral Ti sites in these frameworks. On the other hand, TS-1 is not known to catalyze the oxidation of acetone at rates that are competitive with benzene hydroxylation, yet Ti-HMS catalyzes acetone oxidation readily.

In the case of V-HMS and V-MCM-41, the siting of the V=O moiety was distinguishable on the basis of NMR and FTIR spectroscopy [14]. Nevertheless, both derivatives were effective catalysts for hydroxylation of 2,6-DTBP to the corresponding benzoquinone with exceptionally high conversion and peroxide efficiency. V-HMS materials also catalyze the oxidation of phenol, naphthalene and cyclododecanal with peroxide or tert-butyl hydroperoxide as the oxidant. However, depending on the nature of the substrate, solvent and oxidant, vanadium leaching from the framework was observed [15].

Kaliaguine and his co-workers [16] have compared the catalytic reactivity of Ti-HMS and Ti-MCM-41 silicas for the epoxidation of α-pinene with tert-butylhydroperoxide as the oxidant. The conversions over Ti-HMS were somewhat lower than Ti-MCM-41, although epoxide selectivities were similar. Also, Gontier and Tuel [17,18] prepared a series of Ti-HMS using reaction times as short as 15 min at ambient temperature. Tetrahedral Ti loadings up to 2 wt % and thermal stabilities up to 650°C in air were possible without forming extraframework titania [17]. However, they subsequently found no substantial difference in catalytic reactivity between Ti-HMS and Ti-MCM-41 for the oxidation of aniline [18]. These results suggest that the Ti-HMS derivatives used in these studies did not possess the high textural porosity needed for facile access to framework Ti sites. In fact, the absence of textural mesoporosity in the Ti-HMS catalyst used by Gontier and Tuel was confirmed by N_2 adsorption studies [19]. These reports verify that the interparticle mesoporosity of Ti-HMS and HMS molecular sieves, in general, is dependent on the reaction conditions used for framework assembly. In general, textural porosity is formed when the assembly process is carried out in a water-rich solvent. Alcohol-rich solvents tend to eliminate textural porosity. Enhanced reactivity for large molecule conversions over HMS catalysts can be expected only when the textural and framework mesoporosities are comparable in magnitude.

4. RING OPENING POLYMERIZATION

The small crystallite domain sizes and high textural mesoporosity that can be achieved for HMS derivatives through S°I° assembly may be especially beneficial in promoting polymerizations and other bulky conversions where diffusion effects can limit the catalytic effectiveness of larger particle mesostructures. In order to demonstrate the potential utility of mesostructures for polymerization reactions, we have shown [20] that Sn(IV)-substituted HMS is remarkably effective for the ring opening polymerization of *l*-lactide dimer to poly (*l*-lactic acid), abbreviated PLA, with a high molecular weight and low polydispersity.

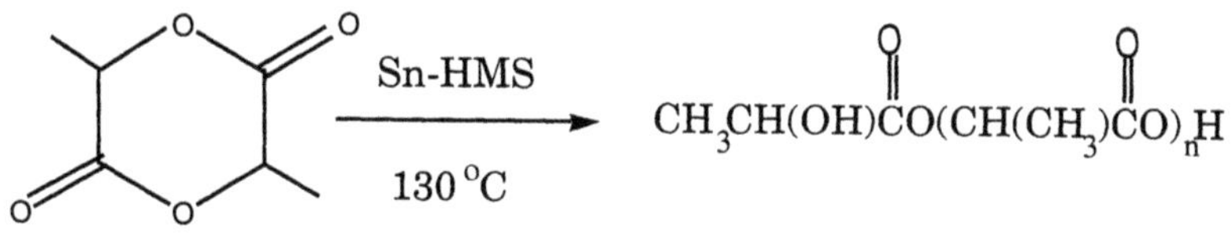

Lactide Dimer PLA

Sn-HMS containing 1 mol % tin(IV) was prepared at ambient temperature by S°I° assembly in ethanol:water (3:1V/V) using a 100:1 molar mixture of $Si(OC_2H_5)_4$ and Sn(*iso*-$C_3H_7)_4$ as the inorganic precursors and dodecylamine as the structure directing surfactant. The N_2 adsorption isotherm for the calcined (550°C) mesostructure indicated a BET surface area of 886 m^2/g, an average framework pore size of 2.7 nm, and a textural (interparticle) mesoporosity in excess of the framework mesoporosity.

Table 3 reports the conversions of *L*-lactide dimer to PLA at 130°C. Included for comparison purposes are the conversions for Ti-, Fe-, Y-, and La-HMS derivatives, pure HMS silica, a Sn-doped silica gel (1.0 mol% Sn) and pure SnO_2. The latter two catalysts were prepared by hydrolysis of the corresponding alkoxides under condition analogous to those used to form Sn-HMS.

The polymerization product obtained from Sn-HMS exhibited the highest conversion as determined by ^{1}HNMR (82%), the largest average molecular mass (36000) and a low polydispersity (1.1). In the case of pure tin oxide as the catalyst, the conversion was substantial (73%), but the polymerization product had a much lower molecular mass (17800) and a high polydispersity (1.7). The low activity in the case of Fe-, Y- and La-HMS and the Sn-doped silica gel may indicate the lack of a suitable dispersion of active metal centres in the host silica. Sn-HMS, however, clearly combines the reactivity of tin Lewis-acid sites with the selectivity of a regular mesopore structure in affording PLA with a reasonably high molecular mass and low polydispersity. It appears that the ordered pore structure improves average molecular mass and polydispersity values in comparison to homogeneous catalysts by imposing steric constraints on the propagating PLA chains and minimizing 'back-biting' and intermolecular transesterification reactions.

Table 3
Lactide polymerization over heterogeneous catalysts[a]

Catalyst	Conversion (%)	PLA Molecular mass
Sn-HMS	82	36 000
Ti-HMS	15	808
Fe-HMS	2	583
Y-HMS	0	--
La-HMS	0	--
HMS	0	--
Sn-doped Silica	22	3 200
SnO_2	73	17 800

[a]Reaction conditions: 2.00 g (13.9 mmol) lactide dimer; 0.1 g catalyst (except for SnO_2, where 0.001 g was used); T = 130°C; reaction time 72 h. All catalysts were calcined at 550°C prior to use.

5. SELECTIVE CATALYTIC REDUCTION (SCR)

Yang et al. [7], recently investigated the activity of Fe^{3+} exchanged forms of Al-MCM-41 and Al-HMS for the selective catalytic reduction (SCR) of NO by NH_3. In order to minimize the number of factors which may cause significant differences in the catalytic properties between Al-MCM-41 and Al-HMS, the aluminum contents of both molecular sieves were controlled to around 8%. Also, the pore size distribution for the two supports was controlled to around 28 Å, the BET surface areas were in the range 800-850 m^2/g, and the framework pore volumes were similar (0.62 and 0.75 cm^3/g for Al-HMS and Al-MCM-41, respectively). However, the textural mesoporosity of Al-HMS was comparable to the framework mesoporosity, whereas little or no textural porosity was present for Al-MCM-41.

Although HMS and MCM-41 are similar both in terms of chemical composition, surface area, and pore volume, Fe/Al-HMS showed considerably higher activities than Fe/Al-MCM-41. The main differences between HMS and MCM-41 are the crystal domain sizes and the unique textural porosity of HMS. The crystal domain size for HMS was of the order 150Å, whereas that of MCM-41 was larger by two orders of magnitude, as judged by XRD line widths and TEM.

The comparison of the estimated apparent and intrinsic rate constants given in Table 4 show that the reaction for the Fe/Al-MCM-41 sample was severely limited by pore diffusion, but not for HMS molecular sieve. The Thiele effectiveness factor for the HMS was nearly 1.0, whereas that for the

MCM-41 were 0.53 at 350°C and 0.44 at 400°C. An overall activation energy of 6.5 kcal/mol for Fe-HMS was in the range for SCR without diffusion limitation. The high effectiveness factor for the HMS catalyst is attributable to the small domain size and short diffusion path in the framework channels. Thus, the advantage for the Fe^{3+} exchanged mesoporous molecular sieves, in particular Al-HMS, for the SCR reaction is clearly demonstrated.

Table 4
Comparison of the apparent and intrinsic first-order rate constants and the over all diffusivity for NO SCR reaction over Fe^{3+} exchanged Al-HMS and Al-MCM-41*

Catalyst	Apparent k (s^{-1})		Intrinsic k (s^{-1})[a]		Diffusivity (cm^2/s)[a]	
	350°C	400°C	350°C	400°C	350°C	400°C
Fe/Al-HMS	56.5	87.5				
			56.6	87.6	3.95×10^{-7}	4.0×10^{-7}
Fe/Al-MCM-41	30.0	38.5				

[a]The intrinsic rate constant and diffusivity are assumed to be the same for both mesoporous catalysts.

6. ACID CATALYSIS

The replacement of silicon by trivalent elements in HMS materials has been investigated by several groups in an effort to improve Bronsted and Lewis acidity for catalytic applications [21-26]. Although Al-MCM-41 derivatives can be prepared directly by electrostatic assembly pathways, the as-synthesized materials are structurally sensitive to calcination [27]. Also, if sodium silicate is used in the synthesis a post-synthesis treatment with NH_4NO_3 is needed to remove residual Na^+ ions from the exchange sites [30]. It has been reported by Corma and his co-workers [29] that Al-MCM-41 undergoes framework dealumination upon calcination to remove the surfactant. Hitz and Prins [30] verified the dealumination of surfactant-intercalated Al-MCM-41 upon calcination, but they showed that dealumination could be minimized by first removing up to 73% of the surfactant by proton exchange in ethanol.

In contrast to the electrostatic assembly pathways to Al-MCM-41, the S°I° pathway to Al-HMS derivatives offers a convenient route to acidic mesoporous molecular sieves with retention of framework aluminum and other trivalent ions. As noted earlier, the replacement of Si^{4+} by Al^{3+} requires the protonation of one equivalent of structure-directing amine surfactant for every equivalent of Al^{3+} incorporated into the framework. However, because only up to 15 mole % of the silicon can be replaced at tetrahedral sites, the majority of the framework is assembled through H bonding interactions of the amine surfactant with the silica framework. Consequently, most of the surfactant (>90%) can be efficiently removed by simple solvent extraction and this preserves the tetrahedral siting of aluminum centers in the framework.

Tuel and Gontier [22] have recognized the potential importance of the S°I° assembly pathway for the preparation of HMS derivatives containing trivalent framework elements of catalytic significance. They found that Al^{3+}, Ga^{3+}, Fe^{3+} and B^{3+} all could be incorporated into the framework of HMS silica at Si/M^{3+} ratios as low as 10. More importantly, the neutral amine surfactant could be removed by solvent extraction without damaging the framework. Also, the remaining small concentration of charge balancing primary alkylammonium ions associated with the aluminum sites could be removed either by calcination or by ion exchange without collapsing the framework or removing the aluminum from the framework. Similar results were obtained for other M^{3+}-substituted derivatives. Thus, the structural degradation that normally occurs upon calcining M^{3+}-substituted mesostructured silicas to remove surfactant can be avoided through S°I° assembly and the subsequent solvent extraction of the vast majority of the surfactant. The heat of combustion associated with the removal of the remaining protonated surfactant is not sufficient to significantly alter the framework structure.

Mokaya and Jones [23,24] also have recognized the advantages of preparing mesoporous silica and aluminosilicate molecular sieves through a S°I° assembly pathway. Using a synthetic methodology equivalent to that originally reported for HMS synthesis (i.e., alkyl amines as the surfactant, alkoxides as the inorganic precursors, ethanol/water as a solvent, ambient temperature assembly, and solvent extraction of the surfactant), they prepared a series of derivatives with XRD and textural properties identical to HMS materials. However, they designated their products as MMS rather than HMS derivatives. Interestingly, their Al^{3+}-substituted products with Si/Al ratios in the range 5-40 exhibited greater Bronsted acidity and catalytic activity for cumene cracking compared to Al-MCM-41, amorphous aluminosilicate gel and zeolite HY of similar aluminum content [25]. In addition, the Al-HMS catalysts were less prone to deactivation. It is unlikely that the siting and intrinsic acidity of tetrahedral aluminum centers in Al-HMS, Al-MCM-41, and aluminosilicate gels differ significantly, because the framework walls in all three materials are amorphous. However, *access* to the acidic sites in these materials may differ greatly. What is particularly distinctive of most Al-HMS materials compared to Al-MCM-41 and amorphous gels is the uniformity of the framework pores.

Table 5
Cumene conversions (%) at 300°C

Catalyst	20 min.	3 hr.
Al-MCM-41	14.7	13.6
PNA-1	41.3	37.5
Al-HMS	24.8	26.8
PNA-2	47.6	42.4
ZSM-5	95.1	93.7

The short intersecting framework pores, in addition to the small crystallite domain size, may well facilitate access to the active sites in HMS structures. This improved access may be the main reason for the superior reactivity of Al-HMS materials as acid catalysts relative to Al-MCM-41, which have very long channel lengths, and amorphous gels with irregular, highly constrained pore-structures.

Very recently, van Bekkum and his co-workers [26] have reported modifying Al-HMS and Al-MCM-41 (Si/Al ≅ 30) to contain entrapped unit cells of ZSM-5. These modified mesostructures were prepared by ion exchanging the aluminosilicates with tetrapropylammonium cations as MFI structure directors and subsequently digesting the mesostructures in glycerol at 120°C for 24 h. The resulting nanoporous aluminosilicate products, designated NPA-1 and NPA-2, respectively, exhibited XRD patterns characteristic of Al-MCM-41 and Al-HMS, but reflections characteristic of ZSM-5 were absent. An FTIR band at 550-560 cm^{-1} in both products was considered to be indicative of unit cells of ZSM-5 highly dispersed in the mesostructure framework. Further evidence for the presence of entrapped ZSM-5 units cells was provided by the acid catalytic activity of NPA-1 and NPA-2 toward cumene cracking. As can be seen from the cumene conversions listed in Table 5, the acid catalytic activity of the modified mesostructures is substantially greater than the parent mesostructures. It is especially noteworthy that PNA-2, obtained by modification of Al-HMS, is more active than the PNA-1 product derived from Al-MCM-41. The superior activity of Al-HMS and NPA-2 relative to Al-MCM-41 and NPA-1 may again be attributed to the intrinsically more efficient access to framework catalytic sites for HMS derivatives.

REFERENCES

1. J.S. Beck, J.C. Vartuli, W.J. Roth, M.E. Leonowicz, C.T. Kresge, K.D. Schmit, C.T.-W. Chu, D.H. Olson, E.W. Sheppard, S.B. McCullen, J.B. Higgins and J.L. Schlenker, J. Am. Chem. Soc., 114 (1992) 10834.
2. S. Inagaki, Y. Fukushima, K. Kuroda, J. Chem. Soc., Chem. Commun., 680 (1993).
3. Q. Huo, D.I. Margolese, U. Ciesla, P. Feng, T.E. Gier, P. Sieger, R. Leon, P.M. Petroff, F. Schüth, and G.D. Stucky, Nature, 368 (1994) 317.
4. P.T. Tanev, M. Chibwe and T.J. Pinnavaia, Nature, 368 (1994) 321.
5. P.T. Tanev and T.J. Pinnavaia, Science, 267 (1995) 865.

6. S.A. Bagshaw, E. Prouzet and T.J. Pinnavaia, Science, 269 (1995) 1242.
7. D.M. Antonelli and J.Y. Ying, Angew. Chem., Int. Ed. Engl., 35 (1996) 426.
8. W. Zhang, T.R. Pauly and T.J. Pinnavaia, Chem. Mater., 9 (1997) 2491.
9. W. Zhang, M. Fröba, J. Wang, P.T. Tanev, J. Wong and T.J. Pinnavaia, J. Amer. Chem. Soc., 118 (1996) 9164.
10. W. Zhang, J. Wang, P.T. Tanev and T.J. Pinnavaia, J. Chem. Soc. Chem. Commun., 979 (1996).
11. B. Notari, Adv. Catal., 41 (1996) 253.
12. K.M. Reddy, I. Moudrakovski, A. Sayari, J. Chem. Soc. Chem. Commun., 1059 (1994).
13. J.S. Reddy and A.J. Sayari, Chem. Soc. Chem. Commun., 2231 (1995).
14. A. Sayari, Chem. Mater., 8 (1996) 1840.
15. J.S. Reddy, P. Liu and A. Sayari, Appl. Catal. A, 148 (1996) 7.
16. T.D. On, M.P. Kapoor, P.N. Joshi, L. Bonneviot, S. Kaliaguine, Catal. Lett., 44 (1997) 171.
17. S. Gontier, A. Tuel, Zeolites, 15 (1995) 601.
18. S. Gontier, A. Tuel, J. Catal., 157 (1995) 124.
19. A. Tuel and S. Gontier, Chem. Mater., 8 (1996) 114.
20. T.M. Abdel-Fattah and T.J. Pinnavaia, J. Chem. Soc. Chem. Commun., 665 (1996).
21. R.T. Yang, T.J. Pinnavaia, W. Li and W. Zhang, J. Catal., 172 (1997) 488.
22. A. Tuel and S. Gontier, Chem. Mater., 8 (1996) 114.
23. R. Mokaya and W. Jones, Chem. Commun., 981 (1996).
24. R. Mokaya and W. Jones, Chem. Commun., 983 (1996).
25. R. Mokaya and W. Jones, J. Catal., 172 (1997) 211.
26. K.R. Kloetstra, H. van Bekkum, J.C. Jansen, J. Chem. Soc. Chem. Commun., 2281 (1997).
27. A. Corma, Chem. Rev., 97 (1997) 2373.
28. Z. Luan, C.F. Cheng, H. He and Klinowski, J. Phys. Chem., 99 (1995) 10590.
29. A. Corma, V. Fornés, M.T. Navarro and Pérez-Pariente, J. Catal., 148 (1994) 569.
30. S. Hitz and R. Prins, J. Catal., 168 (1997) 194.

MESOPOROUS MOLECULAR SIEVES 1998
Studies in Surface Science and Catalysis, Vol. 117
L. Bonneviot, F. Béland, C. Danumah, S. Giasson and S. Kaliaguine (Editors)

Structure simulation of mesoporous molecular sieves

K. Kleestorfer, H. Vinek and A. Jentys*

Vienna University of Technology, Institute of Physical Chemistry
Getreidemarkt 9/156, A-1060 Wien, Austria.

The structural properties of mesoporous molecular sieves with MCM-41 like structure were studied by Molecular Dynamics simulations. The ratio between the wall thickness and the pore diameter was varied and the stabilization energy of the lattice was determined. With decreasing wall thickness an increase of the stabilization energy was observed. Structures with partially interrupted walls were found to be energetically highly unstable.

1. INTRODUCTION

Mesoporous molecular sieves like MCM-41 [1, 2] have become a matter of wide spread interest [3, 4], nevertheless, their structural properties on an atomistic level are still under discussion. The questions being debated include the shape and possible irregularities such as narrow sections or interconnections of the pores, the thickness of the walls and if the walls are amorphous or partially crystalline. While numerous experimental studies of the structural properties of MCM-41 exist, e.g. see reviews [5, 6], only a few publications are describing mesoporous molecular sieves on the basis of a structure simulation up to now [7 - 9]. This might result from the large number of atoms necessary to describe the MCM-41 structure without applying symmetrical constraints, i.e. ~3000 - 5000 atoms compared to 200 - 500 atoms usually needed for microporous molecular sieves and zeolites. Additionally, the question to which extent the walls are crystalline influences the selection of the best starting geometry for the simulation.

In this paper we describe the results of a Molecular Dynamics study on the structural properties of MCM-41 type materials. Geometries with different pore sizes were analysed and the question to which extent the thickness of the walls between the pores influences the structural stability of MCM-41 was investigated.

---

* Author, to whom correspondence should be addressed (e-mail: jentys@tuwien.ac.at)

2. THEORETICAL METHODS

2.1. Construction of the initial structures for simulation

The structure of MCM-41, a mesoporous molecular sieve with a hexagonal arrangement of one-dimensional cylindrical pores in the range between 25 and 100 Å [1], was generated from α-quartz by removing the atoms inside the pores of the chosen diameter. A hexagonal unit cell with a size of 49.1 × 49.1 × 27.0 Å was used to ensure that the pore size could be varied over the range between 15 and 50 Å, without the necessity to generate a new unit cell structure for each model studied. An infinite lattice of the MCM-41 like structure was generated using periodical boundary conditions on the unit cell applying only P1 symmetry. After removing the silicon and oxygen atoms inside the pores, at first the vacant bonds of the silicon atoms were saturated with oxygen atoms and subsequently, all vacant bonds of the oxygen atoms were saturated with hydrogen atoms. The procedure of generating the starting structure is depicted in Figure 1.

The pores were initially circular and in the structure with the largest pore diameter studied the walls were built up at their thinnest section from only one silicon atom. Additionally one structure was studied where the pore diameter was equal to the unit cell size, thus generating a discontinuity in the walls.

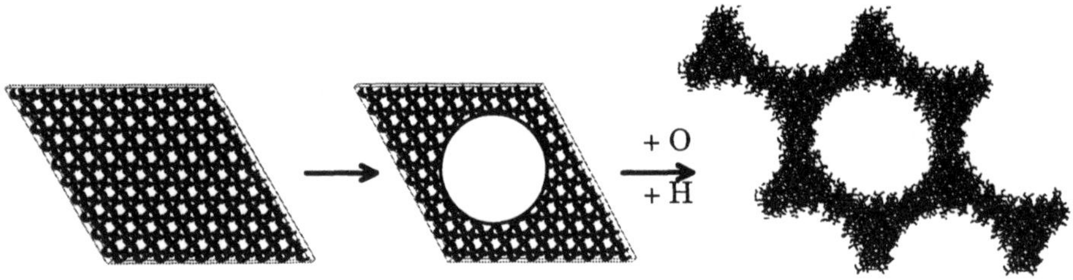

Figure 1. Construction of the initial structures

2.2. Simulation techniques

Due to the large number of atoms in the unit cell (~ 2500 - 5000 depending on the pore diameter) the cff91/czeol force field [10] with parameters derived by Sauer *et al.* was used for the simulations. The energies were calculated from Molecular Dynamics simulations using Discover 2.9.8 [10]. The simulations were carried out at a temperature of 700 K, using a time step of 1 fs and a total simulation time between 1000 and 10000 fs. Constant volume and constant temperature calculations were performed using an NVT ensemble.

2.3. Determination of the stabilization energy

The stabilization energy (E_{stab}) of the lattice was calculated from the following energy contributions (Eq. 1).

$$E_{tot} = n_{Si} \cdot E_{Si-O} + n_H \cdot E_{H-O} + E_{stab} \tag{1}$$

Each model structure consisted of a different number of atoms, therefore, the energy was normalized to the composition of the corresponding unit cell. The energy contribution of a silicon atom bond to an oxygen atom (E_{Si-O}) was determined from the initial quartz structure by dividing the lattice energy of SiO_2 (E^0_{tot}) by the number of silicon atoms (n_{Si}) (Eq. 2).

$$E_{Si-O} = \frac{E^0_{tot}}{n_{Si}} \tag{2}$$

The energy needed for the formation of a hydroxyl group (E_{H-O}), was determined from a homologous series of $H\text{-}[OSi(OH)_2]_n\text{-}OH$ molecules (with n = 1 - 3).

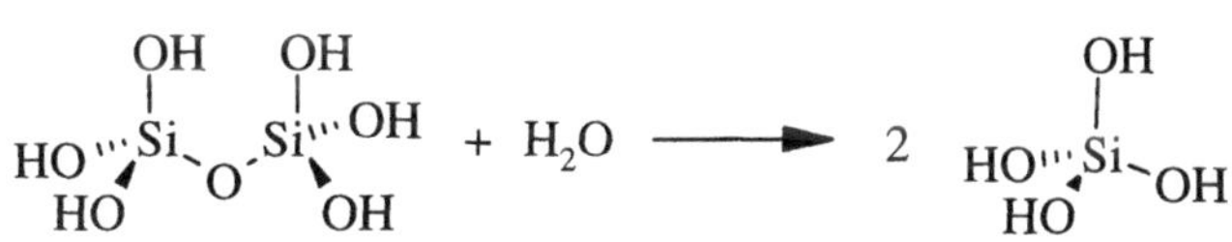

ΔE = 50.21 kcal·mol^{-1}

$$E_{H-O} = \frac{1}{2} \cdot \left(2 \cdot E_{Si(OH)_4} - E_{Si_2O_7H_6} - E_{H_2O}\right) \tag{3}$$

An energy of E_{H-O} = 25.11 kcal·mol^{-1} was determined for a hydroxyl group being formed (Eq. 3).

The stabilization energy of the mesoporous molecular sieves was calculated from these energy contributions using the energy balance given in Eq. 1.

$$E_{stab} = \frac{E_{tot} - n_{Si} \cdot E_{Si-O} - n_H \cdot E_{H-O}}{n_{Si}} \tag{4}$$

3. RESULTS AND DISCUSSION

The total energies, the concentration of hydroxyl groups and the stabilization energies of the molecular sieves as a function of their pore diameter are summarized in Table 1.

Table 1
Energies and concentrations of OH groups

Pore radius [Å]	OH loading [$mmol \cdot g^{-1}$]	Stabilization energy per silicon atom [$kcal \cdot mol^{-1}$]
0.0	0.0	——
15.0	1.5	56.5
27.3	2.7	-6.8
31.1	4.1	-5.0
34.7	4.9	-25.3
39.5	7.7	-22.2
41.9	9.0	-25.8
42.6	10.0	-29.4
49.1	10.1	16.7

The concentration of hydroxyl groups and the stabilization energies (Eq. 4), both as a function of the pore diameter are shown Figure 2 and 3, respectively.

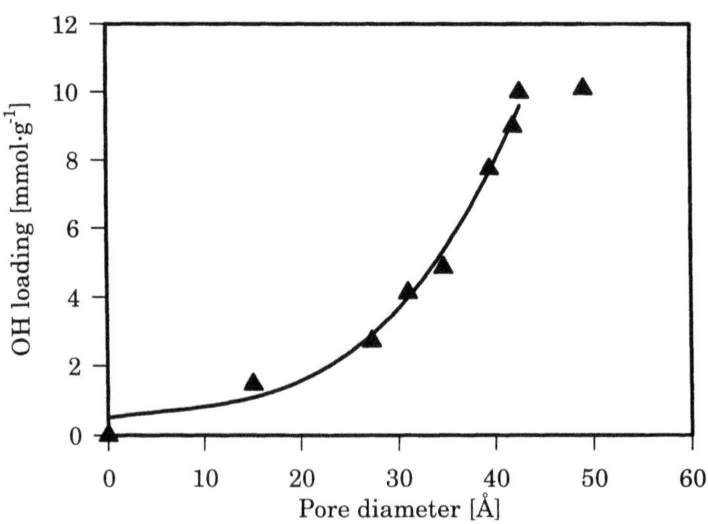

Figure 2. Concentration of hydroxyl groups as function of pore diameter

The concentration of OH groups of the structure with 49.1 Å pore diameter deviates from the trend observed for the other structures as the walls are already partially interrupted.

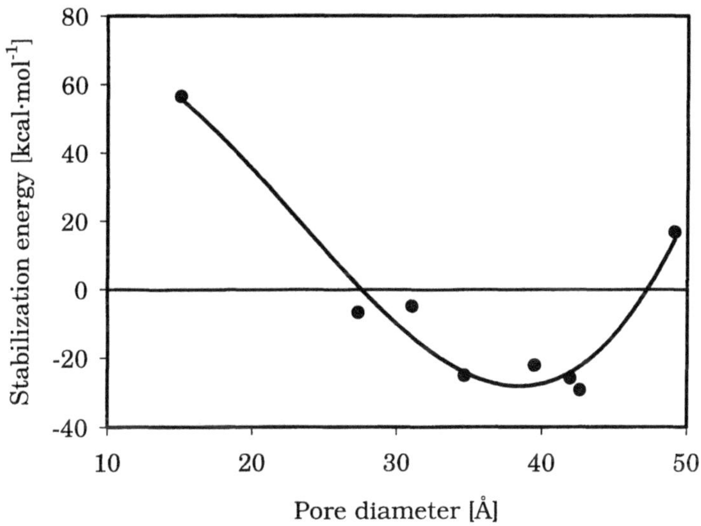

Figure 3. Stabilization energy as a function of pore diameter

The stabilization energy of the lattice showed a broad minimum at pore diameters between 35 and 42 Å, while structures with smaller pores and with pores that are not completely formed were found to be energetically unfavorable. This results also confirm that a dense phase is the thermodynamically most stable structure of SiO_2 and that isolated $Si(OH)_4$ species are energetically not stable.

The wall thickness of the most stable structures was in the range between 8 and 15 Å, a value which is in perfect agreement with the experimentally determined ones between 7 and 15 Å by XRD, N_2-BET and TEM [2, 11, 12]. However, it should be noted that all experimental methods mentioned only allow the indirect determination of the wall thickness from the difference of the pore diameter and the size of the unit cell. The concentration of hydroxyl groups increased with increasing pore diameter. This results from the use of a constant unit cell size and might be an artifact in this simulation, because the internal surface increased with increasing pore diameter, while in parallel, the wall thickness and consequently the total mass decreased. At this point we would like to mention that we determined the concentration of hydroxyl groups on the internal surface of MCM-41, prepared with C_{12}TMABr and C_{16}TMABr templates by pyridine adsorption followed by IR-spectroscopy [13], and it was in the range of 0.8 $mmol.g^{-1}$. Although there is not a perfect agreement between the experimentally determined concentrations and those resulting from the simulation, the numbers are in the same order of magnitude. Additionally it has to be considered that the experimentally studied materials had smaller pores than the structures reported here.

Two examples of the final structures resulting for the simulations are presented in Figures 4 and 5.

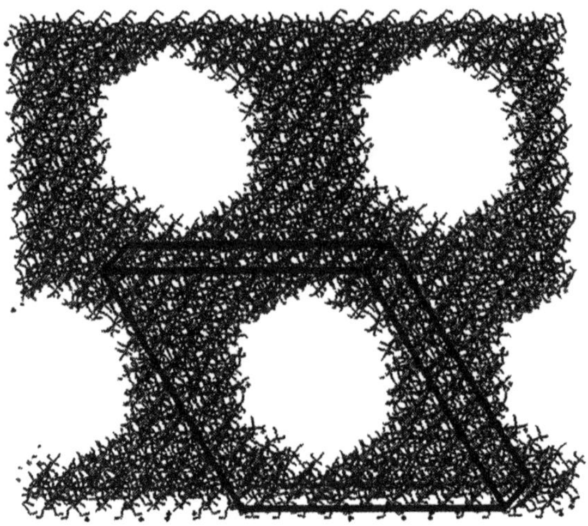

Figure 4. Structure with a pore diameter of 31 Å

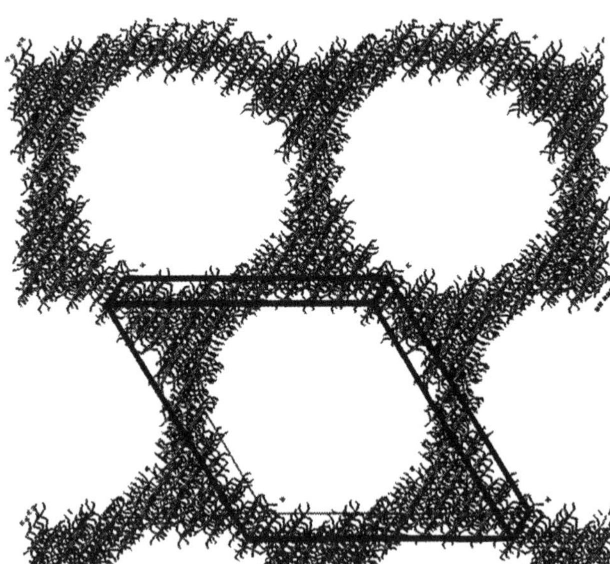

Figure 5. Structure with a pore diameter of 42 Å

For both structures the simulation indicated that the walls a preferably amorphous and the pores are only slightly hexagonal.

4. CONCLUSIONS

The results of the simulation indicate that for mesoporous molecular sieves of the MCM-41 structure type an optimum range for the pore size and wall thickness exist. We could show that with decreasing wall thickness the lattice energy decreases, i.e., the structure becomes more stable, until a pore size is reached where the walls become partially disconnected at their thinnest sections. The optimum wall thickness for MCM-41 is predicted to be in the range between 8 to 15 Å. On the basis of these simulations we would further like to speculate that the walls between the pores of MCM-41 are as thin as possible, but the formation of interconnection between the pores is energetically highly unfavorable.

ACKNOWLEDGMENT

The work was supported by the "Fonds zur Förderung der Wissenschaftlichen Forschung" under project P10874 CHE.

REFERENCES

1. C. T. Kresge, M.E. Leonowicz, W. J. Roth, J.C. Vartuli and J. S. Beck, Nature, 359 (1992) 710.
2. J.S. Beck, J.C. Vartuli, W.J. Roth, M.E. Leonowicz, C.T. Kresge, K.D. Schmitt, C.T.W. Chu, D.H. Olson, E.W. Sheppard, S.B. McCullen, J.B. Higgins and J.L. Schlenker, J. Am. Chem. Soc., 114 (1992) 10834.
3. F. Schüth, Ber. Bunsenges., 99 (1995) 1306.
4. A. Corma, Chem. Rev., 97 (1997) 2373.
5. A. Sayari, Stud. Surf. Sci. Catal., 102 (1996) 1.
6. P. Behrens, Angew. Chem. Int. Ed. Engl., 35 (1996) 515.
7. B. P. Feuston and J. B. Higgins, J. Phys. Chem., 98 (1994) 4459.
8. M. W. Maddox, J. P. Olivier and K. E. Gubbins, Langmuir 13, (1997), 1737.
9. R. D. Oldroyd, J. M. Thomas and G. Sankar, J. Chem. Soc., Chem. Commun., (1997) 2025.
10. Discover® 2.9.8, September 1996. San Diego: Molecular Simulations, 1996.
11. C. N. Wu, T. S. Tsai, C. N. Liao and K. J. Chao, Microp. Mater., 7 (1996) 173.
12. K. J. Edler, P. A. Reynolds, J. W. White and D. Cookson, J. Chem. Soc., Faraday Trans., 93 (1997) 199.
13. A. Jentys, K. Kleestorfer and H. Vinek, Microp. Mater, submitted (1998).

MESOPOROUS MOLECULAR SIEVES 1998
Studies in Surface Science and Catalysis, Vol. 117
L. Bonneviot, F. Béland, C. Danumah, S. Giasson and S. Kaliaguine (Editors)

In-situ XRD Study of the Initial Stages of Formation of MCM-41 in a Tubular reactor

M. Lindén[a,b], S. Schunk[a] and F. Schüth[a]

[a]Institut für Anorganische Chemie, Johann Wolfgang Goethe-Universität Frankfurt, Marie-Curie Straße 11, 60439 Frankfurt, Germany

[b]Department of Physical Chemistry, Abo Akademi University, Porthansgatan 3-5, 20500 Abo, Finland

1. INTRODUCTION

Since the discovery of MCM-41 and related materials in the early nineties [1,2] many publications on the synthesis, formation mechanism, properties, and possible applications of such materials have appeared. However, only little attention has been paid to the very early stages of the formation of these materials, due to the fact that this regime is very difficult to analyze. All syntheses reported so far have been performed in batch systems, where solutions containing the silica source, the surfactants and other ingredients are mixed and then reacted to completion.

A different approach, i.e. the use of a tubular reactor, seemed an attractive alternative for three reasons. Firstly, if commercial bulk applications for ordered mesoporous oxides can be developed, it would be highly attractive run a high throughput, continuous synthesis process. For this purpose, a tubular reactor system with continuous operation seems to be ideal. Secondly, the quality of MCM-41 type materials synthesized in different labs and even in the same lab still differs substantially with respect to perfection of the structure, sorption capacity, and hydrothermal stability. One of the factors which is difficult to control, but might influence the quality of the resulting product, is the mixing of the reagents. A tubular reactor allows highly controlled mixing, the intensity of which can be varied by changing mixing elements. Thirdly, and most important in the context of our present investigation, any position in a tubular reactor corresponds to a certain reaction time in a batch system. Ideally, there is thus no change in the state of a reaction at a constant position along the tubular reactor. This enables us to study fast processes occuring during the synthesis of MCM-41 with slow techniques, if an analytical cell is placed at a certain position in the tubular reactor. In principle, it is possible to use any analytical technique suitable for the study of the system under investigation, i.e. X-ray diffraction, Raman spectroscopy, IR-spectroscopy, or UV/Vis-spectroscopy. Even NMR spectroscopic *in-situ* measurements seem to be feasible.

In this contribution we report results concerning the early stages of the formation of siliceous MCM-41 as well as the formation of zirconia- and titania-based mesostructured materials using X-ray diffraction on a conventional diffractometer with a time resolution of less than a second.

2. EXPERIMENTAL

2.1 Systems investigated

Two different systems were investigated in the tubular reactor, the synthesis of siliceous MCM-41 [3] and the synthesis of mesostructured titania. Zirconia, following the work of Ciesla et al. [4], was studied in some experiments as well. However, due to the high absorption of zirconium and thus the bad signal to noise ratio from the X-ray experiment in solution, the data were of poor quality. The behavior of this system seemed to be similar to the titania.

The composition of the silica system is close to the one reported by Grün et al.[5]. However, in order to stabilize the emulsion, small amounts of 1-butanol were added to the system. This does not dramatically influence the nature of the products formed, as a comparison of materials synthesized with and without butanol showed. The overall composition (by weight) of the reaction mixture was 400 H_2O : 40 NH_3 (25 wt %) : 8 CTAB : 2 1-butanol : 31 TEOS. One of the stock solutions was TEOS, the other solution contained all other components dissolved in water. All experiments were carried out at a temperature of 303 ± 2 K, since this reaction system allows fast mesophase formation already at room temperature. The pH immediately after addition of the TEOS was 10.3.

For the preparation of mesostructured titania, titanium-oxosulfate ($TiOSO_4 \cdot xH_2SO_4 \cdot xH_2O$, Aldrich) was used as the titanium source. Two solutions were prepared, one containing 2.0 g of CTAB dissolved in 100 g of water and one containing 2 g of titanium-oxosulfate and 2 g of HCl (37 %) dissolved in 100 g of water. After combining these solutions, a white solid immediately precipitates. The reaction was performed at room temperature in the tubular reactor and analyzed at very short residence times of down to 300 ms, as well as in flasks, either at room temperature or at 90°C, for comparison. Most reactions in flasks were performed at about fourfold higher concentrations in order to obtain more product. This did not affect the mesophase formation, as checked in a series of experiments. High concentrations are not suitable for the investigations in the tubular reactor, since this leads to strong absorption losses.

2.2 Reactor and XRD-cell

The tubular reactor simply consists of a teflon or polyethylene tube. In most experiments a tube with an inner diameter of 3 mm was used. Other diameters are possible if a different range of residence times is desired. The maximum length of the tube was 50 m, but most of the experiments were performed with substantially shorter tubes. The reaction solutions were fed from beakers containing the stock solutions into the tube by two peristaltic pumps (Ismatec) via an ultrasonic bath and static mixing elements (Sulzer) to efficiently mix the TEOS with the aqueous phase for the preparation of siliceous materials. The residence time of the solution in the mixing unit (ultrasonic bath and mixing elements) was 5 to 10 s, depending on the total flow rates. In order to minimize backmixing, air bubbles were introduced periodically after the mixer, resulting in approximately 20 cm long trains of reaction solution separated by 3 cm long bubbles of air. This technique had been developed in K. Unger's group for the continuous preparation of amorphous silica spheres [6].

In the case of the titania, where the structure develops very fast, the mixing time of 5 to 10 s would be too long to detect the early stages of structure formation. In addition, since here the solutions are miscible, the ultrasonic step is not necessary. Therefore, the different precursor solutions were directly mixed without special mixing elements in a T-shaped piece of tube.

Mixing is certainly not perfect in this case, but this arrangement allows analysis at about 300 ms after contacting the stock solutions with each other.

The heart of the setup is the X-ray diffraction cell (fig. 1). This cell is mounted on the goniometer of a STOE Stadi P transmission X-ray diffractometer, and can be fitted into the reactor tube at any point. The reaction solution passes through a thin slit (about 2.5 mm) between two polyetheretherketone films (Goodfellow) which are sealed with teflon against the cell body. The slit dimensions are a compromise between good flow conditions and little peak broadening, which is due to part of the sample being out of focus. The slit with the reaction solution is placed at the focus of the diffractometer by two high precision adjustment screws. The reaction solution leaves the X-ray cell at the top and is then collected to analyze the final sample quality.

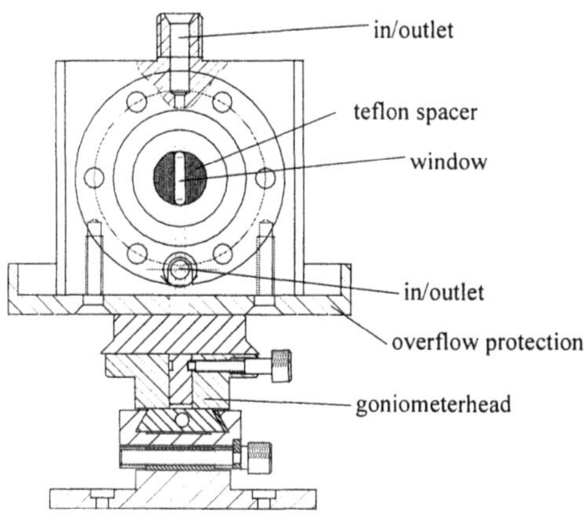

Figure 1: X-ray diffraction cell for *in-situ* experiments

In some cases accumulation of solid was observed on the walls of the tube and on the windows of the XRD-cell. In order to avoid any misinterpretation, the cell was only operated for 4 to 10 min at a time. After this the feed was switched to water and a background diffraction pattern was accumulated which was subtracted from the sample diffractogram. This overestimates the influence of the background, since at the beginning of each run no precipitate is deposited on the windows. The fouling was most pronounced for the silica at short residence times below 2 min, where the TEOS is only partly hydrolyzed, but never was higher than about 15 % of the total signal. In the titania case, fouling was no problem at all. Prior to the next run, reactor and the cell were cleaned by flushing with 0.2 M NaOH followed by large amounts of distilled water.

3. RESULTS AND DISCUSSION

3.1 Silica

As already described by Grün et al. [5], the MCM-41 synthesis with TEOS and ammonia is very fast already at room temperature. How fast the assembly of the mesophase is, can be seen with the *in-situ* X-ray experiment. Fig. 2 shows the X-ray patterns obtained during synthesis runs as described in the experimental. In this set of experiments, the cell was placed at different positions along the tubular reactor after the mixer, corresponding to different residence times and thus different reaction times.

No intensity is recorded, if only the surfactant containing solution without TEOS is passed through the cell (bottom trace). However, if TEOS is added, low angle intensity develops rapidly. A peak around 2° (2 theta) starts to grow after times as short as 15 s. At a reaction time of 90 s, a well developed peak at 2° is clearly visible, and after 180 s the MCM-41 structure is fully developed with even higher order peaks being discernible. The diffraction patterns are certainly not as well resolved as such recorded on dry samples. One has to bear in mind, though, that the system is highly diluted and the patterns are those of a dispersion containing less than 10 wt% of solid.

The material present after 3 min in solution has somewhat different properties than a material which is isolated after this reaction time, filtered, and dried (fig. 3). The d-spacing is shifted to higher angles in the dried material, corresponding to a shrinking of the unit cell (from 4.4 nm to 4 nm). This is due to the condensation process in the siliceous walls, which occurs on a much longer time scale than the self assembly in this system, and/or removal of water from the silicate surfactant system which leads to a deswelling. Calcination at 500°C results in another decrease of the d-spacing (fig. 3) to 3.7 nm, a fact which has been repeatedly described in the literature. The properties of the materials after calcination with respect to sorption behavior and X-ray pattern are essentially identical for samples isolated after 3 min from the tubular reactor and samples synthesized in a batch experiment. The pore filling step in the nitrogen isotherm lies between p/p_0 = 0.2 - 0.3, which means that the pores have a diameter of approximately 2 - 3 nm. The total volume adsorbed is about 450 cm^3(STP)/g, regardless, whether the samples were prepared in a batch reaction or in the tubular reactor. This suggests, that the basic features of the structure are already fully developed after very short reaction times.

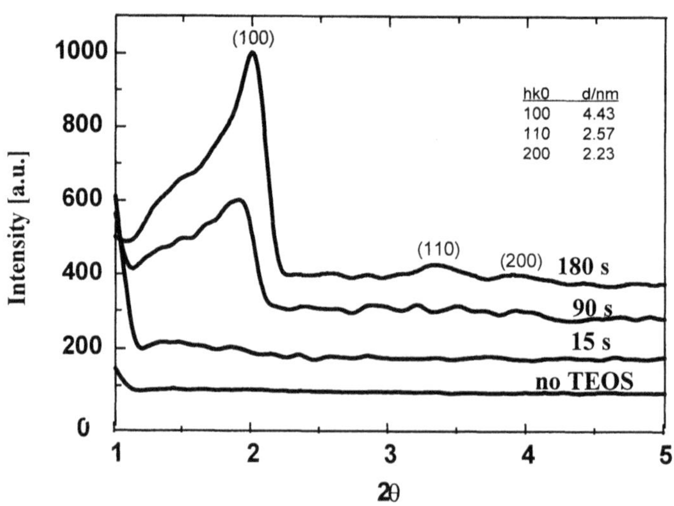

Figure 2: *In-situ* X-ray diffraction patterns recorded for different residence time in the tubular reactor for the TEOS/CTAB system

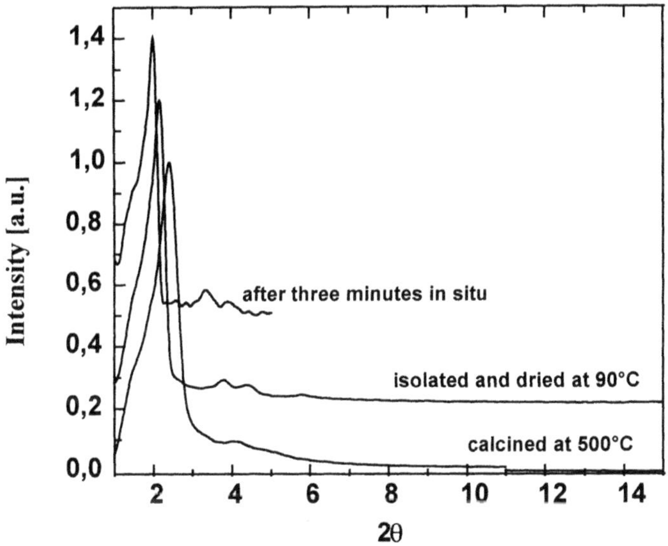

Figure 3: Comparison of XRD patterns recorded *in-situ*, after drying, and after calcination

In our experiments we did not observe a lamellar intermediate, as described before for a synthesis system with Cab-O-Sil and sodiumsilicate as the silica source at a higher pH [7]. This is most probably due to the fact that the charges on the silica species are relatively low under the conditions investigated in the study presented here. The species, which interact with the surfactant molecules resulting from hydrolysis of TEOS need not be fully hydrolized, which reduces their charge compared to fully hydrolized silicate monomers or oligomers. In addition, at lower pH the silica species are not highly charged. Such conditions would directly lead to a composite with a higher curvature than the lamellar phase, i.e. the hexagonal phase as observed here. Investigations of a system resembling the one which was studied in [7] are in progress.

3.2 Titania

Titania based ordered mesoporous materials have first been prepared by Antonelli and Ying [8] using an anionic surfactant and a titanium alkoxide. In our group we developed a synthesis procedure for a mesoporous zirconia which uses a cationic surfactant as in the conventional MCM-41 synthesis [4]. We first tried to study the assembly process in this system with the tubular reactor setup. However, since zirconia is a very strong absorber, only little intensity reached the detector after passing through approximately 2.5 mm of a relatively highly concentrated zirconia solution. The diffraction patterns thus had a very bad signal to noise ratio. It could be seen even from these bad data, that the assembly process is extremely rapid in this system, because low angle intensity was recorded already after the shortest residence time which can be realized in our setup.

To overcome the absorption problem, we chose to investigate the related titania system as well, in which also a mesostructured material, albeit less ordered after calcination, can be prepared [9]. Although also the titanium is a stronger absorber than silicon, reliable data could be collected. Fig. 4 shows the results of an experiment with a residence time of nominally 300 ms. For such short times the precision in determining the residence time is not very high and can only be calculated from the volumes involved and the flow. However, it is certain that the reaction time was less than a second.

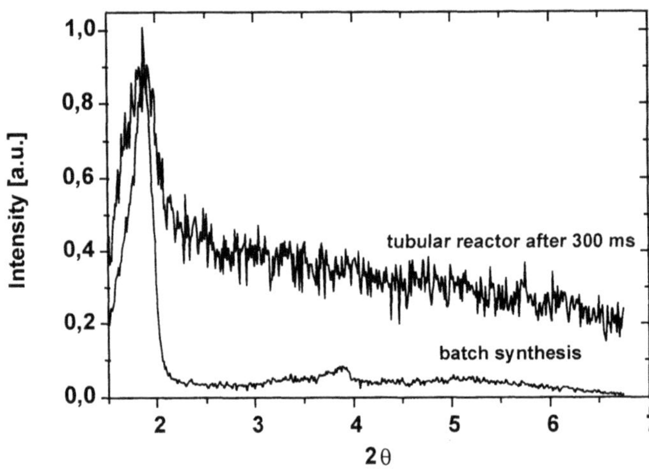

Figure 4: XRD pattern of mesostructured titania recorded *in-situ* in the tubular reactor, and wet material recovered from a batch synthesis

One can clearly see that even after a reaction time of less than a second the mesostructure has formed. The low angle reflection exactly coincides with the (100) reflection of a material which was recovered from a batch synthesis 2 min after mixing the reagents. Higher order reflections can not be discriminated in the tubular reactor setup, because the noise is too large. These data suggest that the driving force for the formation of a composite organic/inorganic liquid crystal is indeed very high and the relatively slow kinetics often observed for silica based systems might be attributed to the need for the formation of the right oligomers. This can take some time in silicate solutions, depending on the starting materials: Colloidal or fumed silica needs to depolymerize, silicate solutions need to oligomerize after changes of pH, in alkoxysilanes hydrolysis and subsequent oligomerization are necessary. These processes, rather than the assembly to the composite liquid crystal might be the rate determining step in the formation of the mesostructure.

However, if the mesostructure has assembled, it is certainly not complete, since the inorganic framework is not fully condensed. Condensation in the silica after assembly can be inferred from ^{29}Si NMR spectroscopy which shows that Q^2 and Q^3 silicon atoms are converted to fully condensed Q^4 units. In addition, the reduction of the unit cell size discussed above is a strong indication that additional condensation occurs in the silica framework.

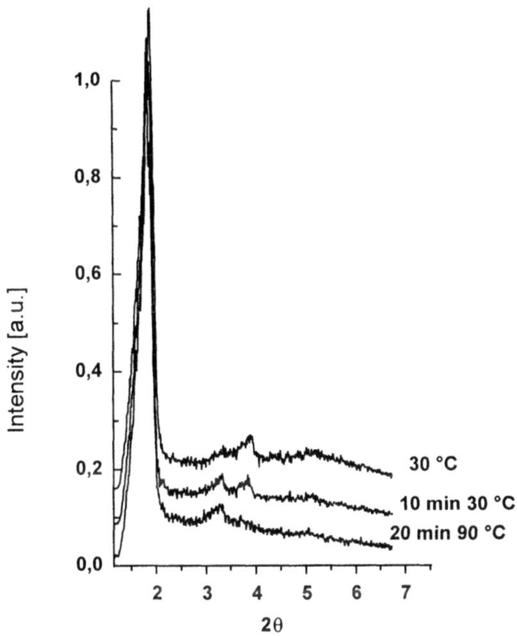

Figure 5: XRD patterns for titanias at different stages of the condensation of the inorganic framework

Also for the titania materials there is additional evidence that the inorganic framework undergoes still substantial changes after formation of the mesophase. In fig. 4 the (200) peak in the pattern of the material from the batch synthesis has a substantially higher intensity than the (110) peak. This, however, changes when the samples are heated to 90 °C for prolonged periods of time in the synthetic solution (figure 5). The intensity of the (110) peak increases on expense of the (200) peak. A similar behavior was also observed for the zirconias, the only difference being the slower change of the diffraction patterns in case of the zirconia. This change in intensities is most probably not due to the fact that a layered phase is transformed to a hex phase, since then the d-spacings should change. The reason for this intensity change could rather lie in the fact that upon condensation of the inorganic framework the density along the (110) direction increases while it remains essentially unchanged in the (100) direction. This would lead to a redistribution of intensity between the differently indexed reflections in the X-ray pattern as observed experimentally. An X-ray model for this process was recently developed in our group, which will be discussed in more detail in a forthcoming publication[10].

4. CONCLUSION

Using the newly developed tubular reactor concept which allows the analysis of fast reactions, such as the formation of a solid from solution, with slow analytical techniques, we could demonstrate, that MCM-41 type materials can form very rapidly under certain conditions. Siliceous MCM-41 prepared from TEOS starts to develop in less than a minute, the structure formation is complete after about 3 min. From then on only condensation processes in the silica framework occur. The rate limiting step in this reaction might in fact not be the assembly but rather the hydrolysis of the alkoxide.

For the mesostructured titania material synthesized with titanium oxosulfate as the precursor, the formation of the mesostructure is even more rapid and faster than the shortest time detectable with our present instrumentation. A diffraction peak at 2° (2 theta) is clearly discernible even after less than a second. Also in this case, though, further condensation takes place in the inorganic framework subsequently.

The results of this study clearly demonstrate the power of the method used. Future experiments will be directed to decreasing the minimum reaction time which can be analyzed to 30 ms and to the investigation of systems where phase changes occur during formation of the mesophase.

5. ACKNOWLEDGEMENT

This research was funded by the EC-TMR program under grant number ERB-FMRX-CT96-0084.

6. REFERENCES

1. C.T. Kresge, M.E. Leonowicz, W.J. Roth, J.C. Vartuli and J.S. Beck, Nature, 359 (1992) 710.
2. T. Yanagisawa, T. Shimizu, K. Kuroda and C. Kato, Bull. Chem. Soc. Jpn., 63 (1990) 988.
3. M. Lindén, S. Schunk and F. Schüth, Angew.Chem., in print
4. U. Ciesla, S. Schacht, G.D. Stucky, K.K. Unger and F. Schüth, Angew. Chem., 108 (1996) 597; Angew. Chem. Int. Ed. Engl., 35 (1996) 541.
5. M. Grün, I. Lauer and K.K. Unger, Adv. Mater., 9 (1997) 254.
6. C. Kaiser, Diploma Thesis, Mainz 1992
7. A. Monnier, F. Schüth, Q. Huo, D. Kumar, D. Margolese, R.S. Maxwell, G.D. Stucky, M. Krishnamurty, P. Petroff, A. Firouzi, M. Janicke and B.F. Chmelka, Science, 261 (1993) 1299.
8. D.M. Antonelli and J.Y. Ying, Angew. Chem., 107 (1995) 2202; Angew. Chem. Int. Ed. Engl., 34 (1995) 2014.
9. M. Thieme and F. Schüth, Microporous Mater., submitted
10. M. Lindèn, S. Schacht, S. Schunk and F. Schüth, in preparation

MESOPOROUS MOLECULAR SIEVES 1998
Studies in Surface Science and Catalysis, Vol. 117
L. Bonneviot, F. Béland, C. Danumah, S. Giasson and S. Kaliaguine (Editors)

Internal Modification of Ordered Mesoporous Hosts

K. Moller and T. Bein

Department of Chemistry, Purdue University, West Lafayette, IN 47907, USA

This review provides an overview on different aspects of inclusion chemistry in ordered mesoporous host materials such as MCM-41 or MCM-48, with a focus on grafting of silane coupling agents, grafting of reactive metal complexes, and polymerization in the channels. Co-condensation of reactive species during the mesopore synthesis is discussed as a method to incorporate functionality into the walls of the channel system. Applications of these modified and functionalized systems are in heterogenous catalysis and photocatalysis involving bulky catalysts and/or large substrates. Other applications include removal of heavy metals, stabilization of dyes, and polymer composites.

1. INTRODUCTION

The discovery of surfactant-templated mesoporous materials with variable, controlled pore diameters and topologies including both hexagonal and cubic systems has opened the way to do intriguing experiments *inside* the resulting channel structures. In utilizing the chemical reactivity of the mesoporous host, a number of functional groups have been covalently anchored to the channel walls, including ligands intended for the attachment of metal complexes. The reactivity of intra-channel sorbates even includes polymerization, thus leading to confined filaments of common polymers as well as conducting materials such as carbon.

A high level of synthetic control and optimization is required for the most demanding approach of grafting a functional molecule into the mesoporous host: Co-condensation of framework precursors with metal sources that are covalently bonded to a functional group. Several of these strategies have been demonstrated in recent reports. Intriguing applications of these systems include catalysis with large substrate molecules, formation of novel nanocomposites, and separations.

An overview on inclusion chemistry in mesoporous hosts will be provided on the covalent grafting of functional groups, co-condensation reactions, and polymerization in mesoporous systems.

2. COVALENT GRAFTING OF LIGANDS AND FUNCTIONAL GROUPS

Surface reactions exploiting terminal silanol groups have long been performed with amorphous supports or crystalline zeolites for applications in catalysis, separations or sensing. The mesoporous MCM materials add a new dimension due to their large and variable pore diameters. The internal surface composition is comparable to that of amorphous supports such as silica. The introduction of functional organic groups is usually performed through attachment of silane-coupling agents to the mesoporous walls of the previously synthesized and calcined MCM materials. The functional group is either directly incorporated in the silane-coupling agent or it is grafted onto it in a second or further reaction steps.

The versatility of the MCM materials was recognized in two patents shortly after their invention. Thus, silanol groups were functionalized with main group and transition metal halides, hydrides, alkoxides or acetates for applications in sorption and catalysis,[1] or for sensing through the addition of detecting components to the pore walls.[2]

2.1. Silane coupling agents

Detailed reports describing surface modifications, including the anchoring of functional groups and metal complexes to mesoporous MCM supports, have been published in the recent past. A comprehensive study of the surface chemistry of siliceous MCM-41[3] reveals three different types of surface silanols, e.g., single, hydrogen bonded and geminal hydroxyls. Only the single and geminal hydroxyl groups were shown to be reactive towards chlorotrimethylsilane. Surface silylation with trimethylsilyl groups was shown to enhance the stability of MCM-48 to moisture and compression.[4]

Direct introduction of functionality through reaction of silanol groups in mesoporous MCM-41 with 36 Å or 55 Å pore size has been achieved with tris(methoxy)mercapto-propylsilane.[5] Anchoring of the thiol groups in the internal channels was evident from reduction of the pore size from 36 to 27 Å. In a related report, the concentration of functional groups was varied and reached up to 76% of the full surface coverage (Figure 1).[6] The SH-moieties were subsequently used in both studies for removing mercury from waste streams. The highly efficient materials were also stable after regeneration by washing with concentrated HCl.

Brunel and coworkers first reported the anchoring of catalytic functionalities into MCM-supports. They used 3-amino- and 3-chloropropyl-triethoxysilane as an intermediate for the grafting of primary and tertiary aminofunctions to MCM-41.[7] The dehydrated mesoporous support was refluxed in toluene containing the respective aminopropyl- or chloropropyl-silane coupling agent.

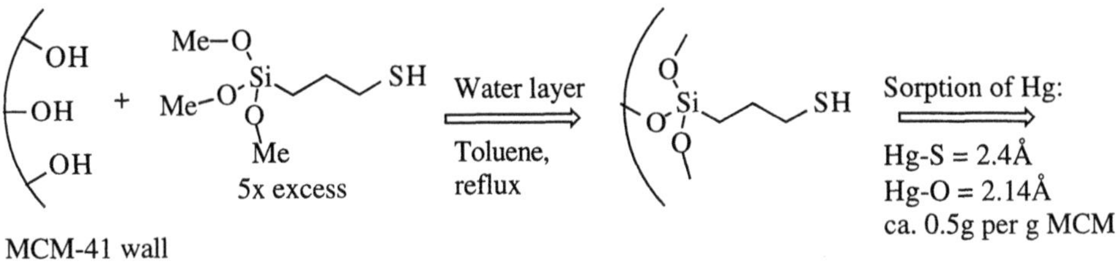

Figure 1. Attachment of thiol functions into MCM-41.[6]

Subsequent exposure of the latter sample to a solution of piperidine resulted in the grafted piperidinopropyl functionalities of the support. The authors showed that these new heterogeneous catalysts allow the highly regioselective formation of alpha monoglycerides in the epoxide ring-opening reaction of glycidol with fatty acids. In related work[8] they grafted MCM-41 with various functional alkoxysilanes, e. g., $(RO)_3Si(CH_2)_3X$ with X = Cl, $NH(CH_2)_2NH_2$ and NHC(O)N-Salpr (salpr = 3-[N,N'-bis(3,5-di-tert-butylsalicylidenamino)-propyl]-amine). The functional groups of the first two attached organic moieties were further converted into 2-($NHCH_2$)Pyr and 4-($NH(CH_2)_2NHCO$)Pyr.

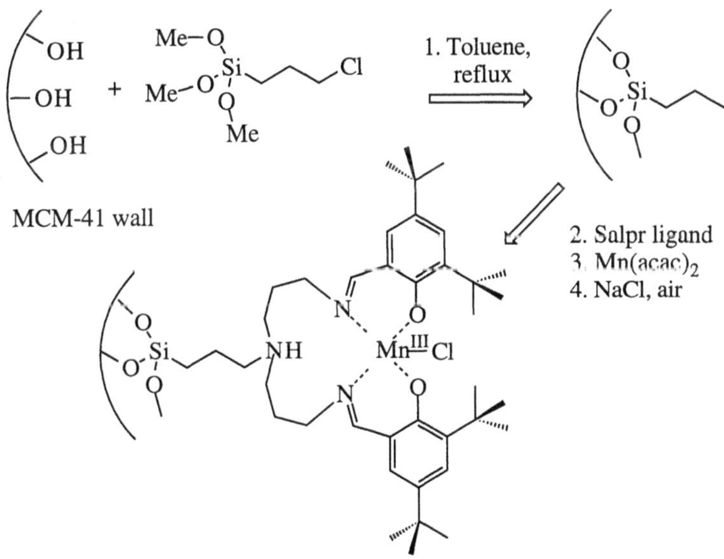

Figure 2. Synthesis of ClMn(salpr) complexes grafted to MCM-41.[9]

A similar route was used by the same authors to prepare MCM-bonded manganese(III) Schiff-base complexes (Figure 2).[9] Pentadentate ligands, 3-[N,N'-bis(3,5-di-tert-butylsalicylidenamino)propyl]amine (salpr), were anchored to the surface of siliceous MCM-41 through the silane coupling agent 3-chloropropyl-triethoxysilane, and transformed into Mn(III)(salpr) complexes by ligand exchange without loss of the regular porous structure of the support.

The reaction of the amine functions of triazacyclononane (TACN) with the epoxy groups of MCM-41-grafted (3-glycidyloxypropyl)-trimethoxysilane was used to anchor 2-hydroxyalkyl-substituted TACN ligands to the host (Figure 3).[10] Manganese complexes of the grafted ligands were shown to be active in the epoxidation of styrene and cyclohexene with hydrogen peroxide.

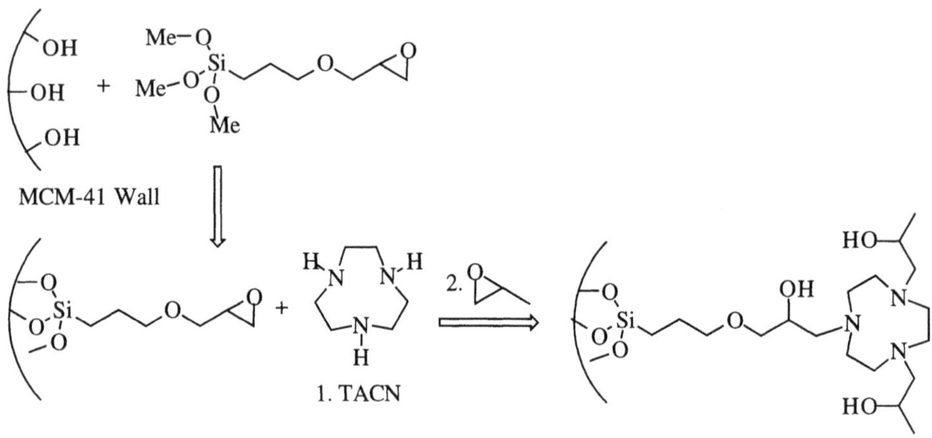

Figure 3. Heterogenization of 2-hydroxyalkyl-substituted TACN ligands in the mesoporous host.[10]

In another approach, immobilization of other cobalt complexes[11] was performed via methoxysilyl-propylamines to yield surface bound chelate ligands including ethylenediamine (ED), diethylenetriamine (DET) and ethylenediaminetriacetic acid salt (EDT). They were subsequently reacted with Co(II), and the resulting materials showed some reactivity towards oxygen adduct formation with grafted ED and DET. Similarly, 3-aminopropyl-triethoxysilane attached to the MCM wall was further modified with [Ru(II)*meso*-tetrakis(4-chlorophenyl)porphyrin]. Attachment of the porphyrin complex is attributed to exchange of an ethanol ligand on the precursor porphyrin complex with the amino group on the coupling agent. The complex showed good catalytic activity for alkene oxidations with t-butyl hydroxyperoxide.[12]

Ferrocenyl groups were attached by O'Hare to the walls of MCM-41 *via* ring opening of (1,1′-ferrocenediyl)dimethylsilane. The samples were characterized by EXAFS and solid state NMR.[13]

2.2. Grafting of metal complexes.

Surface attachment of transition metal oxides and metal or bimetal complexes without the use of intermediate silane coupling agents was achieved by direct reaction of surface hydroxyl groups with reactive species. Titanocene dichloride was anchored to MCM-41 by Maschmeyer et al.[14] The complex was converted to titanium-oxo species through calcination in air. EXAFS spectroscopy revealed the presence of surface bound, four-coordinated titanium species that showed a higher catalytic activity for the epoxidation of alkenes than other supports containing Ti in their framework. Maschmeyer et al.[15] also reported the inclusion of the cobalt complex $[Co_3(\mu_3\text{-}O)(OAc)_5(\mu_2\text{-}OH)(py)_3]PF_6$ that was studied in the oxidation of cyclohexane to cyclohexanone and cyclohexanol. In this case, the hydroxyl group of the complex was expected to react with the surface silanols. The catalyst lifetime was less than 48 hours when the complex was introduced via imbibition. The lifetime was improved when the MCM surface was previously functionalized with Me_2SiCl_2 and $Cl_3Si(CH_2)_3Br$ which was further reacted with glycine such as to immobilize the catalyst.

The reactivity of chloride ligands towards hydroxyl groups of the MCM-41 host was exploited by several researchers[16] for the anchoring of the stereospecific catalyst ethylenebis(indenyl) zirconium dichloride, which showed high activity for the production of isotactic polypropene when activated with methyl-alumoxane. It was emphasized that the cylindrical pores of the host suppress the formation of inactive complexes between catalyst molecules, thus increasing reactivity. EXAFS studies showed retention of the Zr-indenyl framework and removal of chloride ligands on anchoring. In order to synthesize another prospective stereoselective catalyst, Anwander et al.[17] used lanthanide silylamide precursors that were grafted via protolytic exchange reactions to the support. In previous work, we introduced the bimetallic complex $Me_3SnMo(CO)_3Cp$ into the MCM-41 pores; this complex was anchored *via* the oxophilic trimethyl tin component to the mesoporous walls (Figure 4).[18] EXAFS spectroscopy revealed the formation of small bimetallic clusters upon heating in vacuum.

Metal chlorides and acetates were also reacted with calcined siliceous MCM hosts for a different purpose[19], i.e., to establish a general route for the implantation of metal atoms into the channel walls. Calcination subsequent to surface anchoring of $AlCl_3$, $SnCl_2$ $Zn(O_2CMe)_2$ or $Mn(O_2CMe)_2$ led to mesoporous materials with ion-exchange capacity, high stability and catalytic activity.

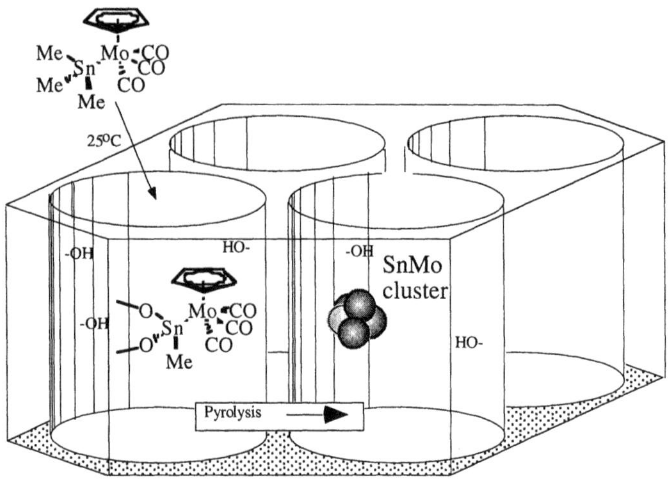

Figure 4. Grafting of the bimetallic complex $Me_3SnMo(CO)_3Cp$ into the pores of MCM-41.[18]

3. HYBRID MATERIALS OBTAINED BY IN SITU CO-CONDENSATION

As described in the previous sections, MCM hosts have been subjected to a variety of chemical modifications to create reactive surfaces. The grafting of active species is usually performed after MCM synthesis and template removal. However, since the preparation of MCM-materials is a template-assisted sol-gel process, it was recognized that the co-condensation of siloxane and organosiloxane precursors during MCM-formation can result in truly new hybrid materials. Under certain conditions, the condensation of the MCM walls proceeds already at room temperature, thus allowing even the co-condensation of organic moieties that would not withstand the rigorous conditions of the 'classical' hydrothermal synthesis. Thus, hybrid materials can be prepared that have, in contrast to the above-mentioned grafted MCM hosts, functional groups directly incorporated into the walls of the host. The connectivity between functional group and host can be anticipated as one with [Si]-C bonds reaching out of the walls instead of [Si]-O-Si-C when using grafting methods (where [Si] designates silicon atoms in the wall). This different connectivity is expected to result in greater hydrolytic stability of the hybrid materials.

The co-condensation route was pioneered by Mann et al.[20] who documented the possibility of reacting the silica precursor TEOS with phenyl- and octyl-triethoxysilane in the presence of hexadecyltrimethylammonium bromide template to result in hybrid mesoporous MCM-type materials. In a later study, Mann et al. extended their research to a number of functional groups including thiol, amine, epoxide, imidazole and alkyl-organosiloxanes.[21] Incorporation of all these groups except octyl-triethoxysilane gave high quality MCM-materials. In the original synthesis of MCM-type materials, removal of the surfactant template was typically performed through calcination in N_2/air at 550 °C. However, the fragility of the organic moieties in the hybrid structures requires extraction of the template at moderate temperatures. This process was frequently found to disrupt the structure of some of the hybrid materials, especially at high concentrations of the functional groups.[21] However, Mann's phenyl, thiol and amino-functionalized MCM materials were stable upon extraction in HCl/EtOH.

Functionalized MCM hosts that are apparently stable to ethanol extraction of the template were prepared by Macquarrie et al..[22] They used neutral surfactant templates together with aminopropyl- and cyanoethyl-trialkoxysilanes. An example of an incorporated functional group showing chemical reactivity was reported by Stein et al..[23] They included vinyl functions into MCM-41 (by co-condensation of TEOS with vinyltriethoxysilane) which were stable to an acid extraction process (Figure 5). The authors report the complete consumption of the vinyl functional groups, as followed by FTIR and NMR spectroscopy, after up to six days in the presence of bromine. In a later paper Stein[24] reported also on the co-condensation of TMOS and mercaptopropyl groups, which showed a slightly higher uptake of mercury than the thio-grafted MCM described by Feng et al..[6] Stein also oxidized the thiol groups to sulfonic acid groups, thus creating a solid acid with micropores of about 14 Å. This new hybrid was active in the conversion of 3,4-dihydro-2H-pyran to 2-ethoxytetrahydropyran.

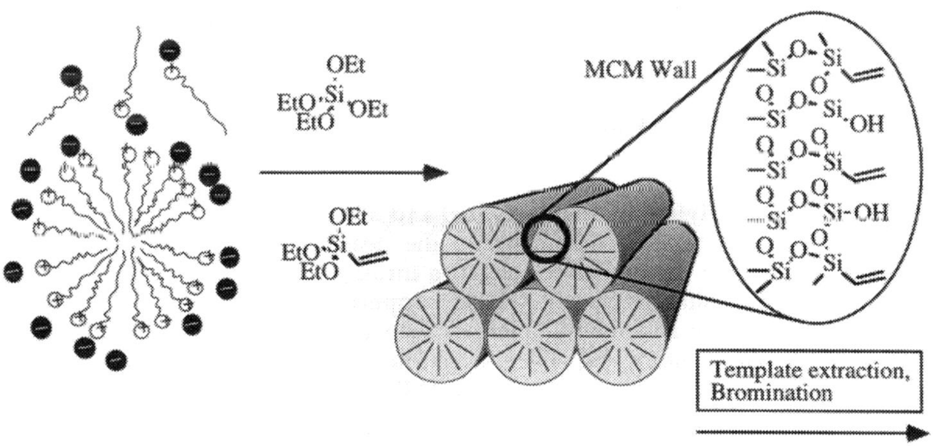

Figure 5. Inclusion of vinyl functionalities into MCM hosts by co-condensation of TEOS with vinyltriethoxysilane.[23]

A functionalized mesoporous hybrid material based on the MCM-41 structure was synthesized by us[25] through room temperature co-condensation of tetramethylorthosilicate (TMOS) and 3(trimethoxysilyl)propyl methacrylate

(TMSiPMA) in the presence of cetyltrimethylammonium chloride. A gradual decrease of the d(100) spacing from 40 to 36 Å and loss of outer diffraction peaks was observed upon increasing the degree of methacrylate inclusion, ranging from 0 : 1 up to 0.33 : 1 molar ratios of TMSiPMA : TMOS. Successful template removal through acid extraction resulted in highly crystalline mesoporous materials. Nitrogen sorption shows a reduced pore volume in the functionalized MCM materials as compared to the parent MCM host. The methacrylate groups are easily monitored with IR spectroscopy. It was demonstrated that the C=C double bonds are readily accessible for bromine addition; total consumption was observed within two hours.

Finally, a unique type of co-condensation involving the surfactant itself was presented by Ying et al..[26] The reaction between $Nb(OEt)_5$ and tetradecylamine gave a niobium complex that on treatment with water resulted in a hexagonal mesostructure similar to MCM-41. Mesoporous niobium oxide was obtained after removal of the surfactant.

4. POLYMERIZATION IN MESOPOROUS CHANNELS

The large pores in MCM materials offer not only new perspectives in areas such as catalysis but are of great interest also for the preparation of new composite materials. It has been found that the mechanical strength of polymers can be greatly enhanced by the addition of inorganic fillers. A review on this topic was given by Frisch et al..[27] While amorphous silica and to a smaller extent zeolites have been used in polymer composites, only few articles describe the utilization of MCM materials for the preparation of nanocomposites. The main purpose of adding meso- or microporous instead of compact glassy fillers to a polymer matrix is to achieve a maximum in interconnectivity between both phases by channeling polymers through the pores and crosslinking them around the filler particles.

4.1. Inclusion of poly(methylmethacrylate) and other polymers

Several recent reports demonstrate the feasibility of forming polymer networks within the internal pore structure through adsorption and subsequent polymerization of monomers in porous supports. Llewellyn et al.[28] adsorbed styrene, vinyl acetate and methyl methacrylate into MCM hosts with different pore sizes in order to study the effect of confinement on polymer growth. They observed an increase in chain length of PMMA with decreasing pore size of MCM-41. This effect was associated with inhibited termination processes in the channels. We have studied the polymerization of methyl methacrylate (MMA) in the microporous channels of zeolites NaY/X and ZSM-5 as well as in the mesoporous hosts MCM-41 and MCM-48.[29] To avoid the formation of polymer layers on the external surface, adsorption was carefully performed on previously dehydrated and evacuated hosts *via* gas phase adsorption. A nearly complete polymerization *via* radical initiation resulted in nanocomposites in which the polymer content was shown to correlate with the internal pore volume of the different supports. The encapsulation of the polymer results in strong host-guest interactions that suppress the glass-transitions normally found in bulk polymers. This is true for the microporous as well as the mesoporous supports.

The adsorption of methyl methacrylate in MCM-41 was also performed by Tatsumi et al.[30] They observed that PMMA grown in the mesoporous channels of MCM-41 exceeds the bulk molecular weight by one order of magnitude and found that the molecular weight could be manipulated by varying the monomer-to-initiator ratio.

4.2. Conducting polymers and carbon

Polyaniline filaments in MCM-41. Conducting polymers such as polyaniline or polyacetylene have attracted much attention due to the unique opportunities involved in creating lightweight conductors and semiconductors, as well as luminescent materials. As efforts continue to decrease the size of electronic circuitry in order to achieve higher storage density and faster data processing speeds, ideas have been developed to build electronic circuits from *molecular* components. In contrast to the lithographic "top down" methods that are reaching diffraction limits in optical pattern transfer, this strategy could be considered a "bottom up" approach. Due to the presence of directional bonds in organic conducting polymers, these materials have been prime candidates for the design of molecular electronics components. The inclusion chemistry of conducting materials in inorganic hosts has been reviewed.[31]

In an effort to stabilize polymer conductors at nanometer scales, we have developed a synthetic protocol for the formation of conducting filaments of polyaniline in the 3 nanometer wide channels of the aluminosilicate MCM-41 (Figure 6).[32] In this approach, aniline vapor was diffused into the dehydrated channels of the host at room temperature, followed by immersion into a solution of oxidant. Host-encapsulated polyaniline filaments in the conducting emeraldine salt form were produced in this reaction. Chromatography of polymer recovered from the host (with dilute HF) showed polymer chain lengths of several hundred aniline rings. Nitrogen sorption was used to detect a significant drop in pore volume, confirming the presence of encapsulated polyaniline.

Contactless microwave absorption at 2.6 GHz was used to determine the conductivity of the encapsulated filaments to be about a quarter of bulk polyaniline. This result shows that conjugated polymers can be encapsulated in nanometer channels and still support mobile charge carriers.

In an additional study, the stabilization of conducting polyaniline filaments in the 3 nm channels of Cu(II) or Fe(III)-containing aluminosilicate MCM-41 was achieved.[33] In this case, adsorption of aniline vapor into the dehydrated, metal-containing host produced radical cations or short oligomers that can "lock in" the aniline species in the host channels. Exposure of these systems to oxidant under exclusion of air led to the formation of conducting emeraldine salt in the MCM channels.

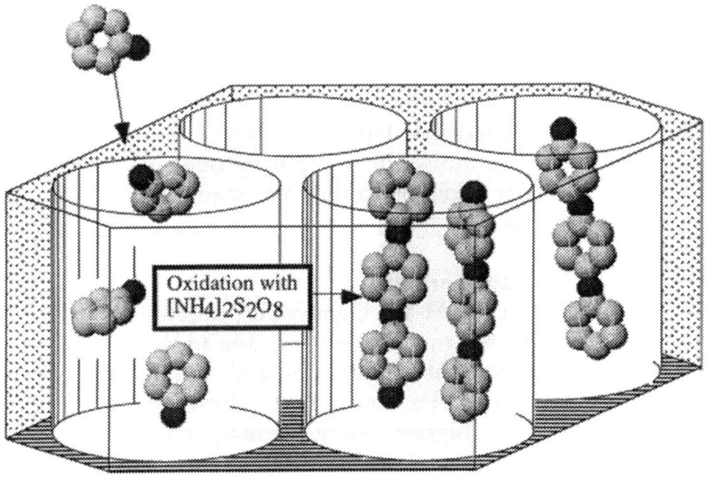

Figure 6. Adsorption and polymerization of aniline in MCM-41.[32]

Conducting carbon wires encapsulated in the MCM-41 host. As a candidate for the design of conductive nanostructures, graphitic carbon is attractive because it is chemically stable and exhibits high conductivity. We have studied the formation of carbon "wires" in the regular channels of MCM-41.[34] The precursor molecule acrylonitrile was introduced into the host through vapor or solution transfer, and polymerized in the channels with radical initiators. The recovered polyacrylonitrile (PAN) shows a chain length at the order of about 1000 monomers, enough to extend through significant parts of the MCM-41 particles.

As expected, pyrolysis of the included polyacrylonitrile led to loss of nitrogen, and the formation of carbonized material in the channels of the host. Raman data were used to estimate the average domain size in the pyrolized PAN-MCM samples, and it was found that the carbon phase in the MCM-41 host is consistently more ordered than in bulk samples treated similarly. This shows that the MCM host plays an important role in structuring the growing graphitic carbon sheets, possibly through the parallel alignment of the precursor polymer chains. Microwave measurements showed that the conductivity of dry pyrolized PAN-MCM increased with pyrolysis temperature. For treatment temperatures of 800 and 1000 °C, the conductivity was about ten times greater than that of bulk material treated similarly. This surprising result was explained with the greater order of the graphitic material in the MCM channels.

In a recent report, the preparation of carbon nanotubes from CVD reactions of hydrocarbon gas in mesoporous supports was described.[35] The phonon properties of the nanotubes were characterized with Raman scattering.

5. SUMMARY

In this brief review it has become obvious that the newly available ordered mesoporous solids offer numerous opportunities for creative chemistry in the interior of nanometer-size pores. The archetype host in this context is the aluminosilicate or siliceous hexagonal MCM-41 material, but recently additional systems such as the cubic MCM-48 and transition-metal containing ordered mesoporous materials have also been explored as hosts in this context. One of the major applications envisioned in this field is, not surprisingly, heterogeneous catalysis, where the large pores (compared to zeolites) enable the heterogenization and stabilization of a wide variety of catalytic centers. Both the catalyst and the reactant molecules entering the channels can be quite large. Many other applications are being envisioned, for example removal of heavy metals, stabilization of dyes, and polymer composites. The inclusion chemistry needed to achieve these goals ranges from various grafting techniques using the intrachannel hydroxyls, co-condensation with the walls during synthesis, and intrachannel polymerization.

ACKNOWLEDGEMENT

The authors gratefully acknowledge generous funding from the U.S. Department of Energy and from the German Ministry of Education and Research (BMBF).

REFERENCES

---

1 J. S. Beck, D. C. Calabro, S. B. McCullen, B. P. Pelrine, K. D. Schmitt, and J. C. Vartuli, US 5,145,816, Sept. 8, 1992.

2 D. H. Olson, G. D. Stucky, and J. C. Vartuli, US 5,364,797, Nov. 15, 1994.

3 X. S. Zhao, G. Q. Lu, A. K. Whittaker, G. J. Millar, and H. Y. Zhu, J. Phys. Chem. B, 101 (1997), 6525.

4 K. A. Koyano, T. Tatsumi, Y. Tanaka, and S. Nakata, J. Phys. Chem. B, 101 (1997), 9436.

5 L. Mercier, and T. J. Pinnavaia, Adv. Mater., 9 (1997), 500.

6 X. Feng, G. E. Fryxell, L.-Q. Wang, A. Y. Kim, J. Liu, and K. M. Kemner, Science, 276 (1997), 923.

7 A. Cauvel, G. Renard, and D. Brunel, J. Org. Chem., 62 (1997), 749.

8 D. Brunel, A. Cauvel, F. Fajula, and F. DiRenzo, Stud. Surf. Sci. Catal., 97 (1995), 173.

9 P. Sutra, and D. Brunel, Chem. Commun. (Cambridge), (1996), 2485.

10 Y. V. Subba Rao, D. E. DeVos, T. Bein and P. A. Jacobs, Chem. Commun. (1997), 355.

11 J. F. Diaz, K. J. Balkus, Jr., F. Bedioui, V. Kurshev, and L. Kevan, Chem. Mater., 9 (1997), 61.

12 C.-J. Liu, S.-G. Li, W.-Q. Pang, and C.-M. Che, Chem. Commun. (1997), 65.

13 S. O'Brien, J. Tudor, S. Barlow, M. J. Drewitt, S. J. Heyes, and D. O'Hare, Chem. Commun. (Cambridge), (1997), 641.

14 T. Maschmeyer, F. Rey, G. Sankar, and J. M. Thomas, Nature, 378 (1995), 159.

15 T. Maschmeyer, R. D. Oldroyd, G. Sankar, J. M. Thomas, I. J. Shannon, J. A. Klepetko, A. F. Masters, J. K. Beattie, and C. R. A. Catlow, Angew. Chem., Int. Ed. Engl., 36 (1997), 1639.

16 (a) Y. S. Ko, T. K. Han, J. W. Park, and S. I. Woo, Macromol. Rapid Commun., 17 (1996), 749. (b) J. Tudor and D. O'Hare, Chem. Commun. (Cambridge), (1997), 603. (c) S. O'Brien, J. Tudor, T. Maschmeyer, and D. O'Hare, Chem. Commun., (1997), 1905.

17 R. Anwander, and R. Roesky, J. Chem. Soc., Dalton Trans., (1997), 137.

18 C. Huber, K. Moller, and T. Bein, J. Chem. Soc., Chem. Commun., (1994), 2619.

19 R. Ryoo, S. Jun, J. M. Kim, and M. J. Kim, J. Chem. Soc., Chem. Commun., (1997), 2225.

20 S. L. Burkett, S. D. Sims, and S. Mann, Chem. Commun., (1996), 1367.

21 C. E. Fowler, S. L. Burkett, and S. Mann, Chem. Commun., (1997), 1769.

22 D. J. Macquarrie, Chem. Commun., (1996), 1961.

23 M. H. Lim, C. F. Blanford, and A. Stein, J. Am. Chem. Soc., 119 (1997), 4090.

24 M. H. Lim, C. F. Blanford, and A. Stein, Chem. Mater. , 10 (1998), 467.

25 K. Moller, T. Bein, and R. X. Fischer, to be submitted.

26 D. M. Antonelli and J. Y. Ying, Angew. Chem., Int. Ed. Engl., 35 (1996), 426.

27 H. L. Frisch, and J. E. Mark, Chem. Mater., 8 (1996), 1735.

28 P. L. Llewellyn, U. Ciesla, H. Decher, R. Stadler, F. Schueth, and K. K. Unger, Stud. Surf. Sci. Catal., 84 (1994), 2013.

29 K. Moller, T. Bein, and R. X. Fischer, Chem. Mater., in press.

30 S. M. Ng, S. Ogino, T. Aida, K. A. Koyano, and T. Tatsumi, Macromol. Rapid Commun. 18, (1997), 991.

31 T. Bein, Stud. Surf. Sci. Catal., 102 (1996), 295.

32 C.-G. Wu and T. Bein, Science, 264 (1994) 1757.

33 C.-G. Wu and T. Bein, Chem. Mater., 6 (1994) 1109.

34 C.-G. Wu and T. Bein, Science, 266 (1994) 1013.

35 S. Xie, W. Li, C. Wang, L. Xu, H. Zhang, Y. Zhang, and L. Qian, Sci. China, Ser. A: Math. Phys. Astron., 40 (1997), 971 (CA 1997: 643 028).

MESOPOROUS MOLECULAR SIEVES 1998
Studies in Surface Science and Catalysis, Vol. 117
L. Bonneviot, F. Béland, C. Danumah, S. Giasson and S. Kaliaguine (Editors)

Mesoporous materials derived from layered silicates and the adsorption properties

S. Inagaki, S. Ogata, Y. Goto, Y. Fukushima
Toyota Central R&D Labs., Co., Nagakute, Aichi, 480-1192, Japan

The form and structure of silica source as a starting material affected the structure of the produced mesoporous material. Mesoporous silicas derived from a layer-formed silica possessing a structure similar to a phyllosilicate had characters of small particle size, thin pore-wall thickness and high regularity of pore arrangement. Easy dissolving character and many nucleation sites for crystal growth would attributed to the characters. Adsorption isotherms of water vapor revealed the change of hydrophobicity of mesoporous material by contact with water vapor. Mesoporous material adsorbed the largest amount of surfactant which was used for the synthesis as a template. High density mesoporous silica showed high separation ability of both polar and non-polar molecules in each normal- and reversed-phases high-performance liquid chromatography. Mesoporous silica including porphyrin molecules in the channels showed adsorption property of oxygen and the evalution of oxygen was responded to the UV-light.

1. INTRODUCTION

Porous solids are important materials due to their wide applications in various separation, purification and catalytic processes. Although we have many kinds of amorphous porous materials which offer micropore, mesopore and macropore, we had no crystalline material

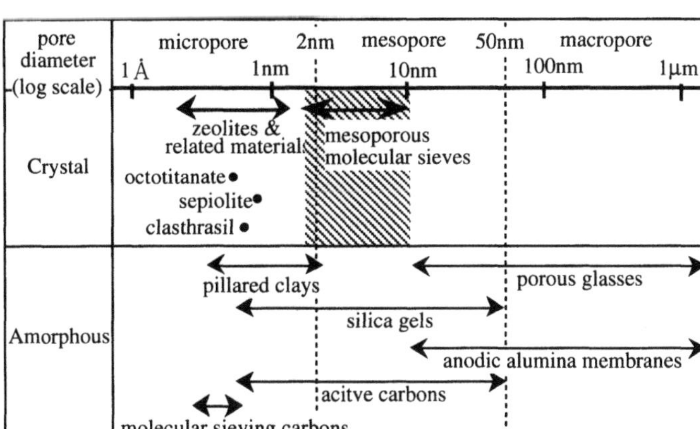

Figure 1 Classification of porous materials.

possessing mesopore until 1992 (Figure 1). Zeolites and zeolite-like materials such as aluminophosphate materials have only micropores with diameters of about 0.3-1.4 nm. The mesoporous molecular sieves such as MCM-41 and FSM-16 are first materials possessing well-defined pores whose sizes are larger than those in zeolites. The mesoporous molecular sieves would bring about a progress in an understanding of the unique properties of nanospaces.

Zeolites are synthesized typically by hydrothermal treatments of silica- and alumna-gels in high pH condition at 100-200°C. The organic reagents such as a tetrapropylammonium $[(C_3H_7)_4N^+]$, which are called as templates or structure directing reagents, play an important role to crystallize zeolites. The synthetic method using an organic reagent have given us many types of zeolite and the related microporous crystals. Recent trend is synthesizing zeolites with large pore size. However, new synthetic procedure was necessary to be developed for a synthesis of porous crystal with pore diameter over 1.4 nm .

The mesoporous molecular sieves were synthesized by using a new template concept, that is molecular assembly templating. The use of surfactant aggregations as the templates is one of the significant difference to zeolite syntheses, in which separated organic molecules were used as templates.

The first report to synthesize the mesoporous materials by using surfactant as template was that of Yanagisawa et al[1]. in 1990. The mesoporous silicas with pore diameter from 1.8 to 3.2 nm were prepared by heating a layered polysilicate kanemite in a aqueous solution of alkyltrimethylammonium with different alkyl-chain lengths. In 1992, researcher in Mobil Corporation reported a family of mesoporous materials (M41S)[2] with hexagonal MCM-41 , cubic (MCM-48) and lamellar (MCM-50) by liquid-crystal templating method. And, we reported ordered mesoporous material (FSM-16)[3] from kanemite by modifying the Yanagisawa's method in 1993.

A "folding sheet" mechanism was proposed[4] for the formation of FSM-16 with a three-dimensional silicate framework from the layered (two-dimensional) silicate. The noble topochemical formation mechanism was studied by dividing the formation process into two elemental stages, i) folding of silicate layers and ii) interlayer cross-linking[5]. The dividing of formation process also enable us to optimize the synthetic conditions of highly ordered and thermally stable FSM-16. An aluminosilicate FSM-16 with framework aluminum was synthesized from a layered silicate containing aluminum[6]. The mesoporous aluminosilicate showed a large amount of acidity and 70% of incorporated aluminum served as acid sites. Pore-wall structure of FSM-16, which was important for understanding properties as catalysts and adsorbents, was examined by X-ray diffraction (XRD), transmission electron micrograph (TEM), physisorption, and XRD pattern simulations[7]. A pore-wall thickness and pore diameter of FSM-16, which could not be determined by crystallographic data because of the amorphous-like wall structure, were estimated by combining the results of various methods and a simulation. Adsorption isotherms of nitrogen, benzene and water vapor on FSM-16 showed unusual sudden increases in adsorption due to capillary condensations in the uniform mesopores[8]. The relationship between the adsorption hysteresis and pore size of FSM-16 was also discussed.

Here, we report some topics of our recent researches on syntheses of mesoporous molecular sieves from clay minerals and the some adsorption properties.

2. SYNTHESES OF MESOPOROUS SILICAS FROM CLAY MINERALS

In the folding sheet mechanism, an ion exchange with cationic surfactant such as hexadecyltrimethylammonium for Na^+ cation in the interlayer of a polysilicate kanemite ($NaHSi_2O_5 \cdot 3H_2O$) folded the sheet and formed an organic/silica complex with hexagonal symmetry of several nanometer periods[5)]. The structure was kept even after removal of the organic surfactant by a calcination to produce FSM-16. The FSM-16 has an advantage of higher heat and steam resistance compared with MCM-41[9)]. Silicon alkoxides such as tetraethoxysilane or tetramethoxysilane, sodium silicate aqueous solution or amorphous silica were used as a silica source for the synthesis of MCM-41. The silica monomers or origomers were polymerized with organic template to form a amorphous silica walls. On the other hands, local sheet structure of the crystalline kanemite was preserved in the silica walls of FSM-16. The high heat resistivity of FSM-16 would be attributable to the regular atomic arrangement in the silica wall.

A shape and size of an individual particles of the mesoporous materials should be an important characters for using them as an adsorbents or catalyses. Some ways for forming the mesoporous silica with film[10)], balloon[11)], or fiber[12)], and other shapes have been reported. Polymerization reactions of silica are again an important process for the shape control. A fine powder of A-type zeolite suitable for additives of synthetic detergents could be made from silica skeletons made from clay minerals, from which acid aqueous solutions leaches other metal cations. Both the atomic structure and the macroscopic shape of the source silica should be keys for making useful porous materials.

Although the mesoporous silica materials have been attracted great deal of attention and countless studies on the synthesis and properties of the mesoporous materials were reported, the atomic structure of silica walls and influence of the silica source on them have been paid only a little attentions. We made mesoporous silicas from various clay minerals and studied the structure and the morphology of them, and discussed the effects of the silica source on the wall structure and shapes of the mesoporous silica.

Clay minerals, sepiolite, hectorite, vermiculite, kaolinite and chrysotile were leached by 3-5 N HCl or H_2SO_4 aqueous solution at 90-100 °C for 5 -19 h to remove aluminum, magnesium or iron in octahedral-sites of the phyllosilicates and we got the leached silica powders. TEM observation showed that the leaching treatment did not break the outward forms of the original clay minerals. Relatively sharp Q^3 and Q^4 NMR peaks for the leached sepiolite, chrysotile and vermiculite, suggested a little distortion of local arrangement around the silicon atoms in the silica layers during the leaching process.

The leached silica powders were dispersed in 50-300 cm^3 of 1N NaOH aqueous solutions and mixed with hexadecyltrimethylammonium chloride aqueous solutions. After heating the mixture at 70 °C for a few hours the pH was adjusted to 8.5 by adding HCl aqueous solution. The mesoporous silica powders were obtained by the calcination at 650 °C for 4 h in air. An aerosil powder, an amorphous silica-gel fine powder, was also used as a silica source for a comparison.

Two set of peaks were overlapped in the XRD patterns for the mesoporous silica. One is a set of broad peaks whose d-value of the main peak is about 5 nm. The other is a set of peaks indexed hexagonal regularity whose d(100) value is about 3.5 nm. The d-values are listed in Table 1. The intensity of the broads peak decreased and that of the sharp hexagonal sharp peaks increased with increasing the amount of the NaOH aqueous solution additive. A small plate-like particles whose mean diameter was about 0.5 μm were observed by TEM. The lattice image of hexagonal arrangement of mesopore was observed in TEM photographs. An part of

Table 1 Results of the characterizations of the mesoporous silicas.

Silica source	Hexagonal peak $(2/\sqrt{3})$xd(100) nm	Broad peak $(2/\sqrt{3})$xd nm	BET surface area $m^2 \cdot g^{-1}$	Pore diameter nm
Aerosil	4.68	5.66	760	2.4
Leached Sepiolite	4.25	-	1280	2.7
Leached Hectorite	4.25	5.36	970	2.7
Leached Vermiculite	4.25	5.99	860	-
Leached Kaolinite	4.25	5.99	870	2.7
Leached Chrysotile	4.08	6.36	860	2.7

the irregular pore arrangement were also observed around the each plates. A large area of the irregular parts was observed for the sample with higher intensity of broad peak in XRD patterns.

BET surface area and pore diameter are also listed in Table 1. Although the d-values of XRD peaks for the mesoporous silica from clay minerals were smaller than that from aerosil, the values of the pore diameter and the surface area larger than that from aerosil. These results suggested the thinner silica wall in the mesoporous materials from clay minerals. Increasing of NaOH addition before the silica/surfactant complexes caused significant changes in XRD patterns, but a little change in the BET surface area.

The mesoporous silica from clay minerals was plate-like fine powder with highly ordered hexagonal symmetry, which were surrounded with fringes of irregular mesoporous arrangement. The result of N_2 adsorption isotherms showed that the diameter of the pores in higher ordered and the irregular ordered region was resemble to each other. However, distances between the pores were smaller in regular parts than that in irregular parts, that means thinner silica wall thickness in the highly ordered part. The irregular pore arrangement with thicker wall was resemble to that in the mesoporous materials, MSU, synthesized with neutral surfactants[13)]. The broad peaks in the XRD obtained in this work and the TEM images in the fringe parts were resemble to those for the MSU.

The result of NMR for aqueous suspension at high pH suggested that solid fractions of silica with ordered atomic structure resemble to phyllosilicate and dissolved silica monomers and oligomers are existed in the solutions during the forming of the silica/surfactant complexes with nanometer scale regularity. The mesostructure forming reaction with surfactant was caused simultaneously with the dissolving the silica, which was promoted by OH^- anions. The silica monomers or oligomers with small molecular weight were dissolved from the leached clay minerals, and the layered arrangement of the silica surface also plays as a substrates for the aggregation of the mesostructured silica/surfactant complex. Such a substrate effect of a two dimensional regular arrangements on the surface of clay minerals have been already reported in mesoporous film formation[10)].

The easy dissolving character and many nucleation sites for crystal growth would attributed to the characters of the mesoporous silica made from clay minerals, that are the small particle size, thin wall thickness and high regularity of pore arrangement.

3. ADSORPTION PROPERTIES

Since the mesoporous materials are good models for mesoporous adsorbents because of their narrow pore size distribution, some investigations have focused on adsorption-desorption behavior of gases and vapors on those mesoporous materials. There have been many reports on adsorption isotherms of nitrogen[14-18], argon[14,16], oxygen[14] and organic vapors[19] on MCM-41 and FSM-16. A behavior of hysteresis loop on the isotherms was discussed in relation to pore size[15,16] and temperature[19]. Determination of the precise pore size of MCM-41 and FSM-16 from the isotherms have been also attempted[7,20,22]. Here, we report some our recent researches on adsorption proprieties of the mesoporous materials directed toward future applications.

3.1 Hydrophobicity[23]

Siliceous MCM-41 and FSM-16 materials have been reported to show a hydrophobic property[24]. Although the surface properties affect the catalysis and adsorption behavior directly, we have had few report on a study of the evaluation of hydrophobicity of the mesoporous materials. We make a quantitative estimation of hydrophobicity of FSM-16 from adsorption isotherms of water vapor by using the Kelvin equation.

The adsorption isotherms of water vapor repeatedly measured for the same siliceous FSM-16 sample with pore diameter of 2.7 nm are shown in Figure 2. The first adsorption isotherm of water vapor showed the type V isotherm, which was similar isotherm observed for the siliceous MCM-41. An adsorbed amount of water was small in lower P/Po region under 0.54, whose P/Po region was that of multilayer adsorption of water on the surface. It suggests a weak interaction between the surface and water, that is some hydrophobic feature of FSM-16. The adsorbed density of water on the surface was 3.4 molecules nm^{-2} at P/Po=0.54 and was in good agreement with the density of silanol groups[25] of FSM-16. The steep increase in adsorption at P/Po above 0.54 is due to a capillary condensation of water in the mesopores. The second and third adsorption isotherms, which were measured for the same sample after evacuations at 25 °C, show different features from the first isotherm. The P/Po position, which the capillary condensation occurred shifted to 0.42 and 0.40 for the second and third isotherms. The adsorbed densities at multilayer adsorption region increased to 6.4 and 6.8 molecules•nm^{-2} at P/Po=0.42 and 0.40 respectively.

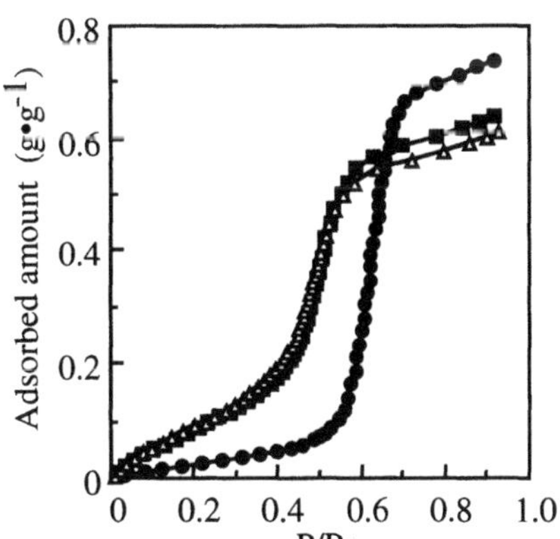

Figure 2 Adsorption isotherms of water vapor for FSM-16 (2.7 nm). First (●), second (■) and third (▲) isotherms.

These changes of isotherm indicate that the hydrophobicity of FSM-16 decreases during the measurements of the isotherms. The decrease in the hydrophobicity would be attributable to surface hydration of FSM-16 with water vapor during the measurements of the isotherms. The surface hydration was also confirmed by infrared spectroscopy of FSM-16, that is increase in silanol density after contact with water vapor and evacuation.

We can estimate the hydrophobicity from the isotherms by using the Kelvin equation[26] (1) as follows:

$$\ln(P/Po) = -2\gamma V_L \cos\theta / (r_k R T) \quad (1)$$

where P/Po is the relative vapor pressure at capillary condensation, γ and V_L are surface tension and molar volume of liquid water, θ is contact angle between surface and water, r_k is the Kelvin radius, R is gas constant and T is absolute temperature. The contact angle θ is a parameter to reflect the hydrophobicity of the surface and might be one of hydrophobicity index. In order to calculating the precise contact angles by the Kelvin equation, we must determine the thickness of multilayer adsorption of water t_m because the capillary condensation occur in the inner core of multilayer adsorption of water on the pore wall[26]. The Kelvin radius r_k was determined by r_p-t_m, where r_p represents the pore radius of the mesoporous material. The t_m were calculated from the adsorbed density of water at beginning of capillary condensation as shown above. The calculated r_k were 1.25, 1.15 and 1.16 nm for the first, second and third isotherms, respectively. The bulk values of γ and V_L of water at 25 °C were used for the calculation. The calculated contact angle decreased from 47 to 27 and 19 degrees in the ordered of first, second and third isotherms. These values also confirm the change in the surface hydrophobicity of the FSM-16 materials.

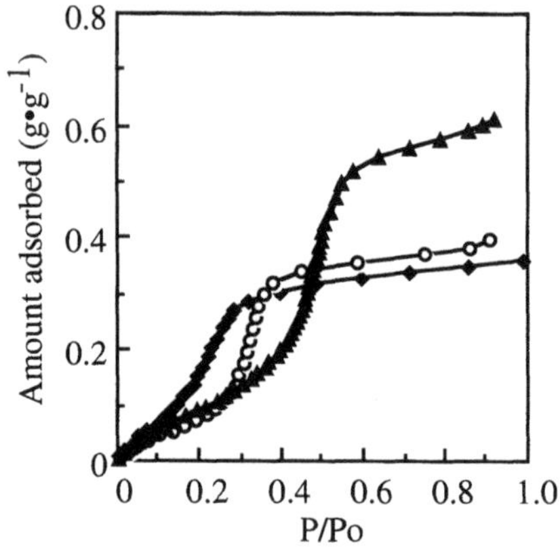

Figure 3 Adsorption isotherms of water vapor for the fully hydrated FSM-16 with pore diameters of 2.7 (▲), 1.7 (○) and 1.4 nm (◆).

The P/Po position at capillary condensation shifted to lower P/Po with decrease of pore size of FSM-16 as shown in Figure 3. As the isotherms were measured for the fully hydrated FSM-16 samples, the same contact angles of 19 degree could be employed. The P/Po positions at capillary condensation for the FSM-16 with pore diameters of 1.7 and 1.4 nm were calculated by using the contact angle of 19 degree and their pore sizes. They were in good agreement with the experimental results and this supported the reasonable application of the Kelvin equation to the calculation of contact angles.

The degree of hydrophobicity of FSM-16 would be almost same to that of MCM-41, but smaller than those of high silica zeolites and aluminophosphate molecular sieves. Because the mesoporous materials have silanol groups on the inner and outer surfaces of mesopores, while zeolites have almost no silanol groups on the pore wall. As the pore-wall of the mesoporous materials have not well-defined structure, the hydrophobicity would change in accordance with preparation methods and the conditions. This method would be useful an estimation of the degree of hydrophobicity of the mesoporous materials.

3.2 Liquid-Phase Adsorption of Surfactants[27]

There are few report on a liquid-phase adsorption for the mesoporous materials although they are considered to be suitable for the adsorption of relatively large molecules. Especially, adsorption of surfactants in a solution for the mesoporous materials is interesting because their pore structure are formed by templating of surfactant micelles.

FSM-16 samples were dispersed in surfactant aqueous solutions and stirred at appropriate temperature for 5 h. The FSM-16 samples were filtered, washed with deionized water and dried. Adsorbed amounts of surfactant were determined by thermal gravimetric measurements of the FSM-16 samples. Adsorption equilibriums were reached in a hour at 70 °C.

FSM-16, which was prepared by using alkyltrimethylammonium as template, adsorbed all types of cationic, anionic and nonionic surfactants, while silica gel adsorbed no cationic and anionic surfactants as shown in Figure 4. Especially, large amount of alkyltrimethylammonium were adsorbed for FSM-16. The maximum adsorbed amount reached ca. 60 % of alkyltrimethylammonium existing in the as-synthesized FSM-16 material as template. The adsorbed amount of alkyltrimethylammonium for FSM-16 increased steeply at the critical micelle concentration (cmc) of the surfactant. It shows that FSM-16 has a strong adsorption force to the micelle-like surfactant but weak force to molecular one. The size fit between pore opening of FSM-16 and micelles would enhance the adsorption potential

This was also supported by the result on adsorption of various alkyltrimethylammonium with different alkyl-chain length (carbon number, n = 8, 10, 12, 14, 16 and 18) on FSM-16 with pore diameter of 2.7nm. When the same surfactant concentration of 0.1 M were employed for each adsorption experiments, adsorbed amount of alkyltrimethylammonium with n=12,14,16 and 18 (ca. 0.74 mmol•g^{-1}) were larger than those with n=8 and 10 (ca. 0.43 mmol•g^{-1}). The cmc of alkyltrimethylammonium increases with decrease in the alkyl-chain length of the surfactant. The surfactant concentration employed here (0.1 M) were larger than cmc of the surfactants with n=12, 14, 16 and 18, but smaller than that of the surfactants with n=8 and 10. It indicates that the micelle formation promotes an adsorption of surfactants for FSM-16.

We also measured adsorbed amount of hexadecyltrimethylammonium (HDTMA, n=16) on

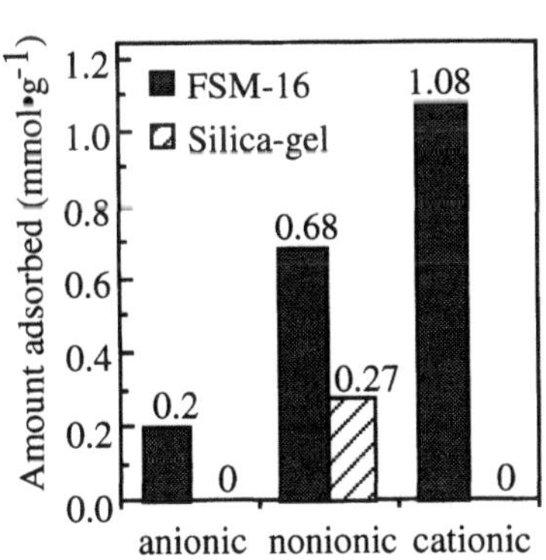

Figure 4 Adsorbed amount of anionic (Sodium laurylsulfonate), nonioni (polyoxyethyleneoctylphenylether) an cationic (hexadecyltrimethylammoniun chloride) surfactants for FSM-16 (2.7 nm) and silica-gel.

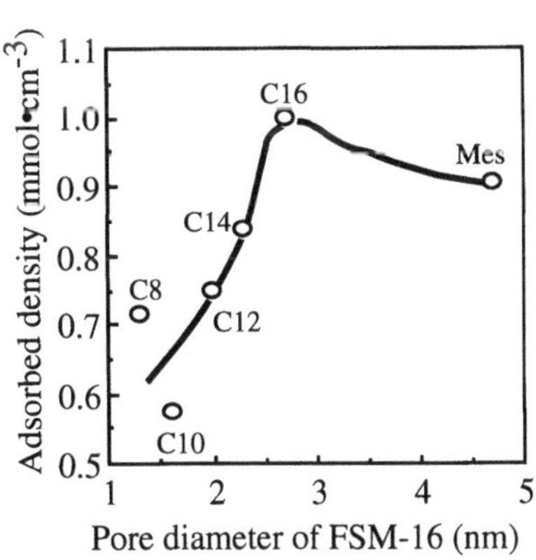

Figure 5 Adsorbed density of surfactant (n=16) in a pore of FSM-16 prepared by using surfactant with n=8-16 (C10-C16) or mesitylene (Mes).

various FSM-16 samples with pore diameters from 1.2 to 4.7 nm, which were synthesized by using various alkyltrimethylammonium with n=8-16 or auxiliary mesitylene. Adsorbed densities of HDTMA in the pore of FSM-16 were calculated by dividing the adsorbed amounts by pore volumes of FSM-16 and plotted against the pore diameter in Figure 5. The adsorbed density of HDTMA in the pore was largest for the FSM-16 with pore diameter of 2.7 nm. The 2.7nm-FSM-16 material was prepared by using the same HDTMA as template. FSM-16 showed the efficient adsorption of the same surfactant which was used for the synthesis as template. The efficient packing of surfactant in the pore would be attributable to size fit between surfactant micelle and pore diameter.

3.3 Separation Performance as Packing Material[28)]

High-performance liquid chromatography (HPLC) is one of useful separation methods. Silica gel is typical packing material for the chromatography and is usually used in normal-phase HPLC. For the separation of non-polar molecules, silica is modified with organic pendants and used in reversed-phase HPLC. Here, we study a chromatographic performance of high density mesoporous silica prepared by using surfactant as template and a comparison to the performance of silica gel.

MCM-41 and FSM-16 have been usually obtained as fine particle and have too poor mechanical strength to be used for column packing material. So, we have prepared high density mesoporous silica (HDMS) with high mechanical strength by condensation of tetramethoxysilane in a presence of cationic surfactant (hexadecyltrimethylammonium chloride). Small amount of water and hydrochloric acid were added to the tetramethoxysilane to catalyze the condensation. We can get transparent bulk materials by drying the gel and

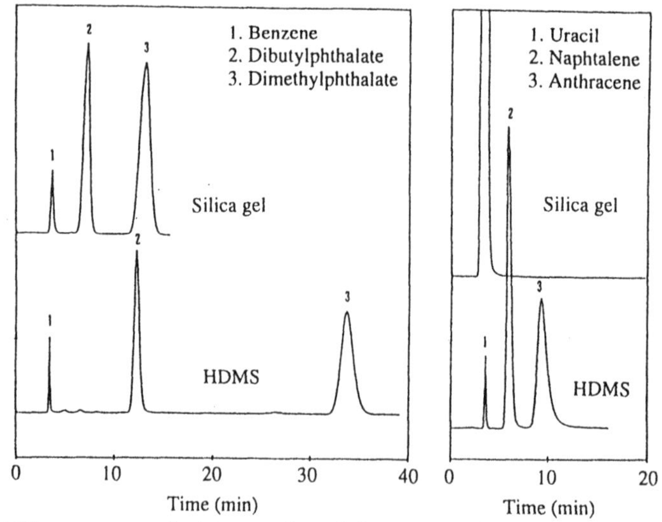

Figure 6 Chromatogram in (a) normal- and (b) reverse-phases high-perfomance liquid chromatography (HPLC) on FSM-16 and silica-gel. 1% methanol/n-hexane for normal-phase and 40% methanol/water for the reversed-phase were employed. Column : 250 x 4 mm ID. Flow rate: 1ml/min.

calcining at 550 °C. They were milled and classified into powder in a 15 micrometers diameter. HDMS had a mean pore diameter of 2.3 nm, surface area of 952 $m^2{\bullet}g^{-1}$ and pore volume of 0.60 $cm^3{\bullet}g^{-1}$. The powder was packed in 250 x 4 mm stainless steel column. The packing density of the powder was 0.56 $g{\bullet}cm^{-3}$. A mixture of methanol-n-hexane (1:99. v/v) was used as eluent for normal-phase and methanol-water (40:60, v/v) was used for reversed-phase HPLC.

Figure 6 shows chromatograms of relatively large-polar phthalate esters in normal-phase chromatography and non-polar aromatic compounds in reversed-phase chromatography by HDMS and silica-gel. HDMS showed longer retention time (k') for dibutylphthalate (k'=2.60) and dimethylphthalate (k'=8.91) than silica gel (k'=0.94, 2.59, respectively). The large retention time would be attributable to the large surface area and uniformity of mesopore of HDMS. HDMS also shows high separation performance of aromatic compounds in reversed-phase chromatography, while silica gel show no separation ability in the reversed-phase chromatography.

HDMS shows separation performance in both normal- and reversed-phases chromatography with no surface modification, that is quite different feature from that of silica gel. The difference would be attributable to not only different silanol densities but also different surface properties due to their distinct pore structures between HDMS and silica gel. The no necessity of the surface treatment of HDMS when used in reversed-phase chromatography is an advantage to be used actually because of no contamination with eluated surface agent in the separated product and recyclable use of the packing material by calcination

3.4 Oxygen Adsorption by Porphyrin/FSM-16 complex[29)]

Porphyrin molecules play important roles in vivo such as oxygen carrier in hemoglobin and active site for photosynthesis in chlorophyll. Although porphyrin holds oxygen reversibly to form oxygen-porphyrin complex in the vital systems, porphyrin in a molecular form do not adsorb oxygen because of the formation of stable dimers. It is interesting to make an artificial hemoglobin or chlorophyll by fixing the porphyrin molecules in some substrate. We incorporated porphyrin molecules in the channels of FSM-16 and examined the adsorption property of oxygen.

Porphyrin molecules were introduced in the mesopores of FSM-16 by sublimation in a vacuumed sealed glass tube at 350-500 °C. Figure 7 shows a schematic model of FSM-16 channel containing porphyrin molecules.

The porphyrin/FSM-16 complex was put in a glass cell equipped with pressure gage. Oxygen gas was introduced in the cell after evacuation and the cell was closed with a cock. Oxygen pressure in the cell was monitored with the pressure gage. Large decrease in the pressure were

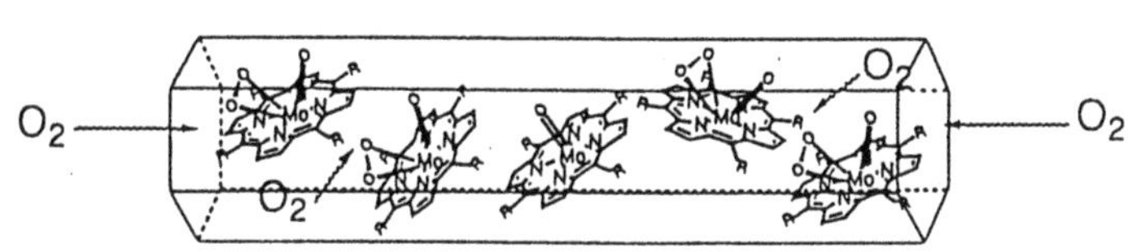

Figure 7 Schematic model of the channels of FSM-16 containing porphyrin molecules.

observed for porphyrin/FSM-16 complexes, while almost no decrease in the pressure were observed for only FSM-16 sample and only porphyrin molecule. Table 2 lists the adsorbed amount of oxygen for several porphyrin/FSM-16 complexes calculated from the pressure decrease of oxygen. The Mo^{IV}-tetraphenylporphyrin(tpp)/FSM-16 complex with pore diameter of 2.7 nm had larger adsorbed amount of oxygen than that with pore diameter of 4.7 nm. The larger adsorbed amount for smaller pore FSM-16 was also observed for Fe^{II}-tpp/FSM-16 complexes. It would be attributable to dimer formation of porphyrin in the channel of 4.7 nm because the pore size is too large to disperse porphyrin in isolated-molecular form. Mo^{IV}-tpp fixed on silica (aerosil) showed only small adsorbed amount of oxygen. Ru^{II}-tpp/FSM-16 with pore diameter of 2.7 nm had the largest adsorbed amount of oxygen, in which 88 % of Ru in the porphyrin complex adsorbed oxygen molecules.

Table 2 Adsorbed amount of oxygen for porphyrin/FSM-16 complex.

Porphyrin Complex	Pore Diameter nm	O_2/Metal mol%
$Mo^{IV}O$(tpp)/FSM-16	27	65
$Mo^{IV}O$(tpp)/FSM-16	47	34
$Mo^{IV}O$(tpp)/Aerosil	-	10
Fe^{II}(tpp)/FSM-16	27	66
Fe^{II}(tpp)/FSM-16	47	44
Ru^{II}(tpp)/FSM-16	27	88

tpp:tetraphenylporphyrin

Moreover, UV-light irradiation on the Mo^{IV}-tpp/FSM-16 complex adsorbing oxygen promoted a desorption of oxygen from the complex. Turning off the light stopped the desorption. The photosensitive adsorption behavior of oxygen would be related to the photosynthesis by chlorophyll in plants.

4. SCOPE

Figure 8 shows the advances in the synthesis of mesostructure materials by using self-assembly in a surfactant/inorganic system in the past 5 years and scope in a future. Variation of the mesostructure material was extended in terms of composition, form and structure. Composition of the mesoporous materials have been extended from SiO_2 or SiO_2-Al_2O_3 to metallosilicates containing Ti, V, etc.[30] and recently extended to non-silicon transition metal oxide such as Al_2O_3[31-33], TiO_2[34-36], ZrO_2[37-41], Ta_2O_5[42], Nb_2O_5[43], SnO_2[35,37] and HfO_2[44] etc. However, these transition metal oxide mesoporous materials have poor thermal stability and structural regularity. The synthetic method of thermally stable and highly regular mesoporous non-silicon transition metal oxide should be established in a near future because of their variety of applications such as devises and catalysts. The synthesis of mesoporous material with a framework of organic/inorganic hybrid or only organic are also important for applications to gas storage and adsorbents.

Recently, we have had many reports on the control of the morphology and form of the mesoporous materials. There have been many efforts to make mesoporous films such as supported film[10], self-standing film[45], transparent film[46] and oriented film[47]. These efforts are important to be used separation membranes and devices such as high density memory mediums. Spherical mesoporous materials have been developed by a few groups[48-50] and applied to chromatographic separations. Most recently, patterning[51] on the substrate in micrometer size

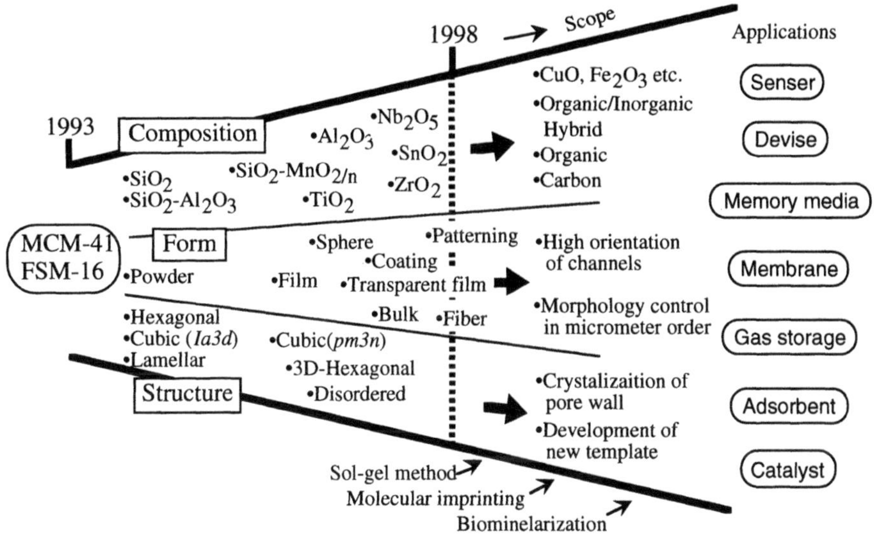

Figure 8 Advances in the synthesis method of mesostructure materials by using a self-assembly in organic/inorganic system.

and fibrous mesoporous material with oriented channels[52] have been reported. High orientation of the channels and morphology control in micrometer order[53] will be important in a near future. The relationship between the formation of well-defined micrometer-sized particles and biomineralization is also an interesting.

Crystallization of pore wall of mesoporous material is important for improvement of heat and steam stability and appearance of strong acidity similar to zeolites. Development of nw templates other than surfactants would open for interesting extensions of this strategy.

Acknowledgments

We thank Prof. K. Kuroda, Waseda Univ., Prof. M. Ichikawa, Hokkaido Univ. and Dr. K. Nobuhara, Fuji Silysia Chemical Ltd., for discussions. A part of this study was supported by the Proposal-Based New Industry Creative Type Technology R&D Promotion Program from the New Energy and Industrial Technology Development Organization (NEDO) of Japan. We wish to thank Toyota Central R&D Labs., Inc. for the permission to publish this work.

REFERENCES

1) Yanagisawa, T. et al.: Bull. Chem. Soc., Jpn., 63(1990),988.
2) Kresge, C. T. et al.: Nature, 359(1992), 710 ; Beck, J. S. et al.: J. Am. Chem. Soc., 114(1992), 10834.
3) Inagaki, S. et al.: J. Chem. Soc., Chem. Commun., (1993), 680.
4) Inagaki, S. et al. : Studies in Surface Science and Catalysis, 84(1994),125.
5) Inagaki, S. et al. : Bull. Chem. Soc., Jpn., 69(1996), 1449.
6) Inagaki, S. et al. : Studies in Surface Science and Catalysis., 105(1996), 109.
7) Inagaki, S. et al.: Chem. Mater., 8(1996), 2089.

8) Inagaki, S. et al.: J. Colloid Interface Sci., 180(1996), 623.
9) Chen, C. Y. et al.: Microporous Mater., 4(1993), 1.
10) Yang, H. et al.: Nature, 379(1996), 703.
11) Schacht, S. et al.: Science, 173(1996), 768.
12) Davis, S. A. et al.: Nature, 385(1997), 420.
13) Bagshaw, S. A. et al.: Science, 29(1995), 1242.
14) Branton, P. J. et al. : J. Chem. Soc., Faraday Trans., 90(1994),2965.
15) Rathousky, J. et al. : J. Chem. Soc., Faraday Trans., 90(1994),2821.
16) Llewellyn, P. L. et al. : Microporous Mater., 3(1994),345.
17) Schmidt, R. et al. : Microporous Mater., 3(1994),443.
18) Branton, P. J. et al. : Langmuir, 12(1996),599.
19) Rathousky, J. et al. : J. Chem. Soc., Fraday Trans., 91(1995),937.
20) Neimark, A. V. et al. : Langmuir in press.
21) Kruk, M. and Jaroniec, M.: J. Phys. Chem. B, 101(1997),583
22) Naono, H. et al. : J. Colloid Interface Sci., in press.
23) Inagaki, S. et al.: Microporous and Mesoporous Mater., in press.
24) Chen, C.-Y. et al.: Microporous Mater., 2(19930, 17.
25) Ishikawa, T. et al.: J. Chem Soc., Faraday Trans., 92(1996), 1985.
26) Gregg, S. J. and Sing, K. S. W.:"Adsorption, Surface Area and Porosity," 2nd. ed., Academic Press, New York, 1982.
27) Inagaki, S. et al.: 11th. Meeting of the Japan Society on Adsorption, Yamanashi, October, 1997, Abstr., pp.68, P-42.
28) Nobuhara, K. et al.: Chromatography, 117(1996), 282.
29) Goto, Y. et al.: 13th. Meeting of the Japan Association of Zeolite, Nagasaki, Nobember, 1997, Abst., pp.77, B3.
30) Sayari, A.: Chem. Mater., 8(1996), 1840.
31) Bagshaw, S. A. et al. : Angew. Chem. Int. Ed. Engl., 35(1996), 1102.
32) Yada, M. et al.: J. Chem. Soc., Chem. Commun., (1996), 769.
33) Vaudry, F. et al. : Chem. Mater., 8(1996), 1451.
34) Antonelli, D. M. and Ying, J. Y. : Angew, Chem., Intl. Ed. Engl., 34(1995), 2014.
35) Ulagappan, N. and Rao, C. N. R.: Chem. Commun., (1996), 1685.
36) Ulagappan, N. and Rao, C. N. R.: J. Serb. Chem. Soc., 61(1996), 1107.
37) Knowles, J. A. and Hudson, M. J. : J. Chem. Soc., Chem. Commun, (1995), 2083.
38) Kim, A. Y. et al. : Mat. Res. Soc. Symp. Proc. 435(1996), 131.
39) Larsen, G. et al. : J. Cat., 164(1996), 246.
40) Kim, A. Y. et al. : Chem. Commun., (1997), 161.
41) Pacheco, G. et al. : Chem. Commun., (1997), 491.
42) Antonelli, D. M. and Ying, J. Y.: Chem. Mater., 8(1996), 874.
43) Antonelli, D. M. et al. : Inorg. Chem., 35(1996), 3126.
44) Liu, P. et al. : Chem. Commun., (1997), 577.
45) Yang, H. et al.: Nature, 381(1996), 589.
46) Ogawa, M. : J. Chem Soc., 116(1994), 7941.
47) Tolbert, S. H. et al.: Science, 278(1997), 264.
48) Liu, J. et al.: J. Porous Mater., 2(1996), 201.
49) Grun, M. et al.: Adv. Mater., 9(1997), 254.
50) Huo, Q. et al.: Chem. Mater., 9(1997), 14.
51) Trau, M. et al.: Nature, 390(1997), 674.
52) Tolbert, S. H. et al., Science, 278(1997) , 264.
53) Yang, H. et al.: Nature, 386(1997), 692.

MESOPOROUS MOLECULAR SIEVES 1998
Studies in Surface Science and Catalysis, Vol. 117
L. Bonneviot, F. Béland, C. Danumah, S. Giasson and S. Kaliaguine (Editors)

Adsorption characterization of mesoporous molecular sieves

Peter I. Ravikovitch,[♣] Gary L. Haller,[♣] and Alexander V. Neimark[♠,♣,*]

[♣]Chemical Engineering Department, Yale University, New Haven, CT 06520, USA

[♠]TRI/Princeton, Princeton, NJ 08542-0625, USA

Low-temperature nitrogen and argon adsorption is one of the most informative experimental techniques for structural characterization of mesoporous molecular sieves (MMS). Although adsorption methods are well documented in the literature, the problem of quantitative interpretation of adsorption isotherms in terms of pore size distributions is quite topical. Conventional methods, such as the BJH and HK methods, are not adequate in the nanometer range of pore sizes typical for MMS. We discuss foundations and applications of the non-local density functional theory (NLDFT) for characterization of MMS. NLDFT is a molecular level model capable of constructing the adsorption isotherms in pores of different geometry and predicting the capillary condensation transitions in mesopores based on the potentials of intermolecular interactions. On the example of novel MCM-41 catalysts, we show that the NLDFT model is a reliable method for calculating the pore size distributions from adsorption isotherms, and in combination with XRD, for assessing the pore wall thickness in MMS of uniform structure.

1. INTRODUCTION

Low temperature nitrogen and argon adsorption isotherms are widely used for assessing the quality of mesoporous molecular sieves of M41S type and other nanoporous materials. In many cases, the isotherms exhibit prominent swings related to the capillary condensation transition in nanopores. The swing shape serves as a fingerprint of geometrical specifics of the pore structure. The swing position determines the characteristic pore size. Its amplitude determines the pore volume. The sharpness determines the pore size heterogeneity, which is expressed in terms of pore size distributions. Conventional adsorption models, which are currently used for calculations of pore structure parameters of MMS from experimental isotherms, are proven to have substantial limitations and shortcomings. As a rational alternative, we have recently developed a new method for interpretation of adsorption isotherms and pore structure characterization in the range of nanometers. The method developed is based on the non-local density functional theory (NLDFT) model for physical adsorption of nitrogen and argon in model cylindrical pores. In this paper, we discuss the NLDFT model on the example of the pore structure characterization of MMS of MCM-41 type. The nitrogen

* address for correspondence: aneimark@triprinceton.org

and argon isotherms were measured on newly synthesized samples of mesoporous catalysts prepared at different conditions and studied by XRD. We show that the NLDFT model provides a rigorous method for calculating the pore volume, specific surface area, pore size distribution, and pore wall thickness.

2. LIMITATIONS OF CONVENTIONAL METHODS FOR PORE SIZE ANALYSIS IN THE RANGE OF NANOPORES

Several theoretical models are routinely used for the characterization of porous materials. The most commonly used methods include the BET and comparison (or α-S) methods for calculating the specific surface area [1], and the Barrett-Joyner-Halenda (BJH) and Horvath-Kawazoe (HK) methods for calculating the pore size distribution [2, 3]. While these methods are incorporated in the software packages of automated commercial adsorption instruments, it has been well documented in the literature that their accuracy is limited, especially in the nanometer range of pore sizes, typical for M41S and other recently discovered MMS. The pore sizes in these materials vary from ca. 7 to 20 molecular diameters. In such confinements, the adsorbate molecules experience strong interactions with the pore walls. The properties of the adsorbate confined in nanopores differ appreciably from those of the bulk liquid. The Kelvin equation corrected for the multilayer adsorption on the pore walls, which is the basis of the BJH method, does not account for the adsorbate-adsorbent interactions correctly. The Kelvin equation presents a rigorous asymptotic which is valid in the limit of large pores but becomes inaccurate in nanopores [4]. Derjaguin [5] proposed an effective macroscopic approach to account for the solid-fluid interactions in terms of disjoining pressure acting in the adsorbed film. This refined thermodynamic treatment leads to significant corrections to the Kelvin equation [6], however, it is rarely used.

A more rigorous and accurate description of adsorption in nanopores can be obtained from molecular level models based on statistical mechanics of confined fluids. The HK method [3] is a simplified approach in which the change of the free energy during adsorption is equated to the average potential field arising from both adsorbate-adsorbate and adsorbate-adsorbent interactions. This method has been developed for nitrogen adsorption in carbon slit-shaped micropores and is not grounded for capillary condensation in pores of MMS. The density functional theory provides a rational model of confined fluids [7]. It was proven particularly useful for unified quantitative description of adsorption and capillary condensation in nanopores and calculating the pore size distributions [8-12].

3. NON-LOCAL DENSITY FUNCTIONAL THEORY OF ADSORPTION IN PORES

In the DFT approach [7], the local fluid density $\rho(\mathbf{r})$ of adsorbate confined in a pore at given chemical potential μ and temperature T is determined by minimization of the grand thermodynamic potential (GP). The GP is expressed as a functional of the local fluid density $\rho(\mathbf{r})$:

$$\Omega[\rho(\mathbf{r})] = F[\rho(\mathbf{r})] - \int d\mathbf{r}\rho(\mathbf{r})[\mu - U_{ext}(\mathbf{r})] \qquad (1)$$

where $F[\rho(\mathbf{r})]$ is the intrinsic Helmholtz free energy functional, and $U_{ext}(\mathbf{r})$ is the potential imposed by the pore walls. In a perturbation fashion, the Helmholtz free energy $F[\rho(\mathbf{r})]$ is divided into a contribution from the reference system of hard spheres, $F_{HS}[\rho(\mathbf{r})]$, and a contribution from attractive interactions which are treated in a mean-field approximation:

$$F[\rho(\mathbf{r})] = F_{HS}[\rho(\mathbf{r})] + \frac{1}{2}\iint d\mathbf{r}d\mathbf{r}'\,\rho(\mathbf{r})\rho(\mathbf{r}')\Phi_{attr}(|\mathbf{r}-\mathbf{r}'|) \qquad (2)$$

Here, $\Phi_{attr}(r)$ is the attractive potential. In this work we use the non-local density functional for hard sphere fluid developed by Tarazona [13]. The attractive interactions are modeled by the Lennard-Jones potential using the Weeks-Chandler-Andersen prescription [14].

The cylindrical pore model is used to approximate geometry of pore channels. The solid-fluid interactions are modeled with the LJ potential integrated over the cylindrical surface. Equilibrium density profiles, $\rho(r)$, are obtained by minimization of the GP by the method of Indeterminate Lagrange Multipliers [15]. In the cylindrical pores, the excess adsorption per unit pore area is expressed as:

$$N_S(P/P_0) = \frac{2}{D-\sigma_{ss}}\int_0^{(D-\sigma_{ss})/2} r dr\left[\rho(r) - \rho_g(P/P_0)\right] \qquad (3)$$

Here, ρ_g is the bulk gas density, D is the pore diameter measured between the centers of the atoms in the first layer of the pore wall, and $\sigma_{ss} = 2.76$ Å is the Van der Waals diameter of oxygen atom in the pore wall. Here, adsorption is defined as the excess quantity with respect to the equilibrium density in the bulk, and the reference fluid volume is defined as the internal volume of the cylindrical pore of diameter $D_{in} = D - \sigma_{ss}$.

Parameters of the fluid-fluid interactions have been chosen to reproduce the bulk thermodynamic properties of nitrogen and argon at low temperatures [8,10]. These include liquid-gas coexistence densities, saturation pressure and the surface tension of the free interface. Parameters of the solid-fluid interactions were fitted to reproduce the standard nitrogen and argon isotherms on non-porous oxides [16]. More detailed description of the model and parameters involved can be found in Refs. [8,10,12]. Note that the solid-fluid interaction parameters used in this work differ somewhat from the parameters which we used in previous calculations of nitrogen adsorption on MCM-41 [8]. The updated parameters were chosen to provide a better agreement with the standard nitrogen adsorption isotherm.

In Figure 1 we present NLDFT isotherms in comparison with the standard nitrogen isotherm on non-porous solids [16]. To demonstrate how the strength of solid-fluid interactions affects capillary condensation in nanopores we varied the Lennard-Jones solid-fluid interaction parameter, ε_{sf}. Figure 1 shows that in a pore of 13 molecular diameters, σ_{ff}, (43.7Å), 20% decrease or increase of ε_{sf} shifts the equilibrium transition pressure by at most 0.04 P/P_0. In the range of MCM-41 pores, such variations in the transition pressure are equivalent to only ca. 3Å change in the pore diameter. At the same time,

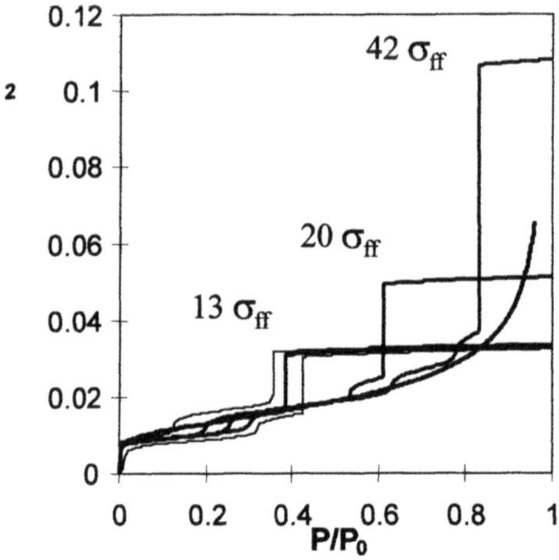

Figure 1. NLDFT isotherms in pores and standard nitrogen adsorption isotherm on non-porous solid. Isotherms in a $13\sigma_{ff}$ pore obtained by increasing and decreasing ε_{sf} by 20% (thin lines).

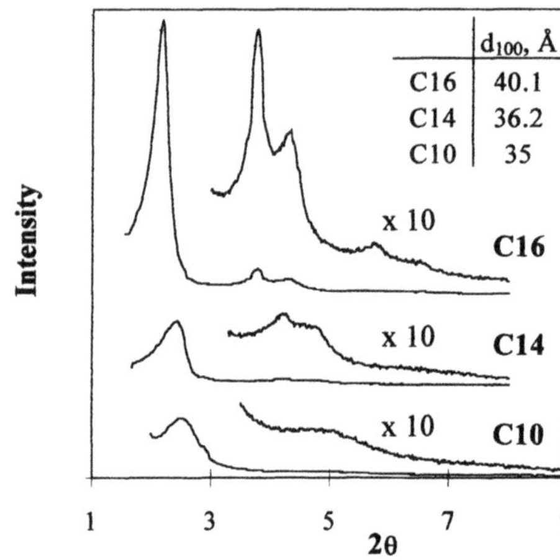

Figure 2. X-ray diffraction patterns the of MCM-41 samples.

monolayer and multilayer adsorption are strongly influenced by the solid-fluid interactions - the monolayer peak shifts by one order of magnitude in P/P_0. In a wider pore ($20\sigma_{ff}$, 68.7 Å), the capillary condensation transition is less influenced by the solid-fluid interactions and 20% variation of ε_{sf} shifts the transition by only ca. 0.01-0.02 P/P_0. It is worth noticing that changes in the strength of the solid-fluid interactions also worsen the agreement with the standard nitrogen isotherm on a non-porous surface (Figure 1). We conclude that the fluid-fluid interactions are the most important factor determining the capillary condensation pressure, and that the NLDFT predictions are quite accurate despite the simplified treatment of the solid-fluid interactions which neglects the heterogeneity of the surface. Incorporation of such effects has been recently made in the MC simulations by Maddox *et. al.* [17]. The MC predictions for the capillary condensation transition pressures in the pores of MCM-41 were very close to the NLDFT model down to the most narrow mesopores (within 0.02 P/P_0 for pores < 40 Å [12]).

4. CALCULATION OF STRUCTURAL PARAMETERS FROM THE NLDFT MODEL

To calculate the pore structure parameters, the experimental isotherm can be represented as a combination of the isotherms in individual pores. This approach implies that adsorption in a given pore is not influenced by adsorption in other pores which is well-justified in the case of high-quality MCM-41 materials. The hysteresis effects in ideal cylindrical pores is caused by metastability of the adsorption branch of the isotherm and thus, the experimental desorption branch is best suited for the pore size analysis [8, 9]. The experimental isotherm is written as

$$N_{\exp}(P/P_0) = \int_{D_{\min}}^{D_{\max}} \varphi(D_{in}) N_S(D_{in}, P/P_0) dD_{in} + S_{ext} N_{std}^S(P/P_0) \tag{4}$$

where $N_S(D_{in}, P/P_0)$ is the theoretical adsorption in a pore of size D_{in}, $\varphi_S(D_{in})$ is the pore size distribution, S_{ext} is the surface area of macropores, and $N_{std}^S(P/P_0)$ is the standard adsorption per unit area on a non-porous surface. Here we account for the fact that real materials may contain significant amounts of amorphous, essentially non-porous phase, which contributes to the total isotherm increasing the specific surface area. Unlike the BJH method, Eq. (4) also accounts for variations of the adsorbate density in filled pores, which is somewhat greater than the bulk-liquid density.

While an average pore diameter of MCM-41 can be easily estimated from the inflection point of the isotherm swing (using, e.g., the NLDFT dependence of the capillary condensation pressure versus the pore diameter [12]), an accurate determination of the cumulative pore surface area and the pore volume requires a rational multilayer adsorption isotherm. The NLDFT model gives reliable values of the capillary condensation pressure and the adsorbate density in filled pores, however the model isotherms exhibit pronounced steps in the pre-transition region which are not observed in the experiments (Figure 1). This limitation can be overcame by using a well-established experimental fact that nitrogen and argon adsorption isotherms on MCM-41 materials in the multilayer region can be reduced to the corresponding standard isotherms on a non-porous surface [11,12]. This experimental observation holds true even for very narrow-pore samples [11]. Thus, we substituted the theoretical isotherms in the multilayer adsorption region with a standard isotherm on non-porous surface with corrections on the curvature of cylindrical pores.

With the above correction, equation (4) is converted into the linear matrix equation and is solved numerically using Tikhonov regularization method [18]. More details on the numerical method will be presented elsewhere.

In Figures 3 and 4, we show nitrogen and argon adsorption isotherms at 77 K on three MCM-41 samples prepared with surfactants of different chain length (C10, C14 and C16). The C14 sample contains 2% of Sn. Preparation procedures will be described elsewhere. All isotherms exhibit characteristic capillary condensation swings which are more pronounced for Ar then for N_2 because of lower reduced critical temperature of Ar. Although the temperature 77 K is below the triple point of bulk Ar, fluid in such narrow pores remains liquid [19]. Therefore, we used the saturation pressure of the supercooled liquid Ar at 77 K while constructing the isotherms. The C16 sample has a higher quality than the other two samples. This is supported by the XRD pattern (Figure 2) exhibiting 5 peaks which can be indexed on the hexagonal unit cell.

The pore size distributions are presented in Figure 5 and the pore structure parameters are summarized in Table 1. We note a very good agreement between the average pore sizes obtained from N_2 and Ar desorption isotherms that supports the consistency of our approach. For comparison, we included the pore size distribution calculated from the adsorption branch of Ar isotherm on C16 sample. In this case, the pore size distribution peak is shifted significantly to larger pores (Figure 5).

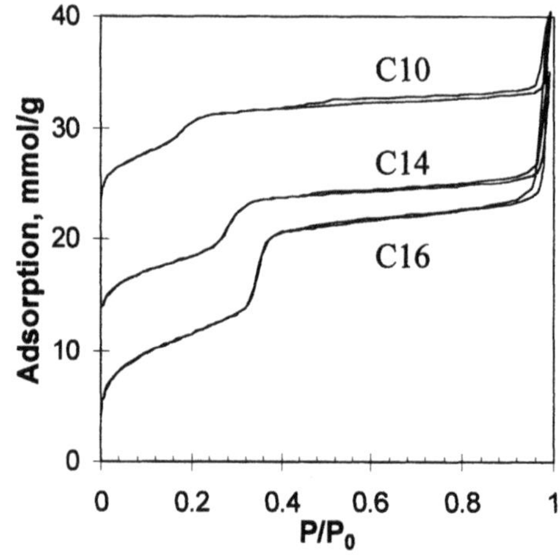

Figure 3. Nitrogen adsorption isotherms at 77.4 K. The scale is shifted by 10 mmol/g for sample C14 and by 20 mmol/g for C10 sample.

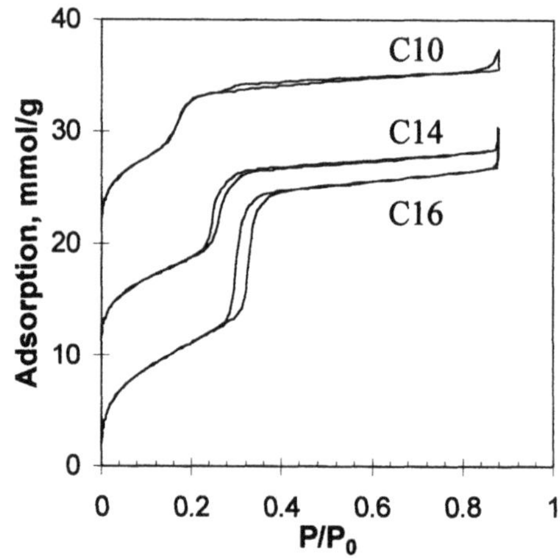

Figure 4. Argon adsorption isotherms at 77.4 K.

This supports our earlier conclusion that namely the desorption branch of the isotherm should be employed for the pore size distribution analysis in the MCM-41 type of MMS. When the pore channels are non-uniform and are interconnected into a pore network, the analysis of adsorption and desorption isotherms requires a refinement. Network effects cause a wider hysteresis loop with abrupt desorption at a certain relative pressure corresponding to evaporation from the blocking pore constrictions. Accounting for network effects by use of percolation methods [19] is desirable.

The cumulative pore volumes and surface areas of primary mesopores (Table 1) were determined for pores smaller than D_{max}=60 Å. Larger pores were treated as an open surface characterized by the external surface area, S_{ext}. Within intrinsic uncertainties introduced by our choice of the multilayer adsorption isotherm and the ill-posed nature of equation (4), the results obtained from N_2 and Ar isotherms are in a good agreement. The calculated external surface areas were ca. 50% larger for N_2 than for Ar. This effect may be attributed to possible freezing of Ar in wide mesopores at 77 K. The experimental studies have shown that Ar at 77 K remains liquid in the MCM-41 of size ca. 25 Å [20], and that Ar freezes in ca. 36 Å pores of Vycor glass at 75.55 K, and in ca. 51 Å pores of xerogel at 72.3 K [21]. Thus, it appears that Ar remains liquid in the primary mesopores of our MCM-41 materials, while it may freeze in larger mesopores. This is also supported by the absence of a noticeable increase of Ar adsorption in the vicinity of saturation, while the increase of N_2 adsorption at $P/P_0 \rightarrow 1$ clearly shows that the samples under consideration contained an appreciable amount of macropores. It should be noted that the NLDFT model does not show the onset of freezing for Ar at 77 K even in sufficiently wide pores.

Assuming the cylindrical pore model, the pore wall thickness, t_{cyl}, in a hexagonal array of pores is calculated as $t_{cyl} = 2/\sqrt{3}d_{100} - D_{in} = a_0 - D_{in}$. The cylindrical pore model implies the minimum wall thickness among other possible cross-sections with the same area. For example, if we assume the hexagonal cross-section of real pores, and that D_{in} is the equivalent pore diameter which yields the same cross-section area, the pore wall thickness in a hexagonal array of hexagonal pores equals

$$t_{hex} = a_0 - D_{in}\sqrt{\pi\cos(\pi/6)/3} \approx a_0 - 0.9523 D_{in}, \quad (5)$$

which is somewhat larger than t_{cyl} (Table 1).

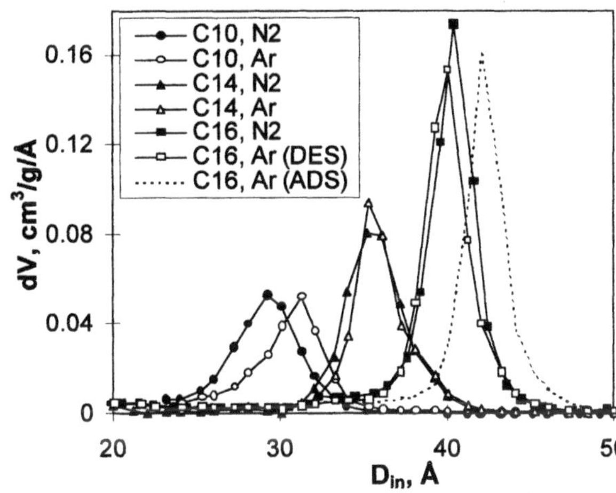

Figure 5. Pore volume distributions calculated from the NLDFT model.

Table 1
Pore structure parameters of MCM-41 samples

Sample	C10		C14		C16	
	N_2	Ar	N_2	Ar	N_2	Ar
V_{meso}, [cm^3/g]	0.34±0.01	0.34±0.01	0.41±0.01	0.41±0.01	0.64±0.01	0.62±0.01
S_{meso}, [m^2/g]	530±15	530±15	500±15	520±15	700±15	710±15
S_{ext}, [m^2/g]	80±15	50±15	100±15	70±15	120±15	80±15
D_{in}, [Å]	29.2	31.3	35.2	35.3	40.3	40.1
t_{cyl}, [Å]	11.2	9.1	6.5	6.5	6.0	6.2
t_{hex}, [Å]	12.6	10.6	8.2	8.2	8.0	8.1

5. CONCLUSIONS

On the example of the pore structure characterization of novel mesoporous catalysts of MCM-41 type, we have demonstrated that the NLDFT model provides a rigorous approach to interpreting the adsorption isotherms and calculating the pore structure parameters of MMS. The samples with average pore diameter of 3, 3.5, and 4 nm were prepared at different conditions and studied by low temperature nitrogen and argon adsorption, and XRD. The structural parameters, such the pore volume, specific surface area, pore size distribution, and pore wall thickness, calculated independently from the nitrogen and argon isotherms are in a satisfactory agreement.

ACKNOWLEDGEMENT

We thank Naisheng Yao for the preparation of the samples and Sangyun Lim for the XRD measurements. This work was supported, in part, by Grant DE-FG02-88ER13836 from DOE, Office of Basic Energy Sciences, and the TRI/Princeton exploratory research program.

REFERENCES

1. S.J. Gregg, K.S.W. Sing, Adsorption, Surface Area and Porosity, Academic Press: New York, 1982.
2. E.P. Barrett, L.G. Joyner, P.H. Halenda, J. Am. Chem. Soc., 73 (1951) 373.
3. J. Horvath, K. Kawazoe, J. Chem. Eng. Japan, 16 (1983) 470.
4. R. Evans, U. Marini Bettolo Marconi, P. Tarazona, J.Chem.Soc.Faraday Trans. 2, 82 (1986) 1763.
5. B.V. Derjaguin, Acta Phys. Chem URSS, 12 (1940) 181.
6. J.C.P. Broekhoff, J.H. de Boer, J.Catalysis, 9 (1967) 8; 10 (1968) 368.
7. R. Evans, in D. Henderson, (ed.), Fundamentals of Inhomogeneous Fluids, Marcel Dekker, New York, 1992, Chapter 5.
8. P.I. Ravikovitch, S.C. Ó Domhnaill, A.V. Neimark, F. Schüth and K.K. Unger, Langmuir, 11 (1995) 4765.
9. A.V. Neimark, P.I. Ravikovitch, S.C. Ó Domhnaill, F. Schüth and K.K. Unger, in M.D. LeVan (ed.), Fundamentals of Adsorption, Kluwer Academic Publishers, Boston, MA, 1996, p.667.
10. A.V. Neimark, P.I Ravikovitch, M. Grün, F. Schüth, K.K. Unger, Characterization of Porous Solids IV, 15-18 September 1996, Bath, UK.
11. P.I. Ravikovitch, D. Wei, W.T. Chueh, G.L. Haller, A.V. Neimark, J. Phys. Chem. B, 101 (1997) 3671.
12. P.I. Ravikovitch, G.L. Haller, A.V. Neimark, TRI/Princeton Workshop "Characterization of Porous Materials: from Angstroms to Millimeters", June 18-20, 1997, Princeton, NJ; Advances in Colloid and Interface Science (in press).
13. P. Tarazona, Phys.Rev.A, 31 (1985) 2672.
14. J.D. Weeks, D. Chandler, H.C. Andersen, J. Chem. Phys., 54 (1971) 5237.
15. A.V. Neimark, Langmuir, 11 (1995) 4183.
16. J.H. de Boer, B.G. Linsen, Th.J. Osinga, J. Catalysis, 4 (1965) 643.
17. M.W. Maddox, J.P. Olivier, K.E. Gubbins, 13 (1997) 1737.
18. C.L. Lawson, R.J. Hanson, Solving Least Squares Problems, SIAM: Philadelphia, 1995.
19. A.V. Neimark, Studies in Surf. Sci. & Catal., 62 (1991) 67.
20. P.L. Llewellyn,Y. Grillet, J. Rouquerol, C. Martin, J.-P. Coulomb, Surf. Sci., 352-354 (1996) 468.
21. E. Molz, A.P.Y. Wong, M.H.W. Chan, J.R. Beamish, Phys. Rev. E, 48 (1993) 5741.

MESOPOROUS MOLECULAR SIEVES 1998
Studies in Surface Science and Catalysis, Vol. 117
L. Bonneviot, F. Béland, C. Danumah, S. Giasson and S. Kaliaguine (Editors)

Synthesis and Applications of Nanoporous Materials

Jackie Y. Ying

Department of Chemical Engineering, Massachusetts Institute of Technology, Cambridge, MA 02139-4307, USA

Nanoporous systems are characterized by the molecular assembly of structures consisting of nanometer-sized cavities or pores. Conventional porous structures with long-range crystalline order are typically limited to aluminosilicates (e.g. zeolites) and phosphates with pore openings of less than 15 Å. The inflexibility in the composition and pore size of such materials does not address the needs of the variety of catalytic reactions and gas adsorption applications that exist in chemical industries. Our ability to process nanoporous materials with greater control in compositional and pore structure variation will open up tremendous opportunities in advancing catalysis and separation technologies.

In 1992, Mobil researchers reported that a family of aluminosilicates (termed M41S) with pores larger than 20 Å could be synthesized in an ordered packing through a liquid crystal route [1,2]. Of particular interest is MCM-41, which has hexagonally-packed cylindrical pore channels containing surface areas greater than 1200 m^2/g and uniform pore sizes that can be tailored from 20 to 100 Å in diameter, making for attractive heterogeneous catalysts, catalyst supports, and nanocomposite host materials for a wide range of novel applications [3-6]. The ability to synthesize these materials and to control the structural characteristics on a mesoscopic scale stems from the unique synthesis route of using surfactant solutions to template the inorganic precursors into a continuous solid framework. The formation of mesoporous materials with a variety of crystallographically well-defined frameworks [7] has been made possible via a generalized "liquid-crystal templating" (LCT) mechanism [8,9].

To create materials with superior structural and surface chemical properties, much work has been done to expand upon the LCT synthesis to include non-aluminum metal dopants within the silicate MCM-41 framework [10,11], and to derive non-siliceous MCM-41-type of mesostructures [8,12]. The ability to introduce compositional variation into the supramolecular templating of nanoporous structure will present exciting opportunities for attaining a great variety of molecular sieves well beyond the current synthetic flexibility of the zeolitic systems. For example, transition metal oxides play an important role as industrial catalysts in petrochemical production, pollution control, and pharmaceutical and fine chemical syntheses, but they are typically available with relatively low surface areas and poorly-defined pore structures. Improved performance can be achieved by developing transition metal oxide materials with the highly desirable microstructural characteristics of MCM-41. A number of non-siliceous M41S-like materials have been reported for the oxides of such transition metals as antimony, iron, molybdenum, tin, and vanadium; however, these mesostructured materials collapsed upon surfactant removal and did not yield mesoporous molecular sieves [8,13-15]. These attempts at synthesizing transition metal oxide mesostructures were based on the LCT mechanism involving Coulombic charge interplay between charged surfactant head groups, metal precursor species and counterions in solution.

We have found that to successfully synthesize transition metal oxide mesoporous materials, it is critical to establish a strong interaction between the surfactant head groups and the inorganic precursor prior to micellar self-assembly [16]. Based on this idea, our laboratory has created new organometallic compounds that consist of transition metal alkoxides ligated with a variety of surfactants through covalent bonding. This precursor design proves to be ideal for creating transition metal oxides with a hexagonally-packed cylindrical mesoporous structure. The resulting family of transition metal oxides, termed TMS or *Tech Molecular Sieves*, is characterized by uniform pore diameters that can be flexibly altered through variations in the surfactant chain lengths and the addition of swelling agents [17-21]. Consequently, we are able to achieve mesoporous transition metal oxides (TMS1) with surface areas in excess of 600 m^2/g, and pore sizes tailored in the range of 20-100 Å. Also of interest to the development of nanoporous materials is the extension of supramolecular templating synthesis to the pore diameter range below 20 Å. We have found that using the ligand-assisted approach, supramolecular templating can be applied to attain microporous transition metal oxides (TMS5) using short-chain templating molecules [22]. Such systematic bridging of the microporous and mesoporous regimes would be very useful in overcoming the current limitations with existing zeolitic pore structures for selective petrochemical and fine chemical processing. The transition metal-based composition further offers the fundamental chemical flexibility in tailoring mixed oxidation states, electronic interaction, surface modification, and thermal stability. A wide variety of reactions and processes may profit from the dual microstructural and chemical functionalities of the TMS family of molecular sieves.

Research on the synthesis of advanced nanoporous materials would be most fruitful when targeted towards specific catalytic and separation applications of interest. The uniformly large pore openings of mesoporous materials overcome the configurational diffusion limitations of zeolitic systems. Such materials offer unique opportunities to provide shape and size selectivity for molecules too large to react within the framework of conventional microporous zeolitic structures. We have examined the chemical vapor grafting approach as a means to achieve high Pd dispersion on the pore walls of MCM-41. The resulting Pd-TMS11 catalyst has successfully been used in carbon-carbon coupling reactions [23]. Remarkable activity, easy accessibility and exceptional stability of this system provides an excellent example for a new generation of simple and efficient heterogeneous Heck catalysts that even rivals some of the best homogeneous catalysts. Transition metal-doped mesoporous materials may also serve as stable catalytic supports for fixating bulky organometallic catalytic complexes, such as metalloporphyrins [24,25]. The resulting heterogenized catalyst systems have the advantage of ease of handling and separation in continuous operations over homogeneous catalysts, while retaining the high activity and selectivity associated with designed organometallic ligands.

REFERENCES

1. C. T. Kresge, M. E. Leonowicz, W. J. Roth, J. C. Vartuli, and J. S. Beck, Nature, 359 (1992) 710.
2. J. S. Beck, J. C. Vartuli, W. J. Roth, M. E. Leonowicz, C. T. Kresge, K. D. Schmitt, C. T.-W. Chu, D. H. Olsen, E. W. Sheppard, S. B. McCullen, J. B. Higgins, and J. L. Schlenker, J. Am. Chem. Soc., 114 (1992) 10834.
3. C.-G. Wu and T. Bein, Chem. Mater., 6 (1994) 1109.
4. C. L. Bowes, A. Malek, and G. A. Ozin, Chem. Vap. Deposition, 2 (1996) 97.
5. A. Sayari, Chem. Mater., 8 (1996) 1840.
6. C. J. Brinker, Curr. Opin. Solid State Mater. Sci., 1 (1996) 798.
7. Q. Huo, D. I. Margolese, and G. D. Stucky, Chem. Mater., 8 (1996) 1147.
8. Q. Huo, D. I. Margolese, U. Ciesla, P. Feng, T. E. Gier, P. Sieger, R. Leon, P. M. Petroff, F. Schüth, and G. D. Stucky, Nature, 368 (1994) 317.
9. C.-Y. Chen, S. L. Burkett, H.-X. Li, and M. E. Davis, Microporous Mater., 2 (1993) 27.
10. P.T. Tanev, M. Chibwe, and T. J. Pinnavaia, Nature, 368 (1994) 321.
11. A. Corma, M. T. Navarro, and J. Pérez Pariente, J. Chem. Soc., Chem. Commun., (1994) 147.
12. P. V. Braun, P. Osenar, and S. I. Stupp, Nature, 380 (1996) 325.
13. A. Stein, M. Fendorf, T. P. Jarvie, K. T. Mueller, A. J. Benesi, and T. E. Mallouk, Chem. Mater., 7 (1995) 304.
14. G. G. Janauer, A. Dobley, J. Guo, P. Zavalij, and M. S. Whittingham, Chem. Mater., 8 (1996) 2096.
15. N. Ulagappan and C. N. R. Rao, Chem. Commun., (1996) 1685.
16. D. M. Antonelli and J. Y. Ying, Curr. Opin. Coll. Interf. Sci., 1 (1996) 523.
17. D. M. Antonelli and J. Y. Ying, Angew. Chem. Int. Ed. Engl., 35 (1996) 426.
18. D. M. Antonelli, A. Nakahira, and J. Y. Ying, Inorg. Chem., 35 (1996) 3126.
19. D. M. Antonelli and J. Y. Ying, Angew. Chem. Int. Ed. Engl., 34 (1995) 2014.
20. D. M. Antonelli and J. Y. Ying, Chem. Mater., 8 (1996) 874.
21. M. S. Wong, D. M. Antonelli, and J. Y. Ying, Nanostr. Mater., 9 (1997) 165.
22. T. Sun and J. Y. Ying, Nature, 389 (1997) 704.
23. C. P. Mehnert and J. Y. Ying, Chem. Commun., (1997) 2215.
24. L. Zhang and J. Y. Ying, AIChE J., 43 (1997) 2793.
25. L. Zhang, T. Sun, and J. Y. Ying, submitted to Chem. Commun.

MESOPOROUS MOLECULAR SIEVES 1998
Studies in Surface Science and Catalysis, Vol. 117
L. Bonneviot, F. Béland, C. Danumah, S. Giasson and S. Kaliaguine (Editors)

The Preparation of Mesoporous Metals from Preformed Surfactant Assemblies

G. S. Attard, N. R. B. Coleman, J. M. Elliott

Department of Chemistry, University of Southampton,
Southampton SO17 1BJ, United Kingdom.

ABSTRACT

Mixtures of surfactant molecules and water form intricate three-dimensional structures at high surfactant concentrations. These phases can be utilised as molds for the formation of highly ordered mesoporous materials such as inorganic oxides and metals. It has already been shown that non-ionic surfactant assemblies can be employed in the synthesis of mesoporous silicas [1]. By applying the same principles to metals, we have produced a new family of mesoporous materials.

The hexagonal phases formed by a number of ternary non-ionic surfactant + water + metal precursor systems have enabled us to prepare mesoporous metals such as platinum. The hexagonal pore structure of mesoporous platinum, as observed by TEM, was a direct cast of the surfactant phase from which it was formed [2]. By changing the chain length of the surfactant, the pore size could be altered, and surface area measurements of the platinum materials were greater than those of commercially available platinum black. These novel, nanostructured, metals are of considerable interest for applications in catalysis and electronic devices.

1. INTRODUCTION

The use of surfactants in the synthesis of mesoporous silicates has been well documented since the initial work of Beck and co-workers [3,4]. In 1992 they prepared a family of mesoporous silicates, known as M41S, from reaction mixtures containing a micellar solution of an ionic surfactant and a silicate precursor. Interestingly, the pore morphology, of the resulting silicates, resembled the nanostructure associated with aggregates formed at higher concentrations of surfactant. The existence of these structures at the relatively low concentrations of surfactant, used in the synthesis of M41S materials, was explained in terms of an ionic interaction between the surfactant and silicate [5]. It was proposed that aggregation of the micelles, induced by the presence of the negatively charged silicate species, led to the more complicated structures that were observed. Later work confirmed that the presence of micelles was critical in producing an ordered mesoporous structure [6].

In 1995 it was shown that mesoporous silicates, with pore morphologies identical to the M41S materials (Figure 1), could be produced from aggregates formed by non-ionic polyoxyethylene surfactants [1]. The major difference between this method of preparing mesoporous materials and the M41S route was the concentration of surfactant, which in this case was much higher than in the M41S method. A higher concentration of surfactant resulted in the formation of surfactant aggregates that formed liquid crystal phases and molding of the silica network was determined by the structure of the aqueous domains in the pre-formed liquid crystal phase. This contrasted with the M41S route where assembly of the micelles, during the formation of the silicate network, was important in determining the final mesopore structure.

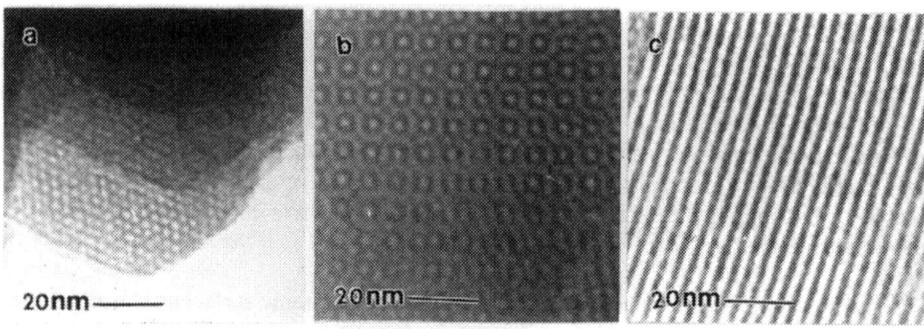

Figure 1. TEM images of mesoporous silicas. a) Hexagonal pore morphology (hexagonal array of cylindrical holes), b) Cubic pore morphology (complex interconnecting channel system with the Ia3d space group), c) Lamellar pore morphology (sheets of silica).

There are a number of advantages of the direct liquid crystal templating of mesoporous materials over the M41S route. Firstly, the silicates that are formed have millimetre dimensions whereas, M41S materials are micron sized powders. Secondly, this approach allows a high degree of control and predictability over the nanostructure of the material through choice of phase and surfactant. Finally, the range of materials that can be produced by direct liquid crystal templating extends beyond inorganic oxides to semiconductors [7], metals [8] and polymers [9].

Nanostructured metals are of considerable interest for applications in catalysis, batteries, fuel cells, capacitors and sensors [10-13]. By extending the direct liquid crystal templating of mesoporous silicates to purely metal systems we have produced novel nanostructured metals. Recently we reported on the chemical synthesis of mesoporous platinum by the reduction of a platinum precursor in the aqueous domains of an hexagonal (H_I) phase. Transmission electron microscopy (TEM) of the resulting granular H_I platinum showed it to have the characteristic hexagonal pore morphology. In addition, extended X-ray absorption fine structure (EXAFS) indicated that the metal, in the walls of the material, was chemically pure, and had a face-centred cubic structure [8]. Following on from the work on chemically produced mesoporous platinum we reported on the electrodeposition of H_I platinum [2]. The prepared mesoporous platinum films were shown to have high specific surface areas and correspondingly high capacitances.

Here we compare and contrast the electrochemically-produced mesoporous, platinum, films (H_I-ePt) [2] and the chemically produced powders (H_I-Pt) [2]. Control over the pore size, in the case of the electrochemically deposited films, has been achieved by changing the type of surfactant and by the addition of a hydrophobic species to swell the pores.

2. EXPERIMENTAL

The H_I phase, containing a platinum precursor, was made from a mixture containing 42wt% of a nonionic surfactant, 29wt% of deionised water and 29wt% of platinic acid. octaethyleneglycol monohexadecyl ether ($C_{16}EO_8$) and octaethyleneglycol monododecyl ether ($C_{12}EO_8$) were chosen as the nonionic surfactants as they form H_I phases which are stable over a large compositional range. Also, quaternary mixtures containing $C_{16}EO_8$ and *n*-heptane in the molar ratios 2:1 and 1:1 were used for electrodeposition experiments.

Characterisation of the phase behaviour of these mixtures was carried out by optical polarising microscopy. Our electrochemical plating baths consisted of a polished gold working electrode, large-area platinum gauze counter electrode and a saturated calomel reference electrode (SCE) with the platinic acid containing, H_I phase as the electrolyte. Electrodeposition of mesoporous platinum from the H_I phase of the surfactant onto the gold electrodes was achieved by stepping the potential from +0.6V versus SCE to –0.1V versus SCE. Films were grown at temperatures between 25 and 65°C. After deposition, the gold electrodes were washed with water for 2 days to remove the surfactant. Confirmation that all the surfactant had been removed was achieved by energy dispersive analysis of x-rays.

H_I-Pt powders were prepared from mixtures containing $C_{16}EO_8$, water and platinic acid in similar proportions to those used in the electrodeposition experiments. Chemical reduction was achieved by placing a thin layer of the phase onto a sheet of steel for ca. 4 hours. The resulting black mixture was removed from the steel surface and the H_I-Pt extracted by washing successively with acetone, water, 6M hydrochloric acid, water, and acetone.

Transmission electron microscopy (TEM) provided evidence for a nano-structure and scanning electron microscopy (SEM) was used to study the thickness of the H_I-ePt films. Surface area analysis was performed electrochemically by cyclic voltammetry of the H_I-ePt films in 2M sulfuric acid [2]. Brunauer, Emmett and Teller (BET) nitrogen surface area measurements for H_I-Pt powders and commercially available platinum black were also made for comparison.

3. RESULTS AND DISCUSSION

The presence of platinic acid in the non-ionic surfactant/water mixtures did not destroy the liquid crystalline phases of the surfactant. In the case of the ternary $C_{16}EO_8$ + platinic acid + water mixture the optical texture for the H_I phase was observed at temperatures up to and above 85°C. H_I phases were also formed in the quaternary mixtures containing *n*-heptane. Electrodeposition on gold gave a uniform coherent film of H_I-ePt that was highly reflective and metallic in appearance. This contrasted with the H_I-Pt, which was a micron sized powder and almost matt black in appearance. Previous Scanning tunnelling microscopy (STM) studies of the electrodeposited films had shown them to be remarkably flat (surface roughness: ±10nm over $1cm^2$) and this accounted for the high reflectance of the films [2].

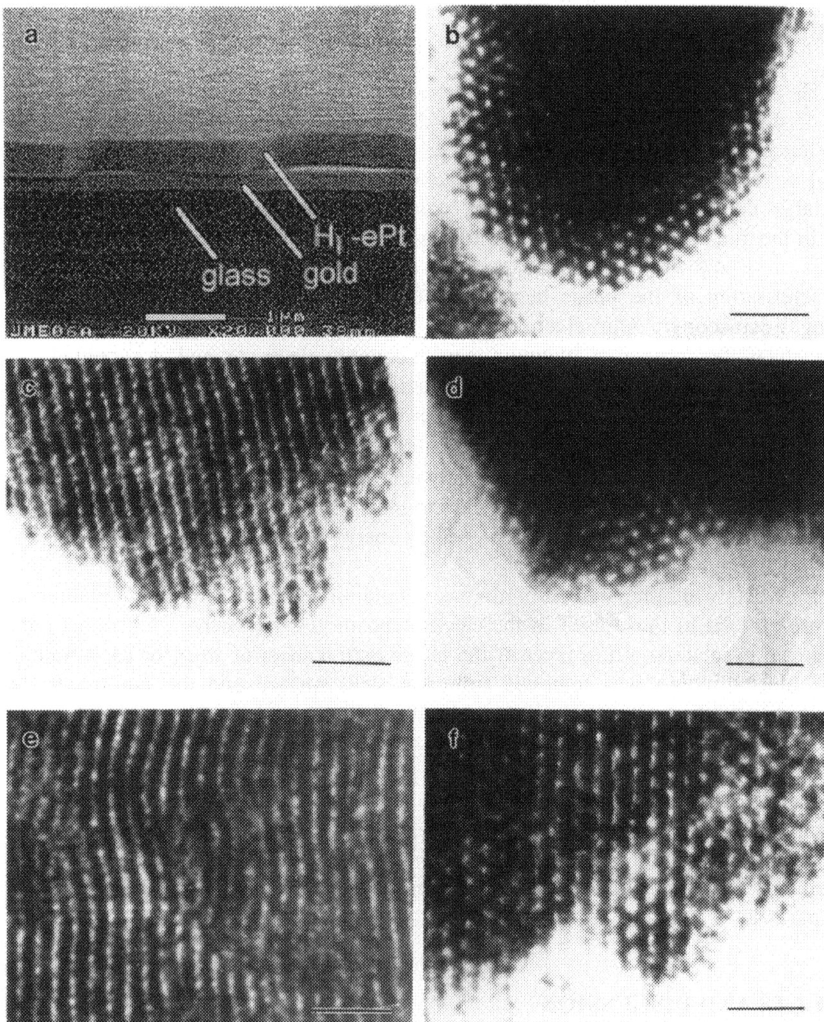

Figure 2. a) SEM micrograph of H_I-ePt deposited on a gold-coated glass slide (scale bar: 1μm). Transmission electron micrographs showing the hexagonal nanostructure of: b) chemically produced platinum from a $C_{16}EO_8$ H_I phase, electrodeposited platinum from a $C_{16}EO_8$ H_I phase c) side on view of pores, d) end on view of pores, electrodeposited platinum from a $C_{16}EO_8$ H_I phase containing 1 mole equivalent of *n*-heptane e) side on view of pores, f) end on view of pores (scale bars: 25nm).

The thicknesses of the H_I-ePt films, from SEM, varied depending on the charge density but most were less than 0.5μm across (Figure 2a). TEM studies of H_I-ePt revealed a highly mesoporous structure consisting of cylindrical holes arranged on an hexagonal lattice (Figure 2c-f). The platinum wall thickness was found to be ~2.5(±0.2)nm for all the samples, but the diameter of the holes varied depending on the chain length of the surfactant and the proportion of *n*-heptane (Table 1). Hole size was determined by the hydrophobic sections of the H_I phase as platinum was not formed in these domains. The shorter chained, $C_{12}EO_8$, surfactant formed smaller hydrophobic domains than $C_{16}EO_8$ and thus the corresponding hole size was smaller. Addition of *n*-heptane to the $C_{16}EO_8$ mixtures increased the diameter of the holes, by up to 1.5nm, through swelling of the hydrophobic domains. Similar results of pore swelling, by the inclusion of a hydrophobic additive, have been reported in the preparation of M41S materials [14]. For comparison, the H_I-Pt powders, prepared from mixtures containing $C_{16}EO_8$, were similar in terms of pore size (~2.5nm) and wall thickness (~2.5nm) compared to the corresponding H_I-ePt films (Figure 2b).

Table 1.
Change in pore size with chain length of surfactant and quantity of *n*-heptane

Surfactant	Mole ratio surfactant:*n*-heptane	Pore diameter /nm ±0.2
$C_{12}EO_8$	No alkane	1.7
$C_{16}EO_8$	No alkane	2.5
$C_{16}EO_8$	2:1	3.5
$C_{16}EO_8$	1:1	4.0

Surface areas of the H_I-ePt films were estimated to be 22(±2) m^2g^{-1} from the cyclic voltammetry measurements. Platinum deposited in the absence of surfactant but under similar potential and charge density conditions was found to have a much smaller surface area of 4.5(±0.5) m^2g^{-1}. BET nitrogen surface area measurements for the chemically produced H_I-Pt were larger than H_I-ePt at 60(±6) m^2g^{-1} and more than platinum black (20-30m^2g^{-1}). In both the H_I-ePt films and H_I-Pt powders the increased surface area could be accounted for by the highly porous nanostrucure of the materials.

4. CONCLUSION

Our experiments demonstrate that liquid crystal phases of non-ionic surfactants provide well defined structures for the formation of mesoporous platinum. Quantitative control over the pore size in H_I platinum can be achieved by either the use of a hydrophobic additive or by changing the chain length of the surfactant. While H_I-Pt powders have potential applications in catalysis the H_I-ePt films, that have been produced, are of particular relevance for use in batteries, fuel cells, sensors, and electrochemical capacitors.

REFERENCES

1. G. S. Attard, J. C. Glyde and C. G. Goltner, Nature, 378 (1995) 366-368.

2. G. S. Attard, P. N. Bartlett, N. R. B. Coleman, J. M. Elliott, J. R. Owen and J. H. Wang, Science, 278 (1997) 838.
3. J. S. Beck, J. C. Vartuli, W. J. Roth, M. E. Leonowicz, C. T. Kresge, K. D. Schmitt, C. T-W. Chu, D. H. Olson, E. W. Sheppard, S. B. McCullen, J. B. Higgins and J. L. Schlenker, J. Am. Chem. Soc., 114 (1992) 10834.
4. C. T. Kresge, M. E. Leonowicz, W. J. Roth, J. C. Vartuli and J. S. Beck, Nature, 359 (1992) 710.
5. A. Monnier, F. Schuth, Q. Huo, D. Kumar, D. Margolese, R. S. Maxwell, G. D. Stucky, M. Krishnamurty, P. Petroff, A. Firouzi, M. Janicke and B. F. Chimelka, Science, 261 (1993) 1299.
6. C-F. Cheng, Z. Luan and J. Klinowski, Langmuir, 11 (1995) 2815.
7. P. V. Braun, P. Osenar and S. I. Stupp, Nature, 380 (1996) 325.
8. G. S. Attard, J. M. Corker, C. G. Goltner, S. Henke and R. H. Templer, Angew. Chem. Intl. Ed. Engl., 36 (1997) 1315.
9. G. S. Attard, S. Leclerc and C. G. Goltner, unpublished data.
10. F. Leroux, B. E. Koene and L. F. Nazar, J. Electrochem. Soc., 143 (1996) L181.
11. S. Ye, A. K. Vijh and L. H. Dao, ibid, p. L7.
12. W. G. Pell and B. E. Conway, J. Power Sources, 63 (1996) 255.
13. J. Wang and L. Anges, Anal. Chem., 64 (1992) 456.
14. N. Ulagappan and C. N. R. Rao, J. Chem. Soc. Chem. Commun., (1996) 2759.

The authors would like to thank Dr. Barbara Cressey for her help with electron microscopy, the Engineering and Physical Sciences Research Council of the United Kingdom and British Nuclear Fuels plc who supported this work.

MESOPOROUS MOLECULAR SIEVES 1998
Studies in Surface Science and Catalysis, Vol. 117
L. Bonneviot, F. Béland, C. Danumah, S. Giasson and S. Kaliaguine (Editors)

Preparation of novel mixed [Cu, Zn, Al] hexagonal and lamellar mesophases

S. Valange and Z. Gabelica*

ENSCMulhouse, Université de Haute Alsace, Laboratoire de Matériaux Minéraux, UPESA 7016, 3 rue Alfred Werner, F-68093 Mulhouse Cedex, France

1. INTRODUCTION

CuO-ZnO (mixed) oxides supported on various carriers like alumina or chromia are important catalysts for various oxidation or hydrogenation reactions, including the conversion of syngas to methanol [1, 2]. Their composition, structure and structural modification through various post-preparation treatments were shown to dramatically influence the nature and structure of the final phases, the composition of the active centers and hence their overall catalytic performances [2]. For example, in the case of selective hydrogenation of either CO or CO_2 into methanol or methane for which metallic copper was recognized to be the main active phase [2, 3], the unique activity of these catalysts was ascribed to some synergy between Cu and ZnO, while Al_2O_3 is believed to have primary a structural role thereby minimizing or preventing sintering of Cu particles [1, 3].

Huo et al [4] demonstrated that the now well addressed method to obtain periodic mesoporous silicates could be extended to the synthesis of non silica based mesostructures, in particular to oxides of a variety of metals. Both cationic and anionic surfactants were used along with water soluble metallic salts to stabilize the corresponding inorganic-organic mesophases, including Pb, Mg, Al, Ga, Mn, Fe, Ni, Co, and Zn lamellar oxides, Sb and W hexagonal oxides, Sb cubic oxide and a lamellar Zn phosphate [4, 5]. Fe, W, Pb and Sb also formed hexagonal mesophases but they did not exhibit any porosity after calcination [4]. Besides this pioneering work, an ever wider series of more recent attempts to synthesize non silica based single phase- or binary mesostructured materials has been recently reviewed [6-8]. Most of these lamellar metallic oxide composites were supposed to form through the classical S^-I^+ mechanism. They readily collapse after calcination, giving dense, in most cases amorphous metal oxides. This was supposed to be caused most probably by the low degree of condensation of the oligomeric ions constituting the mesoporous walls, although the redox instability of the metallic cations was not excluded to play a fragilizing role.

The incorporation of several transition metal ions including Cu^{2+} and/or Zn^{2+} was reported to be successfully achieved for low metal amounts (Si/[Cu + Zn] molar ratios above 35) by direct synthesis into mesostructured silicas [9, 10]. Despite the fact that in such compounds the (neutral) silica is not expected to behave as the most efficient support for Cu and Zn species [11], (Cu, Zn)-MCM-41 prepared through direct synthesis showed promising catalytic properties, for example in CO_2 hydrogenation [9] or in selective reduction of NO with propene

[12]. Moreover, the synergetic effect of ZnO in reducing Cu^{2+} also proved effective in Cu and Zn bearing MCM-48 [10].

No attempts were reported to support (incorporate) these ions on (in) other mesoporous carriers like Al_2O_3 for which several preparation recipes were recently proposed [13-17]. On the other hand, while the synthesis of a pure Cu bearing mesophase was never mentioned, lamellar, thermally unstable Zn based oxidic mesostructures could be prepared by heating at 100°C for 5 days Zn nitrate fully hydrolyzed as cationic species at pH = 3 and admixed with dodecylphosphate template (anionic surfactant) [4].

In a preliminary work [13], we have succeeded in preparing and stabilizing high specific surface area pseudo-hexagonal and thermostable alumina mesophases. Such materials are by no means excellent candidates to accommodate into their porous structure CuO and/or ZnO oxidic species as guests, admixed upon post synthesis treatments. Even if they collapse upon calcination, the resulting high surface Al_2O_3 amorphous phase could also provide a support for the metallic oxides, possibly more efficient than the conventional low surface area Al_2O_3.

The present study deals with another kind of approach to prepare dispersed mixed oxides, namely the use of precursors that involve a mesostructured framework exclusively composed of the 3 oxidic species. We describe for the first time the preparation and characterization of a series of such compounds.

2. EXPERIMENTAL SECTION

2.1. Chemicals

The following surfactants were used for the various syntheses: (i) cationic surfactant such as the quaternary alkylammonium salts C_{14}- and C_{16}TMABr (Fluka) and (ii) anionic surfactants, namely sodium dodecyl sulfate, $C_{12}H_{25}OSO_3Na$ (SDS) and long alkyl chain carboxylic acids and/or their sodium salts, $C_nH_{2n-1}O_2H/Na$ (Fluka), with n = 14 and 16. The aluminum-containing reactants were aluminum nitrate and the aluminum Keggin polycation (Al_{13}), synthesized by a slow adding of a NaOH solution to an aqueous $Al(NO_3)_3.9H_2O$ solution, according to well established procedures [18]. Two different concentrations of the Keggin salt, namely 0.1M Al (diluted Al_{13}) and 1.25 M Al (concentrated Al_{13}) were used. $Cu(NO_3)_2.\ 3H_2O$ and $Zn(NO_3)_2.\ 6H_2O$ (Fluka) were used as metallic sources.

2.2. Synthesis procedure

The general procedure adopted for the preparation of mesoporous Cu-Zn-Al phases was as follows: Solution A was achieved by dissolving in a small amount of water appropriate quantities of specific surfactants, namely palmitic (C_{16}) or myristic (C_{14}) acid (partly or totally neutralized with NaOH) and C_{16}- or C_{14}TMABr respectively, with C_xOO^-/C_xTMA^+ ratio of 4, so as to achieve an anionic surfactant medium. This solution was added along with a NaOH solution to an aqueous solution containing the inorganic source (Al, Cu and Zn) (solution B). The composition range of the reaction mixture was: w Al_2O_3 x CuO : y ZnO : 0.1 to 1 S : z NaOH : 1 to 12 H_2O (S being the micelle; w/2 + x + y = 1, z being variable so as to adjust the appropriate final pH). All syntheses, achieved at ambient temperature, yielded instantaneously blue colored (when Cu present) or white (without Cu) solids and colorless mother liquors, provided the pH was carefully controlled and monitored during the simultaneous or successive addition of the reactants. The resulting suspension was maintained under stirring for at least 15 h at ambient temperature. The solid was recovered by filtration, washed with de-ionized water and dried at 60°C in an oven overnight.

2.3. Characterization

Calcination was carried out by heating the as-synthesized material to 450°C for 3h under nitrogen flow at a heating rate of 2.5 °C/min, followed by a 2h isotherm at 450°C. XRD patterns were recorded on a Philips PW 1050 diffractometer using the Cu Kα radiation. Bulk chemical analysis for C, H, O, N, Al, Cu and Zn was achieved by atomic adsorption. N_2 isotherms were obtained on a Micromeritics ASAP 2010 instrument. Pore size distributions were calculated using the BJH method on the adsorption branch while BET surface areas were evaluated from the linear part of the BET plots.

3. RESULTS AND DISCUSSION

3.1 Sample composition selection

Table 1 gives the synthesis conditions and some characteristics of a series of samples involving variable Cu and Zn contents.

Table 1

Sample	Starting materials				Synthesis Conditions		Characteristics		
	Al source	Composition (% atom)			pH	Phase	Surface area (m^2/g)	Pore size (Å)	Mesopore volume (cm^3/g)
		Cu	Zn	Al					
1	Al^{3+}	0	0	100	7.5-8	PsH	526	51	0.88
2	Al_{13d}	0	0	100	6.5	PsH	810	28	0.61
3	Al_{13c}	0	0	100	6	PsH	670	17	0.34
4	Al^{3+}	3	7	90	6.5	PsH	656	60	0.49
5	Al_{13d}	3	7	90	6.5	PsH	662	27	0.46
6	Al_{13c}	3	7	90	6.5	PsH	773	24	0.38
7	Al^{3+}	12	28	60	6.5	PsH	370	64	0.77
8	Al_{13d}	12	28	60	6.5	PsH	480	38	0.56
9	Al_{13d}	25	15	60	6.5	PsL	558	42	0.60
10	Al^{3+}	50	30	20	6.5	PsL	167	31	0.09
11	Al^{3+}	50	30	20	7	PsL	208	27	0.10
12	Al_{13d}	50	30	20	7	PsL	189	29	0.10
13	Al_{13d}	50	30	20	8	PsL	118	34	0.08
14	Al_{13d}	50	30	20	11.5	PsL	134	33	0.09

Samples 1 to 3 are pure alumina mesophases prepared through our earlier optimized procedure involving the use of either concentrated and diluted aluminum Keggin polycations $[AlO_4Al_{12}(OH)_{24}(H_2O)_{12}]^{7+}$($Al_{13}$) or monomeric $Al(H_2O)_6^{3+}$ (Al^{3+}) cations and an admixture of appropriate amounts of palmitic acid or its Na salt and $C_{16}TMABr$, with the C_{16}^-/C_{16}^+ ratio arbitrary adjusted to 4 so as to maintain a global negative charge of the resulting "mixed micellar system" [13]. Samples 4 to 6 are comparable to their respective analogs 1 to 3 except that they involve 10 at. % of (Cu + Zn) with the Cu/Zn ratio adjusted to 3/7, as optimized by Herman et al. for the classical ICI CuO-ZnO/alumina methanol synthesis catalysts [19]. Samples 7 and 8 are similar to their analogs 4 and 5 but they involve 4 times as much Cu and Zn in the same atomic ratio, with the resulting Al content of 60 at. %. Sample 9 has the same Al concentration as sample 8 but its Cu/Zn ratio was adjusted to 5/3, as optimized in the case of selective hydrogenation of aldehydes over our (Cu, Zn, Al) catalysts [20]. Samples 10 to 14 involve twice as much Cu and Zn in the same ratio and only 20 at. % Al. Samples 15 to 20 involve frameworks only composed of Cu and/or Zn (Table 2).

Table 2

Sample	Starting materials				Synthesis conditions		XRD
	Al source	Cu (% atom)	Zn (% atom)	Al (% atom)	Surfactant	pH	Phase	d_{100} (Å)
15	-	100	-	-	$C_{14}OONa$	4.5	L	37
16	-	100	-	-	$C_{14}OONa$	4-4.5	L	42
17	-	100	-	-	$C_{16}OOH$	4.5	L	42
18	-	-	100	-	$C_{14}OONa$	5-5.5	L	34
19	-	50	50	-	$C_{14}OONa$	4.5	L	44
20	-	50	50	-	SDS	5.5-6	PsL	30

3.2. Mesopore characteristics of the various Cu-Zn-Al materials

A structure was considered here as PsH when at least the d_{110} and d_{210} peaks are observable next to the intense d_{100} peak and when the d_{100}/d_{110} ratio was close to 1.73. A structure was considered as PsL or lamellar (L) when at least the d_{200} and d_{300} peaks are observable (even if not well resolved) next to the intense d_{100} peak in the as-synthesized phase and when the d_{100}/d_{200} and d_{200}/d_{300} ratios were close to 2.

Pure Al and Al-rich [(Cu + Zn) ≤ 10] mesoporous materials prepared in the presence of Al^{3+} and diluted Al_{13} (Al_{13d}) exhibit a PsH arrangement with XRD d_{100} values close to 55 Å (samples 1 to 6), provided the pH of the synthesis batch was higher than 5. Quite high mesoporous volumes and surface areas (between 500 and 810 m^2/g), as well as narrow pore size distribution observed in the case of phases prepared from the Al_{13} precursors (not shown), argue for a regular (probably essentially alumina-based) mesostructure of good quality. The overall shape of the corresponding isotherms with a very weak hysteresis loop, similar to those characterizing classical Si-MCM-41 materials, confirmed the regular organization of the mesopores [13]. Bulk chemical analysis (not reported) indicates that the C_{16}^-/C_{16}^+ ratio was maintained within both the as synthesized alumina [13] and the (Cu + Zn)-poor phases (see below).

The increase in pore diameter and pore volume in samples 1 and 4 with respect to samples 2 and 5 (Table 1) is explained by considering that materials prepared through polycondensation of monomeric Al precursors involve less well organized and more hydrated walls, as well as a distribution of pores having a relatively large diameter. After the thermal treatment responsible of the rapid release of water, the walls get thinner, thereby generating pores with even wider diameters. In contrast, using oligomeric Al pre-structured species would favor the stabilization of well organized as synthesized solids that are expected to further exhibit an enhanced thermal stability.

This trend was enhanced upon using more concentrated Al_{13} (Al_{13c}) precursors. Indeed, samples 3 [13] and 6 show a surprisingly high surface area (more than 770 m^2/g for the Cu and Zn bearing sample 6), a medium pore volume (Table 1) and a relatively narrow pore size distribution [21]. These observations are consistent with the formation of rather thick walls composing these as-made Al-rich phases and explains their increased resistance upon calcination with respect to the analog phases prepared from Al^{3+} and Al_{13d} precursors.

Despite the large amounts of Cu and Zn present in compounds prepared from Al_{13d}, samples 8 and 9 still exhibit an adsorption-desorption isotherm typical of mesostructured solids showing a narrow pore distribution (Figures 1 and 2).

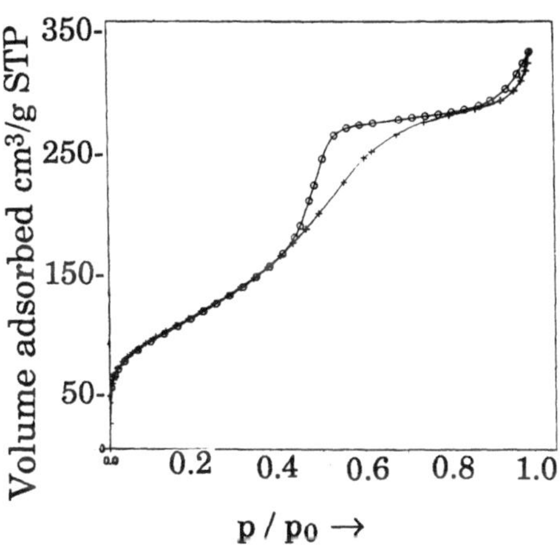

Figure 1. N_2 isotherm of sample 9 [25 Cu - 15 Zn - 60 Al]

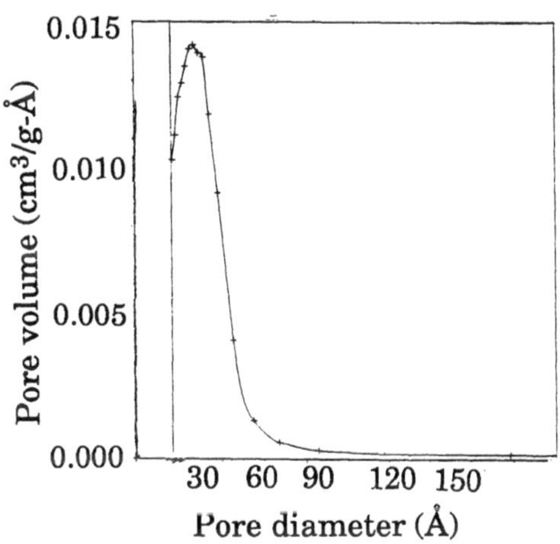

Figure 2. Corresponding BJH pore size distribution determined from the adsorption branch

Sample 7 synthesized in the presence of Al^{3+}, exhibits a different isotherm, still with a marked hysteresis loop but with the adsorption step occurring at higher p/p_o values [21] which is related to a very wide pore distribution [21] As all the Al-rich samples 1 to 9 show similar characteristics, we can conclude that amounts of (Cu + Zn) as high as 40 at. % do not fundamentally modify the sorption properties of the mixed phases. The presence of important amounts of divalent metallic ions does probably not result in the formation of a separate (lamellar) phase that would simply admix with the initial alumina network. Even if these elements use to form separate layered hydroxynitrate phases (see § 3.3), in our case they probably easily polycondense along with the alumina network and result in a rather homogeneous Cu-Zn-Al intergrown network in which Cu and Zn ions partly substitute for Al. Further studies are being under way to confirm this hypothesis.

The Cu and Zn-rich samples 10 to 14 show a pseudo-lamellar structure that is at least partly thermostable at 450°C, as suggested by the presence of a medium intense broad XRD peak around d = 45-50 A in the calcined phase [21]. This was substantiated by adsorption data showing that these phases still involve a medium surface area, a limited amount of (large) mesopores probably resulting from major textural porosity, and a very small total pore volume (Table 1).

3.3. Chemical composition of the as synthesized phases

Table 3 compares the framework metallic composition (at. %) of a series of as-made samples, with the theoretical composition calculated from the reactants.

Table 3

Sample.	Cu		Zn		Al		pH
	Theor.	Found	Theor.	Found	Theor.	Found	
4	3	3.0	7	5.9	90	91.1	6.5
5	3	3.1	7	7.2	90	89.7	6.5
8	12	12.1	28	27.1	60	60.8	6.5
12	50	51.8	30	28.8	20	19.4	7.0
14	50	51.9	30	30.3	20	17.3	11.5

The results indicate that in each case, the initial metallic ratios were maintained in the final solids thanks to a careful pH-monitored and surfactant-assisted selective precipitation in aqueous solution. Only sample 14 was found partly depleted in Al, due to the too high pH.

The full chemical composition of samples 5 and 8, determined from combined bulk chemical and thermal analyses, very accurately matches their theoretical composition expected from the ingredients and assuming that the excess of positive charges must be compensated by the anions present in the medium, namely hydroxyls and nitrates.

Sample 5

Theor: $Cu_{3.0}$ - $Zn_{7.0}$ - $Al_{90.0}$ - $(C_{16}TMA^+)_{6.7}$ - $(C_{16}COO^-)_{26.7}$ - $(OH^- + NO_3^-)_{270}$

Found: $Cu_{3.1}$ - $Zn_{7.1}$ - $Al_{89.7}$ - $(C_{16}TMA^+)_{6.7}$ - $(C_{16}COO^-)_{27.5}$ - $(OH^-)_{264}$ - $(NO_3^-)_{4.8}$

Sample 8

Theor: $Cu_{12.0}$ - $Zn_{28.0}$ - $Al_{60.0}$ - $(C_{16}TMA^+)_{6.}$ - $(C_{16}COO^-)_{26.}$ - $(OH^- + NO_3^-)_{244}$

Found: $Cu_{12.1}$ - $Zn_{27.1}$ - $Al_{60.8}$ - $(C_{16}TMA^+)_{6.}$ - $(C_{16}COO^-)_{27.2}$ - $(OH^-)_{217}$ - $(NO_3^-)_{16.2}$

The amount of $C_{16}TMA^+$ and $C_{16}COO^-$ species in the final solid was maintained as well, suggesting that an optimized selection of the initial surfactant amounts was achieved for a ready interaction with the metallic ions and a complete precipitation of the ingredients as a mesophase. Bearing in mind that no nitrate ions were found in the pure mesoporous aluminas [13], the above results suggest that nitrates along with OH^- species preferably neutralize the Cu^{2+} and Zn^{2+}species. These ions are indeed known to form an extraordinary variety of hydroxy-oxysalts among which the most representative hydroxy-nitrates are $Cu_2(OH)_3NO_3$ and $Zn_5(OH)_8NO_3)_2$ [22]. Both compounds involve polymeric sheets composed respectively of $Cu_2(OH)_3O$ and $Zn_5(OH)_8$ layers interlinked through nitrate ions that are hydrogen bonded to the OH^- ions of the adjacent layers, and in which both Zn and Cu occupy octahedral sites [22]. The total amount of nitrates found in samples 5 and 8 (respectively 4.8 and 16.2) fairly well fit the above nitrate formulas. EPR data confirmed the exclusive presence of distorted octahedral Cu^{2+} species in the as synthesized phases, similar to those involved in $Cu_2(OH)_3(NO_3)$ [23].

It is speculated that in the pH range of 6-7, similar pre-polycondensed Cu and Zn basic nitrates can also readily form in the presence of Al hydroxy-oxy species that are already pre-structured into a PsH phase through specific interactions with the surfactant micelles. The basic nitrates could then undergo some structural rearrangement so as to get accommodated within the alumina mesophase, in which they would stay interlinked through the (partial) participation of the residual nitrates and eventually form the mixed Cu-Zn-Al phase. When the actual amount of Al is high, it will favor the formation of a PsH type structure. In contrast, for low Al contents or when only Cu and/or Zn ions are present, their hydroxy-nitrate polymeric layered structure (L or PsL) will prevail and optionally accommodate some residual Al species.

3.4 Influence of the Cu and Zn contents on the phase structure.

Samples 4 to 9 are essentially based on an Al rich framework (Al at.% ≥ 60) and, except sample 8, essentially exhibit a pseudo-hexagonal structure while samples 10 to 14 show a XRD pattern better corresponding to a pseudo lamellar structure. Samples 15 to 19 that only contain Cu or Zn, exhibit a pure lamellar structure.

A PsL structure would still retain a porosity after calcination, as illustrated by at least one XRD peak still visible in the very low angle region. This may be explained by considering that our so called PsL structure consists in (not necessarily regularly) folded Al-based flexible sheets partly cross-linked through the condensation of M-OH groups (M being mainly Al) of the adjacent layers. The folding and cross-linking could occur through a mechanism implying an early winding of the partly pre-structured Al based sheets around the intercalated surfactant micelle as already suggested by Inagaki et al. [24] or Monnier et al. [5] for various silica mesophases. Al-free Cu and/or Zn based phases (samples 15-19, Table 2) present a XRD pattern characteristic of a pure lamellar mesostructure. It would readily collapse upon calcination, as ascertained by the absence of any X-ray diffraction peak.

Obviously the overall Al content does influence the structure. The Al-rich Cu and Zn bearing phases retain their hexagonal or pseudo-hexagonal structure characteristic of the pure alumina analog, suggesting that the minor amount of Cu and Zn atoms probably either stay well dispersed as separate phases within (onto) the alumina based network, or are truly substituting lattice Al. The fact that samples 7 and 8, relatively rich in (Cu + Zn), still involve large pore sizes and mesopore volumes after calcination, suggests that their mesopores are not plugged by any extra framework Zn or Cu oxide phase and favors the true substitution hypothesis.

Al-free samples involving a pure Cu, Zn or (Cu + Zn) framework are possibly composed of $Cu_x(OH)_y(NO_3)_z$ relatively pre-condensed but quasi unfolded neutral sheets that readily collapse upon calcination. Such phases are still well mesostructured probably through hydrogen type bonding between the metal bearing sheets and the nearly neutral carboxylic acid micelles that are hardly dissociated in the 4-6 pH range ($S^\circ I^\circ$ type predominant mechanism). Sample 20, probably involving only still (partly) positively charged Cu and Zn species that interact with the pure anionic surfactant SDS would involve a different type of structure, possibly stabilized through a S^-I^+ type mechanism.

REFERENCES

1. K. Klier, Adv. Catal., 31 (1981) 243
2.. G.C. Chinchen, P.J. Denny, J.R. Jennings, M.S. Spencer and K.C. Waugh, Appl. Catal., 36 (1988) 1
3. J.L.G. Fierro, in: Proc. 8° Seminario Brasileiro de Catalise, IBP (eds.), Anais, Rio de Janeiro, Brazil, 1995, vol. II, p 544
4. Q. Huo, D.I. Margolese, U. Ciesla, G.D. Demuth, P. Feng, T.E. Gier, P. Sieger, A. Firouzi, B.F. Chmelka, F. Schüth and G.D. Stucky, Chem. Mater., (1994) 1176
5. A. Monnier, F. Schüth, Q. Huo, D. Kumar, D. Margolese, R.S. Maxwell, G.D. Stucky, M. Krishnamurty, P. Petroff, A. Firouzi, M. Janicke and B.Chmelka, Science, 261 (1993) 1299
6. P. Behrens, Angew. Chem. Int. Ed. Engl., 35 (1996) 515
7. A. Corma, Chem. Rev., 97 (1997) 2373
8. A. Sayari and P. Liu, Microporous Mater., 12 (1997) 283
9. S. Valange, Z. Gabelica, M. Lopez Granados, S. Rojas and J.L.G. Fierro, in: Discussions on Zeolite and Microporous Materials, H. Chon and Y.S. Uh, (eds), Hanrimwon Publ. Co., Seoul, Korea, 1997, p 203
10. M. Hartmann, S. Racouchot and C. Bischof, J. Chem. Soc. Chem. Commun., (1997) 2367
11. T. Tagawa, N. Nomura, M. Shimakage and S. Goto, Res. Chem. Intermed. 21 (1995) 193
12. S. Valange, Z. Gabelica, C. Torre-Abreu and M.F. Ribeiro, in: "Discussions on Zeolite and Microporous Materials", (H. Chon and Y.S. Uh, Eds), Hanrimwon Publ. Co., Seoul, Korea, 1997, p 331
13. S. Valange, J.L. Guth and Z. Gabelica, accepted for presentation at this meeting
14. M. Yada, H. Hiyoshi, K. Ohe, M. Machida and T. Kijima, Inorg. Chem., 36 (1997) 5565.
15. A. Stein and B. Holland, J. Porous Mater., 3 (1996) 83.
16. S.A. Bagshaw and T.J. Pinnavaia, Angew. Chem. Int. Ed. Engl., 35 (1996) 1102.
17. F. Vaudry, S. Khodabandeh and M.E. Davis, Chem. Mater., 8 (1996) 1451.
18. J.W. Akitt and J.M. Elders, J. Chem. Soc. Dalton Trans., (1988) 1347.
19. R.G. Herman, K. Klier, G.W. Simmons, B.P. Finn, J.B. Bulko and T.P. Kobylinski, J. Catal., 56 (1979) 407
20. S. Valange, Z. Gabelica, J. M. Clacens, A. Derouault and J. Barrault, to be presented at the 12th IZC, Baltimore, July 1998
21. Figures shown at the oral presentation
22. A.F. Wells, Structural Inorganic Chemistry, 5th ed., Clarendon Press, Oxford, 1984, pp 649 and 1141
23. K.J. Chao, personal communication
24. S. Inagaki, Y. Fukushima and K. Kuroda, J. Chem. Soc. Chem. Commun. 1993, 680

MESOPOROUS MOLECULAR SIEVES 1998
Studies in Surface Science and Catalysis, Vol. 117
L. Bonneviot, F. Béland, C. Danumah, S. Giasson and S. Kaliaguine (Editors)

Tubular aluminophosphate mesoporous materials containing framework silicon, vanadium and manganese

Z. Luan,[a] D. Zhao,[a] H. He,[b] J. Klinowski[b] and L. Kevan[a]

[a] Department of Chemistry, University of Houston, Houston, Texas 77204-5641, U. S. A.
[b] Department of Chemistry, University of Cambridge, Lensfield Road, Cambridge CB2 1EW, U. K.

Aluminophosphate-based tubular mesoporous materials containing framework silicon, vanadium and manganese have been successfully synthesized. X-ray diffraction and transmission electron microscopy reveal that these solid products possess disordered arrays of tubular mesoporous channels which mimic the hexagonal MCM-41 structure. N_2 adsorption further suggests that these materials are relatively thermally stable, have an average free mesopore diameter of 35 Å and have high BET specific surface areas up to 700 m^2/g. Electron probe microanalysis reveals that the overall P/Al ratio of these materials is less than unity and ^{27}Al and ^{31}P MAS NMR shows that both tetrahedral and octahedral aluminum occur as well as tetrahedral phosphorous and hydroxylated phosphorus species, indicating a nonideal aluminophosphate framework. Incorporation of silicon results in a less disordered mesoporous structure and substantial increases in the relative amounts of tetrahedral aluminum and tetrahedral phosphorus. Transition metal ions can be incorporated up to V/Al = 0.10 for vanadium and Mn/Al = 0.20 for manganese. Electron spin resonance and diffuse reflectance ultraviolet-visible spectroscopies reveal that both vanadium and manganese are present as isolated species and suggest framework locations for these ions.

1. INTRODUCTION

Aluminophosphate molecular sieves ($AlPO_4$-n), which are constructed by aluminum and phosphorus tetrahedra, are extensively studied crystalline microporous materials [1]. As a sequel to the successful synthesis of the M41S family of silica-based mesoporous molecular sieves [2,3], several research groups have attempted to prepare aluminophosphate-based mesoporous materials, but these attempts led to a lamellar phase with poor thermal stability [4-7]. A tubular mesoporous aluminophosphate phase UHM-1, which is structurally similar to MCM-41 and has improved thermal stability, has been reported more recently by the Houston laboratory [8] and from some other groups [9-11]. In contrast to previous approaches, these latter studies use a cationic alkylamine surfactant as a structure-directing-agent and conduct the synthesis at room temperature and in basic aqueous solution.

Most of the catalytic applications of microporous aluminophosphate $AlPO_4$-n molecular sieves involve acid sites or oxidation-reduction sites, which are associated with the incorporation of framework silicon or transition metal ions. In this work we extend our original synthesis of aluminophosphate UHM-1 [8] to its silicon-substituted analogy UHM-3 and vanadium or manganese incorporated UHM-1, VUHM-1 or MnUHM-1, and present a comprehensive characterization of these materials.

2. EXPERIMENTAL SECTION

2.1. Synthesis.

UHM-1 was synthesized in the presence of tetramethylammonium hydroxide (TMAOH) using cetyltrimethylammonium chloride as the structure-directing-agent, and aluminum hydroxide and phosphoric acid as the aluminum and phosphorus sources. A detailed synthesis procedure has been reported [8]. For the synthesis of UHM-3 with various Si/Al ratios, the same procedure was employed except that the synthesis was performed at a P/Al ratio of 2.00 and a desired amount of tetraethylorthosilicate (Aldrich) was added with stirring into the gel mixture before the addition of TMAOH. VUHM-1 or MnUHM-1 were also synthesized with the same procedure for UHM-1, but an amount of vanadyl sulfate trihydrate (Aldrich) or 0.2 M $MnCl_2$ (Aldrich) solution for the desired V/Al or Mn/Al ratio was added to the gel mixture with stirring prior to the addition of TMAOH. The solid product was filtered, repeatedly washed with deionized water, and dried in air. The organic structure-directing-agent was removed by heating a sample at increasing temperature steps between 200 ~ 500 °C for 5 h at each step in flowing nitrogen, and then at 500 °C for 6 h in flowing oxygen.

For comparison, a Mn^{2+} ion-exchanged UHM-1 sample, Mn-UHM-1, was prepared with calcined UHM-1 via a standard ion-exchange procedure.

2.2. Characterization.

Electron probe microanalysis (EPMA) was performed on a Jeol JXA-8600 spectrometer. Powder X-ray diffraction (XRD) patterns were collected using a Philips 1840 powder diffractometer with Cu Kα radiation (40 kV, 25 mA). N_2 adsorption isotherms were measured at 77 K using a Micromeritics Gemini 2375 analyzer. The specific surface area, A_{BET}, was determined from the linear part of the BET equation. The calculation of the pore size distribution and the cumulative mesopore (17 ~ 100 Å) surface area, A_{BJH}, were performed using the desorption branches of the N_2 adsorption isotherms and the Barrett-Joyner-Halenda (BJH) formula. Solid-state NMR spectra were recorded at 9.4 T using a Chemagnetics CMX-400 spectrometer. Electron spin resonance (ESR) spectra were recorded at X-band at 293 K or 77 K on a Bruker ESP 300 spectrometer. Diffuse reflectance ultraviolet-visible (UV-vis) spectra were measured with a Perkin-Elmer 330 spectrophotometer equipped with a 60 mm Hitachi integrating sphere accessory.

3. RESULTS

3.1. Powder X-ray diffraction

As-synthesized UHM-1 gives a well-defined XRD pattern with a prominent peak at 2.2° 2θ and some broad but clearly present peaks in the 2θ range of 3.0 ~ 8.0° (Figure 1a). This XRD pattern is similar to that of siliceous MCM-41 material [2,3] and can be indexed to a hexagonal lattice with a d(100) spacing of about 40 Å. However, as compared to siliceous MCM-41, the diffraction peaks are broadened as previously observed for disordered mesoporous silica materials MSU-1 [12] and KIT-1 [13]. The pore structure of UHM-1 is believed to be a three-dimensional, disordered network of short, wormlike, tubular channels while the channel widths are uniform, as further supported by transmission electron microscopy [8].

XRD patterns of as-synthesized UHM-3 samples with variable Si/Al ratio up to unity are similar to that of UHM-1, indicating that the aluminophosphate framework of UHM-1 does not deteriorate with the incorporation of silicon. This is also the case for as-synthesized VUHM-1 and MnUHM-1 with low metal content, around 0.025 for a V/Al ratio in VUHM-1 (Figure 1b) or 0.05 for a Mn/Al ratio in MnUHM-1. But, with further increase of the metal content, the resolution of the XRD patterns deteriorates, especially when the V/Al ratio in VUHM-1 or the Mn/Al ratio in MnUHM-1 approaches 0.1 or 0.2 (not shown). Thus, the highest vanadium or manganese incorporation in UHM-1 is limited but is comparable to that in microporous materials [14]. Calcined UHM-1, UHM-3 and MnUHM-1 with low manganese loading show less resolved XRD patterns than as-synthesized ones (Figure 1a'), indicating a decreasing order of the mesoporous structure. However, the integrity of the tubular mesoporous structures is retained. VUHM-1 shows surprisingly less thermally stability, its structure collapses after thermal treatment in flowing N_2 at 300 °C (Figure 1b').

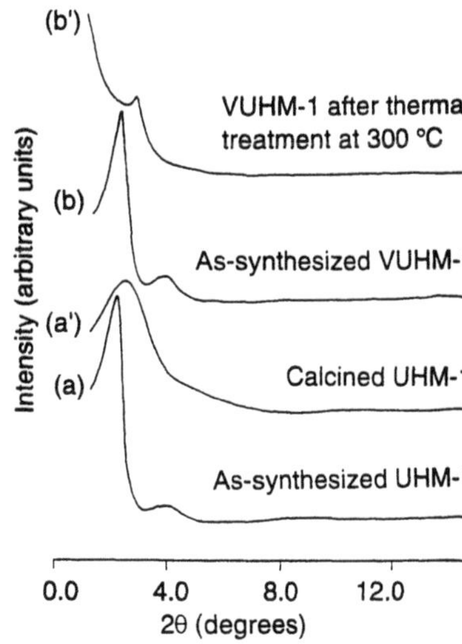

Figure 1. Powder XRD patterns of (a) as-synthesized UHM-1, (a') calcined UHM-1, (b) as-synthesized VUHM-1 with V/Al ratio 0.025 and (b') (b) after subjected to heat treatment at 300 °C for 5 h.

3.2. Electron probe microanalysis

EPMA analysis shows that both as-synthesized and calcined samples have a unique composition; therefore, the formation of either pure phase Al_2O_3 or P_2O_5 can be excluded. However, the overall P/Al ratios of all samples are between 0.5 ~ 0.8, which is less than unity and suggests that aluminophosphate-based mesoporous materials possess a nonideal aluminophosphate framework different from microporous aluminophosphate molecular sieves [1].

Analysis of UHM-3, VUHM-1 and MnUHM-1 confirms silicon, vanadium and manganese are homogeneously distributed, and that the overall Si/Al ratio in UHM-3 and the V/Al or Mn/Al ratio in VUHM-1 or MnUHM-1 are comparable with those in the initial gel mixtures.

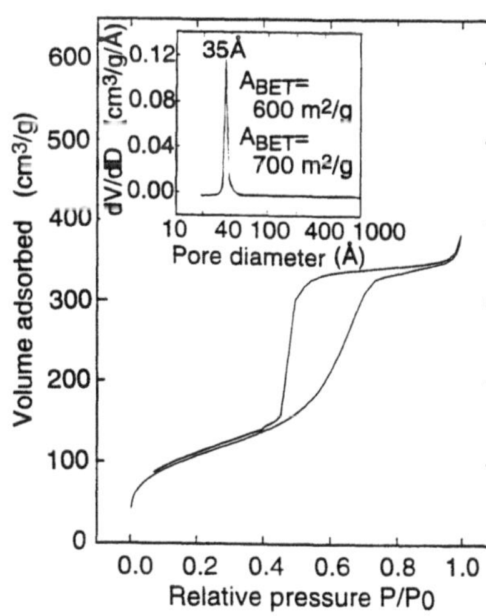

Figure 2. N_2-adsorption isotherm of calcined UHM-3 with Si/Al ratio 0.10. The inset shows the corresponding pore size distribution curve.

3.3. N_2 Adsorption measurement

Calcined UHM-1, UHM-3 and MnUHM-1

samples give an irreversible type IV adsorption isotherm with an H1 hysteresis loop, as typically shown in Figure 2, which exhibits sharp inflection in the P/P_0 range from 0.45 to 0.75 characteristic of capillary condensation within uniform mesopores [15]. The broad hysteresis loop in the isotherm reflects some mesopore disorder in shape. A BJH plot leads to a relatively narrow pore size distribution with a pore size of 35 Å and a high mesopore surface area A_{BJH} over 600 m^2/g (Figure 2). These results directly confirm the presence of tubular mesoporous channels in these aluminophosphate materials. The wall thickness can be estimated from a_o - pore size as 11 Å (unit cell parameter $a_o = 2d(100)/\sqrt{3}$, where $d(100) =$ 40 Å for typical samples).

3.4. ^{27}Al MAS NMR

The ^{27}Al MAS NMR spectrum of as-synthesized UHM-1 gives two resolved lines at 43 and 1 ppm (Figure 3a). The line at 43 ppm can be assigned to aluminum in a tetrahedral environment, in which aluminum is covalently bound to four P atoms via oxygen bridges, and the chemical shift at 1 ppm can be assigned to octahedral aluminum [16,17]. This clearly shows the simultaneous presence of both AlO_4 and AlO_6 structural units in UHM-1 corresponding to the deviation of the overall P/Al ratio from unity. Upon incorporation of silicon (UHM-3), the 43 ppm line increases in intensity with increasing silicon incorporation, whereas the 1 ppm line decreases (Figure 3b to 3c). As the silicon loading approaches a Si/Al ratio of 1.01 only the 43 ppm line is resolved (Figure 3d). This indicates that the incorporation of silicon promotes a more ordered arrangement of the aluminophosphate framework.

3.5. ^{31}P MAS NMR

The ^{31}P MAS NMR spectra of a UHM-3 sample with a low Si/Al ratio of 0.10 exhibits two intense lines at -19 and 0 ppm, and two other weak lines at -6 and -11 ppm (Figure 4b). The occurrence of the ^{31}P chemical shift at -19 ppm is due to tetrahedral phosphorus bonded to four aluminums via oxygen bridges, whereas the lines

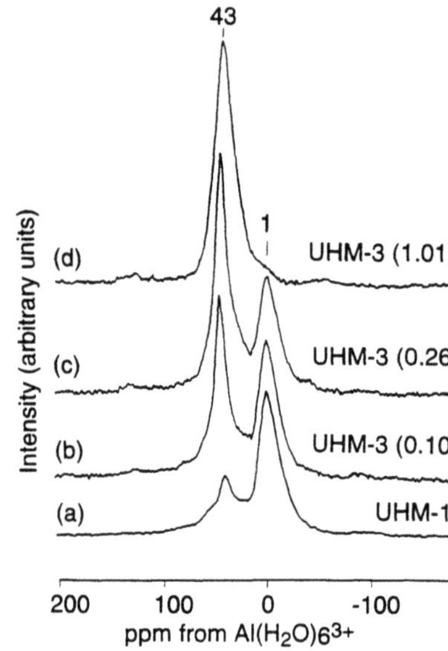

Figure 3. ^{27}Al MAS NMR spectra of as-synthesized UHM-1 and UHM-3. The number indicates the Si/Al ratio.

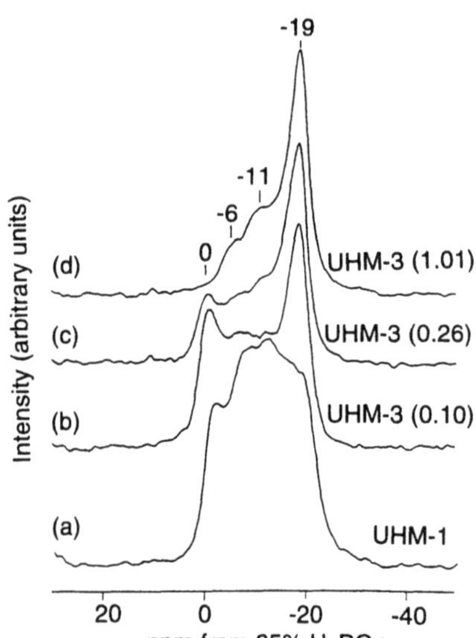

Figure 4. ^{31}P MAS NMR spectra of as-synthesized UHM-1 and UHM-3. The number indicates the Si/Al ratio.

at -11, -6 and 0 ppm are assigned to phosphorus atoms with mixed coordination $P(OH)_x(OAl)_{4-x}$, where x is 1 to 3 [5,16-18]. Without substituted silicon, UHM-1 gives a broad ^{31}P MAS NMR spectrum over the range from 0 to -20 ppm, but four lines at 0, -6, -11 and -19 ppm still can be identified (Figure 4a). The first three lines show larger intensity as compared to UHM-3. This suggests that there is no strict aluminum and phosphorus ordering of alternating AlO_4 and PO_4 structural units in UHM-1 and UHM-3 with low silicon loading. However, with increasing silicon incorporation in UHM-3, the three lines at 0, -6 and -11 ppm corresponding to phosphorus with variable numbers of hydroxyl ligands show less intensity relative to the -19 ppm line due to the $P(OAl)_4$ unit (Figure 4b to 4c). As the silicon loading approaches a Si/Al ratio of unity, the ^{31}P MAS NMR spectrum becomes dominated by the line at -19 ppm (Figure 4d). The incorporation of silicon into UHM-1 seems to facilitate the formation of a more ordered aluminophosphate-based mesoporous structure.

3.6. ^{29}Si MAS NMR

The ^{29}Si MAS NMR spectrum of UHM-3 with a low Si/Al ratio shows only one resolved line around -87 ppm (Figure 5a). At increasing Si/Al ratio a second line around -94 ppm becomes resolvable (Figure 5b). The intensities of both lines increase with increasing Si/Al ratio, but the -94 ppm line increases relatively more and dominates the spectrum at a Si/Al ratio of 1.01 (Figure 5b to 5d). Considering the excess of aluminum over phosphorus in the UHM-3 materials and the high amount of substituted silicon, the -87 ppm line can be assigned to silicon coordinated through oxygens to four aluminums as in $Si(OAl)_4$ structural units, and the -94 ppm line to silicon in $Si(OAl)_3OSi$ units in which there exists a Si-O-Si linkage [18,19]. For comparison, Figure 5e shows a typical ^{29}Si MAS NMR spectrum from silicate MCM-41 materials in which the intense line around -110 ppm comes from $Si(OSi)_4$ structural units and the shoulder around -100 ppm from silicons in $Si(OSi)_3OH$ units [15]. The lack of these lines in the ^{29}Si MAS NMR spectra of UHM-3 materials suggests that islands of pure silicate MCM-41 phases can be ruled out. But the broad ^{29}Si MAS NMR line shape of UHM-3 similar to that of siliceous MCM-41 suggests that the O-T-O angle varies somewhat in these UHM-3 materials, where T represents an Al, P, or Si tetrahedral atom.

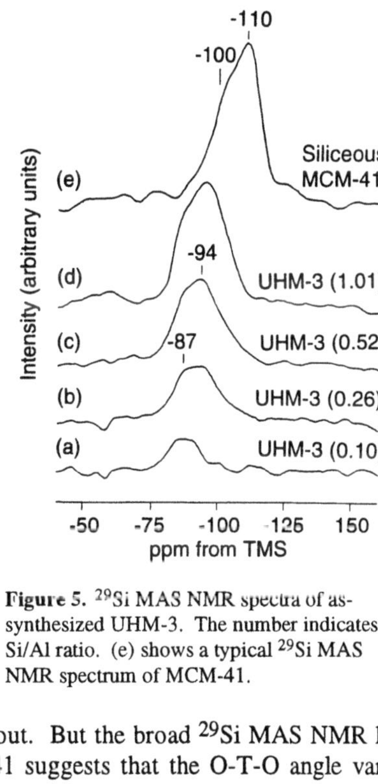

Figure 5. ^{29}Si MAS NMR spectra of as-synthesized UHM-3. The number indicates the Si/Al ratio. (e) shows a typical ^{29}Si MAS NMR spectrum of MCM-41.

3.7. Diffuse reflectance UV-vis spectroscopy

As-synthesized VUHM-1 shows a strong UV-vis absorption band near 285 nm and a weak shoulder near 350 nm, the intensity of which increases with increasing vanadium incorporation (Figure 6). No additional absorption bands are seen in the visible region. This spectrum is typical for framework V^{5+} ions in zeolites due to charge-transfer transitions between tetrahedral oxygen ligands and central V^{5+} ions [20,21]. This indicates that most of the V^{4+} ions initially added to the gel are oxidized to V^{5+} ions during synthesis. The 285 and

350 nm bands can be assigned to two different tetrahedral environments, one inside and one on the surface of the tubular walls [21].

3.8. Electron spin resonance spectroscopy

VUHM-1. As-synthesized VUHM-1 gives an ESR spectrum with strong anisotropy (Figure 7), which is characterized by an axially symmetric set of eight lines originating from V^{4+} coupled to its own nuclear spin (^{51}V, I_n=7/2, natural abundance 99.8%) and described by the spin Hamiltonian parameters $g_{||}$ = 1.937, $A_{||}$ = 192 G and $g_{\perp}$ = 1.984, $A_{\perp}$ = 69 G, where A is the hyperfine coupling. No superimposed broad singlet due to a high local spin concentration of VO^{2+} ions is observed. These g values and hyperfine coupling constants are strikingly different from those of tetrahedral V^{4+} ions [22] and are typical of VO^{2+} complexes with square pyramidal coordination in vanadium-substituted molecular sieves [21].

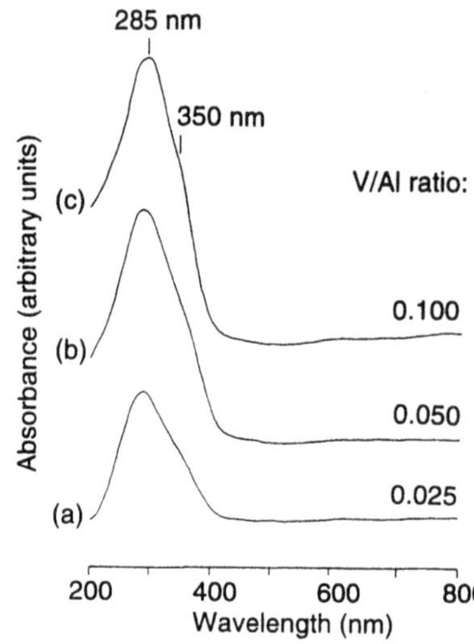

Figure 6. Diffuse reflectance UV-vis spectra of as-synthesized VUHM-1 with variable V/Al ratios.

MnUHM-1. MnUHM-1 with a low Mn/Al ratio of 3 x 10^{-4} exhibits a typical Mn^{2+} ESR spectrum with a six line hyperfine coupling of ~ 82 G and g = 2.008 [23]. The small peaks in between the six major hyperfine lines represent forbidden transitions involving $\Delta m_1 = \pm 1$ and indicate that the zero field splitting is nonnegligible [23]. This is consistent with an immobile Mn^{2+} species [24]. With increasing Mn/Al ratio up to 0.1 MnUHM-1 give similar ESR spectra but with less resolved hyperfine lines. As the Mn/Al ratio approaches 0.2, the spectra loses all hyperfine resolution. A broad line at g = 4.30 with a linewidth of about 530 G is observed for all MnUHM-1 samples (Figure 8Aa). The interpretation of this g = 4.30 peak is somewhat controversial although it can be ascribed to transitions of the $M_S = \pm 3/2$ doublet for a high spin S = 3/2 system in a rhombic ligand field [25].

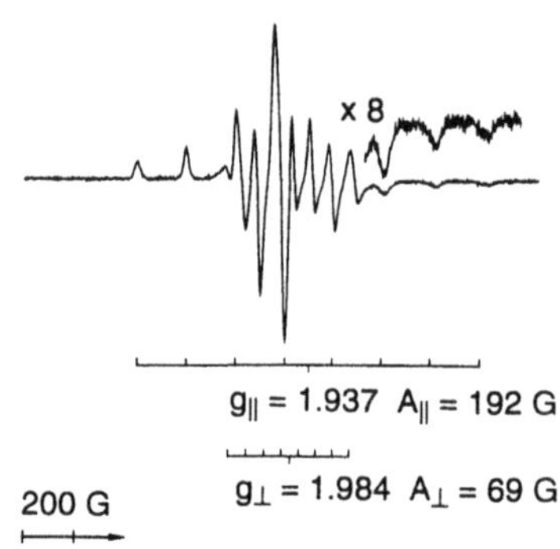

Figure 7. Room temperature ESR spectrum of as-synthesized VUHM-1 with V/Al ratio 0.050.

Upon dehydration, MnUHM-1 gives a spectrum with poorer hyperfine resolution (Figure 8Ab). After adsorption of ammonia (ND_3), the ESR spectrum remains similar to the dehydrated sample except that the relative intensity of the peak at g = 4.30 is increased (Figure 8Ac). This is also the case for pyridine (C_5D_5N) adsorption (Figure 8Ad). For comparison, the corresponding ESR spectra of Mn^{2+} ion-exchanged into UHM-1 (Mn-UHM-1) are shown. Fresh Mn-UHM-1 shows a similar ESR spectrum to MnUHM-1, but the forbidden transitions are weaker, indicating mobility of the Mn^{2+} species (Figure 8Ba). Evacuation increases the peak at g = 4.30 (Figure 8Ba). After adsorption of ND_3 or C_5D_5N, the g = 4.30 peak increases substantially (Figure 8Bc to 8Bd). These differences from MnUHM-1 indicate that the Mn^{2+} species in Mn-UHM-1 is more accessible to adsorbates.

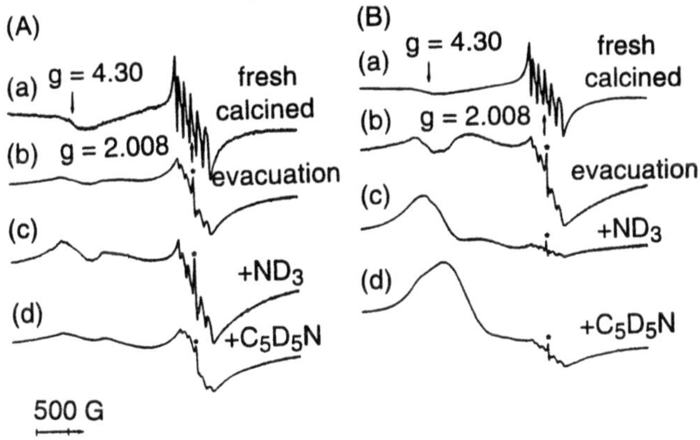

Figure 8. X-band ESR spectra at 77 K of (A) calcined MnUHM-1 and (B) Mn-UHM-1. (a) fresh calcined, and after (b) evacuation at 250 °C overnight, (c) adsorption of ammonia (ND_3), or (d) adsorption of pyridine (C_5D_5N). The asterisk indicates an unknown paramagnetic center.

4. DISCUSSION

4.1. Mesoporous structures

All the aluminophosphate materials exhibit a similar XRD pattern with one intense peak with a d-spacing about 40 Å similar to those from disordered silica materials MSU-1 and KIT-1 with three-dimensional, short wormlike, tubular channels. These materials are different from but mimic hexagonal MCM-41 mesopore materials. This conclusion is well supported by TEM images which show an array of disordered tubular channels about 40 Å in diameter. N_2-adsorption reveals that calcined UHM-1, UHM-3 and MnUHM-1 possess an average mesopore size of about 35 Å and a mesopore surface area up to 600 m^2/g, which together with the XRD results indicate 11 Å thick pore walls.

4.2. Framework atomic arrangements

The ^{27}Al and ^{31}P MAS NMR spectra reveal that UHM-1 is composed of AlO_4, AlO_6 and PO_4 structural units and hydroxylated phosphorus species. Such mixed coordination environments indicates that the condensation of either aluminum or phosphorus sources is incomplete and that UHM-1 has more disorder than siliceous MCM-41. The incorporation of silicon into UHM-1 to form UHM-3 causes a substantial increase of tetrahedral aluminum and tetrahedral phosphorus even at very low silicon loading. Virtually all the aluminums are tetrahedral and all the phosphorus is present as $P(OAl)_4$ structural units after the silicon loading approaches a Si/Al ratio of unity. Since the ^{29}Si MAS NMR shows that the incorporated silicon is atomically dispersed, it is suggested that the incorporation of silicon into UHM-3 promotes the formation of a more ordered framework similar to that in microporous silicoaluminophosphate molecular sieves.

4.3. Oxidation states and coordinations of incorporated transition-metal ions

ESR and UV-vis reveal that vanadium exists in VUHM-1 as monoatomically dispersed vanadyl VO^{2+} species in square pyramidal symmetry and also as V^{5+} ions in two different tetrahedral environments inside the tubular walls and on surface of the tubular walls. The V^{4+}

ions initially added to the gel are oxidized to V^{5+} ions. Like in vanadium-substituted MCM-41 [23], the spectroscopic parameters of these vanadium species in VUHM-1 suggest that vanadium is in a tetrahedral framework site.

ESR reveals that Mn^{2+} in both as-synthesized and calcined MnUHM-1 is immobile and inaccessible to adsorbates such as water, ammonia and pyridine, which is strikingly different from the results from ion-exchanged Mn-UHM-1. This suggests that Mn^{2+} is located in the mesoporous aluminophosphate framework in MnUHM-1.

5. CONCLUSIONS

Relatively thermally stable aluminophosphate mesoporous materials UHM-1, UHM-3, VUHM-1 and MnUHM-1, which possess disordered tubular channels of about 40 Å in diameter, have been successfully synthesized. The spectroscopic characterization reveals that these materials have a nonideal aluminophosphate framework due to incomplete condensation of phosphorus species, but the incorporation of silicon into UHM-1 to form UHM-3 promotes more complete condensation of phosphorus. Spectroscopic data also suggest a framework location for incorporated silicon, vanadium and manganese in the aluminophosphate network.

ACKNOWLEDGMENT

This research was supported by the Robert A. Welch Foundation, the University of Houston Energy Laboratory and the National Science Foundation.

REFERENCES

1. S.T. Wilson, B.M. Lok, C.A. Messina, T.R. Cannan, E.M. Flanigen, J. Am. Chem. Soc., 104 (1982) 1146.
2. C.T. Kresge, M.E. Leonowicz, W.J. Roth, J.C. Vartuli, J.S. Beck, Nature, 359 (1992) 710.
3. A. Sayari, Chem. Mater., 8 (1996) 1840.
4. S. Oliver, A. Kuperman, N. Coombs, A. Louth, G.A. Ozin, Nature, 378 (1995) 47.
5. A. Sayari, I.L. Moudrakovski, J.S. Reddy, Chem. Mater., 8 (1996) 2080.
6. Q. Gao, J. Chen, E. Xu, Y. Yue, Chem. Mater., 9 (1997) 457.
7. S. Cheng, J.-N. Tzeng, B.-Y. Hsu, Chem. Mater., 9 (1997) 1788.
8. D. Zhao, Z. Luan, L. Kevan, J. Chem. Soc.,Chem. Commun., (1997) 1009.
9. T. Kimura, Y. Sugahara, K. Kuroda, 11th Intl. Zeolite Conference Abstract RP45; Seoul, Korea, 1996.
10. B. Chakraborty, A.C. Pulikottil, S. Das, B. Viswanathan, J. Chem. Soc., Chem. Commun., (1997) 911.
11. P. Feng, Y. Xia, J. Feng, X. Bu, G.D. Stucky, J. Chem. Soc., Chem. Commun., (1997) 949.
12. S.A. Bagshaw, E. Prouzet, T.J. Pinnavaia, Science, 269 (1995) 1242.
13. R. Ryoo, J.M. Kim, C.H. Ko, C.H. Shin, J. Phys. Chem., 100 (1996) 17718.
14. G. Bellussi, M. S. Rigutto, in Advanced Zeolite Science and Applications; J.C. Jansen, M. Stöcker, H.G. Karge, J. Weitkamp, Eds.; Studies in Surface Science and Catalysis, Vol. 85; Elsevier: Amsterdam, 1994; pp 177 - 213.
15. Z. Luan, H. He, W. Zhou, C.-F.Cheng, J. Klinowski, J. Chem. Soc., Faraday Trans., 91 (1995) 2955.
16. D. Müller, J.B. Fahlke, G. Ladwig, Zeolites, 5 (1985) 53.
17. J. Rocha, W. Kolodziejski, H. He, J. Klinowski, J. Am. Chem. Soc., 114 (1992) 4884.
18. L.S. de Saldarriaga, C. Saldarriaga, M.E. Davis, J. Am. Chem. Soc., 109 (1987) 2686.
19. B.M. Lok, C.A. Messina, R.L. Patton, E.M. Flanigen, J. Amer. Chem. Soc., 106 (1984) 6092.
20. M. Morey, A. Davidson, H. Eckert, G. Stucky, Chem. Mater., 8 (1996) 486.
21. Z. Luan, J. Xu, H. He, J. Klinowski, L. Kevan, J. Phys. Chem., 100 (1996) 19595.
22. M. Narayana, L. Kevan, J. Phys. C: Solid State Phys., 16 (1983) L863.
23. G. Brouet, X. Chen, C.W. Lee, L. Kevan, J. Am. Chem. Soc., 114 (1992) 3720.
24. D. Zhao, D. Goldfarb, J. Chem. Soc., Chem. Commun., (1995) 875.
25. J. Kliava, Phys. Stat. Sol. (b), 134 (1986) 411.

MESOPOROUS MOLECULAR SIEVES 1998
Studies in Surface Science and Catalysis, Vol. 117
L. Bonneviot, F. Béland, C. Danumah, S. Giasson and S. Kaliaguine (Editors)

Mesoporous molecular sieve thin films

M. E. Gimon-Kinsel and K. J. Balkus, Jr.*

Department of Chemistry, The University of Texas at Dallas, PO Box 830688, Richardson, Texas 75083-0688 United States

Mesoporous films of MCM-41 and transition metal oxides (Me-TMS1) were generated using pulsed laser deposition followed by a brief post hydrothermal treatment. In particular, the Nb-TMS1 molecular sieves possess semiconductor properties and thin films of Nb-TMS1 lend themselves to applications such a capacitive-type chemical sensors made using the Nb-TMS1 films as the dielectric phase. The capacitance change of such mesoporous molecular sieve based chemical sensors is sensitive to the relative moisture level in their environment making them useful as potential humidity sensors. These sensors were also evaluated for the detection of volatile organic compounds such as methanol and acetone.

1. INTRODUCTION

Mesoporous molecular sieves are of great interest to the materials community because their pore structure, catalytic, conductive and magnetic properties can be somewhat tailored for specific applications. These high surface area materials may find utility in areas ranging from catalysis to chemical sensors. Many of these applications could benefit from thin film configurations of the mesoporous materials. There has been recent progress in this area for mesoporous silicate and aluminosilicate materials [1]. Ozin et al. [1b] have synthesized oriented mesoporous silicate films at the air/water interface which can be lifted from the water and show a resilience to bending. Aksay et al. [1c] have grown mesoporous silicate films on a variety of interfaces including water/mica, water/graphite and water/silicon. The nanostructure of the resultant film was found to depend on the specific interactions of the organic surfactant with the substrate. Ogawa [1d] has used a rapid spin coating procedure to produce mesoporous silica-surfactant composite films which are regarded as unique due to their transparency. Ganguli et al. [1e] have reported the formation of mesostructured silica films using a dip-coating method. However, it is unclear how universally applicable these methods will be. We have employed pulsed laser deposition (PLD) for the fabrication of mesoporous films. PLD has proven to be an effective method for producing thin films of other low density molecular sieves including microporous zeolites such as UTD-1 [2] as well as aluminophosphate molecular sieves including $AlPO_4$-5 and MeAPO-5 (Me = Co, Mn, Mg and V) [3]. The laser deposited molecular sieve films generally have low crystallinity but can be reorganized (crystallinity improved) by a brief hydrothermal treatment. In this paper we report the generation of mesoporous films of MCM-41 and the transition metal oxide mesoporous molecular sieve Nb-TMS1 [4] by PLD/post hydrothermal treatment. This is the first time

* Author to whom correspondence should be addressed

mesoporous films have been made using this laser deposition technique. The laser deposition of silicate and aluminosilicate MCM-41 molecular sieves were found to require a sensitizer (UV absorbing molecule) to assist in the ablation process. We refer to this modification as guest assisted laser ablation (GALA). In contrast, the Nb-TMS1 could be ablated without modification. The films of Nb-TMS1 lend themselves to applications such as capacitive-type chemical sensors made using the Nb-TMS1 films as the dielectric phase. We have found that the capacitance change of such mesoporous molecular sieve based chemical sensors is sensitive to the relative moisture level in their environment. Additionally, the response of Nb-TMS1 chemical sensors was compared to that of analogous sensors made using dense phase Nb_2O_5 as the dielectric phase.

2. EXPERIMENTAL

2.1. Molecular Sieve Synthesis

The aluminosilicate MCM-41 was synthesized according to the published procedure [5]. The white MCM-41 powder was prepared by mixing the reagents in the following molar ratios, 1 SiO_2: 0.1 $Al_2(SO_4)_3$: 0.26 $(TMA)_2O$: 0.27 $(CTMA)_2O$ where TMA and CTMA are the trimethylammonium ion and cetyltrimethylammonium ion, respectively. The source of silica was Cab-O-Sil M5 fumed silica (Cabot). The resulting gel was heated under static conditions at 150°C for 48 hours. The molecular sieve product had a Si/Al ratio of ~13 with >85% of the Al in tetrahedral sites as determined by ^{27}Al MAS NMR. Organometallic modified aluminosilicate MCM-41 was prepared by first calcining at 540°C followed by stirring the template-free molecular sieve in an aqueous solution containing the bis(pentamethylcyclopentadienyl)cobaltocium ($Cp_2{}^*Co^+$) ion overnight. The yellow aluminosilicate MCM-41 containing the $Cp_2{}^*Co^+$ ions was collected and washed with deionized water.

The transition metal oxide mesoporous molecular sieve Nb-TMS1 (pore size 22Å) was synthesized according to the published procedure [4]. A typical preparation of the synthesis gel involved the mixing of niobium (V) ethoxide (CHEMAT Technology, Inc.) and dodecylamine (Aldrich) with the molar ratio 1:1.25. The resulting mixture was heated at 50°C for approximately 1 minute, then 10 mL ethanol followed by 10 mL deionized water were added which resulted in the precipitation of a white solid. The solid was aged in the supernatant for 72 hours. The white solid and supernatant were loaded into a 23 mL Teflon lined stainless steel autoclave and heated for 24 hours at 80°C, 24 hours at 100°C, followed by 7 days at 180°C. The white product was isolated by suction filtration, washed with deionized water, ethanol and diethyl ether and dried overnight at room temperature.

2.2. Pulsed Laser Deposition and Post Hydrothermal Treatment of Films

In general, a pressed pellet of the molecular sieve was placed in a controlled-atmosphere chamber and irradiated using a Lumonics HyperEx-400 excimer laser (248 nm (KrF*), pulse length 14 ns, repetition rate 10 Hz). The laser energy was measured using a Scientech pyroelectric head (Model 380402) and found to range from 90 to 100 mJ/pulse. A computer controlled restoring mirror (Oriel) was used to turn the laser beam 90° and move the beam across the pellet surface. A focusing lens was employed to decrease the laser beam to a spot size of ca. 0.001 cm^2. Films were deposited by placing a substrate such as titanium nitride

(TiN) coated silicon on the heated substrate holder located approximately 2.5 cm in front of the pellet. Typical film deposition conditions are as follows: laser power density, 100 MW/cm^2; substrate temperature, 150-200°C; background oxygen pressure, 250 mTorr; deposition time, 20 min. Scanning electron microscopy (SEM) was used to examine the morphology and determine the thickness of the deposited films.

The crystallinity of the Nb-TMS1 thin film deposited as described above was improved by a brief post hydrothermal treatment. The film coated TiN/silicon substrate was positioned in a 23 mL Teflon lined stainless steel autoclave at approximately 60° angle with the film side facing down. The melted niobium (V) ethoxide and dodecylamine (molar ratio 1:1.25) mixture was added to the Teflon liner followed by 20 mL ethanol. The resulting mixture was allowed to age for 72 hours at room temperature, 4 hours at 90°C, 4 hours at 140°C and for 24 hours to 3 days at 180°C.

2.3. Chemical Sensor Fabrication

A capacitive-type chemical sensor was prepared from a hydrothermally treated/template-free Nb-TMS1 thin film which had been deposited onto a TiN/silicon substrate. The organic template was removed by washing in a solution of 3:1 isopropanol: water acidified with nitric acid to pH 1.75 for 3 hours followed by washing in ethanol. A Au/Pd alloy was patterned onto the film surface using a shadow mask. The device was than heated to 200°C to desorb water and immediately transferred to an environmental chamber under a positive pressure of high purity nitrogen where it was allow to cool to room temperature (24°C).

The environmental chamber was equipped with two probes by which film plus adsorbate capacitance was measured. One probe made electrical contact with the bare TiN surface and the other probe made contact with a 0.184 mm^2 Au/Pd electrode. A voltage was applied and the capacitance signal was received by a WayneKerr Precision Component Analyzer 6425. A virtual instrument program written in-house using LabVIEW for Windows 3.0.1 allowed the applied voltage to be stepped from 0V to 0.300V and the measured capacitance plotted as a function of applied voltage.

3. RESULTS AND DISCUSSION

We have previously shown that excimer irradiation of all silica MCM-41 results in the generation of defect sites and associated photoluminescence [6]. The as-synthesized aluminosilicate MCM-41 (containing framework mostly aluminum) was strongly photoluminescent prior to excimer irradiation and material was not deposited by PLD. It would appear that films of these materials could not be generated using PLD because the MCM-41 absorbs the excimer energy which generates defect sites. However, we found that if the surfactant template was removed and the organometallic $Cp_2{*}Co^+$ was exchanged in the mesoporous structure, the PLD of MCM-41 films was possible. Figure 1 shows a SEM of laser-deposited modified aluminosilicate MCM-41 containing $Cp_2{*}Co^+$ ions on a polished silicon substrate. The experimental conditions used to generate the film were 100 MW/cm^2, 240 mTorr, 155°C substrate temperature and 30 minutes deposition time. The particle size of the deposited material is on average less than 500 nm. Examination of the laser irradiated pellet revealed no increase in photoluminescent intensity in the 248-nm irradiated areas.

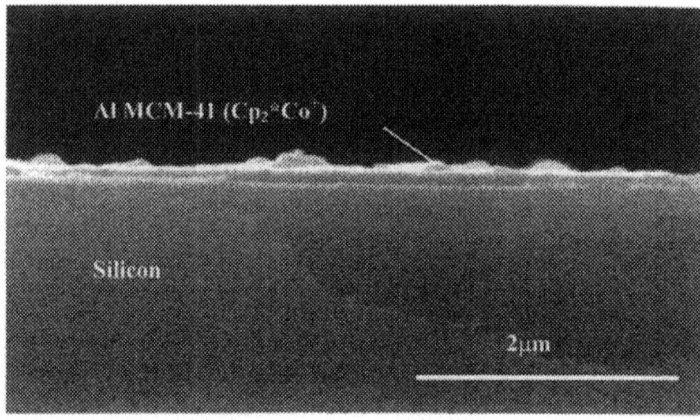

Figure 1. SEM showing laser-deposited modified aluminosilicate MCM-41 film cross-section (170 nm).

The orientation and ordering of this film are under evaluation. We refer to this modification of the PLD technique as guest assisted laser ablation (GALA). From our initial observations, it appears that the $Cp_2{*}Co^+$ molecules absorb the laser irradiation preventing the generating of defects in the molecular sieve framework. We are currently evaluating a number of small UV absorbing molecules as possible guest molecules. With the discovery of GALA, laser deposited films could be generated from a wide variety of all silica and aluminosilicate molecular sieves.

Table 1
Experimental conditions for pulsed laser deposition of Nb-TMS1 films on TiN/silicon

Example	Laser Power Density (MW/cm^2)	Pressure (mTorr)	Substrate Temperature (°C)	Deposition Time (min)	Film Thicknes (nm)
1A	100	300	150	80	180
1B	180	150	130	20	120
1C	230	240	100	14	90
1D	260	230	300	20	240
1E	290	240	120	14	130
1F	300	250	300	20	300
1G	470	250	200	20	383
1H	500	260	300	20	400
1I	500	250	100	20	403
1J	600	250	120	14	400

Selected deposition conditions and the resultant film thicknesses for Nb-TMS1 are shown in Table 1. All the TiN/silicon substrates, with the exception of 1C, 1E, 1I and 1J were observed to be coated with a uniform continuous layer of nanometer scale particles having diameters ranging from about 30 nm to 250 nm. The film morphology of Examples 1C, 1E, 1I and 1J was observed to be nonuniform or "cratered" in appearance. SEM images of the surface morphology and cross-section for Example 3D film are shown in Figure 2.

A

B

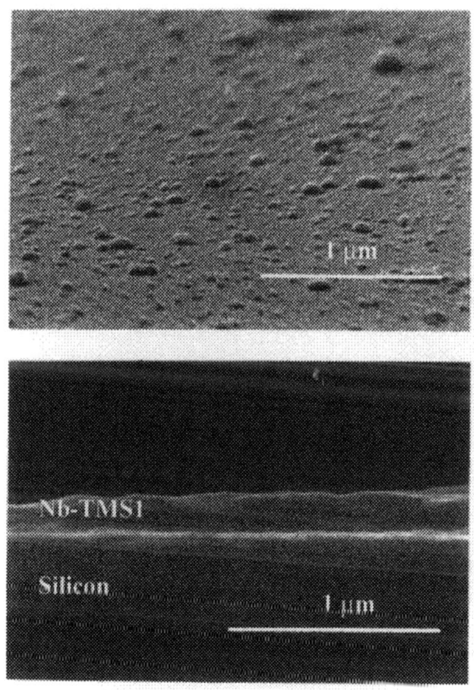

Figure 2. (A) SEM image of the surface morphology and (B) cross-section (240 nm) of a niobium oxide film deposited by pulsed laser deposition using a Nb-TMS1 pellet.

Following a brief post hydrothermal treatment (see Table 2), the deposited films were transformed into well adhered Nb-TMS1-like films. By "Nb-TMS1-like" it is meant that the films exhibit the X-ray diffraction patterns that would be expected for the Nb-TMS1 bulk material. Interestingly, little growth occurred on the uncoated portions of the TiN/silicon substrate surface also exposed to the Nb-TMS1 synthesis mixture. The X-ray diffraction patterns of Examples 2C-2E included a strong relative intensity line at 27.3 +/- 0.3Å d-spacing. The position of this reflection does not change dramatically (within error) following template removal using acid wash. Figure 3 shows a SEM image of Example 2A hydrothermally treated film.

Table 2
Experimental conditions for hydrothermal treatment of laser deposited Nb-TMS1 films.

Example	Days Aging at 180°C	Initial Film Thickness (nm)	Final Film Thickness (nm)	d_{100} present?
2A	2.5	300	330	no
2B	2.5	240	300	no
2C	3	400	10000	yes
2D	3	383	7000	yes
2E	3	403	13000	yes

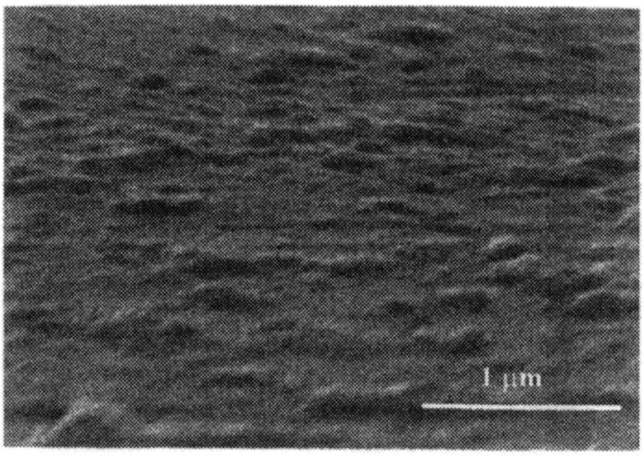

Figure 3. SEM image of a hydrothermally treated Nb-TMS1 film.

Capacitive-type chemical sensors were fabricated using the hydrothermally treated/template-free Nb-TMS1 films as the dielectric phase, the TiN substrate as an electrode and patterned Au/Pd as the other electrode. By applying a potential between the electrodes, changes in the capacitance were recorded upon absorption of analyte molecules into the molecular sieve film. We found that the Nb-TMS1 films were particularly sensitive to the relative level of moisture in the environmental chamber. When the Nb-TMS1 based sensor was subjected to a flow a nitrogen containing water vapor, the capacitance measured across the hydrothermally treated film increased as the water was adsorbed. The background capacitance/area measured in flowing high purity nitrogen was on the order of 0.30 nF/mm^2. The capacitance/area increased to 5.2 nF/mm^2 when water was introduced into the flowing nitrogen. Because water can move freely into and out of the pores of the hydrothermally treated film, the capacitance response can be monitored over a period of time. For example, Figure 3 illustrates the reversible response to water of sensors based on mesoporous films which are approximately 300 nm thick. A repeat cycle (x3) of 90% humidity followed by a flush of high purity nitrogen was used to determine the reproducibility of the capacitance response. During each 10 min flush with nitrogen, the capacitance returned to its baseline value within 1 min. The average capacitance change varied no more that +/- 2.5% for the three sequential cycles.

In order to develop a humidity sensor using the hydrothermally treated/template-free Nb-TMS1 films, we would need to know the specificity of the capacitance response to water. During the course of these experiments, gaseous molecules such as carbon dioxide (100% concentration) did not produce a measurable increase in capacitance above the background value. The vapors of dry methanol and acetone produced a minimal capacitance change and therefore would not interfere in the detection of water. The effect of other potential analytes such as ammonia is the subject of current studies.

Niobium oxide films consisting of Nb_2O_5 have been shown to function as humidity sensors [7] .The mechanism by which the mesoporous Nb-TMS1 films act as a humidity sensor may parallel that of the dense phase material. Laser deposited films of Nb_2O_5 were generated and determined to be crystalline using XRD. The capacitance response of chemical sensors based on the mesoporous Nb-TMS1 films was compared to the response of sensors fabricated from laser-deposited dense phase Nb_2O_5 films of similar thicknesses (results not shown). The absolute capacitance response measure for the sensors based on the Nb_2O_5 films is lower by 3 orders of magnitude than that measured for the Nb-TMS1 based devices. However, the relative sensor response (response to water/response to nitrogen) calculated for the Nb-TMS1 and Nb_2O_5 based sensors are identical within experimental error. Thus, we are led to speculate at this point that the larger absolute capacitance response of the Nb-TMS1 based sensors is due the larger surface area/increased absorption capacity of the mesoporous films.

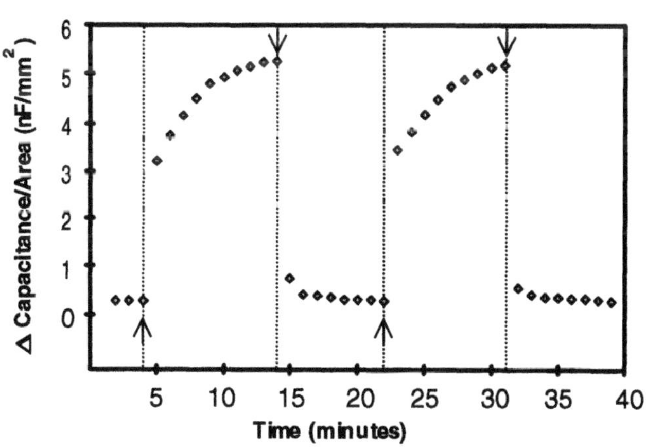

Figure 4. Average reversible response of a capacitive-type Nb-TMS1 chemical sensor to the presence of water vapor.

4. CONCLUSIONS

Continuous films of aluminosilicate MCM-41 and Nb-TMS1 mesoporous molecular sieves have been prepared using pulsed laser deposition. Ablation of the aluminosilicate MCM-41 requires the presence of a sensitizer (i.e. UV absorbing molecule) which is a phenomenon we refer to as GALA. PLD followed by a brief post hydrothermal treatment can be used to generate films of mesoporous Nb-TMS1 molecular sieves. Capacitive-type chemical sensors can then be fabricated using these films as the dielectric phase. It was shown that such chemical sensors are sensitive to the relative level of moisture in the environment which makes them useful as potential humidity sensors. We anticipate that the compositional variances available for mesoporous molecular sieves will lead to more selective sensors as well as other applications for these thin films.

ACKNOWLEDGEMENTS

We thank the Texas Advanced Technology Program for funding this research.

REFERENCES

1. (a) G. Ozin, Nature, 379 (1996) 703. (b) H. Yang, N. Coombs, Igor. Sokolov, G. A. Ozin, Nature, 382 (1996) 589. (c) I. A. Aksay, M. Trau, S. Manne, I. Honma, N. Yao, L. Zhou, P. Fenter, P. M. Eisenberger, S. M. Gruner, Science, 273 (1996) 892. (d) M. Ogawa, Chem Commun. (1996) 1149. (e) R. Ganguli, Y. Lu, M. T. Anderson, C. A. Drewien, C. J. Brinker, H. Soyez, B. Dunn, M. H. Huang, J. I. Zink, Nature, 389 (1997) 364.
2. K. J. Balkus, Jr., T. Muñoz, Jr., M. E. Gimon-Kinsel Chem. Mater. (in press).
3. (a) K. J. Balkus, Jr., L. J. Sottile, S. J. Riley, B. E. Gnade, Thin Solid Films 260 (1995) 1183. (b) K. J. Balkus, Jr., L. J. Sottile, H. Nguyen, S. J. Riley, B. E. Gnade, Mater. Res. Soc. Symp. Proc., 375 (1994) 33. (c) L. J. Sottile, K. J. Balkus, Jr., S. J. Riley, B. E. Gnade, Mater. Res. Soc. Symp. Proc., 351 (1994) 263. (d) K. J. Balkus, Jr., S. J. Riley, B. E. Gnade, Mater. Res. Soc. Symp. Proc., 351 (1994) 437.
4. D. M. Antonelli, A. Nakahira, J. Ying, Inorg. Chem., 35 (1996) 3126.
5. Z. Luan, C.-F. Cheng, W. Zhou, J. Klinowski, J. Phys. Chem., 99 (1995) 1018.
6. M. E. Gimon-Kinsel, K. Groothuis, K. J. Balkus, Jr., Micropor. Mater. (in press).
7. N. Kurioka, D. Watanabe, M. Haneda, T. Shimanouchi, T. Mizushima, N. Kakuta, A. Ueno, T. Hanaoka, Y. Sugi, Catalysis Today, 16 (1993) 495.

MESOPOROUS MOLECULAR SIEVES 1998
Studies in Surface Science and Catalysis, Vol. 117
L. Bonneviot, F. Béland, C. Danumah, S. Giasson and S. Kaliaguine (Editors)

Nucleation, growth and form of mesoporous silica: role of defects and a language of shape

Geoffrey A. Ozin,*[a] Charles T. Kresge[b] and Hong Yang[a]

[a]Materials Chemistry Research Group, Lash Miller Chemical Laboratories, 80 St. George Street, University of Toronto, Toronto, Ontario, Canada M5S 3H6.

[b]Mobil Technology Company, P.O. Box 480, Paulsboro, New Jersey, 08066-0480, USA

Topological defects in a silicate liquid crystal seed are considered to play a central part in directing the growth of hexagonal mesoporous silica towards specific morphologies. Below the silica isoelectric range, pH = 1-3, silicate is protonated and the seed carries a sheath of positive surface charge, that can be tuned by adjusting the acidity of the growth medium. Surface anchoring interactions at the polar interface between the seed and surroundings are believed to favor certain types of defects and patterns of director fields in the seed, while van der Waals attractive and electrical double layer repulsive forces between colloidal dimension seeds and rod micelles promote particular kinds of accretive growth processes between them. A survey of the relation between synthesis composition and shape reveals that high acidity and large electrical double layer repulsions favor end-to-end (homeotropic anchoring) seed-rod growth, leading to fiber-based morphologies bearing different extents of curvature. Lowering the acidity facilitates side-to-side (planar anchoring) seed-rod growth resulting in higher curvature gyroid and spiral shapes. As the isoelectric point is approached, a further reduction in electrical double layer repulsive interactions induces a switch in the growth process to one involving seed-seed flocculation, culminating in sphere shapes. A defect initiated growth model is able to rationalize the origin of curved mesoporous silica shapes.

1. INTRODUCTION

Mesoporous silica represents a new class of materials containing periodic arrays of channels and cavities, that are tunable over the range 20-100 Å.[1] Synthetic strategies have recently been developed that provide hexagonal mesoporous silica with a collection of fascinating and highly curved shapes, including, fibers, toroids, discoids, gyroids, spirals and spheres.[2-6] Herein we describe the results of new experiments that identify growth nuclei in the formation of mesoporous silica under dilute aqueous acidic conditions, delineate the relation between synthesis conditions and shape, and define the connection between mesostructure and morphology. These observations enable a model to be formulated that can rationalize the nucleation, growth and form of these curved mesoporous silica shapes which is based upon concepts borrowed from surfactant-sol-gel chemistry, colloid science, liquid crystal physics and crystal growth.

2. RESULTS AND DISCUSSION

From an initially homogeneous synthesis solution (xHCl : yCTACl : zTEOS : $100H_2O$ at RT and 80°C), thermal fluctuations cause silicate micelles to assemble into a critical size nucleus that can sustain further growth. In preparations that favor predominantly fibers, gyroids or spheres, growth nuclei have been deposited from clear synthesis solutions onto freshly cleaved graphite and copper grids, and the early stage probed by tapping mode AFM and TEM. The smallest object that has been observed is *ca.* 500 Å in size and has a droplet appearance, **Figure 1**. At slightly longer reaction times these nuclei enlarge to form recognizable mesoporous silica shapes which continue to evolve over time to the final morphology. High acidity favors the fibers while the gyroids and spheres appear sequentially as the acidity is progressively decreased towards the isoelectric range of silica, pH = 1-3. The growth process that yields these shapes is accelerated at higher concentrations of CTACl and TEOS. A synthesis composition-shape ternary diagram summarizing some of these results is shown in **Figure 2**.

Imaging studies of these shapes, using SEM and TEM, define the relation between the acidity of a synthesis, the emergence of a particular shape, and the connection of mesostructure to morphology.[2-6] It is found that well-ordered channels run down the body length of the fibers and whirl concentrically around the rotation axis of the gyroids, however, the channels are not as well-ordered in the spheres and the surface texture shows a granular appearance. POM studies of these mesoporous silica shapes confirm the association of channel structure to overall morphology.[2-6]

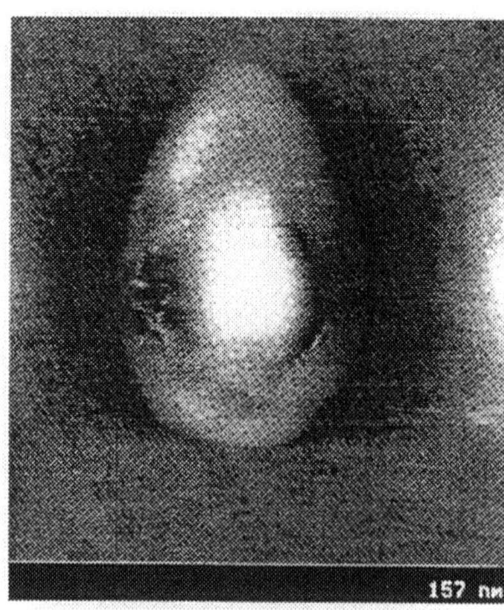

Figure 1. AFM image (height mode), of the early stage growth nucleus that has been observed in a mesoporous silica fiber preparation.

It is especially noteworthy, that POM reveals optical birefringence patterns that are characteristic of the types of line defects usually found in hexagonal lyotropic liquid crystals, but now they are rigidified and preserved in mesoporous silica.[7,8] This strongly suggests that topological defects exist in silicate liquid crystal growth nuclei and are responsible for initiating a particular kind of growth process that is manifest spatially in the form of the contour surfaces of the resulting mesoporous silica morphology.

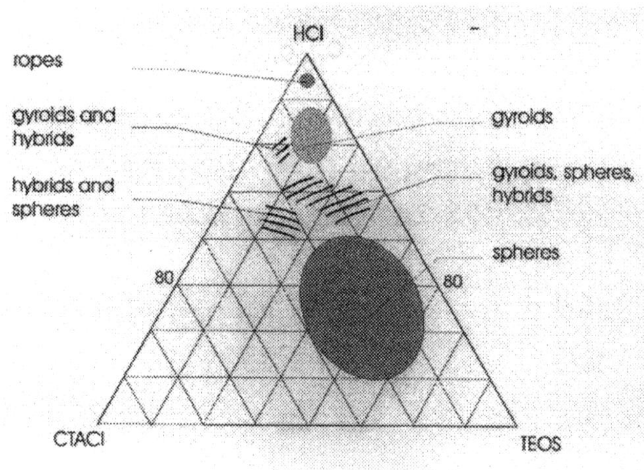

Figure 2. Illustration of synthesis composition-shape ternary diagram for mesoporous silica morphologies: xHCl : yCTACl : zTEOS : $100H_2O$, 80°C.

The problem of growth and form in mesoporous silica therefore appears to depend upon a unique combination of physico-chemical phenomena, taken from surfactant-sol-gel chemistry, colloid science, liquid crystal physics and crystal growth. First, it is important to define the factors that control the formation of particular kinds of defects and director field configurations in a silicate liquid nucleus, because these likely play a pivotal role in the initiation and progress of a particular growth process. Within an elastic medium definition of the growth nucleus, it is the bulk and surface elastic constants and surface anchoring (i.e., homeotropic or planar) strengths at the polar interface, that will determine defects and stable director field patterns. These are expected to be sensitive to the dielectric and ionic strength of the growth medium.[9] Second, it is necessary to understand the kinds of intermolecular forces involved in the assembly of the critical size nucleus and its subsequent growth, because these processes involve interactions between colloidal dimension particles, where the main part of the free energy of interaction is from the sum of van der Waals attractions and electrical double layer repulsions.[10] In view of the observed sensitivity of mesoporous silica shape on acidity, particular attention is drawn to the way in which the double layer repulsion between colloidal dimension particles depends on the dielectric and ionic strength of the intervening medium, which likely directs a specific growth process.

Point, line and planar defects, intrinsic to regular solid state materials, are of central importance to crystal nucleation, growth and morphology.[11] Furthermore, the energy required for stabilizing defects in liquid crystals is far less than in crystalline solids. Thus, it should therefore be of no surprise that topological defects can be easily formed in a silicate liquid crystal seed which can profoundly affect growth and form of mesoporous silica materials. Assuming that the packing geometry of flexible rod-like micelles in the seed is more-or-less hexagonal, **Figure 3**, then the Volterra method enables translational topological defects to be defined.[7,8,12]

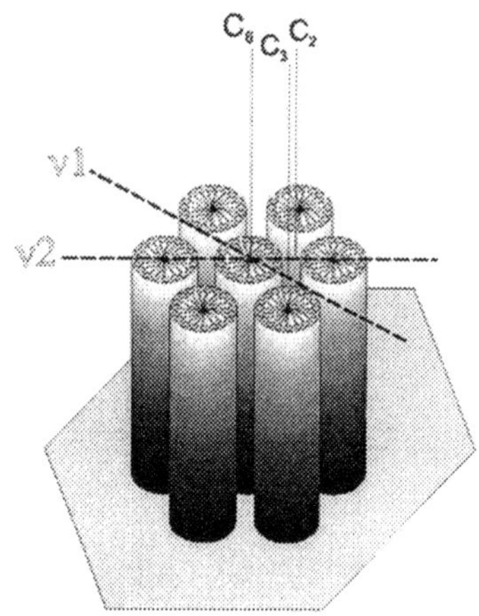

Figure 3. Symmetry elements in a hexagonal close packed rod micelle arrangement

The defects pertinent to the hexagonal silicate liquid crystal seed are transverse and longitudinal edge dislocations, transverse screw dislocations, and disclinations normal and parallel to the axes of the micellar rods.[6,12] Although the cause and effect of particular defects in the seed has yet to be established, it is reasonable to expect that curved morphologies can originate from longitudinal and transverse disclinations, with the rotation vector along the six-fold (C_6, $\pm\pi/3$), three-fold (C_3, $\pm 2\pi/3$), two-fold (C_2, $\pm\pi$), (ν_1, or ν_2, $+\pi$ or $+2\pi$) axes to produce for example, curved fiber, gyroids, and toroid morphologies.

Substantial numbers of the simple mesoporous silica shapes have a remarkable similarity to the contour surfaces predicted from defect initiated growth.[12] A comparison between some of the theoretically predicted and observed synthetic shapes is presented in **Figure 4**. Most of these morphologies could arise from disclination defects. The simplest morphologies are the discoids and flat toroids which may be initiated from a $+2\pi$ disclination rotated along the transverse axis in the *ab*-plane, **Figures 4A**, and single point spirals from different extents of a $+2\pi$ screw dislocation, **Figure 4B**. Bent fibers probably emerge from a $+\pi$ disclination rotated along the transverse axis, and the twist feature observed in hexagonal fibers can be thought of as the consequence of $+\pi/3$ or $-\pi/3$ disclinations around the longitudinal axis C_6, **Figure 4C.** TEM images that define mesostructure-morphology relations in mesoporous silica gyroid shapes provide compelling evidence for the proposed defect structures.[2-6] Several types of defect structures in non-descript mesoporous silica have been described.[13]

Figure 4. Representative morphologies that may arise from simple defects; left: SEM images, right: theoretical models.

Hybrid structures based upon a combination of two defects can lead to a variety of mesoporous morphologies, **Figure 5**. Three major combinations, involving two $+\pi$ screw dislocations, two $+\pi$ disclinations, and a $-\pi/3$ disclination normal to and a $+2\pi$ disclination orthogonal to the main rotation axis, have been identified. If growth is initiated by a parallel pair of $+\pi$ screw dislocations, a dual center spiral can be generated, **Figure 5A**. In the case of growth initiated by two $+\pi/+\pi$ disclinations having their transverse axes at different angles then either bagel or mussel shapes can evolve, **Figures 5B**. Similarly an irregular bagel shape appears when the two axes are at an angle other than 0° or 90°, **Figure 5C**. A third variety involves fusion of a $-\pi/3$ disclination normal and $+2\pi$ disclination orthogonal to the main rotation axis and this defect could initiate development of a twisted toroid, **Figure 5D**.

These defect-channel plan-shape relations manifest themselves as characteristic birefringence patterns when viewed between crossed polarizers in an optical microscope. The optical textures of discoid ($+2\pi$ disclination), ribbon ($+\pi$ disclination), elongated discoid (two parallel $+\pi$ disclinations), and bagel (two orthogonal $+\pi$ disclinations) shapes with at least one transverse axis parallel to the optic axis of the hexagonal mesoporous silica have been identified.[6] Clearly all of these mesoporous silica optical textures bear a close resemblance to those observed for defect structures and patterns of director fields in organic liquid crystals.

Acidity and temperature in a mesoporous silica synthesis will affect surface anchoring interactions at the interface between the silicate liquid crystal seed and silicate rod micelles in the surrounding medium. These interactions will also influence which topological defect(s) is stabilized in the seed and how it directs the growth of the seed to a related mesoporous silica shape.

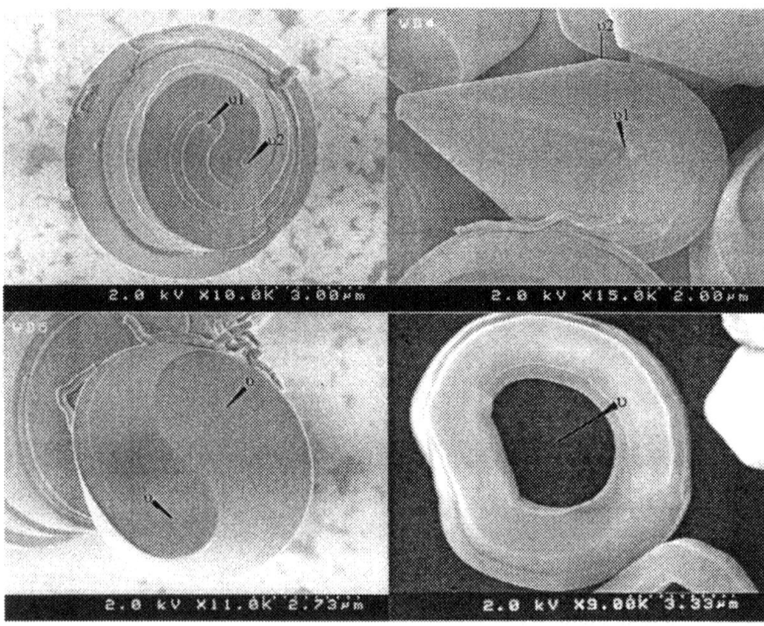

Figure 5. Representative morphologies that may arise from combined defects.

The fiber, gyroid and sphere shapes all emerge from acidic syntheses with pH values increasing from around -0.6 to +0.1, which is below the isoelectric range of silica, pH = 1-3.[14] The silicate liquid crystal seed is therefore expected to be comprised of a CTA^+Cl^- $SilicateH_x^{x+}xCl^-$ assembly with hydrogen-bonded, protonated and polymerized silicate moieties located at the polar interface, and exhibits pH tunable surface charge. The seed forms from silicate rod micelles and grows in size by continued accretion of silicate micelles. Both the silicate seed and rod micelles are colloidal dimension particles with a sheath of positive surface charge, hence the attachment of one to the other will be controlled by the minimum free energy pathway. It is anticipated that the two major sites of deposition of rods on the seed are end-to-end (homeotropic anchoring) and side-to-side (planar anchoring). Furthermore, it is logical that this process will be controlled, to a large extent, by a compromise between van der

Waals attractive and electrical double layer repulsive interactions between the rod and seed, which will depend on the magnitude and thickness of surface positive charge.

At high acidity and high surface charge, particular defects are favored in the seed and it is to be expected that electrical double layer repulsion will dominate, leading to end-to-end aggregation between rod and seed, resulting in rapid axial growth, **Figure 6**. Under these conditions, defects will direct the growth towards mesoporous fiber morphologies. As the acidity and surface charge are decreased, the defect structure in the seed is likely to change and van der Waals attractions will begin to take over, favoring side-to-side growth of rod and seed. Under these conditions defects direct the growth towards mesoporous silica gyroid, spiral, and bagel morphologies. Interestingly, as the acidity moves even lower and towards the isoelectric point of silica, mesoporous sphere shapes begin to emerge with a surface texture that appears to be an agglomeration of grains (RMSR ~ 500 Å). It appears that in this region of diminished surface charge the electrical double layer repulsion has decreased to the point that flocculation of seeds is the favored growth pathway rather than homeotropic or planar anchoring of rods to seeds.

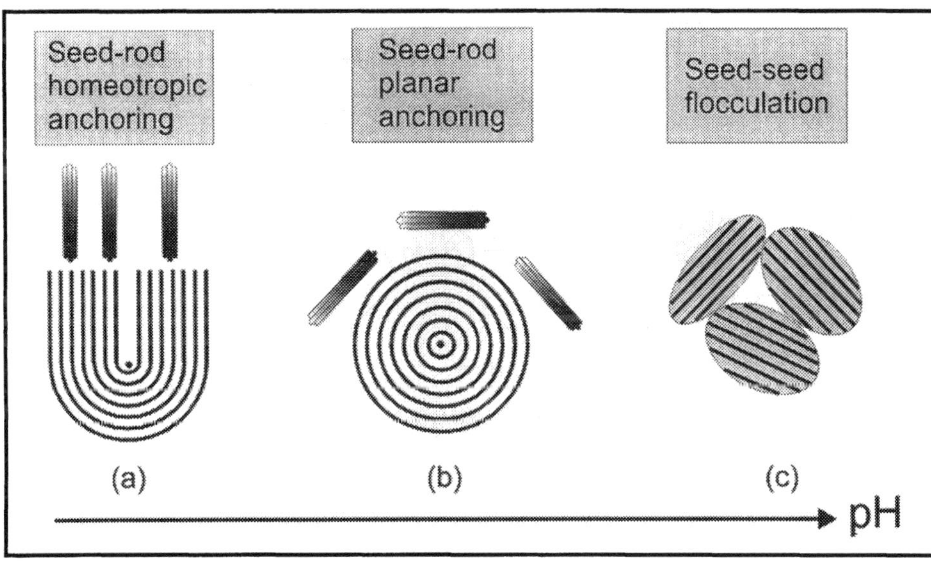

Figure 6. Defect and director field patterns, and growth of silicate liquid crystal seeds, in the synthesis of mesoporous silica with different curved shapes.

Some pivotal experiments that shed further light on the proposed growth mechanism of mesoporous silica, involve the effect of added electrolytes and co-solvents to syntheses that favor the production of fiber and gyroid shapes. For example, the addition of increasing amounts of NaCl up to saturation has a negligible effect on the production of these shapes. In stark contrast, the addition of increasing increments of ethanol causes a gradual transformation of the fiber to gyroid to sphere shapes. As to be expected, there is a concomitant reduction in the mesopore diameter of the fibers and gyroids as ethanol is known to reduce silicate head group repulsions and thus the diameter of the silicate rod micelles. These observations provide strong support for the colloidal growth pathway from silicate rod micelles and silicate liquid crystal seeds. Clearly the ionic strength of the HCl in a typical synthesis has reduced the thickness of the electrical double layer to the point that added NaCl has little effect on the energetics and dynamics of the seed-rod aggregation pathway. On the other hand, the dielectric screening of the surface charge by the less polar solvent ethanol and concomitant reduction of the electrical double layer repulsion, is sufficiently pronounced to favor seed-seed flocculation and growth of sphere shapes, instead of rod-seed assembly to fiber and gyroid shapes. The observation that added EtOH sequentially transforms fibers to gyroids to spheres and therefore has the same outcome as decreasing the acidity in a synthesis, provides credence for the colloid growth pathway. The proposed colloidal based assembly-growth mechanism is summarized in Figure 7.

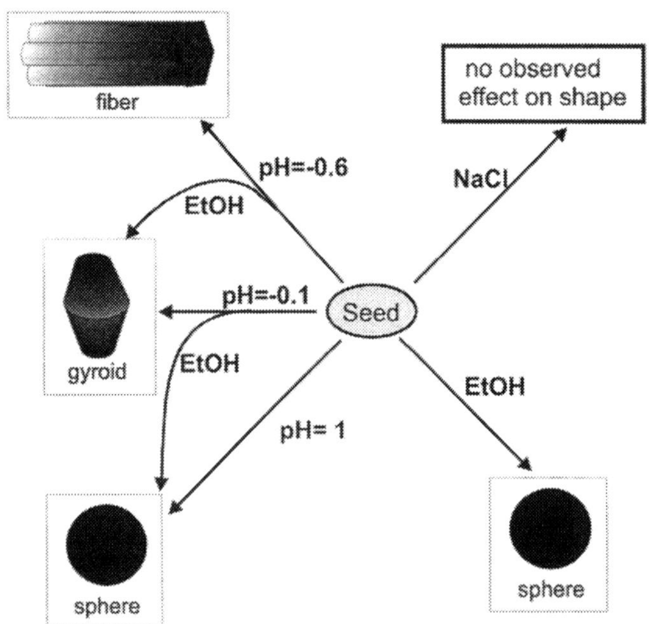

Figure 7. Summary of proposed colloid-based growth mechanism

3. CONCLUSIONS

Concepts have been drawn from surfactant-sol-gel chemistry, colloid science, liquid crystal physics and crystal growth in order to develop a model for the nucleation, growth and form of low and high curvature hexagonal mesoporous silica morphologies. Attention is directed to synthesis composition-shape and mesostructure-morphology relations, confined defect and director field patterns in liquid crystal seeds, van der Waals and electrical double layer forces, ionic strength and dielectric medium effects on the interaction between colloidal dimension silicate rod micelles and silicate liquid crystal seeds. Within an initially clear and homogeneous synthesis solution, Brownian motion induces collisions between co-assembled silicate rod micelles and enables spontaneous formation of *ca.* 500 Å diameter silicate liquid crystal growth nuclei. The pH, electrolyte concentration and dielectric strength of the synthesis medium pre-determine surface charge, electrical double layer thickness, type of defect and director field in the seed. Subsequent seed-rod accretive steps are therefore pre-destined to emerge as specific mesoporous silica morphologies simply by virtue of the shape-directing capability of the director field patterns and electrical double layer repulsions in the seed.

ACKNOWLEDGEMENTS

We are deeply indebted to Mobil Technology Company for financial support of this research. HY is grateful for an Ontario Graduate Scholarship held during this research.

REFERENCES

1. C. T. Kresge, M. Leonowicz, W. J. Roth, J. C. Vartuli and J. C. Beck, Nature (London), 359 (1992) 710.
2. H. Yang, N. Coombs and G. A. Ozin, Nature (London), 386 (1997) 692.
3. G. A. Ozin, H. Yang, I. Sokolov and N. Coombs, Adv. Mater., 9 (1997) 662.
4. N. Coombs, D. Khushalani, G. A. Ozin, S. Oliver, G. C. Shen, I. Sokolov and H. Yang, Dalton Trans., (1997) 3941.
5. G. A. Ozin, H. Yang, I. Sokolov, and N. Coombs, J. Mater. Chem., (1998), in press.
6. H. Yang, G. A. Ozin, and C. T. Kresge, Adv. Mater., 10 (1998), in press.
7. (a) M. Kléman, Points, Lines, and Walls: in Liquid Crystals, Magnetic Systems and Various Ordered Media, John Wiley & Sons, Chichester, 1983; (b) H. Zocher, M. Kléman, P. G. de Gennes, in Liquid Crystals & Plastic Crystals, Volume 1, G. W. Gray and P. A. Windsor (eds), Ellis Horwood Ltd., New York, 1974.
8. D. Demus and L. Richter, Textures of Liquid Crystals, Verlag Chemie, Weinheim, 1978.
9. G. P. Crawford and S. Zumer (eds.), Liquid Crystals in Complex Geometries, Taylor and Francis, London, 1996.
10. D. H. Everett, Basic Principles of Colloid Science, Royal Society of Chemistry Paperbacks, Royal Society of Chemistry, Cambridge, 1994.
11. W. A. Tiller, The Science of Crystallization, Cambridge University Press, Cambridge 1991.
12. Y. Bouligand, J. Phys., 41 (1980) 1297; *ibid*, 1307.
13. J. Feng, Q. Huo, P. M. Petroff and G. D. Stucky, Appl. Phys. Lett., 71 (1997) 620.
14. C. J. Brinker and G. W. Scherer, Sol-Gel Science: The Physics and Chemistry of Sol-Gel Processing, Academic Press, San Diego, 1990.

MESOPOROUS MOLECULAR SIEVES 1998
Studies in Surface Science and Catalysis, Vol. 117
L. Bonneviot, F. Béland, C. Danumah, S. Giasson and S. Kaliaguine (Editors)

Structure and dynamic properties of surfactant systems

Dominique Langevin

Laboratoire de Physique des Solides, Université Paris-sud, 91405 Orsay, France

Surfactant molecules self-assemble in solution to give rise to a large variety of structures: micelles, vesicles, sponge phases, liquid crystal lyotropic phases (hexagonal, cubic, lamellar) among others. Micelles and lamellar phases were perhaps the most extensively studied systems, but the variety of structures observed with surfactants also attracted considerable interest, and recent progresses allowed to rationalize the physical chemistry principles of their molecular organization. In this short review, we will recall the main ideas allowing to understand the origin of the self-assembled structures. We will also discuss briefly their dynamic properties which play important roles in many circumstances. Indeed, it should not be forgotten that the self-assembled structures are transient because of the exchanges of the surfactant molecules between the aggregates and the solvent. These dynamic properties processes are certainly important in the elaboration of mesoporous materials with surfactant systems as templates. We will mainly focus here on aqueous surfactant solutions, but surfactant self-assembly also occurs in other solvents. Surfactants also help to stabilize dispersions of water-insoluble liquids such as oils in water (emulsions, microemulsions). Many excellent books have been written on surfactant self-assembly. For more details, the reader can usefully consult reference 1, which is one of the most recent.

1. DILUTE SURFACTANT SOLUTIONS

1.1. Surfactant Aggregation

Surfactant molecules are dispersed as monomers in water when their concentration c is very low. At some point the solubility limit is reached, and either the excess surfactant precipitates or it forms aggregates in which the hydrophobic part of the molecule is hidden in the interior, to minimize contact with water. These aggregates are in equilibrium with monomers, and the monomer concentration remains close to the solubility limit called "cac" (for critical aggregation concentration). When the hydrophobic part of the surfactant is made of CH_2 chains, one can evaluate simply the difference in chemical potential of the surfactant in the aggregates and in the bulk solution (as a monomer) from the free enthalpy gain Δg when a CH_2 group is removed from water and incorporated in an alkane environment ($\Delta g \sim kT$ per CH_2 group). This leads to the well known relation [2]:

$$\ln cac = a - bn \qquad (1)$$

where a and b are constants and n is the number of CH_2 units ($b=\Delta g/kT$). Since b ~1, an increase of n by two units leads approximately to a decrease of the cac by a factor 10.

When n is too large, aggregates are not formed, and the excess surfactant precipitates : the solubilization limit is called the Krafft point. This is the case of CTAB (cetyl trimethyl ammonium bromide) at room temperature. CTAB micelles can however be obtained at higher temperatures, above 25°C. The limit temperature above which aggregation in bulk is observed is called Krafft temperature. The Krafft temperature depends mainly on the number of CH_2 units, but also on the nature of the polar part of the surfactant molecules: for instance sulfates have smaller Krafft temperatures than sulfonates with the same chain.

1.2. Shape of the aggregates

Let us now discuss the nature of the aggregates formed. This important issue was addressed first by Tanford [2] and later on by Israelachvili, Mitchell and Ninham [3]. The idea is to compare the relative bulkiness of the polar and non-polar parts of the surfactant molecule. If V is the volume of the non-polar part, l its length and A the area occupied by the molecule at the hydrocarbon-water interface, one can introduce a surfactant "parameter" V/Al. When $V/Al < 1/3$, the relative bulkiness of the polar part dominates : when the molecules start to pack together to hide the non-polar part from water, the aggregates show a strong curvature towards water, and spherical micelles are formed. If the non-polar part is made of CH_2 chains, V and l can be calculated from known values of CH_2 groups dimensions. For DTAB (dodecyl trimethyl ammonium bromide) for instance, $A \sim 50Å^2$ and $V/Al \sim 1/3$: spherical micelles are indeed formed above the cac, called in this case "cmc" (for critical micellar concentration). When $1/2 > V/Al > 1/3$, cylindrical aggregates are favoured and for $2 > V/Al > 1/2$, flat aggregates are favoured. This last case corresponds frequently to double chain lipids: above the cac, the solution becomes turbid, because there is not enough surfactant to form a lamellar phase and the portions of lamellae close up into vesicles. Vesicles are larger than micelles and give rise to an appreciable scattering, even close to cac. For $V/Al > 2$, the formation of aggregates into non-polar solvents is favoured (reversed micelles, for instance).

Some precautions are needed when using the notion of surfactant parameter. The molecular parameters V, A and l are not intrinsic to the molecule, they rather refer to distances between adjacent molecules in the aggregates. For instance, A is frequently determined by the repulsion between the polar parts; for a ionic surfactant, the repulsion is electrostatic and A decreases when salt is added to the solution because of the screening by the ions. In some cases, for instance with CTAB, the addition of salt can transform spherical micelles into cylindrical ones. For nonionic surfactants, such as polyethylene glycol ethers, A decreases with increasing temperature because the hydration water of the polar parts goes away. The curvature of the aggregates eventually goes through zero (V/Al =1) around the so called "cloud point temperature" and reverses at high temperature. When oil is present, this oil can be solubilized in the chains, so that V can increase, l remaining constant. The surfactant parameter needs therefore to be used with caution, and provided it is done so, a very good qualitative understanding of the nature of the aggregates can be achieved.

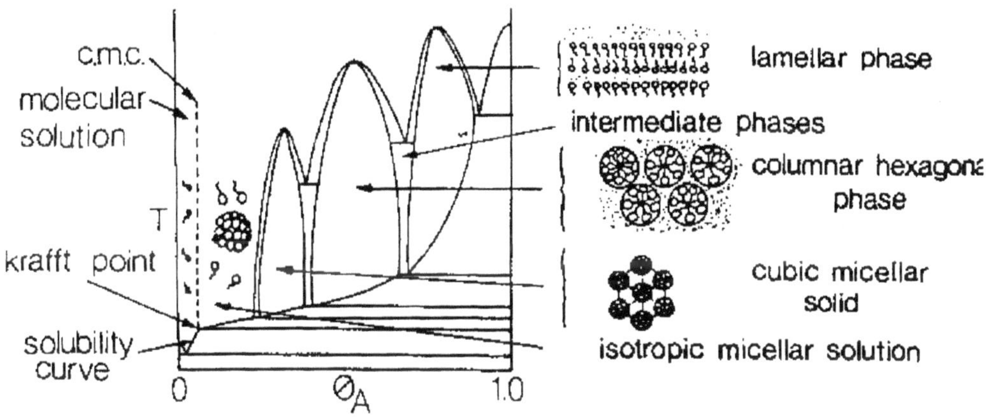

Figure 1. Schematic phase diagram; ϕ_A is the surfactant volume fraction (after reference 1, page 162)

1.3. Surfactant layers bending energy

Recently, this notion of "spontaneous" curvature of surfactant layers have been generalized by Helfrich to include the notion of curvature elasticity [4]. If the surfactant layer is replaced by an mathematical surface (the neutral surface) of principal curvatures C_1 and C_2, one can write the elastic energy per unit area of the system as :

$$F = \frac{1}{2} K (C_1 + C_2 - 2C_o)^2 + \bar{K} C_1 C_2 \qquad (2)$$

where C_0 is the spontaneous curvature of the layer, by convention positive if the curvature is against water (V/Al<1); flat aggregates correspond to $C_0=0$ (V/Al=1).

The elastic constants K and K are related respectively to deviations to mean curvature and to gaussian curvature. This energy allows to explain the behaviour of the systems for which C_0~0. When the elastic constants are large compared to kT, flat lamellae are formed; when they are comparable to kT, the layers are undulated because of thermal fluctuations. The lamellar phase is no longer stable and a "sponge" phase, with no long range order is obtained. Whereas the meaning of K is rather straightforward, the role of K is more difficult to assess. If K is negative, the second term of (2) is smaller if C_1 and C_2 have the same sign, i.e. favouring vesicles. If K is positive, structures with saddles (regions where C_1 and C_2 have opposite signs) are favoured : these are the "sponge" phases. The actual structure is determined not only by the two terms of equation 2, but also by the many other contributions to the free energy of the system : dispersion entropy, interaction between aggregates, etc. In particular, dilute lamellar phases are only stable if there are long range repulsions between the surfactant bilayers. A more complete discussion of the influence of K and K on the phase diagrams can be found for instance in reference 5.

It must be stressed that the description in terms of equation 2 is only valid for surfactant layer thicknesses much smaller than the inverse curvatures and it is not suitable for spherical micelles.

2. CONCENTRATED SURFACTANT SOLUTIONS

The interactions between aggregates can play an important role in more concentrated solutions. Interactions can affect both size and polydispersity of the micelles [6]. They can also promote transitions from spherical to cylindrical micelles in ionic surfactant solutions, because when c is increased, the ionic concentration is increased and the spontaneous curvature is decreased (see §1.2). The growth of cylindrical micelles with surfactant concentration can be extremely fast if particular ions are added to CTAB solutions, for instance salicylate ions. When the cylinders are long enough, the micelles can entangle, and the solutions become viscoelastic as semi-dilute polymer solutions [7].

In somewhat more concentrated solutions, the excluded volume interactions among others, can promote the appearance of liquid crystalline phases. They occur generally in the order hexagonal->lamellar, with sometimes cubic phases before and/or after the hexagonal phase. The cubic phase can be either made of isolated micelles in a cubic lattice, or have a bicontinous structure (such as those described by the Schwartz surface) [8]. Nematic phases have also been observed, for instance with mixtures of sodium decyl sulfate and decyl alcohol. Although disc-like aggregates are not favoured in solutions with only one surfactant, here the formation of discs is allowed: due to molecular packing considerations, the alcohol sits in the flat parts of the discs and the ionic sulfate in the curved ones; the discs then pack into discotic arrangements. Rectangular phases, where cylinders are flattened and exhibit elliptical sections, have also been observed. In general, it can be said that in each phase, the packing conditions are only optimal for a given concentration c. In the range of existence of this phase and away from this concentration, there is some degree of geometrical frustation which is released by the production of topological defects. The amount of defects is obviously larger at the phase boundaries and could promote the transition between the different phases.

3. DYNAMIC PROPERTIES

Surfactant aggregates are transient because the surfactant constantly exchanges between the aggregates and the solvent. The exchange time τ is related to the solubility of the surfactant in the solvent. When the non-polar part is made of CH_2 groups, this solubility is related to the free enthalpy cost Δg of transfer of a CH_2 group from an alkane environment to an aqueous environment. As the cac, the exchange time τ decreases by a factor 10 when 2 CH_2 groups are added to the chain [9]. For surfactants with 16 carbon chains, $\tau \sim 10^{-5}$ s. For lipids with two carbon chains, τ is much larger, up to hours.

The rheological behaviour of these systems is strongly affected by the exchanges. An extreme case is the sponge phase which viscosity is only slightly larger than that of water although the structure is connected over a macroscopy scale. The entangled solutions of giant micelles are also less viscous than expected from their equilibrium length. The main control parameter is the scission energy found close to 30 kT in most systems. This could seem surprising, but it should be remarked that single chain surfactants leading to known giant micelles systems have all 16 carbon chains (as CTAB). The scission energy is probably less for 14 carbon chains, and because the micelle length is an exponential function of the scission energy, the length of 14 carbons chain surfactant micelles could be much smaller than for surfactants with 16 carbons. The 18 carbon chain surfactants have larger Krafft temperatures and do not give micelles close to room temperature. The situation is somewhat different for the so called "gemini" surfactants, which are made of two cationic surfactant molecules with the polar parts linked by a spacer made of several CH_2 groups. These surfactants form cylindrical micelles more easily, at least without salt added [10].

4. POLYMERIZATION IN SURFACTANT SYSTEMS

Many attempts to use surfactant structures as templates for the synthesis of small colloidal particles and mesoporous materials have been made in the past. There is an extensive literature on the subject [11]. The main problem encountered is that first, the monomers introduced in the systems might aborb at the surface of the aggregates and change therefore their curvature. This is probably one of the reasons why polymerization in bicontinous microemulsion phases ($C_0 \sim 0$) very frequently leads to spherical polymers. The change of spontaneous curvature can also be produced by the polymer, once formed. The polymer also strongly affects the elastic bending moduli of the surfactant layers. Changes in spontaneous curvature and in bending moduli produced by polymers are dependent on the way in which the polymer is connected to the layer : anchored, adsorbed, etc [12]. Electrostatic interactions also play an important role in the complexation of surfactant aggregates and polyelectrolytes with charges of opposite signs [13].

5. CONCLUSION

In this short review, we have briefly described several key ideas to the understanding of the structure and the dynamic properties of surfactant solutions. The study of aqueous mixed solutions of polymer and surfactant is presently a very active field of research [14], which could give useful ideas to better understand the kinetics of formation of mesoporous materials using the surfactant microstructures as templates.

REFERENCES

1. W. M. Gelbart, A. Ben-Shaul, D. Roux, (eds.), Micelles, Membranes, Microemulsions and Monolayers, Springer, N. Y., 1994.
2. C.Tanford,The Hydrophobic Effect, Wiley, N. Y., 1973.
3. J. N. Israelachvili, D. J. Mitchell and B. W. Ninham, J. Chem. Soc. Faraday Trans. II, 72 (1976) 1525.
4. W. Helfrich, Z. Naturforsch. Teil C, 28 (1973) 693.
5. D. A. Huse, S. Leibler, J. Physique, 49 , 605 (1988).
6. W. M. Gelbart, W. E. Mc Mullen, A. Ben Shaul, Mol. Cryst. Liq. Cryst., 132 (1986) 325.
7. M. E. Cates, J. Candau, J. Phys. Condens. Mater., 2 (1990) 6869.
8. J. Charvolin, J. F. Sadoc, J. Physique, 49 (1988) 521.
9. J. Lang, R. Zana in "Surfactant Science Series", ed. R. Zana, Dekker, N.Y., 22 (1987) 405.
10. R. Zana, Current Opinion in Colloid & Interface Sci., 1 (1996) 667.
11. S. D. Desai, R. D. Gordon, A. M. Gronda, E. L. Cussler, Current Opinion in Colloid & Interface Sci., 1 (1996) 519.
12. C. Hiegeist, R. Lipowski, J. Phys. II France, 6 (1996) 1465.
13. M. Antonietti, S. Henke, A. Thuneman, Adv. Mater., 8 (1996) 41.
14. E. D. Goddard, K. P. Ananthapadmanabhan, Interaction of surfactants with polymers and proteins, CRC press, Boca Raton, 1993.

MESOPOROUS MOLECULAR SIEVES 1998
Studies in Surface Science and Catalysis, Vol. 117
L. Bonneviot, F. Béland, C. Danumah, S. Giasson and S. Kaliaguine (Editors)

Silazane-silylation of mesoporous silicates: towards tailor-made support materials

R. Anwander*, C. Palm, J. Stelzer, O. Groeger and G. Engelhardt

Institut für Technische Chemie I, Universität Stuttgart, Pfaffenwaldring 55, D-70569 Stuttgart, Germany

Mesoporous silicate MCM-41 was modified by disilazane agents of type $HN(SiR^1R^2_2)_2$ with $R^{1,2}$ = H, Me, Ph, Vin (Vin = vinyl). Calculations from elemental analysis and spectroscopic studies (FTIR, ^{1}H MAS NMR) revealed that the degree of silylation and hence the surface hydroxyl consumption depend on the steric bulk/shape of groups R. The formation of various siloxy functionalities $[OSiR^1R^2_2]$ is reflected in the pore characteristics and implies effective pore size engineering as evidenced by nitrogen physisorption measurements. Competitive H_2O/toluene adsorption revealed that the sorption capacity and hydrophobicity of the silylated MCM-41 materials are significantly affected by the nature of the attached $[OSiR^1R^2_2]$ group. Fully silylated materials carrying reactive "Si-Vin" moieties were functionalized by hydroboration with BH_3(THF) and 9-BBN.

1. INTRODUCTION

Silylation reactions are of fundamental importance for the manipulation of siliceous support materials including zeolites [1,2]. Silylating agents such as chlorosilanes, alkoxysilanes and silylamines have found widespread application and their reaction with surface silanol functionalities has been reported in detail (Scheme 1). We are particularly interested in liquid phase disilazane modifications of mesoporous silicates in order to exploit their potential as structural probes and to tailor the oxidic support for catalytic applications [3]. Disilazane agents are known to work under favorable reaction conditions, producing the most tolerable co-products [4].

$$\equiv Si{-}OH \; + \; Cl{-}SiR_3 \xrightarrow[25\,^{\circ}C]{base} \equiv Si{-}OSiR_3 \; + \; HCl{\cdot}base \qquad (1)$$

$$\equiv Si{-}OH \; + \; RO{-}SiR_3 \xrightarrow[100\,^{\circ}C]{} \equiv Si{-}OSiR_3 \; + \; HOR \qquad (2)$$

$$2 \equiv Si{-}OH \; + \; HN(SiR_3)_2 \xrightarrow[25\,^{\circ}C]{} 2 \equiv Si{-}OSiR_3 \; + \; NH_3 \qquad (3)$$

Scheme 1. Commonly applied liquid phase silylation procedures for siliceous materials.

---

* Corresponding author, present address: Anorganisch-chemisches Institut, Technische Universität München, Lichtenbergstr. 4, D-85747 Garching, Germany, E-mail: anwander@arthur.anorg.chemie.tu-muenchen.de.

2. EXPERIMENTAL

2.1. General

Purely siliceous MCM-41 material was synthesized according to the literature, employing $C_{16}H_{33}N(CH_3)_3Br$ as a templating agent [5,6]. After calcination (N_2: 540 °C, 5h, heating rate 1.5 °C min^{-1}; air: 540 °C, 5h) and dehydration (280 °C, 10^{-5} torr, 4 h, heating rate 1 °C min^{-1}) the material was stored in a glovebox (MB Braun MB150B-G-II; <1 ppm O_2, <1 ppm H_2O) where successively all silylation reactions and product manipulations were performed. *n*-Hexane and THF were distilled from Na/K alloy under nitrogen. The disilazane agents purchased from Aldrich and Gelest were degassed (freeze-pump-thaw) prior to use. Toluene, *tert*-butanol, and the borane agents BH_3(THF) (1 M in THF) and 9-BBN (9-borabicyclo-[3.3.1]nonane, 0.5 M in THF) were used as received from Aldrich.

2.2. Syntheses

The silylation reactions were carried out as reported earlier [7] by adding a solution of 5-6 mmol (excess) disilazane in *n*-hexane to a suspension of approximately 1.0 g of dehydrated MCM-41 in *n*-hexane (20 mL). After stirring the reaction mixture for 20 h at ambient temperature, unreacted silylamine was removed by several washings with *n*-hexane (50 mL). The silylated materials were dried *in vacuo* for at least 5 h at ambient temperature. The hydroboration reactions were carried out by contacting a suspension of [MCM-41]$SiMe_2$Vin (0.50 g) in 10 mL THF precooled to -35 °C with 1.5 mmol (slight excess based on the reactive vinylic sites) of borane agent. After stirring the reaction mixture for 20 h at ambient temperature, unreacted borane agent was removed by several washings with THF (20 mL). The resulting hybrid materials were dried *in vacuo* for at least 5 h at ambient temperature.

2.3. Characterization

Elemental analyses were performed on an Elementar VarioEL. FTIR spectra were recorded on a Perkin Elmer FTIR spectrometer 1760X, using Nujol mulls between CsI plates. ^1H MAS NMR experiments were carried out at 400.13 MHz on a Bruker MSL-400 spectrometer equipped with a standard 4 mm MAS probe. Single pulse excitation has been applied (pulse repetition 5s, $\pi/2$ pulse, spinning speed 10 KHz) and the spectra were referenced to $Si(CH_3)_4$. Nitrogen physisorption measurements were performed on an ASAP 2010 volumetric adsorption apparatus (Micromeritics) at 77 K for relative pressures from 10^{-2} to 0.60 [$a_m(N_2$, 77 K) = 0.162 nm^2]. A schematic drawing of the experimental set-up applied for the competitive H_2O/toluene and single component *tert*-butanol adsorption measurements is given in ref. 8. The flow-apparatus consists of a saturator filled with chromosorb P-NAW (60/80 mesh, Machery Nagel), the liquid (solid) adsorptives and a fixed-bed adsorber (bed height h = 30-50 mm). Nitrogen was used as a carrier gas. The silylated MCM-41 materials (0.40-0.80 g) were used unpressed and mixed with seasand (Fluka) to decrease the pressure drop. The adsorption capacities of seasand were negligible. The materials were activated at 523 K in flowing nitrogen for 4 h. The adsorption temperature was varied between 303 K and 363 K. The saturator was held at 293 K (p_W = 2.34 kPa, p_{Tol} = 2.92 kPa) for the competitive adsorption measurements and at 277 K (p_{tBuOH} = 1.26 kPa) for the single component adsorption. The gas stream was analyzed in intervals of 3 minutes, using a capillary gas chromatograph. The final loadings X_i were obtained from the breakthrough curves. The Hydrophobicity Indices (HI) were calculated as HI = X_{Tol}/X_W [9].

3. RESULTS AND DISCUSSION

Post-synthetic modification of structurally highly ordered MCM-41 materials offers an alternative route to the co-condensation "one-pot" reactions of differently functionalized silanes in the presence of a templating agent [10]. The silylated MCM-41 materials **2-7** were synthesized according to Scheme 1/(3), applying for the first time a variety of silylamines.

Methyl and phenyl substituted disilazanes were chosen to probe the impact of differently sized/shaped groups on surface coverage [11]. The more reactive Si-H and Si-Vin moieties constitute excellent spectroscopic probes and proved to be stable under the prevailing reaction conditions and subsequent thermal treatment. The silylated materials were treated at 250 °C/ 10^{-3} torr prior to elemental and physico-chemical characterization. The SiR_3 surface coverage was derived from carbon elemental analysis, ranging from 0.97 to 1.53 SiR_3/nm^2 for materials **2, 3, 5** and **6** (Table 1). $SiMePh_2$-silylated material **4** ([MCM-41]$SiMePh_2$) displayed a significantly lower surface coverage of 0.58 SiR_3/nm^2 probably due to the presence of two bulky phenyl substituents per silyl group.

Table 1
Analytical data, surface area, pore volume, and pore diameter

Material	Silylating agent(s)	Elemental analysis wt% C	SiR_3/nm^2	a_s(BET)[c] / m^2g^{-1} (C)	V_p[d] / cm^3g^{-1}	d_p[e] / nm
1[a]	–	–	–	1139 (80)	0.96	2.8
2[a]	$HN(SiMe_3)_2$	8.49	1.25	851 (26)	0.58	2.3
3[a]	$HN(SiMe_2Ph)_2$	17.70	0.97	765 (23)	0.48	2.1
4[a]	$HN(SiMePh_2)_2$	17.26	0.58	836 (36)	0.57	2.25
5[a]	$HN(SiHMe_2)_2$	6.95	1.53	847 (28)	0.62	2.2
6[a]	$HN(SiMe_2Vin)_2$	11.97	1.32	764 (30)	0.54	2.1
7[a]	$HN(SiHMe_2)_2/HN(SiMe_2Vin)_2$	7.21	–	812 (28)	0.60	2.2
8[b]	$HN(SiMe_2Vin)_2 + BH_3(THF)$	11.22	–	733 (37)	0.51	2.15
9[b]	$HN(SiMe_2Vin)_2$ + 9-BBN	26.76	–	582 (25)	0.20	< 1.5

[a] Pretreatment temperature 250°C, 3h, 10^{-3} torr. [b] Pretreatment temperature 25°C, 3h, 10^{-3} torr. [c] Specific BET surface area (C = BET constant). [d] BJH desorption cumulative pore volume of pores between 1.5 and 3.5 nm diameter. [e] Pore diameter according to the maximum of the BJH pore size distribution calculated from the desorption branch.

3.1. Spectroscopic studies

The infrared spectra of the silylated materials **2, 3, 5** and **6** reveal the consumption of all isolated silanol groups which are characterized by a sharp and intensive vibration at 3695 cm^{-1}. A small amount of hydrogen-bonded SiOH moieties [ν(OH) ca. 3660 cm^{-1}] seems to be not accessible to these silylating agents. However, the sterically less demanding tetramethyldisilazane consumed all isolated and hydrogen-bonded silanol functionalities to yield material **5**. This reaction behavior is in accord with the aforementioned highest surface coverage of the "$SiHMe_2$" group. A more peculiar behavior was observed for material [MCM-41]$SiMePh_2$ (**4**). The original sharp ν(OH) mode at 3695 cm^{-1} did not completely disappear and another sharp OH stretch vibration appeared at 3616 cm^{-1} indicative of an interaction of residual silanol groups with silicon-attached phenyl groups.

These findings could be further clarified by ^1H MAS NMR spectroscopy (Fig. 1). Spectrum A shows the proton resonance of the isolated SiOH groups of dehydrated MCM-41 material **1** which is located at 1.8 ppm. The ^1H MAS NMR spectra of the silylation products **2** (B), **3** (C), **5** (E) and **6** (F) reveal the consumption of all isolated silanol groups. The chemical shifts of the silyl groups can be assigned to Si-CH_3 (0.1 ppm), Si-H (4.6 ppm), Si-C_2H_3 (5.7 ppm) and Si-C_6H_5 groups (7.1 ppm). Peak intensities confirm the survival of the more reactive Si-H and Si-Vin functionalities under the prevailing reaction conditions. The ^1H MAS NMR spectrum of material **4** is more complex (Fig. 1, D). At least three resonances located at

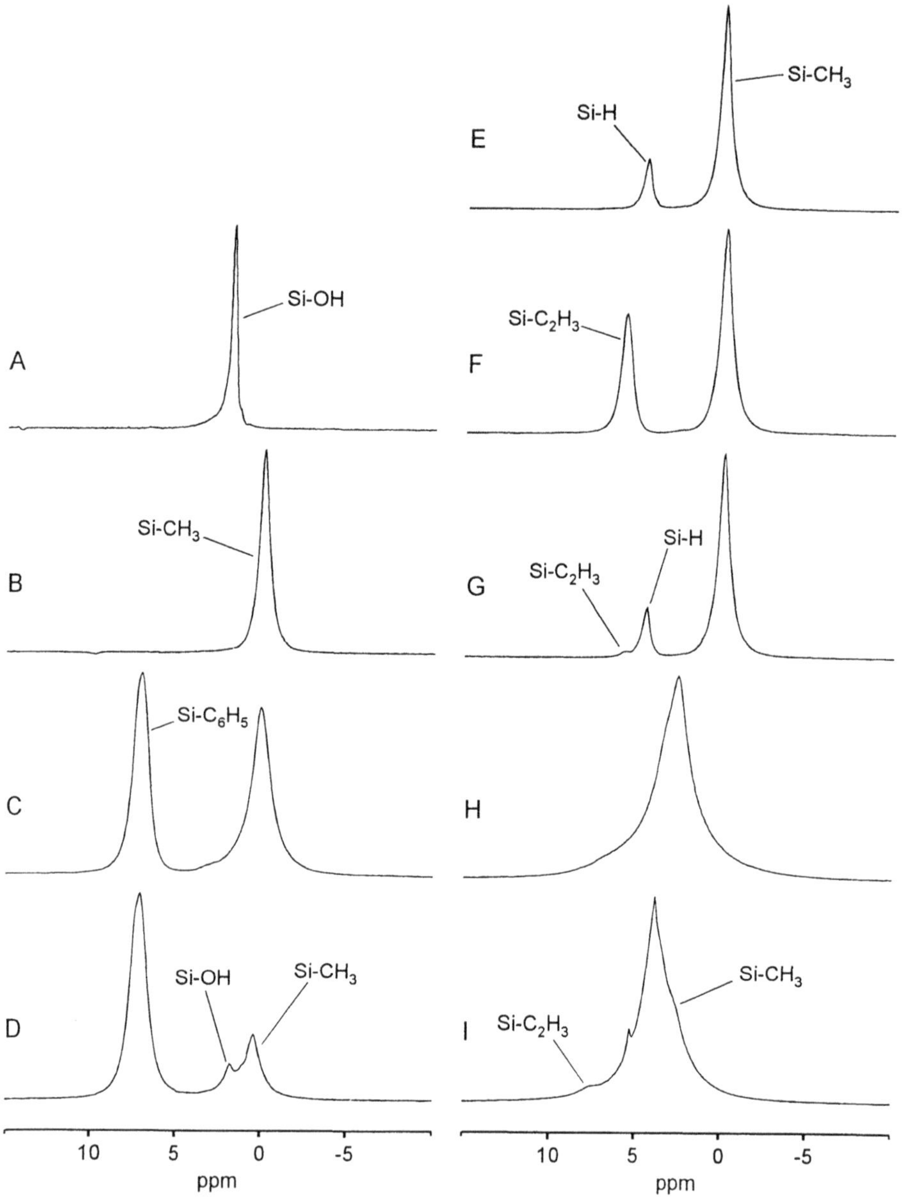

Figure 1. ^1H MAS NMR spectra of unmodified and silylated samples: (A) MCM-41 dehydrated at 280 °C/10^{-5} torr (**1**); (B) [MCM-41]$SiMe_3$ (**2**); (C) [MCM-41]$SiMe_2Ph$ (**3**); (D) [MCM-41]$SiMePh_2$ (**4**); (E) [MCM-41]$SiHMe_2$ (**5**); (F) [MCM-41]$SiMe_2$Vin (**6**); (G) [MCM-41]$SiHMe_2$/$SiMe_2$Vin (**7**); (H) [MCM-41]$SiMe_2$Vin + BH_3(THF) (**8**); (I) [MCM-41]$SiMe_2$Vin + 9-BBN (**9**); all silylated samples were measured after pretreatment at 250 °C/10^{-3} torr.

0.5, 1.8, and 7.2 ppm with relative signal intensities of 2.7 : 1.0 : 9.4 are observed. The signal at 1.8 ppm reflects a considerable amount of residual silanol groups. The steric bulk/shape of silyl groups carrying two phenyl rings seem to markedly affect the surface reaction and hence the extent of silanol group consumption. A competitive silylation reaction demonstrating the potential of the Si-H and Si-Vin functionalities as spectroscopic probes revealed an approximately 35 times higher reaction rate for the sterically less demanding $HN(SiHMe_2)_2$ agent (spectrum G).

3.2. Adsorptive properties

Nitrogen physisorption (Table 1). All silylated materials display adsorption/desorption isotherms of type IV according to the IUPAC nomenclature [12]. The isotherms are characterized by a sharp inflection point at relative pressures p/p_o, ranging from 0.18 to 0.25, as shown for materials **2-4** in Figure 2. The BJH differential pore size distribution and the effective mean pore diameter (2.1-2.3 nm), respectively, reveal effective pore size engineering. Also the BET surface areas were considerably decreased upon silylation. The isotherm of material [MCM-41]$SiMePh_2$ (**4**) corresponds to a comparatively high pore volume, which refers to the distinct reaction behavior of the corresponding silylamine.

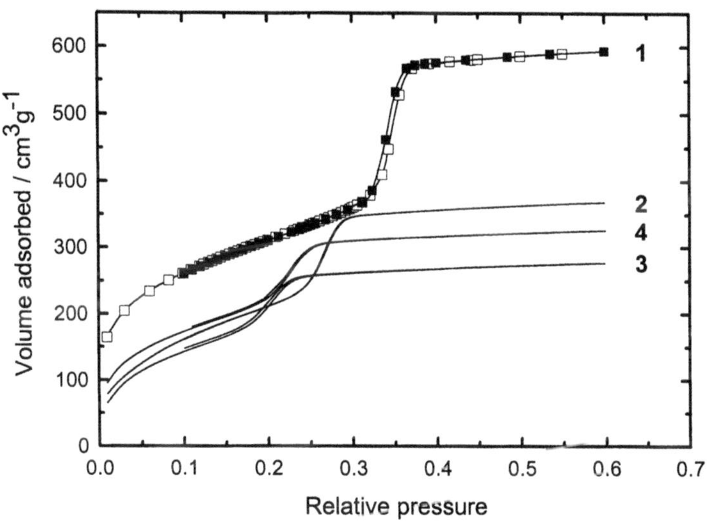

Figure 2. Nitrogen adsorption/desorption isotherms at 77.4 K (compare to Table 1; material **1**: filled symbols denote desorption).

Competitive H_2O/toluene adsorption – Hydrophobicity Index HI. Figure 3 representatively shows the temperature-dependent breakthrough curves of the parent MCM-41 material **1**, featuring (i) a pronounced step at the lower adsorption temperatures of 303 K and 348 K, respectively, due to mesopore filling, (ii) that water breaks through first, and (iii) successive replacement of water by toluene. Table 2 summarizes the water/toluene loadings and the resulting Hydrophobicity Indices of the parent and the Si(Me,Ph)-silylated materials **2-4**. The unsilylated material **1** adsorbed the largest amount of both toluene and water independent of the adsorption temperature. This behavior corresponds to the largest pore volume, the largest surface area, and certainly the most polarizing surface (high number of silanol groups). As expected, the silylated materials exhibited very low adsorption capacities for water, tentatively reflecting the amount of unsilylated silanol functionalities (see FTIR, ^{1}H MAS NMR spectroscopy). The hydrophobic nature of the silylated surfaces combined with high toluene loadings due to mesopore condensation accomplish high HI numbers at 303 K, ranging from 20 to 71. A distinct adsorption behavior was observed for the phenyl substituted materials **3** and **4** at conditions where no mesopore condensation occurs (363 K). Although the SiR_3 surface coverage of the $SiMe_2Ph$-silylated material **3** is only about 80 % of that of the $SiMe_3$-silylated material **2**, no pronounced preference for toluene adsorption is detected. This suggests that the phenyl groups of the $SiMe_2Ph$ moieties adopt a surface arrangement which hampers the interaction with the adsorptive toluene. In contrast, material **4** carrying two phenyl substituents per silyl group shows the proposed interaction with toluene, affording about 50 % of the toluene loading of the unsilylated material **1**. The presence of a considerable amount of surface silanol groups due to low $SiMePh_2$ surface coverage also increases the water capacity of material **4**. The relatively small pore volume of **3** obtained by nitrogen physisorption cannot be accounted for this behavior because the toluene capacity is the same for material **3** (43.2 wt%) and **4** (42.4 wt%) at 303 K.

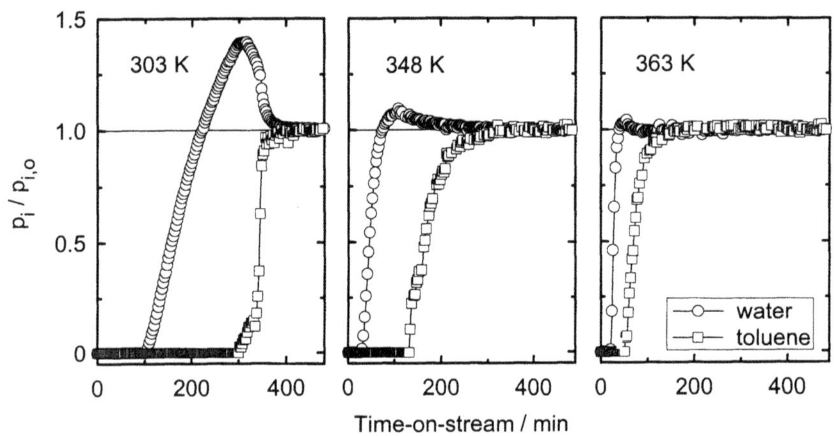

Figure 3. Competitive H_2O/toluene adsorption on activated, unmodified MCM-41 material **1** at different adsorption temperatures (compare to Table 2).

Table 2
Substrate loadings and Hydrophobicity Indices[a]

Material	Loadings (303 K, wt%) Water	Toluene	HI	Loadings (363 K, wt%) Water	Toluene	HI	Loading *t*-BuOH (353 K, wt%)
1	5.4	94.6	17.5	0.8	13.0	16.3	26.5
2	0.8	56.8	71	0.3	3.8	12.7	2.6
3	1.0	43.2	43.2	0.3	3.4	11.3	2.3
4	2.1	42.4	20.2	0.6	7.3	12.2	8.0

[a] Sample preparation according to the experimental part.

A single component adsorption of *tert*-butanol on the silylated materials **2-4** emphasized the results obtained for the competitive water/toluene adsorption, clearly showing the amount and accessibility of silanol functionalities in the order MCM-41 (**1**) » [MCM-41]$SiMePh_2$ (**4**) > [MCM-41]$SiMe_3$ (**2**) ≈ [MCM-41]$SiMe_2Ph$ (**3**).

3.3. Hydroboration of vinyl-functionalized material 6

Preliminary hydroboration reactions were performed by reaction of material **6** with BH_3(THF) and 9-BBN in THF at ambient temperature (Scheme 2). 9-BBN is commonly employed to direct the regioselectivity of the addition process, generating B-C bonds at the sterically less encumbered carbon atom (isomer **A**) [13]. Siliceous materials modified with boron compounds are discussed as promising Lewis acidic catalysts [1].

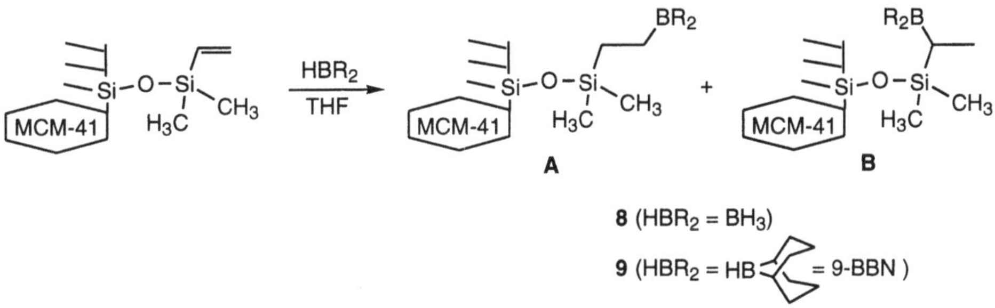

Scheme 2. Hydroboration of vinyl-functionalized silicate MCM-41.

Thus modified materials **8** and **9** were characterized by elemental analysis (Table 1), FTIR and ^{1}H MAS NMR spectroscopy. The characteristic C=C stretching frequency at 1597 cm^{-1} [14] was completely consumed upon addition of BH_3(THF). New vibration modes appeared at 2507 and 1558 cm^{-1} which are assignable to B-H and B-H-B bridge stretching frequencies [15]. Formation of the BH_3(THF) addition product **8** was confirmed by the disappearance of the vinylic protons at 5.7 ppm in the ^{1}H MAS NMR spectrum. A broad resonance centered at 0.1 ppm appeared which originates from overlapping resonances of Si-CH_3 and B-H protons (Fig.1, H). The broadening of the signal is probably due to the coexistence of different addition products "R-BH_2", including the stereoisomers **A** and **B** (Scheme 2). The formation of Lewis acid/base derived surface species of type "(Si_2O)→(BH_3)" cannot be ruled out. The clear appearance of the resonance at 5.7 ppm in the ^{1}H MAS NMR spectrum of material **9** (Fig.1, I) reveals that the reaction with 9-BBN renders a significant amount of vinylic protons unmodified. The limited reaction is probably due to steric restrictions at the surface, supporting an anti-Markovnikov addition of the borane to produce isomer **A** (Scheme 2). The peaks at 3.4 ppm and 1.7 ppm result from the CH and CH_2 protons of the immobilized 9-BBN group which are superimposed by the broader resonances of Si-CH_2-CH_2-BR_2 and Si-CH_3 protons, the maximum of the latter being pronounced by the shoulder at about 0 ppm.

The steric bulk of the 9-BBN moiety considerably decreases the pore volume of the hybrid material **9** as evidenced by nitrogen physisorption (Table 1). The resulting isotherm is of type I and points to a microporous solid with a relatively small external surface. Addition of the smaller BH_3 group implies only a slight drop in pore volume and effective mean pore diameter, respectively, the original type-IV isotherm is retained.

4. CONCLUSIONS

Disilazane agents represent an effective "silylation pool" to study the surface morphology of mesoporous silicate materials. As shown by spectroscopic studies and nitrogen physisorption measurements the silylation reactions involving surface silanol groups are directed by size/shape effects of the organic substituent of the introduced silyl moiety. E.g., disilazane agents carrying sterically less demanding silyl groups such as "$SiHMe_2$" accomplish the highest surface coverage. Hydrophobicity Indices derived from competitive water/toluene adsorption prove to be complex numbers markedly affected by SiR_3 surface coverage and nature of the silyl substituents. Competing silylation reactions allow the controlled introduction of reactive surface components as shown for vinyl moieties which provide a unique platform for subsequent modification reactions as verified by the hydroboration reaction.

ACKNOWLEDGEMENT

We thank the Deutsche Forschungsgemeinschaft for generous support.

REFERENCES

1. E.F. Vansant, P. Van Der Voort and K.C. Vrancken (eds.), Characterization and Chemical Modification of the Silica Surface (Stud. Surf. Sci. Catal., Volume 93), Elsevier, Amsterdam, 1995.
2. A. Choplin, J. Mol. Catal., 86 (1994) 501.
3. See for chlorosilane silylation of MCM-41 silicates: (a) J. Chen, Q. Li, R. Xu and F. Xiao, Angew. Chem. Int. Ed. Engl., 34 (1995) 2898; (b) X.S. Zhao, G.Q. Lu, A.K. Whittaker, G.J. Millar and H.Y. Zhu, J. Phys. Chem., 101 (1997) 6525.
4. See for leading references: (a) W. Hertl and M.L. Hair, J. Phys. Chem., 75 (1971) 2181; (b) D.W. Sindorf and G.E. Maciel, J. Phys. Chem., 86 (1982) 5208; (c) S. Haukka and A. Root, J. Phys. Chem., 98 (1994) 1695.
5. J.S. Beck, J.C. Vartuli, W.J. Roth, M.E. Leonowicz, C.T. Kresge, K.D. Schmitt, C. T.-W. Chu, D.H. Olson, E.W. Sheppard, S.B. McCullen, J.B. Higgins and J.L. Schlenker, J. Am. Chem. Soc., 114 (1992) 10834.
6. T. Boger, R. Roesky, R. Gläser, S. Ernst, G. Eigenberger and J. Weitkamp, Microporous Mater., 8 (1997) 79.
7. R. Anwander, O. Runte, J. Eppinger, G. Gerstberger, E. Herdtweck and M. Spiegler, in press.
8. S. Ernst and J. Weitkamp, GIT Special Chromatography International, (1996) 82.
9. C.H. Berke, A. Kiss, P. Kleinschmitt and J. Weitkamp, Chem.-Ing.-Tech., 63 (1991) 623.
10. See for examples: (a) S.L. Burkett, S.D. Sims and S. Mann, Chem. Commun., (1996) 1367; (b) D.J. Macquarrie, Chem. Commun., (1996) 1961; (c) M.H. Lim, C.F. Blanford and A. Stein, J. Am. Chem. Soc., 119 (1997) 4090.
11. D.-W. Hua and D.M. Smith, Langmuir, 8 (1992) 2753.
12. K.S.W. Sing, D.H. Everett, R.A.W. Haul, L. Moscou, R.A. Pierotti, J. Rouquérol and T. Siemieniewska, Pure Appl. Chem., 57 (1985) 603.
13. J.A. Soderquist and H.C. Brown, J. Org. Chem., 45 (1980) 3571.
14. C. Bonhomme, P. Toledano, J. Maquet, J. Livage and L. Bonhomme-Coury, J. Chem. Soc., Dalton Trans., (1997) 1617.
15. B.A. Morrow and R.A. McFarlane, Langmuir, 2 (1986) 315.

MESOPOROUS MOLECULAR SIEVES 1998
Studies in Surface Science and Catalysis, Vol. 117
L. Bonneviot, F. Béland, C. Danumah, S. Giasson and S. Kaliaguine (Editors)

Stabilization of M41S Materials by Trimethylsilylation

T. Tatsumi[a], K. A. Koyano[a], Y. Tanaka[b], and S. Nakata[b]

[a]Engineering Research Institute, School of Engineering, The University of Tokyo, Yayoi, Bunkyo-ku, Tokyo 113, Japan

[b]Chiyoda Corporation, Moriya-cho, Kanagawa-ku, Yokohama 221, Japan

Trimethylsilylation of M41S mesoporous molecular sieves, MCM-48 and MCM-41, results in a distinct improvement in their stability to moisture and compression. This stabilizing effect is ascribed to the increase in hydrophobicity. The resultant stability is retained after removal of the methyl groups by calcination, although the calcined material is no longer hydrophobic. This stabilizing method is also applicable to Ti-containing M41S materials.

1. INTRODUCTION

The M41S mesoporous molecular sieves are highly promising as adsorbents, supports, and catalysts [1,2]. However, the range of application has been rather limited by the instability of the structure of the mesoporous materials. It has been reported that MCM-41 [3] as well as FSM-16 [4], similar material synthesized from kanemite, is mechanically unstable compared to other materials such as alumina and silica gels. We have found that the structure collapse of mesoporous molecular sieves MCM-48 and MCM-41 materials by mechanical compression occurs mechanochemically through the hydrolysis of Si-O-Si bond in the presence of adsorbed water [5]. Although pure-silica and Al-containing MCM-41 have high thermal and hydrothermal stability and can be heated to at least 1123 K in dry air or 1073 K in air with low water vapor pressure before the collapse of the structure [6,7], Kim *et al.* claimed that the structure of MCM-41 was lost completely upon boiling in water for 2 days due to silicate hydrolysis [8]. Furthermore, the regular structure of FSM-16, from which the surfactant was removed by HCl treatment, disappeared on exposure to water vapor even at room temperature [9]. We also have observed that the structure of MCM-41 is not stable in the long term when exposed to water vapor over a saturated aqueous solution of NH_4Cl at room temperature. Particularly, the stability of Ti-MCM-41 was found low compared to pure-silica MCM-41; the substitution of Ti for Si in the silica framework of MCM-41 greatly reduces the stability of the structure [10]. Rhee and Lee also reported that the Ti-MCM-41 structure collapsed and titanium was leached out of the silicate framework by hydrolysis under mild reaction conditions for phenol hydroxylation with H_2O_2 [11]. Coustel *et al.* suggested that the local hydrolysis of the Si-O-Al bond occurred by thermal treatment and that the increase in the wall thickness is effective in improving the stability of Al-containing MCM-41 [12].

Since the instability of the structure of the mesoporous materials seems to be due to the adsorbed water, we expected that the stability will increase by increasing the hydrophobicity of the mesoporous structures. We report that trimethylsilylation of mesoporous materials is effective in enhancing their stability; furthermore, the stabilizing effect was unexpectedly maintained after removal of the methyl groups by calcination. A preliminary report has been published elsewhere [13].

2. EXPERIMENTAL SECTION

2.1. Preparation of Materials

For the preparation of pure-silica MCM-48, an aqueous solution of cetyltrimethylammonium chloride / hydroxide (CTMACl/OH, Cl/OH = 70/30, Tokyo Chemical Industry (TCI)) was added dropwise to TEOS (TCI) under vigorous stirring at 278 K. After stirring for 1 h, the mixture was heated at 358 K for 4 h to remove the ethanol produced in the hydrolysis of the TEOS. The resultant gel was poured into a Teflon bottle, and heated under static conditions at 373 K for 10 days. After hydrothermal treatment, the solids were filtered, washed with distilled water, and dried at 333 K overnight. The template was removed by calcination in air at 813 K for 6 h. The molar composition of the gels subjected to hydrothermal synthesis was as follows: $SiO_2 \cdot CTMA \cdot 46.5H_2O$.

For the preparation of pure-silica MCM-41, fumed silica (Aerosil 200, Nippon Aerosil) and dodecyltrimethylammonium bromide (DTMABr, TCI) were used as a Si source and template, respectively. The procedure was based on the method reported by Reddy *et al* [14]. Fumed silica was dissolved in a NaOH solution and stirred for 30 min. Then DTMABr was added to the mixture, which was stirred for 1 h. The resultant mixture was hydrothermally treated at 373 K for 6 days. The following steps were the same as described above for the synthesis of pure-silica MCM-48. The molar composition of the gels subjected to hydrothermal synthesis was as follows: $SiO_2 \cdot 0.5DTMA \cdot 0.17NaOH \cdot 30H_2O$.

Ti-MCM-41 was prepared as follows [10]: to a 25% aqueous solution of DTMABr, TEOS and tetramethylammonium hydroxide (TMAOH, TCI) were added dropwise in turn to partly hydrolyze TEOS at 278 K. After 30 min, a solution of TBOT in 2-propanol was added to this resultant mixture under vigorous stirring. The mixture was then stirred for 30 minutes, when the residual amount of H_2O was added. After the heating at 358 K for 4 h, the resultant gels were hydrothermally treated in a static Teflon bottle at 373 K for 10 days. The molar composition of the gels subjected to hydrothermal synthesis was as follows: $SiO_2 \cdot 0.01TiO_2 \cdot 0.6DTMABr \cdot 0.3TMAOH \cdot 60H_2O$ or $Si \cdot 0.0125Ti \cdot 0.6CTMABr \cdot 0.36TMAOH \cdot 75H_2O$. Ti-MCM-48 was similarly synthesized by using CTMACl/OH (Cl/OH = 70/30) as a template [15]. The molar composition of the gels was as follows: $SiO_2 \cdot 0.02\ TiO_2 \cdot CTMA \cdot 46.5\ H_2O$.

2.2. Trimethylsilylation

Trimethylsilylation was performed according to the procedure reported by Mobil [2]; 0.5 g samples, 10 g $(CH_3)_3SiCl$ (TMSCl), and 15 g $((CH_3)_3Si)_2O$ (HMDS) were refluxed for 16 h under N_2. The HCl produced was stripped off on a rotary evaporator, and the dry powder was washed 4 times with 10 ml dehydrated acetone with centrifuging and drying at 393 K for 3 h. The samples before and after trimethylsilylation are named Si-MCM-48(non-sil) and Si-MCM-48(sil), respectively. The Si-MCM-48(sil) sample was calcined at 823 K for 6 h to remove methyl groups to give MCM-48(sil-cal).

2.3. Characterization

X-ray powder diffraction patterns were collected on a Rigaku Rint 2400 diffractometer using Cu*K*α radiation. IR spectra were taken on a Perkin-Elmer 1600 FT-IR spectrometer as KBr pellets containing 1 wt% sample. Nitrogen adsorption/desorption measurements were performed on a Japan Bel Belsorp 28SA analyzer. The solid-state ^{29}Si NMR measurements were performed on a JEOL JNM-GX270 spectrometer equipped with a magic angle spinning (MAS) unit. Measurement conditions were as follows: resonance frequency, 53.54 MHz; pulse width, 6 μs; pulse delay 7 s; spinning rate, 3.5 kHz; standard, $(CH_3)_4Si$.

2.4. Stability Measurement

Mechanical stability of the mesoporous samples was investigated as follows: all the samples were dried in air at 393 K for 3 h immediately before compression. The samples (0.1 g) were pressed in a steel die of 20 mm diameter by a pressing apparatus under ambient air for

10 min. The structure of pressed samples was examined by XRD and N_2 adsorption. For the purpose of estimating the stability against water, samples were exposed to moisture over a saturated aqueous solution of NH_4Cl at room temperature for varying times. This procedure is denoted 'moisture treatment'. The samples were dried at 473 K for 2 h before XRD measurements.

3. RESULTS AND DISCUSSION

3.1. Preparation of Samples

The XRD patterns of synthesized Si-MCM-48(non-sil) showed eight peaks indexed on a cubic structure as reported by Kresge *et al.* [1] The nitrogen physisorption isotherms of the Si-MCM-48(non-sil) were characteristic of a mesoporous material with a BET surface area above 1000 m^2/g and a uniform pore size of 25.2 Å as determined by the Dolimore-Heal (D-H) method. There was no significant change in the XRD patterns and *d* spacings upon trimethylsilylation and the subsequent calcination, as shown in Figure 1. In contrast, the N_2 adsorption isotherm was significantly changed upon trimethylsilylation. The N_2 adsorption

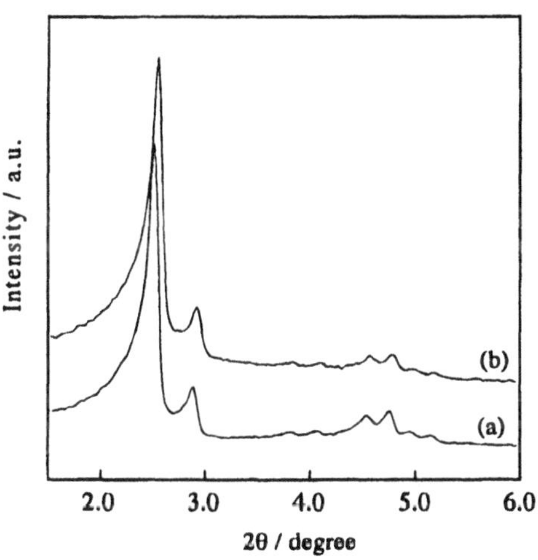

Figure 1. XRD patterns of (a) Si-MCM-48 (non-sil) and (b) Si-MCM-48(sil).

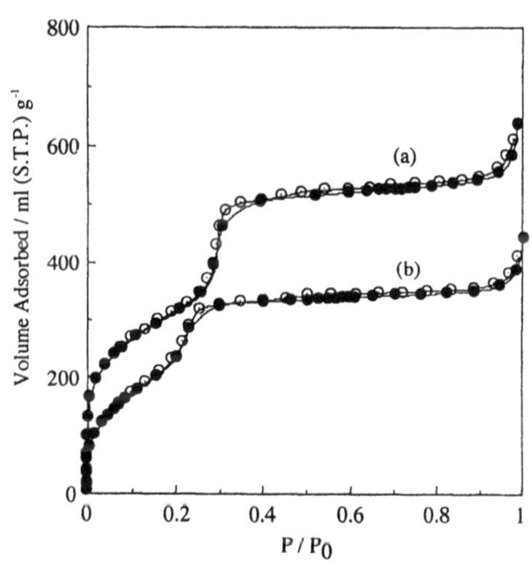

Figure 2. N_2 adsorption isotherms of (a) Si-MCM-48(non-sil) and (b) Si-MCM-48(sil).

Table 1
Physical Properties of Si-MCM-48

Samples	d_{211} [a] (Å)	BET surface area (m^2/g)	Pore diameter[b] (Å)	Pore volume (ml/g)
Si-MCM-48(non-sil)	35.2	1074	2.5	0.89
Si-MCM-48(sil)	35.3	736	2.1	0.62
Si-MCM-48(sil-cal)	35.2	866	2.3	0.68

[a] d_{211} Bragg spacing from XRD measurement. [b] Pore diameter calculated by the D-H method.

isotherms before and after trimethylsilylation are compared in Figure 2. Table 1 summarizes the pore diameter determined by the D-H method, the BET surface area, and the pore volume obtained from the N_2 adsorption data. These observations could be accounted for by assuming that the silanol groups inside the pore were trimethylsilylated. The BET surface area, pore size, and pore volume decreased by trimethylsilylation, and was partly restored by the subsequent calcination.

The ^{29}Si MAS NMR spectrum of Si-MCM-48(non-sil) exhibited two broad resonances, at -110 ppm for a Q^4 environment and at -100 ppm for a Q^3 environment together with a shoulder at -90 ppm ascribable to Q^2 species. The spectrum Si-MCM-48(sil) showed two peaks at 14.3 and -108.5 ppm ascribed to the silicon atoms of trimethylsilyl groups and Q^4 silicate, respectively. Because of the lack of symmetry in the peak for the Q^4 environment, it was suggested that the peak for the Q^3 environment was superimposed on that for the Q^4 environment. As shown in Figure 3, this hypothesis was substantiated by the enhancement of the peak for the Q^3 environment in ^1H-^{29}Si CP MAS NMR spectra, indicating that a portion of the silanol groups was not trimethylsilylated. The spectrum of Si-MCM-48(sil-cal) obtained by calcination of Si-MCM-48(sil) showed that the Q^2, Q^3, and Q^4 species were increased at the expense of trimethylsilyl groups. Thus silanol groups, which should have been formed by calcination of trimethylsilyl groups, condensed to a considerable extent.

The results are summarized in Table 2, in which the amount of Si in Si-MCM-48(non-sil) was set as a standard. The content of silanol groups was 0.40 mol/mol-Si (for Si-MCM-48(non-sil)) before trimethylsilylation, and that of trimethylsilyl groups was 0.24 mol/mol-Si (for Si-MCM-48(non-sil)) after the treatment, indicating that the degree of modification was about 60 %. The incomplete trimethylsilylation is considered to be due to the bulkiness of the trimethylsilyl group (*ca.* 6.0 Å). Nevertheless, Si-MCM-48 (sil) was highly hydrophobic. Thermogravimetry measurements showed that Si-MCM-48(sil) adsorbed *ca.* 1 wt% of H_2O, being much more hydrophobic than Si-MCM-48(non-sil) and Si-MCM-48(sil-cal), which adsorbed 30 and 43 wt% of H_2O, respectively. Probably trimethylsilyl groups cover nearby silanol groups, preventing water from approaching the silanol groups.

As shown in Table 3, trimethylsilylation of Si-MCM-41 also resulted in similar change in the N_2 adsorption behavior. The BET surface

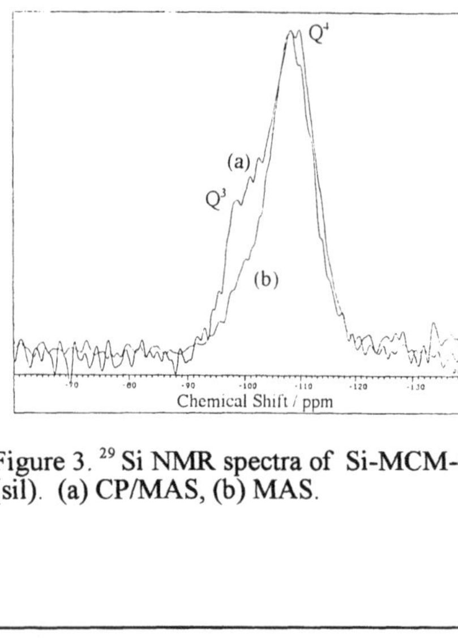

Figure 3. ^{29}Si NMR spectra of Si-MCM-48 (sil). (a) CP/MAS, (b) MAS.

Table 2
^{29}Si MAS NMR data for pure-silica MCM-48

samples	Si[a] (mol/mol-Si $\times 10^2$) Q^2	Q^3	Q^4	Si (TMS)	total
Si-MCM-48(non-sil)	3.4	33	64	-	100
Si-MCM-48(sil)	0.0	13	87	24	124
Si-MCM-48(sil-cal)	4.4	27	92	0.0	123

[a]The amount of Si in Si-MCM-48(non-sil) was set as a standard.

Table 3
Physical Properties of Si-MCM-41

Samples	d_{100} [a] (Å)	BET surface area (m^2/g)	Pore diameter[b] (Å)	Pore volume (ml/g)
Si-MCM-41(non-sil)	33.4	1111	2.1	0.81
Si-MCM-41(sil)	33.6	753	1.9	0.49
Si-MCM-41(sil-cal)	32.9	903	1.9	0.60

[a] d_{100} Bragg spacing from XRD measurement. [b] Pore diameter calculated by theD-H method.

area, pore size, and pore volume decreased by trimethylsilylation, and was partly restored by the following calcination. However, the XRD patterns were practically unchanged upon trimethylsilylation and the subsequent calcination.

3.2. Stability to Moisture

The stability to water/moisture is one of the most important factors in the application of mesoporous materials. Upon exposure to water vapor, over a saturated aqueous solution of NH_4Cl at room temperature, Si-MCM-48(non-sil) sample lost its regularity with time as shown in Table 4. The collapse of the structure was accompanied by a remarkable increase in the Q^3 and Q^2 silicate peaks in the ^{29}Si MAS NMR spectra; the molar ratios of Q^2 /Q^3 /Q^4 was 7.6 / 46 / 46 after 10-day moisture treatment (see Table 2 for before moisture treatment). Thus presumably Si-O-Si bonds were hydrolyzed, resulting in the decrease in the regularity. However, the XRD patterns and ^{29}Si MAS NMR spectra of Si-MCM-48(sil) were unchanged by moisture treatment. This is interpreted in terms of hydrophobicity; Si-MCM-48(sil) was much more hydrophobic than Si-MCM-48(non-sil) and adsorbed little water, being hardly affected by moisture.

Noteworthy is that the XRD peak intensity of Si-MCM-48(sil-cal), d_{211} intensity, in particular, hardly decreased by 30-day moisture treatment, as shown in Table 4. The Q^3 and

Table 4
XRD data for Si-MCM-48 exposed to water vapor

Exposure	Si-MCM-48(non-sil)		Si-MCM-48(sil)		Si-MCM-48(sil-cal)	
Time (days)	d_{211} (Å)	P-H[a] (%)	d_{211} (Å)	P-H[a] (%)	d_{211} (Å)	P-H[a] (%)
0	34.9	100	35.8	100	34.9	100
1	35.3	90	36.1	98	-	-
5	35.2	86	36.0	99	34.0	102
10	34.2	86	36.0	97	33.4	95
20	33.7	64	-	-	-	-
30	-	-	36.0	100	33.7	98
90	31.3	35	-	-	-	-

[a] height of d_{211} peak in X-Ray diffractograms.

Q^2 silicate peaks in ^{29}Si MAS NMR spectra, however, increased; the molar ratios of Q^2 /Q^3 /Q^4 were 3.6 / 22 / 75, and 6.5 / 36 / 57 before and after 10-day moisture treatment, respectively. Similarly to Si-MCM-48(non-sil), the hydrolysis of the Si-O-Si bonds obviously occurred on exposure to moisture. As shown in Table 4, the long-range order of Si-MCM-48(sil-cal) observed by XRD was kept on long-time exposure to moisture at room temperature despite the hydrolytic cleavage of the Si-O-Si bonds.

The stability of Si-MCM-41(non-sil) to moisture was also investigated. On exposure to moisture for 30 and 90 days, the d_{100} peak height of Si-MCM-41(non-sil) decreased by 6 and 41 %, respectively,. Changes in the ^{29}Si MAS NMR spectra of Si-MCM-41(non-sil) were similar to those of Si-MCM-48(non-sil); the Q^3 and Q^2 silicate peaks increased remarkably on exposure to moisture. The molar ratios of Q^2 /Q^3 /Q^4 were 0.0 / 21 / 79, and 6.4 / 38 / 56 before and after 10-day moisture treatment. Thus it is indicated that Si-O-Si bonds were hydrolyzed, resulting in the decrease in the regularity, in a similar manner to Si-MCM-48(non-sil). The stability of Si-MCM-41(non-sil) was also greatly improved by trimethylsilylation and trimethylsilylation followed by calcination; the XRD patterns of Si-MCM-41(sil) and Si-MCM-41(sil-cal) hardly changed by moisture treatment for 30 days, as demonstrated in Figure 4 for the latter.

3.3. Mechanical Stability

Figure 5 shows the N_2 adsorption isotherms of Si-MCM-48 samples after compression at various pressures. The results of XRD and N_2 adsorption measurements are summarized as follows. In the case of Si-MCM-48(non-sil), the intensities of the characteristic d_{211} diffraction peaks were markedly decreased with increasing pressure. Simultaneously the steep increase in adsorption at p/p_0 around 0.3 due to the ordered mesoporous structure faded and the amount of adsorbed N_2 reduced. These results are similar to those reported for MCM-41 [3] and of FSM-16 [4]. In

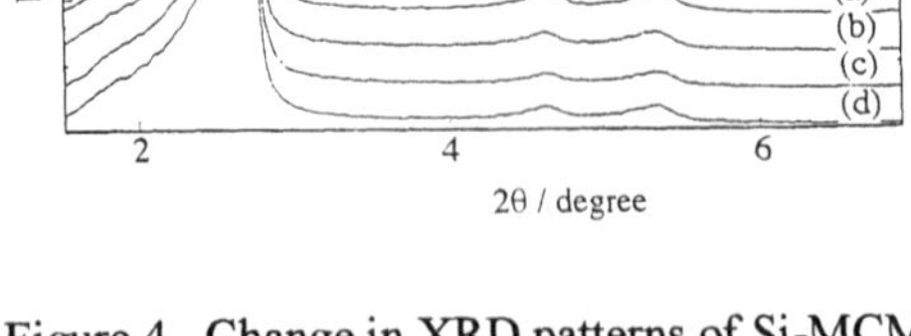

Figure 4. Change in XRD patterns of Si-MCM -41 (sil-cal) upon exposure to moisture for (a) 0 day; (b) 5 days; (c) 10 days; (d) 30 days.

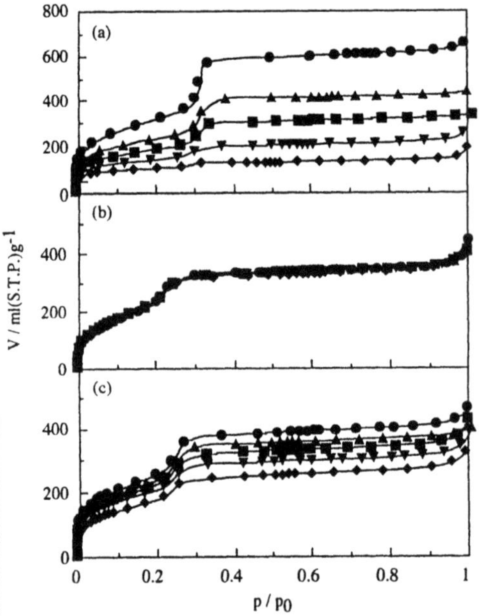

Figure 5. N_2 Adsorption Isotherms of Si-MCM-48 Samples after Treatment at Varoius Pressures. (a) Si-MCM-48(non-sil); (b) Si-MCM-48(sil); (c) Si-MCM-48(sil-cal); ●, 0 kg/cm^2; ▲, 441 kg/cm^2; ■, 882 kg/cm^2; ▼, 1323 kg/cm^2; ◆, 1764 kg/cm^2.

contrast, Si-MCM-48(sil) proved structurally more stable to pressure; the compression led to no change in the N_2 adsorption isotherm, although the d_{211} peak height was slightly decreased with increasing pressure. As-synthesized Si-MCM-48 also withstood the compression. This is probably due to the non-porosity of Si-MCM-48(non-sil) as a result of the presence of template molecules in the pore. In contrast, Si-MCM-48(sil) has a sizeable void, as shown in Table 1. We have revealed that Si-MCM-48(non-sil) mechanochemically collapsed by adsorbed water because of its hydrophilicity [5]. Since Si-MCM-48(sil) was hydrophobic, the difference between Si-MCM-48(non-sil) and Si-MCM-48(sil) in the mechanical stability can be explained by the hydrophobicity / hydrophilicity of the samples. Contrary to what would be expected, however, Si-MCM-48(sil-cal) also exhibited high mechanical stability compared to Si-MCM-48(non-sil). This improvement in mechanical stability upon silylation and silylation followed by calcination was also observed for Si-MCM-41.

3.4. Stability of Ti-MCM-41

The stability of Ti-MCM-41(non-sil) to moisture was extremely low compared to Si-MCM-41(non-sil) and Si-MCM-48(non-sil). Although the reason is not clear, it is conceivable that the Si-O-Ti linkage is more susceptible to hydrolysis than the Si-O-Si linkage. The stability of Ti-MCM-41, however, was remarkably increased by trimethylsilylation and was also retained after the subsequent calcination. Changes in the ^{29}Si MAS NMR spectra of Ti-MCM-41(sil-cal) were similar to those of Si-MCM-48(sil-cal) and Si-MCM-41(sil-cal); the Q^3 and Q^2 silicate peaks increased on exposure to moisture.

Trimethylsilylation of Ti-containing mesoporous molecular sieves led to another advantage, Trimethylsilylated Ti-MCM-41 and Ti-MCM-48 materials, being hydrophobic, proved to be highly active in oxidation of alkenes and alkanes with H_2O_2 compared to the corresponding non-trimethylsilylated samples [16].

3.5. Cause of the Stability Improvement

It is conceivable that water adsorbed onto silanol groups causes the hydrolysis of nearby Si-O-Si bonds, resulting in the collapse of the ordered mesoporous structure. The trimethylsilyl groups capping the hydrophilic silanol groups will protect those Si-O-Si bonds from hydrolysis, resulting in an increase in stability.

One might wonder that the difference in the stability to water and compression between non-silylated samples and silylated-calcined samples, both of which are purely inorganic and rather hydrophilic, is partly due to the increased wall thickness. In the case of Si-MCM-48 samples, the increment of wall thickness is estimated at *ca.* 2 Å by N_2 adsorption data, as shown in Table 1. The number of Si atoms increased by 23% based on ^{29}Si MAS NMR measurements (Table 2). Thus we evaluate the wall thickness of Si-MCM-48(sil-cal) at *ca.* 12 Å, assuming that of Si-MCM-48(non-sil) is *ca.* 10 Å Coustel *et al.* reported that there was little difference in the hydrothermal stability between MCM-41 with 10 Å wall thickness and that with 12 Å [12]. Hence the remarkable stability of silylated-calcined samples should be mainly due to the factors other than the wall thickness.

We presume that Si-O-Si bonds in the silylated-calcined samples are classified into three types with different stability against hydrolysis; (1) Si-O-Si bonds originally present and constituting the wall, (2) those which were formed by the trimethylsilylation, and (3) those derived from condensation of trimethylsilyl groups or a trimethylsilyl group and a silanol group on calcination. The hydrolysis of type (3) bonds would not result in the collapse of the structure since they did not originally constitute the wall. In contrast, the hydrolysis of the type (1) bonds causes the collapse of the structure. The type (2) bonds derived from externally introduced trimethylsilyl groups should be rather free from strain and not susceptible to hydrolysis. If this is the case, few silanol groups would be present near the type (1) bonds and hence the type (1) bonds will hardly chance to be hydrolyzed, resulting in the stabilization of the structure of the silylated-calcined samples.

4. CONCLUSIONS

Although Si-MCM-48 samples have low stability against moisture and compression, the stability can be improved drastically by trimethylsilylation; no hydrolytic cleavage of Si-O-Si bonds occurred on exposure to water vapor over a saturated aqueous solution of NH_4Cl at room temperature for 30 days, and the structure did not collapse upon applying external pressure of 1764 kg/cm^2. The high stability was attributed to the high hydrophobicity of Si-MCM-48(sil). Moreover, the stability was retained after removal of methyl groups by calcination, although the silylated-calcined material was no longer hydrophobic. The stabilization procedure is also applicable to Si-MCM-41 and Ti-containing mesoporous molecular sieves. The stability of Ti-MCM-41, which was much less resistant to moisture than pure-silica MCM-41, was drastically increased by similar treatment.

ACKNOWLEDGMENTS

This work was supported in part by Grant-in-aid for Scientific Research on Priority Areas (No. 0742104) from the Ministry of Education, Science and Culture, Japan.

REFERENCES

1. C. T. Kresge, M. E. Leonowicz, W. J. Roth, J. C. Vartuli and J. S. Beck, Nature, 359 (1992) 710.
2. J.S. Beck, J.C. Vartuli, W.J. Roth, M.E. Leonowicz, C.T. Kresge, K.D. Schmitt, C.T-U. Chu, D.H. Olson, E.W. Sheppard, S. B. McCullen, J.B. Higgins and J.L. Schlenker, J. Am. Chem. Soc., 114 (1992) 10834.
3. V.Y. Gusev, X. Feng, Z. Bu, G.L. Haller and J.A. O'Brien, J. Phys. Chem., 100 (1996) 1989.
4. T. Ishikawa, M. Matsuda, A. Yasukawa, K. Kandori, S. Inagaki, T. Fukushima and S. Kondo, J. Chem. Soc., Faraday Trans., 92 (1996) 1985.
5. T. Tatsumi, K.A. Koyano, Y. Tanaka and S. Nakata, Chem. Lett. (1997) 469.
6. C.Y. Chen, H.X. Li and M.E. Davis, Microporous Mater., 2 (1993) 17.
7. J.M. Kim, J.H. Kwak, S. Jun and R. Ryoo, J. Phys. Chem., 99 (1995) 16742.
8. R. Ryoo, J.M. Kim, C.H. Ko and C.H. Shin, J. Phys. Chem., 100 (1996) 17718.
9. S. Inagaki, Y. Sakamoto, Y. Fukushima and O. Terasaki, Chem. Mater., 8 (1996) 2089.
10. K.A. Koyano and T. Tatsumi, Microporous Mater., 10 (1997) 259.
11. C.H. Rhee and J.S. Lee, Catal. Lett., 40 (1996) 261.
12. N. Coustel, F.D. Renzo and F. Fajula, J. Chem. Soc., Chem. Commun., (1994) 967.
13. T. Tatsumi, K.A. Koyano, Y. Tanaka and S. Nakata, J. Phys. Chem., 101 (1997) 9436.
14. K.M. Reddy, I. Moudrakovski and A. Sayari, J. Chem. Soc., Chem. Commun., (1994) 1059.
15. K. A. Koyano and T. Tatsumi, Chem. Commun., (1996) 145; Microporous Mater., 10 (1997) 259.
16. T. Tatsumi, K.A. Koyano and N. Igarashi, Chem. Commun., (1998) 325.

MESOPOROUS MOLECULAR SIEVES 1998
Studies in Surface Science and Catalysis, Vol. 117
L. Bonneviot, F. Béland, C. Danumah, S. Giasson and S. Kaliaguine (Editors)

Improvement of structural integrity of mesoporous molecular sieves for practical applications

R. Ryoo, J. M. Kim and C. H. Ko

Materials Chemistry Laboratory, Department of Chemistry and Center for Molecular Science, Korea Advanced Institute of Science and Technology, Taejon, 305-701, Korea

The synthesis and post-synthetic metal grafting of mesoporous silica molecular sieves designated as MCM-41 (P6m hexagonal), MCM-48 (Ia3d cubic), KIT-1 (disordered network), SBA-1 (Pm3n cubic) and MSU-1 (disordered sponge) has been studied in order to improve the integrity of the mesoporous structures. Our results show that disordered materials are obtained in the form of irregular particles with uniform pore size distribution. On the other hand, the ordered materials can be synthesised in the form of single crystallites with excellent structural integrity. The hydrothermal stability of the materials can be improved by the addition of salts during synthesis. The mesoporous silica can be converted to the corresponding metallosilicate analogues by grafting metal elements onto the silica frameworks using non-aqueous solutions of aluminium chloride, zinc acetate, copper acetate and iron acetate. The resultant metallosilicates, which maintain the high structural integrity of the pure silica forms, are suitable for applications in catalysis, ion exchange and adsorption.

1. INTRODUCTION

In recent years, the synthesis route to porous silica using surfactant micelles as the structure-directing agent has led to the discovery of various mesoporous molecular sieves such as MCM-41 (hexagonal structure with P6m symmetry) [1,2], MCM-48 (cubic Ia3d) [1,2], SBA-1 (cubic Pm3n) [3], MSU-1 (sponge-like disordered) [4] and KIT-1 (disordered network) [5]. The mesoporous materials attract particular attention in catalysis and separation technology due to their precisely controllable pore diameters, in the range of 1.6 - 32 nm [6] far exceeding that of conventional zeolites. Various applications are considered in chromatography, catalytic cracking and reforming of heavy oil, and shape-selective synthesis of bulky organic compounds for which the micropores of conventional zeolites are not suitable.

However, practical applications of the mesoporous materials have been inhibited by low hydrothermal stability, mechanical stability, ion exchange capacity, acidity and catalytic activity compared with those of ordinary zeolites. This has stimulated a great deal of efforts for improving the physicochemical properties of the mesoporous materials [7,8]. For example, it is reported that the structures of MCM-41 and MCM-48 disintegrated during the impregnation of metal clusters using aqueous solutions although they showed excellent thermal stability in dry air up to 1000 K [9-11]. The low hydrothermal stability was due to hydrolysis of the silica frameworks. The frameworks of MCM-41 and MCM-48 were reported to collapse during the attempt to pelletization with compression above 1000 kg cm^{-1} [12,13]. The hydrothermal and mechanical stability was improved by decreasing the concentration of silanol groups of the frameworks using post-synthetic silylation techniques

[2,13]. The addition of salts during synthesis also resulted in a significant improvement of the hydrothermal stability [14].

The incorporation of Al and transition metal elements within the silica frameworks has been considered as a means of increasing the acidity, ion exchange capacity and catalytic activity of the mesoporous silica molecular sieves [7,8]. A large number of papers report the metal incorporation by adding metal sources prior to the hydrothermal reaction. However, the methods incorporating metal precursors into the surfactant-silicate gel require specialised synthesis conditions depending on the respective structures of the materials being synthesised, incorporating metals and silica sources. Besides, the synthesis is hindered by a severe decrease in the structural order as the metal content increased. X-ray diffraction (XRD) patterns of MCM-41 obtained by the synthesis method with the Si/Al ratios less than 30 show poor resolution of lines. Until now, the direct synthesis of MCM-48 [15] has not been successful for the Si/Al ratios below 22. To overcome the synthesis difficulty, a post-synthetic method of grafting metal sources onto the silica frameworks has been developed from our laboratory as a generalised means to the preparation of mesoporous metallosilicates [16]. The method is simply to slurry the mesoporous silica materials in non-aqueous solutions of various anhydrous compounds such as $AlCl_3$, $Al(NO_3)_3$, $SnCl_2$, Zn acetate, Mn acetate, Cu acetate and Fe acetate before calcination in air. The resultant metallosilicate materials exhibit cation exchange capacity that can be attributed to the metal incorporation with the oxidation states probably less than 4. The samples show superior structural integrity, acidity and catalytic activity for n-butane cracking, compared to mesoporous materials obtained by the direct synthesis methods which incorporate metal precursors into the surfactant-silicate gel.

Physicochemical properties of the mesoporous materials such as the pore arrangement, pore size distribution, hydrothermal stability, mechanical stability and silanol group content depend very much on synthesis methods. The silica frameworks are not completely cross-linked, unlike those of conventional microporous zeolites. Studies with magic angle spinning (MAS) ^{29}Si NMR spectroscopy on MCM-41 indicates that the silica frameworks of the mesoporous materials have various degrees of the silicate cross-linking with terminal silanol groups, depending on the synthesis methods [17]. The degree of the silicate cross-linking becomes very similar after calcination at 773 K, giving rise to a silanol/Si molar ratio of approximately 0.3. These results indicate that additional cross-linking occurs during the calcination, more or less depending on the degree of the cross-linking achieved during the synthesis. The additional cross-linking causes the lattice parameters to contract by 3 - 30% depending on the various synthesis methods [2,17,18]. In general, materials with much contraction have a poorly ordered arrangement of mesopores and a poor hydrothermal stability. The structural integrity of the mesoporous materials depends on details of synthesis conditions, compared with those of crystalline zeolites. The synthesis of the mesoporous silica has been extensively investigated to achieve high structural integrity [18-20].

Considering the significance of the synthesis conditions, we present our recent works which have been carried out for the improvement of structural integrity of various mesoporous silica molecular sieves.

2. EFFECT OF pH ADJUSTMENT FOR MCM-41 SYNTHESIS UNDER BASIC CONDITIONS

A hydrothermal procedure with repeated pH adjustment has been developed in order to obtain MCM-41 of high structural order [18]. The colloidal silica, Ludox HS40 (39.5 wt% SiO_2, 0.4 wt% Na_2O and 60.1 wt% H_2O, Du Pont) was added, with magnetic stirring, to 1 M aqueous solution of NaOH after preheating to 343 - 353 K. The resulting gel mixture was continuously stirred for 2 h at the same temperature, to obtain a clear sodium silicate solution with a Na/Si ratio of 0.5 (2.4 wt% Na_2O, 9.2 wt% SiO_2 and 88.4 wt% H_2O). This solution was cooled to room temperature and slowly added to a hexadecyltrimethylammonium chloride solution (24.9 wt% HTACl, 0.4 wt% NH_3 and 74.7 wt% H_2O) with vigorous magnetic stirring at room temperature. The resultant HTA-silicate gel mixture, which has a molar composition of 4 SiO_2 : 1 HTACl : 1 Na_2O : 0.15 $(NH_4)_2O$: 200 H_2O, was contained in a polypropylene bottle and heated without stirring to 370 K in an oven for one day (the first heating period). The mixture was then cooled to room temperature. The pH of the reaction mixture was adjusted to 10 by dropwise addition of 30 wt% acetic acid with vigorous stirring. The bottle was heated again to 370 K for one day (the second heating period). This procedure for pH adjustment and subsequent heating was repeated twice (i.e., carried out three times in all). The precipitated product, MCM-41 containing HTA surfactant, was filtered, washed with doubly distilled water and dried in an oven at 393 K. More than 95% of the surfactant in the dried product was removed by washing with stirring in an ethanol-HCl mixture ($\leq$ 1 g sample with a mixture of 0.5 ml of 37% HCl and 20 ml of 95% ethanol). The product after the surfactant removal was calcined easily in air under static conditions using a muffle furnace. The calcination temperature was increased to 823 K over 5 h and maintained there for 5 h.

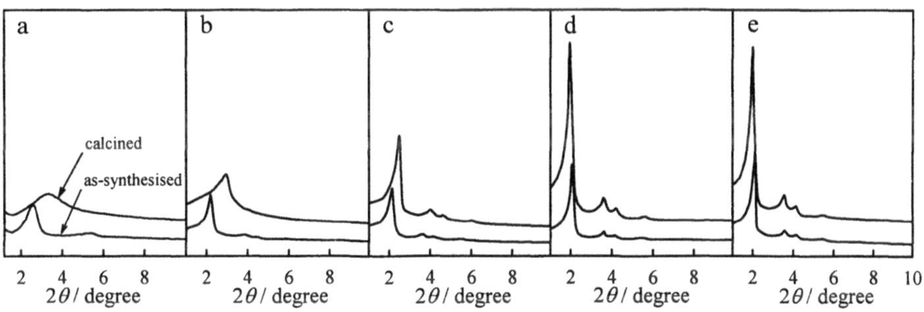

Figure 1. XRD patterns for MCM-41 synthesised (a) at room temp. for 5 h, and at 373 K (b) without pH adjustment, (c) with pH adjustment to 10, (d) by repeating pH adjustment twice, and (e) three times [18].

MCM-41 samples were also synthesised by decreasing the number of the pH adjustment and heating steps. The XRD patterns in Figure 1 show that the formation of the surfactant-silicate mesostructure takes place at room temperature under the present synthesis conditions. However, the mesostructure formed at room temperature has a relatively low diffraction intensity and resolution. The mesostructure obtained after heating the reaction mixture for one day at 370 K has four Bragg diffraction peaks with high resolution and intensity. However, the mesostructure thus obtained has a large number of excess silanol groups, due to a low degree of silicate cross-linking in the framework. The excess silanol groups are condensed during calcination, leading to lattice contraction as much as 25%. Loss of the structural order occurs during the lattice contraction. The lattice contraction and the loss of the structural order are prevented by repeating the pH adjustment and heating. Five well-resolved XRD peaks are obtained from pure silica MCM-41 sample obtained after alternating the pH adjustment and heating three subsequent times. The XRD line width changes very little during the calcination. The lattice contraction occurs less than 3% during the calcination. The pore-size distribution for the calcined sample shows a peak with a width less than 0.3 nm at half the full height, in the mesopore range [21]. These effects resulting from the pH adjustment during the hydrothermal synthesis can be attributed to shifts in the reaction equilibrium toward the silicate condensation while the framework is supported by the surfactant [18].

3. DISORDERED MESOPOROUS SILICA WITH 3-DIMENSIONAL CHANNEL NETWORK

The structure of MCM-41 consists of hexagonal packing of one-dimensional channels that are susceptible to pore blockage due to coke formation, structure damage and sintering of supported metal particles. Mesoporous materials with three-dimensional channel structures have an advantage from a catalytic and diffusional point of view. Such a material designated as KIT-1 has been synthesised in our laboratory using organic acid salts (ethylenediaminetetraacetic acid tetrasodium salt and adipic acid disodium salt) [5] as a micelle disrupter under synthesis conditions similar to that for MCM-41 in Section 2. Typical gel compositions for the synthesis are 5 SiO_2 : 1.25 Na_2O : 1 HTACl : 0.15 $(NH_4)_2O$: 4 ~ 8 $EDTANa_4$: 350 H_2O. The $EDTANa_4$ is dissolved in the surfactant solution, prior to the mixing with the silicate solution. The remainder of the synthesis procedure is the same as that for MCM-41 in Section 2, except for the difference in the gel composition.

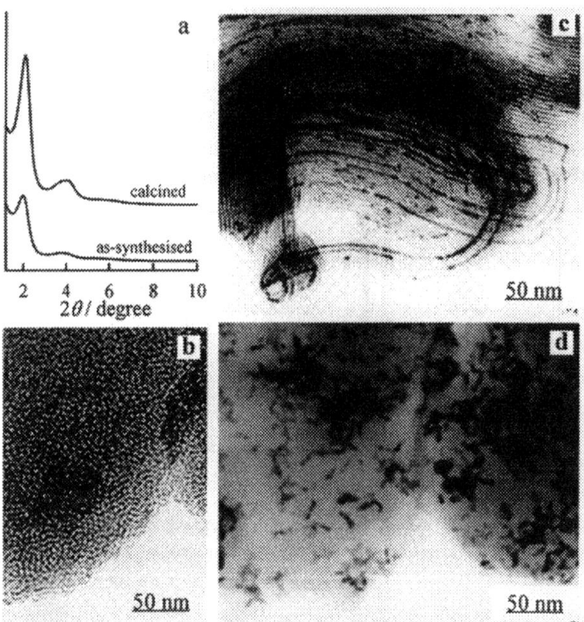

Figure 2. (a) XRD patterns and (b) TEM image for KIT-1 and TEM images for (c) MCM-41 and (d) KIT-1 obtained after supporting Pt nanowires [5].

XRD pattern for KIT-1 is shown in Figure 2(a). The XRD pattern consists of three broad peaks that may be arbitrarily indexed to (100), (200) and (300) diffraction with d_{100} = 4.0 - 4.2 nm. Calcination does not affect the XRD pattern, except that the intensity increases 3 times and the d_{100} spacing decreases slightly ($\leq$ 0.1 nm). The presence of the three-line XRD pattern after the calcination suggests that the KIT-1 does not have a layered structure, but the broad diffraction peaks come from its structural disorder with very uniform pore sizes. In order to investigate the structure of the KIT-1 further, we have compared transmission electron microscopy (TEM) images for the KIT-1 obtained after supporting Pt nanowires inside the mesoporous channels with those for MCM-41. The technique for supporting the Pt nanowires is described elsewhere [22]. The microscopy images in Figure 2 show that the structure of the KIT-1 is constructed by a three-dimensional disordered network of short worm-like channels, whereas the MCM-41 channels are one-dimensional.

The pore-size distribution for the KIT-1 has been analysed from N_2 adsorption isotherm at liquid N_2 temperature following the Horvath-Kawazoe analysis. The analysis gives a narrow peak centred at the pore diameter of 3.4 nm, with the peak width of 0.3 nm at half the full height. It is remarkable that, despite the disordered channel arrangement, the pore size for the KIT-1 is as uniform as that for the ordered MCM-41. The specific surface area of the KIT-1 obtained by the BET method is 1000 m^2g^{-1}, similar to that of MCM-41. Besides, the structure of the KIT-1 exhibits outstanding hydrothermal stability and thermal stability, compared with that of the MCM-41 sample whose XRD pattern is shown in Figure 1(e). The structure of the KIT-1 is stable in boiling water for 24 h whereas the MCM-41 structure is lost completely [5].

4. SALT EFFECTS FOR THE IMPROVEMENT OF HYDROTHERMAL STABILITY

The high hydrothermal stability of the disordered KIT-1 is very significant, considering that the improvement of hydrothermal stability has been a critical problem for ultimate successful commercialisation of the MCM-41 type materials. In order to understand the cause of the high hydrothermal stability of the KIT-1 and to apply the result to the synthesis of other materials, we have synthesised similarly disordered materials with changing the silica-to-surfactant molar ratios from that of the MCM-41 synthesis in Section 2. However, such disordered silica materials show no differences in the hydrothermal stability from the ordered MCM-41. It is therefore reasonable that the hydrothermal stability of the KIT-1 can be due to the addition of salts during synthesis. Since the structure is disordered if salts are combined with the surfactant solution before the silica source is added, we have added salts into the reaction mixture following the second heating period in Section 2. The MCM-41 synthesis and calcination have been carried out in the same ways described in Section 2, except for the use of various amounts of salts (KCl, NaCl, Na acetate, Na_4EDTA, K_4EDTA, etc.) [14]. The hydrothermal stability of the synthesised materials has been determined from the XRD intensity after heating in boiling water [5,10].

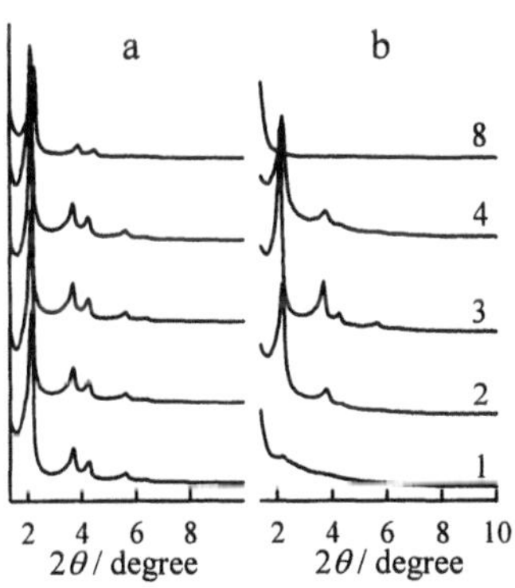

Figure 3. XRD patterns of MCM-41 displayed relative to the amount of NaCl added to the synthesis gel mixture: (a) calcined and (b) heated in boiling water for 12 h. Numbers denote the NaCl-to-HTA molar ratio [14].

The XRD patterns in Figure 3 show that the MCM-41 materials obtained using suitable amounts of salts are stable for at least 12 h in boiling water under reflux conditions. The salt effects depend very dramatically on the concentration and the nature of the salts. In the present condition, the addition up to 3 moles of NaCl per HTA, 2 moles of KCl, 4 moles of sodium acetate, and 8 moles of Na_4EDTA leads to an excellent improvement of hydrothermal stability. The salt effects can be due to moderation of the electrostatic interaction between cationic surfactant micelles and surrounding silicate anions. While the surfactant-silicate mesostructure is formed, the salt can attenuate the electrostatic interaction between the surface of the surfactant micelles and the silicates. It may be speculated that the silicates can be freed from strong electrostatic binding with the surfactant under such circumstances. Consequently, the degree of silicate polymerisation (i.e., silanol group condensation) is expected to increase compared to a situation without such electrostatic attenuation, which is confirmed using MAS ^{29}Si NMR.

5. SYNTHESIS OF MCM-48 SINGLE CRYSTALS

The structure of MCM-48, consisting of two independent and intricately interwoven networks of mesoporous channels [23,24] is preferred to the one-dimensional channel structure of MCM-41 from a diffusional and catalytic point of view. Besides, the structure of the MCM-48 is isotropic. It is reasonable that the anisotropic structure of MCM-41 is anisotropically affected by 'frustration' or 'stress' in the micelles which comes from mismatch between the palisade region and head groups. The consequence is curvatures in the channel alignment. However, the isotropic structure of the MCM-48 may not be affected by the micelle 'frustration'. If this is the case, it is likely to obtain *single* crystal-like morphology with the MCM-48, and the perfectly ordered MCM-48 structure would be suitable as a nanoporous host. However, the synthesis conditions for MCM-48 are known to be very particular, compared with those for MCM-41. The synthesis difficulty is believed to have inhibited researches on the MCM-48. We have therefore developed an easily reproducible, hydrothermal synthesis procedure to obtain the MCM-48 using a sodium silicate solution as the silica source, HTABr as the structure-directing agent, and various kinds of alcohol such as methanol, ethanol and t-butanol as an additive for the mesophase control [20].

The silica source is a sodium silicate solution which is prepared in the same way as described in Section 2, except that the molar composition is 0.25 Na_2O : 1.0 SiO_2 : 12.5 H_2O in the solution. A surfactant solution was prepared in a Teflon bottle by dissolving HTABr in a mixture of doubly distilled water and ethanol. The sodium silicate solution was added into the Teflon bottle containing the surfactant solution with vigorously magnetic stirring at room temperature. The resulting gel mixture, with a molar composition of 1.4 SiO_2 : 1.0 HTABr : 0.35 Na_2O : 5.0 ethanol : 140 H_2O, was heated for 16 h at 413 K (or 4 d at 373 K) in an autoclave under static conditions and subsequently cooled to around 340 K. The supernatant liquid in the reaction mixture was quickly separated from the precipitated MCM-48 phase by decantation and filtration before further cooling. The cooling resulted in the formation of insoluble silicate gel dispersed in the liquid, making it difficult to separate the silicate gel from the MCM-48 phase afterward. The separated MCM-48 powder was washed quickly with hot distilled water before further cooling. The product was then slurried in an ethanol-HCl mixture, filtered, washed with ethanol and calcined in the same ways used for the MCM-41 synthesis in Section 2.

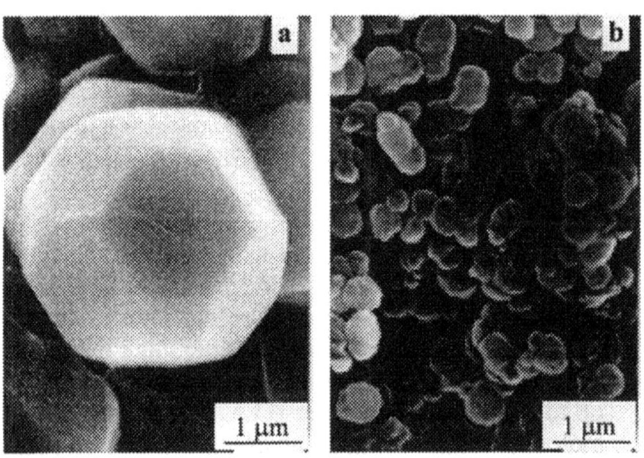

Figure 4. Field emission scanning electron microscopy images for MCM-48 obtained (a) 413 K and (b) 373 K [20].

The field emission scanning electron microscopy images in Figure 4 show that the MCM-48 sample synthesised at 413 K exhibits not only excellent structural order but also crystal-like morphology of truncated rhombic dodecahedron. The same high quality MCM-48 can be obtained by substituting 5.0 ethanol in the above procedure with 3.0 methanol. The synthesis can also be performed with 2.0 isopropanol, 0.5 sec-butanol or 0.5 t-butanol. All MCM-48 materials obtained with the alcohol showed the similar BET surface areas in the range of 1000±50 m^2g^{-1}. However, the structural order decreased with the use of the higher alcohols. The role of the alcohol is to control the surface curvature of the surfactant micelles so that the cubic surfactant-silicate mesostructure is favoured over the formation of hexagonal mesostructures [19]. Loss of alcohol during the hydrothermal reaction results in disordering of the structure. Our results indicate that the synthesis of the MCM-48 under the present conditions is not the direct transcription of the Ia3d surfactant phase [20]. Instead, a disordered surfactant-silicate mesostructure is formed during the initial period of the hydrothermal reaction. The cubic MCM-48 crystals grow from the disordered particles, subsequently transforming to a lamellar phase through a typical dissolution process. The activation energy is 14 kcal mol^{-1} for this kinetic-controlled crystallisation process for MCM-48. All these results show remarkable similarities between the formation of the MCM-48 particles under the present hydrothermal conditions and hydrothermal growth of other conventional zeolite crystals. It is very interesting that the synergistic self-assembly of the silica framework with the surfactant micelles leads to the formation of the single crystals, in spite of the amorphous nature of the silicate framework. The building blocks, that is, the lattice points of the MCM-48 crystals are not atoms or molecules but large ensembles of amorphous silica.

6. MESOPOROUS METALLOSILICATES BY POST-SYNTHETIC METAL GRAFTING

Pure silica forms of MCM-41, KIT-1, MCM-48, MSU-1 and SBA-1 were obtained following the hydrothermal synthesis procedures described in Sections 2, 3 and 5, and methods reported in the literature [3,4], respectively. The as-synthesised materials were washed with slurrying in an ethanol-HCl mixture at room temp, in the same way used for MCM-41 in Section 2. The samples were immediately dried in an oven at 413 K. The washing resulted in the removal of more than 95% of the surfactant, which was confirmed by infrared absorption spectroscopy. The mesoporous materials after the removal of the surfactant were slurried in absolute ethanol solutions containing the anhydrous form of various metal compounds such as $AlCl_3$, $Al(NO_3)_3$, $SnCl_2$, Zn acetate $[Zn(O_2CMe)_2]$, $Mn(O_2CMe)_2$, $Cu(O_2CMe)_2$ and $Fe(O_2CMe)_2$, with magnetic stirring for 1 h at 298 - 333 K [16]. The mesoporous materials were then washed with absolute ethanol, dried and calcined in air using a muffle furnace. Calcination temperature was increased to 773-823 K (depending on metals) over 5 h or less and maintained there for 2 h.

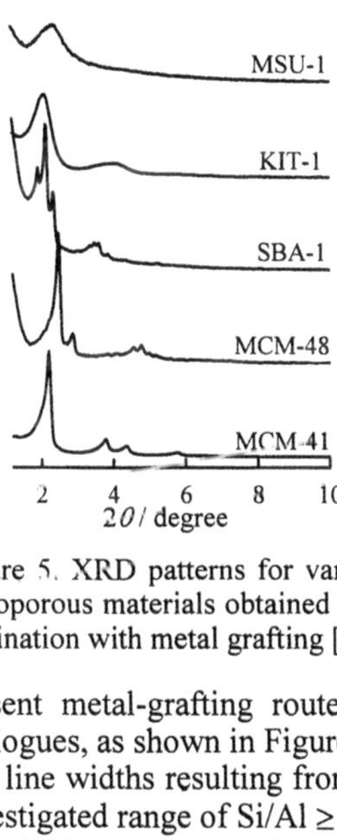

Figure 5. XRD patterns for various mesoporous materials obtained after calcination with metal grafting [16].

All the mesoporous metallosilicates obtained by the present metal-grafting route give the same XRD patterns as those for their pure-silica analogues, as shown in Figure 5. There are no significant changes in the XRD peak intensity and line widths resulting from the implantation of Al, Sn, Zn, Mn, Cu and Fe throughout the investigated range of $Si/Al \geq 8$, $Si/Sn \geq 20$, $Si/Zn \geq 10$, $Si/Mn \geq 20$, $Si/Cu \geq 30$ and $Si/Fe \geq 30$. No additional XRD peaks appeared beyond the 2θ region shown in Figure 5, indicating that the grafted metal elements are well dispersed. The BET areas of the various samples range from 900 to 1200 m^2g^{-1}, and show no more decreases. All the materials exhibit significant cation exchange capacity, which indicates that the metals are incorporated within the silicate frameworks with the oxidation states less than +4. In the case of aluminium incorporation, MAS ^{27}Al NMR spectra show much higher intensity for aluminium with tetrahedral co-ordination than that for octahedral aluminium. Temperature programmed desorption (TPD) of ammonia showed that the present method gives, in general, solids with stronger acidity than those synthesised by the incorporation of Al sources during hydrothermal synthesis [16]. The metal-grafted samples exhibit higher catalytic activity for butane cracking at 800 K. This can be due to the presence of a relatively large amount of aluminium on the framework surface of the grafted samples, while much aluminium in the other samples is embedded within the pore wall.

7. PERSPECTIVES

The structural integrity of mesoporous molecular sieves has been remarkably improved in recent years. If progress is made at this rate, it will be possible to have the mesoporous materials in the form of fairly large uniform single crystals that are suitable for chromatographic applications and single crystal X-ray studies in the near future.

REFERENCES

1. C. T. Kresge, M. E. Leonowicz, W. J. Roth, J. C. Vartuli and J. S. Beck, Nature, 359 (1992) 710.
2. J. S. Beck, J. C. Vartuli, W. J. Roth, M. E. Leonowicz, C. T. Kresge, K. D. Schmitt, C. T.-W. Chu, D. H. Olson, E. W. Sheppard, S. B. McCullen, J. B. Higgins and J. L. Schlenker, J. Am. Chem. Soc., 114 (1992) 10834.
3. Q. Huo, R. Leon, P. M. Petroff and G. D. Stucky, Science, 268 (1995) 1324.
4. S. A. Bagshaw, E. Prouzet and T. J. Pinnavaia, Science, 269 (1995) 1242.
5. R. Ryoo, J. M. Kim, C. H. Ko and C. H. Shin, J. Phys. Chem., 100 (1996) 17718.
6. D. Zhao, J. Feng, Q. Huo, N. Melosh, G. H. Fredrickson, B. F. Chmelka and G. D. Stucky, Science, 279 (1998) 548.
7. A. Sayari, Stud. Surf. Sci. Catal., 102 (1996) 1.
8. A. Corma, Chem. Rev., 97 (1997) 2373.
9. J. M. Kim, J. H. Kwak, S. Jun and R. Ryoo, J. Phys. Chem., 99 (1995) 16742.
10. J. M. Kim and R. Ryoo, Bull. Korean Chem. Soc., 17 (1996) 66.
11. R. Ryoo, C. H. Ko, J. M. Kim and R. Howe, Catal. Lett., 37 (1996) 29.
12. P. I. Ravikovitch, D. Wei, W. T. Chueh, G. L. Haller and A. V. Neimark, J. Phys. Chem. B 101 (1997) 3671.
13. K. A. Koyano, T. Tatsumi, Y. Tanaka and S. Nakata, J. Phys. Chem. B, 101 (1997) 9436.
14. R. Ryoo and S. Jun, J. Phys. Chem. B, 101 (1997) 317.
15. R. Schmidt, H. Junggreen and M. Stocker, Chem. Commun., (1996) 875.
16. R. Ryoo, S. Jun, J. M. Kim and M. J. Kim, Chem. Commun., (1997) 2225.
17. C.-Y. Chen, H.-X. Li and M. E. Davis, Microporous Mater., 2 (1993) 17.
18. R. Ryoo and J. M. Kim, J. Chem. Soc., Chem. Commun., (1995) 711.
19. Q. Huo, D. I. Margolese and G. D. Stucky, Chem. Mater., 8 (1996) 1147.
20. J. M. Kim, S. K. Kim and R. Ryoo, Chem. Commun., (1998) 259.
21. M. Kruk, M. Jaroniec, R. Ryoo and J. M. Kim, Microporous Mater., 12 (1997) 93.
22. R. Ryoo and C. H. Ko, Chem. Commun. (1996) 2467.
23. M. W. Anderson, Zeolites, 19 (1997) 220.
24. V. Alfredsson, M. W. Anderson, T. Ohsuna, O. Terasaki, M. Jacob and M. Bojrup, Chem. Mater. 9 (1997) 2066.

MESOPOROUS MOLECULAR SIEVES 1998
Studies in Surface Science and Catalysis, Vol. 117
L. Bonneviot, F. Béland, C. Danumah, S. Giasson and S. Kaliaguine (Editors)

Transition metal-modified mesoporous silicas as catalysts for oxidation reactions

Alain TUEL

Institut de Recherches sur la Catalyse. CNRS. 2, avenue A. Einstein 69626 Villeurbanne Cedex France

1. INTRODUCTION

Selective oxidation is one of the key reactions for the preparation of intermediates and fine chemicals. At an industrial scale, most of these reactions are still stoechiometric and lead to the formation of millions of tons of non recyclable and polluting by-products. Therefore, the development of new processes associating a stable, recyclable heterogeneous catalyst with a clean oxidant like H_2O_2 represents a challenge for the forthcoming decade. As far as catalysts are concerned, transition metal-substituted zeolites like TS-1 have shown excellent performances in the liquid phase oxidation of small organic substrates with aqueous hydrogen peroxide at low temperature [1]. The activity of these catalysts have been attributed to isolated tetrahedrally coordinated metal species in the hydrophobic environment of the zeolite channels [2]. However, their use is limited due to the small pore opening of zeolitic structures and the necessity to develop large pore or mesoporous materials capable of catalyzing the oxidation of bulky molecules is thus of prime importance. As for zeolites, incorporation of heteroelements in the amorphous framework of mesoporous silicas can provide the latter with potential applications in catalysis. This can be achieved by different procedures, including direct synthesis in the presence of the metallic precursor or post-synthesis treatments. Various types of materials can be prepared depending on the synthesis conditions, principally the nature of the surfactant molecule. While modified-MCM-41 samples are usually obtained using long chain quaternary ammonium cations as surfactants [3], the use of neutral molecules like primary amines in C_8 to C_{18} leads to HMS (Hexagonal Mesoporous Silica) [4]. HMS slightly differs from MCM-41 by the lack of a long range order of the mesopores and thicker framework walls [5]. However, despite the presence of an array of regular mesopores, these materials cannot be considered as "very large pore" zeolites but rather as ordered mixed oxides. Indeed, they are amorphous and the internal surface of the mesopores is covered with silanol groups, which drastically increases their hydrophilic character as compared to most of the conventional zeolites [6]. Therefore, they cannot be used with aqueous H_2O_2 in reactions which require the total absence of water like the oxidation of alkanes or the hydroxylation of substituted aromatic compounds. Nevertheless, transition metal-modified mesoporous silicas can be a good alternative to zeolitic materials in the oxidation of bulky substrates for which microporous molecular sieves are totally inactive.

This presentation reviews the most recent advances on the use of various transition metal-modified mesoporous silicas for the catalytic oxidation of organic substrates. Most of the data available in the literature concern the liquid phase oxidation of molecules with aqueous hydrogen peroxide or organic hydroperoxides. Unfortunately, it is sometimes difficult to evaluate the catalyst efficiency when reactions are illustrated by a few examples of experimental data. Moreover, since fundamental aspects like the leaching of active metal species under reaction conditions are rarely reported, some of the activities and selectivities have to be considered with great caution.

2. CATALYSTS

Liquid phase oxidations require catalysts with isolated active centers in an inorganic support, generally silica [7]. These centers can be transition metal ions like titanium or vanadium, which are known to provide microporous zeolites with specific properties [8]. The synthesis and characterization of transition metal-modified mesoporous silicas has been widely reported in recent reviews [9-11]. Ti-MCM-41 and Ti-HMS can be relatively easily synthesized with metal contents up to ca. 7 wt %, corresponding to Si/Ti > 10 [12]. The amount of metal incorporated is high compared to that introduced in zeolitic framework, due to the amorphous nature of the solids and the absence of strains which are present in crystalline materials. For both MCM-41 and HMS derivatives, incorporation of titanium results in an increase of the d_{100} spacing in XRD patterns and of the pore wall thickness. Characterization of these solids by means of IR, UV-Vis and XANES spectroscopies indicated that, in calcined materials, Ti species are isolated in the silica framework with a predominant tetrahedral coordination along with rehydrated five and six coordinated species [13]. The higher coordination of Ti species as compared to TS-1, where Ti is essentially tetrahedrally coordinated, is a direct consequence of the hydrophilic character of mesoporous silicas. An important difference between Ti-MCM-41 or Ti-HMS and Ti-containing zeolites is that some of the active sites in mesoporous silicas can be located inside the silica walls, and are thus inaccessible to substrate or oxidant molecules. In order to circumvent this problem, Oldroyd et al. [14] have prepared Ti-supported MCM-41 by reaction of a pure silica material with titanocene dichloride. Characterization of the calcined material by means of X-ray absorption spectroscopy revealed that Ti species were tetrahedrally coordinated and could readily change their coordination in the presence of moisture. The catalytic activity of the material in the epoxidation of cyclohexene with TBHP was considerably increased with respect to a conventional Ti-MCM-41 material.

Vanadium-modified mesoporous silicas V-MCM-41 and V-HMS have also been prepared by several groups [15,16]. Characterization of calcined materials showed that the environment of V(V) species was different in V-HMS and V-MCM-41, particularly after rehydration of the samples. Gontier et al. [16] have reported that V(IV) species in as-prepared V-HMS materials were probably occluded in the mesopores along with surfactant molecules and that oxidation and grafting on the silica surface took place upon calcination. Thus, the coordination of V(V) species in calcined V-HMS samples is similar to that observed in V-MCM-48 materials prepared by reaction of a pure silica material with $VO(OiPr)_3$ [17]. In particular, it changes very rapidly from tetrahedral to octahedral with exposure to ambient atmosphere, as evidenced by UV-Vis and NMR spectroscopies.

Other cations such as Zr^{4+} have also been reported to be active and selective catalysts in homogeneous systems. However, incorporation of the large Zr^{4+} cations in zeolites is very difficult due to the strains imposed by the crystalline framework. On the other hand, we have reported in our group that Zr^{4+} could be introduced in the amorphous framework of HMS materials [18]. Samples were synthesized following a method adapted from that used to prepare Ti-HMS Solids with Zr loadings up to ca. 8 wt % could be obtained without formation of segregated ZrO_2 particles, as evidenced by UV-Vis spectroscopy. As it will be discussed later, Zr-containing mesoporous silicas are efficient catalysts for the oxidation and epoxidation of a variety of organic substrates with H_2O_2 and TBHP.

Sn-modified mesoporous silicas were also prepared by cationic and neutral assembly pathways [19,20]. More recently, several groups have reported on the synthesis of Cr-, Mn and Mo-modified mesoporous silicas and their application as catalysts in the oxidation of organic substrates like benzene, phenol, 1-naphthol or aniline [21,22]. However, catalytic data were often illustrated by a single example and the advanced characterization of the catalysts in terms of dispersion, coordination and oxidation state of the metal was not reported.

Another class of catalysts consists in supporting active species on the surface of mesoporous silicas. This can be achieved by grafting an organometallic complex onto the walls of the mesoporous support or by preliminary functionalization of the silica surface followed by anchoring the complex. The example of Ti-grafted MCM-41 using titanocene complexes has already been mentioned above [14]. More recently various complexes of manganese or cobalt have been anchored to the silica walls of MCM-41 and used in selective oxidations [23,24].

3. LIQUID PHASE OXIDATIONS

As mentioned in the introduction, all these materials were essentially used in the liquid phase oxidation of organic substrates with H_2O_2 or organic hydroperoxides. Due to the presence of mesopores in the catalysts, special attention has been given to the oxidation of bulky molecules, which cannot penetrate the channels of microporous zeolitic systems. However, modified mesoporous silicas can also be used to enhance the catalytic activity when diffusion limitations exist over microporous solids. A summary of the application of these materials for catalyzing oxidation, hydroxylation and epoxidation reactions is given below.

3.1. Oxidation of primary amines

Ti-MCM-41 shows very little activity in the conversion of primary aliphatic amines to the corresponding hydroxylamines and oximes compared to TS-1 [25]. On the other hand, Ti- and Zr-HMS materials are very active catalysts in the liquid phase oxidation of arylamines with H_2O_2 or TBHP [26]. With H_2O_2 as oxidant, azoxy compounds can be obtained with selectivities up to 95 % when the reaction is performed at relatively high temperature. Decreasing the temperature of the reaction to 0°C modifies the product selectivity and nitroso compounds are the major products formed. In fact, modified mesoporous solids catalyze the oxidation of the amine to nitroso and nitro derivatives. Azoxy and azo compounds are further formed by bimolecular reactions between products and/or unreacted amine. Decreasing the reaction temperature thus diminishes the probability of bimolecular reactions and increases the selectivity in nitroso compounds. No differences in activity and selectivity were observed between Ti- and Zr-containing materials. Even though the non-substituted aniline could be selectively oxidized over the microporous TS-1, turnover numbers observed on mesoporous catalysts were 10 to 50 time higher. The presence of nitroso compounds in the reaction

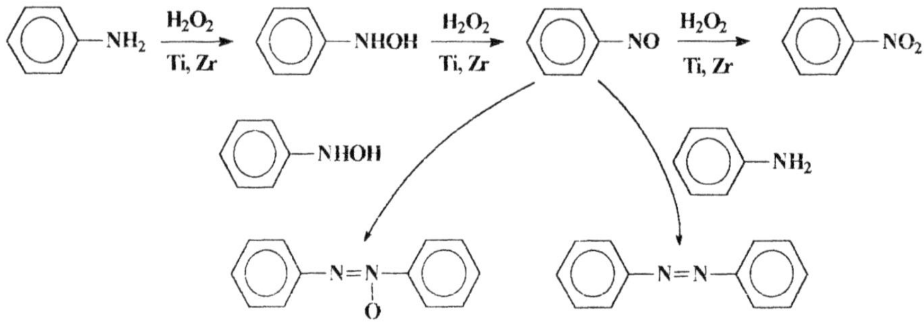

mixture was in favor of an electrophilic attack of the amine lone pair by peroxo species formed by reaction of the metal center with H_2O_2. By contrast, TBHP leads to significant amounts of azo compounds, thus supporting a radical mechanism. The advantage of mesoporous systems with respect to zeolites was illustrated by data on the oxidation of 3-methylaniline : the corresponding azoxy compound was formed within less than 30 min over Ti- or Zr-HMS materials with 90 % selectivity and a complete consumption of H_2O_2 (Table 1). The same experiment performed over TS-1 led to trace amounts of oxygenated compounds after several hours reaction.

Table 1
Oxidation of 3-methylaniline with H_2O_2 over Ti and Zr-modified mesoporous silicas

Catalyst[a]	T (°C)	Time (h)	H_2O_2 eff (%)[b]	Prod. sel. (%)		
				Nitroso	Azo	Azy
Ti-HMS	70	0.5	99	3	-	97
	0	3	68	71	-	29
Zr-HMS	70	0.5	95	5	-	95
	0	3	62	68	-	32
TS-1	70	3	12	17	-	83

[a]Si/Ti = 108 and Si/Zr = 98 in mesoporous silicas. TS-1 was prepared with Si/Ti = 79.
[b]H_2O_2 efficiency = (moles of H_2O_2 utilized for the formation of oxygenated compounds/moles of H_2O_2 introduced) x 100.
Reaction conditions : 0.5 g catalyst, 0.1 mole 3-methylaniline, 0.03 mole H_2O_2, solvent = acetonitrile.

V-HMS also catalyzes the oxidation of substituted anilines with TBHP [26]. In this case, nitro compounds are obtained with 100 % selectivity and no azoxy or azo derivatives are observed. However, a reaction performed with the substrate after removal of the catalyst resulted in a similar conversion, thus suggesting the presence of active V species in solution. Leaching of V species from V-containing mesoporous silicas under reaction conditions was further reported by different groups for various substrates and oxidants.

3.2. Oxidation of organic sulfides

Corma et al. [27] have reported that Ti-MCM-41 could catalyze the oxidation of bulky thioethers to the corresponding sulfoxides and sulfones :

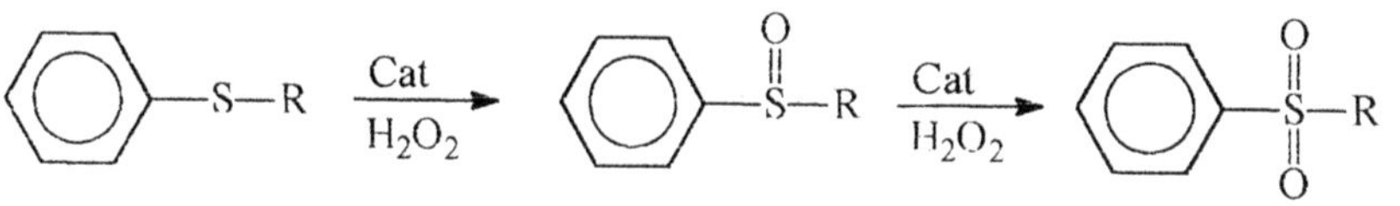

In the case of methyl-phenyl sulfide (R = CH_3), Ti-MCM-41 was less active than the large pore Ti-Beta zeolite, in agreement with the higher intrinsic activity expected for Ti sites in the zeolite. Nevertheless, the order of reactivity was reversed in the case of the bulky isopentyl-phenyl sulfide, which diffuses more slowly in the zeolite pores. For both substrates, it was observed that the sulfoxide selectivity was higher when the reaction was performed over Ti-MCM-41.

3.3. Hydroxylation of aromatic compounds.

Benzene, phenol, 1-naphthol as well as bulkier aromatic compounds like 2,6-di-*tert*-butyl phenol (DTBP) have been oxidized in the presence of H_2O_2 or TBHP over Ti, V and Sn-modified mesoporous silicas. Das et al. [19] have reported that Sn-MCM-41 could catalyze the hydroxylation of phenol with H_2O_2. No activity in this reaction was observed with a pure silica material and the phenol conversion increased with the Sn content in the catalysts. Moreover, a negligible conversion was observed with Sn-impregnated MCM-41, showing that active species are isolated Sn^{4+} cations incorporated to the silica walls. By contrast to the Ti-containing zeolite TS-1, the 1,2 - dihydroxybenzene was the major product formed along with the para isomer and para-benzoquinone. The product selectivity was not strongly affected by the Sn content in the materials.

V-HMS has also been used to catalyze the hydroxylation of phenol to dihydroxybenzenes [15]. In this case, no quinone was detected and the para/ortho ratio was close to 0.75. However, leaching of V species in the presence of H_2O_2 and homogeneous reaction by dissolved species could explain, at least partially, the observed activity.

Tanev et al. [4] reported an exceptional activity of Ti-HMS materials in the hydroxylation of benzene with H_2O_2 in the presence of acetone as the solvent. However, following papers reported that products were actually due to the oxidation of acetone, and not separated from phenol by gas chromatography [21]. In fact, Ti, V, Mn, Cr and Mo-modified MCM-41 have shown interesting activities in the benzene hydroxylation in the absence of solvent. However, leaching of active species during the reaction and the possibility of an homogeneous catalysis were not mentioned.

Larger aromatic compounds like naphthalene and 1-naphthol can also be hydroxylated with Sn or V-MCM-41 to 1,4 - naphthoquinone. Selectivities up to 90 % in the quinone were observed over Sn-MCM-41 [15,19].

Oxidation of the bulky 2,6-di-*tert*-butyl phenol (DTBP) to the corresponding quinone and diphenoquinone has also been widely reported.

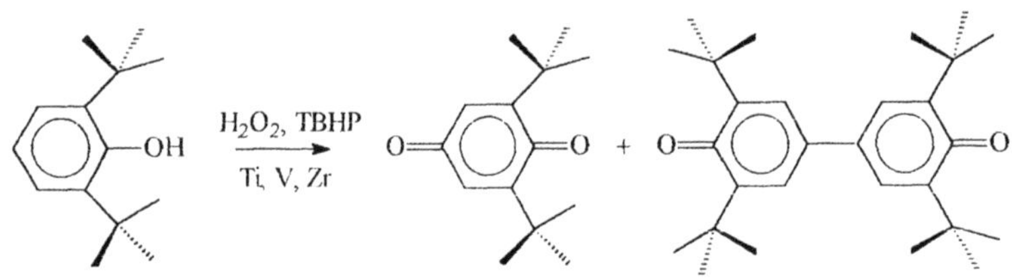

Pinnavaia et al. [28] were the first to show that Ti-modified mesoporous silicas could catalyze the hydroxylation of such a bulky substrate in the presence of an excess of H_2O_2 (H_2O_2/DTBP = 6). Ti-HMS and Ti-MCM-41 were much more active in this reaction than TS-1, which was explained by the small pore opening of the zeolite. The authors also observed that Ti-HMS materials were much more active than the corresponding Ti-MCM-41. The difference was attributed to the complementary textural mesoporosity in Ti-HMS, thus allowing a better access of the reagent molecules to the active sites. However, the possibility of differences in Ti sitting between the two materials was also suggested.

Zr-HMS materials show similar activities and selectivities than Ti-containing materials in this reaction and high quinone selectivities up to 90 % have been obtained [18]. V-HMS also exhibit exceptional activity and H_2O_2 efficiency in this reaction, by far superior to that of its titanium analog [15]. However, as for the hydroxylation of phenol, the activity was essentially due to dissolved species resulting from the leaching of V species in the presence of H_2O_2.

3.4. Epoxidation of olefins

Several transition metal-modified mesoporous silicas have been reported to catalyze the epoxidation of linear and cyclic olefins with H_2O_2 and organic hydroperoxides. The activity of Ti-MCM-41 in the epoxidation of linear olefins like 1-hexene was found to be lower than when using Ti-containing zeolites like TS-1 or Ti-Beta. Since diffusion limitations are negligible with such a substrate, the superior activity of zeolites was attributed to a higher intrinsic activity of Ti species in the latter. However, mesoporous molecular sieves become much more active than zeolites when cyclic or bulky olefins are used as substrates. For example, the selective oxidation of methyl methacrylate to methyl pyruvate proceeds 2-2.5 times faster over mesoporous Ti-containing catalysts than with TS-1 [13].

Great differences in activity are also observed in the oxidation of styrene with H_2O_2. Ti-HMS is approximately 4 times more active than TS-1 and the major product formed is acetaldehyde, resulting of the cleavage of the C-C bond [13]. The selectivity in acetaldehyde is higher than 70 %, the other products formed being styrene oxide and the corresponding diol. However, relatively high styrene oxide selectivities can be obtained in anhydrous media. Kim et al. [28] have performed the reaction with iodosylbenzene over $[Mn(bpy)_2]^{2+}$ complexes immobilized by electrostatic interactions in the mesopores of Al-MCM-41. In this case, similar amounts of styrene oxide and phenylacetaldehyde were formed. Even though catalysts did not show any loss in activity when recycled, the possibility of leaching of the complex, which was not chemically bonded to the support, was not mentioned.

Epoxidation of norbornene to norbornene oxide has also been carried out using Ti and Zr-modified mesoporous silicas [18]. The oxidation of norbornene can lead to the exo and endo epoxides, which can further give the alcohol by ring opening in the presence of acid sites. The exo epoxide is the major product formed over Ti-MCM-41 [29]. The activity is higher than that observed over the Ti-Beta zeolite, which leads to the preferential formation of the alcohol, catalyzed by protons associated with the framework Al sites. Comparison between Zr and Ti-HMS in this reaction shows that both types of materials possess similar activities with H_2O_2 as oxidizing agent in this reaction but that Zr-HMS leads to the selective formation of the alcohol (Table 2). This was attributed to the strong Lewis acidity of the Zr^{4+} cation in the silica matrix, as evidenced by ammoniac adsorption [18].

Table 2
Epoxidation of norbornene with H_2O_2 over Ti and Zr-HMS (from ref. [18]).

Catalyst[a]	Substrate conv. (%)	Epoxide select. (%)	Alcohol select. (%)	H_2O_2 conv. (%)
Ti-HMS	6.5	70	30	82
Zr-HMS	7.5	20	80	74

[a]Si/Ti = 51 and Si/Zr = 47 in Ti-HMS and Zr-HMS, respectively.

We have also reported that Ti and Zr-HMS samples were efficient catalysts for the oxidation of cyclohexene with aqueous H_2O_2 and TBHP (Table 3) [11].

Table 3
Oxidation of cyclohexene over a series of Ti and Zr-modified mesoporous silicas

Catalyst	Si/Me	Oxidant	Oxid. Conv. (%)	Select. (%)[a] Epox.	Alc.	Diol
Ti-HMS	52	TBHP	92	94	2	4
Ti-HMS	184	H_2O_2	85	50(86)	35(4)	15(10)
Ti-HMS	85	H_2O_2	97	52(82)	30(6)	18(14)
Ti-HMS	52	H_2O_2	95	48	30	22
Ti-HMS	19.5	H_2O_2	97	41	33	26
Zr-HMS	125	H_2O_2	90	48(72)	27(6)	25(22)
Zr-HMS	47	H_2O_2	92	32	23	45
Zr-HMS	26	H_2O_2	97	30	22	48

[a]selectivities in cyclohexene oxide (Epox.), cyclohexenol (Alc.) and cyclohexanediol (Diol). Values in parentheses correspond to ethanol extracted samples.
Reaction conditions : 0.5 g catalyst, 0.1 mole cyclohexene, Oxidant/cyclohexene = 0.2, T = 70°C, solvent = acetonitrile. Data obtained after 3 h reaction.

When Ti-modified mesoporous silicas and anhydrous organic hydroperoxides are used, cyclohexene is selectively converted to cyclohexene oxide. The activity can be considerably increased by using Ti-grafted MCM-41 instead of a conventional catalyst where some of the Ti species can be located inside the silica walls [14]. By contrast, the use of H_2O_2 as oxidizing agent leads to the formation of cyclohexene oxide along with 2-cyclohexene-1-ol and cyclohexanediol. Over calcined Ti-HMS materials, the epoxide and alcohol are formed in similar amounts, suggesting a radical mechanism involving the participation of cyclohexenyl hydroperoxide as the oxidant :

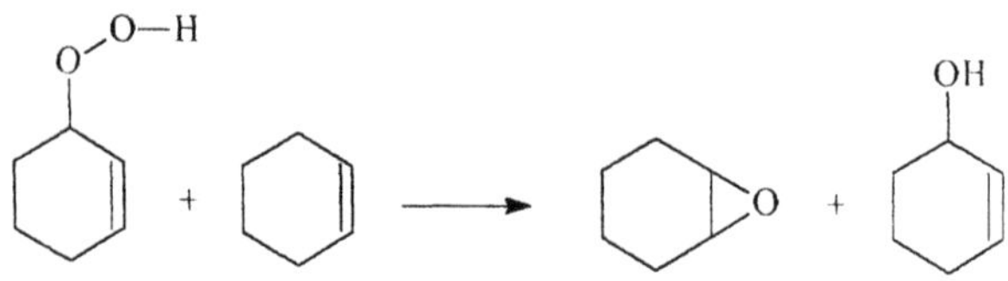

This mechanism has already been proposed for the epoxidation of cyclohexene over.Ti/SiO_2 catalysts [30]. Therefore, two different routes can be envisaged during the epoxidation of cyclohexene over Ti-modified mesoporous silicas.
Route (1) predominates in hydrophobic systems while traces of water direct the reaction to the formation of cyclohexenyl hydroperoxide (route (2)).

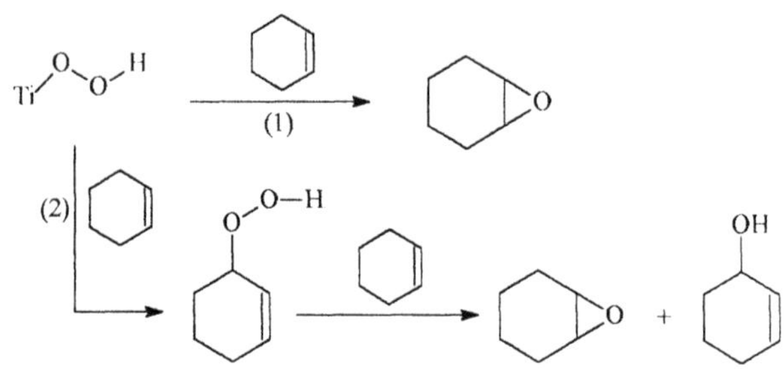

As far as the Ti content in Ti-HMS does not exceed 2 wt %, only traces of cyclohexanediol are detected. The diol selectivity increases with the Ti content for higher loadings due to the formation of Brönsted sites, resulting from a lower dispersion of Ti species in the catalysts.

Removing the surfactant molecules from the mesopores of Ti-HMS by ethanol extraction without further calcination considerably increases the epoxide selectivity. We have observed that a partial esterification of the internal surface of the mesopores took place upon solvent extraction. The Si-O-C_2H_5 moieties thus formed probably reinforce the hydrophobic character of Ti-HMS which favors pathway (1) in the above mechanism. As for norbornene, the selectivity in cyclohexanediol is higher when Zr-HMS are used as catalysts, due to the stronger Lewis acidity.

Corma et al. [29] took advantage of the presence of acid sites in Al-Ti-MCM-41 to perform directly the two-step oxidation of α-terpinol following the scheme :

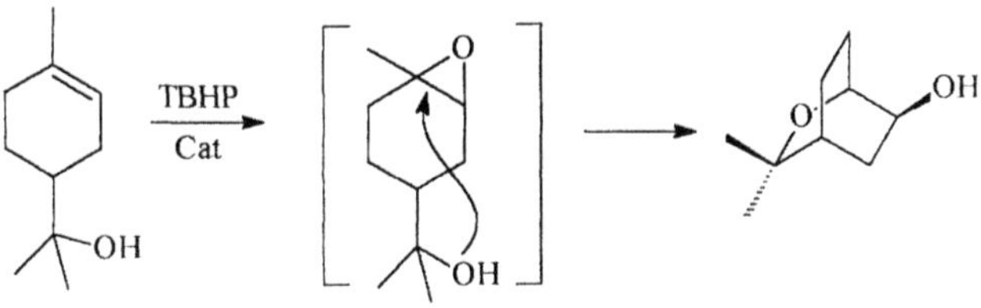

(1) (2) (3)

Titanium sites catalyze the epoxidation of α-terpinol to the corresponding epoxide (2), which further reacts with acid sites to give product (3), which is interesting as an additive. The activity of Ti-Al-MCM-41 in this reaction was by far superior to that of Ti-Al-Beta zeolite, and selectivities as high as 80 % in product (3) were obtained after 8 h reaction.

In a similar manner, Ti-Al-MCM-41 catalyzes the multistep oxidation of linalool to cyclic furan and pyran hydroxy ethers with 100 % selectivity [29].

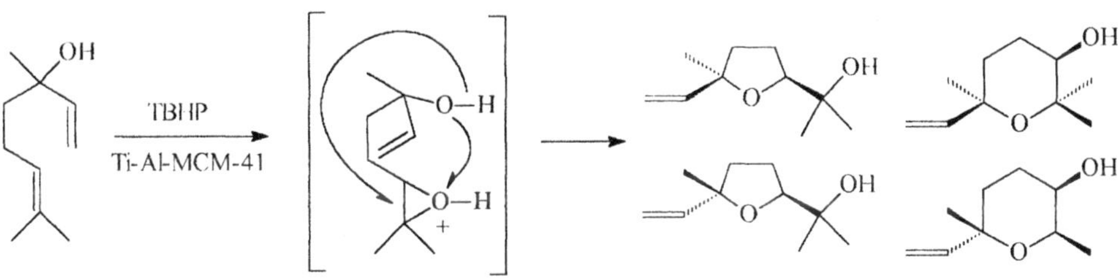

For this reaction, TBHP was used as the oxidant and conversions of 80 % were obtained. As shown in the reaction scheme, the epoxidation of the double bond catalyzed by Ti centers is followed by the acid catalyzed intramolecular ring opening. Furan to pyran ratios of about 0.9 were obtained, comparable to those observed with epoxidase enzyme as catalyst.

3.5. Oxidation of alkanes

The literature concerning the oxidation of alkanes over transition metal-modified mesoporous silicas is scarce. One of the reason is that Ti-containing catalysts are almost inactive in this reaction. Up to date, only V-MCM-41 and V-HMS materials have been reported to catalyze the selective oxidation of cyclic alkanes and cyclic alcohols to the corresponding ketones [15,31]. Over V-HMS, cyclododecane is converted to cyclododecanone (67 %) and cyclododecanol (27 %) with 34.2 % conversion and 21 % H_2O_2 selectivity. Oxidation of cyclododecanol leads to the selective formation of cyclododecanone (selectivity > 90 %) in acetonitrile using either H_2O_2 or TBHP as oxidizing agent. However, dissolution of vanadium species was observed with both types of oxidants, which makes that the observed activities more likely result from homogeneous reactions. It was also the conclusion of Neumann et al. [32] who studied the oxidation of a series of cyclic alkanes over V-MCM-41 using various oxidants. They claimed that vanadium was almost entirely leached from the heterogeneous catalyst in the presence of H_2O_2. Cyclic alkanes gave the corresponding ketones with high selectivities at reasonable conversions when TBHP was used as oxidizing agent. For all substrates, cycloalkyl-*tert*-butyl peroxides were formed as secondary products, thus supporting a radical mechanism. By measuring the amount of dissolved vanadium in the solution after reaction, the authors came to the conclusion that approx. 0.1 % of the total vanadium was leached during the catalytic process. Surprisingly, they observed that these V species were inactive in the oxidation of cyclic alkanes and they concluded that the reaction was truly heterogeneous. To circumvent problems of leaching of active species and to increase the selectivity in ketones, reactions were carried out using the isobutyraldehyde (IBA)/O_2 system, which is known to form in situ the anhydrous peracid $Me_2CHC(O)OOH$ and/or the peroxo radical $Me_2CHC(O)OO^{\cdot}$. Exceptionally high ketone selectivities were observed in the room temperature oxidation of cyclooctane. V-MCM-41 was found to be more active than V_2O_5-SiO_2 catalysts and could be recycled without loss in activity. A mechanism has been proposed to explain the exclusive formation of cyclooctanone from cyclooctane in the presence of IBA/O_2 without formation of the intermediate alcohol.

This mechanism shows that quite large amounts of isobutyric acid, alkyl peracid and acetone are formed. Under such conditions, it is difficult to believe that V-MCM-41 catalysts are stable and that no leaching of V species are observed, as claimed by the authors.

Recently, Maschmeyer et al. [23] have reported the selective low temperature oxidation of cyclohexane to cyclohexanol and cyclohexanone over a cobalt complex immobilized on MCM-41. When MCM-41 was functionalized with glycine prior to anchoring the complex, high TOF's of 216 (mol cyclohexane/mol catalyst.h) were observed for long reaction times and the total selectivity in alcohol and ketone exceeded 95 %. The reaction was truly heterogeneous since no leaching of the complex was detected.

4. GAS PHASE OXIDATIONS

Despite a relatively abundant literature concerning the application of modified mesoporous silicas in liquid phase oxidations, very few studies have reported on the use of such materials for the gas phase oxidation of organic molecules. This can appear surprising if we consider the numerous advantages that possess mesoporous silicas with respect to conventional supports like amorphous SiO_2 to prepare metal-supported catalysts. They are stable at high temperatures and their pore dimension combined with very high surface areas make that it is possible to prepare accessible highly dispersed metal species. Moreover, since some of the problems associated with the liquid phase like the leaching of active species is not expected to occur in the gas phase, materials like V or Cr-modified silicas might be used. As an example, Centi et al. [33] have studied the oxidation of toluene with oxygen over a series of various vanadium-containing micro and mesoporous molecular sieves. The reaction led to principally benzaldehyde but phenol was also formed over mesoporous catalysts. A maximum in benzaldehyde selectivity was observed for low V contents in V-HMS (approx. 2 wt % V) but the phenol selectivity increased with the vanadium loading. A mechanism was proposed to explain the formation of phenol via the formation of large intermediates like benzophenone and phenylbenzoate. These intermediates are too bulky to form inside the micropores of V-containing zeolites like VS-1, which explains why phenol is only detected at trace levels over these catalysts.

Another recent work concerns the total oxidation of propene to carbon dioxide over manganese-oxo species grafted onto the surface of MCM-41 mesopores [34]. In this case, MCM-41 was treated with $Mn_2(CO)_{10}$, followed by a calcination in air for 2 h at 300°C. The material showed a high activity in the conversion of propene compared to a conventional supported manganese catalyst, which was attributed to highly dispersed MnO_x species in the mesopores.

5. CONCLUSION

With the discovery of mesoporous silicas, a new generation of "redox" catalysts has been obtained by expanding the area of microporous redox molecular sieves to the mesopore range. Moreover, their amorphous character afforded the possibility to incorporate large cations like Zr^{4+}, which could not be introduced in crystalline microporous structures.

Some of these materials, particularly those containing Ti and Zr, have shown to be promising catalysts for the oxidation of bulky substrates like cyclic olefins and polyaromatics with aqueous H_2O_2 or TBHP. However, despite the advantages related with the presence of mesopores, the intrinsic activity (per active site) is usually lower than that observed on zeolites. The difference is probably due to the amorphous nature of the framework along with a relatively high hydrophilic character. Moreover, their applications are limited since they do not catalyze reactions such as the oxidation of linear aliphatic compounds or the selective hydroxylation of substituted aromatics to the corresponding di-hydroxy molecules with H_2O_2. Another aspect which is often neglected is the stability of these materials under reactions conditions. If leaching of active species in the presence of the substrate, the solvent or the oxidizing agent is sometimes reported, the study of the stability of the mesopore structure upon long exposure to solvents and water is lacking. This point is however particularly important if processes have to be developed in industry, where catalysts are often working for several days or weeks before recycling.

Even though applications of these solids in the gas phase oxidation of organic substrates is still limited, the possibility of preparing highly dispersed metal species opens new opportunities in the field and there is no doubt that many new processes will be developed in the near future.

Finally, a very exciting field of research concerns the heterogeneization of homogeneous systems by anchoring metal complexes to the surface of the mesopores. These complexes usually possess specific properties as homogeneous catalysts which depend on the nature of the metal itself and of the ligands, and which cannot be reproduced with metal-incorporated mesoporous silicas. Therefore, new families of heterogeneous catalysts have to be developed, which will considerably enlarge the possibilities of such materials in selective oxidation reactions.

REFERENCES

1. B. Notari, Catalysis Today, 18 (1993) 163
2. G. Perego, G. Bellussi, C. Cordo, M. Taramasso, F. Buonomo and A. Esposito, Stud. Surf. Sci. Catal., 28 (1986) 129.
3. J.S. Beck, J.C. Vartuli, W.J. Roth, M.E. Leonowicz, CT Kresge, K.T. Schmitt, C.T-W Chu, D.H. Olson, E.W. Sheppard, S.B. McCullen, J.B. Higgins and J.L. Schlenker, J. Am. Chem. Soc., 114 (1992) 10834.
4. P.T. Tanev, M. Chibwe and T.J. Pinnavaia, Nature, 368 (1994) 321.

5. P.T. Tanev and T.J. Pinnavaia, Science, 267 (1995) 865.
6. P.L. Llewellyn, F. Schüth, Y. Grillet, F. Rouquerol and K.K. Unger, Langmuir, 11 (1995) 574.
7. I.W.C.E. Arends, R.A. Sheldon, M. Wallau and U. Schuchardt, Angew. Chem. Int. Ed. Engl., 36 (1997) 1144.
8. A.V. Ramaswamy and S. Sivasanker, Catal. Lett., 22 (1993) 239.
9. A. Sayari, Chem. Mater., 8 (1996) 1840.
10. A. Corma, Chem. Rev., 97 (1997) 2373.
11. A. Tuel, Microporous and Mesoporous Materials, (in press).
12. S. Gontier and A. Tuel, Zeolites, 15 (1995) 601.
13. W. Zhang, M. Fröba. J. Wang, P.T. Tanev, J. Wong and T.J. Pinnavaia, J. Am. Chem. Soc., 118 (1996) 9164.
14. R.D. Oldroyd, J.M. Thomas, T. Maschmeyer, P.A. MacFaul, D.W. Snelgrove, K.U. Ingold and D.D.M. Wayner, Angew. Chem. Int. Ed. Engl., 35 (1996) 2787.
15. J.S. Reddy, P. Liu and A. Sayari, Appl. Catal. A : General, 148 (1996) 7.
16. S. Gontier and A. Tuel, Microporous Mater., 5 (1995) 161.
17. M. Morey, A. Davidson, H. Eckert and G. Stucky, Chem. mater., 8 (1996) 486.
18. S. Gontier and A. Tuel, Appl. Catal. A : General, 143 (1996) 125.
19. T.K. Das, K. Chandary, A.J. Chandwadkar and S. Sivasanker, J. Chem. Soc., Chem. Commun., (1995) 2495.
20. T.M. Abdel-Fattah and T.J. Pinnavaia, J. Chem. Soc., Chem. Commun., (1996) 665.
21. W. Zhang, J. Wang, P.T. Tanev and T.J. Pinnavaia, J. Chem. Soc., Chem. Commun., (1996) 979.
22. N. Ulagapan and C.N.R. Rao, J. Chem. Soc., Chem. Commun., (1996) 1047.
23. T. Maschmeyer, R.D. Oldroyd, G. Sankar, J.M. Thomas, I.J. Shannon, J.A. Klepetko, A.F. Masters, J.K. Beattie and C.R.A. Catlow, Angew. Chem. Int. Ed. Engl., 36 (1997) 1639.
24. J.F. Diaz, K.J. Balkus Jr., F. Bedioni, V. Kurshev and L. Kevan, Chem. Mater., 9 (1997) 61.
25. J.S. Reddy and A. Sayari, Appl. Catal. A : General, 128 (1995) 231.
26. S. Gontier and A. Tuel, J. catal., 157 (1995) 124.
27. A. Corma, M. Iglesias and F. Sanchez, Catal. Lett., 39 (1996) 153.
28. S-S. Kim, W. Zhang and T.J. Pinnavaia, , Catal. Lett., 43 (1997) 149.
29. A. Corma, M.T. Navarro, J. Perez-Pariente and F. Sanchez, Stud. Surf. Sci. Catal., 84 (1994) 69.
30. E. Jorda, A. Tuel, R. Teissier, and J. Kervennal, J. Catal. (in press).
31. K.M. Reddy, I. Moudrakowski and A. Sayari, J. Chem. Soc., Chem. Commun., (1994) 1059.
32. R. Neumann and A.M. Khenkin, J. Chem. Soc., Chem. Commun., (1996) 2643.
33. G. Centi, F. Fazzini, L. Canesson and A. Tuel, Stud. Surf. Sci. Catal., 110 (1997) 893.
34. R. Burch, N. Cruise, D. Gleeson and S-C. Tsang, J. Chem. Soc., Chem. Commun., (1996) 951.

MESOPOROUS MOLECULAR SIEVES 1998
Studies in Surface Science and Catalysis, Vol. 117
L. Bonneviot, F. Béland, C. Danumah, S. Giasson and S. Kaliaguine (Editors)

New organic chemical conversions over MCM-41-type materials

H. van Bekkum and K.R. Kloetstra

Laboratory of Organic Chemistry and Catalysis, Delft University of Technology,
Julianalaan 136, 2628 BL Delft, The Netherlands

Mesoporous MCM-41 type materials offer a rich spectrum of opportunities to serve as catalysts in organic chemicals syntheses.

1. INTRODUCTION

The discovery of the mesoporous MCM-41 type materials [1] and the related HMS materials [2], with pore diameters of 2-8 nm, have enlarged the accessibility window of zeolitic materials [3] to perform organic syntheses greatly.

Recently the preparation of even more spacious ordered silica materials (up to 30 nm) was reported [4].

Some differences between MCM-41 type and the less-ordered HMS materials include (i) the way and ease of formation, (ii) the pore wall thickness, and connected with that the exposure of the Al units, (iii) the possibility of extraction of the template molecules. As to the use in organic conversions the accent was so far on the MCM-41 type materials.

MCM-41 type materials have been modified by
- ion exchange
- adsorption or other introduction of catalytic entitities
- grafting of the pore walls with organic functions
- build-in of redox centers
- coating of the walls with inorganic oxides
- partial recrystallization

2.1. Ion-exchanged MCM-41 materials

Al-containing MCM-41 materials can be ion-exchanged to the mildly acidic [5] H-forms, to the mildly basic Na- or Cs-forms [6], or to Lewis acid type cationic forms, e.g. the Zn-form [7].

Various organic reactions require just mild acid catalysis and the MCM-41 materials in the H-form meet the requirements, while offering larger surfaces [3] than conventional amorphous silica-aluminas.

For example, MCM-41 systems are excellent acetalisation catalysts. In this field the Corma group studied [8] acetalisation of aldehydes, showing that bulky reactants, like diphenylacetaldehyde, can be successfully acetalised on MCM-41. With respect to activity and re-use MCM-41 materials behaved favourably compared to amorphous silica-alumina and the zeolites H-Y and H-Beta.

Our group showed [7] that in fructose alkylation (another acetalisation) MCM-41 is the catalyst of choice. With lower alcohols quantitative conversions are obtained, whereas with long chain alcohols a two-step process is preferred. With strong acid catalysts, like sulfonic acids, homogeneous or as a resin, or with H-zeolites, yields of alkyl fructosides are low and dehydration to 2-hydroxymethylfurfural prevails. H-MCM-41's mild acidity suffices to activate fructose, especially its furanose, five-membered forms (cf. Fig. 1). Consequently in the initial reaction mixture the alkylated furanose forms (2 and 3) dominate. The effect of the Si/Al ratio of the MCM-41 material was small within the range Si/Al = 30-100.

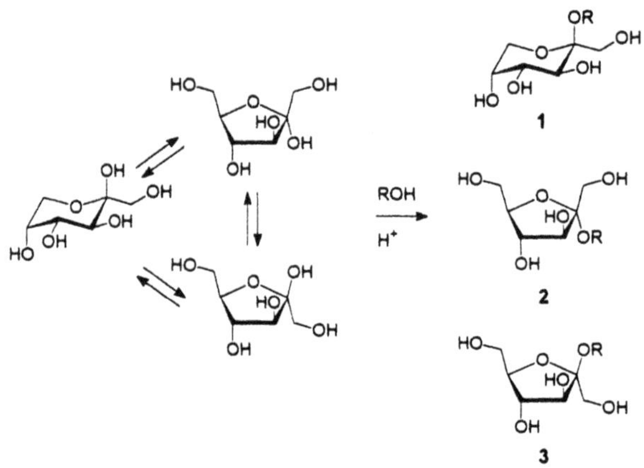

Figure 1. The reaction of D-fructose with fatty alcohols.

It is to be noted that glucose is not alkylated in the presence of H-MCM-41. Here the stronger acidic H-zeolites [10] or sulfonic acids are required as catalysts. The different reactivities of fructose and glucose seem related to the stabilities of the intermediate oxocarbenium ions formed.

Also fructose-containing disaccharides such as leucrose, isomaltulose and lactulose can be alkylated [9] over H-MCM-41. Importantly, the glycosidic bond connecting the two sugar units is not broken.

H-MCM-41 is also a mild and efficient catalyst in the reaction of alcohols and phenols with the activated double bond of dihydropyran [12]. This reaction is often used for temporary OH-protection. The mesoporous channels allow the conversion of bulky reactants such as cholesterol.

An interesting aromatic alkylation coupled with a ring closure over H-MCM-41 was reported [13] by the Corma group: cinnamyl alcohol and 2,4-di-t-butylphenol were found to react to a dihydrobenzopyran system. Within a series of acidic catalysts H-MCM-41 gave the highest conversion and yield.

An example of direct aromatic acylation catalysed by H-MCM-41 is the reaction of 1-naphthol with 2,6-dimethylbenzoic acid [14] (Fig. 2). In a solvent-free procedure an excellent yield was obtained. First the ester is formed which undergoes Fries-rearrangement to the ketone.

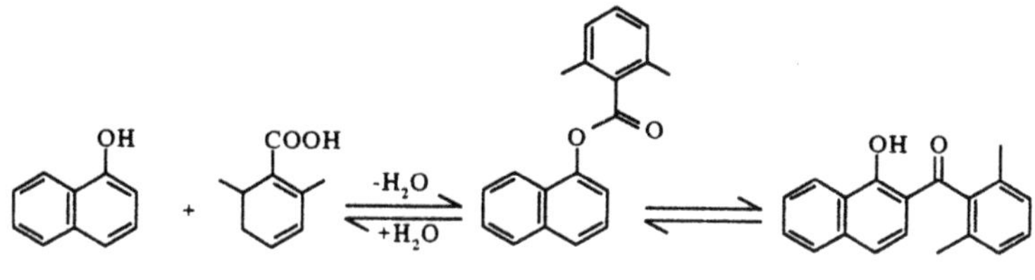

Figure 2. Acylation of 1-naphthol with 2,6-dimethylbenzoic acid.

Similarly, resorcinol and 2,6-dimethylbenzoic acid could be converted over H-MCM-41 to the corresponding benzophenone in > 80% yield. The MCM-41 catalyst was recycled without significant loss of activity.

In the two above acylations zeolite H-Beta behaves poorly because the product molecules are too bulky to be formed inside Beta.

Sodium and cesium cation-exchanged MCM-41 materials were prepared by treatment of Na,H-MCM-41 with 0.1 M aqueous NaCl and 0.1 M CsCl, respectively. The materials were characterized by CO_2 TPD, ^{23}Na NMR and ^{133}Cs NMR.

Na- and Cs-MCM-41 turned out to be good and water-stable basic catalysts in the reaction of benzaldehyde and ethyl cyanoacetate (Fig. 3). In this so-called Knoevenagel condensation the first step is thought to be the deprotonation of ethyl cyanoacetate ($pK_a \approx 12$) on a basic oxygen of MCM-41. The small difference in activities between Na- and Cs-MCM-41 suggests that the type of extra framework cation has little influence on the intrinsic basicity of MCM-41.

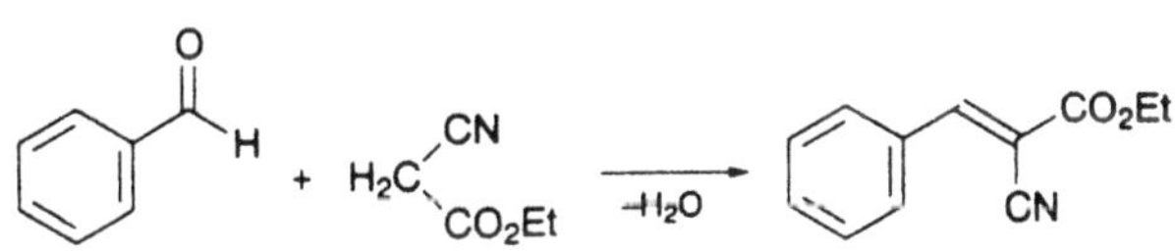

Figure 3. Knoevenagel condensation.

When executing the Knoevenagel reaction in ethanol as the solvent acetalisation towards benzaldehyde diethyl acetal prevailed, showing the presence of residual acidity in Na-MCM-41.

Na-MCM-41 also catalysed the aldol condensations of benzaldehyde with acetophenone, with 2-hydroxyacetophenone (Fig. 4), and with α-tetralone. Applying a solventless procedure at 150 °C with 5 wt % catalyst good results were obtained. Thus benzaldehyde and α-tetralone showed 60% conversion with 98% selectivity to 2-benzylidene-1-tetralone after 6 h of reaction [6].

In the reaction of benzaldehyde with 2-hydroxyacetophenone (Fig. 4, R = OH) intramolecular Michael addition was observed as a consecutive reaction, which, however, did not reach completion in our experiments.

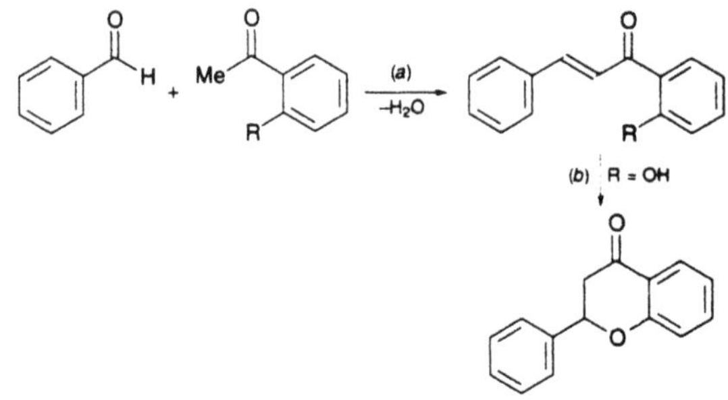

Figure 4. Aldol condensation of benzaldehyde.

The acidity of the α-C-H bonds of acetophenone and α-tetralone (pK_a ~ 17) would seem too low for deprotonation on the MCM-41 materials, considering the fact that diethyl malonate (pK_a 13.3) hardly showed reaction in the Knoevenagel condensation with benzaldehyde. However, the acidity of the α-C-H bonds might be substantially increased [15] by coordination of the carbonyl oxygen of the ketones to Al-sites in MCM-41.

Finally, we mention in this paragraph the use of Zn-exchanged MCM-41 material in aromatic acylation [7], and the use of Cr-exchanged MCM-41 in 1-decene oligomerization [16].

2.2. MCM-41 Materials as catalyst supports

MCM-41 type materials have been applied as supports for many types of catalytic entities from noble metal particles [17] to chiral complexes [18].

Some interesting recent approaches are the adsorption of a Ti-containing silsesquioxane [19] and the immobilization of enzymes [20] into MCM-41.

In the work of the van Santen group $(c\text{-}C_6H_{11})_7Si_7O_{12}Ti$ (η^5-C_5H_5) is adsorbed on all-silica MCM-41, which material has been made hydrophobic by reaction with diphenylsilyl chloride [19]. In this way leaching is prevented. The system is active in cyclooctene epoxidation with t-butyl hydroperoxide and can be recycled.

The Balkus group succeeded in immobilizing the enzymes cytochrome C (d. 3.0 nm), papain (d. 3.6 nm), and trypsin (d. 3.8 nm) into all-silica MCM-41 [20]. The enzymes were introduced by adsorption and the problem of leaching was tackled by subsequently reacting surface silanol groups with 3-aminopropyltriethoxysilane. Further work in > 4 nm pore systems is programmed to explore advantages of the immobilization.

Our group has been engaged - working in close cooperation with Prof. Kozhevnikov from the Boreskov Institute - with the preparation and application [21, 22] of heteropoly acid (HPA), $H_3PW_{12}O_{40}$ (PW), supported on MCM-41, aiming at strong acidity within the channels. Independently this system was studied by Mobil [23] and by Exxon [24] workers.

Heteropoly acid (ø 1.2 nm)
loaded MCM-41

HPA on MCM-41 catalysts were prepared [22] by impregnation with solutions of PW in methanol towards loadings from 10-50 wt %. The Keggin structure of the PW is fully retained as shown by ^{31}P NMR. Up to 50% loading no HPA crystal phase is developed. TEM shows the PW mainly to be located in the channels; amorphisation of the dominant MCM-41 structure was applied to make the inner HPA visible.

When the impregnation is performed with aqueous $H_3PW_{12}O_{40}$ an extra species - assumed [22] to be $H_6P_2W_{18}O_{62}$ or $H_6P_2W_{21}O_{71}$ - is observed by ^{31}P NMR for the lower loadings.

PW/MCM-41 exhibits higher catalytic activity than H_2SO_4 or bulk PW in liquid-phase alkylation of 4-t-butylphenol (TBP) by isobutene and styrene. In the alkylation of TBP by styrene, PW/MCM-41 shows a size selectivity compared to bulk PW and PW/SiO_2, providing higher yields of a 2-(1-phenylethyl)-4-t-butylphenol, at the expense of the more bulky 2,6-bis(1-phenylethyl)-4-t-butylphenol (see Table 1).

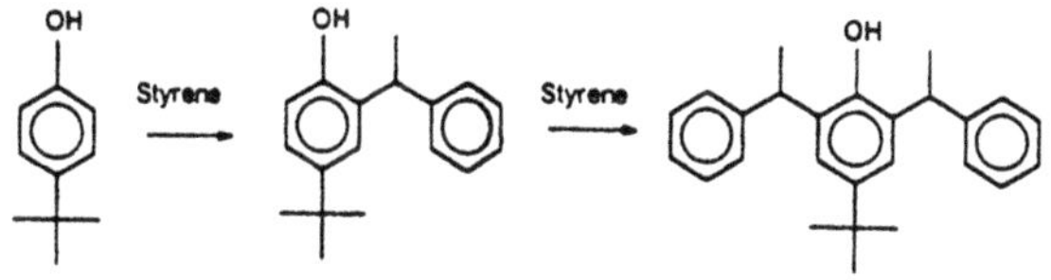

I II

Table 1
Alkylation of TBP with styrene in n-octane (150 °C, TBP/styrene 1:1.2 mol/mol, TBP/n-octane 1:2.5 wt/wt, 4.0 wt % catalyst, 1.5-2 h)

Catalyst[a]	Selectivity[b] (mol %)		
	I	II	other products[c]
PW	23 (17)	68 (73)	9 (10)
40% PW/MCM/41	66 (27)	21 (60)	13 (13)
40% PW/SiO_2	24 (20)	64 (69)	12 (11)

[a] PW catalysts were prepared by impregnation, dried at room temperature in vacuum and stored over P_2O_5.

[b] The selectivity based on TBP at 90% TBP conversion; in parentheses at 99% TBP conversion.

[c] Di-t-butylphenol and 6-(1-phenylethyl)-2,4-di-t-butylphenol.

Another consecutive reaction is the t-butylation of TBP with isobutene to give 2,4-di-t-butylphenol (BTBP) followed by 2,4,6-tri-t-butylphenol (TTBP).

Fig. 5 shows typical kinetic curves for the homogeneous (with H_2SO_4) and heterogeneous (with a PW based catalyst) alkylation of TBP. In the homogeneous reaction, both the first and the second step are fairly fast, yielding 90% TTBP. In the heterogeneous reaction, the first step is also fast, but the second one goes rather slowly, probably due to a significant steric hindrance for the conversion of bulky DTBP on the catalyst surface. This results in a significantly higher selectivity to DTBP at the expense of TTBP in the heterogeneous process.

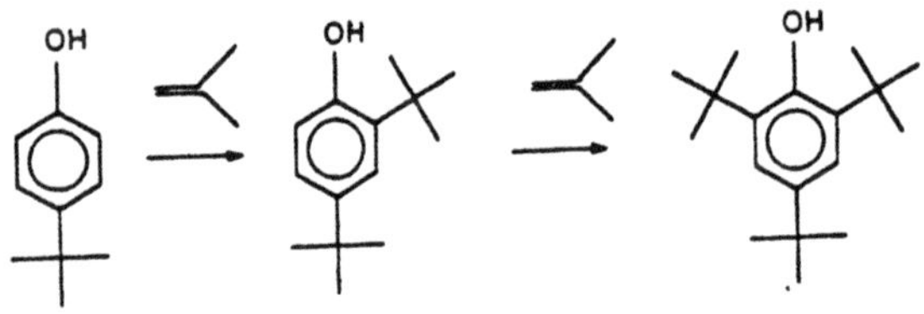

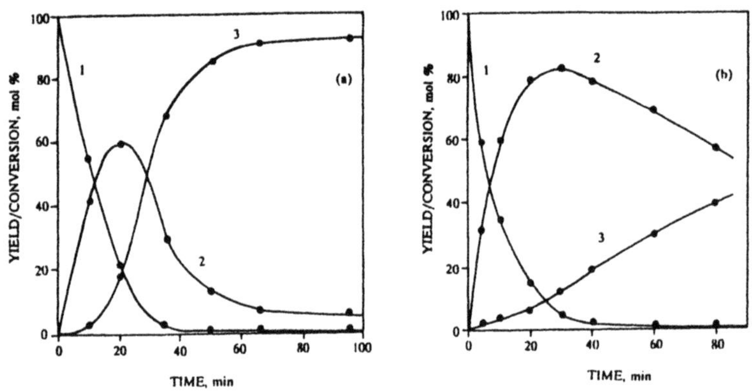

Figure 5. Plot of yield/conversion versus time for the TBP alkylation with isobutene catalyzed by (a) H_2SO_4 (5.0 wt %) and (b) wt % PW/MCM-41 (2.5 wt %) in benzene (70 °C, TBP/C_6H_6 50:50 wt/wt): (1) TBP, (2) DTBP, (3) TTBP.

In the alkylation of benzene with tetradecene PW-loaded MCM-41 catalysts were shown [23] to give higher benzene conversion and mono-alkylate selectivity than silica-alumina and H-MCM-41. All-silica MCM-41 materials were inactive.

Recently, Chinese workers [25] compared Al-containing MCM-41 and all-silica HMS materials as supports for $H_4SiW_{12}O_{40}$ (SiW). The test reaction was the gas phase esterification of 1-butanol and acetic acid over the temperature range 110-150 °C. The HMS-supported catalysts showed somewhat higher activity than the MCM-41 systems. Moreover, the authors observed some HPA-leaching.

In our group liquid and gas phase esterification over PW on MCM-41 were studied [26]. The liquid phase experiments were conducted in toluene with azeotropical removal of the water formed. Here, severe clustering of the HPA units towards large clusters situated at the outer MCM-41 surface was seen in TEM pictures of the spent catalyst.

Apparently, the relatively polar reactants (1-propanol and hexanoic acid) and the product water enhance the inner mobility of the HPA by solvation. Upon arrival at the pore mouth the HPA is - in view of its low solubility in toluene - deposited on the outer surface of the particle. There is remaining but lower esterification activity of the system.

In the gas phase experiments this sintering phenomenon was also observed [26], though less pronounced.

In conclusion, the SiW/- and PW/MCM-41 compositions, having strong acid sites and a regular mesoporous system, are promising catalysts for the acid-type conversion and formation of organic compounds of large molecular size, provided the reactants are nonpolar and no water is formed.

2.3. Grafting of the MCM-41 pore walls with organic functions

The availability of various trialkoxysilyl functional compounds opened the way to attach organic functional groups - with catalytic potential - via a spacer chain to the inner pore walls of MCM-41 and HMS materials.

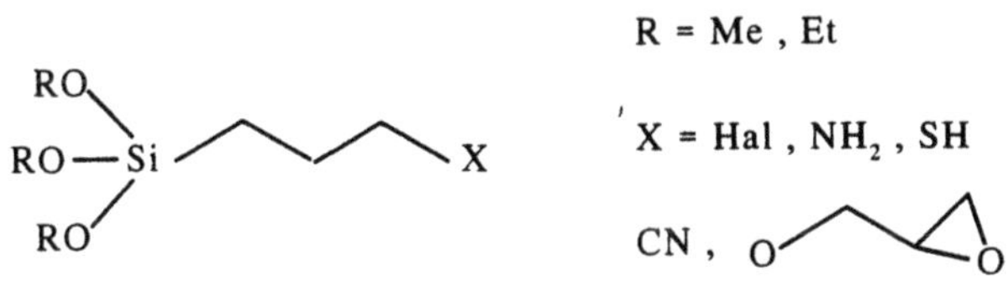

Two approaches can be considered and have been applied: (i) by reactive coating of the pore walls which are known to contain many SiOH groups (post-synthesis); (ii) introduction during synthesis (co-condensation).

The Jacobs group succeeded in preparing [27, 28] strongly acidic MCM-41 and HMS-based materials by introducing sulfonic acid groups by both methods. As precursor reagent 3-mercaptopropyltrimethoxysilane was applied. The SH-group of the precursor is easily converted into a SO_3H-group by oxidation with H_2O_2.

The sulfonated mesoporous materials showed good activity in furan hydroxyalkylation/-condensation and show promise in polyol esterification [27, 28].

It should be mentioned that Degussa AG recently announced [29] - under the trade name Deloxan - solid acid catalysts consisting of polysiloxanes bearing alkylsulfonic acid groups. The materials are prepared by the co-condensation method. A comparison of these amorphous systems with the MCM-41-based catalysts would seem worthwhile.

The Brunel group attached [30, 31] amine functions (amino, piperidino) through a halogen-containing precursor to MCM-41 materials. The mesoporous basic systems obtained catalysed the addition of lauric acid to glycidol [30] and the Knoevenagel condensation of benzaldehyde and ethyl cyanoacetate [31]. In the epoxide ring opening the piperidinopropyl-MCM-41 was the most active catalyst whereas in the Knoevenagel condensation the aminopropyl system was more active. The latter observation is explained by assuming the amino group to form an imino intermediate in the reaction.

Aminopropyl- (and cyanoethyl-) modified mesoporous materials of the HMS-type have also been prepared [32] by co-condensation.

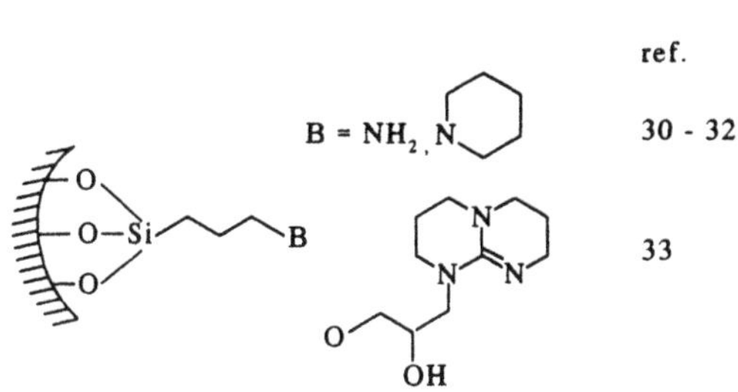

A more strongly basic MCM-41 based catalyst was designed by the Jacobs group [33]. Here a guanidine base, 1,5,7-triazabicyclo[4.4.0]dec-5-ene (TBD), is connected via a spacer (see Formula) to the mesopore wall.

This system catalyzes under mild conditions (40-80 °C) the Michael addition of ethyl cyanoacetate to various acceptor molecules (e.g. ethyl acrylate). Also the less acidic diethyl malonate reacts. When comparing the MCM-TBP catalyst with TBD on amorphous silica, prepared analogously, then the latter system is more active for the somewhat larger reactants (cyclohexenone/diethyl malonate) indicating a reduced space in MCM-41 due to the catalyst introduction.

The MCM-41/TBP system turned out [33] to be also an active catalyst in the Knoevenagel condensation of benzaldehydes and in the epoxidation of conjugated enones with H_2O_2 (due to its ionization towards HOO^-).

In connection with heterogenisation of guanidine bases, we mention here the entrapment [34] of N',N',N-tricyclohexylguanidine in dealuminated zeolite Y. This catalyst converted benzaldehyde and acetone into 4-phenyl-4-hydroxybutan-2-one, which is halfway the aldol condensation.

2.4. Mesoporous redox catalysts

In the search for new low-waste oxidation techniques much attention is being paid to the design and development of MCM-41- and HMS-based redox catalysts [35]. Without aiming at completeness some examples follow.

Using the (3-glycidyloxypropyl)trimethoxysilane reagent the Jacobs group attached 1,4,7-triazacyclononane ligands to the MCM-41 pore wall. Equipped with a central Mn atom this system proved to be a selective epoxidation catalyst with H_2O_2 as the oxidant [36]. Another attached Mn-complex was reported by Sutra and Brunel [37].

In view of the versatile Ti-containing zeolites TS-1 and Ti-Beta an obvious focus is on Ti-containing mesoporous materials. In addition Ti systems are expected to be less vulnerable to leaching than other redox metal loaded materials.

Co-condensation techniques [38, 39] as well as post-synthesis methods [40] have been applied to introduce Ti into MCM-41 and HMS materials. As post-synthesis reagent $TiCp_2Cl_2$ was applied by the Thomas group [40]. The Ti centres are believed to be attached to the support via three Ti-O-Si linkages. Upon calcination the residual Cp ligand is oxidized and the catalytic Ti site formed. The epoxidation ability of the system was further improved [41] by first grafting the MCM-41 surface with Ge(IV) using Bu_4Ge.

Reported reactions catalysed by Ti-MCM-41 and Ti-HMS include epoxidation, benzene [42] and phenol hydroxylation, 2,6-dialkylphenol oxidation to the corresponding quinone, oxidative aromatic bromination [43], and carbohydrate oxidation [44].

The thermal and gas phase reaction stability of Ti-MCM-41 have been reported [45] as excellent. The stability under liquid phase oxidation conditions has to be proven, however [45, 46].

The above has also aroused renewed interest in Ti-containing inexpensive silica forms.

By applying co-condensation of zirconium isopropoxide the Tuel group was able [47] to introduce Zr into HMS material. The catalysts obtained behaved similarly to Ti-HMS in a number of oxidation reactions.

In the Meerwein-Ponndorf-Verley reduction of ketones, Zr-HMS was more active than Ti-HMS. Moreover, the Zr system remained active in the presence of amines [48], allowing the reduction of tropinone (Fig. 6).

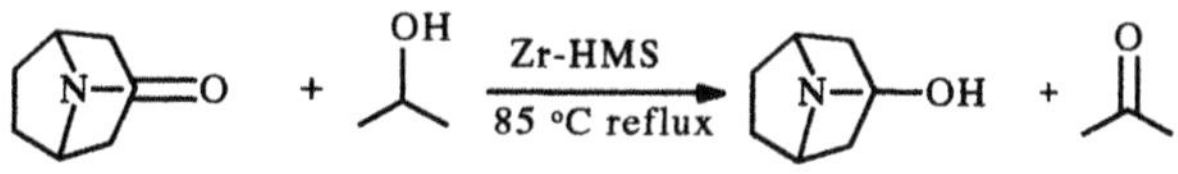

Figure 6. MPV-reduction of tropinone.

2.5. Coating of the pore walls with inorganic oxide

Following our work on Cs-exchanged MCM-41 materials, basic catalysts were prepared [49] by impregnation of MCM-41 with solutions of cesium acetate, so as to obtain - after calcination - cesium oxide inside the pores.

A material containing 20 wt % Cs_2O was an excellent catalyst in the Michael addition of the weakly acidic diethyl malonate (pK_a 13.3) to the bulky chalcone (Fig. 7). Applying these reactants in a molar ratio of 1:1.3 in the absence of solvent at 150 °C with 2 wt % catalyst a conversion of 87% with a selectivity of 91% was achieved within 30 min.

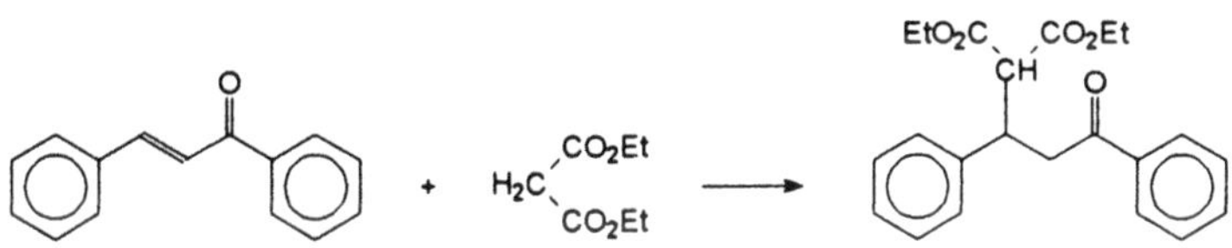

Figure 7. Michael addition of chalcone with diethyl malonate.

In a Michael reaction no water is formed. When this is the case, as in aldol or Knoevenagel condensations, Cs_2O-MCM-41 is not a suitable catalyst. Moreover, the regenerability of the materials is relatively poor.

Recently, our group developed [50] thermally and water-stable catalysts by impregnating equimolar amounts of cesium acetate and lanthanum nitrate onto MCM-41 and HMS materials followed by calcination. These catalysts are assumed to contain a $CsLaO_x$ layer onto the MCM-41 wall and performed well in Knoevenagel and Michael reactions, while being fully regenerable.

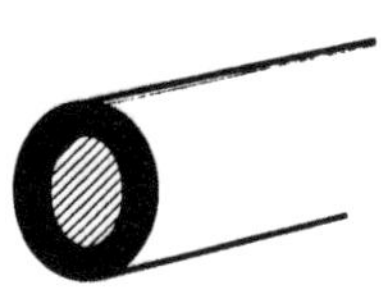

$CsLaO_2$-Coated MCM-41
Water-Stable Basic Catalyst

Reactions studied include Knoevenagel condensations of benzaldehyde and the two-step Michael addition of ethyl cyanoacetate to ethyl acrylate [50].

Furthermore, a new reaction was found [51]: the isomerization of ω-phenylalkanals towards phenyl alkyl ketones (selectivities up to 83%). The mechanism of this new rearrangement (Fig. 8) is as yet unsolved.

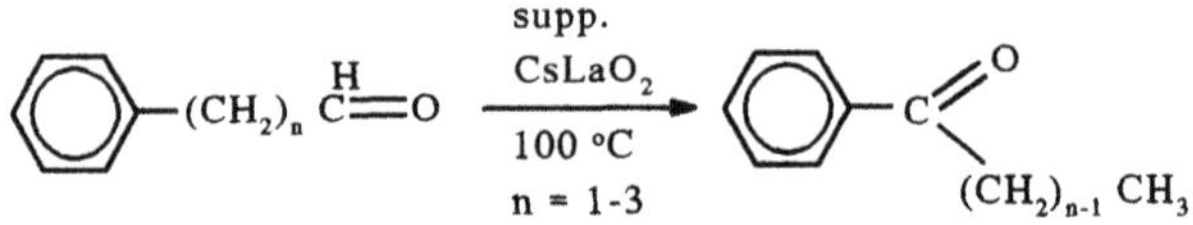

Figure 8. Rearrangement of ω-phenylalkanals.

2.6. Partial recrystallization of mesoporous systems

Finally, we mention the partial crystallization of MCM-41 and HMS pore walls towards more acidic zeolitic domains [52]. The approach is to introduce the appropriate template by ion exchange and to allow subsequently a gradual partial crystallization. In the first attempts tetrapropylammonium was positioned inside the pores so as to induce formation of ZSM-5 structures.

In conclusion, mesoporous materials - as such or after modification - offer many opportunities to serve as catalysts in organic conversions.

REFERENCES

1. J.S. Beck, J.C. Vartuli, W.J. Roth, M.E. Leonowicz, C.T. Kresge, K.D. Schmitt, C.T.W. Chu, D.H. Olson, E.W. Sheppard, S.B. McCullen, J.B. Higgins and J.L. Schlenker, J. Am. Chem. Soc., 114 (1992) 10834.
2. P.T. Tanev and T.J. Pinnavaia, Science, 267 (1995) 856.
3. Reviews: a) J.L. Casci, Stud. Surf. Sci. Catal., 85 (1994) 329; b) J.S. Beck and J.C. Vartuli, Curr. Opin. Solid State Mater. Sci., 1 (1996) 76; c) X.S. Zhao, G.Q.M. Lu and G.J. Millar, Ind. Eng. Chem. Res., 35 (1996) 2075.
4. D. Zhao, J. Feng, Q. Huo, N. Melosh, G.H. Frederickson, B.F. Chmelka and G.D. Stucky, Science, 279 (1998) 548.
5. R. Mokaya, W. Jones, Z. Luan, M.D. Alba and J. Klinowski, Catal. Lett. 37 (1996) 113. Cf. also A. Corma, M.S. Grande, V. Gonzalez-Alfaro and A.V. Orchilles, J. Catal., 159 (1996) 375.
6. K.R. Kloetstra and H. van Bekkum, J. Chem. Soc., Chem. Commun., 1995, 1005.
7. E.A. Gunnewegh, S.S. Gopie and H. van Bekkum, J. Mol. Catal., 106 (1996) 151.
8. M.J. Climent, A. Corma, S. Iborra, M.C. Navarro and J. Primo, J. Catal., 161 (1996) 783.
9. A.T.J.W. de Goede, I.G. van der Leij, A.M. van der Heijden, F. van Rantwijk and H. van Bekkum, PCT Int. Appl., WO 96 136640 (1996).
10. A. Corma, S. Iborra, S. Miquel and J. Primo, J. Catal., 161 (1996) 713.
11. A.T.J.W. de Goede, M.P.J. van Deurzen, I.G. van der Leij, A.M. van der Heijden, J.M.A. Baas, F. van Rantwijk and H. van Bekkum, J. Carbohydr. Chem., 15 (1996) 331.
12. K.R. Kloetstra and H. van Bekkum, J. Chem. Res. (S), (1995) 26.
13. E. Armengol, M.L. Cano, A. Corma, H. Garcia and M.T. Navarro, J. Chem. Soc., Chem. Commun., 1995, 519.

14. H. van Bekkum, A.J. Hoefnagel, M.A. van Koten, E.A. Gunnewegh, A.H.G. Vogt and H.W. Kouwenhoven, Stud. Surf. Sci. Catal., 83 (1994) 379.
15. K.R. Kloetstra, PhD Thesis, Delft University of Technology, 1997.
16. B.P. Pelrine, K.D. Schmitt and J.C. Vartuli, US Patent 5.105.051 (1992).
17. e.g. U. Junges, W. Jacobs, I. Voigt-Martin, B. Krutzsch and F. Schüth, J. Chem. Soc., Chem. Commun., 1995, 2283.
18. L. Frunza, H. Kosslick, H. Landmesser, E. Höft and R. Fricke, J. Mol. Catal. A: Chemical, 123 (1997) 179.
19. S. Krijnen, H.C.L. Abbenhuis, R.W.J.M. Hanssen, J.H.C. van Hooff and R.A. van Santen, Angew. Chem., in the press.
20. J. Felipe Diaz and K.J. Balkus Jr., J. Mol. Catal. B: Enzymatic, 2 (1996) 115.
21. I.V. Kozhevnikov, A. Sinnema, R.J.J. Jansen, K. Pamin and H. van Bekkum, Catal. Lett., 30 (1995) 241.
22. I.V. Kozhevnikov, K.R. Kloetstra, A. Sinnema, H.W. Zandbergen and H. van Bekkum, J. Mol. Catal. A: Chemical, 114 (1996) 287.
23. C.T. Kresge, D.D. Marler, G.S. Rao and B.H. Rose, US Patents 5.324.881 and 5.366.945 (1994).
24. S.L. Soled, J.G. Da Costa Paes, A. Gutierrez, S. Miseo, W.E. Gates and K.L. Riley, PCT Int. Appl., WO 95 13.869 (1995).
25. W. Chu, X. Yang, Y. Shan, X. Ye and Y. Wu, Catal. Lett., 42 (1996) 201.
26. M.J. Verhoef, P.J. Kooyman, J. A. Peters and H. van Bekkum, submitted to Microporous Mesoporous Mater.
27. W.M. van Rhijn, D.E. De Vos, B.F. Sels, W.D. Bossaert and P.A. Jacobs, J. Chem. Soc., Chem. Commun., 1998, in the press.
28. W.M. van Rhijn, D.E. De Vos, W.D. Bossaert, J.Bullen, B. Wouters, P. Grobet and P.A. Jacobs, contribution to the Baltimore Symposium 1998.
29. S. Wieland and P. Panster, Stud. Surf. Sci. Catal., 108 (1997) 67.
30. A. Cauvel, G. Renard and D. Brunel, J. Org. Chem., 62 (1997) 749.
31. M. Lasperas, T. Lloret, L. Chaves, I. Rodriguez, A. Cauvel and D. Brunel, Stud. Surf. Sci. Catal., 108 (1997) 75.
32. D.J. Macquarrie, J. Chem. Soc., Chem. Commun., 1996, 1961.
33. Y.V. Subba Rao, D.E. De Vos and P.A. Jacobs, Angew. Chem., 36 (1997) 2661.
34. R. Sercheli, A.L.B. Ferreira, M.C. Guerreiro, R.M. Vargas, R.A. Sheldon and U. Schuchardt, Tetrahedron Lett., 38 (1997) 1325.
35. For a review on redox molecular sieves see: I.W.C.E. Arends, R.A. Sheldon, M. Wallau and U. Schuchardt, Angew. Chem. Int. Ed. Engl., 36 (1997) 1144.
36. Y.V. Subba Rao, D.E. De Vos, T. Bein and P.A. Jacobs, J. Chem. Soc., Chem. Commun., 1997, 355.
37. P. Sutra and D. Brunel, J. Chem. Soc., Chem. Commun., 1996, 2485.
38. A. Corma, M.T. Navarro and J. Perez-Pariente, J. Chem. Soc., Chem. Commun., 1994, 147.
39. P.T. Tanev, M. Chibwe and T.J. Pinnavaia, Nature, 368 (1994) 321.
40. T. Maschmeyer, F. Rey, G. Sankar and J.M. Thomas, Nature, 378 (1995) 159.
41. R.D. Oldroyd, J.M. Thomas and G. Sankar, J. Chem. Soc., Chem. Commun., 1997, 2025.
42. W. Zhang, J. Wang, P.T. Tanev and T.J. Pinnavaia, J. Chem. Soc., Chem. Commun., 1996, 979.

43. J.V. Walker, M. Morey, H. Carlsson, A. Davidson, G.D. Stucky and A. Butler, J. Am. Chem. Soc., 119 (1997) 6921.
44. E.J.M. Mombarg, S.J.M. Osnabrug, F. van Rantwijk and H. van Bekkum, Stud. Surf. Sci. Catal., 108 (1997) 385.
45. C.H. Rhee and J.S. Lee, Catal. Lett., 40 (1996) 261.
46. D. Trong On, M.P. Kapoor, P.N. Joshi, L. Bonneviot and S. Kaliaguine, Catal. Lett., 44 (1997) 171.
47. A. Tuel, S. Gontier and R. Teissier, J. Chem. Soc., Chem. Commun., 1996, 651.
48. J.C. van der Waal et al., unpublished results.
49. K.R. Kloetstra and H. van Bekkum, Stud. Surf. Sci. Catal., 105 (1997) 431.
50. K.R. Kloetstra, M. van Laren and H. van Bekkum, J. Chem. Soc., Faraday Trans., 93 (1997) 1211.
51. K.R. Kloetstra, J. van den Broek and H. van Bekkum, Catal. Lett., 47 (1997) 235.
52. K.R. Kloetstra, H. van Bekkum and J.C. Jansen, J. Chem. Soc., Chem. Commun., 1997, 2281.

MESOPOROUS MOLECULAR SIEVES 1998
Studies in Surface Science and Catalysis, Vol. 117
L. Bonneviot, F. Béland, C. Danumah, S. Giasson and S. Kaliaguine (Editors)

Sulfonic acid bearing mesoporous materials as catalysts in furan and polyol derivatization*

W. Van Rhijn, D. De Vos, W. Bossaert, J. Bullen, B. Wouters, P. Grobet and P. Jacobs*

Center for Surface Science and Catalysis, KULeuven, Kard. Mercierlaan 92, B-3001 Heverlee, Belgium

Fax. 32 16 32 19 98, E-mail: pierre.jacobs@agr.kuleuven.ac.be

We have designed new acid catalysts by incorporation of propylsulfonic acid groups into MCM and HMS type structures, with HS-$(CH_2)_3$-Si(OR')$_3$ as the building block. Material synthesis was either primary, *via* sol-gel processing of TEOS and HS-$(CH_2)_3$-Si(OMe)$_3$ in the presence of an ionic or neutral template, or secondary, *via* silylation of a pre-synthesized MCM support. The eventual surface modification towards -$(CH_2)_3$-SO_3H was assessed with ^{13}C-NMR, IR, sorption measurements and TGA. In the catalytic work, two reaction types, which are inherently difficult even with optimal acid zeolites are discussed. First, 2,2-[bis-(5-methylfuryl)propane is formed by condensation of two methylfurans (MF) and acetone. With propylsulfonic acid MCMs, the furan oligomerization is largely circumvented and furan-based yields are increased from 40% (for H-β) to over 80%. In the esterification of sorbitol with long fatty acids, zeolites are hardly active; however, at the hydrophobic surface of the propylsulfonic acid MCMs, both reactant phases are brought into intimate contact and this results in considerable catalytic activity.

1. INTRODUCTION

In spite of the success of zeolites as solid acid catalysts, there are relatively few condensed phase processes in which they have been implemented. This is undoubtedly due to specific problems of the liquid phase, *e.g.* phase separation, or strong sorption when molecules with heteroatoms are the reactants. Reactions in which zeolites are of little practical use are for instance the condensation of furans, or the esterification of polyols. Nevertheless, both processes lead to economically and environmentally attractive products. Hydroxyalkylation of furan derivatives yields to bisfurylalkanes [1]; esters of polyols are widely applied as detergents, emulsifiers or low-caloric fats [2].
In the present contribution a new group of solid acids is introduced, based on the covalent anchoring of sulfonic acids in a mesoporous environment, and the advantages of this new design for liquid phase acid catalysis are illustrated. Preliminary results of this work have been published [3].

---

* This work was supported in a IUAP-PAI program of the Belgian Federal Government. W.V.R. acknowledges IWT and D.D.V. FWO for fellowships.

2. EXPERIMENTAL

2.1. Catalyst preparation

The procedures were designed based on the synthesis recipes for purely siliceous mesoporous materials [4]. Recent examples of synthesis of other hybrid mesoporous materials can be found in literature [5-7].

Preparation of MCM-SH and HMS-SH via co-condensation (Scheme 1, route 1) [5]. MCM-41 type materials were synthesized from a mixture of MPTS and $Si(OEt)_4$ (TEOS). The molar gel composition was 0.12 $C_{16}NMe_3Br$: 0.5 NaOH : 0.8 TEOS : 0.2 MPTS : 130 H_2O (24 h, 293 K). The ionic surfactant was removed by washing with 0.1 N HCl. Mercaptopropyl-modified HMS was obtained from following gel: 0.8 TEOS : 0.2 MPTS : 0.275 n-C_{12}-amine : 8.9 EtOH : 29.4 H_2O (24 h, 293 K). n-C_{12}-amine was extracted with boiling EtOH (12 h, 3 times).

Surface modification of an MCM material via silylation (Scheme 1, route 2a) [6]. MCM-41 was synthesized from Ludox. After calcination and evacuation at 393 K, an excess MPTS was added in dry toluene. After 4 h heating under reflux in a water-free atmosphere, the material was collected and soxhlet extracted with Et_2O/CH_2Cl_2 (50:50).

Surface modification via coating (route 2b) [7]. A calcined MCM-41 sample (3.0 g) was refluxed in water for 4 h and filtered. The wet filter cake was suspended in toluene (300 ml). Part of the solvent (100, 150 or 200 ml of the toluene-water mixture) was removed in an azeotropic distillation with a Dean-Stark trap. Finally, the filtered material was extracted with Et_2O and CH_2Cl_2.

Immobilized mercaptopropyl groups were oxidized with H_2O_2. Several parameters were varied, *e.g.* time, pH, and oxidant excess. The best results were obtained in a 24 h treatment with a 6-fold excess of non-neutralized H_2O_2. After washing with H_2O and EtOH, and acidification in 0.1N H_2SO_4, the powder was rinsed thoroughly with distilled H_2O and dried at 333 K.

2.2. Catalytic procedures

Hydroxyalkylation-condensation of methylfuran and acetone. The MCM-SO_3H catalyst (0.18 g) was saturated at ambient temperature with acetone (3.2 g) for 15 min, before addition of the highly reactive methylfuran (1.8 g). This suspension was heated for 24 hours (323 K), and the filtrate was analyzed by GC using a CP-Sil-5 column (Chrompack). Conversion and selectivity were determined using naphthalene as the internal standard. Reaction and side products were identified via GCMS.

Esterification of D-sorbitol with lauric acid. 3.64 g D-sorbitol and 24.0 g lauric acid (molar ratio = 1:6) together with 0.36 g catalyst were loaded into a stirred 100 ml-reactor at 385 K. Quantitative analyses were carried out with size-exclusion HPLC and ^{1}H-^{13}C-NMR. Size-exclusion HPLC was performed on a HP 1090 liquid chromatograph with a Microgel-3 column from Chrompack. After dilution (5 wt% in THF), samples (10 μl) were injected into the mobile phase (THF, flow rate 1 ml/min). The RI-detector and the column were held at 313 K while the column pressure was kept at 52 bar. ^{1}H-^{13}C-NMR spectra were measured on a Bruker AMX-300-spectrometer. The samples were diluted in $CDCl_3$ (10-50 %). COSY- and HETCOR-NMR-spectroscopy was used to determine the structure of the products, at a proton frequency of 500 MHz.

3. RESULTS AND DISCUSSION

3.1. Synthesis and characterization of immobilized sulfonic acid groups

Mercaptopropyl groups were immobilized on mesoporous supports *via* the different synthetic routes outlined in Scheme 1. In this scheme, Si-$(OR)_3$, X and A represent respectively the hydrolizable Si-$(OMe)_3$ moiety, the hydrolytically stable propyl spacer and the oxidizable -SH group in MPTS. First, with an ionic or non-ionic aqueous surfactant (*viz.* hexadecyltrimethylammonium bromide and n-dodecylamine) co-condensation of MPTS and tetraethyl orthosilicate (TEOS) results in straightforward mercaptopropyl encapsulation *(route 1)*. In this micelle-templated synthesis the mercaptopropyl part of MPTS is probably oriented along the apolar tails of the surfactant micelles. Alternatively, in *route 2a* the high surface area of surfactant-free calcined MCM was used as a substrate for silylation of Si-OH and Si-O-Si groups. In *route 2b,* the H_2O content of the MCM support is controlled to direct the coverage of the surface with -SH groups. Well-defined volumes of water/toluene (100, 150, 200 ml) were withdrawn from a wet MCM support in dry toluene *via* Dean Stark destillation, and addition of the organosilane to the finely dispersed MCM resulted in materials covered with an MPTS layer. Thus three materials with different surface loadings were obtained.

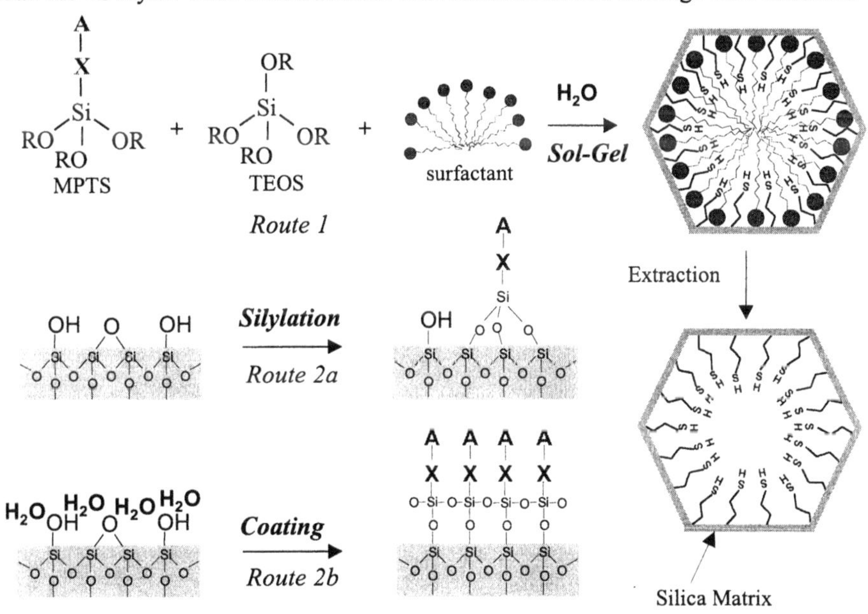

Scheme 1. Mercaptopropyl encapsulation into mesoporous ordered materials.

All synthesis routes result in materials functionalized with mostly intact -$(CH_2)_3$-SH groups, as is shown by an IR $\nu_{S\text{-}H}$ vibration of medium intensity at 2575 cm^{-1} and the IR $\nu_{C\text{-}H}$ stretchings at 2972, 2894, and 2867 cm^{-1} (Figure 1). The solid state ^{13}C-NMR chemical shifts (δ=11 ppm (C^1), 27 ppm (C^2,C^3)) (Figure 2a) are in accordance with the shifts of liquid MPTS

(δ=8.23 ppm (C^1); 27.46, 27.55 ppm (C^2,C^3)). The signal of the C^1 atom next to Si is shifted downfield to 11 ppm. Signals of unhydrolized CH_3O- groups (around 50 ppm) were usually not detected.

The thiol-functionalized materials were also studied with sorption measurements and TGA. Typical loadings, as measured by thermogravimetry, vary between 0.5 and 4.5 meq.g^{-1} and depend strongly on the synthesis route (Table 1). As demonstrated by the measurement of the N_2 sorption isotherm, the pore volume with a radius larger than 2 nm decreases by the anchoring of the $(CH_2)_3SH$ units [3]. Moreover, the average pore diameter shows also a decreasing tendency, suggesting that the environment for organic reactions is spatially more constrained after surface modification.

Next, the SH groups are oxidized with H_2O_2 to sulfonic acid groups. As can be seen from Figure 2a and 2b, mercaptans are easily oxidized to disulfides (δ= 41, 23 ppm).

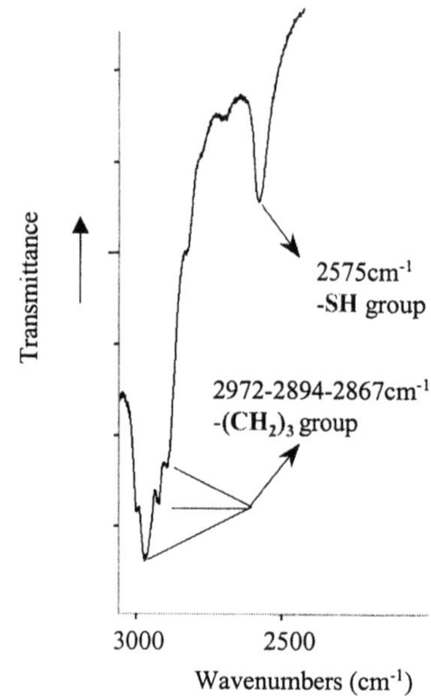

Figure 1. IR-spectrum of MCM-SH, prepared via surface coating.

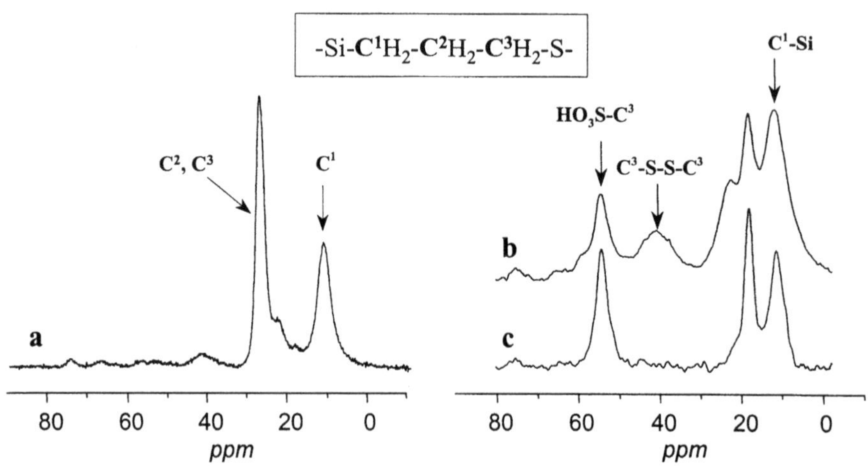

Figure 2. Solid state ^{13}C-CP-NMR spectra of MCM materials obtained via surface coating. (a) $(CH_2)_3$-SH functionalized material, (b) = (a) after a sub-optimal oxidation, with both disulfide and $(CH_2)_3$-SO_3H groups, (c) = (a) after optimized oxidation, with exclusively $(CH_2)_3$-SO_3H groups.

Therefore the oxidation conditions must be well tuned to achieve selective sulfonic acid formation. Figures 2b and c show the NMR spectra of a material with residual disulfide contamination and a material with exclusively sulfonic acid groups. Relatively long oxidation times (12-24 h) and a six-fold excess of H_2O_2 are recommendable. In these conditions, the signals of disulfide totally disappear and only signals at δ = 54, 18 and 11 ppm of the covalently linked $-(CH_2)_3SO_3H$ remain. Finally, acidification and thorough washing to remove all liquid acid yields the heterogeneous propylsulfonic acid.

Table 1
Concentration of covalently linked $-(CH_2)_3-SH$ or $-(CH_2)_3-SO_3H$ groups (meq/g) for materials obtained via different synthesis routes.

Catalyst	Synthesis Route	Loading (meq/g)
HMS-$(CH_2)_3-SO_3H$	Co-condensation (*1a*)	1.6
MCM-$(CH_2)_3$-SH	Silylation (*2a)*	0.5
MCM-$(CH_2)_3$-SH (100 ml)[a]	Coating (*2b*)	4.5
MCM-$(CH_2)_3$-SH (150 ml)[a]	Coating (*2b*)	3.8
MCM-$(CH_2)_3$-SH (200 ml)[a]	Coating (*2b*)	2.7

[a] Volume of toluene/water removed *via* Dean-Stark distillation.

3.2. Hydroxyalkylation-condensation of methylfuran with acetone

The catalytic properties of the new materials were tested in the synthesis of 2,2-bis(5-methylfuryl)propane (DMP). DMP is produced from 2-methylfuran (MF) and acetone in a strong acid catalyzed hydroxyalkylation reaction, followed by a condensation reaction with a second MF:

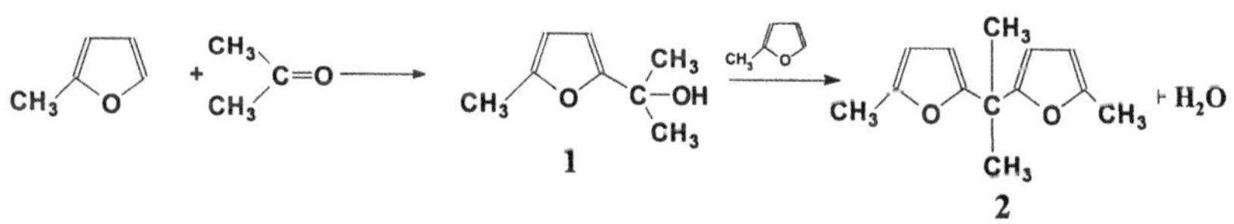

All MCM-$(CH_2)_3SO_3H$ catalysts combine a remarkable DMP selectivity (often up to 95%) with elevated MF conversion [3]. Despite the acetone excess (2.5:1), acid-catalyzed aldol condensation of acetone into mesityloxide was not observed; in our reaction conditions, the electrophilic aromatic substitution of the highly reactive MF with acetone prevails. The tertiary carbinol (**1**) was not directly detected. In acid conditions, **1** is protonated and dehydrated to form a carbenium ion, which immediately reacts further with MF to yield DMP (**2**). Therefore the hydroxyalkylation of MF with acetone is the rate determining step. Further hydroxyalkylation of DMP or formation of 5-isopropenyl-2-methylfuran from **1** are the principal side reactions (<10%).

For this furan reaction, the materials obtained via the coating procedure (Scheme 1, 2b) are catalytically superior to those obtained via other methods (routes 1a, 1b and 2a). The highest

DMP yield on MF basis was a remarkable 82% for coated MCM-$(CH_2)_3SO_3H$. Therefore several parameters were altered in the synthesis of coated materials, and the catalytic effects of these variations were examined in more detail.

First the –SH oxidation procedure was optimized and evaluated in function of the DMP synthesis. A material with exclusively SO_3H groups was compared to a catalyst with residual disulfide (*cf.* Figure 2 b-c). As can be seen in Figure 3, a thorough oxidation is important to maximize the activity.

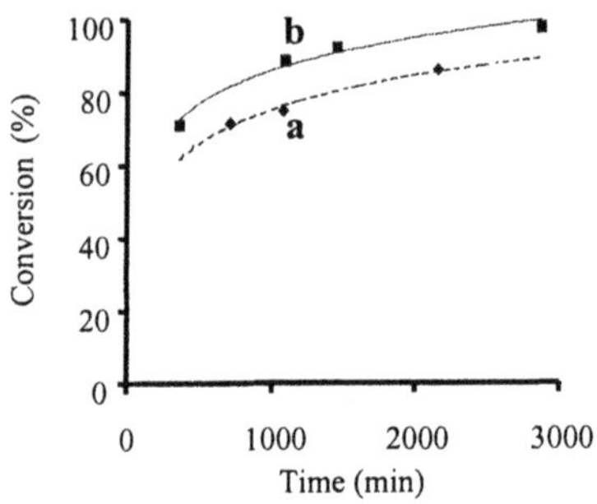

Figure 3. Methylfuran (MF) conversion vs. time in the reaction of MF with acetone (1:2.5). Catalysts: (a) an MCM with both -S-S- and -SO_3H, (b) an MCM with exclusively -SO_3H functional groups.

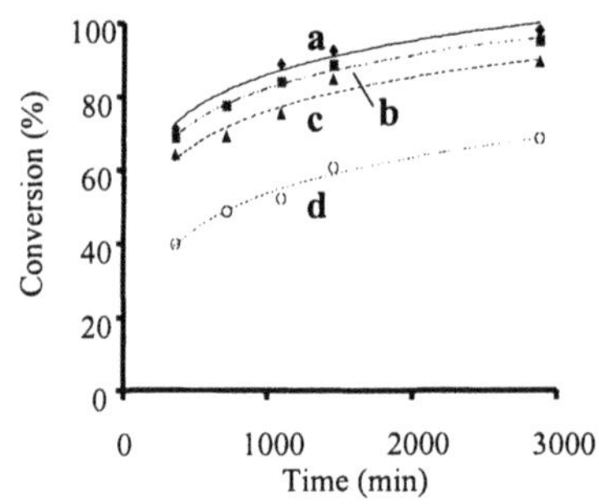

Figure 4. Methylfuran (MF) conversion vs. time in the reaction of MF with acetone (1:2.5). MCM-SO_3H Catalysts: (a) 4.5 meq/g, (b) 3.8 meq/g, (c) 2.7 meq/g. The catalyst in (d) was obtained by coating of an amorphous SiO_2 (kieselgel 60) as for (a).

Next different materials were synthesized by altering the water content on the calcined MCM *via* Dean Stark destillation. The data of Table 1 (entries 3-5) prove that this is a flexible method to change the loading with propylsulfonic acid groups. The catalytic results (Figure 4) show that the highest catalytic activity is observed for the material with the highest surface concentration of catalytic groups. For the sake of comparison, an amorphous silica was coated via the standard procedure, with removal of 100 ml water/toluene. Due to the lower surface area of kieselgel 60, a lower loading and consequently, a slower conversion of MF is observed.

3.3. Esterification of D-sorbitol

Heating of the biphasic sorbitol/lauric acid mixture in the presence of zeolites (e.g. H-β) does not result in appreciable lauric acid conversion (Figure 5). Apparently, the strong adsorption of the polyol on the zeolite blocks the access of the fatty acid to the acid sites. A brown residue resulting from sorbitol degradation is formed in the polyol layer.

At contrast, the fatty acid reacts easily in the presence of an HMS-SO_3H material, synthesized via co-condensation of MPTS and TEOS (20:80) in the presence of n-dodecylamine (route 1b). Conversions based on the HPLC data are given in Figure 5. Alternatively, ^{13}C-NMR spectra were integrated to calculate the molar ratio of free acid (δ=180 ppm) over esters (δ = 172-174 ppm). Agreement between both methods was satisfactory.

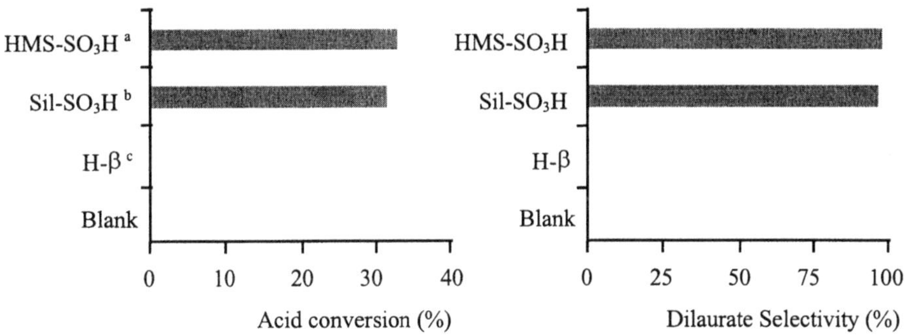

Figure 5. Conversion of lauric acid and selectivity for dilauryl isosorbide in the esterification of sorbitol (reaction time: 24 h). [a] catalyst prepared by co-condensation of MPTS and TEOS in the presence of dodecylamine. [b] amorphous silica, prepared from MPTS and TEOS in the absence of a template. [c] Si:Al = 12.5.

The dominant product at completion of the reaction is isosorbide dilaurate (**3**). The product NMR spectrum matches that of the diester synthesized by acylation of isosorbide **4** with lauroyl chloride in the presence of pyridine [8]. Moreover, the ^1H-^1H and ^{13}C-^1H coupling constants in the reaction product were identified via correlated NMR-spectroscopy (COSY, HETCOR) and are consistent with the bicyclic structure of **3**. Isosorbide can be viewed as the double water elimination product of sorbitol and retains only two alcohol groups for esterification. As the molar ratio of acid to polyol was 6, a lauric acid conversion of close to 33 % with a selectivity over 95 % for **3** (as in Figure 5) corresponds to an essentially full conversion of the sorbitol.

The high esterification activity of the MCM-SO_3H materials in comparison with zeolites is likely due to the hydrophobicity of the former. This favors approach of the lauric acid to the catalytic surface. Moreover, once mono-esters are formed, the detergency characteristics of these compounds promote an easy mixing of the polyol and fatty acid into a monophasic system. Visually, a sudden homogenisation of the reaction system is observed after 6 h.

At an intermediate lauric acid conversion, the apolar layer contains isosorbide, isosorbide monolaurate and isosorbide dilaurate (Figure 6). This suggests that the primary reaction is the dehydration of the polyol. The apolar isosorbide subsequently migrates towards the apolar lauric acid phase and is esterified. Such dehydrations mostly require fairly drastic conditions. For instance, the commercial synthesis of anhydrosorbitol esters is performed by heating a sorbitol/fatty acid-mixture at temperatures of at least 498 K [9]; Bock *et al.* used 3 M H_2SO_4 and long reaction times (40 h) to produce isosorbide from sorbitol at 377 K [10]. The dehydration observed in the present reaction (385 K, 24 h) thus illustrates the strong acid nature of the heterogenized sulfonic acid groups.

Both the dehydrating and esterifying activity of mesoporous sulfonic acids can be understood in terms of their hydrophobic nature. This favors expulsion of liberated H_2O out of the reaction mixture which drives dehydration and esterification towards completion in a Fisher-type reaction [11].

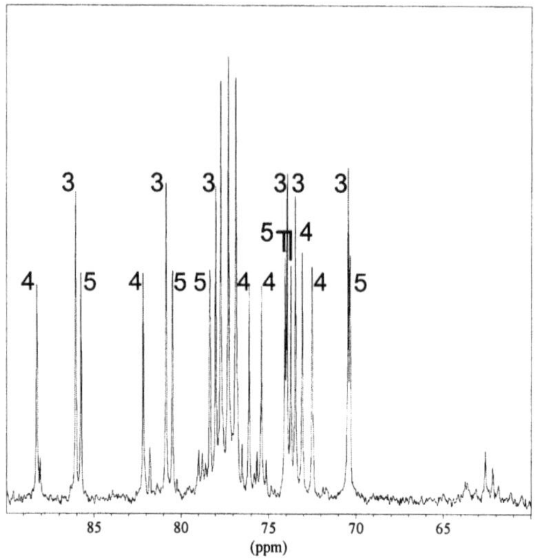

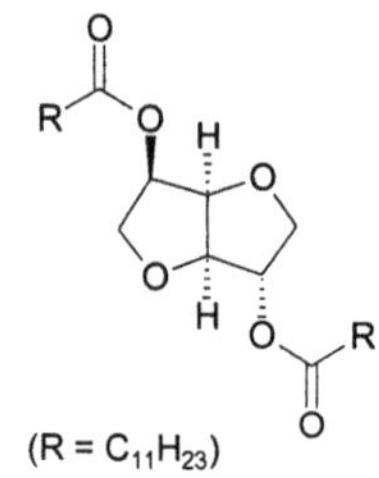

3

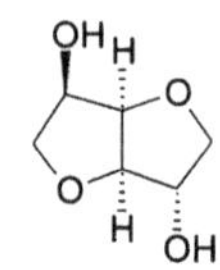

4

Figure 6. Partial ^{13}C-NMR spectrum of the apolar layer in the reaction of sorbitol and lauric acid (acid conversion: 17 %). **3** = isosorbide dilaurate, **4** = isosorbide, **5** = isosorbide monolaurate. The weakness of the signals between 60-65 ppm proves that most primary alcohol groups have been etherified.

REFERENCES

1. A. Riad, Z. Mouloungui, M. Delmas and A. Gaset, Synthetic Communications, 19 (1989) 3169. F. Algarra, A. Corma, H. Garcia and J. Primo, Appl. Catal. A, 128 (1995) 126. I. Iovel, Y. Gol'dberg and M. Shimanskaya, Chem. of Heterocyclic Cmpnds, 25 (1989) 613.
2. D. Swern (ed.), Bailey's Industrial Oil and Fat Products, Wiley-Interscience, New York, 1982.
3. W.M. Van Rhijn, D.E. De Vos, B.F. Sels, W.D. Bossaert and P.A. Jacobs, Chem. Commun. (1998) 317.
4. J. S. Beck, J. C. Vartuli, W. J. Roth, M. E. Leonowicz, C. T. Kresge, K. D. Schmitt, C. T. Chu, D. H. Olson, E. W. Sheppard, S. B. McCullen, J. B. Higgins and J. L. Schlenker, J. Am. Chem. Soc., 114 (1992) 10834. C. T. Kresge, M. E. Leonowicz, W. J. Roth, J. C. Vartuli and J. S. Beck, Nature, 359 (1992) 710. X.S. Zhao, G.Q. Lu and G.J. Millar, Ind. Eng. Chem. Res., 35 (1996) 2075. P.T. Tanev and T.J. Pinnavaia, Science, 267 (1995) 865.
5. S.L. Burkett, S.D. Sims and S. Mann, Chem. Commun. (1996) 1367. D.J. Macquarrie, Chem. Commun. (1996) 1961. L. Mercier and T.J. Pinnavaia, Adv. Mater., 9 (1997) 500.
6. A. Cauvel, G. Renard and D. Brunel, J. Org. Chem., 62 (1997) 749.
7. X. Feng, G.E. Fryxell, L.-Q. Wang, A.Y. Kim, J. Liu and K.M. Kemner, Science, 276 (1997) 923.
8. K. Hess and E. Messmer, Chem. Ber., 54 (1921) 499.
9. Kirk and Othmer's Encyclopaedia of Chemical Technology, Wiley-Interscience, New York, 1991.
10. K. Bock, C. Pedersen and H. Thøgersen, Acta Chem. Scand., 35 (1981) 441.
11. J. March, Advanced Organic Chemistry, Wiley-Interscience, New York, 4th edition ,1992.

MESOPOROUS MOLECULAR SIEVES 1998
Studies in Surface Science and Catalysis, Vol. 117
L. Bonneviot, F. Béland, C. Danumah, S. Giasson and S. Kaliaguine (Editors)

Titanium containing MCM-41 molecular sieves prepared by secondary treatment

Anke Hagen[a], Di Wei[b] and Gary L. Haller[b]

[a]Carl von Ossietzky University Oldenburg, FB9/Institut of Technical Chemistry, PF 2503, D-26111 Oldenburg, Germany

[b]Yale University, Department of Chemical Engineering, P.O. Box 208286, New Haven, CT 06520-8286, USA

Secondary treatment of Al-MCM-41 with $Ti(OEt)_4$ led to Ti anchored to the walls giving samples with high activities in cyclohexene oxidation producing mainly cyclohexenyl hydroperoxide and cyclohexene oxide. Ion exchange with K_2TiF_6 as well as hydrothermal treatment in the presence of TiX_4 (X=F, Cl, Br) left Ti oxide species inside the channels which performed poorly in cyclohexene oxidation. In addition, the latter treatment caused a removal of Al from the structure of Al-MCM-41 with no indication of Ti insertion into the vacant positions of the structure.

1. INTRODUCTION

The so called 'direct synthesis' or 'primary synthesis' is a method commonly used to isomorphously substitute lattice Si^{4+} of zeolites or molecular sieves by foreign ions. The procedures generally involve direct addition of a foreign metal source in the solution used for hydrothermal synthesis of the silica based molecular sieves. In contrast, the 'secondary synthesis' was first developed and used by Skeels and Breck [1] to produce Si enriched frameworks without structural defects through a reaction of zeolites with an aqueous fluorosilicate solution under relatively mild conditions. Thereafter, Skeels and Flanigen [2] described the substitution of Fe and Ti for Al in a number of zeolite frameworks through secondary synthesis. Characterization showed a successful substitution of Fe or Ti into the framework tetrahedral positions. Similarly, Kraushaar and Van Hooff [3] reported an alternative route to synthesize titanium silicate (TS-1) by treating the dealuminated ZSM-5 with saturated $TiCl_4$ vapor. It is concluded that acid leaching of ZSM-5 results in highly siliceous product with vacant T-atom positions in the lattice which are available for the insertion of Ti ions.

The method of secondary synthesis is a versatile approach for incorporating various atoms into the structure of molecular sieves. One advantage lies in the usually easily obtainable starting Al containing structure. Furthermore, once a procedure is found, it is applicable to a variety of lattice ions and a search for specific synthesis conditions is not necessary as in the case for direct hydrothermal preparation.

So far, crystalline zeolites are restricted to structures having pore diameters smaller than about 13 Å ($AlPO_4$-8, VPI-5, cloverite [4]). The desire and need of fine chemical producers to subject larger molecules to heterogeneous shape selective catalysis has driven extended research and led to the discovery of a new family of mesoporous molecular sieves designated as MCM-41 [5,6]. Thereafter, considerable effort has been devoted to obtain oxidation catalysts by placing transition metals into the MCM-41 structure. Titanium containing mesoporous molecular sieves have been prepared by electrostatic (MCM-41) and neutral assembly pathways (HMS) [7,8]. Additionally, in a special approach, Thomas et al. [9] used a MCM-41 matrix to attach Ti as titanocene onto the walls of the channels. The complex is anchored to the silanol groups and highly dispersed Ti sites were formed after calcination (6.76 wt.% Ti). The thereby achieved accessibility of all Ti sites is claimed to be responsible for the high turnover frequencies obtained in epoxidation of olefins being one order of magnitude larger than on conventionally prepared Ti-MCM-41.

In the present paper the potential of secondary incorporation of Ti into MCM-41 is discussed. Procedures include hydrothermal treatment, ion exchange and impregnation starting from Al containing MCM-41. Samples were characterized with respect to their structure by small angle XRD and to the location and coordination of Al and Ti with ^{27}Al MAS NMR, XANES and DRUV spectroscopies. Catalytic performance of the samples was tested in both the gas and liquid phase reactions. A comparison between physicochemical properties of Ti-MCM-41 prepared by different synthesis methods will also be presented.

2. EXPERIMENTAL

2.1. Secondary synthesis

Chemicals: Silica sources were HiSil-233 (Pittsburgh Plate Glass); tetramethylammonium silicate (SACHEM, 10- wt% silica). The Ti sources were $Ti(OC_2H_5)_4$ (Aldrich, 20-wt% Ti in ethanol), $TiCl_4$ (Fluka, 99%), and K_2TiF_6 (Fluka, 98%). The Al source was sodium aluminate (EM Science). The quaternary ammonium surfactant $C_{14}H_{29}(CH_3)_3NBr$ was purchased from Sigma Co. The $C_{14}H_{29}(CH_3)_3NOH$ solution was prepared by ion-exchanging the 29-wt% of $C_{14}H_{29}(CH_3)_3NBr$ aqueous solution with equal molar exchanging capacity of IRA-400(OH) ion exchange resin (Aldrich) by batch mixing.

2.1.1. Synthesis of the parent Al-MCM-41

The Na forms of Al-MCM-41 were synthesized with Si/Al *ratios* (synthesis mixture) of 15, 60 and 120 by using $C_{14}H_{29}(CH_3)_3NOH$ according to Beck et al. [1], method C. After the hydrothermal synthesis, samples were filtered, washed and dried at room temperature. Subsequently, the pre-dried solid was loaded into a glass reactor which had a ceramic distributor for the inlet gas. It was heated to 540°C in helium flow (~100 ml/min) within 20 hr and calcined at this temperature in helium for 1 hr followed by oxygen (~80 ml/min) for 5 hr in order to remove the template. The weight loss during this treatment amounted to about 40 wt%.

2.1.2. Incorporation of Ti

After the template removal, one of the following methods was applied to the Al-MCM-41 samples:

Method 1 (hydrothermal treatment, *Ti,Al-MCM-41)* consisted of addition of TiX_4 (X=F, Cl, Br) with Ti/Al=3 in a single step under hydrothermal conditions thereby modifying a process introduced by Kraushaar and Van Hooff [3] using ZSM-5 zeolites. The corresponding acid is formed with water present in the samples. Autoclave reaction was performed at 120°C for 14 or 24 hours. The samples were thoroughly washed with diluted HCl, NH_4Cl or ethanol after the treatment to remove non-reacted TiX_4 and dried at room temperature.

Method 2 (ion exchange, *Ti-Al-MCM-41*) was based on results obtained by Skeels and Flanigen [2] studying the incorporation of transition metals into microporous zeolites. Al-MCM-41 was stirred under reflux with an aqueous solution of K_2TiF_6 (Ti/Al=1) at 95°C for 10 hr, filtered, washed with water and dried at room temperature.

In Method 3 (impregnation, *Ti/Al-MCM-41*), Al-MCM-41 was impregnated with $Ti(OEt)_4$ in ethanolic solution and dried at room temperature. One part of the samples was heated in air flow to 300°C within 1 hr and kept at this temperature for 1hr.

2.2. Direct synthesis

The $C_{14}H_{29}(CH_3)_3NOH$ solution was combined with tetramethylammonium silicate and HiSil-233, respectively, with stirring. $Ti(OC_2H_5)_4$ was then added to the above mixtures, stirred and placed into an autoclave at 120°C for six days. The resulting solid was recovered by filtering, washing and drying in air at ambient temperature. It was then heated from room temperature to 540°C in the flow of helium for a period of 20 hr and followed 1 hr in helium and 5 hr in a flow of oxygen at 540°C.

2.3. Characterization

The experimental details of small angle XRD measurements have been described elsewhere [10].

The solid state MAS ^{27}Al NMR spectra were recorded at room temperature using an Am-500 Bruker spectrometer (11.7 T) at 130.32 MHz, equipped with multinuclear MAS probe from Doty Scientific Inc. A 5 mm standard zirconia rotor was driven by air to spin at about 5 KHz. Chemical shifts are referenced against an aqueous solution of $Al(NO_3)_3$. The spectra were obtained with a pulse length of 1.0 μs equivalent to about a 90 degree pulse. Relaxation delay was 0.5 s; spectral width was chosen as 25 kHz and 1 K data points were used.

Diffuse reflectance spectra were recorded in the UV-visible region on a HP 8452A Diode Array Spectrometer. The spectra were recorded in air at room temperature. The sample was pressed at 100 kg cm^{-2} into a thick wafer with an effective thickness of about 1g cm^{-2}.

X-Ray absorption measurements were performed at NSLS Brookhaven National Laboratory (2.5 GeV storage ring, station X19A, Si(111) monochromator crystal) and at Cornell High Energy Synchrotron Source (CHESS). Samples were pressed into self supporting wafers and placed into a stainless steel cell, equipped with water cooled kapton windows, gas in- and outlets and heating system allowing in situ treatment/reaction. Selected samples were subjected to in situ dehydration by heating to 300°C in 1 hr and holding at this temperature for 1 hr while purging the cell with an oxygen/helium mixture.

2.4. Catalytic reactions

The ethanol oxidation was carried out in an isothermal fixed-bed downflow reactor operating at atmospheric pressure. Helium was bubbled through an ethanol saturator and were combined with oxygen to give a reactant mixture of O_2 (5%), ethanol vapor (5%) and helium (balance). The products were analyzed by an on-line gas chromatograph (HP-5700A) equipped with TCD and Porapack column. Blank runs were performed on the empty Pyrex reactor without any detectable conversions. Pretreatment of the catalyst samples were performed at 773 K for 0.5 hr in a stream of 5% oxygen / helium prior to each run. The oxidative dehydrogenation of ethanol was then carried out over a 50 K temperature range (563 K- 603 K).

Liquid phase reactions were carried out in a three-neck flask under reflux, heated in a mineral oil bath. Solvent (13 ml methanol-MeOH or acetonitrile-ACN), oxidant (7 mmol of 35 % hydrogen peroxide or of 90 % *tert-butylhydroperoxide*-TBHP) and 35 mmol substrate were brought to the reaction temperature of 60°C and 100 mg catalyst added marking *t*=0. Samples of 0.01 ml were taken at selected reaction times, filtered and analyzed by GC on a 320 μm 30 m HP5 capillary column connected to a FID.

3. RESULTS AND DISCUSSION

3.1. Mesoporous structure

Figure 1 shows XRD spectra of the parent Al-MCM-41 samples. Parameters determined from XRD measurements are summarized in Table 1. An intense signal is obtained for as-synthesized Al-MCM-41 (Figure 1). The d spacing (d_{100}) is slightly decreased from Si/Al=120 and 60 to 15. Thus, a shrinkage of the unit cell dimensions with increasing Al

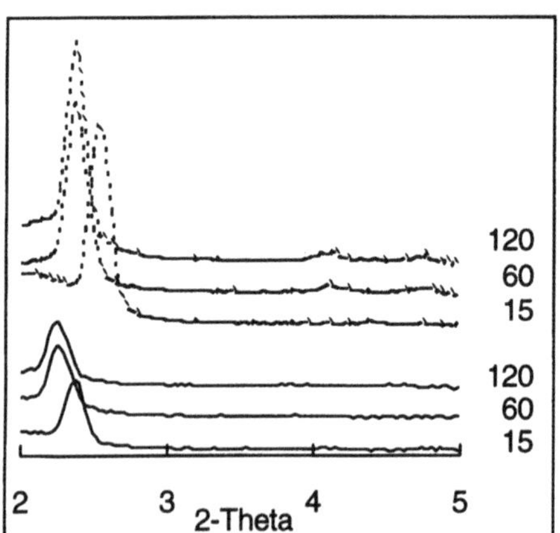

Figure 1. Small angle XRD spectra of Al-MCM-41 of as-synthesized (solid lines) and after calcination (dotted lines). The Si/Al ratio is indicated in the figure.

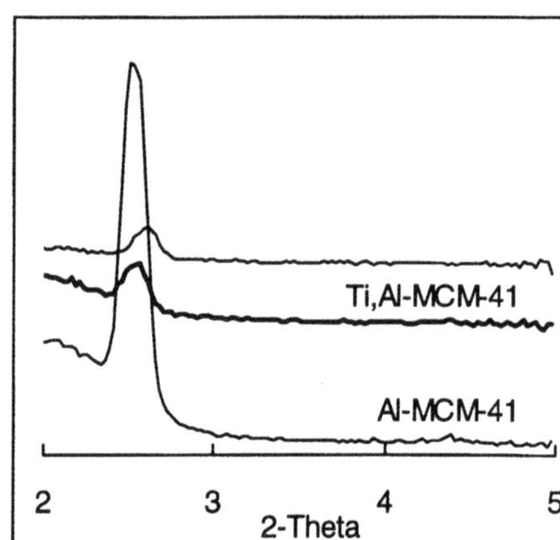

Figure 2. Small angle XRD spectra of the parent Al-MCM-41 (Si/Al=15) and after treatment with $TiCl_4$ for 14h (dark line) and 24h.

content was observed, despite of the longer bond distance of Al-O compared to Si-O. After calcination to remove the template, a framework contraction also occurs, as already reported by several groups (e.g. [8,11]). It is more prominent at higher aluminum content and thus an even more pronounced decrease in unit cell dimensions occurs. Furthermore, the intensity of the d_{100} peak strongly increases, a second (d_{110}) peak and for Si/Al=60 and 120, a third (200) peak appears in the XRD spectrum after calcination (Figure 1). This finding indicates the importance of template removal for the formation/stabilization of the MCM-41 structure and the higher structural regularity which can be obtained at lower Al content.

Figure 2 shows XRD spectra of titanium containing Al-MCM-41 prepared by the hydrothermal treatment. Parameters of all samples are also listed in Table 1. After treatment of Al-MCM-41 (Si/Al=15) with TiX_4 (X=F, Cl and Br) under hydrothermal conditions, the intensity of the first peak in the XRD spectrum decreases strongly indicating severe structural damage. The application of TiF_4 for 24 h causes a complete breakdown of the structure. Also ion exchange with K_2TiF_6 causes a slight destruction of the material's structure whereas impregnation gives almost the same spectrum as the parent Al-MCM-41.

Table 1
The d_{100} spacings and unit cell parameters ($a_0 = 2d_{100}/\sqrt{3}$) for Al-MCM-41 before and after modification.

Al-MCM-41	As-synthesized			Calcined		
Si/Al ratio	15	60	120	15	60	120
d_{100} / Å	37.4	39.0	39.0	35.2	37.4	37.4
a_0 / Å	43.2	45.0	45.0	40.6	43.2	43.2
Ti-MCM-41 From Si/Al=15	$TiCl_4$ 14h	24h	TiF_4 14h	K_2TiF_6 24h	$Ti(OEt)_4$ Impreg*	TiC16 -DS*
d_{100} / Å	34.5	33.9	34.2	34.8	34.8	36.3
a_0 / Å	39.8	39.1	39.5	40.2	40.2	41.9

* Impreg: Impregnated Ti-MCM-41 sample using $Ti(OEt)_4$ precursor.
TiC16-DS: Ti-MCM-41 sample (TiC16) from direct synthesis.

Analyzing the d_{100} parameter for the TiX_4 treated samples (Table 1), a decrease is observed, which is more pronounced at longer treatment times. Thus, the d_{100} spacings change from 35.2 through 34.5 to 33.9 Å for the calcined Al-MCM-41 (Si/Al=15) and Ti,Al-MCM-41 treated for 14 and 24 hours with $TiCl_4$, respectively. Together with the observations of Al-MCM-41 with varying Si/Al ratios it appears, that the unit cell parameters increase in the order Ti < Al < Si in MCM-41 although the crystal radii changes in the opposite direction being 0.68 (Ti^{4+}), 0.50 (Al^{3+}) and 0.41 Å (Si^{4+}). This observation may similarly be explained in terms of a strong deformation of the O-T-O bond angles to balance the effect of the T-O bond distances and would require a sufficient flexibility of the MCM-41 framework structure which is not usually observed in zeolite systems where the unit cell parameters can be directly related to the incorporation of T-atoms in MFI structures [12,13]. All other treatments (ion exchange, impregnation) did not alter the cell parameters significantly (Table 1). The largest unit cell parameters were obtained on Ti-MCM-41 prepared by direct synthesis.

3.2. Local Environment of Ti

3.2.1. ^{27}Al MAS NMR spectroscopy

In Figure 3, ^{27}Al MAS NMR spectra of MCM-41 after several treatments are presented. Both, the spectrum of as-synthesized as well as calcined Al-MCM-41 contain a peak at 52.4 ppm. After calcination, a small peak at -4 ppm appears indicating a partial transformation of aluminum from tetrahedral to octahedral positions [4,5,7].

After treatment of the sample with $TiCl_4$, the peak around 52.4 ppm diminishes dramatically accompanied by a growth of the peak at -1.3 ppm. This observation leads to the conclusion that Al is partially removed from tetrahedral framework sites.

After ion exchange with K_2TiF_6, no peak around 0 ppm can be detected (Figure 3) indicating no severe dealumination leading to octahedral extraframework aluminum. The spectrum of Al-MCM-41 does not change after impregnation with $Ti(OEt)_4$, as would be expected.

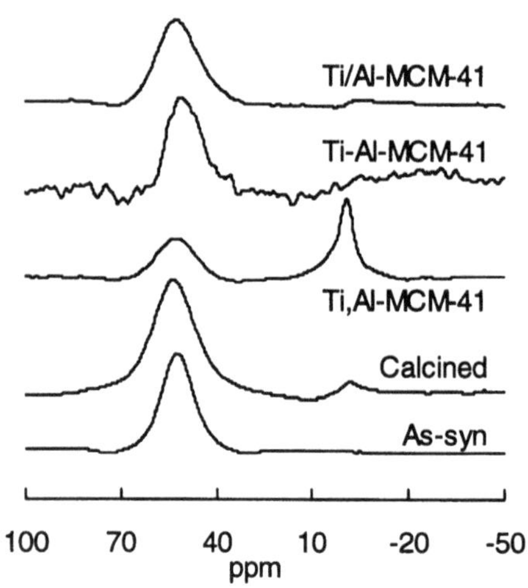

Figure 3. ^{27}Al MAS NMR spectra of Al-MCM-41 (Si/Al=15) before and after modification with Ti: $TiCl_4$ for 14h (*Ti,Al-MCM-41*), K_2TiF_6 (*Ti-Al-MCM-41*) and $Ti(OEt)_4$ (impregnated *Ti/Al-MCM-41*)

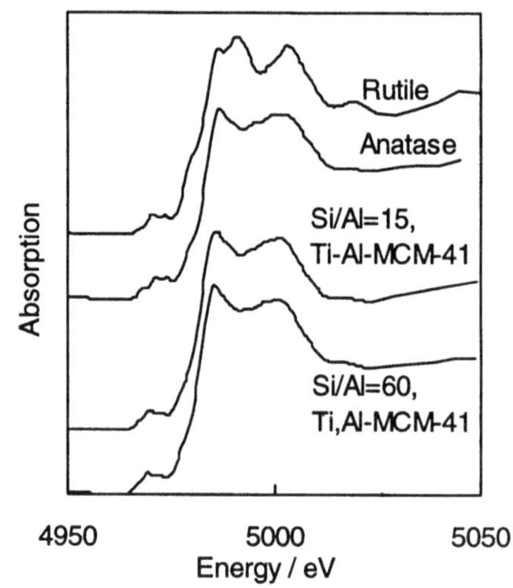

Figure 4. XANES spectra of Ti containing Al-MCM-41, Si/Al=15 (K_2TiF_6, *Ti-Al-MCM-41*), Si/Al=60 ($TiCl_4$, *Ti,Al-MCM-41*) and TiO_2

3.2.2. X-ray Absorption Near Edge Structure (XANES)

The XANES spectra of Ti-MCM-41 samples are shown in Figure 4 and 5 along with titanium reference compounds. Intensities and positions of pre-edge features of the samples and references are listed in Table 2. The anatase and rutile contain three small pre-edge peaks. With increasing distortion of the octahedral environment, the intensity of the center peak increases. Upon transition to tetrahedral coordination, one intense pre-edge peak at lower energy values is to be observed.

Features of the spectra of Al-MCM-41 treated with $TiCl_4$ and K_2TiF_6 closely resemble those obtained with titania (Figure 4, Table 2). In situ dehydration of $TiCl_4$ and K_2TiF_6 treated Al-MCM-41 samples at 300°C did not change the XANES characteristics. This indicates that none of the applied methods lead to an isomorphous incorporation of Ti into the framework. Indeed, extraframework oxide species are formed.

In contrast, the XANES of impregnated Al-MCM-41 gave different results. A single pre-edge peak rather than three is observed at room temperature (Figure 5). Upon heating, the intensity of the peak increased, the position, however, does not change significantly (Table 2). These features are retained after cooling down under oxygen/helium flow. Results can be similarly interpreted in terms of a distorted octahedral coordination of titanium in hydrated sample, probably involving water molecules. Dehydration causes a decrease in the coordination number. The intensity observed on tetrahedral references is not completely reached, indicating the presence of a mixture of both, tetrahedrally and octahedrally coordinated titanium species. On the other hand, the increase lies in the same range of Ti-MCM-41 (dehydrated at the same temperature) prepared by direct synthesis (Table 2). Chemically, the transformation of the octahedrally coordinated Ti in $Ti(OEt)_4$ in the course of the preparation could be explained by an alcoholysis reaction taking place between the ethyl group and the silanol groups of the MCM-41. Such alcohol interchange reactions are known to proceed on alkoxy derivatives of several atoms [14].

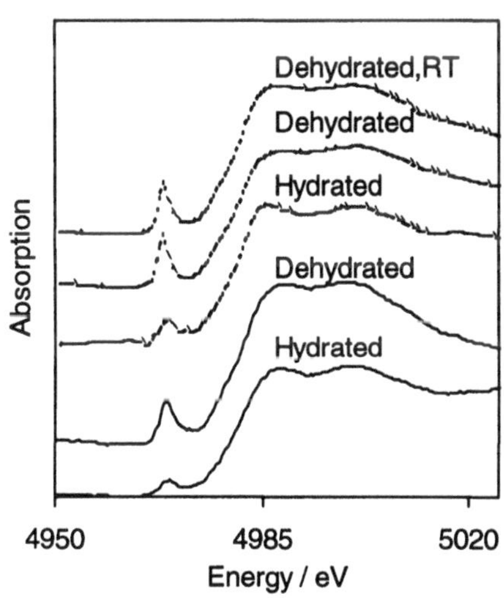

Figure 5. XANES spectra of Ti impregnated Al-MCM-41 before and after in situ recalcination at 300°C and subsequent cooling to room temperature in helium flow, Si/Al=15 (solid line), Si/Al=60 (dotted line)

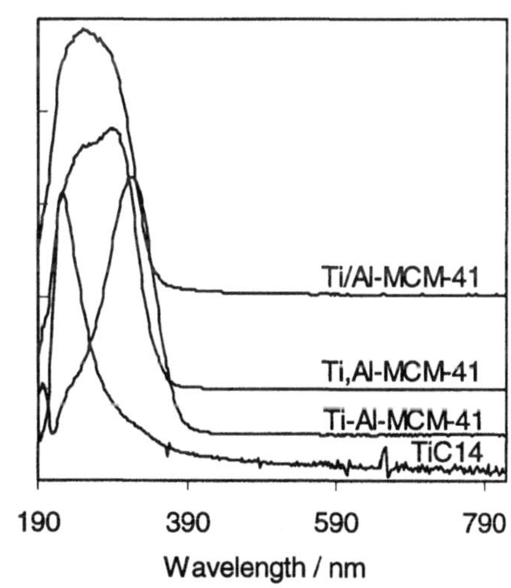

Figure 6. DRUV spectra of TiC14 prepared by direct synthesis (*Ti-MCM-41*), secondary treatment of Al-MCM-41 with $TiCl_4$ for 14 h (Si/Al=60, *Ti,Al-MCM-41*), K_2TiF_6 (Si/Al=60, *Ti-Al-MCM-41*) and $Ti(OEt)_4$ (Si/Al=15, *Ti/Al-MCM-41*)

3.2.3. Diffuse Reflectance UV-visible spectroscopy (DRUV)

Figure 6 presents results of DRUV spectroscopy for Ti containing MCM-41 prepared by secondary synthesis and for comparison, one sample from direct synthesis. The ion exchanged sample shows a pronounced peak around 300 nm indicating the presence of TiO_2, preferentially. In contrast, the Ti-MCM-41 prepared by direct synthesis shows a sharp peak around 220 nm and, thus in the expected region for Ti in sixfold coordination including two water molecules in hydrated state. The samples treated with $TiCl_4$ and impregnated with $Ti(OEt)_4$ show intermediate behavior. A slight blue shift of the 300 nm peak and the appearance of a second structure at lower wavelengths is observed which could be explained by the presence of species with polymeric Ti-O-Ti linkages.

Table 2
XANES pre-edge peak positions (relative to the edge of Ti foil) and intensities. Selected samples were studied before and after in situ dehydration at 300°C with subsequent cooling to room temperature in oxygen/helium.

Sample (Si/Al)	Treatment	Intensity / a.u.
Ti,Al-MCM-41 (60)	hydrated	0.17
Ti,Al-MCM-41 (60)	300 °C	0.19
Ti/Al-MCM-41 (15)	hydrated	0.15
Ti/Al-MCM-41 (15)	300 °C	0.39
Ti/Al-MCM-41 (60)*	hydrated	0.27
Ti/Al-MCM-41 (60)*	300 °C	0.51
Ti/Al-MCM-41 (60)*	300 °C, cooling	0.46
Ti/Al-MCM-41 (120)*	hydrated	0.20
Ti/Al-MCM-41 (120)*	300 °C	0.44
Ti/Al-MCM-41 (120)*	300 °C, cooling	0.47
TiC16#*	hydrated	0.27
TiC16#*	300 °C	0.47
TiC16#*	300 °C, cooling	0.44
TiO_2 anatase		0.16
TiO_2 rutile		0.15
$Ti(OEt)_4$		0.20

* Samples were measured at CHESS. Others at NSLS.
# TiC16 is the Ti-MCM-41 sample from 'direct synthesis'.

4. CATALYTIC BEHAVIOR

4.1. Ethanol oxidation

The ethanol oxidation has been carried out on the Ti-MCM-41 samples obtained from a secondary synthesis. A comparison between the catalytic activities (ethanol total turnover frequency normalized by the metal loading) of Ti-MCM-41 from direct synthesis and Ti-MCM-41 prepared by secondary synthesis is presented in Figure 7. The Ti-MCM-41 from secondary synthesis has comparable or even higher activity than Ti-MCM-41 from direct synthesis. It indicates that in the gas phase reaction, the catalytic performance is not sensitive to the extraframework oxidic Ti species because it has been shown by DRUV characterization that Ti-MCM-41 catalyst from secondary synthesis contains most of Ti sites sitting outside the lattice. Indeed, the supported/unsupported Ti and V oxides are active catalysts for gas phase oxidations [15]. Additionally, the selectivity to acetaldehyde is observed to be very low with the secondary synthesized Ti-MCM-41 samples because the small amount of Al^{3+} left in the framework during the synthesis would create strong Brønsted acidity, which greatly facilitated the side reaction to diethyl ether. This is also a reason for the higher total activity of secondary synthesized samples, compared to the samples from direct synthesis since now they (secondary synthesized) may have to be considered as a kind of bifunctional catalyst.

4.2. Oxidation of cyclohexene

A comparison of catalytic activities (cyclohexene conversion at the same conditions normalized by the metal loading) of cyclohexene oxidation is also made between the different MCM-41 catalysts systems (as shown in Figure 8). The activities of Ti-MCM-41 catalyst from secondary synthesis are now much lower than that of Ti-MCM-41 samples prepared by direct synthesis. This suggests that in the liquid phase oxidations, the randomly isolated, multiple bonded framework Ti species play an important role while the extraframework oxidic species has a high activity for the decomposition of the H_2O_2. The multiple bonds of framework Ti effectively prevent the complete hydrolysis of the active sites. Moreover, in the case of secondary synthesized samples, the Al^{3+} left in the framework during the synthesis would also facilitate the decomposition of H_2O_2. Additionally, it is noticed that among secondary samples, the impregnated Ti-MCM-41 sample has the best performance. This higher activity may possibly be ascribed to the still isolated Ti sites which are grafted to the channels by an alcoholysis reaction between the ethyl group of $Ti(OEt)_4$ and the silanol groups of MCM-41 during impregnation.

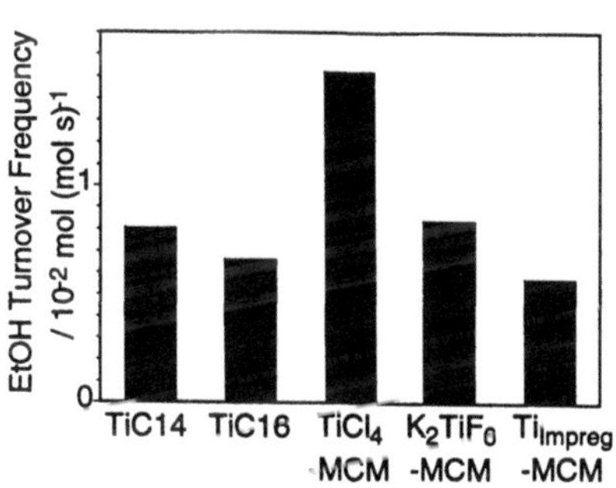

Figure 7. A comparison of total turnover frequencies (normalized by the metal loading) of ethanol oxidation between MCM-41 catalysts from different synthesis methods. Reaction conditions: O_2 (5%), EtOH (5%) and He (balance), T = 320°C.

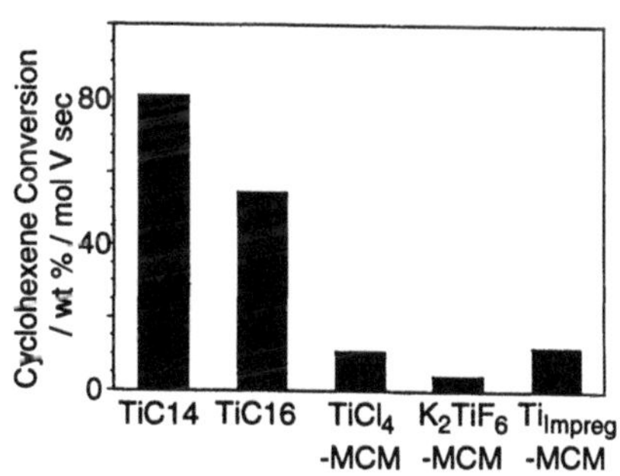

Figure 8. A comparison of cyclohexene conversions between MCM-41 catalysts from different synthesis methods. Reaction conditions: solvent: methanol; cyclohexene / H_2O_2 = 5; reaction time: 5hr; temperature: 60°c; sample weight: 200mg.

5. SUMMARY

Hydrothermal treatment of Al-MCM-41 with TiX_4/HX (X=F, Cl, Br) causes a removal of aluminum from framework positions, however, the samples experience severe structure damage at the same time and there is no indication of an insertion of Ti into framework positions, left by Al. Oxidic Ti species inside the channels and at the outer surface are formed. Consequently, the catalytic performance in olefin epoxidation is poor.

Ion exchange with K_2TiF_6 similarly leads to oxidic Ti species inside the channels and there is no significant removal of Al from framework positions.

Samples impregnated with $Ti(OEt)_4$ possess completely different characteristics. Aluminum in the framework is not affected by treatment. The $Ti(OEt)_4$ is probably already at room temperature anchored to silanol groups pointing inside the channels. Calcination should strengthen the grafting of the complex onto the channels via alcoholysis with silanol groups. Finally, Ti should be coordinated to 4 oxygen neighbors with 2 water molecules completing the 6-fold coordination, presumably. In the course of catalytic oxidations, these water molecules can be replaced forming a titanium hydroperoxide complex which expresses itself in the yellow color of the reaction mixture. The obtained samples show an excellent activity in cyclohexene oxidation which is comparable or even superior to samples prepared by direct synthesis.

REFERENCES

1. G. W. Skeels and D. W. Breck, Proceedings of the Sixth International Zeolite Conference, Edited by D. Olson and A. Bisio, Butterworths Ltd, UK (1984) 87-96.
2. G. W. Skeels and E. M. Flanigen, ACS Symp. Series, 398 (1989) 420.
3. B. Kraushaar and J. H. C. Van Hooff, Catal. Lett., 1 (1988) 81.
4. M. Janicke, D. Kumar, G. D. Stucky and B. F. Chmelka, Stud. Surf. Sci. Catal., 84 (1994) 243.
5. R. Schmidt, E. W. Hansen, M. Stöcker, D. Akporiaye and O. H. Ellestad, J. Am. Chem. Soc., 117 (1995) 4049.
6. G. Perego, G. Bellussi, C. Corno, M. Taramasso, F. Buonomo and A. Esposito, Stud. Surf. Sci. Catal., 28 (1986) 129.
7. B. Kraushaar-Czarnetzki and J. H. C. Van Hooff, Catal. Lett., 2 (1989) 43.
8. D. C. Bradley, R. C. Mehrotra and D. P. Gaur, Metal Alkoxides, Academic Press, New York, 1978, pp. 27.
9. Maschmeyer, T.; Rey, F.; Sankar, G.; J. M. Thomas, Nature 378 (1995) 159.
10. P. I. Ravikovitch, D. Wei, W. Chueh, G. L. Haller and A. V. Neimark, J. Phys. Chem., B101 (1997) 3671.
11. G. Centi, F. Trifiro, J. R. Ebner and V. Franchetti, Chem. Rev., 28 (1989) 400.
12. G. C. Bond, A. J. Sakany and G. D. Parfitt, J. Catal., 57 (1979) 476.
13. G. C. Bond and P. König, J. Catal., 77 (1982) 309.
14. P. J. Gellings, Catalysis, 7 (1985) 105.
15. S. T. Oyama and G. A. Somorjai, J. Phys. Chem., 94 (1990) 5022.

MESOPOROUS MOLECULAR SIEVES 1998
Studies in Surface Science and Catalysis, Vol. 117
L. Bonneviot, F. Béland, C. Danumah, S. Giasson and S. Kaliaguine (Editors)

Possibilities of Mesoporous Materials in Catalysis

A. Corma and D. Kumar

Instituto de Tecnología Química, UPV-CSIC, Universidad Politécnica de Valencia, Avda. de los Naranjos s/n, 46022 Valencia, Spain.

The possibilities of mesoporous molecular sieves (MMS) in catalysis has been reviewed. It is shown that due to their mild acidity Al containing MCM-41 has potential for use in bifunctional catalysis such as hydroisomerization and hydrocracking, as well as for the production of fine chemicals, where strong acid sites are not required. When containing transition metals in the wall, they can be used as selective oxidation catalysts using H_2O_2 and organic hydroperoxides as oxidants. Finally, owing to their high surface area and porosity they can be used as supports for acid, base and redox catalysts including transition metal complexes. The major drawbacks of MMS are their low hydrothermal stability and low stability in the presence of water. It will be shown that these limitations can be overcome.

1. THE NEED FOR NEW MESOPOROUS CATALYSTS

Crystalline materials with pore sizes in the microporous range have shown a wide range of applications . Zeolites, and in general microporous molecular sieves, are good examples of such materials. The reason for the success of zeolites in catalytic processes is related to the following specific features of these materials.

- They have a very high surface area and adsorption capacity.
- The adsorption properties of the zeolites can be controlled (tunable pore size) and they can be varied from hydrophobic to hydrophilic type materials.
- Active sites, such as acid sites can be generated in the framework and their strength and concentration can be tailored for a particular application.
- The sizes of their channels and cavities are in the range typical for many molecules of interest (5-12 Å).
- The intricate channel structure allows the zeolites to present different types of shape selectivity.
- Zeolites can be activated to produce very stable materials that are both heat and chemical attack resistant.

Although they possess these catalytically desirable properties, zeolites are not of use when reactants of sizes larger than the pore dimensions are processed. Therefore, logically one would like to employ zeolites with the same porous structure but with an increase in pore ...ely, only materials employing Al and P or Ga and P as framework elements were successfully synthesized. These metallophosphates include APO_4-8, ULM-5, VPI-5, cloverite, cacoxenite and JDF-20 with 14-, 16-, 18-, and 20-membered rings respectively [1-12]. UTD-1, a 14-member ring unidirectional aluminosilicate zeolite was discovered recently [12,13]. The Co organometallic template used for its synthesis can be removed such that the Brönsted acidity of the framework is sufficient for the cracking of paraffins.

Due to a number of reasons, however, including thermal instability (cacoxenite), diffusion restriction due to protruding hydroxyl groups (cloverite), stacking disorder (VPI-5), difficulty of Co template removal (UTD-1) these materials are limited in their use as catalysts..

Thus, a gap has been bridged by the discovery of both the M41S family of materials and FSM-16, which have opened up new possibilities for preparing catalysts with uniform pores in the mesoporous region that can be easily accessed by bulky molecules that are present in crude oils and fine chemical production [14-27].

Following on from their discovery in 1992 these materials have inspired scientists to carry out a great deal of creative research in the design, synthesis and characterisation with the aim of application in many fields, including adsorption, gas separation, guest-host chemistry, ion exchange and catalysis [28].

The family of materials known as M41S consists of an hexagonal phase (p6m) referred to as MCM-41, a cubic phase (Ia3d) known as MCM-48 and a lamellar phase known as MCM-50 [29]. The synthesis of these novel materials all employ organic surfactant molecules which alter the inorganic structure and can be cationic, anionic, gemini or neutral surfactants [30-32]. Since the organic molecule is removed post-synthetically the organic framework left behind can find application in a wide range of fields as mentioned earlier. Due to the deep understanding and flexibility of the synthesis conditions one is now able to tune the pore size from ~ 20 Å-100 Å as well as dictate exactly the framework composition in terms of pore silicates or metallosilicates (such as Al, V, or Ti-substituted silicates) as well as non silica materials such as Nb, Al, Zr, Hf, Ta, Zr-P and Ti-P oxides, sulphates and aluminophosphates [15,33-39].

The possibility to introduce different atoms in these mesoporous networks to generate catalytically active centres together with long range order and high surface areas (~ 700 m^2/g) and tunable pore size and narrow pore size distribution enabled one to use these materials in acid, base and redox catalysis, as well as high surface area supports for acids, bases, metal oxides, and transition metal complexes.

2. ACID CATALYSIS

2.1. Brönsted and Lewis acid sites in MCM-41

Research has been thoroughly carried out on the stability and acidity of these mesoporous materials. Meziani et al. [40] have shown that the number of Lewis acidic-basic sites in MCM-41 samples increases with the degree of Al incorporation into the framework. The acidity of Al-containing MCM-41 samples has been measured by adsorption/desorption of bases such as pyridine and NH_3 [41-45]. Corma et al. [41] have shown by pyridine adsorption that calcined aluminosilicate samples have both Brönsted and Lewis acidity, wherein the former increases with an increase in aluminium content of the sample. Spectroscopic characterization into the strength, types and stability of the Brönsted and Lewis sites on these mesoporous aluminosilicate catalysts has been conducted by various research groups [46-50]. Feng et al. illustrated that the acidity and selectivity of these materials, according to the isomerization selectivity of 2-methyl-2-pentene, are affected by both the Si/Al ratio and the pore size at fixed composition [50]. Thus, it can be concluded that it is important to fully characterise these novel materials in order to take advantage of their potential catalytic properties [51-54]. Figure 1 illustrates the TPD of NH_3 spectra of USY, calcined aluminosilicate MCM-41 (denoted as Si) and amorphous silica/alumina [41] indicate that MCM-41 aluminosilicate has a similar acid strength to amorphous silica/alumina, but has the advantage of possessing regular pores.

It is interesting to observe that we have already reached the stage where methods are being innovated to enhance acidity and catalytic cracking activity of these mesoporous catalysts.

Table 1 [55] illustrates the elemental composition, d-spacing and textural properties and acidity of siliceous MCM-41 materials containing grafted aluminium. In fact, the authors claim that the resulting materials retain their hexagonal order and exhibit higher Brönsted acid content with respect to Al-MCM-41 prepared by direct hydrothermal synthesis.

Table 1.
Elemental composition, *d* spacing and textural properties and acidity of the study materials (CHA = cyclohexylamine).

	SAMPLE				
	PSMCM	Al-PSMCM40	Al-PSMCM20	Al-PSMCM10	Al-PSMCM5
Si/Al		38.5	19.8	11.2	6.1
d_{100}/Å	33.8	34.8	36.4	36.0	35.7
Surface Area/$m^2\ g^{-1}$	1110	1000	1004	996	980
Pore volume /$cm^3\ g^{-1}$	0.93	0.82	0.81	0.78	0.77
APD^a/Å	26.3	26.8	26.7	26.6	26.5
$a_o{}^b$/Å	39.0	40.2	42.0	41.6	41.2
Wall thickness/Å	12.7	13.4	15.3	14.9	14.7
Acidity/mmol CHA g^{-1}		0.16	0.29	0.48	0.51

^{a}APD = Average pore diameter (determined using BJH analysis). $^b a_o$ = Lattice parameter from XRD data using the formula $a_o = 2d_{100}/\sqrt{3}$.

Kloestra and his co-workers [56] report the design of an MCM-41 catalyst by inducing an increase in the local crystallinity of the amorphous framework by ion-exchange with tetrapropylammonium template. The resultant effect is an upgrading of the catalytic properties of the materials.

Finally, Medina-Valtierra and his co-workers developed MCM-41/gamma-Al_2O_3 catalysts that contained mainly Lewis acid sites and could be employed successfully for the para-selective alkylation of toluene with isopropanol [57]. The physico-chemical and catalytic properties of Al-, Ga-, and Fe-substituted materials related to MCM-41 have also been studied during the last two years [58-63], such that this field has really expanded to exciting frontiers.

But it was the incorporation of tetrahedrally coordinated aluminium species into the silicate walls that was the starting point for a wide range of applications.

2.2. Catalytic cracking and hydrocracking by Al-MCM-41

It was one year after its discovery that Aufdembrink et al. reported the substitution of Al-MCM-41 (35 wt%) in a silica alumina-kaolin clay matrix [64] and applied this mixture to cracking processes. Al-MCM-41 actually showed a higher selectivity towards liquid fuels than USY zeolite.

Table 2
First-order kinetic rate constants for n-heptane cracking

Sample	First-order kinetic rate (h^{-1})[a]	K Al(Al + Si)
MCM-41.1	0.6211	8.873
Amorphous Si-Al	3.20	35.56
USY1	12.37	1237

[a] Calculated from plot stop of $-\ln(1-x)$vs $WHSV^{-1}$

Table 2 indicates that Al-MCM-41 has been used to crack n-heptane as well as polyethylene and a vacuum gasoil [65,66]. In such a MAT (Microactivity Test) experiments, it has been found that USY zeolite is 139 times more active than Al-MCM-41 with n-heptane, due to the larger amount of Brönsted acid sites on the USY zeolite as well as stronger acid sites. In the case of gasoil, MCM-41 activity was closer to that of USY zeolite than amorphous silica-alumina. In terms of gasoil cracking Al-MCM-41 was seen to produce a higher amount of liquid fuels and less gases and coke than amorphous silica/alumina. However, it was seen to be more selective to diesel formation and yielded less gasoline and higher coke levels when compared to USY zeolite. Mesoporous silica/aluminas are generally used prior to USY [67] zeolites, in order to precrack large feed molecules. Therefore, developing on from these observations Schipper et al. [68] prepared a cracking catalyst whose core consisted of USY, ZSM-5 zeolites and whose shell consisted of MCM-41. It was Chen et al [69] who observed that Al-MCM-41 materials exhibit high activity for n-C-16 cracking and a good selectivity for producing low olefins, particularly i-$C_4^{=}$. They attributed this to the large surface areas, low density of acid sites and low activities of hydrogen transfer reactions. Roos et al. [70] investigated the catalytic properties of Al-MCM-41 using MAT n-hexadecane and 1,3,5-triisopropylbenzene and demonstrated that the acidity and accessibility of the internal active sites influenced the activity and selectivity.

It was found that [71] when MCM-41, amorphous silica-alumina and USY zeolite were steamed at 750°C the loss in surface area of MCM-41 was the largest of all 3 samples. This indicates that MCM-41 does not resist hydrothermal deactivation in the regenerator of the FCC unit (which is at 800°C). This means that one must not only consider the initial activity of the catalyst [44,45,71] but also its hydrothermal stability. It was also observed that Al-MCM-41 was more hydrothermally stable when it contained a lower number of acid sites (i.e. lower aluminium content) but then was less catalytically active. Therefore, it was logical to measure the gasoil cracking activity of steamed MCM-41, USY zeolite and amorphous silica-alumina such that it was found that steamed Al-MCM-41 was the least active. Although Borade et al. found [72] that Al-MCM-41 materials were inactive for n-hexane cracking (due to the requirement of stronger acid sites, with respect to cracking of hexadecane reported by Chen et al. [69] which requires weaker acid sites) Al-MCM-41 does exhibit promising applications for the cracking of vacuum gasoil and straight run naphtha [73]. It is obvious however, that more work has to be carried out on the stability of MCM-41 [50,74-77]. One such method would be to increase the stability of the pore walls by increasing their thickness [76-80].

Although a relatively expensive process, Aguado et al. reported the conversion of low density polyethylene into hydrocarbon feedstocks [65,81]. Refinary processes such as hydroisomerization, hydrocracking, demetallisation and olefin oligomerization require weakly acidic catalysts that do not necessarily need to be extremely hydrothermally stable.

Employing Al-MCM-41 as a mildly acidic support onto which NiO and Mo_2O_3 were incorporated proved successful for the hydrocracking and hydrodesulphurization of feed. This MHC (mild hydrocracking process) was well suited to Al-MCM-41 because it has a high surface area for achieving good dispersion of the transition metal oxides and at the same time the aluminium form of the Al-MCM-41 possesses mildly acidic character [82-84]. The MHC performance of a NiMo/MCM-41 catalyst was compared to that of amorphous silica-alumina and USY, containing the same amount of Ni and Mo [83]. The results indicate that the USY zeolite contained the strongest and highest number of Brönsted acid sites and MCM-41 had the same acidity as amorphous silica-alumina. Nevertheless, the MCM-41 catalyst gave a better performance in terms of HDS, HDN and hydrocracking conversion when vacuum gasoil containing 2.53 wt% of sulphur and 2900 ppm of N_2 were hydrotreated and tested for HDS, HDN (hydrodenitrification) and hydrocracking performance. Thus, NiMo/MCM-41 bifunctional catalysts may be used for MHC catalysts owing to their high surface area (which favours a high dispersion of the active oxide species, while increasing the accessibility of the large molecules of the gas oil feed containing heteroatoms to the catalytically active sites). Following on from this research, HDS has been carried out on extensively characterised Ni, Mo-supported MCM-41 catalysts. Cui et al. [85] concluded from the analysis of adsorption isotherms that the dispersion of Ni and Mo in MCM-41 is more active than in NaY supported zeolite.

It is noteworthy to mention that [86] a new disordered mesoporous molecular sieve, KIT-1, was used as a support for MoO_3 and NiO and employed in the hydrodesulfurization of thiophene reaction. Interestingly, it was seen to exhibit a higher catalytic activity than MCM-41 and NaY zeolite supposedly due to its 3-D disordered network of short channels which apparently reduce the risk of blockage in the catalysts and facilitate the transport of reaction and product molecules. Song and Reddy [87,88] successfully supported Co-Mo catalysts on Al-MCM-41 for the HDS of dibenzothiophene in distillate fuels as well as for the HDS of petroleum resids.

Al-MCM-41 can be used to isomerize normal paraffins using the bifunctional catalysts Pt/MCM-41 [89,90], and comparison with the conventional Pt-silica-alumina catalysts Pt/MCM-41 was seen to produce less cracked products which was probably due to a greater dispersion of Pt.

Marler et al. [91] combined hydrocracking, and hydroisomerization processes in order to convert wax feeds to high viscosity index lubricants. This 2 step process involved the

Pt/MCM-41 catalysts which produced less waxy branched paraffins. More recently Girgis et al. [90] used the dual function catalyst Pt/MCM-41 to investigate the reaction pathways and kinetics of n-hexadecane hydroisomerization and hydrocracking,.

Corma et al. [92] studied the hydrogenation of aromatics in diesel fuels on Pt/MCM-41 catalysts. The fact that stringent environmental regulations directed lower hazard emissions from vehicle exhausts and also the growing demand for high quality diesel fuels this group was prompted to employ Pt/MCM-41 for studying the hydrogenation of naphthalene (see scheme) (as a reduction in aromatics content is a new diesel fuel specification). The conventional noble metal catalysts used were unfortunately readily poisoned by the small amounts of sulphur and nitrogeneous organic compounds present in the feed. Pt-MCM-41, however, owing to the high metal dispersion and the interaction of small Pt clusters with the mildly acidic centres in Al-MCM-41 exhibited a high sulphur tolerance, and a superior hydrogenation activity with respect to Pt/USY.

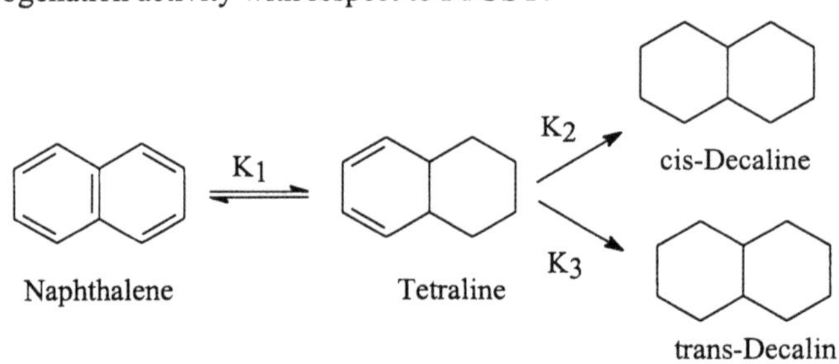

Scheme for the hydrogenation of naphthalene

Pt and/or Pd containing MCM-41 were claimed as catalysts for the low temperature hydrogenation of benzene [93], for the hydrogenation of aromatics in diesel and kerosene feeds [94] and geraniol [95].

2.3. Al-MCM-41 for the upgrading of olefins, disproportionation and oligomerization of olefins

H-MCM-41 has been used for upgrading olefins in several different processes [96]. The large sorption capacity and mild acidity rendered this material useful. One of the routes investigated was the oligomerization of C_3, C_4 or C_5 olefins to C_6-C_{18}^+ hydrocarbons that were then recycled and cracked in a multistage process to form C_3-C_5 range olefins. The same researchers also employed H-MCM-41 to produce isobutylene and isoamylenes via the disproportionation of olefins. The products are generally used in the production of MTBE and TAME. Various groups have researched into the production of fuels and lubricants from the oligomerization of olefins [97-101] on metal containing MCM-41. It appears, therefore, that MCM-41 type catalysts are suitable for the oligomerization of larger olefins or generally to produce large lubricant molecules. Figure 1 indicates the oligomerization of propylene on different acid catalysts.

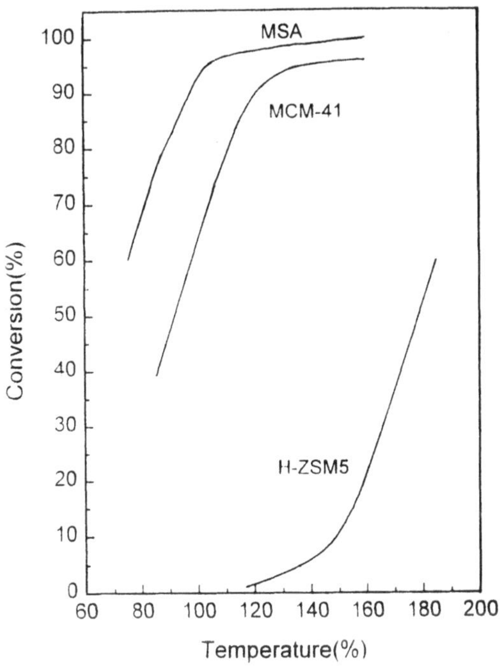

2.4. Al-MCM-41 applied for fine chemical synthesis

A great application of these novel mesoporous materials is to take advantage of the large regular pores which will allow the diffusion of bulky reactants and bulky products, thus, minimizing unwanted consecutive reactions and catalyst decay that are usually encountered in the acid catalysed organic reactions of fine chemicals. Fine chemical synthesis involves the liquid phase that consists of bulky reactants and products that can easily diffuse in and out of the pores [102]. The special combination of large pores and mild acidity in H-MCM-41 has been exploited for carrying out reactions such as acetalyzations [103], Beckmann rearrangements [104], glicosidation [105] and aldol condensation [106].

Keeping this is mind, Al-MCM-41 has been used successfully in Friedel-Crafts alkylations and acylations [107-113]. An example in which the shape selectivity of MCM-41 produces good results is for the alkylation of 2,4-di-tert-butphenol with cinnamyl alcohol [107]. Similar pore size effects were observed during the tertbutylation of antharacene, naphthalene and tianthrene [109]. Acylation reactions are normally used for the production of fine chemicals and stoichiometric amounts of $AlCl_3$ are generally used to catalyse such reactions. However, the disposal of $AlCl_3$ residues has posed a problem in the past. Looking towards the future Rhone-Poulenc have sensibly claimed the use of acid zeolites for the Friedel-Crafts acylation of aromatics [114], and furthermore Spagnol et al. explored the possibility of using MCM-41 for the acylation of bulky reactants.

Aromatic acylation requires mild acidity and is an important step for the synthesis of aromatic ketones which are used in the preparation of synthetic fragances and pharmaceuticals. Quite recently Gunnewegh et al. [111] reported the acylation of 2-methoxy naphthalene with acetic, benzoic and isobutyric anhydride to yield a product which was acylated in the 6-position, which is a precursor for the production of Naproxen, an anti-inflammatory drug. It was observed that although at temperatures higher than 100°C selectivity to acylation in the 6 position was always below 10% on the H-MCM-41 catalyst, the H-MCM-41 catalyst was easily regenerated.

The catalytic preparation of acetals is of importance in both the pharmaceutical industry [106,115] and for fragances [116]. Table 3 illustrates the rates of acetalization of bulky aldehydes of various sizes; n-heptanal, 2-phenylpropanal and diphenylacetaldehyde with trimethylorthoformiate on β-zeolite, mesoporous Si-Al and Al-MCM-41. Concluding the results it can be seen that when the reactants are too bulky to diffuse in and out of the β-zeolite pores Al-MCM-41 is the best catalyst as no diffusional restrictions are observed.

Table 3
Influence of the catalyst pore size on the rate and conversion in the acetalization reaction of 1, 2, and 3.

Catalyst	r_0 (mol.h^{-1}.g^{-1}) x 10^3		
	1[a]	**2**[b]	**3**[c]
1-MCM-41	2500	2480	2340
2-βH	--	--	180
Si-Al	1500	1600	460

[a] **1** = n-heptanal; [b] **2** = 2-phenylpropanal; [c] **3** = diphenylacetaldehyde

Thus, MCM-41, due to its large pores and channels, was seen not to restrict the diffusion of the products formed within its channels thus preventing blocking, deactivation and poisoning of the catalyst as is sometimes the case with acetalization on HY zeolite.

The Beckman rearrangement of cyclohexanone oxime [117-121] has been carried out on medium and large pore zeolites to yield the conclusion [119] that even weak acid sites are able to carry out this reaction. Although MCM-41 possesses the weak acidity required the yields and selectivity were lower than those obtained on conventional zeolites. Work carried out very recently by Semmer et al. [122] studied the Beckman rearrangement of oximes of different sizes on Al-MCM-41. High selectivities to the corresponding lactames indicated the benefit of MCM-41 when working with larger oximes. De Goede et al. [105] studied a process for the preparation of alkyl glycosides by reaction of an alcohol with a saccharide or a lower-alkyl glycoside in the presence of MCM-41.

Some very interesting work [106] using Al-MCM-41 catalyst was the synthesis of α-n-amyl cinnamaldehyde (jasminaldehyde), a traditional fragance of violet like scent, which entailed the condensation of 1-heptanal (1) with benzaldehyde (2) to form (3) jasminaldehyde. The mesoporous catalyst exhibited a high degree of selectivity even when using low ratios of benzaldehyde/heptanal.

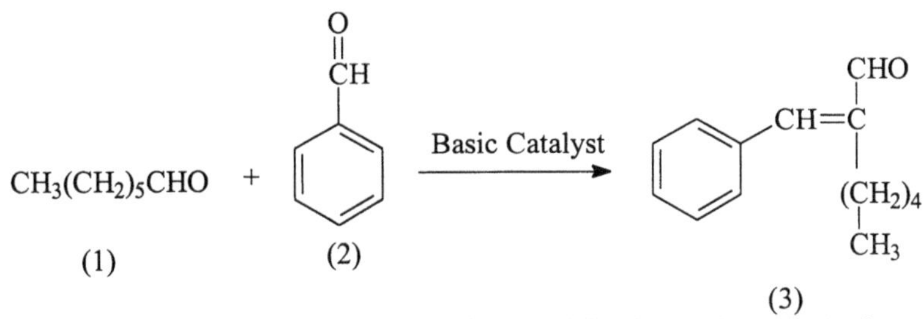

Lourvanij et al. [123] developed a kinetic model for the partial dehydration of glucose to organic acids by micro- and mesoporous aluminosilicate catalysts. Catalysts studied included HY-zeolite, MCM-20 and MCM-41.

3. BASE CATALYSTS

It has been [124,125] reported that basic catalytic character was exhibited when the negative charge of the aluminium atom in microporous aluminosilicates was compensated by alkaline cations. Stronger basicity was observed when the charge to radius ratio of the charge balancing cation was decreased. The negative charge of the tetrahedrally coordinated aluminium atom in Al-MCM-41 was balanced by Na^+ and Cs^+, and the resulting materials were active and selective for the base-catalyzed Knoevenagel condensation of benzaldehyde with ethyl cyanoacetate [126] as seen in the scheme.

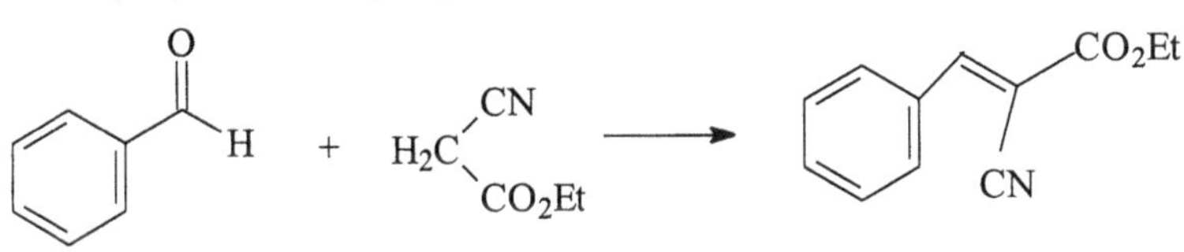

Both the Cs^+- and Na^+-exchanged samples could perform the reaction, however, the former material exhibited a higher activity. One observation was that when more than the required amount of alkaline species was introduced in order to ion exchange the MCM-41, Na_2O and Cs_2O particles were formed. These resultant materials were seen to possess a stronger basicity than cesium- and sodium-exchanged MCM-41.

Following on from this observation Kloetstra et al. developed a range of CsLa or Cs_2O supported Al-MCM-41 catalysts. MCM-41 containing intraporous cesium oxide (which was prepared by the impregnation of aqueous cesium acetate and subsequent calcination) showed a high regioselectivity for the Michael addition of diethyl malonate to neopentyl glycol diacrylate. The advantage in supporting the Cs_2O on MCM-41 was apparent by the lack of regioselectivity of bulk cesium oxide [127]. Likewise, binary cesium-lanthanium oxide supported on MCM-41 [128] produced a heterogeneous basic catalyst that was product selective in the liquid phase Michael addition of ethyl cyanoacetate to ethyl acrylate. The Knoevenagel addition of enolates to benzaldehyde in aqueous media was also catalyzed by the CsLa-oxide/MCM-41 system. Furthermore, the catalyst was reusable after regeneration at high temperatures without loss of activity.

The activities of cesium oxide or binary cesium-lanthanum oxide supported on Al-MCM-41, Al-HMS, amorphous silica-alumina and gamma-alumina in the novel isomerization of omega phenylalkanals to phenyl alkyl ketones were also compared [129].

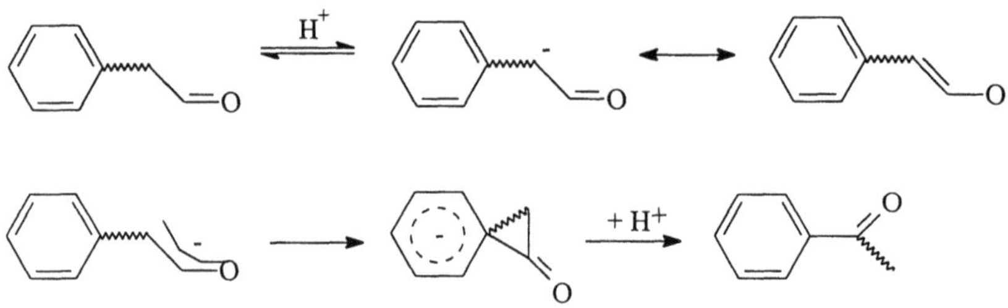

4. REDOX CATALYSIS

4.1. Titanium substituted MCM-41

It was logical to compare the activity of Ti-MCM-41 to that of Ti-silicalite [130-133] and the large pore Ti-β-zeolite [134-136] which were originally used for the selective oxidation of paraffins, olefins and alcohols. Ti-MCM-41 was first reported in 1994 [137] and epoxidised selectively olefins to epoxides using H_2O_2 as the oxidising agent. Although its activity in epoxidizing small linear olefins with H_2O_2 was lower than Ti-silicalite and Ti-β, it was especially useful for the oxidation of large molecules such as d-terpineol and norbornene at 70°C [130] using tert-butyl hydroperoxide as the oxidant. Another mesoporous material called Ti-HMS [138,139] showed quite a low activity for the hydroxylation of phenol with respect to Ti-MCM-41. Many reports have developed this observation further [140-142]. Ti-HMS (which has been prepared with a neutral surfactant (S^oI^o)) was nevertheless seen to exhibit catalytic activity for the liquid-phase peroxide oxidations of methyl methacrylate [143].

Inevitably much research has been conducted to characterize the Ti-MCM-41 material. Rhee et al. [144] employed XRD and XAFS in order to elucidate the coordination environment of Ti in the Si framework. Likewise, Alba et al. [145] indicated with XANES and ^{29}Si MAS NMR that the titanium is 4-coordinate and that Ti(IV) does not form a regular TiO_4 tetrahedron due to its larger size when compared to S(IV) in the silica matrix. Very recently the detailed structural and surface properties of siliceous and titanium modified HMS molecular sieves were elucidated. It was shown that at high titanium loadings, a significant decrease in the primary mesopore volume, specific surface area and pore size uniformity took place [146].

Camblor et al. were able to epoxidise methyl oleate using hydrogen peroxide or tert-butyl hydroperoxide as oxygen donors with high conversion and epoxide selectivities using Ti-MCM.41 [147]. Koyano et al. [148] were able to synthesize titanium containing MCM-41 which was applicable to the oxidation of bulky alkenes with H_2O_2 or as with Camblor et al. [147] with t-butyl hydroperoxide as the oxidant. However, the stability of Ti-MCM-41 to water was low with respect to pure silica MCM-41.

The groups of Zhang [149] and Morey [150] reported the synthesis of Ti-containing mesoporous material, isomorphous to cubic MCM-48. The high pH synthesis of Ti-MCM-48 was reported using (pH=11) NaOH. A well known fact is that the presence of alkaline metal hydroxides should be avoided as they hinder the incorporation of Ti into the microporous and mesoporous pore walls by favouring the formation of alkaline metal titanates. Thus, according to Zhang et al. [149] an increase in the unit cell size of only 0.3 Å may be due to the fact that not all the titanium present in the synthesis is eventually tetrahedrally coordinated. However, it is quite remarkable that Ti-MCM-48 is nevertheless able to selectively oxidise methyl methacrylate and styrene. Both Ti-MCM-41 and Ti-MCM-48 were used for the photocatalytic reduction of CO_2 with H_2O at 328 K. Both mesoporous materials exhibited a high photocatalytic reactivity for the formation of CH_3OH, more so, in fact, than bulk TiO_2 [151].

The chemical and pharmaceutical industries are interested in the oxidation of amines [152,153]. This has been successfully attempted on Ti-MCM-41 and Ti-HMS mesoporous materials [154,155]. Ti-HMS was active for the oxidation of arylamines in the liquid phase due to the fact that it contained larger pores and the possibility of employing organic peroxides as oxidants. Ti-MCM-41 oxidized bulky sulfides to sulfoxides and sulfones better than the large pore Ti-β zeolite [156]. A novel piece of research was conducted by Vankelecom et al. [157] who incorporated Ti-MCM-41 catalyst and TS-1 catalyst in polydimethylsiloxone membranes which were subsequently able to oxidise hexane and epoxidise cis-cyclooctene with high conversion under solvent-free reaction conditions.

Bifunctional acid/oxidation catalysts which are able to perform two different reaction steps on the same catalyst were synthesized by Corma et al. in 1995 [158].

MCM-41 structures that contained both Al and Ti atoms in the silicate framework contained both Ti^{IV} oxidation sites and H^+ ions associated with the Al^{IV} acid sites; and were thus able to carry out the multistep oxidation of linalool to cyclic furan and pyran hydroxyethers with 100% selectivity and a ratio of 0.89.

Similarly, On et al. [159] synthesized bifunctional titanium mesoporous molecular sieves containing B^{3+}, Al^{3+} or Fe^{3+} trivalent ions which were very active for the epoxidation of bulky olefins and highly selective in diol formation. Alkoxide sol-gel processes are used to prepare mesoporous titania-silica aerogels with highly dispersed titanium. Ti-MCM-41, however, possesses long range order and requires surfactant molecules in order to be synthesized.

Thus, in order to justify the use of the more expensive Ti-MCM-41 catalyst the aerogel and Ti-MCM-41 samples were compared catalytically [160-165]. Optimized TiO_2-SiO_2 aerogel [161] contains a high density of active Ti-O-Si centres and allows accessibility to bulky olefins and organic hydroperoxides. However, optimized Ti-MCM-41 which contains only 2 wt% TiO_2 (w.r.t. 20 wt% in the aerogel) exhibits a higher olefin epoxidation per Ti atom [166]. It is also important to note that H_2O_2 oxidant can only be used on the Ti-MCM-41 sample. So for specific reasons it is worthwhile to research further into the preparation of titania-silica aerogels by the sol-gel route as this synthesis procedure avoids the use of organic templates and larger amounts of titanium can be introduced into the sample.

Following on from the successful incorporation and subsequent application of Ti in MCM-41 Fe, Cr, V and Mn were also studied in the same manner [167]. X-ray absorption spectroscopy studies indicated that the stability of the heteroatoms to remain in the framework after calcination was Ti > Fe > V > Cr, where the chromium was present as a highly dispersed chromium oxide species. The nature and location of the active sites are of utmost importance when catalytic activity is considered.

Reddy et al. [168] observed that V-MCM-41 could be used for the hydroxylation of 2,6-di-tert-butylphenol (2,6-TBP) into 2,6-di-tert-butyl-1,4-benzoquinone with high conversion selectivity and efficient H_2O_2 utilization. Although V_2O_5-SiO_2 aerogel gave a good conversion the catalytic activity stopped due to the formation of V_2O_5 after the regeneration process which involved calcination This once again indicates that the possibility of regenerating the catalyst by a calcination process in the absence of air has to be looked into further.

An ingenious idea reported by Luan et al. in 1997 [169] was to load vanadium onto calcined Ti-MCM-41 or Zr-MCM-41 supports by incipient wetness impregnation of a dilute (0.001 - 0.017 M) aqueous vanadyl sulfate solution. ESR and diffuse reflectance UV spectroscopy indicate that the surface titanium or zirconium centers inmobilize these vanadium species and promote the oxidation of VO^{2+} to V^{5+},. thus indicating a strong interaction between V and surface Ti or Zr centers. Sn and Zr have also been introduced in the framework of MCM-41 and found to be active for formation of lactic acid polymers, and the oxidation of aniline, cyclohexene and norbornylene [170-173].

4.2. Ordered mesoporous materials as supports

The groups on the surface of these materials are readily functionalized and thus acid and base functions, as well as metals, oxides, and transition metal complexes can be dispersed within these high surface area materials.

4.2.1. Supporting Acids and Bases.

Reactions such as paraffin isomerization and isobutane/butene alkylation require strong acid sites. Mesoporous aluminosilicates, unfortunately, possess a mild acidity and to overcome this heteropolyacids, which are strong acids (HPA), were supported on pure silica MCM-41 [174]. This was successfully carried out by Kozheunikov et al., $H_3PW_{12}O_{40}$ retained its Keggin structure and even at HPA loadings as high as 50 wt% it was dispersed without the formation of HPA crystals. The BET surface area indicates that pore blocking occurs; hence a non-uniform superficial distribution of HPA is likely. The HPA supported MCM-41 was surprisingly more active than bulk HPA and H_2SO_4 for the liquid phase alkylation of 4-tert-butylphenol with isobutene [175,176].

Recently we have conducted some experiments [177], in which the results clearly indicate that a better dispersion of $H_3\,PW_{12}\,O_{40}$ is obtained when larger pore MCM-41 is used such that the catalytic alkylation of isobutene/butene is successful. This work aimed at solving two factors that need to be resolved in the future; to achieve a better dispersion of $H_6P_2\,W_{18}\,O_{62}$ and to prevent the leaching of HPA when non polar reactants are used. Sulfonic groups have been generated on the surface of MCM-41 by anchoring thiol containing molecules; followed by oxidation of the thiols to sulfonic groups by H_2O_2. The resultant materials show acidity and catalytic activity [178].

As was discussed previously, Cs_2O crystallites on the surface of MCM-41 represent systems with a strong basicity. Brunel et al. [179] anchored primary and secondary amines to the surface of pure silica MCM-41. These materials were weakly basic and were found to be active for carrying out the Knoevenagel condensation reactions, and also for the preparation of monoglycerides starting from 2,3-epoxy alcohols and fatty acids.

4.2.2. Supporting metals and oxides

Noble metal catalysts were obtained by their dispersion on MCM-41. Schüth et al. [180] have introduced Pt by incipient wetness, or by ion-exchange or directly via the synthesis of MCM-41, and duly compared each method. Up to 80% of incorporation was achieved via the synthesis of MCM-41 in the presence of Pt and highly dispersed platinum crystallites (of uniform 4.0 nm size) were obtained. Incipient wetness impregnation yielded crystallites of 2 and 20 nm. Although all catalysts were highly active for the oxidation of CO with air the samples of Pt clusters of 2nm prepared by incipient wetness impregnation were the most active. Ryoo et al. [181] ion exchanged Al-MCM-41 with $[[Pt\text{-}(NH_3)]^{2+}]$, as is conventionally done with Pt/zeolites. It was observed that employing lower calcination temperatures produced Pt crystallites of 10 nm for 2 wt% loadings of Pt. Highly dispersed Pt/MCM-41 and Pd/Al-MCM-41 catalysts have shown catalytic activity for the hydrogenation of benzene, naphthalene, phenantrene and olefins as well as for the hydrocracking of 1,3,5-triisopropylbenzene [93,182,183]. It is interesting to note that Pt clusters were also supported within the channels of FSM-16 [184] by reductive carbonylation of H_2PtCl_6 and modification with NEt_4Cl. Junges et al. [185] very recently showed that MCM-41 could be loaded with ligand stabilized palladium clusters in the size range of 3-4 nm via different pathways. It was concluded that clusters that were introduced by incipient wetness impregnation were thermally stable up to 513 K in air and showed the highest activity for CO oxidation. Similarly Pd/MCM-41 showed a high activity in the hydrogenation of 1-hexene [186], but did not show any activity in the hydrogenation of benzene. The average Pd particle size in this case was 2-2.5 nm. Inui et al. [187] recently used a variety of novel zeolite catalysts (ZSM-5, H-Fe-silicalite, Ni/SAPO-34) including Pt supported on MCM-41 for a variety of processes for the synthesis of fuels and for denox reactions. In particular the Pt/MCM-41 catalyst was seen to hydrogenate the gasoline fraction selectively into iso-paraffinic gasoline. Since noble metals are poisoned by sulfur containing compounds and diesel streams can contain up to 2000 ppm of sulfur (whereas 200 ppm are present in the diesel pool), the greatest challenge for Pt-supported catalysts in the deep hydrogenation of aromatics is its activity in the presence of sulfur compounds.

The type of support employed for the ethylene and propylene dimerization on nickel and chromium catalysts plays an important role. Ethylene dimerization was observed to take place at 70°C in Ni/MCM-41 and Ni/Al-MCM-41 catalysts [188]. Furthermore [189,190], NO_x catalysts containing Fe^{3+} and Pd on MCM-41 have allowed the control of NO_x, CO, and HC emissions generated by the oxygen-rich combustion process. In situ XANES experiments conducted by Jentys et al. [191] were used to study the changes in the oxidation state of Co, Pt and Rh containing MCM-41. It was observed that the presence of water vapour did not influence the oxidation/reduction of Rh/MCM-41 and Pt/MCM-41, while Co/MCM-41 was oxidised by the presence of H_2O.

4.2.3. Supporting Active species

The silanol groups of MCM-41 can be used to immobilize redox species. This can be done in two different ways; the first by grafting organometallic complexes on the inner walls of MCM-41 and the second by functionalization of silanols with silanes and subsequent anchoring of transition metal complexes to the surface. With respect to the first technique Maschmeyer et al. succeeded in grafting a titanocene complex to the walls of MCM-41. This resulted in a well dispersed, high concentration of Ti active species [192] after which X-ray absorption spectroscopy of the calcined sample indicated that the Ti was tetrahedrally coordinated.

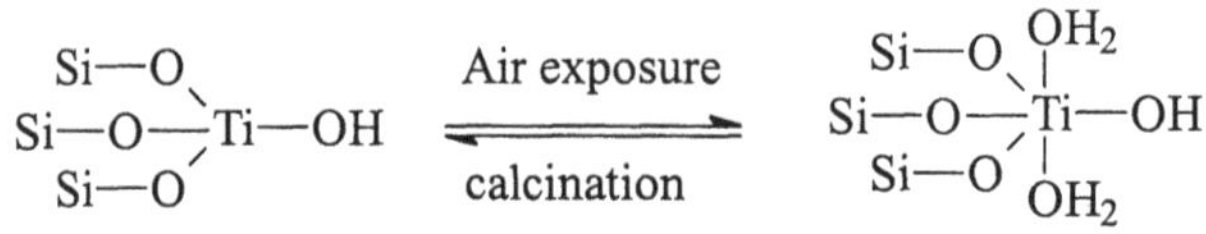

This material was active for the selective epoxidation of cyclohexene and pinene using TBHP as the oxidant, with activities similar to Ti-MCM-41.

Likewise, more recently racethenebis (indenyl) zirconium dichloride catalyst which was used for stereospecific propene polymerization was grafted onto MCM-41; as well as onto methylalumoxane modified MCM-41; where it was used for the production of highly isotactic polypropene [193,194]. Djajanti et al. compared the Mo loading of mesoporous MCM-41 via different methods, including grafting with the pentachloride and MOCV from the hexacarbonyl [195].

Morey et al. reported the immobilization of $O_{32}V{=}O$ centers on the walls of mesoporous materials, probably coordinated by Si-O-V bridges [196]. These transition-metal-ion-grafted mesoporous MCM-41 and MCM-48 materials were used for peroxidative halogenation reactions [197].

A series of non-aqueous solutions including $AlCl_3$, $Al(NO_3)_3$, $SnCl_2$, $Zn(O_2CMe)_2$ and $Mn(O_2CMe)_2$ were used to post-synthetically graft various metal elements onto the siliceous framework of MCM-41, MCM-48, SBA-1 (cubic Pm3n), KIT-1 (disordered) and MSU-1 (disordered). The structural integrity, acidity and catalytic activity was claimed to be superior to those of materials in which the metals were incorporated during the synthesis [198]. The trimethylsilylation of Ti-MCM-41 and Ti-MCM-48 resulted in an increase in their hydrophobicity and thus a remarkable enhancement in their catalytic activity for the oxidation of alkenes and alkanes with H_2O_2 [199].

Balkus et al. [200,201] functionalised the pore surfaces of MCM-41 with silanes to yield surface bound chelate ligands that include ethylenediamine (ED), diethylenetriamine (DET) and ethylenediamine-tetraacetic acid salts (EDTA). These anchored ligands were then used to covalently bind Co(II) complexes which have potential use as oxygen carriers and catalysts. Huber et al. succeeded in encapsulating a $M_3SnMo(CO)_3(\eta\text{-}C_5H_5)$ tin molybdenum complex into MCM-41 [202]. Recently, pentadentate ligands 3-[N,N-bis-3,3-(3,5-di-tert-butylsalicylidenamino) propyl]amine (t-salpr) which were covalently linked via organic moieties to the MCM-41 were further converted into Mn-III (salpr) complexes by ligand exchange. UV-vis diffuse reflectance spectroscopy indicated that the complex integrity before and after was retained. This work obviously opens up the possibility of carrying out enantioselective epoxidation of olefins.

Corma et al. very recently showed that Rh complexes with a chiral ligand that were covalently bound to a modified USY zeolite were able to achieve very high enantioselective excesses (>98%) during the hydrogenation of (Z)-N-acylaminocinnamate derivatives [203]. Also Frunza et al. reported the embedding of an enantioselective homogeneous chiral salen manganese (III) cationic complex into the pores of the MCM-41. Catalytic tests in olefinic epoxidation indicate the maintenance of the catalytic activity and stereoselectivity after embedding in MCM-41 [204]. Similarly, Corma et al. [205] have been able to prepare a Salen Mn complex with a chiral ligand via a "ship in a bottle" approach in a Y zeolite. By using NaOCl as a donor, this catalyst has been used to epoxidize enantioselectively a variety of olefins. When this complex was anchored on the walls of MCM-41 it also gave an enantiomeric excess for the epoxidation of olefins. Yamamoto et al. have also researched into the "ship in a bottle" synthesis of (Pt-15(Co) (30))(2-) encapsulated in the ordered hexagonal mesoporous channels of FSM-16 and its effective catalytic properties in the water-gas shift reaction [206]. Al-MCM-41-immobilized $[Mn(bpy)(2)]^{2+}$ exhibited a higher catalytic activity for styrene oxidation than the corresponding homogeneous catalyst and showed no significant loss of catalytic activity when recycled.

The encapsulation of ruthenium (II) meso-tetrakis (4-chlorophenyl) porphyrin in MCM-41 modified with 3-aminopropyltriethoxysilane yielded a stable catalyst for alkene oxidation by tert-butyl hydroperoxide [207]. Even though other metalloporphyrins have been attached to MCM-41 via ion exchange [208,209] care must be taken due to the relative ease with which leaching may occur under the catalytic conditions employed. Liu et al. have seen that iron (II)-8-quinolinol (I) is supported on MCM-41 successfully and is used to catalyze the hydroxylation of phenol. When compared to non-supported iron (II) it had a higher phenol conversion, better hydroquinone selectivity and better rate of utilization of H_2O_2 [210]. Recently, MCM-41 was used to entrap enzymes such as cytochrome C, papain and trypsin via physisorption. The entrapped trypsin enzyme, for example, was active for the hydrolysis of N-α-benzoyl-DL-arginine-4-nitroaniline [211].

In a similar manner to that discussed previously MnO was supported on high surface area zirconia by grafting from the alkoxide precursor. Although the specific activity of these catalysts and their selectivity towards acetone in isopropanol decomposition were very similar to that of bulk manganese dioxide, the grafted samples showed a high selectivity towards propene [212].

5. CONCLUSIONS

The preparation of these versatile ordered mesoporous materials, has opened new possibilities in most of the fields of catalysis. It is, however, remarkable that most of the work has been done using MCM-41, even though it has a unidirectional structure; such that the 3D MCM-48 has been much less studied. It is, therefore, of interest to be able to synthesize MCM-48 in a simple way, without using alcohols. Moreover, one should also look into the possibility of preparing MCM-48 without alkaline species and with different pore size diameters without using additional organics. This certainly will open up new possibilities for the catalytic use of MCM-48. Finally, a special effort is needed for achieving stability of MCM-41 in aqueous media, as well as under hydrothermal conditions.

Acknowledgments

Financial support by the Spanish MAT97-1016-C02-01 is gratefully acknowledged.

REFERENCES

1. W.M. Meier, D.H. Olson, *"Atlas of Zeolite Structure Types" 3rd Revised Edition*, (Butterworth-Heinemann Ed.), 1992.
2. S.T. Wilson, B.M. Lok, E.M. Flanigen, U.S. Patent 4 310 440 (1982).
3. R.M. Dessau, J.L. Schlenker, J.B. Aiggins, *Zeolites*, 10 (1990) 522.
4. J.W. Richardson Jr, E.T.C. Vogt, *Zeolites,* 12 (1992) 13.
5. M.E. Davies, C. Saldarriaga, C. Montes, J. Garcés, C. Crowder, *Nature* 331 (1988) 698.

6. M.E. Davies, C. Montes, J.M. Garcés, *ACS Symp. Ser.* 398 (1989) 291.
7. M. Esterman, L.B. McCuster, Ch. Baerlocher, A. Merrouche, H. Kessler, *Nature,* 352 (1991) 320.
8. A. Merrouche, J. Patarín, H. Kessler, M. Soulard, L. Delmotze, J.L. Guth, J.F.Jolly, *Zeolites,* 12 (1992) 226.
9. J.L. Guth, H. Kessler, P. Caullet, J. Hazm, A. Mewouche, J. Patarín, *Proc. IX th Int. Zeolite Conference* (R.von Ballmoos et al. Ed.), (1993) 215.
10. R.H. Jones, J.M. Thomas, J. Chen, R. Xu, Q. Huo, S. Li, Z. Ma, A.M. Chippindale, *J. Solid State Chem.*, 102 (1993) 5605.
11. Q. Huo, R. Xu, S. Li, Z. Ma, J.M. Thomas, R.H. Jones, A.M. Chippindale, *J.Chem. Soc., Chem. Commun.,* (1992) 875.
12. C.C. Freyhardt, M. Tsapatsis, R.F. Lobo, K.J. Balkus, Jr., M.E. Davies, *Nature*, 381 (1996) 295.
13. K.J. Balkus, Jr., A.G. Gabrielov, N. Sandler, *Mater. Res. Soc. Symp. Proc.,* 368 (1995) 369.
14. C.T. Kresge, M.E. Leonowicz, W.J. Roth, J.C. Vartulli, U.S. Patent 5 098 684 (1992).
15. C.T. Kresge, M.E. Leonowicz, W.J. Roth, J.C. Vartulli, J.S. Beck, *Nature,* 359 (1992) 710.
16. J.S. Beck, C.T. Chu, I.D. Johnson, C.T. Kresge, M.E. Leonowicz, W.J. Roth, J.C. Vartulli, U.S. Patent 5 108 725 (1992).
17. J.S. Beck, D.C. Calabro, S.B. McCullen, B.P. Pelrine, K.D. Schmitt, J.C. Vartulli, U.S. Patent 5 145 816 (1992).
18. J.S. Beck, C.T. Kresge, M.E. Leonowicz, W.J. Roth, J.C. Vartulli, U.S. Patent 5 264 203 (1993).
19. J.S. Beck, K.D. Smith, J.C. Vartulli, U.S. Patent 5 334 368 (1994); J.S. Beck, C.T. Kresge, S.B. McCullen, W.J. Roth, J.C. Vartulli, U.S. Patent 5 370 785 (1994).
20. J.S. Beck, J.C. Vartulli, W.J. Roth, M.E. Leonowicz, C.T. Kresge, K.D. Schmitt, C.T.-W. Chu, D.H. Olson, E.W. Sheppard, S.B. McCullen, J.B. Higgins, J.L. Schlenker, *J. Am. Chem. Soc.,* 114 (1992) 10834.
21. J.C. Vartulli, C.T. Kresge, M.E. Leonowicz, A.S. Chu, S.B. McCullen, I.D. Johnson, E.W. Sheppard, *Chem. Mater.* 6 (1994) 2070.
22. J.S. Beck, J.C. Vartulli, G.J. Kennedy, C.T. Kresge, W.J. Roth, S.E. Schramm, *Chem. Mater.* 6 (1994) 1816.
23. J.C. Vartulli, K.D. Schmitt, C.T. Kresge, W.J. Roth, M.E. Leonowicz, S.B. McCullen, S.D. Hellring, J.S. Beck, J.L. Schlenker, D.H. Olson, E.W. Sheppard, *Stud. Surf. Sci. Catal.,* 84 (1994) 53.
24. T. Yanagisawa, T. Shimizu, K. Kuroda, C. Kato, *Bull. Chem. Soc. Jpn.,* (1990) 63.
25. S. Inagaki, Y. Fukushima, A.O. Kada, T. Kuranchi, K. Kuroda, C. Kato, 9th *Inter. Zeol. Conf.* , Von Ballmoos et al., Eds., Butterworth-Heinemann, Boston (1993) p. 305.
26. S. 'O' Brien, R.J. Francis, J.S. Price, D. 'O' Hare, S.M. Clark, N. Okazaki, K. Kuroda, *J. Chem. Soc., Chem. Commun.,* (1995) 2423.
27. C.Y. Chen, S.Q. Xiao, M.E. Davies, *Microporous Mater.,* 4 (1995) 1.
28. A. Corma, D. Kumar, "New Trends in Materials Chemistry", R. Catlow and A. Cheetham, Eds., *NATO ASI Series,* Kluwer Academic Publishers, 498 (1997) 403.
29. G.D. Stucky, Q. Huo, A. Firouzi, B.F. Chmelka, S. Schacht, I.G. Voigt-Martin, F. Schüth, *Stud. Surf. Sci. Catal.,* 105 (1996) 3.

30. Q. Huo, D.I. Margolese, U. Ciesla, D.G. Demuth, P. Feng, T.E. Gier, P. Sieger, A. Firouzi, B.F. Chmelka, F. Schüth, G.D. Stucky, *Chem. Mater.*, 6 (1994) 1176.
31. Q. Huo, D.I. Margolese, U. Ciesla, P. Feng, T.E. Gier, P. Sieger, R. Leon, P.M. Petroff, F. Schüth, G.D. Stucky, *Nature*, 368 (1994) 317.
32. A. Sayari, P. Liu, *Microporous Mater.*, 12 (1997) 149.
33. A. Monnier, F. Schüth, Q. Huo, D. Kumar, D. Margolese, R.S. Maxwell, G.D. Stucky, M. Krishnamurty, P. Petroff, A. Firouzi, M. Janicke, B.F. Chmelka, *Science*, 261 (1993) 1299.
34. A. Firouzi, D. Kumar, L.M. Bull, T. Besier, P. Sieger, Q. Huo, S.A. Walker, J.A. Zasadzinski, C. Glinka, J. Nicol, D. Margolese, G.D. Stucky, B.F. Chmelka, *Science*, 267 (1995) 1139.
35. A. Firouzi, F. Atef, A.G. Oertli, G.D. Stucky, B. Chmelka, *J. Am. Chem. Soc.*, 119 (1997) 3596.
36. C. Glinka, J. Nicol, G.D. Stucky, E. Ramli, D. Margolese, Q. Huo, "Advances in Porous Materials"; *Mater. Res. Proc.*, S. Komarneni, D.M. Smith and J.S. Beck, Eds., Materials Research Society, Pittsburgh, P.A. 371 (1995) 47.
37. L.M. Bull, D. Kumar, S.P. Millar, T. Besier, M. Janicke, G.D. Stucky, B.F. Chmelka, *Stud. Surf. Sci. Catal.*, 84 (1994) 429.
38. P.T. Tanev, T.J. Pinnavaia, *Science*, 267 (1995) 865.
39. S.A. Bagshaw, S.A. Prouzet, J.J. Pinnavaia, *Science*, 269 (1995) 1242.
40. M.J. Meziani, J. Zajac, D.J. Jones, J. Roziere, S. Partyka, *Langmuir*, 13 (1997) 5409.
41. A. Corma, V. Fornés, M.T. Navarro, J. Pérez-Pariente, *J. Catal.*, 148 (1994) 569.
42. C.Y. Chen, S.L. Burkell, H.X. Li, M.E. Davies, *Microporous Mater.*, 2 (1993) 27.
43. K.R. Reddy, N. Araki, M. Niwa, *Chem. Lett.*, 7 (1997) 637.
44. A. Liepold, K. Roos, W. Reschetilowski, *Chem. Eng. Sci.*, 51 (1996) 3007.
45. A. Liepold, K. Roos, W. Reschetilowski, A.P. Esculcas, J. Rocha, A. Philippou, M.W. Anderson, *J. Chem. Soc., Faraday Trans.*, 92 (1996) 4623.
46. J. Weglarski, J. Datka, H.Y. He, J. Klinowski, *J. Chem. Soc., Faraday Trans.*, 92 (1996) 5161.
47. S. Viale, E. Garrone, F. Direnzo, B. Chiche, F. Fajula, *Stud. Surf. Sci. Catal.*, 105 (1997) 533.
48. F. Direnzo, B. Chiche, F. Fajula, S. Viale, E. Garrone, *Stud. Surf. Sci. Catal.*, 101 (1996) 851.
49. H. Landmesser, H. Kosslick, W. Storck, R. Fricke, *Solid State Ionics*, 101 (1997) 271
50. X.B. Feng, J.S. Lee, J.W. Lee, J.Y. Lee, D. Wei, G.L. Haller, *Chem. Eng. J.*, 64 (1996) 255.
51. K.M. Reddy, C. Song, *Catal. Lett.*, 36 (1996) 103.
52. Z. Luan, Ch.F. Cheng, W. Zhou, J. Klinowski, *J. Phys. Chem.*, 99 (1995) 1018.
53. R. Schmidt, D. Akporiaye, M. Stocker, O.H. Ellestad, *Stud. Surf. Sci. Catal.*, 84 (1994) 61.
54. M. Janicke, D. Kumar, G.D. Stucky, B.F. Chmelka, *Stud. Surf. Sci. Catal.*, 84 (1994) 243.
55. R. Mokaya, W. Jones, *Chem. Commun.*, 22 (1997) 2185.
56. K.R. Kloetstra, H. Van Bekkum, J.C. Jansen, *Chem. Commun.*, 23 (1997) 2281.

57. J. Medinavaltierra, M.A. Sanchez, J.A. Montoya, J. Navarrete, J.A. Delosreyes, *Appl. Catal. A., General,* 158 (1997) L1-L6.
58. J.B. Kim, T. Inui, *Catal. Lett.,* 36 (1996) 255.
59. H. Kosslick, G. Lischke, G. Walther, W. Storek, A. Martin, R. Fricke, *Microporous Mater.,* 9 (1997) 13.
60. H. Kosslick, H. Landmesser, R. Fricke, *J. Chem. Soc., Faraday Trans.,* 93 (1997) 1849.
61. C.F. Cheng, M.D. Alba, J. Klinowski, *Chem. Phys. Lett.,* 250 (1996) 328.
62. C.F. Cheng, H.Y. He, W.Z. Zhou, J. Klinowski, J.A.S. Goncalves, L.F. Gladden, *J. Phys. Chem.,* 100 (1996) 390.
63. C.F. Cheng, J. Klinowski, *J. Chem. Soc., Faraday Trans.,* 92 (1996) 289.
64. B.A. Aufdembrink, A.W. Chester, J.A. Herbst, C.T. Kresge, U.S. Patent 5 258 114 (1993).
65. J. Aguado, D.P. Serrano, M.D. Romero, J.M. Escola, *J. Chem. Soc. Chem. Commun.* (1996) 725.
66. A. Corma, M.S. Grande, V. Gonzalez-Alfaro, A.V. Orchilles, *J. Catal.* 159 (1996) 375.
67. W.A. Wachter, U.S. Patent 5 221 648 (1993).
68. P.H. Shipper, H. Owen, J.A. Herbst, G.W. Kirker, A. Huss Jr, P. Chu, U.S. Patent 5 179 054 (1993).
69. X.Y. Chen, L.M. Huang, G.Z. Ding, Q.Z. Li, *Catal. Lett.,* 44 (1997) 123.
70. K. Roos, A. Liepold, W. Roschetilowski, R. Schmidt, A. Karlsson, M. Stocker, *Stud. Surf. Sci. Catal.,* 84 (1994) 389.
71. A. Corma, *Chem. Rev.,* 97(6) (1997) 2373.
72. R.B. Borade, A. Clearfield, *Synthesis of Porous Materials,* (1997) 431-449.
73. Q.N. Le, R.T. Thomson, U.S. Patent 5 232 580 (1993).
74. R. Ryoo, S. Jun, *J. Phys. Chem. B.,* 101 (1997) 317.
75. S.B. Mc Cullen, J.C. Vartuli, U.S Patent 5 156 829 (1992).
76. N. Constel, F. Di Renzo, F. Fajula *J. Chem. Soc. Chem. Commun.* (1994) 967.
77. C.Y. Chen, S.Q. Xiao, U. Dahmen, *Book of Abstract 11th Int. Zeolite Conf. Seul RP63,* (1996); R. Ryoo, S. Jun, *J. Phys. Chem. B,* 101 (1997) 317.
78. C.F. Cheng, D.H. Park, J. Klinowski, *J. Chem. Soc., Faraday Trans.,* 93 (1997) 193.
79. F. Direnzo, N. Coustel, M. Mendiboure, H. Cambon, F. Fajula, *Stud. Surf. Sci. Catal.,* 105 (1997) 69.
80. C.F. Cheng, W.Z. Zhou, J. Klinowski, *Chem. Phys. Lett.,* 263 (1996) 247.
81. A.R. Songip, T. Masuda, H. Kuwahara, K. Hashimoto, *Energy Fuels,* 8 (1994) 136.
82. P.J. Nat, E.T.C. Wogt, WO 94/26847 (1994).
83. A. Corma, A. Martínez, V. Martínez-Soria, J.B. Montón, *J. Catal.,* 153 (1995) 25.
84. I.E. Maxwell, *Catal. Today,* 1 (1987) 385.
85. J. Cui, Y.H. Yue, Y. Sun, W.Y. Dong, Z. Gao, *Stud. Surf. Sci. Catal.,* 105 (1997) 687.
86. Y.H. Yue, Y. Sun, Z. Gao, *Catal. Lett.,* 47 (1997) 167.
87. C.S. Song, K.M. Reddy, *Abstracts of Papers of the Am. Chem. Soc.,* 212 (1996) 19.
88. K.M. Reddy, B.L. Wei, C.S. Song, *Abstracts of Papers of the Am. Chem. Soc.,* 213 (1997) 73.
89. K.J. Del Rossi, G.H. Hatzikos, A. Huss, Jr., U.S. Patent 5 256 277 (1993).
90. M.J. Girgis, Y.P. Tsao, *Ind. Eng. Chem. Res.,* 35 (1996) 386.

91. D.O. Marler, D.N. Mazzone, U.S. Patent 5 288 395 (1994).
92. A. Corma, A. Martínez, V. Martinez-Soria, *J. Catal.*, 169 (1997) 480.
93. J.N. Armor, *Appl. Catal.*, 112 (1994) N21.
94. J. Heinerman, E. Vogt, *PCT, Int. Pat. Appl.* WO 94/26846 (1994).
95. D. Tas, R.F. Parton, K. Vercruysse, P.A. Jacobs, *Stud. Surf. Sci. Catal.*, 105 (1997) 1261.
96. Q.N. Le, R.T. Thomson, G.H. Yokomizo, U:S. Patent 5 134 241 (1992).
97. B.P. Pelrine, K.D. Schmidt, J.C. Vartuli, U.S. Patent 5 105 051 (1992).
98. B.P. Pelrine, K.D. Schmidt, J.C. Vartuli, U.S. Patent 5 270 273 (1993).
99. J.B. Kim, T. Imui, *Catal. Lett.* 36 (1996) 255.
100. M. Hartmann, A. Pöppl, L. Kevan, *Stud. Surf. Sci. Catal., 101* (1996) 801.
101. G. Bellusi, C. Perego, A. Caratti, S. Peretello, P.E. Massara, E. Perego, *Stud. Surf. Sci. Catal.*, 85 (1994) 84.
102. A. Barnier, L.F. Admala, P. Grange, *Appl. Catal.*, 77 (1991) 269.
103. M.J. Climent, A. Corma, S. Iborra, M.C. Navarro, J. Primo, *J. Catal.*, 161 (1996) 786.
104. L.X. Dai, R. Hayasaka, Y. Iwaki, T. Tatsumi, *Sogo Shikensho Nenpo (Tokyo Daigaku, Kogakubu),* 55 (1996) 159.
105. A.Th.J.W. De Goede, I.G. Van der Leij, A.M. Van der Heijden, F. Van Rantwijk, H. Van Bekkum, EP 95 201 325 (1995).
106. A. Corma, R. Guil, S. Iborra, J. Primo, *J. Catal.*, (1998) in press.
107. E. Armengol, M.L. Cano, A. Corma, H. García, M.T. Navarro, *J. Chem. Soc. Chem. Commun.*, (1995) 519.
108. E. Armengol, A. Corma, H. García, J. Primo, *Appl. Catal. A. General,* 126 (1995) 391.
109. E. Armengol, A. Corma, H. García, J. Primo, *Appl. Catal. A. General* 149 (1997) 411.
110. Q.N. Le, U.S. Patent 5 191 134 (1993).
111. E.A. Gunnewegh, S.S. Gopie, H. Van Bekkum, *J. Mol. Catal. A. : Chemical* 106 (1996) 151.
112. H. Van Bekkum, A.J. Hoefnagel, M.A. Van Koten, E.A. Gunnewegh, A.H.G. Vogt, H.W. Kouwenhoven, *Stud. Surf. Sci. Catal.* 83 (1994) 379.
113. K.R. Kloestra, H. Van Bekkum, *J. Chem. Res. (S)* 1 (1995) 26.
114. M. Spagnol, L. Gilbert, D. Alby, "The Roots of Organic Development", J.R. Desmurs, S. Ratton, Eds., Elsevier: New York (1996) 29.
115. E.F. Godefroi, J. Meers, U.S. Patent 3 575 999 (1971).
116. K. Bauer, D. Garbe, H. Surburg, *Common Fragances and Flavours Materials,* 2nd Ed., VCH, Weinheim (1990).
117. A. Aucejo, M.C. Burguet, A. Corma, V. Fornés, *Appl. Catal.* 22 (1986) 187.
118. T. Curtin, B.K. Hodnett, *Stud. Surf. Sci. Catal.* 78 (1993) 535.
119. H. Sato, K. Hirose, *Chem. Lett.* (1993) 1765.
120. P.S.Singh, R. Bandyopadhyay, S.G. Hegole, B.S. Rao, *Appl. Catal. A.* 136 (1996) 249.
121. N. Katada, T. Tsubouchi, M. Niwa, Y. Murakami, *Appl. Catat. A.* 124 (1995) 1.
122. V. Semmer, H. García, A. Corma, *to be published.*
123. K. Lourvanij, G.L. Rorrer, *J. Chem. Technol. Biotechnol.*, 69 (1997) 35.
124. A. Corma, V. Fornés, R.M. Martín-Aranda, H. García, J. Primo, *Appl. Catal.*, 59 (1990) 237.
125. D. Barthomeuf, *Catal. Rev. Sci. Eng.*, 38 (1996) 521.

126. K.R. Kloetstra, H. Van Bekkum, *J. Chem. Soc., Chem. Commun.*, (1995) 1005.
127. K.R. Kloetstra, H. Van Bekkum, *Stud. Surf. Sci. Catal.*, 105 (1997) 431.
128. K.R. Kloetstra, M. Van Laren, H. Van Bekkum, *J. Chem. Soc., Faraday Trans.* 93 (1997) 1211.
129. K.R. Kloetstra, H. Van Bekkum, *Catal. Lett.*, 47 (1997) 235.
130. A. Corma, M.T. Navarro, J. Pérez-Pariente, F. Sánchez, *Stud. Surf. Sci. Catal.*, 84 (1994) 69.
131. M.C. Clerici, *Appl. Catal.* 68 (1991) 249.
132. G. Bellusi, A. Carati, M.G. Clerici, G. Meddinelli, R. Millini, *J. Catal.* 133 (1992) 220.
133. D.R.D. Huybrechts, L. De Bruycher, P.A. Jacobs, *Nature*, 345 (1990) 240.
134. M.A. Camblor, A. Corma, A. Martínez, J. Pérez-Pariente, *J. Chem. Soc. Chem. Commun.* (1992) 8.
135. A. Corma, P. Esteve, A. Martínez, S. Valencia, *J. Catal.* 152 (1995) 18.
136. T. Blasco, M.A. Camblor, A. Corma, P. Esteve, A. Martínez, C. Prieto, S. Valencia, *J. Chem. Soc. Chem. Commun,* (1996) 2367.
137. A. Corma, M.T. Navarro, J. Pérez-Pariente, *J. Chem. Soc., Chem. Commun.* (1994) 147.
138. P.T. Tanev, M. Chibwe, T.J. Pinnavaia, N*ature,* 368 (1994) 321.
139. W. Zhang, J. Wang, P.T. Tanev, T.J. Pinnavaia, *J. Chem. Soc., Chem. Commun.*, (1996) 979.
140. A. Sayari, V.R. Karra, J.S. Reddy, Moudrakovski, *Mat. Res. Soc. Symp. Proc.* 371 (1995) 81.
141. K. Kulawik, G. Schulz-Ekkof, J. Rathousky, A. Zukal, *J. Had. Collec. Czech. Chem. Commun.* 60 (1995) 451.
142. S.Gontier, A. Tuel, *Stud. Surf. Sci. and Catal.* 97 (1995) 157.
143. W.H. Zhang, M. Froba, J.L. Wang, P.T. Tanev, J. Wong, T.J. Pinnavaia, *J. Am. Chem. Soc.*, 118 (1996) 9164.
144. C.H. Rhee, J.S. Lee, *Catalysis Today,* 38 (1997) 213.
145. M.D. Alba, Z.H. Luan, J. Klinowski, *J. Phys. Chem.*, 100 (1996) 2178.
146. M. Kruk, M. Jaroniec, A. Sayari, *Microporous Materials,* 9 (1997) 173.
147. M.A. Camblor, A. Corma, P. Esteve, A. Martínez, S. Valencia, *Chem. Commun.*, (1997) 795.
148. K.A. Koyano, T. Tatsumi, *Microporous Materials,* 10 (1997) 259.
149. W. Zhang, T.J. Pinnavaia, *Catal. Lett.*, 38 (1996) 261.
150. M. Morey, A. Davidson, G. Stucky, *Microporous Materials,* 6 (1996) 99.
151. S.G. Zhang, Y. Fujii, K. Yamashita, K. Koyano, T. Tatsumi, M. Anpo, *Chem. Lett.* (1997) 659.
152. S. Tonti, P. Roffia, A. Cesana, M. Mategazza, M. Padovan, Eur. Patent 314,147 (1988).
153. S. Sakane, T. Tsubakino, Y. Nishiyama, Y. Ishii, *J. Org. Chem.*, 58 (1993) 3633.
154. J.S. Reddy, A. Sayari, *Appl. Catal. A, General,* 97 (1995) 157.
155. S. Gontier, A. Tuel, *J. Catal.*, 157 (1995) 124.
156. A. Corma, M. Iglesias, F. Sánchez, *Catal. Lett.*, 39 (1996) 153.
157. I. Vankelecom, K. Vercruysse, N. Moens, R. Parton, J.S. Reddy, P. Jacobs, *Chem. Commun.*, (1997) 137.
158. A. Corma, M. Iglesias, F. Sánchez, *J. Chem. Soc., Chem. Commun.*, (1995) 1635.
159. D.T. On, M.P. Kapoor, P.N. Joshi, L. Bonneviot, S. Kaliaguine, *Catal. Lett.*, 44 (1997) 171.

160. D.C.M. Dutoit, M. Schneider, A. Baiker, *J. Catal.* 153 (1995) 165.
161. R. Hutter, T. Mallat, A. Baiker, *J. Catal.* 153 (1995) 177.
162. R. Hutter, D.C.M. Dutoit, T. Mallat, M. Schneider, A. Baiker, *J. Chem. Soc. Chem. Commun.* (1995) 163.
163. S. Klein, S. Thorimbert, W.F. Maier, *J. Catal.* 163 (1996) 476.
164. Z. Liu, G.M. Crumbaugh, R.J. Davis, *J. Catal.* 159 (1996) 83.
165. S. Klein, J.A. Martens, R. Parton, K. Vercruyse, P.A. Jacobs, W.F. Maier, *Catal. Lett.* 38 (1996) 209.
166. R. Hutler, T. Mallat, D. Dutoit, A. Baiker, *Topics in Catalysis*, 3 (1996) 421.
167. F. Rey, G. Sankar, T. Maschmeyer, J.M. Thomas, R.G. Bell, G.N. Greaves, *Topics in Catalysis,* 3 (1996) 121.
168. J.S. Reddy, A. Sayari, *J. Chem. Soc., Chem. Commun,* (1995) 2231.
169. Z.H. Luan, L. Kevan, *J. Phys. Chem. B,* 101 (1997) 2020.
170. T.M. Abdel-Fattah, T.J. Pinnavaia, *J. Chem. Soc., Chem. Commun.,* (1996) 665.
171. A. Tuel, S. Gontier, *J. Chem. Soc., Chem. Commun.,* (1996) 651.
172. S. Gontier, A. Tuel, *Appl. Catal. A,* 143 (1996) 125.
173. B. Notari, *Catal. Today.,* 18 (1993) 163.
174. Ch.T. Kresge, D.O. Marler, G.S. Rav, B.H. Rose, U.S. Patent 5 366 945 (1994).
175. I.V. Kozheunikov, A. Sinnema, R.J.J. Jansen, K. Pamin, H. Van Bekkum, *Catal. Lett.* 30 (1995) 241.
176. I.V. Kozhevnikov, K.R. Kloetstra, A. Sinnema, H.W. Zandbergen, H. Van Bekkum, *J. Mol. Catal. A.* 114 (1996) 287.
177. A. Corma, J.L. Jorda, A. Martínez, F. Rey, *to be published.*
178. W.M. Van Rhijn, D.E. De Vos, B.F. Sels, W.D. Bossaert, P.A. Jacobs, *Chem. Commun.,* (1998) 317.
179. A. Cauvel, G. Renard, D. Brunel, *J. Org. Chem.,* 62 (1997) 749.
180. U. Junges, W. Jacobs, I. Voigt-Martín, B. Krutzsch, F. Schüth, *J. Chem. Soc. Chem. Commun.* (1995) 2283.
181. R. Ryoo. Ch.H. Ko, J.M. Kim, R. Howe, *Catal. Lett.* 37 (1996) 29.
182. K.M. Reddy, C. Song, *Catal. Today,* 31 (1996) 137.
183. T. Inui, J.B. Kim, M. Seno, *Catal. Lett.,* 29 (1994) 271.
184. T. Shido, T. Yamamoto, M. Harada, M. Ichikawa, *J. Physique IV.,* 7 (1997) 855.
185. U. Junges, F. Schüth, G. Schmid, Y. Uchida, R. Schlogl, *Berichte Der Bunsen Gesellschaft Fur Physikalische Chemie,* 101 (1997) 1631.
186. C.A. Koh, R. Nooney, S. Tahir, *Catal. Lett.,* 47 (1997) 199.
187. T. Inui, *Stud. Surf. Sci. Catal.,* 105 (1997) 1441.
188. N.A. Bhore, I.D. Johnson, K.M. Keville, Q.N. Le, G.H. Yokomizo, U.S. Patent 5,260,501 (1993).
189. P.P. Paul, M.A. Miller, M.J. Heimrich, S.T. Schwab, *Mater. Res. Soc. Symp. Proc. 431 Microporous and Macroporous Materials,* (1996) 117.
190. R.T. Yang, T.J. Pinnavaia, W.B. Li, W.Z. Zhang, *J. Catal.,* 172 (1997) 488.
191. A. Jentys, W. Schiesser, H. Vinek, *Catal. Lett.,* 47 (1997) 193.
192. T. Maschmeyer, F. Rey, G. Sankar, J.M. Thomas, *Nature,* 378 (1995) 159.
193. J. Tudor, D. Ohare, *Chem. Commun.,* (1997) 603.
194. S. O'Brien, J. Tudor, T. Maschmeyer, D. Ohare, *Chem. Commun.,* (1997) 1905.
195. S.D. Djajanti, R.F. Howe, *Stud. Surf. Sci. Catal.,* 105 (1997) 2067.

196. M. Morey, A. Davidson, H. Eckert, G. Stucky, *Chem. Mater.*, 8 (1996) 486.
197. J.V. Walker, M. Morey, H. Carlsson, A. Davidson, G.D. Stucky, A. Buttler, *J. Am. Chem. Soc.*, 119 (1997) 6921.
198. R. Ryoo, S. Jun, J.M. Kim, M.J. Kim, *Chem. Commun.*, 21 (1997) 2225.
199. T. Tatsumi, K.A. Koyano, N. Igarashi, *Chem. Commun.*, (1998) 325.
200. J.F. Diaz, F. Bedioni, E. Briot, J. Devynck, K.J. Balkus Jr, *Mater. Res. Soc. Symp. Proc.* 431 (1996) 89.
201. J.F. Diaz, K.J. Balkus Jr, F. Bedioni, V. Kurshev, L. Kevan, *Chem. Mater.* 9 (1997) 61.
202. C. Huber, K. Moller, T. Bein, *J. Chem. Soc., Chem. Commun.*, (1994) 2619.
203. A. Corma, M. Iglesias, C. Del Pino, F. Sánchez, *J. Chem. Soc., Chem. Commun.*, (1991) 1253.
204. L. Frunza, H. Kosslick, H. Landmesser, E. Hoft, R. Fricke, *J. Mol. Catal. A, Chemical,* 123 (1997) 179.
205. M.J. Sabater, H. García, A. Corma, *J. Chem. Soc., Chem. Commun.*, (1997) in press.
206. T. Yamamoto, T. Shido, S. Inagaki, Y. Fukushima, M. Ichikawa, *J. Am. Chem. Soc.*, 118 (1996) 5810.
207. C.J. Liu, S.G. Li, W.Q. Pang, C.M. Che, *Chem. Commun.*, (1997) 65.
208. L. Graham, R.N. Austin, J. Pham, M. Piech, V. Schunemann, A.X. Trautwein, I. Rietjens, M. Boersma, C. Veeger, D. Mandon, R. Weiss, *Abstracts of the Papers of the Am. Chem. Soc.*, 213 (1997) 154.
209. M. Chibwe, A. Barodawalla, T.J. Pinnavaia, *14th North American Catalysis Meeting of the Catalysis Society* Snowbird, UT, June 1995 paper PB107.
210. C. Liu, Y. Shan, X. Yang, X. Ye, Y. Wu, *Peop. Rep. China J. Catal.*, 168 (1997) 35.
211. K.J. Balkus, J.F. Diaz, *Abstracts of papers of the American Chemical Society* 212 (1996) 231.
212. J. Ma, G.K. Chuah, S. Jaenicke, R. Gopalakrishnan, K.L. Tan, *Berichte Der Bunsen-Gesellschaft-Physical Chemistry Physics*, 100 (1996) 585.

MESOPOROUS MOLECULAR SIEVES 1998
Studies in Surface Science and Catalysis, Vol. 117
L. Bonneviot, F. Béland, C. Danumah, S. Giasson and S. Kaliaguine (Editors)

Synthesis and Characterization of MCM-41with Different Pore Size and Si/Al-ratio

F. P. Matthae [a], D. Genske [a], Ch. Minchev [b] and H. Lechert [a]

[a] Institute of Physical Chemistry , University of Hamburg, Bundesstr. 45, 20146 Hamburg, Germany

[b] Institute of Organic Chemistry, Bulgarian Academy of Sciences, BAN, blok 9, 1113 Sofia, Bulgaria

Samples of MCM-41 have been synthesized using conventional batch compositions, varying the chain length of the surfactant, the temperature, and the content of the co-template to study the possibilities of a controlled influence on the crystallinity of the product and the pore radius distribution. Both parameters have been checked by X-ray and by sorption experiments. The pore radius varies about 0.2 nm for a change of the C-number of the surfactant by 2. The samples prepared at 363 K had a good crystallinity and the narrowest distribution of the pore radii.

An amount of at least about 0.15 mole co-template/SiO_2 was necessary to obtain an optimum crystallinity. Higher co-template contents had no influence.

Further the Si/Al-ratio has been studied at variation of the Si/Al-ratio in the batch. It can be seen that the ratios in the product are generally slightly lower. The distribution of the aluminium is mostly inhomogeneous as can be shown by EDAX. ^{27}Al-MAS-NMR measurements indicate that in the "as-synthesized" samples only tetrahedral Al is present. In the calcined samples a very low amount of octahedral Al can be detected. NH_3-TDS experiments show a broad distribution of acidities.

1. INTRODUCTION

Alumino-silicates of the MCM-41 type are interesting from several points of view. The original interest in developing these substances has been certainly a new cracking catalyst with large pores and a high Si/Al-ratio connected with centres with high acid strength. Precipitating the alumino-silicate in the presence of surfactant micelles with varying length hydrophobic chain, mesoporous substances with narrow pore size distributions and a wide range of pore diameters from about 1 to 10 nm can be obtained.

The available Si/Al-ratio is about 10 up to pure silica with the MCM-41 structure [1, 2, 3, 4]. The walls of the pores are nearly 1 nm thick and amorphous. The synthesis can be carried out in several ways in alkaline and also in weakly acidic solutions using different alumina- or silica-sources.

Depending on the surfactant/SiO_2-ratio in the batch, hexagonal MCM-41, cubic MCM-48, or lamellar MCM-50 can be obtained. Depending on the silica source a certain amount of short chain tetra-alkyl-ammonium-hydroxide (TMAOH, TEAOH or TPAOH) is necessary for the formation of these substances as so called "co-templates".

In the literature a large number of surfactants and their influence on the pore diameter have been studied [5]. Comparatively little is known, however, about the role of the tetra-alkyl-ammonium-ions and their influence on the pore structure. Further, the influence of other synthesis parameters, like e.g. the temperature, has not been studied systematically. The surfactant can be removed by calcination and the remaining alumino-silicate system is stable up to more than 1000 K. The hydrothermal stability is, however poor.

In the present paper, at first, experiments about the synthesis temperature and the quality of the crystallisation products shall be carried out. The quality is checked by X-ray diffraction and by an analysis of the pore size distribution. From the X-ray diagrams the lattice constant of the MCM-41 can be obtained. For an estimation of the pore diameter the thickness of the wall must be taken into account, which is nearly 1 nm.

The pore size distribution has been obtained from measurements of the adsorption-desorption isotherms [6, 7, 8]. The desorption isotherm is analysed automatically with an equation adapted to the Kelvin equation for the peculiarities of the sorption in straight pores taking into account a pore radius corrected by the thickness of the sorption layer. This method runs into problems for narrow pores with diameters below about 1 nm. These problems can be partly overcome by the use of argon as sorption gas.

By comparison of the lattice constants with the pore diameters from the sorption experiments the thickness of the walls of the MCM-41 can be estimated which turns out to be usually near 1 nm. For the low pore radii this value can be used to estimate the pore diameters from the X-ray data.

In the patents and the open literature, temperatures between about ambient temperature and about 500 K are reported. In a series of papers 423 K is recommended. A temperature below 373 K would have advantages for larger series of experiments in the laboratory and also for potential industrial application.

In a second series of experiments the quantity of the tetra-alkyl-ammonium-hydoxide has been varied. The hope of these experiments was to get some insight into the role of these substances for the formation of the MCM-41. Another aspect is, again, the potential industrial application, because these substances are among the most expensive in the synthesis procedure.

Finally, the influence of the batch parameters on the Si/Al ratio has been studied together with some problems of the distribution of the aluminium The Si/Al-ratio of the samples has been analysed by EDAX, and ICP-OES.

The distribution of the Al in the walls has been studied by EDAX-experiments and by MAS-NMR. For the characterization of the acidic properties the TDS-measurements of NH_3 have been carried out for selected samples.

2. EXPERIMENTAL

The batch compositions can be generally described by the formula [1-4].

$$n\ SiO_2 : Al_2O_3 : 1.8\ Na_2O : m\ (TMA)_2O : p\ (C_xTMA)_2O\ q\ (NH_4)_2O : r\ HX : z\ H_2O$$

For the temperature optimization, the parameters were set to the following values:
n = 73.4; m = 11.4; p = 6.5; q = 3.5; r = 13; z = 1420

The values for batch compositions of different Si/Al-ratios were: n = 146/73/25; m = 22/11/4; p = 13/6.5/2 q = 7.5/3.5/1.3; r = 26/13/4.5; z = 2850/1420/482

All experiments have been carried out with surfactants C_x-TMA-Br (e.g. hexadecyl-trimethyl-ammonium-bromide) with chain lengths x of 8,14 and 16.

At first a solution of Na-aluminate in twice the quantity of water has been prepared. This solution was mixed with a 25% wt. aqueous solution of the surfactant. Then the tetra-alkyl-ammonium-hydroxide was added and an amount of a 28% NH_3 aq.-solution, which has been proved to be useful for a partial neutralisation of the HBr which may be expected from the hydrolysis of the surfactant [9]. Mostly $(CH_3)_4NOH$ (TMAOH) was used as co-template. For the surfactant with the shortest chain length C_8 in the surfactant $(C_3H_7)_4NOH$ (TPAOH) proved to be favourable.

To this mixture Cab-osil M-5 (Fluka) was taken as the silica source. The resulting mixture was homogenised for 30 minutes on a shaking machine. Finally, the rest of the water was added. The gel was then placed for 72 hours in an oven at the desired temperature. This reaction time has proved to be optimal in [10]. Experiments with different reaction times are in agreement with these results.

The product was filtered and washed. Then it was dried for 24 hours at 363 K. Different Si/Al-ratios have been adjusted by varying the Na-aluminate content of the batch keeping the silicate/surfactant ratio constant.

The water and the templates can be removed by heating the "as- synthesized" sample for 2 hours at 493 K and by increasing the temperature to 853 K for 8-10 hours. The preparation of the acid form is usually carried out by removal of the Na-ions by triple ion exchange with 2M NH_4NO_3-solution and subsequent activation under vacuum in two steps increasing the temperature to 723 K within about 7 hours. The samples were first characterized by X-ray diffraction. From the shape and the intensity of the peak at $2\theta = 2.26$ deg the quality of the MCM-41 structure can be seen. A second kind of characterization was the inspection of the crystal shape and quality in the scanning microscope.

The determination of the Si/Al-ratio has been carried out with three different methods. To obtain reliable overall data XRF experiments and ICP-OES experiments have been done.

For the XRF-determination a SpectroX-Lab from Spectroanalytical Instruments has been used. The ICP-OES measurements were done with a Perkin-Elmer Plasma-II spectrometer.

In the EDAX measurements has been tried to get information about the distribution of the aluminium in different crystallites and by measurements at different points of the larger crystallites have been carried out. These experiments have been done with a Philips SEM 515/ D 806 V microscope. The pore size distribution has been analysed by measuring the adsorption- and the desorption-isotherm of nitrogen and analysing these curves with the Kelvin-equation used a system from Carlo Erba Instruments, Model Sorptomatic 1900. The distribution of the aluminium to octahedral and tetrahedral sites has been determined from ^{27}Al-MAS-spectra using a BRUKER BKR 500 spectrometer.

3. RESULTS AND DISCUSSION

In Figure 1, the pore size distributions for the standard batch with different surfactants are demonstrated for a reaction temperature of 363 K. For C_8TMACl, TPAOH has proved to be necessary as accompanying template. It can be seen that an increase of the chain length of the surfactant of 2 methylene groups increases with some precision the pore radius to about 0.2 nm. The width of the pore size distribution remains nearly constant for the different surfactants.

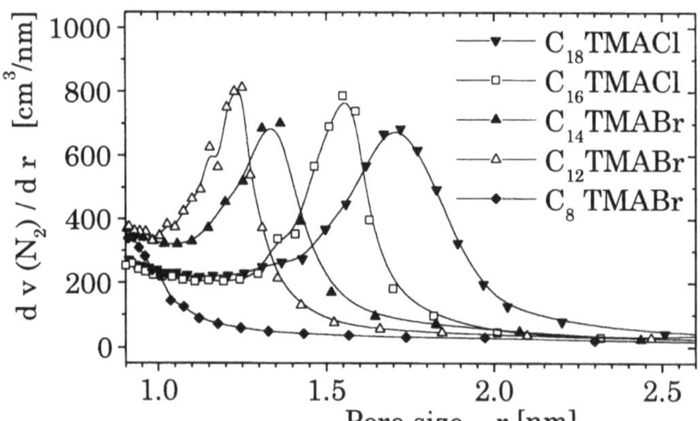

Figure 1. Pore size distribution of MCM-41, synthesized with different surfactants

In Fig. 2a and 2b the pore size distribution for different crystallization temperatures can be seen. In these experiments C_{16}TMACl has been used as surfactant. The diagrams show that the maxima of the distribution curves depend on the temperature rather sensitively.

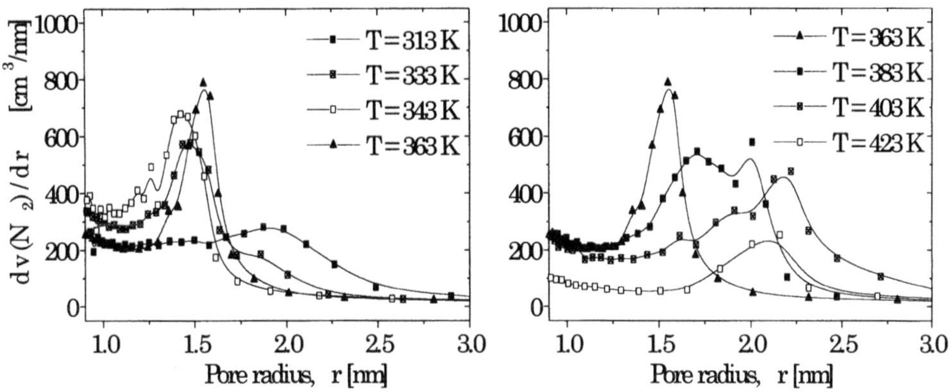

Figure 2. Pore size distribution of MCM-41 at different temperatures

At 313 K this distribution is rather broad with a maximum at about 2 nm. With increasing temperature the distribution narrows. For 363 K the narrowest distribution is obtained with a maximum at 1.5 nm. Increasing the temperature (Fig. 2b) an increase of the most frequent pore radius and a broadening of its distribution can be observed. At 423 K an apparent decrease of the pore volume can be found which is due to a strong broadening of the pore size distribution and no more detectable by our evaluation procedure. From these experiments can be seen that 363 K is the best temperature for a homogeneous pore size distribution and that the temperature offers some possibilities of influencing the characteristics of this distribution.

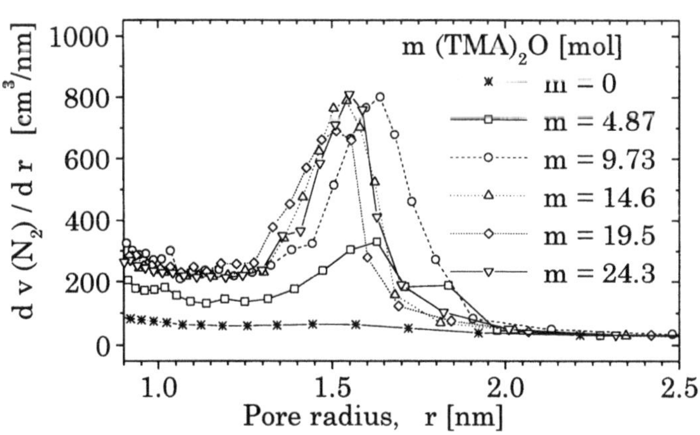

Figure 3. Pore size distribution of MCM-41 with various amounts of co-template TMAOH

Fig. 3 shows some experiments with different amounts of TMAOH in the batch. It can be seen that without any TMAOH no pore structure can be detected and the X-ray pattern shows only an amorphous product.

For $m = 4.967$ TMAOH in the batch the pore structure begins to develop and reaches $m = 9.735$ distribution with a maximum near 1.7 nm. At higher TMAOH contents this distribution does not change significantly.

The X-ray diagrams show similar results. Obviously, a minimum amount of the co-template is necessary for the formation of the porous structure. The influences of a higher content are, however, negligible. The role of the organic cations is not obvious. As is also seen for the syntheses of high silica zeolites, the kinetic control of the formation of a special structure should occur at the surface of the growing crystals. As a rough model may be imagined that the condensation of the silica species is retarded by the co-templates by blocking a part of the SiOH groups at the growing surfaces specifically. In this way the formation of the structure, e.g. by an agglomeration of the micelles may have enough time in comparison with a uncontrolled condensation of the alumino-silicate.

In Table 1 series of experiments with different batch compositions and Si/Al-ratios of about 12, 36, 73 and 115 is summarised.

Table 1.
Physico-chemical characteristics of MCM-41 with different Si/Al-ratios

sample	aluminium source	Si [mg] / 20 mg sample		Al [mg] / 20 mg sample		molar Si/Al-ratios	
		calc.	exp. [a]	calc.	exp. [a]	gel	ICP-OES
MCM-41-12	$NaAlO_2$	8.6	8.1	0.7	0.5	12	16.3
MCM-41-12	$Al_2(SO_4)_3$	32.1	29.7	2.5	3.4	12	8.3
MCM-41-37	$NaAlO_2$	9.1	8.4	0.2	0.3	37	26.7
MCM-41-36	$Al_2(SO_4)_3$	40.9	27.0	1.1	1.2	36	21.0
MCM-41-73	$NaAlO_2$	11.8	9.0	0.2	0.1	73	68.9
MCM-41-58	$Al_2(SO_4)_3$	10.7	6.3	0.2	0.2	58	33.0
MCM-41-115	$Al_2(SO_4)_3$	9.3	7.5	0.1	0.1	115	73.7

[a] determined by ICP-OES-measurements

The ratios obtained in the final products have been analysed by EDAX and by ICP-OES. Experiments with XRF showed that values agreeing with the ICP-OES data can be obtained only for Si/Al = 12. At higher Si/Al-ratios the separation of the Si and Al peaks is no more possible with sufficient accuracy. Generally, the values of the ICP-OES analysis in the products are lower than those in the batch. This is in agreement with the observations in [11].

From these experiments can be seen that the aluminate offered in the batch is almost completely incorporated in the MCM-41. Some of the silicate remains, however, in the batch.

It can be seen that the results of the EDAX and the ICP-OES method are sometimes differing. In the EDAX measurements deviations of the Si/Al-ratio of zeolites grown under similar conditions show that even lower values must be expected.

To homogenize the distribution of the aluminium some experiments have been carried out under stirring. It can be seen that the homogeneity can be slightly improved. Stirring leads, however, to a distinctly wider pore size distribution. For further characterization some samples have been analyzed by ^{27}Al-MAS NMR. These experiments show that the Al in the "as-synthesized" samples is present only in tetrahedral positions. After calcination a few octahedral Al can be detected.

As a final characterization TDS-experiments have been carried out with the ammonium form of selected samples and with samples of the H-form on which ammonia was adsorbed. Both experiments show a release of ammonia over a wide temperature range, indicating centres in a broad range of acidity. A more detailed discussion shall be given in a separate paper.

4. CONCLUSIONS

Some features for a controlled synthesis of MCM-41 materials have been worked out. Following convential batch compositions using tetra-alkyl-ammonium salts as co-templates beside the surfactant, it can be seen that the pore diameter can be adjusted by the chain length of the surfactant with good precision. Asking for narrow distributions of the pore diameters 363 K proved to be an optimal synthesis temperature. At lower and higher temperatures the pore diameters increase and the distribution curves broaden. From 365 K to 400 K an increase to an about 50% higher value can be observed for the most frequent pore diameter showing that this parameter is rather sensitive against the temperature of the synthesis.

For the synthesis a minimum of co-template/ Si = 0.15 is necessary. Higher concentrations of co-template do not improve the formation of the MCM-41 structure.

The Si/Al-ratio of the products is usually lower than in the batch. It could be shown that in the synthesis the silicate is not incorporated completely.

Broad distributions of acidities could be observed in NH_3 TDS experiments.

ACKNOWLEDGEMENTS

We thank the "Deutsche Forschungsgemeinschaft" for financial support of our work.

REFERENCES

1. C.T. Kresge, M.E. Leonowicz, W.J. Roth, J.C. Vartuli and J.S.Beck, Nature, 359 (1992) 710.
2. J.C. Vartuli, W.J. Roth, M.E. Leonowicz, C.T. Kresge, J.S.Beck, K.T. Schmitt, C.T.-W. Chu, D.H. Olson, E.W.Sheppard, Mc Cullen, J.B. Higgins and J.L. Schlenker, J.Amer. Chem Soc. 114, (1994) 10834.
3. J.C. Vartuli, K.T. Schmitt, C.T. Kresge, W.J. Roth, M.E. Leonowicz, Mc Cullen, S.D. Hellring, J.S.Beck, J.L. Schlenker, D.H. Olson and E.W. Schlenker, Stud. Surf. Sci.Catal., 84 (1994) 53.
4. J.C. Vartuli, K.T. Schmitt, C.T. Kresge, W.J. Roth, M.E. Leonowicz, S.B. Mc Cullen, S.D. Hellring, J.S.Beck, J.L. Schlenker, D.H. Olson and E.W. Schlenker, Chem. Mater., 6 (1994) 2317.
5. G.D. Stucky, Q. Huo, A. Firouzi, B.F. Chmelka, S.Schacht, I.G. Voigt-Martin and F. Schüth, Stud. In Surf. Sci.and Catal., 155A (1996) 3.
6. S.J. Gregg and K.S.W Sing, "Adsorption, Surface Area and Porosity", Acad.Press (1991).
7. Pure Appl. Chem. 57 (1985) 603.
8. Pure Appl. Chem. 31 (1985) 578.
9. C.Y. Chen, H.X. Li and M.E. Davis, Microporous Mater 2. (1993) 17.
10. J. Rathousky, A. Zukal, O. Franke and G. Schulz-Ekloff, J. Chem. Soc. Faraday Trans., 90 (1994) 2821.
11. K.M. Reddy and C. Song, Catal. Lett., 36 (1996) 103.

MESOPOROUS MOLECULAR SIEVES 1998
Studies in Surface Science and Catalysis, Vol. 117
L. Bonneviot, F. Béland, C. Danumah, S. Giasson and S. Kaliaguine (Editors)

Improved Synthesis of MSU-X Silica Prepared by Polyethylene Oxide Surfactants

F. Cot[a], P. J. Kooyman[b], A. Larbot[a] and Eric Prouzet[a]

[a]Laboratoire des Matériaux et Procédés Membranaires, CNRS UMR 5635,E.N.S.C.M, 8 rue de l'Ecole Normale, F-34296 Montpellier Cedex 5 (FRANCE) (e-mail:prouzet@cit.enscm.fr)

[b]National Centre for High Resolution Electron Microscopy, Rotterdamseweg 137, 2628 AL Delft (The Netherlands)

Since the discover of the Mobil M41S family of mesoporous molecular sieves [1, 2], the range of potential materials for heterogenous catalysis, supramolecular assembly and filtration processes has greatly expanded. However, up to now, trials for using M41S materials for filtration processes have led either to layers with porosity parallel to the substrat, or to layers only coated on non-porous substrates [3-6].

Among new Micellar Templated Structures (MTS), less ordered structures with wormhole motifs obtained through the N°I° assembly of nonionic polyethylene oxide (PEO) based surfactants (N°) and neutral inorganic precursors (I°) present good potentiality due to the use of low cost and environment friendly surfactants, open air synthesis and easily controlled pore size [7, 8]. Besides, they afford good potentials for supported membranes syntheses since their three dimensional porosity does not preclude of any need for orientation prior to coating and their easy synthesis allows more versatile coating processes.

However, it has been experimented that, whatever the chosen system, syntheses are not always easily reproduced. In our pathway, a typical synthesis of a MSU-1 silica with Tegitol 15-S-12 ($CH_3(CH_2)_{14}(OCH_2CH_2)_{11}OH$) begins with the dropwise addition under sonication of tetraethyl orthosilicate (TEOS: $Si(OCH_2CH_3)_4$) in 100 cc of a 0.02 M aqueous solution of surfactant, in order to obtain a final molar TEOS/Tegitol ratio of 8.0. Sonication is continued for a short period of time to obtain a milky solution. It is allowed to age without agitation to give a clear, colorless solution with a white surnageant that is removed prior to addition of NaF (molar NaF/TEOS ratio of 0.025), a TEOS hydrolysis catalyst. The final product is obtained after a two day reaction at 35°C followed by a six hour calcination at 620°C in air [8].

We show in this report that the initial sonicated mixing of TEOS with Tergitol has a drastic effect on the final structure of the material. Besides, the relationship between the initial hybrid mesophase formed by TEOS molecules trapped in the hydrophilic outer shell of nonionic micelles and the final structure of the material provides a good example of the cooperative formation of inorganic-organic interfaces first demonstrated by Stucky [9, 10]. To this purpose, we undertook several syntheses by only changing the sonication power of a Vibracel 72412 sonication gun (full power: 600 W) mounted with a one square centimeter sonication finger (Table 1).

Table 1
Sonication conditions

Sonication time (min.)	Sonication power (W)
10	6
10	12
5	18
5	24
5	30
2	60
1	120
1	180

1. X-RAY DIFFRACTION

Selected X-ray patterns are given in Figure 1. They have been simulated by fitting the background to an exponential and the diffraction peak - i.e. the correlation peak due to the porous framework - by a gaussian [8]. First, one observes that both low and high power sonication syntheses exhibit similar patterns, with a high intensity background at low angle due to a disordered porous framework. The 6 Watts stirring, for example, only gives a shoulder in the background. This amorphous scatttering fraction has disappeared for the 18-30 W sonication (Fig. 1.c). A crystallinity parameter can be deduced from both the consideration of the half width of the diffraction peak and the area of this peak. Results are displayed in figure 2. Samples prepared with a sonication power between 18 and 30 W have the narrowest diffraction peak (Full Width at Half Maximum ≈ 20 Å). However, they also exhibit the less intense peaks whereas high power stirring leads to larger peak surface areas. This is due to the synthesis of a larger part of a porous silica (higher surface area), but with an ill-ordered porous framework (higher FWHM). Lowest powerful stirrings (6 and 12 W) are poorly structured, which could be due to a too short stirring time to allow a complete homogeneous reaction.

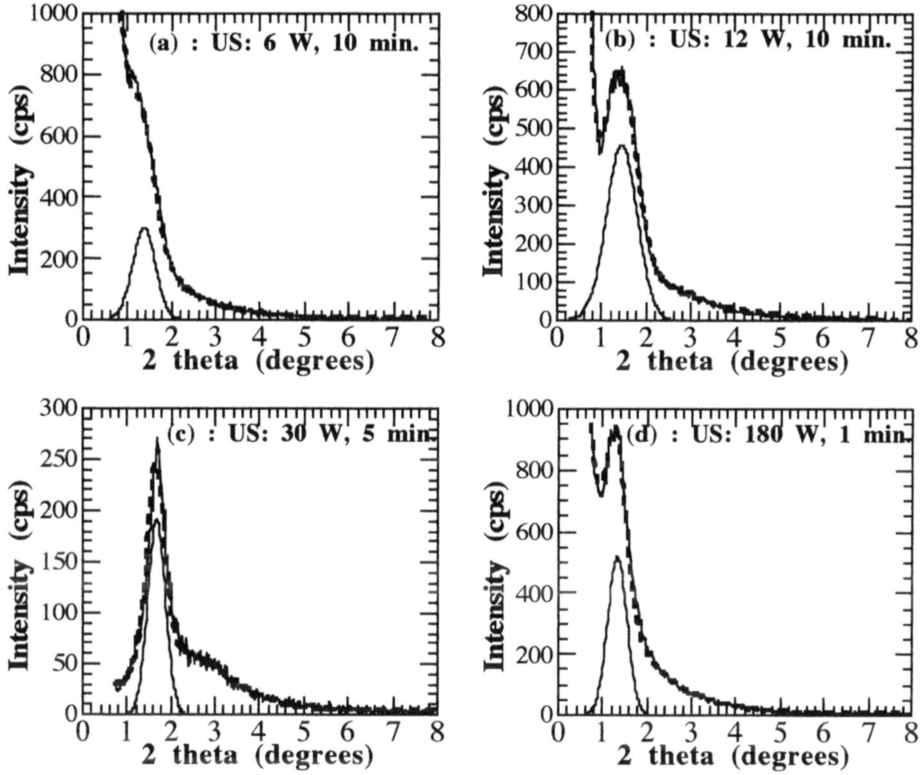

Figure 1. X-ray patterns of MSU-1 synthesized with ultrasonication at (a) 6 W, 10 min., (b) 12 W, 10 min., (c) 30 W, 5 min., (d) 180 W, 1 min. *Continuous line*: exp. pattern; *dashed line*: simulated pattern.

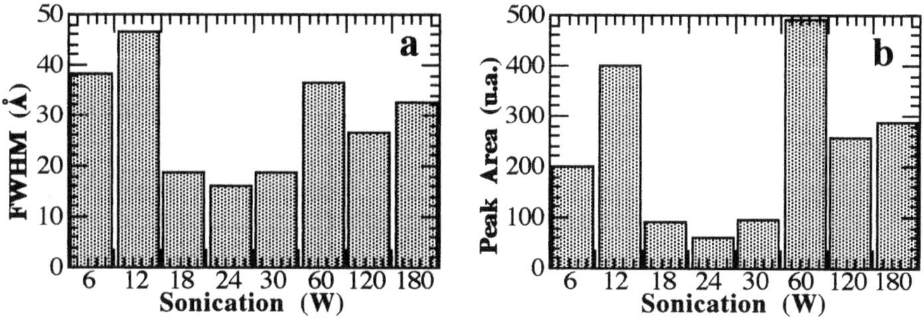

Figure 2. Comparisons of (a) the HWHM and (b) Peak surface areas of the X-ray patterns for the samples synthesized with various sonication powers.

The d-spacing of the final material is also modified by the initial mixing of surfactant and TEOS (Figure 3).

It varies between 49 and 66 Å but the 15-30 W sonication range corresponds to narrow peaks with a d-spacing close to 50 Å. One can also observe that these materials present X-ray patterns with a second broad peak pointing out around 3° (in 2θ) (see Figure 1.c).

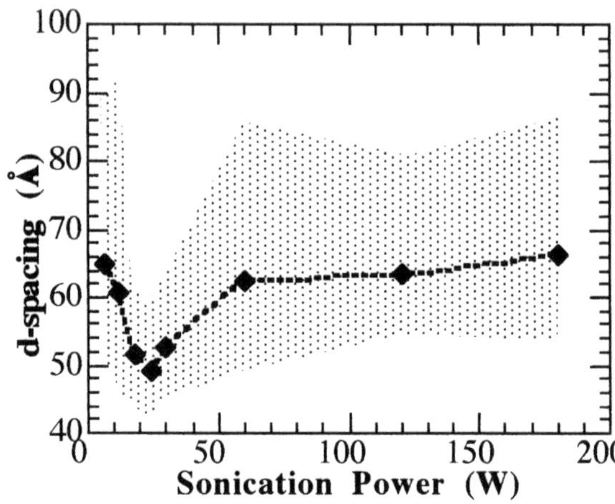

Figure 3. Variation of the d-spacing with the sonication power. Samples prepared in the 15-30 W range exhibit the smaller d-spacing. gray area: limits of the d-spacing range deduced from the FWHM.

2. NITROGEN ADSORPTION

The difference in the structures of materials synthesized with different sonications is totally confirmed by the variations in the porous framework deduced from nitrogen adsorption measurements (see Figure 4 and Table 2). For the 6 W sample, the adsorption/desorption curves give a loop characteristic of a bottle necked porosity. This loop decreases for the 12 W sample and disappears for the materials prepared in the 18-30 W range. For this latter, the Barrett-Joyner-Halender (BJH) model pore size distribution confirms that there is a narrow size distribution centered around 30 Å. The bottle-neck structure is again observed for materials prepared with 60 to 180 W sonications.

Table 2

BET surface areas and porous volume for MSU-1 silica prepared with different sonication powers.

Sonication power (W)	6	12	18	24	30	60	120	180
surface area ($m^2.g^{-1}$)	934	788	868	804	853	913	744	766
Porous volume ($cc.g^{-1}$)	1,02	0,74	0,74	0,68	0,78	0,91	1,05	0,91

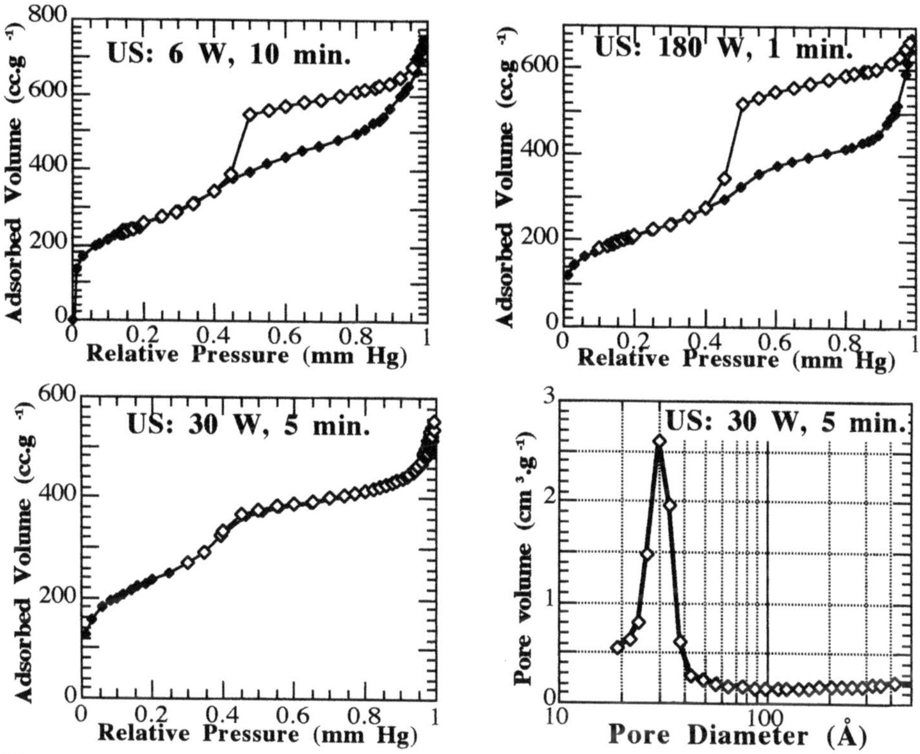

Figure 4. Nitrogen adsorption of calcined samples synthesized after a 6, 30 or a 180 W sonication stirring. The BJH pore size distribution calculated from the desorption curve, is also shown for the 30 W sample.

3. CONCLUSION

The synthesis of nonionic surfactant based mesostructured materials can be achieved through an easy synthesis pathway in open-air containers and at moderate or ambient temperature [7, 8]. However, like other syntheses, it may be hard to reproduce it. This study shows that an initial operation, the mixing of surfactant solution with the silica precursor, is indeed an important step to succeed in the synthesis of well defined materials. It brings a supplementary proof that the initial stage is the formation of a mixed mesophase [9, 10], more or less ordered, that will give the final compound. TEM analyses, presently under progress should confirm X-ray diffraction and nitrogen adsorption results.

REFERENCES

1. C.T. Kresge, M.E. Leonowicz, W.J. Roth, J.C. Vartuli and J.S. Beck, Nature, 359 (1992) 710-712.

2. J.S. Beck *et al.* J.A.C.S. 114 (1992) 10834-10843.
3. H. Yang, A. Kuperman, N. Coombs, S. Mamiche-Afara and G.A. Ozin, Nature, 379 (1996) 703-705.
4. H. Yang, N. Coombs, I. Sokolov and G.A. Ozin, Nature, 381 (1996) 589-592.
5. M.J. Ogawa, Chem. Soc.: Chem. Commu., (1996) 1149-1150.
6. Y. Lu *et al.*, Nature, 389 (1997) 364-368.
7. S.A. Bagshaw, E. Prouzet and T.J. Pinnavaia, Science, 269 (1995) 1242-1244.
8. E. Prouzet and T.J. Pinnavaia, Angewandte Chem. Int. Ed. English, 36 (1997) 516-518.
9. A. Monnier *et al.*, Science, 261 (1993) 1299-1303.
10. A. Firouzi *et al.*, Science, 267 (1995) 1138-1143.

MESOPOROUS MOLECULAR SIEVES 1998
Studies in Surface Science and Catalysis, Vol. 117
L. Bonneviot, F. Béland, C. Danumah, S. Giasson and S. Kaliaguine (Editors)

Thermal decomposition of surfactant occluded in mesoporous MCM-41 type solids.

A. Montes[a], E. Cosenza[a] , G. Giannetto[b] E. Urquieta[c] , R.A. de Melo[c] , N.S. Gnep[d] , and M. Guisnet[d]

[a]Instituto Universitario de Tecnología - Región Capital, Departamento de Química, Apartado 40347, Caracas 1040-A, Venezuela, Fax: 58-2-4617894.
[b]Universidad Central de Venezuela, Facultad de Ingeniería, Los Chaguaramos, Apartado Postal 47100, Caracas 1040-A, Venezuela. Fax 58-2-6054034.
[c]Universidade Federal de São Carlos, Departamento de Engenharia Química. São Carlos, SP, Brasil.
[d]Université de Poitiers, Groupe de Recherches sur la Catalyse, 40, Avenue du Rector Pineau, Poitiers, France.

Mesoporous solids with MCM-41 structure were synthesized. The composition of the gel was varied so as to obtain solids with variable Si/Al ratios (30, 100, ∞). Synthesis was carried out both under atmospheric pressure and autogenous pressure. Characterization of the obtained solids (XRD, STEM, N_2 adsorption) shows that the use of high Si/Al ratios and hydrothermal conditions for the synthesis results in a better ordering of the mesopores hexagonal arrangements, which means higher thermal stability of the solids.

The surfactant decomposition (HTACl) was carried out in a dry air atmosphere. The effects of washing with ethanol and/or the addition of small amounts of platinum previous to the calcination were also studied. To follow the evolution of the surfactant decomposition and elimination, FT-IR, DTA-TGA, ^{27}Al and ^{29}Si NMR-MAS techniques were employed. The favorable effect of the platinum presence during calcination in an oxidizing atmosphere was proven.

1. INTRODUCTION

A considerable increase has been observed over the last decade in the number of publications dealing with synthesis, characterization and evaluation of mesoporous solids in catalysis, adsorption and separation, pollution environmental control and particularly with intra-reticular or intra-zeolitic reactions. Considerable progress has been made in the study of these solids in a number of disciplines, which has resulted in the appearance of a new family of molecular sieves known as M41S [1,2].

The MCM-41 solid is a member of the M41S family, whose mesoporous structure can be controlled by the proper selection of a molding agent, in this case a surfactant, and by adding auxiliary chemical compounds and/or modifying parameters such as temperature and gel composition [3]. The MCM-41 "crystallization" mechanism is different from that of the zeolites, however, the synthesis method is exactly the same: these solids are obtained through hydrothermal reaction of a silica or a silica-aluminum gel in the presence of the structure orienting agent.

The use of these solids as catalysts imposes, even nowadays, developments tending to: i) increase their acidity; ii) increase their thermal and hydrothermal stability; iii) improve the surfactant elimination method, so that part of the solid structure is not damaged during the process. This paper shows the results corresponding to the synthesis of the MCM-41 type solids and discuss those results related to the decomposition and/or elimination of the surfactant used.

2. EXPERIMENTAL

Using the Ryoo method [4,5], slightly modified, solids with MCM-41 structure (Si/Al = 30, 100, ∞) were synthesized. The synthesis kinetics was followed using the reflection intensity [1,0,0] in the XRD pattern. Diffraction patterns were obtained by using a Siemens D-500 diffractometer and CuKα radiation (λ = 1.5418 Å, 40 Kv, 30 mA).

In the characterization by means of Transmission Electronic Microscopy a high resolution Hitachi S500 microscope was used.

Nitrogen adsorption was carried out in a Micromeritics ASAP 2010 sorbometer.

All ^{27}Al and ^{29}Si NMR-MAS spectra were recorded at 20°C on a pulsed Fourier Transform Bruker MSL-400 spectrometer.

The surfactant thermal decomposition (HTACl, hexadecil trimethyl ammonium chloride) using the as synthesized solid was carried out in a dry air atmosphere. The effects of washing with ethanol and/or adding small amounts of platinum before calcination were also studied. In this last case a mixture (90-10) of the MCM-41 solid and the platinum supported over an inert alumina (0.6% Pt/α-Al_2O_3) was prepared. DTA-TGA and FT-IR techniques were used to follow the development of the surfactant decomposition-elimination process up.

Infrared spectra (FT-IR) were recorded by means of a Nicolet MAGNA IR 550 spectrophotometer and the DTA-TGA thermal decomposition diagrams with a TA Instruments S.D.T. 2960 equipment.

3. RESULTS AND DISCUSSION

X-ray diffraction diagrams of the synthesized solids show only the reflections which are characteristic of the MCM-41 type solids [1,6] with regular hexagonal structure and a mean pore diameter of approximately 40 Å. Reflection intensity [1,0,0] in the XRD pattern, compared at the end of the synthesis, is higher as the Si/Al ratio increases. Likewise, this intensity is higher in the syntheses carried out under autogenous pressure as compared with those carried out under atmospheric pressure.

Kinetics of the condensation process of these solids allowed us to determine that this process is completed after 4 or 5 days, in the case of the syntheses under autogenous pressure, whereas under atmospheric pressure at least 10 days are required to complete the process.

Analysis of these solids by means of STEM showed the perfect periodicity of the channels with a diameter of approximately 40 Å. A pore diameter with a similar value could be obtained from the nitrogen adsorption measurements (37 Å) with a surface area amounting to 1000 m^2/g. These last results let us state that the synthesized solid has in fact a regular mesoporous structure, with adsorption parameters corresponding to the values expected for this type of solids [7,8].

The decomposition-elimination process of the surfactant occluded in the channels of the solid was studied by FT-IR following, in particular, the evolution of the bands corresponding to the C-H bonds. These bands are present even in small fragments resulting from the breaking of the original molecule. Stretching bands are in the zone between 2850 and 2960 cm^{-1}[9], as can be observed in Figure 1, whereas the bending bands are in the zone between 1460 and 1490 cm^{-1}[9], as shown in Figure 2. The results show that the surfactant elimination through thermal treatment in dry air atmosphere, using the as synthesized solid, is not complete unless a temperature higher than 400°C is reached, and this temperature varies as a function of the Si/Al ratio (Figure 2). Total elimination of the surfactant requires higher temperatures when the aluminum content of the solid increases. In other words, when its acidity increases.

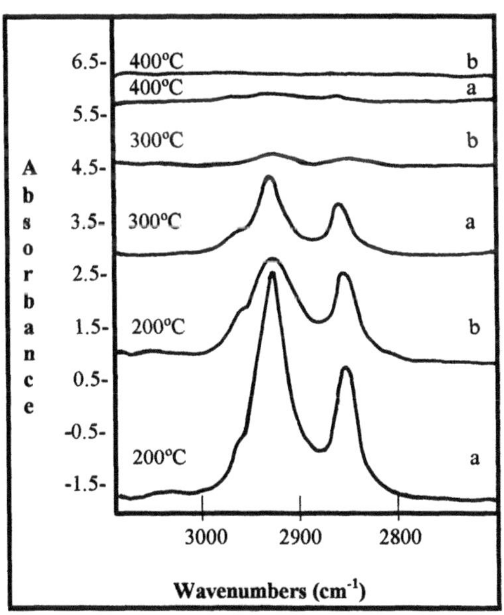

Figure 1. FT-IR spectra of the stretching bands of the surfactant occluded in the channels of the solid (Si/Al = 30) calcinated at different temperatures. a) as synthesized solid, b) as synthesized solid plus 0.6% Pt/αAl_2O_3.

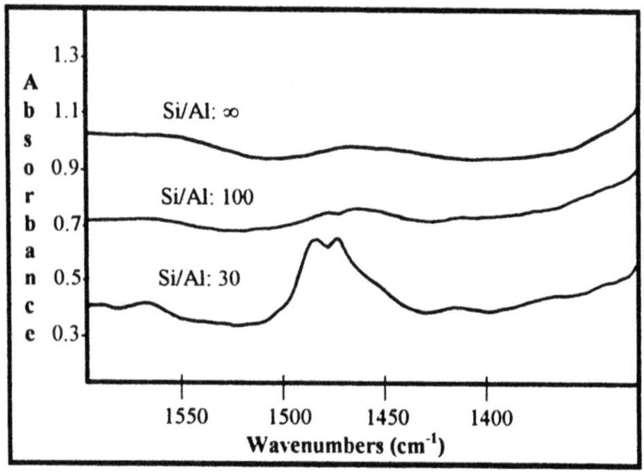

Figure 2. FT-IR spectra of the bending bands of the surfactant occluded in the channels of the MCM-41 type solid with different Si/Al ratios, calcinated at 400°C.

On the other hand, DTA-TGA analysis of the as synthesized solid shows that the calcination effected in a dry air atmosphere results in the decomposition of the surfactant according to a three stage sequence (Figure 3), which corresponds to the results previously observed by Corma et al. [8]: i) a first stage (300-360°C) which must correspond to the primary amine decomposition; ii) a second stage (390-400°C) which must correspond to the breaking of the hydrocarbonated chains; and iii) a last very large stage (550-700°C), which must correspond to the combustion (elimination) of the surfactant fragments.

All of these stages are highly exothermic, which in principle could result in dealumination and deterioration of the solid structure. However, in this case, XRD and ^{27}Al NMR-MAS analyses, carried out before and after calcination at different temperatures allow us to state that the structure of the solid synthesized under autogenous pressure does not suffer major alterations, at least up to 530°C.

On the other hand, it is evident that to minimize the undesired effects which can result from the high energy released during the surfactant decomposition process it is convenient to reduce somehow the amounts of the molding agent to be decomposed, and to decrease the necessary maximum temperature to ensure the complete elimination of this molding agent. To this end a washing with ethanol prior to calcination considerably reduces the amounts of surfactant to be eliminated without provoking changes in its decomposition-elimination pattern (DTA-TGA), that is, without changing the temperatures required for its elimination. Moreover, the addition of small amounts of platinum (max. 0.06% in weight) reduces up to 100°C the temperature at which the complete elimination of the surfactant fragments is produced, which ensures a higher integrity of the structure of the resulting solid. This important effect of the addition of small amounts of platinum, clearly observable by means of FT-IR (Figure 1), can be due to an activation of the molecular oxygen produced by the metal atoms. Besides, these activated species, transferred to the surfactant molecules (spillover effect) facilitate its transformation by means of combustion, which results in a reduction of the temperature required for the process of elimination of the molding agent [10]. This same spillover phenomenon, of oxygen activation and thus reduction of the combustion temperature of the organic compounds (coke), has been observed in other similar systems [11].

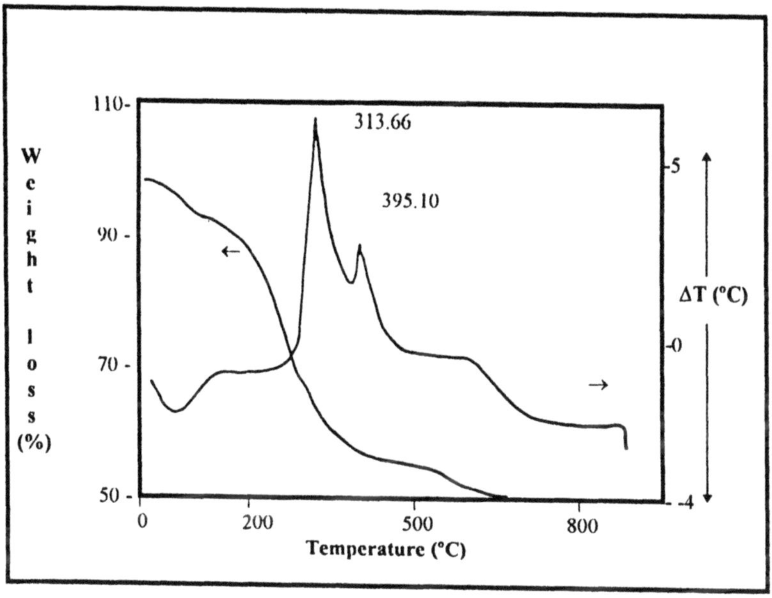

Figure 3. Differential Thermal Analysis (DTA) and Thermal-gravimetric Analysis (TGA) of MCM-41 (Si/Al = 30).

4. CONCLUSIONS

The synthesized materials correspond to mesoporous solids with a regular hexagonal structure, of the MCM-41 type, with pore diameters of approximately 40 Å. The structure of the solids synthesized under autogenous pressure shows good thermal stability.

The amount of surfactant to be eliminated from the structure of the synthesized solid can be considerably reduced with a previous wash with ethanol, whereas the addition of small amounts of platinum reduces up to 100°C the temperature required by the surfactant decomposition-elimination process.

REFERENCES

1. J.S. Beck, J.C. Vartuli, W.J. Roth, M.E. Leonowicz, C.T. Kresge, K.D. Schmitt, G:T-W. Chu, D.H. Olson, E.W. Sheppard, S.B. McCullen, J.B. Higgins and J.L. Schlendker, J. Amer. Chem. Soc., 114 (1992) 10834.
2. C.T. Kresge, M.E. Leonowicz, W.J. Roth, J.C. Vartuli and J.S. Beck, Nature, 359 (1992) 710.
3. J.L. Casci, Stud. Surf. Sci. Catal., 85 (1994) 329.
4. R. Ryoo and J.M. Kim, J. Chem. Soc., Chem. Commun., (1995) 711.
5. J.M. Kim, J.H. Kwak, Sh. Jun and R. Ryoo, J. Phys. Chem., 99 (1995) 16742.
6. A. Monnier, F. Schüth, Q. Huo, D. Kumar, D. Margolese, R.S. Maxwell, G.D. Stucky, M. Krishnamurty, P. Petroff, A. Firouzi, M. Janicke and B.F. Chmelka, Science, 261 (1993) 1299.
7. A. Corma, M.S. Grande, V. González-Alfaro and A.V. Orchiles, J. Catal., 159 (1996) 375.
8. S. Morin, P. Ayrault, S. El Mouahid, N.S. Gnep and M. Guisnet, Appl. Catal. A: General., 159 (1997) 317.
9. A. Corma, V. Fornés, M.T. Navarro and J. Pérez-Pariente, J. Catal., 148 (1994) 569.
10. E. Baumgarten and A. Schuck, Appl. Catal., 37 (1988) 247.
11. S.J. Teichner, in New Aspects of Spillover Effect in Catalysis, (T. Inui et al., Eds.), Elsevier Science Pub., Amsterdam, (1993) 27.

MESOPOROUS MOLECULAR SIEVES 1998
Studies in Surface Science and Catalysis, Vol. 117
L. Bonneviot, F. Béland, C. Danumah, S. Giasson and S. Kaliaguine (Editors)

Hybrid inorganic-organic mesoporous molecular sieves

M. Koya [a] and H. Nakajima [a]

[a] The Noguchi Institute, Kaga, Itabashi, Tokyo, 173-0003, Japan

Hybrid inorganic-organic mesoporous molecular sieves of silica network with various organic groups were prepared with n-alkylamines of various length of carbon chain. Their pore structure, thermal stability and hydrophobic property were studied. The catalytic activity of hybrid titanosilicate mesoporous molecular sieves was studied in the oxidation of cyclooctene with an aqueous hydrogen peroxide solution.

1. INTRODUCTION

Burkett et al.[1] succeeded in the synthesis of hybrid inorganic-organic mesoporous molecular sieves containing phenyl or n-octyl group. Macquarrie[2] also succeeded in the synthesis of them containing aminopropyl or cyanoethyl group, and Lim et al.[3] also did in the synthesis of them containing vinyl group. Concurrently with their studies we independently succeeded in the synthesis of the hybrid inorganic-organic mesoporous molecular sieves containing various organic groups via similar approach[4]. Tatsumi et al. studied the oxidation of alkenes with an aqueous hydrogen peroxide solution by trimethylsilated titanosilicate mesoporous molecular sieves[5] and hybrid titanosilicate mesoporous molecular sieves[6]. We have prepared hybrid mesoporous molecular sieves containing various organic parts with various n-alkylamines, and studied their pore structure, physical property and catalytic activity.

2. EXPERIMENTS

Mesoporous molecular sieves were synthesized according to the method of Tanev et al.[7] A mixture of organotriethoxysilane(alkyl-, alkenyl- or phenyl-triethoxysilane) and tetraethoxysilane was added to an aqueous 44 wt% ethanol solution of n-dodecylamine under stirring at ambient temperature. The molar ratio of organotriethoxysilane to tetraethoxysilane was 1 to 4, and the molar ratio of n-dodecylamine to total silanes was 0.27. The solution was left alone for 20 hrs at ambient temperature after the formation of precipitate had been observed, the precipitate was filtered off, was washed with water, and was dried. n-Dodecylamine was extracted from the precipitate with ethanol at 60 ℃, and the precipitate was again dried at 150 ℃ in a vacuum. n-Decyl-, n-tetradecyl or n-hexadecylamine of a same molar amount was added instead in some other runs. Aluminum isopropoxide or titanium tetraethoxide was added to the solution in the preparation of hybrid aluminosilicate or titanosilicate mesoporous molecular sieves respectively.

The specta of X-ray diffraction, IR, TGA-DTA and ^{29}Si-MASNMR of mesoporous molecular sieves were obtained with Rigaku denki RAD Ⅲ, Perkin-Elmer FTIR-1500, Shimadzu DTA-50/TGA-50 and JEOL NMR-GX270, respectively. The contents of Ti and Si were estimated with XRF. Specific surface area, pore diameter and pore volume of mesoporous molecular sieves were estimated with nitrogen adsorption method with Carlo-Erba SORPTOMATIC 1800, and the amount of the adsorption of water vapor on the mesoporous molecular sieves was estimated as the weight increase of them placed above water in a closed vessel (the partial pressure of water was 0.97) at 30 ℃ for 24 hrs. The catalytic activity of hybrid inorganic-organic titanosilicate mesoporous molecular sieves was estimated in the oxidation of cyclooctene with an aqueous hydrogen peroxide solution at 60 ℃ for 4 hrs. The products were analyzed with GC after the solution had been treated as the following: a cetonitrile was added to the solution and the catalyst was removed from the solution with a centrifuge.

3. RESULTS AND DISCUSSION

The synthesis of hybrid silica mesoporous molecular sieves containing C1 ~ C8 alkyl or phenyl group as their organic part was studied. The results were shown in Table 1.

Table 1
Hybrid inorganic-organic silica mesoporous molecular sieves

Organic part	d* nm	Sp. surface area m^2/g	Pore diameter nm	Combustion temp.*2 ℃
Methyl	3.3	1150	2.4	650
Ethyl	3.3	1180	2.2	500
n-Propyl	2.8	1010	2.2	260
n-Hexyl	2.9	970	2.1	250
n-Octyl	3.5	990	2.1	260
Vinyl	3.2	1200	2.3	400
3-Butenyl	3.5	1100	2.5	300
Phenyl	3.2	1000	2.0	600
-	3.7	1200	3.0	-

* lattice spacing (100 planes)
*2 Combustion temperature was estimated from the exothermic peak in the spectra of DTA (the temperature increased with the rate of 15 ℃/min).

The structure of mesoporous molecular sieves was recognized from their XRD spectra, specific surface area and pore diameter. ^{29}Si-MASNMR spectra of them containing methyl, n-propyl or n-hexyl group showed absorption at the similar chemical shift, -67ppm from the absorption of Si of tetramethyl silane. The absorption was attributed to Si bound to an alkyl group. The chemical shift agreed with that of Burkett et al.[1] observed on the hybrid materials containing n-octyl group. IR spectra of the hybrid mesoporous molecular sieves containing methyl or phenyl group showed absorption at 1,277 and 1,133cm^{-1}, respectively. The absorption was attributed to the bending vibration mode of Si-C of the Si-methyl and Si-phenyl group, respectively. The latter absorption observed on the hybrid ones containing phenyl group agreed with that of Burkett et al[1].

The DTA-TGA analysis of hybrid mesoporous molecular sieves in a stream of air showed an exothermic peak in DTA and a decrease in weight in TGA at about the same temperature. The peak was thought to be attributed to the combustion of their organic part. The exothermic peak of the molecular sieves containing methyl group was observed at the highest temperature of 650 ℃, and the temeratures were as the following decreasing order with the structure of their organic parts; methyl > phenyl > ethyl > vinyl > 3-butenyl > n-propyl > n-hexyl.

Pore sizes of hybrid silica mesoporous molecular sieves prepared with n-dodecylamine were 2.2 ~ 2.5 nm, while that of the ones with no organic group prepared with n-dodecylamine was 3.0 nm. The effects of the length of carbon chain of n-alkylamines were shown in Figure 1. The pore size of the hybrid molecular sieves increased similarly but with less degree as the length of carbon chain of n-alkylamines increased than that of the ones with no organic group increased.

The effects of the content of methyl group of the hybrid mesoporous molecular sieves on the pore size were shown in Table 2. The pore size decreased as the amount of methyltriethoxysilane added in their preparation increased.

The amounts of water adsorbed on the hybrid silica mesoporous molecularsieves were shown in Figure 2. There were less absorbed water than in the mesoporous molecular sieves with no organic group (the first column from the left, the highest one, in Figure 2). The amount adsorbed on the hybrid ones decreased as the length of carbon chain of their organic parts increased. The hybrid mesoporous molecular sieves were more hydrophobic for longer alkyl chains. There was more adsorption on the molecular sieves with phenyl group than with n-hexyl group. A phenyl group gave less hydrophobicity than an alkyl group of the same carbon number to the hybrid mesoporous molecular sieves. A higher water uptake was obtained on the hybrid aluminosilicate mesoporous molecular sieves than on the hybrid titanosilicate, which had almost the same uptake with the hybrid silica ones with the same alkyl group. The insertion of Ti to the silica net work did not change the aqueous hydrophobicity, while that of Al decreased it.

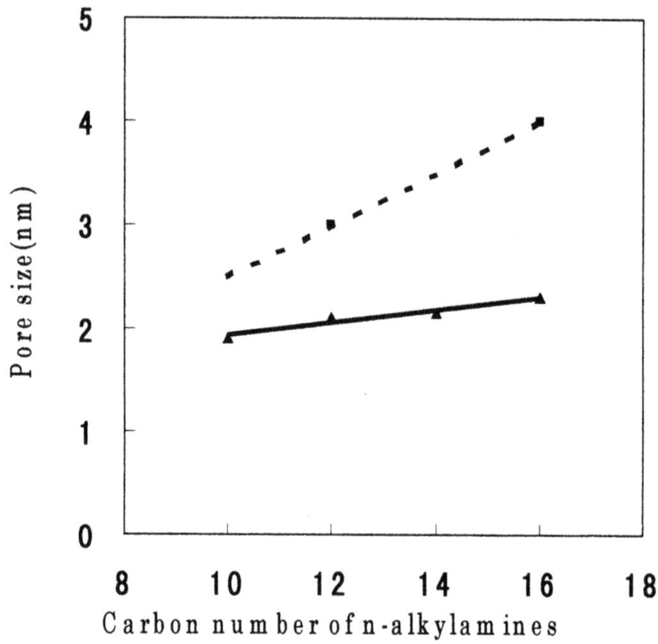

▲;with methyl group,■;with no organic part

Figure 1. Effects of carbon number of n-alkylamines

Table 2
Effects of the content of methyltriethoxysilane in preparation

Ex. No.	(MeSi-/Si-)*	Sp. surface area (m^2/g)	Pore size (nm)
1	1/2	870	2.1
2	1/4	1,130	2.4
3	1/9	1,040	2.6
4	0/1	1,200	3.0

* Molar ratio of methyltriethoxysilane to tetraethoxysilane added in preparation

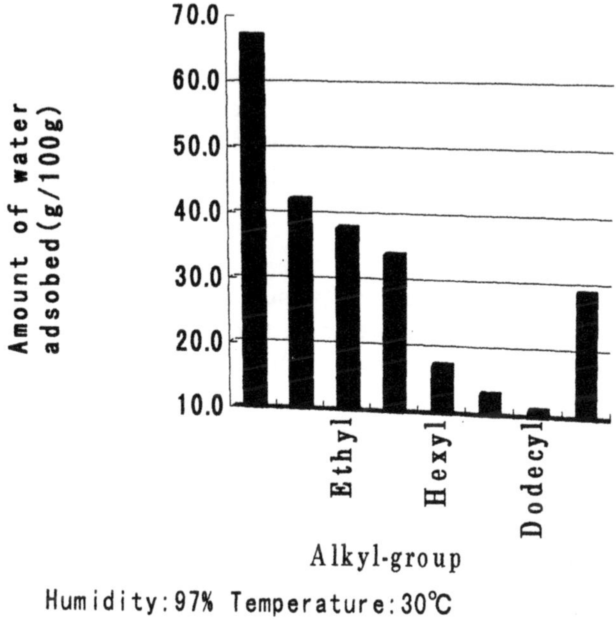

Figure 2 .Adsorption of water on various hybrid mesoporous molecular sieves

Table 3
Oxidation of cyclooctene with an aqeous hydrogen peroxide solution on hybrid titanosilicate mesoporous molecular sieves
Si/Ti:30, catalysts:0.2 g, cyclooctene:50 mmol, H_2O_2:25mmol, temperature: 60 ℃, reaction period: 4 hrs.

Catalysts			Reaction results
Organic part (content)	Sp. surface area m^2/g	Pore diameter nm	Yield of epoxide* (%)
Methyl-Si (1/2)*2	840	2.2	8.6
Phenyl-Si (1/4)*2	970	2.1	8.2
$(Methl)_3$-Si(1/4)*2	920	2.5	7.2
Methyl-Si (1/4)*2	900	2.3	5.8
- (Calcined)	870	3.2	0.3
$(Methyl)_3$-Si*3	710	2.9	3.6

* Yield based on cyclooctene. *2 The molar ratio of organotriethoxysilane to

tetraethoxysilane added in the preparation. *3 Calcined titanosilicate mesoporous molecular sieves were treated with trimethylchlorosilane, the added amount of which was 0.15 mole/Si g-atom in the molecular sieves.

The oxidation of cyclooctene with an aqueous hydrogen peroxide solution by hybrid titanosilicate mesoporous molecular sieves was studied. The results were shown in Table 3. The hybrid titanosilicate mesoporous molecular sieves were more active than the mesoporous molecular sieves with no organic group.

4. CONCLUSIONS

(1) Hybrid inorganic-organic mesoporous molecular sieves containing alkyl, alkenyl or phenyl group were prepared, and their pore sizes were smaller than that of mesoporous molecular sieves with no organic group.
(2) Combustion temperature of the hybrid mesoporous molecular sieves was estimated with DTA, and the temperature of the one with methyl group as an organic group was 650 ℃, which was the highest among those of the hybrid mesoporous molecular sieves with various organic groups.
(3) The amounts of water adsorbed on the hybrid mesoporous molecular sieves were smaller than that on the mesoporous molecular sieves with no organic group, and decreased as the length of alkyl chain of their organic groups increased.
(4) The hybrid titanosilicate mesoporous molecular sieves showed much higher catalytic activity than the ones with no organic group in the oxidation of cyclooctene with an aqueous hydrogen peroxide solution.

ACKNOWLEDGEMENTS

This work was a part of a joint study carried out at Asahi Chemical Industry Co. Ltd. and The Noguchi Institute. The authors are grateful to them for permission to present this paper to this sympodium.

REFERENCES

1. S. L. Burkett, S. D. Sims, S. Mann, Chem. Commun.,1996, 1367.
2. D. J. Macquarre, Chem. Commun.,1996, 1961.
3. M. H. Lim, C. F. Blanford, A. Stein, J. Am. Chem. Soc.,119, 4090(1997).
4. M. Koya, H. Nakajima, I. Aishima, WO97-46743(Asahi Chem. Ind. & Noguchi Inst.)
5. T. Tatsumi, K. A. Koyano, N. Igarashi, Chem. Commun., 1998, 328.
6. S. Kitani, N. Igarashi, T. Tatsumi, Prep.,13th meet'g, Zeolite Soc. Jap., (1997) p83.
7. P. T. Tanev, T. J. Pinnavaia, SCIENCE, 267(1995), 867.

MESOPOROUS MOLECULAR SIEVES 1998
Studies in Surface Science and Catalysis, Vol. 117
L. Bonneviot, F. Béland, C. Danumah, S. Giasson and S. Kaliaguine (Editors)

Synthesis and characterization of aluminum-containing MCM-48

Martin Hartmann and Christian Bischof

Institute of Chemical Technology I, University of Stuttgart, D-70550 Stuttgart, Germany

Pure siliceous and aluminum containing mesoporous MCM-48 materials were synthesized from gels containing C_mTAB (m = 14,16), tetraorthosilicate (TEOS), water and diluted NaOH. The synthesis for the all silica materials, performed in a 5 l autoclave, was monitored by characterization of small samples, taken every two hours, employing X-ray diffraction and ^{29}Si MAS NMR spectroscopy. A number of mesoporous aluminosilicate molecular sieves with MCM-48 structure has been synthesized in 250 ml and 1 l autoclaves using different sources of aluminum and characterized in detail by XRD, ^{29}Si and ^{27}Al MAS NMR and nitrogen adsorption. By using soluble aluminum salts, viz. aluminum nitrate, hydroxide or sulfate, MCM-48 can be prepared with all aluminum in tetrahedral coordination. However, high quality MCM-48 materials can only be obtained to a n_{Si} / n_{Al}-ratio of 80.

1. INTRODUCTION

Since the discovery of ordered mesoporous molecular sieves with pore sizes between 2 and 10 nm, much research work has been devoted to this new class of mesoporous materials, denoted M41S. However, most studies focused on the hexagonal member of this group, MCM-41 [1,2] The same synthesis principles can be applied for the synthesis of another member of this family, the cubic phase MCM-48, although, only little information is available on the synthesis of this material [3]. The incorporation of aluminum into MCM-41, which is interesting with respect to catalytic applications, is discussed in numerous papers, while only Schmidt et al. [4] report on the synthesis of AlMCM-48. Using aluminum sulfate, they succeeded in preparation of an AlMCM-48 ($n_{Si} / n_{Al} = 20$) with aluminum exclusively in tetrahedral coordination. Zhao and Goldfarb have also reported the synthesis of an aluminum-containing MCM-48 ($n_{Si} / n_{Al} = 46$) without, however, stating the details of the synthesis or reporting any characterization data of this material [5]. Furthermore, Ryoo et al [6] described a post synthesis method to incorporate aluminum into all silica MCM-48. Despite of this initial success, aluminum incorporation into MCM-48 yielding high quality AlMCM-48 is still a challenge. In our study, pure siliceous and aluminum-containing MCM-48 materials were synthesized in 250 ml, 1 l and 5 l autoclaves with and without agitation. Different sources of aluminum and variation of other experimental parameters were checked for their potential yielding high quality AlMCM-48 materials.

2. EXPERIMENTAL SECTION

2.1. Synthesis

Mesoporous all silica MCM-48 was synthesized in a 5 l autoclave using hexadecyltrimethylammoniumbromide ($C_{16}TAB$, Aldrich), tetraorthosilicate (TEOS, Aldrich) and diluted NaOH. First, a gel with the molar composition $n_{Si} : n_{CTAB} : n_{Na2O} : n_{H2O} = 1.0 : 0.6 : 0.5 : 63$ was prepared. A typical synthesis procedure is outlined as follows: 840 g of hexadecyltrimethylammoniumbromid ($C_{16}TAB$) were diluted in 2.5 l of distilled water under stirring at 50 °C to give a 25 wt.-% solution. Then 800 g of TEOS were slowly added and the resultant solution was stirred for 5 min. After addition of 1.25 l of a 1 M NaOH solution, the gel was stirred for another 5 min. Thereafter, the resultant synthesis gel was charged into a 5 l autoclave and heated to 100 °C under agitation. To monitor the progress in MCM-48 formation, small samples were taken every 2 h from the autoclave. Each sample was washed with 100 ml of ethanol, 250 ml of destilled water and subsequently dried at 80 °C for 24 h to secure comparable conditions for further characterizations. After 40 h the synthesis was stopped and the resultant material was filtered, washed with ethanol and water and air dried. To remove the surfactant, the as-synthesized material was calcined for 10 h at 540 °C in a flow of nitrogen and air. A similar procedure was used to synthesize MCM-48 with a smaller pore diameter using $C_{14}TAB$ as a surfactant.

For the incorporation of aluminum into MCM-48 materials in tetrahedral coordination, different synthesis strategies and aluminum sources were tested. AlMCM-48 was synthesized using the same general procedure as described above. As sources, aluminum sulfate (Merck) and nitrate (Merck), sodium aluminate, aluminum triisopropylate (Fluka) and aluminum hydroxide were explored. The molar composition, pH-value and temperature of the synthesis gel, the aluminum source and the mixing time was varied in a large number of experiments.

2.2. Characterization

The chemical compositions of the samples were determined by atomic adsorption spectroscopy (AAS). X-ray powder diffraction pattern were recorded after synthesis and template removal on a Siemens D5000 diffractometer using CuK_{α} radiation. After calcination, nitrogen adsorption and desorption isotherms were measured on a Micromeritics ASAP 2010 sorption analyzer. ^{27}Al and ^{29}Si MAS NMR spectra were recorded on a Bruker MSL 400 spectrometer using single pulse excitation with standard 4 mm (^{27}Al) and 7 mm (^{29}Si) rotors. The resonance frequencies were $\omega_0/2\pi$ = 104.31 and 79.49 MHz for ^{27}Al and ^{29}Si, respectively. A 1.0 M solution of aluminum nitrate in water and tetramethylsilane (TMS) were employed as chemical shift references.

3. RESULTS AND DISCUSSION

Figure 1 exhibits the X-ray diffraction patterns for a all silica MCM-48 synthesis as a function of the reaction time. For a better view, the diffraction patterns of samples taken every 4 h were displayed. Hence, it appears that the formation of the MCM-48 structure is completed after 22 h. No further increase in the X-ray diffraction intensity is observed between 22 and 40 h. Between 6 and 18 h of the reaction time a broad reflection is observed at $2\theta = 3.7$ °. With increasing synthesis time, this signal becomes lower in intensity, indicating that the cubic MCM-48 is probably build over an intermediate phase. Probably, this intermediate phase is a disordered three dimensional arrangement and is transformed during the synthesis into a well ordered MCM-48 structure. The ^{29}Si MAS NMR spectra of the as synthesized samples show, that during the synthesis of the MCM-48 condensation of the silanol groups takes place [7]. The Q^4/Q^3-ratio increases with increasing reaction time. The Q^4/Q^3-ratio in the synthesis gel (t = 0) is 0.93 and after a reaction time of 12 h the Q^4/Q^3-ratio increases to 1.58. During this condensation, the structure is probably transformed via a disordered three dimensional phase into the cubic MCM-48. Similar results were found by Huo et al [8], who investigated the formation of MCM-41 using X-ray diffraction and ^{29}Si MAS NMR. In contrast to the formation of the MCM-48 structure, the characteristic pattern of the MCM-41 phase is already recognizable in the synthesis gel. Further crystallization only results in intensity enhancement of the X-ray reflections. This indicates that the hexagonal structure of MCM-41 is already pre-arranged in the synthesis gel.

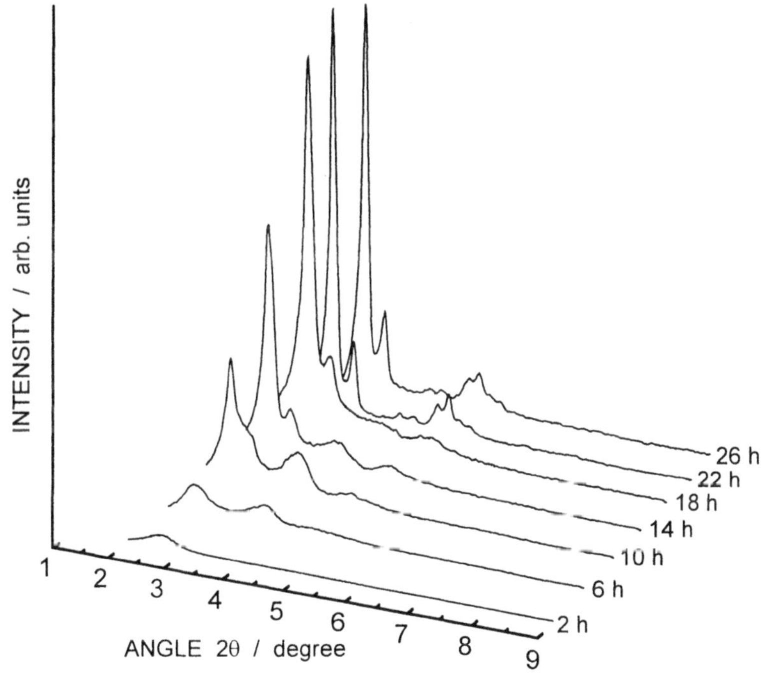

Figure 1. X-ray diffraction patterns for samples taken from an all silica MCM-48 synthesis batch after different reaction times.

While the synthesis and scale up of all silica MCM-48 with C_mTAB (m = 14,16) denoted as MCM-48/14 and MCM-48/16 was successful, the synthesis of MCM-48 with C_mTAB (m = 10,12,18) did not yield in the formation of well ordered MCM-48. A comparison between MCM-48/14 and MCM-48/16 samples is given in Table 1. As expected, the unit cell parameter a_0, the BET-surface, the pore volume and the pore diameter decrease with decreasing length of the alkyl chain.

Table 1
Unit cell parameters and results from the nitrogen adsorption experiments for MCM-48/14 and MCM-48/16

	unit cell parameter / nm		BET-surface /	pore volume[a)] /	pore diameter[a)] /
	as-synthesized	calcined	m^2g^{-1}	cm^3g^{-1}	nm
MCM-48/14	8.43	7.07	1030	0.56	1.7
MCM-48/16	8.91	7.61	1350	0.86	2.0

a) The pore volumes and diameters were calculated from the desorption branch of the adsorption isotherms using the BJH-model.

A collection of the X-ray diffraction patterns of AlMCM-48 materials synthesized with different aluminum sources is shown in Figure 2. In contrast to the results obtained by Schmidt et al. [4], the use of aluminum sulfate does not lead to the formation of AlMCM-48 with a well resolved XRD pattern regardless of the $Al_2(SO_4)_3$ concentration. Only a disordered phase is obtained (Figure 2a), which was also observed by Pu et al. [9] and denoted AlMCM-48 (n_{Si} / n_{Al} = 32). A well ordered cubic MCM-48 phase exhibiting 5 - 8 reflections in the rang of $2\theta = 2 - 9\,°$ is obtained with aluminum nitrate, aluminum hydroxide and aluminum triisopropylate as aluminum sources (Figures 2b-d). In all cases, the sample with the lowest achievable n_{Si} / n_{Al}-ratio leading to a well resolved XRD pattern is displayed.

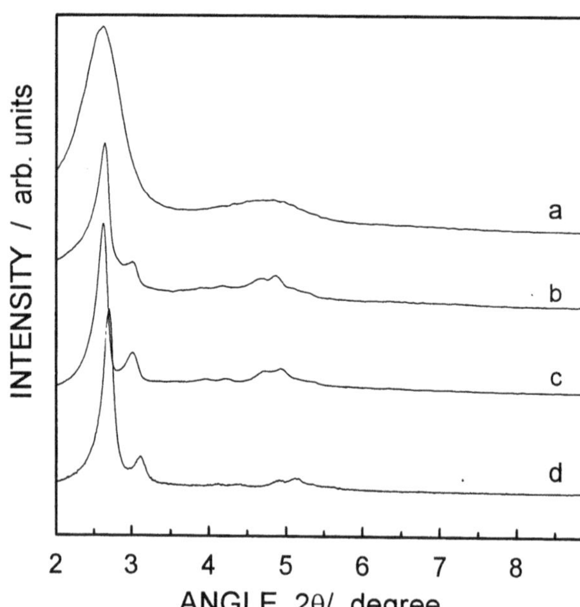

Figure 2. X-ray diffraction patterns of the AlMCM-48 materials synthesized with different aluminum sources: a) aluminum sulfate (n_{Si} / n_{Al} = 17), b) aluminum triisopropylate (n_{Si} / n_{Al} = 5), c) aluminum hydroxide (n_{Si} / n_{Al} = 80), d) aluminum nitrate (n_{Si} / n_{Al} = 100).

Well ordered MCM-48 materials with $n_{Si} / n_{Al} = 100$ and $n_{Si} / n_{Al} = 80$ were synthesized using aluminum nitrate and aluminum hydroxide, respectively. A disordered phase is observed, when the n_{Si} / n_{Al}-ratios were decreased with aluminum nitrate and aluminum hydroxide as aluminum sources in the synthesis gel. In these cases, variation of synthesis parameters did not result in the formation of high quality AlMCM-48. When aluminum triisopropylate is used as an aluminum source, the n_{Si} / n_{Al}-ratio can be reduced to 5.

The nitrogen adsorption and desorption isotherms for the AlMCM-48 materials synthesized with different aluminum sources are displayed in Figure 3. All samples show adsorption isotherms typical for mesoporous materials with small pore size distribution [10]. The results of the nitrogen adsorption experiments and the chemical compositions of these samples are summarized in Table 2.

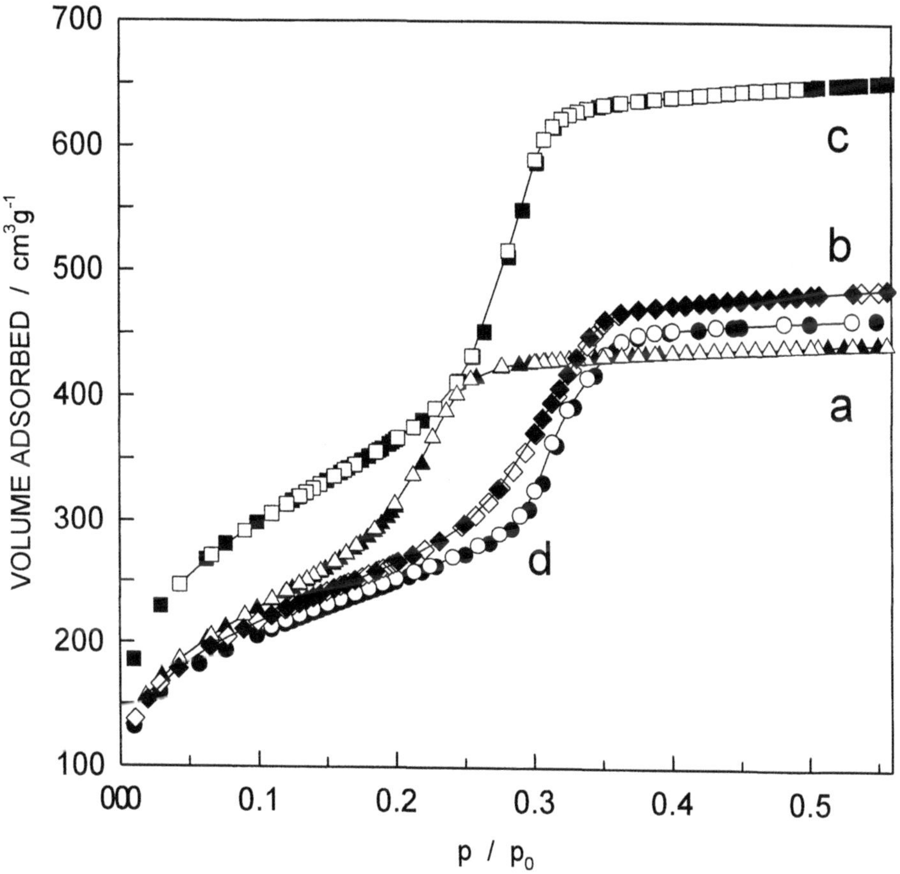

Figure 3. Nitrogen adsorption (open symbols) and desorption isotherms (closed symbols) for AlMCM-48 materials synthesized with different aluminum sources: a) aluminum triisopropylate, b) aluminum sulfate, c) aluminum hydroxide, d) aluminum nitrate.

Table 2
Unit cell parameters, n_{Si}/n_{Al}-ratios and results from the nitrogen adsorption experiments for different AlMCM-48 materials.

Aluminum source	unit cell parameter / nm as-synthesized	unit cell parameter / nm calc.	n_{Si}/n_{Al}	BET-surface / m^2g^{-1}	pore volume[a)] / cm^3g^{-1}	pore diameter[a)] / nm
hydroxide	8.40	8.15	80	1380	1.0	2.2
nitrate	8.45	7.60	100	1080	0.69	1.9
sulfate	-	-	17	1000	0.74	2.3
triisopropylate	9.14	8.08	5	940	0.76	2.6

a) The pore volumes and diameters were calculated from the desorption branch of the adsorption isotherms using the BJH-model.

The low quality of AlMCM-48 synthesized with aluminum sulfate is also reflected in the adsorption data. An unusually broad pore size distribution (not shown) indicates an ordering which is reduced in comparison with all silica MCM-48 and other AlMCM-48 samples. The nitrogen adsorption is significantly enhanced in AlMCM-48 synthesized with aluminum hydroxide in comparison with the material prepared with aluminum nitrate. The pore volume and the surface area of the latter material are substantially lower, although the n_{Si} / n_{Al}-ratio is in the same order of magnitude. AlMCM-48 synthesized with n_{Si} / n_{Al} = 5 using aluminum triisopropylate has a pore diameter of d_p = 2.6 nm, which is larger than d_p of the all silica MCM-48.

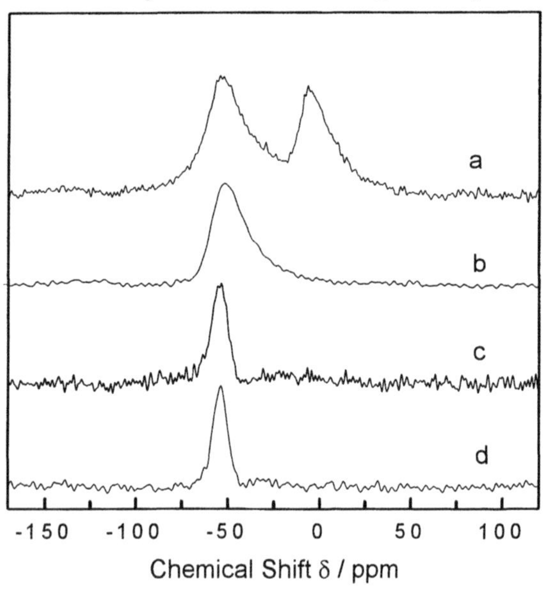

Figure 4. ^{27}Al MAS NMR spectra of the AlMCM-48 materials prepared using a) aluminum triisopropylate (calcined material), b) aluminum sulfate (calcined material), c) aluminum nitrate and d) aluminum hydroxide.

The ^{27}Al MAS NMR spectrum of the sample synthesized with aluminiumtriisopropylate, displayed in Figure 4a, consists of two lines at $\delta = 53$ ppm and $\delta = 0$ ppm. While the latter line corresponds to aluminum in octahedral coordination, the signal at $\delta = 53$ ppm is typical for tetrahedrally coordinated aluminum in mesoporous alumosilicates [11]. A similar ^{27}Al MAS NMR spectrum is observed for AlMCM-41, when aluminum triisopropylate is used as a aluminum source. Samples with aluminum exclusively in tetrahedral coordination are synthesized with water soluble aluminum sources, viz. $Al_2(SO_4)_3$, $Al(NO)_3$ and $Al(OH)_3$, (Figures 4b,c and d). The NMR spectra give a sharp resonance at $\delta = 53$ ppm originating from tetrahedrally coordinated aluminum. Upon template removal, the sharp ^{27}Al resonance broadens, but does not decrease in absolute intensity. The broadening is a result of decreased symmetry of some aluminum sites in the calcined sample. Attempts to synthesize AlMCM-48 with low n_{Si} / n_{Al}-ratios using aluminum nitrate and aluminum hydroxide did not result in the formation of a well ordered cubic phase. In that cases, the resulting X-ray diffraction patterns were similar to the one displayed in Figure 2a. However, the ^{27}Al MAS NMR spectra still only consist of a single line at $\delta = 53$ ppm. The appearance of not well resolved XRD patterns in MCM-41 type materials, when introducing aluminum is explained by Corma [12] not as a result of a less ordered material, but due to the formation of smaller crystallites of MCM-41. However, a literature survey shows that different synthesis strategies and aluminum sources result in materials of widespread quality and properties, which holds also true in for the cubic MCM-48 phase. More work is necessary to understand the influences of synthesis parameters and the nature of the aluminum source.

4. CONCLUSIONS

All silica and aluminum containing MCM-48 samples have been synthesized in 250 ml, 1 l and 5 l autoclaves. The synthesis of a larger quantity and slow agitation was found to have a promoting effect on the synthesis of high quality materials. Formation of the cubic phase is achieved after 22 h via condensation of a less well ordered three dimensional structure. However, the synthesis of well ordered aluminum-containing MCM-48 materials with low n_{Si} / n_{Al}-ratios is still a challenge. An n_{Si} / n_{Al}-ratio as low as 5 can be obtained using aluminum triisopropylate, but not all aluminum is in tetrahedral coordination. Exclusively tetrahedrally coordinated aluminum is found in samples synthesized with water soluble aluminum sources, viz. $Al(OH)_3$, $Al(NO)_3$, and $Al_2(SO_4)_3$. Unfortunately, the synthesis of high quality MCM-48 is limited to $n_{Si} / n_{Al} = 80$ and aluminum hydroxide as aluminum source. Further investigations aiming at a substantial reduction of the n_{Si} / n_{Al}-ratio are in progress.

ACKNOWLEDGMENTS

Financial support from Deutsche Forschungsgemeinschaft (DFG) and Fonds der Chemischen Industrie is gratefully acknowledged. M.H. acknowledges generous support from Prof. J. Weitkamp.

REFERENCES

[1] C.T. Kresge, M.E. Leonowicz, W.J. Roth, J.C. Vartuli and J.S. Beck, Nature, 359, 710 (1992).

[2] J.S. Beck, J.C. Vartuli, W.J. Roth, M.E. Leonowicz, C.T. Kresge, K.D. Schmitt, C. T-W. Chu, D.H. Olson, E.W. Sheppard, S.B. McCullen, J.B. Higgins and J.L. Schlenker, J. Am. Chem. Soc., 114, 10834 (1992).

[3] A.A. Romero, M.D. Alba, W. Zhou, J. Klinowsky, J. Phys. Chem. B, 101, 5294 (1997).

[4] R. Schmidt, H. Junggreen and M. Stöcker, Chem. Commun., 875 (1996).

[5] D. Zhao and D. Goldfarb, in: Zeolites a refined tool for designing catalytic sites, L. Bonneviot and S. Kaliaguine (eds.), Stud. Surf. Sci. Catal. Vol. 97, Elsevier, Amsterdam: 1997, pp. 181-188.

[6] R. Ryoo, S. Jun, J. M. Kim and M.J. Kim, Chem. Commun., 2225 (1997).

[7] C. Bischof and M. Hartmann, in preparation.

[8] Q. Huo, D.I Margolese, U. Ciesla, D.G. Demuth, P. Feng, T.E. Gier, P. Siegler, A. I. Firouzi, B.F. Chmelka, F. Schüth and G.D. Stucky, Chem Mater., 6, 1176 (1994).

[9] S.B. Pu, J.R. Kim, M. Seno and T. Inui, Microporous Mater., 10, 25 (1997).

[10] P.L. Liewellyn, Y. Cillet, F. Schüth, H. Reichert and K.K. Unger, Microporous Mater., 3, 345 (1994).

[11] Z. Luan, , C.F. Cheng, W. Zhou and J. Klinowski, J. Phys. Chem., 99, 1018 (1995).

[12] A. Corma, Chem. Rev., 97, 2373 (1997).

MESOPOROUS MOLECULAR SIEVES 1998
Studies in Surface Science and Catalysis, Vol. 117
L. Bonneviot, F. Béland, C. Danumah, S. Giasson and S. Kaliaguine (Editors)

Preparation of highly ordered MCM-41 with docosyltrimethylammonium chloride (C_{22}TMAC1) as a template and fine control of its pore size

S. Namba, A. Mochizuki, and M. Kito

Department of Materials, Teikyo University of Science & Technology
Uenohara-machi, Kitatsuru-gun, Yamanashi 409-0193, Japan

We succeeded in hydrothermally synthesizing highly ordered silica MCM-41 materials with C_{22}TMACl as a template by reducing the pH (pH=9.0) and the C_{22}TMACl concentration. Moreover the pore sizes of MCM-41 materials were finely controlled by using a template mixture of C_{12}TMABr/C_{16}TMABr or C_{16}TMABr/C_{22}TMACl with various ratios. MCM-41 materials whose pore sizes were finely controlled were also highly ordered. Thus highly ordered MCM-41 materials having *BJH pore size* from 1.8 to 4.2 nm, or intrinsic pore size from *ca.* 2.8 to *ca.* 5.2 nm, were successfully prepared.

1. INTRODUCTION

Recently mesoporous silica molecular sieves have been developed [1-3]. One of them is called "MCM-41", which has a hexagonal structure, uniform one-dimensional mesopores and a very high surface area [2,3]. Pore sizes of MCM-41 materials can be tailored in the range of 1.6-10 nm through the choice of template surfactant and the addition of auxiliary chemicals [1,2]. However, larger pore materials prepared with 1,3,5-trimethylbenzene as an auxiliary chemical display XRD patterns exhibiting a single broad line and also display a broad pore size distribution [2,4]. On the other hand, pore sizes of MCM-41 materials can be increased by increasing the chain length of *n*-alkyl group in *n*-alkyltrimethylammo nium cation used as a template, from C_8 to C_{18}. However, it has been reported that the chain length more than C_{18} gives not a hexagonal but lamellar structure [5].

In our previous paper highly ordered silica MCM-41 could be prepared with hexadecyltrimethylammonium bromide (C_{16}TMABr) as a template by the hydrothermal synthesis and the pore size of highly ordered MCM-41 could be enlarged up to a BJH pore size of 4.7 nm (intrinsic value; *ca*. 6 nm) without loss of structural regularity by addition of 1,3,5-triisopropylbenzene or 1,3,5-trimethylbenzene /1,3,5-triisopropylbenzene mixture as auxiliary chemicals [4].

In this work we tried to prepare MCM-41 materials by using C_{22}TMACI instead of C_{16}TMABr and to finely control the pore size of MCM-41 materials by using a mixture of templates with different alkyl-chain length.

2. EXPERIMENTAL

2.1. Hydrothermal synthesis of MCM-41

MCM-41 materials were synthesized hydrothermally at 373 K by using waterglass (Na_20; 9.5 wt%, SiO_2; 28.6 wt%. Nippon Chemical Industry Co.), sulfuric acid, pure water, and *n*-alkyltrimethylammonium halide as the template. As template, we used C_{12}TMABr (Tokyo Chemical Industry Co.), C_{16}TMABr (Tokyo Chemical Industry Co.), C_{22}TMAC1 (Arquad 22-80, Lion Co.), or their mixtures with various ratios. The procedure was as follows. (1) The template was dissolved in water at 333-343 K and the temperature was held. (2) The waterglass (56.1 g) was mixed with water (120 g). (3) The mixture prepared in (2) was added to the solution prepared in (1). (4) Diluted sulfuric acid (H_2SO_4/H_20=3.6 g/60 g) was added to the mixture and the resulting mixture was stirred vigorously for 0.5 h. (5) The pH was adjusted to the prescribed value by using a small amount of diluted sulfuric acid. (6) Then the mixture was put into a polytetrafluoroethylene vessel and allow to stand for 5-7 days at 373 K. (7) The product was washed with a large amount of warm water (500 ml x 6 times) and dried at 373 K. (8) The dried product was heated to 813 K in air at a rate of 1 K/min and then calcined in air at 813 K for 6 h. The gel composition, pH value, and days for hydrothermal treatment for each run are summarized in Table 1.

2.2. Characterization of MCM-41

MCM-41 materials were characterized by TEM (JEOL JEM-200EX), XRD (JEOL JDX-8030), N_2 adsorption (ASAP-2000) for determination of surface area and pore size distribution (BJH method). Some MCM-41 materials were calcined in air at 973-1273 K for 2 h or treated in a flowing steam (0.5 atm; balance; He) at 1073 K for 2 h to examine thermal and hydrothermal stabilities.

3. RESULTS AND DISCUSSION

3.1. Preparation of MCM-41 with C_{22}TMACl template

MCM-41 was prepared with C_{22}TMACl (MCM-41-22) under the conditions (Run 1 in Table 1) similar to those for MCM-41 prepared with C_{16}TMABr (MCM-41-16) (Run 7 in Table 1). However XRD spectra for the *uncalcined* product

Table 1.
Conditions for hydrothermal synthesis of MCM-41 materials and specific surface area of MCM-41 produced.

Run	Gel composition in molar base							pH	Days of hydrothermal synthesis	Specific surface area / m^2g^{-1}
	SiO_2	C_{22}TMACl	C_{16}TMABr	C_{12}TMABr	H_2SO_4	Na_2O	H_2O			
1	1	0.48			0.14	0.32	82	10	7	667
2	1	0.33			0.14	0.32	82	10	7	818
3	1	0.35			0.14	0.32	82	10	7	978
4	1	0.37			0.14	0.32	82	10	7	1034
5	1	0.35			0.14	0.32	82	9	7	1040
6	1	0.35			0.14	0.32	82	8	7	1018
7	1		0.51		0.14	0.32	75	10	7	1045
8	1	0.17	0.34		0.14	0.32	82	9	5	926
9	1	0.26	0.26		0.14	0.32	82	9	5	995
10	1	0.34	0.17		0.14	0.32	82	9	5	1053
11	1			0.51	0.14	0.32	70	11	5	1584*
12	1		0.17	0.34	0.14	0.32	75	11	5	1194
13	1		0.26	0.26	0.14	0.32	75	11	5	1105
14	1		0.34	0.17	0.14	0.32	75	10	5	1015

* This value may be too large because capillary condensation took place at low P/P_0.

showed only a lamellar phase. The *calcined* product exhibited a extremely weak peak in the XRD spectra and its specific surface area was less than 700 m^2g^{-1} (Table 1). Then we examined the optimum conditions to prepare highly ordered MCM-41-22 materials.

First, the effect of $C_{22}TMACl/SiO_2$ molar ratio was examined (Runs 2-4. Please note that $C_{22}TMACl/SiO_2$ molar ratios used here were much lower than 0.48 in Run 1.). The XRD spectra for the *uncalcined* products are shown in Fig. 1. Both hexagonal and lamellar phases were observed for all samples. However, the fraction of the hexagonal phase is believed to be maximum at $C_{22}TMACl/SiO_2$=0.35. Second, the effect of pH was examined (Runs 3,5,6). The XRD spectra for *uncalcined* samples are shown in Fig. 2. At pH=10 both hexagonal and lamellar phases were observed, while at pH=9 and 8 only hexagonal phase was observed. Moreover, the peak intensity at pH=9 was stronger than the one at pH=8. From these results it is believed that ordered MCM-41-22 can be prepared under the conditions of Run 5 in Table 1.

The BJH pore size distribution for MCM-41-22 prepared under the conditions of Run 5 are shown in Figs. 3(a). The pore size distribution for MCM-41-22 is very sharp as observed for MCM-41-16 in Fig. 4. Moreover, as shown in Fig. 5 the TEM images indicate a regular pore arrangement not only for MCM-41-16 [4,6] but also for the MCM-41-22. The MCM-41-22 thus prepared exhibited a high surface area of 1040 m^2g^{-1} (Table 1). Thus we succeeded in preparing the highly ordered MCM-41-22 with $C_{22}TMACl$ as template under the conditions of Run 5 in Table 1. The BJH pore size of the highly ordered MCM-41-22 prepared here was 4.2 nm. Ciesla *et al*. have reported that the formal application of the BJH method leads to an underestimation of the pore diameter of the MCM-41 materials and the difference between the intrinsic pore diameter and the BJH pore diameter may be *ca*. 1 nm [7]. Therefore the intrinsic pore size of MCM-41-22 is *ca*. 5.2 nm.

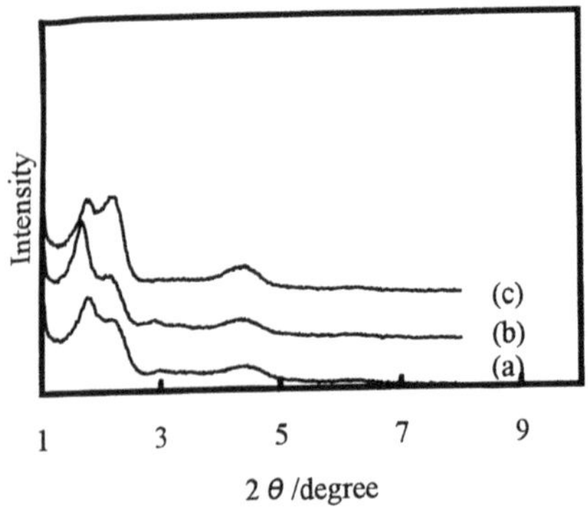

Figure 1. Effect of $C_{22}TMACl/SiO_2$ molar ratio. $C_{22}TMACl/SiO_2$: (a) 0.33, (b) 0.35, (c) 0.37. pH=10.0.

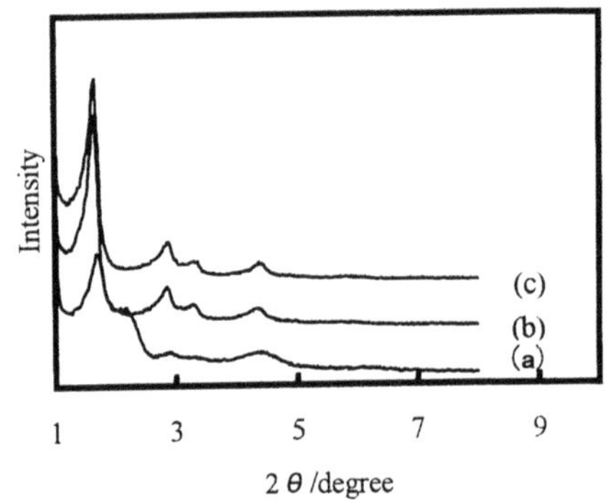

Figure 2. Effect of pH. pH : (a) 10.0, (b) 9.0, (c) 8.0. $C_{22}TMACl/SiO_2$=0.35.

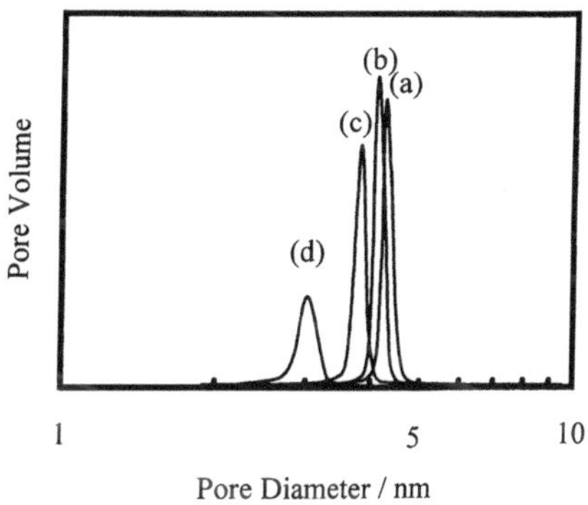

Figure 3. Change in BJH pore size distribution of MCM-41-22 by calcination.
(a) parent, (b) 1073 K, (c) 1173 K, (d) 1273 K.

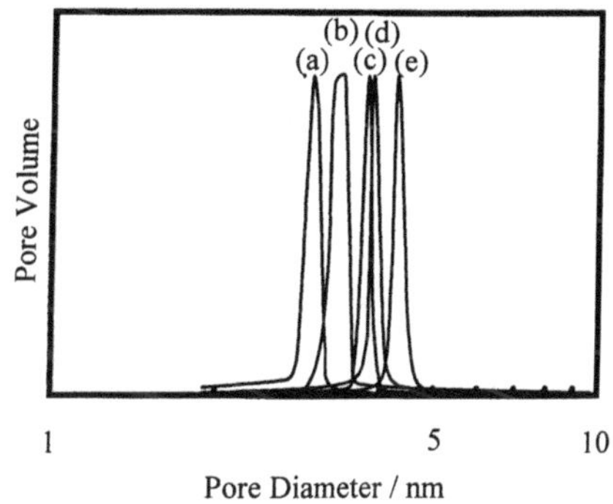

Figure 4. Effect of C_{22}TMACl/C_{16}TMABr ratios on BJH pore size distribution.
C_{22}TMACl/C_{16}TMABr:(a) 0/1, (b) 1/2, (c) 1/1, (d) 2/1, (e) 1/0.

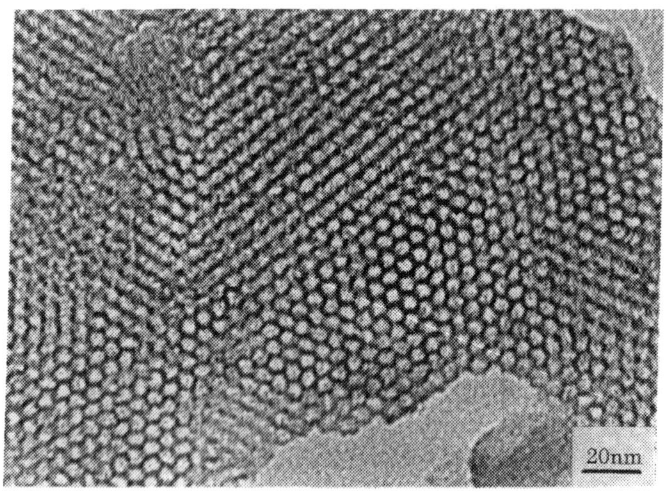

Figure 5. TEM image of MCM-41-22

3.2. Thermal and hydrothermal stabilities of MCM-41-22

In order to establish the thermal stability, MCM-41-22 was calcined at 1073-1273 K and the changes in XRD spectra, BJH pore size distribution and specific surface area upon calcination were measured. Figure 6 shows the change in XRD spectra. The intensity of the four characteristic peaks for the MCM-41-22 calcined up to [illegible]173 K was almost the same as that for the parent MCM-41-22, [illegible] in[illegible]ty for the sample calcined at 1273 K was diminished. The four

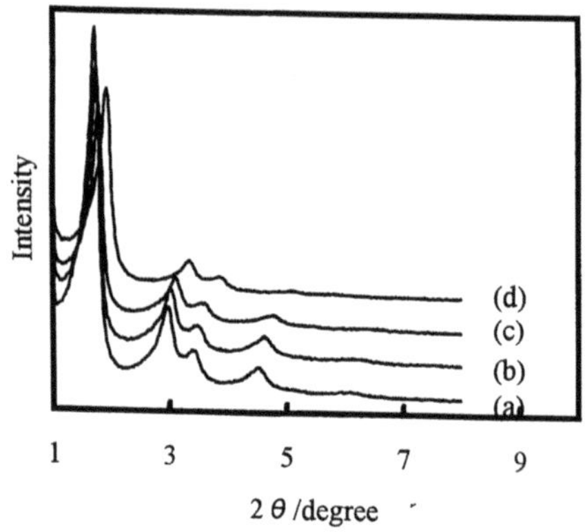

Figure 6. Change in XRD spectra for MCM-41-22 by calcination. (a) parent, (b) 1073 K, (c) 1173 K, (d) 1273K.

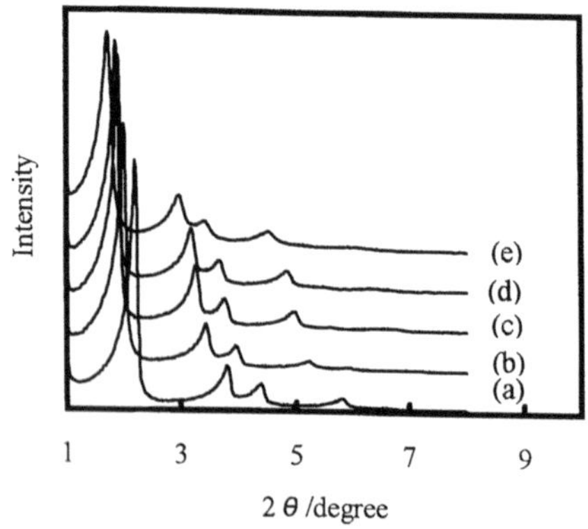

Figure 7. Effect of C_{22}TMACl/C_{16}TMABr ratios on XRD spectra. C_{22}TMACl/C_{16}TMABr:(a)0/1, (b)1/2,(c)1/1,(d)2/1,(e)1/0

peaks were shifted to higher degree at higher calcination temperatures. This fact suggests that as the calcination temperature is raised, more condensation reaction of surface silanol groups proceeds, resulting in smaller pore sizes.

Figure 3 shows the change in BJH pore size distribution upon calcination. The distribution for MCM-41-22 calcined up to 1173 K was as sharp as the one observed for the parent. On the other hand, the distribution for the sample calcined at 1273 K was a little broad and a decrease in pore volume was observed upon calcination. The peak position was shifted to smaller pore sizes with increasing calcination temperatures. This fact also suggests that the condensation reaction of surface silanol groups proceeds.

The surface area of the parent MCM-41-22 was 1040 m^2g^{-1}, while those of the calcined MCM-41-22 at 1073, 1173, and 1273 were 987, 932, and 761 m^2g^{-1}, respectively. The surface area decreased with increasing calcination temperatures. These facts also indicate the condensation reaction of silanol groups. Moreover partial destruction of a hexagonal structure is suggested by the calcination at 1273 K. However the MCM-41-22 calcined at 1273 K still has a hexagonal structure. Thus the highly ordered MCM-41-22 materials is thermally stable up to 1173 K.

The highly ordered MCM-41-22 was treated in a flowing steam at 1073 K for 2 h. The surface area of MCM-41-22 decreased from 1040 to 785 m^2g^{-1} (25 % loss) by steaming. The decrease in surface area by steaming was a little larger than that of MCM-41-16 (16 % loss) [4].

3.3. Fine control of pore size by using mixed templates

When MCM-41 materials are synthesized with a mixture of two kinds of *n*-alkyltrimethylammonium cations with different alkyl-chain length, the question

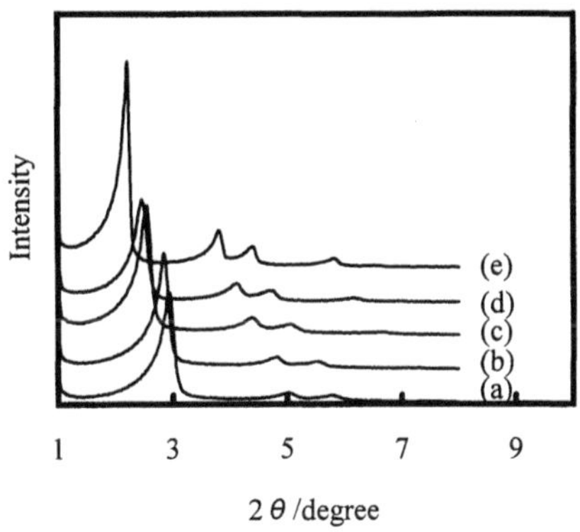

Figure 8. Effect of C_{16}TMABr/C_{12}TMABr ratios on XRD spectra. C_{16}TMABr/C_{12}TMABr: (a) 0/1 (b) 1/2, (c) 1/1, (d) 2/1, (e) 1/0.

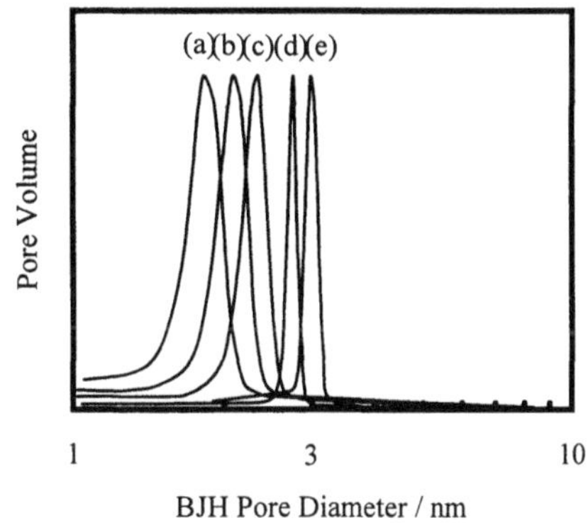

Figure 9. Effect of C_{16}TMABr/C_{12}TMABr ratios on BJH pore size distribution. C_{16}TMABr/C_{12}TMABr: (a) 0/1, (b) 1/2, (c) 1/1, (d) 2/1, (e) 1/0.

may be raised whether the product has uniform or two kinds of pores. The MCM-41 materials were prepared by using the mixture of C_{22}TMACl/C_{16}TMABr with various molar ratios under the conditions of Runs 5, 7-10 (Table 1). The XRD spectra of them are shown in Fig. 7. The characteristic four peaks were observed for every samples. The BJH pore size distributions are shown in Fig. 5. Every distribution was sharp and the pore size decreased with decreasing C_{22}TMACl /C_{16}TMABr ratios. Their specific surface areas are summarized in Table 1. The surface area of every sample was very high. These results show that MCM-41 materials prepared with a mixture of C_{22}TMACl/C_{16}TMABr are highly ordered and their pore size can be finely controlled by changing the C_{22}TMACl/C_{16}TMABr ratios.

MCM-41 materials were also prepared by using a mixture of C_{16}TMABr /C_{12}TMABr with various molar ratios under the conditions of Runs 7,11-14. Figure 8 shows the XRD spectra. The characteristic four peaks were observed for every sample. Figure 9 shows the BJH pore size distributions. Every distribution was sharp and the pore size decreased with decreasing C_{16}TMABr/C_{12}TMABr ratios. The specific surface areas for these materials are summarized in Table 1. The surface area of every sample was very high. These results also indicate that the MCM-41 materials prepared with the mixture of C_{16}TMABr/C_{12}TMABr are highly ordered and the pore size can be finely controlled by changing C_{16}TMABr /C_{12}TMABr ratios.

Thus highly ordered MCM-41 materials having a *BJH pore size* from 1.8 to 4.2 nm, or an intrinsic pore size from 2.8 to 5.2 nm, were successfully prepared.

Acknowledgements

This work has been carried out as a research project of The Japan Petroleum Institute commissioned by the Petroleum Energy Center with the support of the Ministry of International Trade and Industry.

REFERENCES

1. T. Yanagisawa, T. Shimizu, K. Kuroda and C. Kato, Bull. Chem. Soc. Jpn., 63 (1990) 988.
2. C.T. Kresge, M.E. Leonowicz, W.J. Roth, J.C. Vartuli and J.S. Beck, Nature, 359 (1992) 710.
3. J.S. Beck, J.C. Vartuli, W.J. Roth, M.E. Leonowicz, C.T. Kresge, K.D. Schmitt, C.T.-W. Chu, D.H. Olson, E.W. Sheppard, S.B. McCullen, J.B. Higgins and J.L. Schlenker, J. Am. Chem. Soc., 114 (1992) 10834.
4. S. Namba and A. Mochizuki, Res. Chem. Intermed., in press.
5. Q. Huo, D.I. Margolese, and G.D. Stucky, Chem. Mater., 8 (1996) 1147.
6. S. Namba, N. Sugiyama, M. Yamai, I. Shimamura, S. Aoki and J. Izumi, Stud. Surf. Sci. Catal., 105 (1997) 1891.
7. U. Ciesla, M. Grun, T. Isajeva, A.A. Kurganov, A.V. Neimark, P. Ravikovitch, S. Schacht, F. Schuth and K.K. Unger, in Access in Nanoporous Materials, T.J. Pinnavia and M.F. Thorpe (eds.), Plenum Press, New York, 1995, p.231.

MESOPOROUS MOLECULAR SIEVES 1998
Studies in Surface Science and Catalysis, Vol. 117
L. Bonneviot, F. Béland, C. Danumah, S. Giasson and S. Kaliaguine (Editors)

Photoluminescence spectroscopic monitoring in the synthesis of mesoporous materials by microwave-induced heating

S.-E. Park,* D.S. Kim, J.-S. Chang, and W.Y. Kim

Industrial Catalysis Research Laboratory, Korea Research Institute of Chemical Technology (KRICT), Taejon 305-606, Korea

The microwave synthesis of mesoporous MCM-41 material gave a very short period of synthesis time. This was confirmed by monitoring the change of silicate species during microwave synthesis with photoluminescence (PL) spectroscopy. PL spectra of the precursor gel of MCM-41 show an emission band at 367 nm upon the excitation at 220 nm, which is implies the L-center of silicate species in the precursor gels. Attenuated total reflectance (ATR/FTIR) spectra of asymmetric Si-O-Si stretching bands of MCM-41 gel in the region of 1000 - 1250 cm^{-1} are consistent with the PL spectra. Pyrene was used for a luminescence probe in order to monitor the change of dielectric property of the micellar surfactant aggregates. These results support that upon irradiating the microwave the interior of the micellar aggregates becomes more hydrophobic due to the condensation of the silicate surrounding micellar surface. Specially, the ripening of mesoporous silicate network is accelerated by the microwave irradiation of the gel having surfactant due to its microwave-susceptible head group. It is demonstrated that the microwave irradiation not only accelerates the rate of crystallization but also directs the crystallization mechanism like hydrothermal heating.

1. INTRODUCTION

Ever since the liquid crystal templating (LCT) mechanism, in which supramolecular assemblies of cationic alkytrimethylammonium surfactant serve as components of the template for the formation of mesoporous materials [1], so-called M41S family, was proposed to account for their formation [2], modified LCT mechanism [3], cooperative templating model (CTM) [4], ion-exchange mechanism [5], heterogeneous nucleation mechanism (HNM) [6], and worm-like micelle mechanism (WMM) [7] have been proposed from the viewpoint of the organized assemblies of templating surfactants.

Recently, it has been shown that microwave-induced heating technique for the synthesis of zeolites has several fascinating advantages in comparison with the conventional hydrothermal heating [8] such as homogeneous heating throughout a reaction vessel. The microwave technique results in more homogeneous nucleation and shorter crystallization time [9]. The short crystallization time was evidenced in the synthesis of mesoporous MCM-41 material by using microwave irradiation [10,11]. In addition to the shorter synthesis time, microwave irradiation in the synthesis of MCM-41 with the aid of ethylene glycol provided a new way to control its crystallinity and morphology [12].

The role of surfactant chemistry is the key in the formation of mesoporous molecular sieves. The silicate species in solution play an important role in the organization of the surfactant molecules and the electrostatic interaction between the inorganic and surfactant ions determines the morphology of the mesophase [4]. However, there is no report that deals with the relationship between microwave characteristics and surfactant templates or solvent in the synthesis of molecular sieves. But they are eligible to interact with microwaves because surfactant molecules have both a polar head group and a nonpolar end [13].

Several tools such as XRD, ^{14}N and ^{29}Si NMR, EPR, FTIR, and rheological measurements were used for elucidating the synthesis mechanism in the formation of mesoporous materals [3,7,14,15]. Recently, real time, in-situ ATR/FTIR monitoring of M41S-type mesoporous silicate syntheses was used to observe the state of gel phase in synthetic mixture and transformation of silicate structure [16]. In addition, luminescence probe analysis has been extensively studied to characterize microenvironment of surfactant micelles for last two decades [17].

In this work, PL method combined with the probing technique is used as a tool for monitoring the intermediate steps in the synthesis of mesoporous MCM-41 material in addition to ATR/FTIR under the microwave irradiation comparing with hydrothermal conditions.

2. EXPERIMENTAL

Microwave preparation for MCM-41 was carried out in a CEM microwave oven (MDS-2000) with 2450 MHz of microwave frequency. Detailed synthetic procedure was described in our previous report [11]. Myristyl trimethylammonium bromide [MTAB; $C_{14}H_{29}N(CH_3)_3Br$] was used as a quarternary ammonium surfactant. A sodium silicate solution was prepared by mixing aqueous NaOH solution with Ludox HS 40 (39.5 wt% SiO_2, 0.4 wt% Na_2O and 60.1 wt% H_2O, Du Pont) and heating the resultant solution with stirring at 60 °C for 30 min, which was then slowly added to 25 wt% aqueous MTAB solution, with vigorous stirring at room temperature for 1h. The molar composition of resultant gel used in this study was SiO_2: MTAB: NaOH: H_2O = 1.0 : 0.167 : 0.5 : 40.5. The resulting solid product obtained from microwave-induced heating was isolated by filtering, washing with deionized water, and drying in air at 100 °C for 10 h. To remove the organic species occluded in the pores of MCM-41, the as-synthesized samples were calcined at 550 °C for 6 h in air. For comparison, hydrothermal preparation of MCM-41 was carried out at 100 °C for 2 days by using the same gel composition of the microwave synthesis.

In the preparation of MCM-41, the precursor gels and the prepared samples were characterized by ATR/FTIR and PL spectroscopy. ATR/FTIR spectra were recorded at room temperature on a Nicolet FT-IR spectrometer (model Magna 560) using horizontal ATR kit equipped with ZnSe window (transmission window: 650 - 20000 cm^{-1}). PL spectra were measured at room temperature using a Shimazu spectrophotometer (model UV-2501 PC). For ATR/FTIR and PL analyses, the ex-situ monitoring of MCM-41 mesoporous material synthesis has been done by removing small aliquot samples periodically from the ongoing synthetic mixture for characterization. Pyrene as a probe was used for luminescence probing experiments after dissolving in solvents or MCM-41 gel at 2 or 10 μM of pyrene concentration.

3. RESULTS AND DISCUSSION

3.1. Microwave synthesis of molecular sieve MCM-41

The typical preparation of MCM-41 under hydrothermal condition needs 2 days at 100 °C to form crystalline MCM-41, whereas the microwave preparation of MCM-41 takes only 40 min at 100 - 120 °C [11]. The BET surface areas and pore volumes of MCM-41 materials prepared by microwave-induced heating are 700 - 1150 m^2/g and 0.60 - 0.78 cm^3/g, respectively. The d_{100} spacings of products from microwave or hydrothermal heating are 35 - 37 Å. The mesoporous structures of these samples with hexagonal arrangement is also confirmed with transmission electron microscopy (TEM). These results demonstrate that mesoporous material of high crystallinity can be obtained by microwave preparation method.

Using the BJH equation with the N_2 desorption isotherm, the pore diameter of MCM-41 is approximately 21 Å. Combining the results from XRD and N_2 adsorption-desorption, it is confirmed that the thickness of the silicate wall is about 11 Å. In addition to the reduced crystallization time, another feature of the microwave method lies in its ability to provide significant efficiency to form relatively small crystals at higher rates of crystallization compared with conventional hydrothermal procedures. This beneficial effect evidently originates from the transformation of microwave energy into heat through ionic conduction and dipole rotation during the crystallization of precursor gel having dielectric properties.

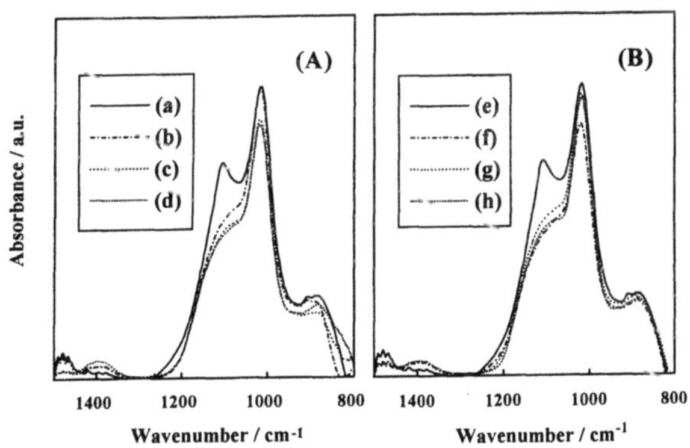

Figure 1. ATR spectra of of MCM-41 precursor gel: (A) (a) before treatment, (b) microwave heating at 120 °C for 5 min, (c) for 10 min under 480 W, and (d) subsequent microwave heating at 100 °C for 20 min under 60 W after treatment (c); (B) (e) before treatment, (f) hydrothermal heating at 100 °C for 4 h, (g) for 16 h, and (h) for 34 h.

3.2. ATR/FTIR analyses of MCM-41 precursor gels

ATR spectrum of aqueous colloidal silica solution containing 39.5 wt% SiO_2 used in this study exhibits a strong absorption at 1150 cm^{-1} with a shoulder at 1050 cm^{-1} (not shown). When NaOH is added to colloidal silica solution, absorption bands at 1020 cm^{-1} and 1108 cm^{-1} are observed. These two bands are different from those of colloidal silica. In this case, former band is dominantly present in comparison with latter band. This suggests the formation of oligomeric silicate species, probably cubic octamer silicate ($Si_8O_{20}^{8-}$), which is ascribed to depolymerization of colloidal silica in the presence of NaOH.

The ATR spectra of MCM-41 precursor and samples obtained from microwave heating for 5, 10 and 30 min, respectively, are measured and compared in Figure 1(A). These spectra include mainly asymmetric Si-O-Si stretching (1000 - 1250 cm^{-1}) and surfactant deformation (1350 - 1500 cm^{-1}) regions. Below 1200 cm^{-1}, the spectrum of the gel containing MTAB surfactant exhibits also two prominent absorption bands at 1020 cm^{-1} and 1108 cm^{-1} with a weak shoulder at 1220 cm^{-1}. Comparing with the mixture of colloidal silica solution and NaOH, bandshape of these two bands is rather sharp and the positions of the bands are slightly shifted. It is assumed that the first band is assigned to the asymmetric Si-O-Si stretching of silicate polyanions and the second one is attributed to silicate oligomers assembled with headgroups of surfactant in micelle. As the gel is heated to 120 °C and subsequently 100 °C with microwave irradiation and heating time increases, the band at 1113 cm^{-1} decreases and then the band at 1018 cm^{-1} becomes a major peak, indicating irreversible polymerization of silicate [16]. In addition, a broad band at around 910 cm^{-1} is present in the gel, which is ascribed to the overlap of the asymmetric (967 cm^{-1}) and symmetric stretching (909 cm^{-1}) bands of CH_3-N^+ in surfactant [16]. However, there is no significant change of the band at 910 cm^{-1}. In general, transmission FT-IR spectrum of solid MCM-41 sample after crystallization diluted with KBr powder shows absorption bands at 1220 and 1020 cm^{-1}. It is assumed that the band at 1220 cm^{-1} is an asymmetric Si-O-Si stretching vibration related to $Si(OSi)_3$ of MCM-41 framework. However, band at 1220 cm^{-1} in ATR spectra of microwave treated samples is weak, which is different from the solid state spectrum of as-synthesized MCM-41. This suggests that ATR spectra reflect the dominant species in the solution and gel phases of the synthetic mixture and thus as-synthesized products in suspension have negligible contribution to the ATR spectra of the mixture. This trend is in good agreement with Calabro *et al.*'s observation monitored in the synthesis of mesoporous silicate by *in-situ* ATR/FTIR study[16]. Figure 1(B) also shows the comparative ATR spectra of MCM-41 precursors with hydrothermal heating as a function of time for 0, 4, 16, and 34 h, respectively. These spectra exhibit the trend similar to those of Figure 1(A) except shortening synthesis time.

In addition, the behavior of ATR spectra in the surfactant region of 1300 -

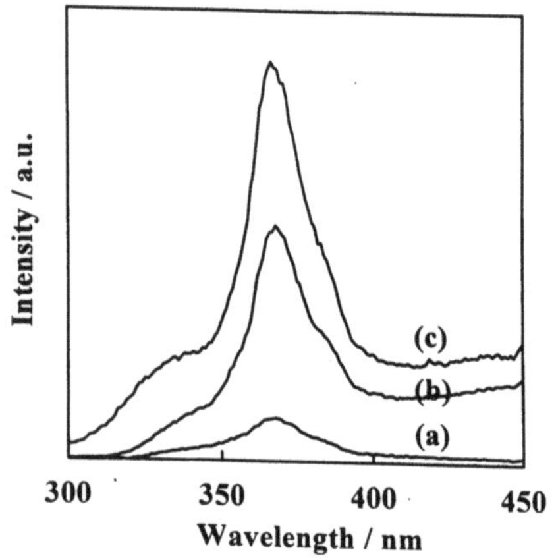

Figure 2. PL spectra of (a) 39.5 wt.% aqueous colloidal SiO_2 solution, (b) the mixture of NaOH and colloidal SiO_2 solution (mol. ratio of SiO_2:NaOH = 1:0.5), and (c) MCM-41 gel contained MTAB surfactant.

1500 cm^{-1} during the synthesis of MCM-41 are also interesting. The features for both microwave-induced heating and hydrothermal heating seem identical. In the region of surfactant deformation between 1350 - 1500 cm^{-1}, ATR spectrum of the precursor gel exhibits relatively sharp peaks at 1490, 1479, and 1467 cm^{-1}. The first two absorption bands are assigned to a split pair of asymmetric methyl (CH_3-N^+) headgroup deformation bands of surfactant molecule. Absorption at 1467 cm^{-1} is assigned to the CH_2 'scissoring' band of surfactant molecule[16]. When the precursor gel is heated hydrothermally or irradiated by microwave for longer time, the absorption bands in the region of 1430 - 1500 cm^{-1} decrease and the bands in the region of 1430 - 1300 cm^{-1} increase gradually. It is known that the absorption band at 1467 cm^{-1} is sensitive to the state of packing of hydrocarbon tail's. A decrease in the intensity of absorption band at 1467 cm^{-1} is expected due to an increase in the *gauche* conformer content of the methylene chains as the hydrophobic region of the micelle adopts a more fully disordered fluid-like state [18]. Therefore, it would indicate the increase in the degree of disorder for stacked methylene chain during nucleation and crystallization.

3.3. PL analyses of MCM-41 precursor gels

PL spectra of colloidal silica in the presence and/or the absence of NaOH and MTAB surfactant are shown in Figure 2. Only weak emission band at 367 nm with shoulders at 340 and 370 nm upon fixing 220 nm of excitation wavelength is observed in colloidal silica solution due to the presence of sodium cation. These bands in sodium silicate solution and the mixture of colloidal silica and MTAB surfactant increase greatly but are not observed in surfactant solution which does not contain Si source. Moreover, the PL intensity emitted at 367 nm is most prominent in the precursor gel containing MTAB. Trukhin has reported that L-center of sodium silicate glass gives emission at 365 nm upon excitation at 207 nm [19,20]. He assigned this band corresponding to emission induced by defect center or structure of [≡Si-O$^-$Na$^+$] species in tetrahedral framework. However, in this work, emission band at 367 nm upon excitation at 220 nm would be assigned favorably to the represented [≡Si-O$^-$Na$^+$] and/or [≡Si-O$^-$C$_{14}$TMA$^+$] species.

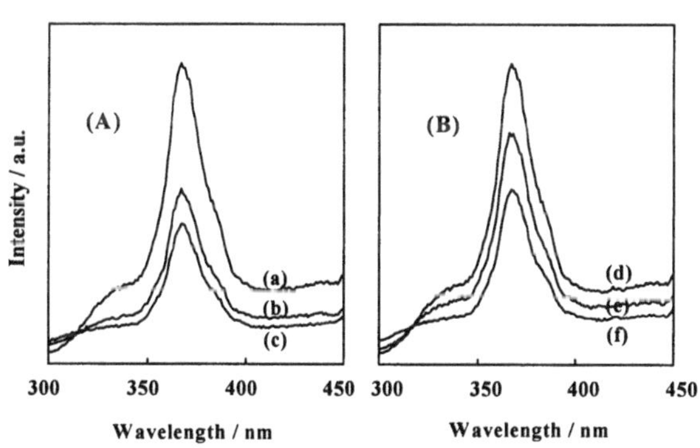

Figure 3. PL spectra of MCM-41 precursor gel: (A) (a) before treatment, (b) microwave heating at 120 °C for 10 min under 480 W of microwave power, and (c) subsequent microwave heating at 100 °C for 20 min under 60 W of microwave power after treatment (b); (B) (d) before treatment, (e) hydrothermal heating at 100 °C for 4 h, and (f) for 12 h: λ_{ex} = 220 nm.

PL spectra (λ_{ex} = 220

nm) of MCM-41 precursor gel solution with microwave irradiation and hydrothermal heating are compared in Figure 3. Intensity of emission band at 367 nm decreases when synthesis temperature and time increases with treatment of microwave irradiation or hydrothermal heating. This evidence implies that during this treatment, anionic silicate species (≡Si-O$^-$) encapsulated around the head group of the surface of rod-like or worm-like micelles continue to condense, leading to the formation of silicate walls and network. The significant feature in the synthesis of mesoporous material by microwave heating is the prominent effect for reducing the synthesis time comparing with hydrothermal heating. The PL intensity obtained by microwave heating only for 10 min is comparable to or less than that from hydrothermal condition for 12 h. This behavior is consistent with the observation of ATR/FTIR investigation.

Pyrene shows a number of photophysical features that make it an attractive fluorophore to probe the microenvironment in micellar aggregates [17]. The relative intensity of the vibrational fine structure peaks in the emission spectrum is sensitive to the probe environment, allowing for the determination of the local polarity at the solubilization site [17,21]. Figure 4 presents PL spectra (λ_{ex}=310 nm) for pyrene dissolved in pure solvents and MCM-41 precursor gel. As reported earlier by Thomas *et al.* [21], these spectra clearly show the five peaks due to the vibrational fine structure of pyrene. However, in these spectra there is no fluorescence of excimer species [22] due to sufficient dilution of pyrene in micellar solution. For the sake of convenience in subsequent discussions, two important peaks among the five predominant peaks are numbered as I and III, and the ratio in the intensity of peak III is discussed with respect to the 0-0 band (peak I). It is known that the intensity ratio of peak III/peak I considers as a measure of solvent polarity [21]. In Figure 4, the III/I ratio of pyrene in water solvent is 0.55, while that in precursor gel solution containing MTAB surfactant is

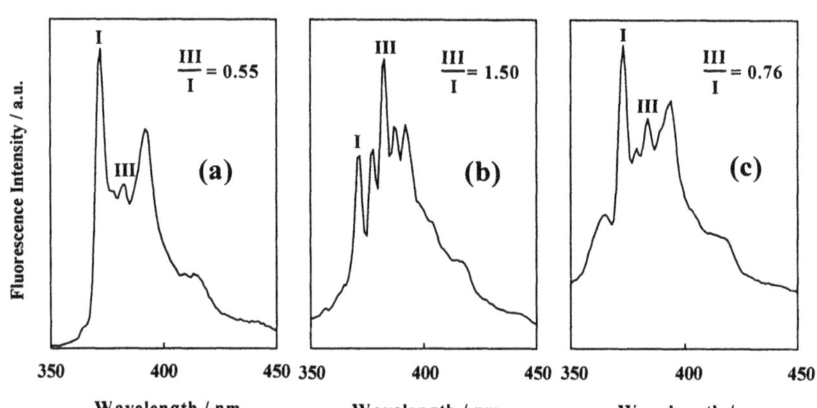

Figure 4. PL spectra of pyrene in dissolved in solvents and MCM-41 precursor gel: (a) pyrene in H_2O, (b) pyrene in n-hexane, and (c) pyrene in MCM-41 precursor gel: [pyrene] = 2 μM for (a) and (c), and [pyrene] = 10 μM for (b); λ_{ex} = 310 nm.

0.76, indicating that pyrene moves into the hydrocabon core region. For normal micellar system, the dielectric constant is low in the micellar core and high at the micellar surface, so that pyrene is preferentially solubilized in the interior hydrophobic regions of micelles since it is a strong hydrophobic probe.

The III/I ratio of pyrene after 20 min of microwave treatment increases up to 0.98. It has been observed that above critical micelle concentration the III/I ratio is quite dependent on the nature of the surfactant head group, but is independent of the surfactant concentration, length of the hydrocarbon chain, or presence of external additives such as hydrocarbon or electrolytes [23]. It is interesting that the III/I ratio of pyrene remains in the range of 0.7 - 1.0 for micellar system regardless of microwave treatment even though pyrene in pure hydrocarbon solvent has a high III/I ratio (e.g., 1.50 for n-hexane). However, it is noted that a pyrene probe is in association-dissociation equilibrium with micelles, and that the observed spectrum of a probe is thus a statistical average over time spent in each location in the micelle. The observed III/I ratio can also be interpreted as the degree of the compactness of the head group region and the extent of surface charge [24]. An alternative explanation can also be proposed in terms of the extent of water penetration into the micellar system. It has been suggested that water can enter the micelles and extend up to several carbons from the head group of the micelle [23]. Therefore, the increase in the III/I ratio of pyrene after microwave heating is regarded as the increase of the compactness of the head group region of the micelle, resulting from polymerization and encapsulation of silicate species around micellar surface. This result also demonstrates that the interior of the micellar aggregates becomes hydrophobic as the mesoporous structure of MCM-41 is formed.

4. CONCLUSIONS

Mesoporous material MCM-41 could be synthesized from the precursor gel containing C_{14} surfactant (MTAB) within a short period of time by the microwave irradiation with 2450 MHz frequency. A shorter period of time compared with hydrothermal method was monitored and confirmed by using PL spectroscopy with the probe technique and ATR/FTIR analysis. From PL and ATR/FTIR spectra according to synthesis condition, the proposed mechanism of MCM-41 formation with microwave irradiation is apparently similar to that with hydrothermal heating. PL spectra with the probe technique suggest that the compactness of the headgroup region of the micellar aggregates and thus the rigidity of surfactant aggregates is significantly enhanced by the microwave treatment of MCM-41 gel. Hence, the enhancement in the crystallization rate of MCM-41 seems to be ascribed to the uneven susceptibility of surfactant head group and silicate species to the microwave irradiation.

ACKNOWLEDGMENTS

This work was supported by a grant from the Ministry of Science and Technology in Korea. We thank to our colleagues, Dr. C.W. Lee and Dr. H.M. Sung-Suh, for helpful discussion.

REFERENCES

1. J.S. Beck, M.C. Vartuli, Current Opinion in Solid State and Mater. Sci., 1 (1996) 76.
2. C.T. Kresge , M.E. Leonowicz, W.J. Roth, J.C. Vartuli and J.S. Beck, Nature, 359 (1992) 710.; J.S. Beck, M.C. Vartuli, W.J. Roth, M.E. Leonowicz, C.T. Kresge, K.O. Schmitt, C.T.W. Chu, D.H. Olson, E.W. Sheppard, S.B. McCullen, J.B. Higgins and J.L. Schlenker, J. Am. Chem. Soc., 114 (1992) 10834.
3. C.Y. Chen, S.L. Burkett, H.X. Li and M.E. Davis, Microporous Mater., 2 (1993) 27.
4. A. Monnier, F. Schuth, Q. Huo, D. Kumar, D. Margolese, R.S. Maxwell, G.D. Stucky, M. Krishnamurthy, P. Petroff, A. Firouzi, M. Janicke and B.F. Chmelka, Science, 262 (1993) 1299.
5. S. Inagaki, Y. Fukshima and K. Kuroda, J. Chem. Soc., Chem. Commun., (1993) 680.
6. J. Liu, A.Y. Kim, J.W. Virden and B.C. Bunker, Langmuir, 11 (1995) 689.
7. Y.S. Lee, D. Surjadi and J.F. Rathman, Langmuir, 12 (1996) 6202.
8. J. C. Jansen, A. Arafat, A. K. Barakat and H. van Bekkum, Synthesis of Microporous Materials, M. L. Occelli and H. E. Robson (eds.), Van Nostrand Reinhold, New York, 1992, Vol. 2, p. 507.
9. U. Lohse, R. Bertram. K. Jancke, I. Kurzawski, B. Parlitz, E. Loeffler and E. Schreier, J. Chem. Soc. Faraday Trans., 91 (1995) 1163.
10. C. -G. Wu and T. Bein, Chem. Commun., (1996) 925.
11. S.-E. Park, D.S. Kim, J.-S. Chang and W.Y. Kim, Catal. Today, in press (1998).
12. D.S. Kim, J.-S. Chang, W.Y. Kim and S.-E. Park, Chem. Commun., submitted (1998).
13. R. Dagani, Chem. Eng. News, Feb. 10, pp. 26-33 (1997).
14. J.Y. Zhang, Z. Luz and D. Goldfarb, J. Phys. Chem. B, 101 (1997) 7087.
15. A.A. Romero, M.D. Alba, W. Zhou and J. Klinowski, J. Phys. Chem. B, 101 (1997) 5294.
16. D. Calabro, E.W. Valyocsik and F.X. Ryan, Microporous Mater., 7 (1996) 243.
17. J.K. Thomas, Chem. Rev., 80 (1980) 283.
18. D.R. Scheuing and J.G. Weers, Colloids Surf., 55 (1991) 41.
19. A.N. Trukhin, J. Non-Cryst. Solids, 123 (1990) 250.
20. A.N. Trukhin, J. Non-Cryst. Solids, 189 (1995) 1.
21. M. Almgreen, F. Grieser and J.K. Thomas, J. Am. Chem. Soc., 102 (1980) 3188.
22. N.J. Turro, M. Gratzel and A.M. Braun, Angew. Chem. Int. Ed. Engl., 19 (1980) 675.
23. K. Kalyanasundaram and J.K. Thomas, J. Am. Chem. Soc., 99 (1977) 2039.
24. H. Itoh, S. Ishido, M. Nomura, T. Hayakawa and S. Mitaku, J. Phys. Chem., 100 (1996) 9047.

MESOPOROUS MOLECULAR SIEVES 1998
Studies in Surface Science and Catalysis, Vol. 117
L. Bonneviot, F. Béland, C. Danumah, S. Giasson and S. Kaliaguine (Editors)

Effect of hydrothermal conditions on the synthesis of siliceous MCM-48 in mixed cationic-anionic surfactants systems

Fengxi Chen, Xuewu Yan and Quanzhi Li

Department of Chemistry, Fudan University, Shanghai 200433, P.R.China

This work examines the effect of specific synthesis variables (i.e., the molar ratio of cationic to anionic surfactant, the carbon number of anionic surfactants, the concentration of the mixed surfactants, the molar ratio of mixed surfactants to SiO_2, crystallization time and temperature) on the synthesis of siliceous MCM-48 (cubic, Ia3d), and obtains its suitable synthesis conditions when the micelles of mixed cetyltrimethylammonium bromide (CTMAB) and sodium carboxylate (SC) are used as templates. These results also show that different M41S-type mesoporous molecular sieves may be synthesized using the micelles of mixed CTMAB-SC as templates, which is complementary to the prevailing electrostatic and neutral templating routes.

1. INTRODUCTION

Since Mobil researchers reported the discovery of a novel family of silica-based mesoporous molecular sieves (M41S) in 1992 [1,2], there has been a growing interest in synthesizing the mesoporous materials with alkyltrimethylammonium surfactant cations ($C_nH_{2n+1}Me_3N^+$, n>6). The M41S family contains MCM-41 (hexagonal, P6m), MCM-48 (cubic, Ia3d) and other species whose structures have not yet been elucidated. MCM-41 has a hexagonal arrangement of unidimensional mesopores. MCM-48 exhibits cubic structure of three dimensional mesopores which can be indexed to an Ia3d unit cell. Among the M41S materials, since the unidimensional mesopores of MCM-41 make it sensitive to blockage and limit its use as catalysts, adsorbents and host matrices, more attention was recently paid to the synthesis and characterization of MCM-48 [3-12]. But the synthesis of MCM-48 requires more specific conditions than that of MCM-41. With the common alkyltrimethylammonium surfactants, a representative synthesis of MCM-48 was presented by Monnier et al [4]. According to their method, MCM-48 was synthesized from a reaction mixture with a molar composition of TEOS:0.25Na_2O:0.65CTMACl:62H_2O at 373K for 3 days. In this system the CTMACl:Si molar ratio and the concentration of CTMACl were 0.65:1 and 13 wt %, respectively. Recently, we reported an improved synthesis of siliceous MCM-48 using mixed cationic-anionic (catanionic) surfactants [i.e., the mixture of cetyltrimethylammonium bromide ($C_{16}H_{33}Me_3NBr$, CTMAB) and sodium carboxylate ($C_nH_{2n+1}COONa$, n=11,13,15,17, SC)] as templates [13]. The improved synthesis of the MCM-48 requires much lower CTMA/silica molar ratio (0.152) and low concentration (5 wt %) of mixed surfactants than those required for a single cationic surfactant system. In this paper, we systematically investigate the effect of hydrothermal conditions on the synthesis of siliceous MCM-48 and obtain its suitable synthesis conditions in the mixed surfactants system.

2. EXPERIMENTAL

2.1. Investigation of the effect of specific synthesis variables

The reactives employed were: Sodium silicate aqueous solution (25.42 wt % SiO_2, 7.40 wt % Na_2O), cetyltrimethylammonium bromide ($C_{16}H_{33}Me_3NBr$, CTMAB), sodium laurate ($C_{11}H_{23}COONa$, SL) and deionized distilled water. A typical synthetic procedure is outlined as follows: The sodium silicate aqueous solution was added dropwise to the mixed solutions of different CTMAB/SL molar ratio according to the gel molar composition described in Table 1. After being stirred at 30^{0}C for 1h, the resultant homogenous gel was crystallized in a static stainless steel autoclave at various temperature for 4 or 8 days. The precipitated product was recovered and calcined as reported in the literature [2]. The similar procedure was used in the mixed solutions containing other carboxylate anionic surfactants with n>11.

According to the above systematic study on the different synthesis variables, we obtain the suitable synthesis conditions of siliceous MCM-48 in mixed catanionic surfactants system, that is, the mixture with molar composition of SiO_2:0.147~0.154CTMAB: 0.011~0.017SC:0.250~0.282Na_2O:58H_2O is crystallized at 100^{0}C for 4 days.

2.2. Characterization

The calcined MCM-48 was characterized by X-ray powder diffraction (XRD), transmission electron microscopy (TEM), N_2 adsorption/desorption isotherm. The XRD pattern was carried out on a Rigaku D/MAX-IIA equipment using the Cu Kα radiation at 40kV and 20mA. The HREM image was obtained with a JEM-200CX transmission electron microscope operated at 200KV from a thin section prepared by ultramicrotomy. The N_2 adsorption and desorption isotherm was carried out at 77K on a ASAP-2000 apparatus and the pore size distribution was calculated from the desorption curve using the BJH model with the Halsey equation for multilayer thickness.

3. RESULTS AND DISCUSSION

Table 1 summarizes the investigation of the effect of specific synthesis variables (i.e., the molar ratio of CTMAB to SL, the concentration of the mixed surfactants, the molar ratio of mixed surfactants to SiO_2, crystallization time, temperature and so on) on the synthesis of siliceous MCM-48 (cubic, Ia3d) in the solutions containing mixed catanionic surfactants.

Table 1
Effect of specific hydrothermal conditions on the synthesis of siliceous MCM-48 in the mixed cationic-anionic surfactants system

Sample number	Gel molar composition SiO_2	:CTMA:	SL:	Na_2O:	H_2O	Crystallization Temperature(^0C)	Crystallization time (days)	Product Obtained
1	1	0.162	0	0.282	58	100	4	MCM-41
2	1	0.154	0.013	0.282	58	100	4	MCM-48
3	1	0.152	0.017	0.282	58	100	4	MCM-48
4	1	0.147	0.025	0.282	58	100	4	MCM-48
5	1	0.135	0.045	0.282	58	100	4	MCM-50
6	1	0.100	0.101	0.282	58	100	4	GEL
7	1	0.152	0.017	0.282	25	100	4	MCM-50
8	1	0.089	0.010	0.282	58	100	4	MCM-48
9	1	0.029	0.003	0.282	58	100	4	MCM-41
10	1	0.614	0.068	0.282	247	100	4	MCM-50
11	1	0.076	0.008	0.282	27	100	4	MCM-41
12	1	0.152	0.017	0.282	58	100	8	MCM-41
13	1	0.152	0.017	0.282	58	20	4	MCM-41
14	1	0.152	0.017	0.282	58	70	4	M.P.*
15	1	0.152	0.017	0.282	58	130	4	MCM-50
16	1	0.152	0.017	0.282	58	170	4	ZSM-5

* M.P. mixed products of MCM-41 and MCM-48.

In the mixed catanionic surfactants system, the assembly process of inorganic and mixed surfactants is dependent on the molar ratio of CTMAB to SL. At the constant concentration (5 wt %) of mixed surfactants and the constant molar ratio (0.168:1) of mixed surfactants to silica, the as-synthesized products were correspondingly MCM-41, MCM-48, MCM-50 (lamellar) and gels when the SL/CTMA molar ratio was 0, 1/12~1/6, 1/6~1/3, 1/3~1, respectively (see Sample 1~6). In classical micelle chemistry, the expected mesophase sequence is a function of the local effective surfactant packing parameter, $g = V/a_0l$, where V is the local volume of the surfactant chains plus any cosolvent organic molecules between the chains, a_0 is the effective head group area at the micelle surface, and l is the kinetic surfactant tail length or curvature elastic energy. According to this treatment, cubic (Pm3n) will form if $g<1/3$, hexagonal (P6m) if $1/3<g<1/2$, cubic (Ia3d) if $1/2<g<2/3$, and lamellar or inverted micelles if $g>1$. Taken the local effective surfactant packing parameter (the value of g) as a molecular structure-directing index to characterize the geometry of the silicate mesophase products [9,14], the above phase transitions show that the value of g apparently increases with the addition of SL. The mechanism for the increase of g value with the addition of SL has been described in our previous work [13]. There are two kinds of force between cationic and anionic surfactant, that is, (1) Van der Waals forces between hydrocarbon chains in the mixed micelles and (2) electrostatic attraction between quaternary ammonium cation and carboxyl anion. The electrostatic attraction at the micelle surface brings about the decrease of a_0 [15] and makes our systems different from the C_{n-3-1}/C_mTMA, $C_{16-12-16}/C_{16-3-1}$ and

CPCl/CTMCl systems reported by Stucky et al [9,14,16] and Khushalani et al [17], respectively, which can be classified into mixed cationic-cationic surfactant systems. As a result of the balance between the two forces, the hydrocarbon chains of SL may not enter the micelle core, but are located at the hydrophobic-hydrophilic portion of the micelles and therefore form a thick palisade layer, which increases V. Since a_0 decreases and V increases, the g value increases in the micelles of mixed CTMA-SL as SL is increasingly added. In this sense SL can be considered as a polar organic co-solvent. We also attempt to investigate the effect of the carbon number of the hydrocarbon chain of sodium carboxylate (SC) on the synthesis of MCM-48, and find that the best molar ratio of CTMAB to SC required for the synthesis of MCM-48 increase from 9 to 14 as the carbon number changes from 11 to 13. Further increasing the carbon number leads to the decrease of the solubility of the mixture of CTMAB and SC at 30^0C in water, which may be overcome by increasing the temperature. This is under investigation in our laboratory. It should be noted that the mixed surfactants/inorganic composite materials can not be formed when the SL/CTMA molar ratio is beyond 1/3 because both the charge density at the micelle interface of the mixed surfactants and counterion binding capacity decrease as the SL/CTMA molar ratio increases [18]. This suggests that the MCM-48 is formed by the electrostatic force between inorganic and mixed surfactants in our synthesis systems, similar to the report by Huo et al [19, 20].

At the fixed molar ratio of CTMAB to SL of 9:1, the composite materials are corresponding to MCM-41, MCM-48 and MCM-50 as the total concentration of mixed surfactants increases from 1 wt %, via 3~5 wt %, to 10 wt % (see sample 3, 7~9). Similar phase transitions are found as the CTMA/silica molar ratio increases from 0.083 to 0.154 at the fixed total concentration of mixed surfactants of 5 wt % and molar ratio of CTMAB to SL of 9:1 (see sample 2~4, 10, 11). Both of these results are consistent with known micellar phase transformations that occur at various CTMAB concentrations in the CTMAB/water binery system [21,22], but it is noteworthy that lower CTMAB concentration and CTMAB /silica molar ratio are required for these two phase transitions, seperately described above, which may be due to the fact that the composite materials are formed via the cooperative organization of inorganic and mixed surfactants [4].

By increasing the crystallization time from 4 days to 8 days, the as-synthesized materials changed from MCM-48 to MCM-41 (see sample 3 and 12). This is another example of a low-to-high curvature phase transition with time, similar to the change from MCM-50 to MCM-41 reported by Monnier et al [4]. According to the charge density matching criterion also suggested by Monnier et al [4], this phase transition may be contributed to the polymerization of the silicate with time. As rearrangement and polymerization of the sillicate species proceed, the density of anionic silanol groups in the silicate wall diminishes, and the optimal head-group area of the surfactant increases (for more details, see ref. 4), so that the head-group area increases toward its optimal value, which increases the surfactant-silicate interface area and brings about a low-to-high curvature phase transition (in this case, from MCM-48 to MCM-41).

Noting the temperature dependence of the products, we find that the temperature is a critical parameter to synthesize MCM-48 at the range of a certain concentration of mixed surfactants (ca. 4~10 wt %) and molar ratio of CTMAB to silica (ca. 0.147~0.154) [23]. Moreover, for a fixed gel molar composition of SiO_2:0.152CTMAB:0.017SL:0.282 Na_2O:58 H_2O, only by changing the crystallization temperature, the different M41S-type mesoporous molecular sieves may be synthesized (see sample 3, 13~16). Under identical gel compositions rather than the absence of SL, only MCM-41 and MCM-50 among the M41S-type molecular sieves can be observed with the variation of the temperature in the single CTMAB system. This implies that the addition of SL makes the assembly process of inorganic and mixed surfactants more sensitive to the crystallization temperature than that of inorganic and single CTMAB surfactant [24]. The difference of temperature sensitivity between the two systems may also be understood in terms of the charge density matching criterion. According to the criterion, one can approximately get $\rho_\varepsilon = N/A$, and thus $dA/d\rho_\varepsilon = -N/\rho_\varepsilon^2$, where ρ_ε is the charge density of the silicate wall, N is the net charge numbers of the surfactant micelles at the surfactant-silicate interface and A is the surfactant-silicate interface area. Since $N_{mixed} < N_{single}$, where the subscripts of mixed and single represent the mixed catanionic surfactants system and the single cationic surfactant system in this paper, respectively, $(dA/d\rho_\varepsilon)_{mixed} > (dA/d\rho_\varepsilon)_{single}$. Therefore, A_{mixed} is more sensitive to the change of ρ_ε than A_{single} (in this case, the change of ρ_ε is brought about by the change of the extent of polymerization of the silicate species with the crystallization temperature). Since the change of the interface area is connected with the phase transitions, one may consider that the phase transitions in the mixed systems are more sensitive to the crystallization temperature than that in the single cationic system. In this sense, the addition of SL enhances the temperature sensitivity of the assembly process of silicate-mixed CTMAB-SL micelles. Moreover, the results of the effect of different parameters on the synthesis of siliceous MCM-48 also show that different M41S-type mesoporous molecular sieves may be synthesized using mixed CTMAB-SC as templating [24], which is complementary to the prevailing electrostatic [19, 20] and neutral [25, 26] templating routes.

A typical XRD powder pattern of siliceous MCM-48 obtained under the suitable synthesis conditions is depicted in Figure 1. The pattern consists of several distinguishable Bragg peaks which can be indexed to different (hkl) reflections. These reflections verify the presence of cubic phase (Ia3d) with unit cell dimension of ca. 94Å. The transmission electron micrographs (TEM) of thin sections of the calcined MCM-48 are shown in Figure 2. The samples were found to be relatively stable in the electron beam, but the use of the electron radiation over a long time made the pore structure more diffuse in some cases (this can be seen by comparison of fig. 2a and 2b). The HREM images of the calcined MCM-48 usually display regular lattice fringes. A variety of two dimensional lattice fringe patterns (e.g., (111) and (110) projections) are observed in these micrographs, which is in good agreement with the results reported in the literature [3,6], while the samples of MCM-41 usually only give their characteristic hexagonal fringe pattern in selected parts of the specimen that happen to be directly aligned in the hk0 projection. These fringe patterns, as well as the XRD powder pattern, suggest that these obtained materials possess observable three dimensional order in

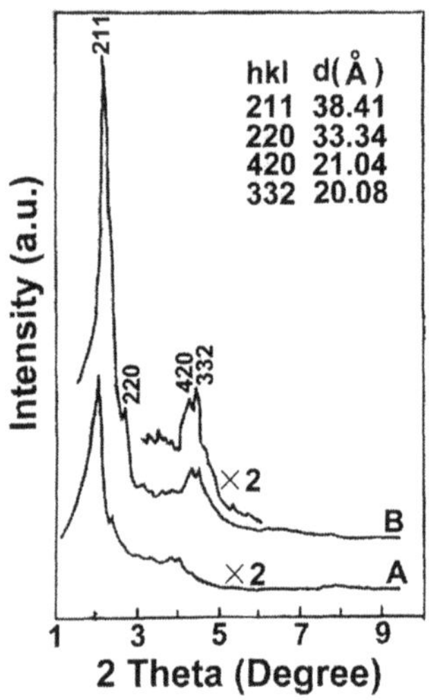

Fig.1. XRD pattern of MCM-48 obtained at 100°C. (A) The as-made MCM-48. (B) The calcined MCM-48.

Fig.2. HREM images of the mesopores on the (111) cubic plane (upper) and the (110) cubic plane (lower) of calcined MCM-48.

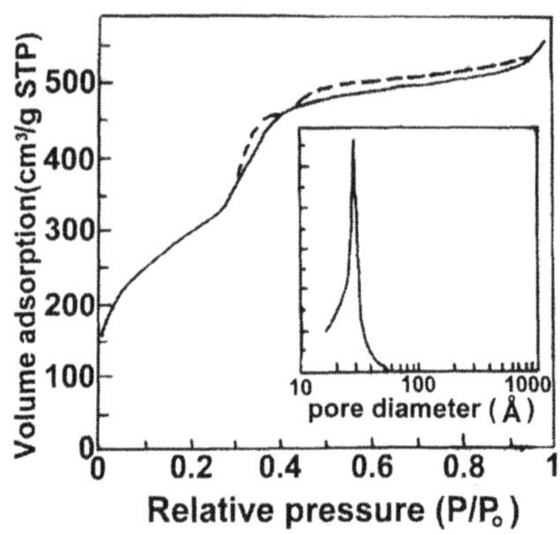

Fig.3. Representative N_2 adsorption (straight) and desorption (dashed) isotherm of calcined MCM-48. Inset is the pore size distribution calculated from the adsorption curve by the Barrett-Joyner-Halenda (BJH) method. The isotherm was obtained on an ASAP-2000 apparatus using nitrogen as adsorbate at 77K.

more than one projection as expected for an analog of the cubic Ia3d liquid-crystal phase. After the TGA/DTA experiment of the calcined MCM-48, the XRD pattern of the residue shows that the cubic structure of the calcined MCM-48 is thermally stable at 1200K. The calcined MCM-48 is further characterized by a N_2 adsorption-desorption isotherm, which gives a BET surface area of 1055.8 m^2/g and pore volume of 0.98 cm^3/g. The pore size distribution, calculated from the desorption curve using BJH model, shows an average pore size of about 29.4Å and half-height width of 4 Å. It corresponds to the hysteresis loop in the 0.3 to 0.4 p/p_0 region of the isotherm, which exhibits the presence of the typical mesoporosity of the calcined MCM-48. The mechanism of the formation of MCM-48 under lower CTMA/silica molar ratio and lower concentration of mixed surfactants may be explained by the local effective surfactant packing parameter, $g = v/a_0l$, as described above. In addition, since the addition of anionic surfactants can increase the surface activity of cationic surfactant aqueous solutions (i.e., lower the surface tension and critical micelle concentration) and thus facilitate the mixed micelle formation [18, 27], SL can be considered as a co-surfactant, as well as a polar organic co-solvent. It's the bifunction of SL that favors the formation of MCM-48 under mild conditions used in this study.

4. CONCLUSIONS

- The molar ratio of CTMAB to SC is a critical parameter to form MCM-48 in the mixed catanionic surfactants system. The molar ratio required increases with the increase of the hydrocarbon chains of SC.

- The phase transformations of the composite materials with the change of the concentration of mixed surfactants or the CTMA/silica molar ratio are consistent with micellar phase transformations that occur at various surfactant concentrations in CTMAB/water binary system, but lower CTMAB surfactants are required for these two phase transformations.

- Increasing the crystallization time leads to phase transition from MCM-48 to MCM-41. The addition of anionic surfactant enhances the temperature sensitivity of the assembly process of inorganic and mixed surfactants.

- Different M41S-type mesoporous molecular sieves may be synthesized using mixed CTMAB-SC as templating, which is complementary to the prevailing electrostatic and neutral templating routes.

ACKNOWLEDGEMENT

This work is financed by National Natural Science Foundation of China (NSFC) under Project 29733070. We also thank Ms. H. Chen for her help in the XRD measurements.

REFERENCE

1. C.T. Kresge, M.E. Leonowicz, W.J. Roth, J.C. Vartuli, J.S. Beck, Nature, 359 (1992) 710.
2. J.S. Beck, J.C. Vartuli, W.J. Roth, M.E. Leonowicz, C.T. Kresge, K.D. Schmitt, C.T.-W. Chu, D.H. Olson, E.W. Sheppard, S.B. McCullen, J.B. Higgins, J.L. Schlenker, J. Am. Chem. Soc., 114 (1992) 10834.
3. J.C. Vartuli, K.D. Schmitt, C.T. Kresge, W.J. Roth, M.E. Leonowicz, S.B. McCullen, S.D. Hellring, J.S. Beck, J.L. Schlenker, D.H. Olson, E.W. Sheppard, Chem. Mater., 6 (1994) 2317.
4. A. Monnier, F. Schüth, Q. Huo, D. Kumar, D. Margolese, R.S. Maxwell, G.D. Stucky, M. Krishnamurty, P. Petroff, A. Firouzi, M. Janicke, B.F. Chmelka, Science 261 (1993) 1299.
5. V. Alfredsson , M.W. Anderson, Chem. Mater., 8 (1996) 1141.
6. R. Schmidt, M. Stöcker, D. Akporiaye, E.H. TΦrstad, A. Olsen, Microporous Mater., 5 (1995) 1.
7. M. Morey, A. Davidson, G. Stucky, Microporous Mater.,6 (1996) 99.
8. W. Zhang , T.J. Pinnavaia, Catalysis Letters, 38 (1996) 261.
9. Q. Huo, D.I. Margolese , G.D. Stucky, Chem. Mater., 8 (1996) 1147.
10. D. Zhao, D. Goldfarb, J. Chem. Soc., Chem. Commun., 8 (1995) 875.
11. J.M. Kim , R. Ryoo, Bull. Korean Chem. Soc. 17(1) (1996) 66.
12. A.A. Romero, M.D. Alba, W. Zhou, J. Klinowski, J. Phys. Chem. B, 101 (1997) 5294.
13. F.X. Chen, L. Huang, Q. Li, Chem. Mater., 9 (1997) 2685.
14. G.D. Stucky, Q. Huo, A. Firouzi, B.F. Chmelka, S. Schacht, I.G. Voigt-Martin, F. Schüth , Stud. Surf. Sci. Catal., 105 (1996) 3.
15. G. Zhao , J. Huang, Acta Phys.-Chim. Sinica, 8(5) (1992) 583.
16. Q. Huo, R. Leon, P.M. Petroff, G.D. Stucky, Science, 268 (1995) 1324.
17. D. Khushalani, A. Kuperman, N. Coombs, G.A. Ozin, Chem. Mater., 8 (1996) 2188.
18. L. Zhang, J. Qiao, H. Ding, G. Zhao, Acta Phys.-Chim. Sinica, 9(4) (1993) 478.
19. Q. Huo, D.I Margolese, U. Ciesla, P. Feng, T.E. Gier, P. Sieger, R. Leon, P.M. Petroff, F. Schüth, G.D. Stucky, Nature, 368 (1994) 317.
20. Q. Huo, D.I Margolese, U. Ciesla, D.G. Demuth, P. Feng, T.E. Gier, P. Sieger, A. Firouzi, B.F. Chmelka, F. Schüth , G. D. Stucky, Chem. Mater., 6 (1994) 1176.
21. X. Auvray, C. Petipas, R. Anthore, J. Phys. Chem., 93 (1989) 7458.
22. T. Wärnhein, A. J. Jönsson, Colloid Interface Sci., 125 (1988) 627.
23. F.X. Chen, H. Chen, X. Yan, Q. Li, CN Patent Application Number 97 1 06615.9.
24. Fengxi Chen, Q. Li, J. Phys. Chem. B (submitted).
25. P.T. Tanev, T.J. Pinnavaia, Science, 267 (1995) 865.
26. S.A. Bagshaw,. E. Prouzet, T.J. Pinnavaia, Science, 269 (1995) 1242.
27. G.X. Zhao, Y. Cheng, J. Ou, B. Tian , Z. Huang, Acta Chimica Sinica, 38(5) (1980) 409.

MESOPOROUS MOLECULAR SIEVES 1998
Studies in Surface Science and Catalysis, Vol. 117
L. Bonneviot, F. Béland, C. Danumah, S. Giasson and S. Kaliaguine (Editors)

Templating effects in the synthesis of MCM-41/MCM-48 phases

Ch. Danumah[1], , S.M.J. Zaidi[1], Normand Voyer[2], S. Giasson[1] and S. Kaliaguine[1*]

[1]Chemical Engineering Department, Université Laval, Ste-Foy, Québec, G1K 7P4 Canada

[2]Chemistry Department, Université Laval, Ste-Foy, Québec, G1K 7P4 Canada

In this study, micellar associations of cetyltrimethylammonium chloride/hydroxide with different Cl/OH exchange ratios and new surfactants having crown-ethers groups as their polar head were used to prepare mesoporous materials of M41S family, with subsequent calcination or CO_2 supercritical fluid extraction to remove the template.

The influences of Cl/OH ratio, the size of the crown-ethers on the final phase and pore size were also investigated. The latter new templates were characterized by ^{1}H NMR and mass spectroscopy and the mesoporous materials by XRD, nitrogen physisorption and solid state ^{29}Si NMR spectroscopy.

1. INTRODUCTION

MCM-41, one of the members of the extensive M41S family of mesoporous sieves, possesses an hexagonal array of uniform mesopores. These pores can be tailored (in the diameter range 16 Å to 100 Å) through an appropriate choice of surfactant, auxiliary chemicals and reaction conditions. The M41S family includes members having pore structures of hexagonal (MCM-41), cubic (MCM-48) and lamellar (MCM-50) symmetry. M41S materials have been synthesized using a variety of liquid crystal templates. These supramolecular assemblies template the condensation of inorganic species and give materials characterized by a very large surface area and a very narrow pore size distribution [1]. The new materials have attracted considerable interest as catalysts [2] and sorbents [3] specially as applied to fine chemicals [4], because of their larger pores than zeolites which allows faster diffusion of bulky molecules.

Over the last few years, a variety of mesoporous materials have been synthesized and given various designations such as FSM-16 [5], MSU [6, 7], HMS [8-10], KIT [11], SBA [12] depending on where they were first prepared, the formation mechanism and the template used. In addition, there has been a lot of papers dealing with potential applications of MCM-41 and yet only a few papers considered MCM-48. This latter material (MCM-48) however having a regular three-dimensional pore system, can be prepared using the same synthesis principles. This scarcity is in part due to the difficulty of synthesizing MCM-48 compared to MCM-41. In spite of the geometry of the porous system, MCM-48 mesoporous materials have similar porosity and thermal stability as MCM-41. In addition, their particles are much better organized than MCM-41 [13].

In this work, we first try to demonstrate, (1) the templating effect via OH concentration on both the polymerization and condensation of the silicate inorganic species, (2) the impact of this parameter on the final mesophase structure and (3) the phase change when increasing the

Cl/OH exchange ratio in the template. Secondly, the use of a new kind of crown ether derived surfactants for the preparation of mesoporous materials is demonstrated.

2 EXPERIMENTAL

2.1 Preparation of template $C_{16}H_{33}N(CH_3)_3OH/Cl$ (CTACl/OH)

The template used in the first part of this work was a micellar association of cetyltrimethylammonium chloride (25wt% in water), the anions of which had been partially substituted for OH^- ions. First a dispersed micellar CTACl/OH solution was prepared by using a solution of cetyltrimethylammonium chloride (CTACl, Aldrich) and subsequently exchanging the chloride with hydroxide ions by contacting the solution with a strong ion-exchange base material (IRA-400, Aldrich). The percent exchange of chloride by hydroxide varied from 18 to 58 percent depending on the amount of resin used.

2.2 MCM-41/MCM-48 silica synthesis

Pure MCM-41 or MCM-48 mesoporous silica were prepared following the procedure of Mobil corporation in the presence of the dispersed micellar CTACl/OH solution by varying the OH percentage from 18 to 58 [14]. To this solution the required amount of tetraethylortosilicate was added. Crystallisation was carried out hydrothermally in open polypropylene bottles at 373 K for 48hr. Then, the product was filtered, washed with deionised water and dried at room temperature. To eliminate the organic template present inside the pores, the samples were calcined at 823 K (heating rate 10 K/min) in air for 6 hr.

2-3 Preparation of template $CH_3(CH_2)_{15}$-O-$CH_2[(CH_2CH_2O)_m]$ (m = 4, 5, 6)

Since their discovery by Pedersen [15], crown ethers have been the object of extensive studies of their preparation and complexing abilities, primarily with metal cations but also with neutral and even anionic species. Moreover, these crown ethers have been used as templates for the formation of zeolite molecular sieves [16].

The surfactants used as templates in this second series of experiments, have a crown-ethers groups as their polar head and an alkyl group as tail.

These new surfactants (Scheme 1) were synthesized by reacting n-hexadecylbromide (Aldrich) with hydroxymethylcrown ethers ($[CH_2CH_2\text{-}O]_mCH_2$-OH, Aldrich) with sodium hydride (NaH, Aldrich) in dimethylformamide (dried over CaH_2). A typical procedure for the synthesis of the 15-crown-5 derivative is as follows:

Hydroxymethyl-15-crown-5 (3.0 g, 12.3 mmol) was stirred with NaH (0.55 g, 18.0 mmol) for 1hr at room temperature in 90 mL of dry dimethylformamide (DMF). Then, 1-bromohexadecane (5.52 g, 18.6 mmol) dissolved in 12-18 mL of DMF was added slowly to the previous mixture under stirring. After reacting for 3 days at room temperature, water (120 mL) was added and the mixture was neutralized and extracted with dichloromethane. After the usual work-up, the hexadecyloxymethyl-15-crown-5 (2.6 g, 45% yield) was obtained as a yellowish oil: ^1H NMR ($CDCl_3$) d 0.90 (t, 3H, CH_3), 1.21 (large s, 28H, $(CH_2)_{14}$), 1.70 (m, 2H, O-CH_2-$\underline{CH_2}$-CH_2), 3.37-3.92 (m, 21H, CH_2O and CH-O); FAB Mass spectrum (thioglycerol): 513 ($M+K^+$), 497 ($M+Na^+$), 475 ($M+H^+$). Hexadecyloxymethyl-12-crown-4 and 18-crown-6 were prepared similarly. These preparations gave 36% of hexadecyloxymethyl-12-crown-4 as a white solid: mp 44 ^{o}C; ^{1}H NMR ($CDCl_3$) d 0.85 (t, 3H, CH_3), 1.23 (large s, 28H, $(CH_2)_{14}$), 1.72 (m, 2H, $O\text{-}CH_2\text{-}C\underline{H_2}\text{-}CH_2$), 3.30-3.90 (m, 19H, CH_2O and CH-O); FAB Mass spectrum (thioglycerol): 563 ($M+Cs^+$), 469 ($M+K^+$), 453 ($M+Na^+$), 431 ($M+H^+$) and 16% yield of hexadecyloxymethyl-18-crown-6 (white solid): mp 44 ^{o}C; ^{1}H NMR ($CDCl_3$) d 0.87 (t, 3H, CH_3), 1.21 (large s, 28H, $(CH_2)_{14}$), 1.65 (m, 2H, $O\text{-}CH_2\text{-}C\underline{H_2}\text{-}CH_2$), 3.40-3.86 (m, 21H, CH_2O and CH-O).

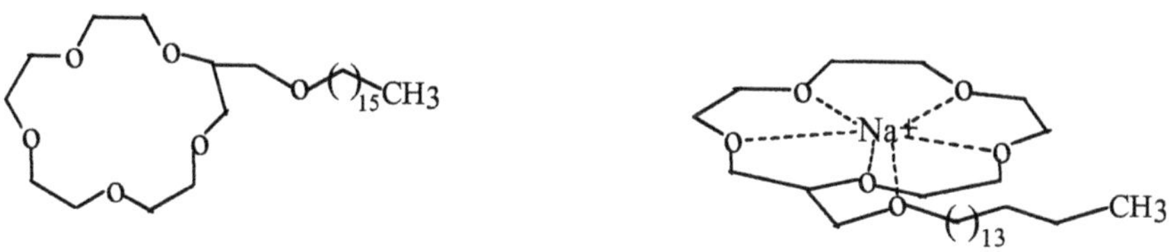

a b

Scheme 1. Structure of the crown ether derived surfactants, hexadecyloxymethyl-15-C-5 ; a) neutral form and b) cation complexed form.

2-4 Synthesis of MCM-41 silica using crown ether templates

Pure MCM-41 mesoporous silica was prepared by mixing these crown ethers as templates with a NaOH solution. The pH was adjusted with HCl after adding the silica source (sodium silicate solution). In some cases the samples were also submitted to post-synthesis treatement in distilled water (pH ~ 7) at 387 K for 24 hr. The products were recovered by filtration, washed with distilled water, dried at room temperature and calcined at 823 K for 5 hr under air, or extracted with supercritical CO_2 at 323 K and 415 atm for various durations.

2-5 Characterization

Solid-State MAS-NMR experiments were conducted at room temperature on a Bruker ASX 300 Spectrometer using 4mm cylindrical Zirconia rotors. ^{29}Si MAS-NMR Spectra were acquired at 59.6 MHz with a 30 ° pulse, duration 7 μs. The repetition times were 30-180 s and chemical shifts were determined relative to tetramethylsilane (TMS). Typical rotor speeds were 9-10.5 kHz and the number of scans varied from 260-1100.

The XRD patterns were recorded with a powder diffractometer Siemens D5000 and Cu Kα radiation (λ = 1.54184 Å).

Surface area measurements were carried out by nitrogen adsorption at 77 K on a Omnisorp sorptometer 100. Before each experiment, the samples were outgassed at 573 K for 2 hours under vacuum (for calcined samples) or at 323 K for 3 hours under vacuum (for CO_2 extracted samples (SFE)).

3. RESULTS AND DISCUSSION

3.1 Pure silica MCM-41/48

The X-ray patterns of as-synthesized and calcined samples (Figure 1a-d) revealed an hexagonal phase at low OH/Cl ratio (18 % OH) and a cubic phase at higher ratios (25-42 % OH) respectively. For the latter, the patterns exibit two peaks at 2θ smaller than 3° and several peaks in range 3.5-8° corresponding to the reflections in the *Ia3*d space group; i.e., consistent with a cubic indexing [14]. In the range of 2θ = 2-8°, at least 8 reflections must be present which are due to the ordering of pores and also seems to be a minimum requirement for a high quality cubic mesoporous structure. For the sample with 18 % OH, we find one relatively broad peak at 2θ smaller than 3° and two broad peaks with very low intensities in the range 2θ = 3-6°. These characteristic features of the peaks indicate a lesser degree of long range order and must be correlated with the presence of more surface silanol groups due to incomplete silica condensation.

The surface area first increases, then decreases whereas the mesopore volume decreases however with increasing OH percentage of the mixture (Table 1). From 18 to 42.1, the characteristic features of the materials N_2 adsorption isotherms are essentially preserved as shown in Figure 2. The materials prepared at high OH/Cl ratios ≤ 45.8 %, collapse partially after template removal indicating the effect of alkalinity on the thermal stability of the material as shown by Di Renzo et al. [17]. Consequently, the surface area and pore volume decrease.

The N_2 adsorption isotherms of all the materials with OH percentage varying from 18 to 42 are of type IV, typical of mesoporous materials [18]. The isotherms show an inflection near p/p_o 0.15-0.3 (34.5 to 42.1 % OH) and 0.3-0.46 (for 18 % OH) indicating capillary condensation within the pores. This inflection becomes oblique for high OH percentage (> 42 %) which indicates a beginning of the non uniformity of the pores. This is consistent with the fact that increasing alkalinity [19-23] corresponds simultaneously to an increasing lamellar phase and a decreasing cubic phase proportions in the final material as shown by X-Ray diffraction patterns. These observations are consistent with the works of Carrazza and al. [24], Voegtlin et al. [25] and Klinowski et al. [26]. These traces of lamellar MCM-50 phase disappear after calcination and are likely responsible for the partial collapse of the framework, generating textural mesoporosity and or macroporosity as shown by an hysteresis loop at higher pressures. This effect is shown in Table 1 by the decrease of the BET surface areas. Recently, Pinnavaia et al. [27, 28] demonstrated that the neutral templating route affords HMS materials with very small scattering domain sizes (less than 17.0 nm) and complementary framework-confined mesopores and textural mesoporosity (at higher $p/p^o > 0.95$). These latter properties are very desirable for accessing the framework-confined mesopores and for improving the performance as adsorbents and catalysts [28, 29]. The N_2 adsorption isotherms of our hexagonal materials (18 % OH) look closely like that of HMS and may therefore have these same characteristic properties. Finally, our work supports partially the results of Pinnavaia et al. [30] which show that ionic templates yield MCM-41 samples that lack appreciable textural mesoporosity.

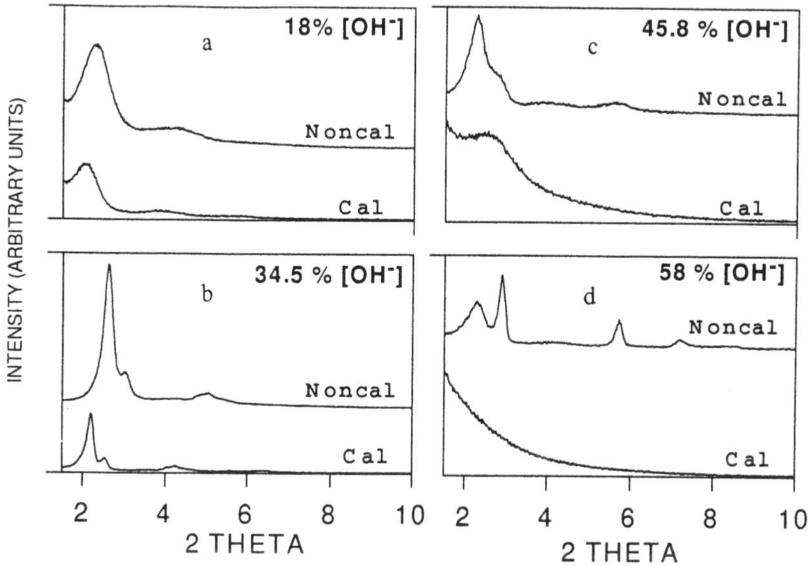

Figure 1. Effect of Cl^-/OH^- ratio on mesophase structure

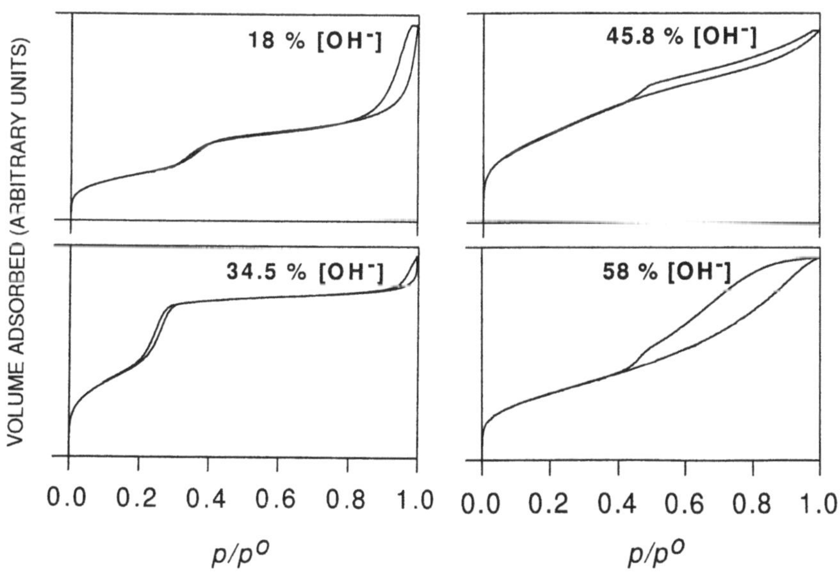

Figure 2. Nitrogen adsorption isotherms at 77 K.

Table 1
Textural parameters deduced from nitrogen sorption isotherms (Calcined samples)

Materials	%OH	d spacing (Å)	BET surface area (m^2/g)	Pore volume (cm^3/g)	BJH Diameter (Å)	Cell Parameter (Å)	Phases
D63	18	42.8	1013	1.8	30	49.5	H
D05	34. 5	33.7	1572	1.4	23	82.5	C
D64	42.1	32	1147	0.9	22	78.1	C
D61	45.8	--	802.2	0.6	36	--	C
D62	51.3		574.9	0.5	38		C + L
D60	58		569.3	0.7	38		C + L

H = Hexagonal (MCM-41); C = Cubic (MCM-48); L = Lamellar (MCM-50) and C + L = Mixed phases.

3.2 Pure Silica templated by crown ethers with side chain alkyl

Before synthesizing the mesoporous materials, the structure of the obtained surfactants was verified by ^{1}H-NMR and MS spectroscopies.

The new family of surfactants proposed in this work has ether groups as its polar head. It may then be compared to recently reported ether surfactants [6, 7, 31]. The surfactants derived from crown ethers (see scheme 1) have several potential advantages as templates in the synthesis of mesoporous molecular sieves. First, the geometrical packing parameter of the molecule may be adjusted by changing both the crown ether size and the length of the alkyl chain. Thus potentially the mesostructure of the bicontinuous phase may be predicted from the packing parameter g [32]. Secondly the fact that both a neutral and charged ion complexing form of the molecule may be obtained (scheme 1, a and b) allows to consider making inorganic mesoporous materials by neutral S^oI^o and charged $S^+X^-I^+$ assemblies using the same surfactant. Conceptually the two mechanisms may even be combined in the same synthesis by adjusting the cation/surfactant ratio.

In the present contribution we report pure silica synthesis results obtained using hexadecyloxymethyl-15-C-5 (D107-D109, 152 and 158) and hexadecyloxymethyl-18-C-6 (D156) in the presence of sodium ions. Thus the expected mechanism is though $S^+X^-I^+$ assemblies in acidic medium.

Figure 3A shows the XRD diagrams of samples D107-D109 both before and after calcination. The as-synthesized samples show clearly the hexagonal symmetry of mesoporous molecular sieves (MCM-41). After calcination however most of this structure is collapsed and the BET surface area of the calcined materials is low (highest value sample D107: 366 m^2/g).

Figure 3B reports XRD patterns for samples D152-156-158. Sample D152 was post-treated at 387 K in distilled water. These three samples were extracted with supercritical CO_2 as the solvant for various lengthes of time. Sample D152 retains some of its hexagonal structure after supercritical extraction but its specific surface area is only 214 m^2/g suggesting incomplete extraction.

The ^{29}Si NMR spectra of our as-synthesized materials are essentially similar to those of MCM-41, MCM-48 and amorphous silica [1]. These spectra exibit three main peaks from Q^4 (framework, $Si[OSi]_4$), Q^3 (silanol, $Si[OSi]_3OH$) and Q^2 (disilanol, $Si[OSi]_2(OH)_2$) with chemical shifts at -91, -101 and -110 ppm respectively. The Q^3 (silanols or silanol nests) intensity is the highest one suggesting an incomplete condensation of the inorganic walls (Q^3/Q^4 = 1.4-1.6).

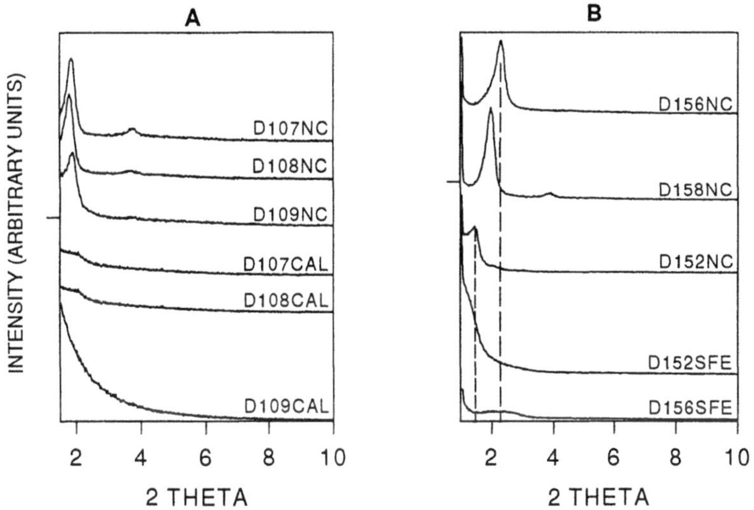

Figure 3. XRD patterns of materials synthesized using hexadecyloxymethyl-15-C-5 surfactants as template (exception D156 hexadecyloxymethyl-18-C-6).

CONCLUSION

It was shown that the variation of OH/Cl ratio as an effect on the final silica mesophase structure and this is consistent with the fact that the solution conditions (ionic strengh, counterion polarisability, surfactant concentration, counterion charge, temperature) affect the liquid crystal structure in surfactant solutions. A phase transition from hexagonal to lamellar via cubic phase can also be observed by varying the ionic strength (increasing the OH percentage) as reported by Beck et al. [1]. It was thus found possible to get the cubic phase MCM-48 by exchanging the halogen counterion of the template $C_{16}H_{33}N(CH_3)_3X$ (X = Br ou Cl) at a certain level instead of increasing the headgroup size or the surfactant to silicium ratio higher than 1 or using gemini surfactants.

We also demonstrate that hexagonal MCM-41 phases can be synthesized using a new type of crown ether based surfactants.

ACKNOWLEDGMENT

We thank Claude Barberis for his technical assistance during the synthesis of crown ether surfactants and Javaid Zaidi for ^{29}Si NMR Results.

REFERENCES

1. (a) C.T. Kresge, M.E. Leonowicz, W.J. Roth, J.C. Vartuli and J.S. Beck, Nature, 359 (1992) 710 ; (b) J.S. Beck, J.C. Vartuli, W.J. Roth, M.E. Leonowicz, C.T. Kresge, K.D. Schmitt, C.T.-W. Chu, D.H. Olson, E.W. Sheppard, S.B. McCullen, J.B. Higgins and J.L. Schlenker, J. Am. Chem. Soc., 114 (1992) 10834.
2. Q.N. Le and R.T. Thomson, Mobil Co. U.S. Patent, US-A 5191144 (1993).
3. P.J. Branton, P.G. Hull and K.S.W. Sing, J. Chem. Soc., Chem. Commun., (1993) 1257.
4. A. Corma, Chem. Rev., 97 (6) (1997) 2373.
5. S. Inagaki, Y. Fukushima and K. Kuroda, J. Chem. Soc., Chem. Commun., (1993) 680.
6. (a) S.A. Bagshaw, E. Prouzet and T.J. Pinnavaia, Science, 269 (1995)1242; (b) S.A. Bagshaw and T.J. Pinnavaia, Angew. Chem. Int. Ed. Engl., 35 (10) (1996) 1102 ;(c) E. Prouzet and T.J. Pinnavaia, Angew. Chem. Int. Ed. Engl., 36 (5) (1997) 516.
7. G.S. Attard, J. C. Glyde and C.G. Göltner, Nature, 378 (1995) 366.
8. P.T. Tanev and T.J. Pinnavaia, Science, 267 (1995) 865.
9. W. Zhang, M. Froba, J. Wang, P.T. Tanev, J. Wong and T.J. Pinnavaia, J. Am. Chem. Soc., 118 (1996) 9164.
10. (a) P.T. Tanev and T.J. Pinnavaia, Chem. Mater., 8 (1996) 2068, (b) A. Tuel and S. Gontier, Chem. Mater., 8 (1996) 114.
11. (a) J.M. Kim, J.H. Kwak, S. Jun and R. Ryoo, J. Phys. Chem., 99 (1995) 16742 ; (b) R. Ryoo and J.M. Kim, J. Chem. Soc., Chem. Commun., (1995) 711.
12. Q. Huo, R. Leon, P.M. Petroff and G.D. Stucky, Science, 268 (1996) 1324.
13. M.W. Anderson, Zeolites, 19 (1997) 220.
14. J.C. Vartuli, K.D. Schmitt, C.T. Kresge, W.J. Roth, M.E. Leonowicz, S.B. McCullen, S.D. Hellring, J.S. Beck, J.L. Schlenker, D.H. Olson et E.W. Sheppard, Chem. Mater., 6 (1994) 2317.
15. (a) C. J. Pedersen, J. Am.. Chem. Soc., 89 (1967) 7017 ; (b) E. Weber; F. Vögtle, Top. Curr. Chem. 98 (1981) 1 ; (c) G.W. Gokel, Monograph in Supramolecular Chemistry: Crown Ethers and Cryptands; The Royal Society of Chemistry, London, 1991. (d) J.-M. Lehn, Nature, 260 (1993) 1762.
16. (a) F. Delprato, L. Delmote, J.L. Guth and L. Huve, Zeolites, 10 (1990) 546; (b) M.W. Anderson, K.S. Pachis, F. Prébin, S.W. Carr, O. Terasaki, T. Ohsuna and V. Alfreddson, J. Chem. Soc., Chem. Commun., (1991) 1660.
17. N. Coustel, F. Di Renzo and F. Fajula, J. Chem. Soc., Chem. Commun. (1994) 967.
18. S.J. Gregg and K.S.W. Sing, Adsorption, Surface Area, and Porosity, 2nd Edition, Academic Press Inc., New York, 1982.
19. M. Ogawa, J. Am. Chem. Soc., 116 (1994) 7941.
20. P.T. Tanev and T.J. Pinnavaia, Science, 271 (1996) 1267.

21. S. Oliver, A. Kuperman, N. Coombs, A. Louph, G.A. Ozin, Nature, 378 (1995) 47.
22. A. Firouzi, D. Kumar, L.M. Bull, T. Besier, P. Sieger, Q. Huo, S.A. Walker, J.A. Zasadzinski, C. Glinka, J. Nicol, D. Margolese, G.D. Stucky and B.F. Chmelka, Science, 267 (1995) 1138.
23. C.A. Fyfe and G. Fu, J. Am. Chem. Soc., 117 (1995) 9709.
24. J. Lujano, Y. Romero and Carrazza, Stud. Surf. Sci. and Catal., 97 (1995) 489.
25. A. Matijasic, A.-C. Voegtlin, J. Patarin, J.L. Guth and L. Huve, Chem. Commun. (1994) 967.
26. A.A. Romero, M.D. Alba, W. Zhou and J. Klinowski, J. Phys. Chem.B, 101 (1997) 5294.
27. P.T. Tanev and T.J. Pinnavaia, Chem. Mater., 8 (1996) 2068.
28. P.T. Tanev, M. Chibwe, and T.J. Pinnavaia, Nature, 368 (1994) 321.
29. N.S. Gnep, P. Roger, P. Cartrand, M. Guisnet, B. Juguin, C. Hamon, Creation of mesopores in monodimensional zeolites. A way of improving their catalytic stability, C. R. Acad. Sci., Ser. 2: 309: 1743 (1989); B. Chauvin, F. Fajula, F. Figueras, C. Gueguen, and J. Bousquet, Sorption properties of zeolite omega, J. Catal., 111 (1988) 94.
30. P.T. Tanev and T.J. Pinnavaia, in Access in Nanoporous Materials, Edited by T.J. Pinnavaia and M.F. Thorpe, Plenum Press, New York, (1995) 13.
31. Zhao, J. Feng, Q. Huo, N. Melosh, G.H. Frederickson, B.F. Chmelka and G. Stucky, Science, 279 (1998) 548.
32. J.N. Israelachvili, D.J. Mitchell and B.W. Ninham, Faraday Transaction II, 72 (1976) 1525.

MESOPOROUS MOLECULAR SIEVES 1998
Studies in Surface Science and Catalysis, Vol. 117
L. Bonneviot, F. Béland, C. Danumah, S. Giasson and S. Kaliaguine (Editors)

Effect of Al Sources on the Synthesis and Acidic Characteristics of Mesoporous Aluminosilicates of MCM-41 Type

Kondam Madhusudan Reddy and Chunshan Song*

Laboratory for Hydrocarbon Process Chemistry and Fuel Science Program, Department of Materials Science and Engineering, Pennsylvania State University, 209 Academic Projects Building, University Park, PA 16802-2303, USA

This paper reports the synthesis of four series of mesoporous aluminosilicates (Al-MCM-41) and their acidic characteristics. Four different Al compounds were examined as Al source in the hydrothermal synthesis of Al-MCM-41, including pseudo boehmite, aluminum sulfate, aluminum isopropoxide, and sodium aluminate. Synthesized samples were characterized by XRD, chemical analysis, nitrogen adsorption, and ^{27}Al MAS NMR. The acidic characteristics of the Al-MCM-41 samples were characterized using temperature-programmed desorption of n-butylamine, and probe hydrocracking reactions of 1,3,5-triisopropylbenzene and 1,3,5-tri-t-butylbenzene. The results clearly indicate that the Al source used for synthesizing Al-MCM-41 samples has a major impact on the Al incorporation in their framework and their acidic properties. Al-MCM-41 prepared with Al isopropoxide and sodium aluminate showed improved acidic properties than those with pseudo boehmite and Al sulfate.

1. INTRODUCTION

The invention by Mobil researchers of a new class of mesoporous molecular sieves has stimulated considerable interest on these materials [1,2]. Because of their attractive properties such as uniform mesopores, high surface area and high hydrocarbon sorption capacity, many studies have been conducted since 1992 on their synthesis, characterization and isomorphous substitution [3-11]. However, only very limited catalytic applications have been reported so far [12-15], probably due to their lack of appropriate acidic characteristics, hydrothermal stability and mechanical strength. It has become an important subject to improve the acidic characteristics of these new materials. We have been working on the synthesis of mesoporous molecular sieves of MCM-41 type in order to improve their acidic characteristics, and apply them as catalytic materials for hydroprocessing of petroleum and coal liquid fractions for producing clean distillate fuels including jet and diesel fuels [16-19]. In this paper we discuss the effects of Al source on the synthesis of Al-MCM-41 and its acidic characteristics.

2. EXPERIMENTAL

Mesoporous molecular sieves (Al-MCM-41) with varying SiO_2/Al_2O_3 ratios were synthesized using different aluminum sources including aluminum isopropoxide (**AI**), pseudo boehmite (**PB**, 75% alumina, Catapal B), aluminum sulfate (**AS**), and sodium aluminate (**SA**). The molar composition of the gel for hydrothermal synthesis is as follows: $100SiO_2$-xAl_2O_3-$8.64Na_2O$-$4.38(TMA)_2O$-31.24CTMABr-$6330H_2O$; where x= 1, 2 and 4. The organic template was removed by a temperature programmed calcination in a tubular furnace at 550°C first in N_2 flow for 1 h and then in an air flow for 5 h. The calcined Al-MCM-41 was exchanged with 0.5 M ammonium nitrate solution at 90°C. The ammonium exchanged Al-MCM-41 was calcined at 460°C for 3 h in an air flow in order to obtain protonated Al-MCM-41. More details of the synthesis procedure may be found elsewhere [16].

* C rresponding author. Email: csong@psu.edu; Fax: 814-865-3075; Tel: 814-863-4466

All the Al-MCM-41 samples were analyzed by ICP-ES for chemical composition, by XRD for d_{100} spacing and crystallinity, by N_2 adsorption and BET for pore size distribution and surface areas of calcined Al-MCM-41 samples. ^{27}Al MAS NMR spectra of as-synthesized Al-MCM-41 samples were recorded on CMX 300 Chemagnetics solid state NMR spectrometer. The chemical shifts for ^{27}Al were measured with respect to $Al(H_2O)_6^{3+}$. SEM observations were made using Philips XL-20 apparatus. The Acidic properties of protonated form of Al-MCM-41 samples were measured by TPD of n-butylamine (n-BA) on TGA and DSC apparatus. The samples were degassed and then saturated with n-butylamine vapor. TPD profiles of the base-covered samples were obtained from 80°C to 700°C at the rate of 10°C/min. Detailed TPD procedure is given elsewhere [20].

Hydrocracking of 1,3,5-triisopropylbenzene (TIPB) and 1,3,5-tri-t-butylbenzene (TTBB) was conducted at 300°C for 30 min using 0.1 g of a proton-form Al-MCM-41, 1.0 g of reactant and 4.0 g of tridecane solvent under H_2 pressure of 1000 psig (cold) in 25-ml stainless-steel batch reactor. For comparison, proton-form Y zeolites (LZ-Y62: SiO_2/Al_2O_3 ~5; CBV740 from Zeolyst: SiO_2/Al_2O_3 ~40), H-mordenite (CBV20A and CBV30A from PQ: SiO_2/Al_2O_3 ~21 and ~38) and an alumina (S-2) were also tested in similar conditions. The products were analyzed by GC-MS and GC-FID; the used catalysts were analyzed by TGA in order to estimate the coke formation, using the procedures described elsewhere [18]. The used catalysts were first heat treated under N_2 flow (200 cc/min) at 30°C to 700°C at the rate of 10°C/min to remove all of the trapped liquid products. Subsequently coke was estimated by TGA from 30°C to 700°C at the rate of 10°C/min in an air flow (100 cc/min). The deactivation due to the coke formation was evaluated by hydrocracking of TTBB using the recovered catalyst from the first run for the second time in similar conditions.

3. RESULTS AND DISCUSSION

3.1. Synthesis and Properties of Al-MCM-41

The four series of aluminosilicates synthesized with the Al sources (**AI**, **PB**, **AS**, and **SA**), all have mesopores and MCM-41 type crystalline phase, as shown in Table 1. Although the analytical data for samples with AI appeared in an earlier communication [18], Table 1 contains the first complete set of data on the physico-chemical characteristics of the four series. All the four Al sources led to synthetic crystalline materials with same XRD pattern of MCM-41 type, but the SiO_2/Al_2O_3 ratios, surface areas and pore diameters were affected considerably by the type of Al source (Table 1). The surface area of the samples with AS was substantially lower; the pore diameters of the samples synthesized with PB are smaller compared to the other series of samples in Table 1. In hydrothermal synthesis of aluminosilicates, the output SiO_2/Al_2O_3 ratios are usually lower than the feed SiO_2/Al_2O_3 ratios. This is generally the case for the synthesis using **AI**, **SA** and **PB**. However, the SiO_2/Al_2O_3 ratios of the Al-MCM-41 synthesized using **AS**, are lower than the corresponding feed ratios, indicating that the Al in this compound was not effectively incorporated into the synthetic Al-MCM-41 samples, probably due to its lower reactivity.

XRD data in Table 1 indicate that the intensity of the 100-reflection peak decreased with decreasing SiO_2/Al_2O_3 feed ratio for all the 4 types of Al sources, indicating that increased Al content in the feed hinders the crystallization process. However the effect of the SiO_2/Al_2O_3 ratio on crystallinity also depends on the Al source. The crystallinity of Al-MCM-41 generally increased in the order of: **AS < SA < AI < PB**. The d-spacing of MCM-41 is related [1,2] to hexagonal unit cell parameter by $a = 2d/\sqrt{3}$. Since the *d*-spacing is directly proportional to unit cell parameter, the d_{100}-spacing values in Table 1 show that the decrease in SiO_2/Al_2O_3 feed ratio can result in unit cell expansion. This unit cell expansion is due to the longer length of Al-O bond compared to that of Si-O bond in the framework. Based on our results, the order of unit cell expansion with respect to Al sources is **SA > AI > AS > PB**, which also represents the order of Al incorporation in the Al-MCM-41 framework. Even for samples with similar

SiO_2/Al_2O_3 output ratios (**AI2, PB2, SA2, AS3**), their XRD data and surface areas can be quite different (Table 1).

SEM photos indicated that the samples of **AI2, PB2, SA2, AS2** are similar in crystallite size, about 0.1-0.2 μm. ^{27}Al MAS NMR spectra of Al-MCM-41 samples clearly illustrated the changes in Al coordination with Al source used for their synthesis. Spectra of the MCM-41 samples may exhibit two peaks at around 50 ppm and 0 ppm, which are attributed to tetrahedral and octahedral Al respectively [18]. If the Al is completely incorporated into the framework one can see a single peak at around 50 ppm, which is generally true for the samples prepared with SiO_2/Al_2O_3 feed ratio of 50 using **SA**, **AI** and **AS**. However, the samples prepared with **PB** exhibit a main peak around 0 ppm corresponding to non-framework Al, indicating that **PB** is not a good source for incorporating Al into Al-MCM-41 framework.

Table 1 Physico-chemical properties of synthesized Al-MCM-41 molecular sieves

Sample[a]	Feed SiO_2/Al_2O_3	Output SiO_2/Al_2O_3	XRD parameters Peak[b] Intensity	XRD parameters d_{100} spacings(Å)	BET Surface Area (m^2/g)	Pore diam. from sorption (Å)
PB1	100	95.5	3642	32.9	1010	21.92
PB2	50	44.3	3216	34.2	1123	23.80
PB3	25	18.4	2722	34.7	1224	25.89
SA1	100	89.6	2842	37.2	1277	28.53
SA2	50	49.3	2635	40.3	1182	28.58
SA3	25	20.2	2204	38.9	1150	28.49
AI1	100	88.4	2967	37.4	1147	27.67
AI2	50	46.8	2906	38.7	1206	28.02
AI3	25	27.4	2272	37.9	1020	28.46
AS1	100	164.6	2511	35.6	834	25.38
AS2	50	87.4	2204	37.7	705	28.88
AS3	25	39.5	583	38.1	610	28.75

a) AI, PB, AS, and SA indicate the type of Al source used in the hydrothermal synthesis.
b) peak due to the 100 reflection.

3.2. Acidic Characteristics of Al-MCM-41

Figures 1 and 2 show the profiles of n-BA TPD conducted on TGA and DSC. The method of n-BA TPD by TGA has been used previously to characterize microporous zeolites [20,21]. Table 2 shows the acidity of Al-MCM-41 samples derived from n-BA TPD in 3 temperature ranges: 100-240, 240-340, and 340-500°C. We have observed that the amount of species desorbed from mesoporous Al-MCM-41 samples between 100-240°C are often more than those from microporous zeolites [20], probably due to the large sorption capacity of MCM-41. From Table 2 and Figure 1, the majority of the n-BA on Al-MCM-41 was desorbed before 240°C. We suspect that the desorption of species at lower temperatures is due partly to physisorbed species, and only partly related to those chemisorbed on weak acid sites on the surface. DSC profiles in Figure 2 support this hypothesis. The total acidity estimated from TPD between 240-500°C appears to be more reproducible and reliable, though the absolute values are smaller than those incorporating TPD between 100-240°C. The acidity data from TPD between 240-500°C indicates that the acidic characteristics of Al-MCM-41 are different for the samples prepared with different Al sources. The variation in acidity may be attributed to differences in Al incorporation and location in the framework with different Al sources. The

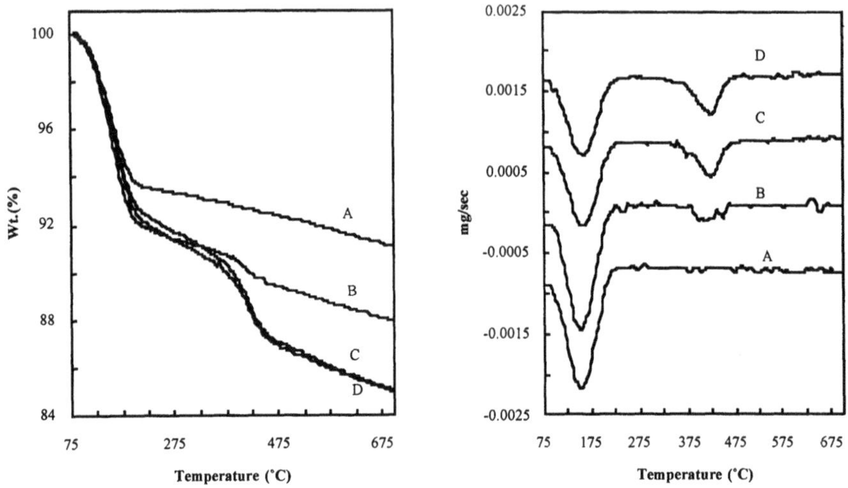

Figure 1. TGA (left) and DTG (right) profiles of n-BA TPD from Al-MCM-41 samples: A) PB2, B) AS2, C) SA2 and D) AI2.

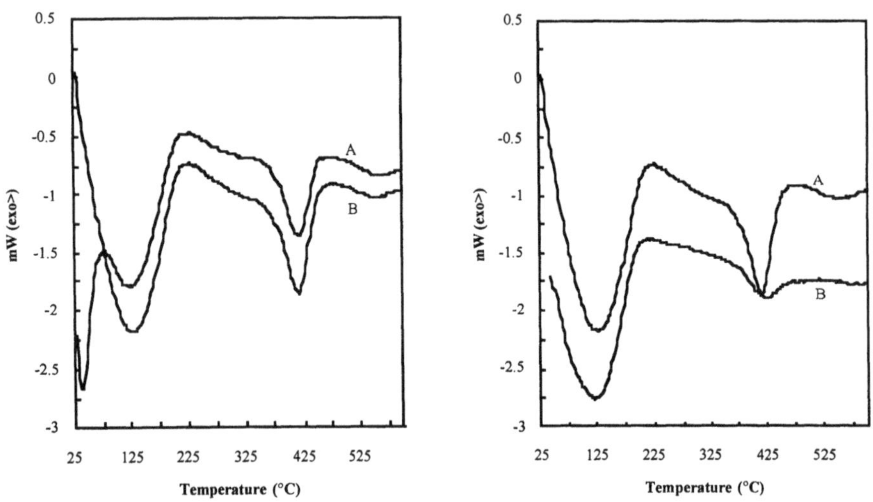

Figure 2. DSC (left) profiles of n-BA TPD from Al-MCM-41 (SA3) without (A) and with (B) initial N_2 flushing a room temperature for 30 min, and DSC (right) profiles of n-BA TPD from Al-MCM-41 samples (A) SA3 (SiO_2/Al_2O_3:25) and (B) SA2 (SiO_2/Al_2O_3:50).

acidity data confirms that at synthetic feed SiO_2/Al_2O_3 ratio of 50, **AI** and **SA** are good Al sources to produce acidic Al-MCM-41, but at the feed ratio SiO_2/Al_2O_3 of 25, the acidity of the final proton-form Al-MCM-41 from **SA** is considerably higher than that from **AI** and **AS**. The decrease in the output SiO_2/Al_2O_3 ratio from 46.8 to 27.4 led to a considerable decrease in acidity from 0.64-0.72 to 0.17-0.20 mmol/g. This is in distinct contrast to the case of H-Y zeolite and H-Mord where increasing the SiO_2/Al_2O_3 ratio decreased the total acidity.

Comparative examination of Table 2 suggests that at moderate feed SiO_2/Al_2O_3 ratio (~50) both **AI** and **SA** are good Al sources for synthesizing acidic materials. Comparing the resulting mesoporous Al-MCM-41 and the well-known Y-zeolite and mordenite at similar SiO_2/Al_2O_3 ratios, one can see similar acidity values. At higher content of aluminum in the synthetic feed (SiO_2/Al_2O_3 ratio =25), **SA** is still a reasonably good aluminum source; but with **AI** as the source, either the aluminum incorporation became much less effective or the incorporated aluminum became structurally less stable upon post-synthesis treatment (calcination, ion-exchange) and the resulting material contained even less acid sites. Acidic characteristics of Al-MCM-41 samples were further examined by hydrocracking of bulky molecules including 1,3,5-TTBB and 1,3,5-TIPB.

Table 2 Acidity of proton-form Al-MCM-41 samples from desorption of n-butylamine

Sample	SiO_2/Al_2O_3	Acidity (mmol/g)			
		Weak[a] (100-240°C)	Intermediate (240-340°C)	Strong (340-500°C)	Total (240-500°C)
MCM-41(AI2)	46.8	1.00	0.18	0.54	0.72
MCM-41(AI2)[b]	46.8	1.01	0.17	0.47	0.64
MCM-41(PB2)	44.3	0.85	0.06	0.11	0.17
MCM-41(AS2)	87.4	1.09	0.09	0.24	0.33
MCM-41(SA2)	49.3	1.04	0.16	0.44	0.60
MCM-41(AI3)	27.4	0.58	0.07	0.13	0.20
MCM-41(AI3)[b]	27.4	0.39	0.07	0.10	0.17
MCM-41(PB3)	18.4	0.50	0.04	0.11	0.15
MCM-41(AS3)	39.5	0.65	0.12	0.24	0.36
MCM-41(SA3)	20.2	1.06	0.19	0.42	0.61
H-Mord(CBV20A)	21.0	0.56	0.16	0.82	1.54
H-Mord(CBV30A)	38.0	0.69	0.13	0.63	1.45
H-Y(LZ-Y62)	4.8	0.79	0.49	1.85	3.12
H-Y(CBV740)	40.0	0.74	0.10	0.44	0.54

a) n-BA desorption between 100-240°C also includes physisorbed n-BA.
b) Experiments were repeated for those particular samples at similar conditions.

Table 3 shows the conversion of a bulky reactant, 1,3,5-TTBB, over three Al-MCM-41 catalysts (SA2, PB2 and AS3) and two microporous catalysts (H-Y with SiO_2/Al_2O_3 of 40, and H-Mord with SiO_2/Al_2O_3 ratio of 38). The major liquid products are di-t-butylbenzene (DTTB), t-butylbenzene (TTB) and benzene. Isobutane and isobutene are the predominant gaseous products, whose yields varied with conversion level. From the results with the fresh catalysts (Run I), the TTBB conversion increased in the order of Al-MCM-41(PB2) < H-Mord < H-Y < Al-MCM-41(AS3) < Al-MCM-41(SA2). These results indicate that properly synthesized mesoporous Al-MCM-41 catalysts are more active than the microporous zeolite Y and mordenite for TTBB hydrocracking. TTBB is a bulky molecule which will not fit in the pores of mordenite and Y zeolite, but it can easily diffuse through the mesopores of Al-MCM-41. However zeolite Y and mordenite did show some activity for hydrocracking of TTBB, which is due to the external acid sites. Among three Al-MCM-41 catalysts PB2 showed very low activity, which corroborate the n-BA TPD and NMR results. Al-MCM-41(SA2) has a higher SiO_2/Al_2O_3 ratio than Al-MCM-41(AS3) catalyst (49 vs 39.5). The former should have lower acidity than the latter if all the aluminium is incorporated into the framework. However, the opposite trend is observed (Table 2), indicating effective incorporation of aluminum from **SA** leading to more catalytically active acid sites (Table 3).

It is of interest to examine the activity of the used (partially coked) mesoporous Al-MCM-41 catalysts. Table 3 also compares the results for hydrocracking of bulky TTBB with the used catalysts (Run II) and the fresh catalysts (Run I). The results of coke analysis by TGA indicated the occurrence of coking reactions during the two consecutive runs, with fresh and then used catalysts, as shown in Figure 3 (left) and Table 4. Table 3 shows that conversion over the used catalysts increased in the order of PB2 ≤ H-Mord < H-Y << AS3 << SA2. More coke was formed on mordenite, followed by H-Y, and the Al-MCM-41. Figure 3 (left) indicates the amounts of coke on the second-used catalysts increased in the order of PB2 < AS3 < SA2 < H-Y < H-Mord. It becomes clear from these results that even after partial coking, the good Al-MCM-41 catalyst prepared using **SA** is still much more active than the ultra-stabilized H-Y and H-Mord for conversion of bulky TTBB. Compared to microporous zeolites, where the TTBB cracking takes place mainly on the external surface which leads to

Table 3 1,3,5-TTBB hydrocracking over fresh (run I) and used (run II) catalysts at 300°C for 30 min

	Proton-form Al-MCM-41						H-Mord[a]		H-Y[b]	
	PB2		AS3		SA2					
	Run I	Run II	Run I	Run II	Run I	Run II	Run I	Run II	Run I	Run II
Conversion (%)	25.31	9.52	77.35	52.02	94.46	74.62	36.67	17.14	33.72	24.17
Products(wt%)										
Gas										
C_1+C_2	0.00	0.00	0.01	0.01	0.01	0.01	0.02	0.01	0.01	0.00
Propane	0.00	0.00	0.01	0.05	0.02	0.02	0.32	0.11	0.01	0.00
Propene	0.02	0.01	0.06	0.01	0.05	0.07	0.02	0.02	0.02	0.03
i-Butane	0.10	0.05	0.96	0.55	1.96	0.09	1.80	0.56	1.02	0.23
n-Butane	0.00	0.00	0.02	0.01	0.02	0.01	0.19	0.05	0.01	0.00
i-Butene	0.38	0.18	0.60	0.70	0.29	0.63	0.07	0.08	0.21	0.80
1-Butene	0.03	0.01	0.12	0.08	0.05	0.08	0.01	0.00	0.02	0.00
C_5^+	0.00	0.01	0.08	0.05	0.08	0.06	0.03	0.07	0.03	0.00
Liquid[c]										
C_4	0.25	0.27	0.10	0.70	1.62	4.40	7.87	2.04	1.65	0.95
C_5	0.11	0.00	0.38	0.41	1.16	0.90	2.41	0.75	0.48	0.17
C_6	0.04	0.00	0.18	0.12	0.49	0.26	0.84	0.37	0.18	0.14
C_7	0.00	0.00	0.52	0.37	0.66	0.53	0.35	0.11	0.23	0.09
C_8	0.69	0.01	4.64	3.07	3.16	4.48	2.01	0.43	0.92	0.46
Benzene	0.19	0.00	0.97	0.66	2.13	0.60	2.99	1.29	1.49	0.07
TBB	1.84	0.11	12.96	7.43	24.61	12.67	1.74	1.32	7.57	3.57
1,3-DTBB	12.37	2.46	33.27	27.09	23.35	29.20	4.54	3.97	8.14	6.90
1,4-DTBB	1.43	1.79	1.53	1.51	1.70	1.36	0.07	0.05	0.35	0.47
1,3,5-TTBB	74.69	90.48	22.65	47.98	5.54	25.38	63.33	82.86	66.28	75.83
1,2,4-TTBB	0.31	0.37	0.28	0.22	0.35	0.14	0.28	0.36	0.27	0.32
Isomers of HC[d]	1.55	1.11	3.64	3.10	5.36	2.42	3.08	2.34	2.06	1.52

a) H-Mordenite (CBV30A); b) H-Y (CBV740); c) In liquid, C_4-C_8 hydrocarbons are mostly olefins; d) Isomers of TBB, DTTB and TTBB.

Table 4 Coke estimation for used catalysts using TGA

Catalyst	Reactant	Run No.	Coke (wt%)[a]
MCM-41 (SA2)	1,3,5-TTBB	I	2.48
MCM-41 (SA2)	1,3,5-TTBB	II	2.88
MCM-41 (PB2)	1,3,5-TTBB	II	1.15
MCM-41 (AS3)	1,3,5-TTBB	II	1.58
H-Mord (CBV30A)	1,3,5-TTBB	II	7.38
H-Y (CBV740)	1,3,5-TTBB	I	5.73
H-Y (CBV740)	1,3,5-TTBB	II	6.06
MCM-41(AI2)	1,3,5-TIPB	I	1.11
MCM-41(AS3)	1,3,5-TIPB	I	1.75
H-Mord (CBV20A)	1,3,5-TIPB	I	6.03
H-Y (LZ-Y62)	1,3,5-TIPB	I	3.41

a) Coke estimated from TGA coke burning profiles above 400°C

faster deactivation by coke deposition, the Al-MCM-41 should be more resistant to coke deactivation due to the presence of mesopores. The first run with fresh catalysts accounts for the majority of the coke accumulated in two consecutive runs, as can be seen from Table 4.

In addition, preliminary results on TIPB hydrocracking were reported earlier [18] and a complete set of data are given in Table 5. The Al-MCM-41 sample PB2 showed very low activity for TIPB hydrocracking, which is similar to that of an amorphous alumina (S-2), details are reported elsewhere [22]. Among the other Al-MCM-41 samples, AI2 and SA2 gave

Table 5 Hydrocracking of 1,3,5-TIPB over molecular sieves catalysts at 300°C for 30 min.

	Proton-form Al-MCM-41					Alumina[a]	H-Mord[b]	H-Y[c]
	AI2	SA2	PB2	AS2	AS3			
Conversion (%)	46.01	47.68	12.52	40.46	56.84	5.44	28.90	49.28
Products(wt%)								
Gas								
C_1+C_2	0.01	0.01	0.00	0.01	0.01	0.01	0.02	0.01
Propane	1.26	1.61	0.06	1.17	1.58	0.02	1.08	2.06
Propylene	1.96	2.99	0.57	2.20	2.82	0.02	0.11	1.80
i-Butane	0.44	0.63	0.01	0.26	0.52	0.00	0.39	0.54
n-Butane	0.02	0.02	0.00	0.01	0.03	0.00	0.08	0.02
C_5^+	0.00	0.03	0.00	0.00	0.08	0.10	0.10	0.06
Liquid								
Benzene	0.07	0.03	0.02	0.02	0.01	0.02	0.68	0.04
Cumene	1.15	1.14	0.03	0.83	0.95	0.04	3.92	1.91
1,3-DIPB	14.21	14.77	3.05	13.80	13.37	0.00	4.00	17.28
1,4-DIPB	4.40	4.57	0.43	4.07	4.04	0.00	1.77	5.85
1,3,5-TIPB	53.99	52.32	87.48	59.54	43.16	94.56	71.10	50.72
1,2,4-TIPB	3.51	3.14	1.85	3.60	2.62	1.06	2.31	3.47
Others	7.73	7.07	0.37	6.28	7.25	0.18	3.76	10.69

a) Alumina (S-2); b) H-Mordenite (CBV20A); c) H-Y (LZ-Y62)

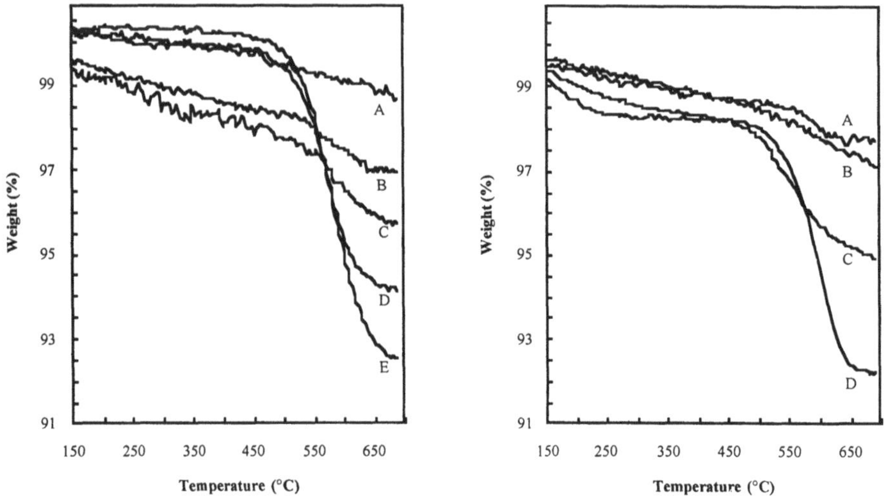

Figure 3. Coke burning profiles (left) of catalysts used twice (run II) in 1,3,5-TTBB hydrocracking: (A) PB2, (B) AS3, (C) SA2, (D) H-Y, and (E) H-Mord; Coke burning profiles (right) of catalysts used in 1,3,5-TIPB hydrocracking: (A) AI2, (B) AS3, (C) H-Y and (D) H-Mord.

similar results. From the data in Table 2, the Al-MCM-41 catalysts prepared with Al isopropoxide (AI2) and sodium aluminate (SA2) have similar acidity values whose level is similar to that of the ultra-stabilized H-Y. Two Al-MCM-41 samples in the AS series were tested; increasing the aluminum content (from AS2 to AS3) increased TIPB conversion (Table 5) which is consistent with the slightly increased acidity (Table 2). The coke analysis of the used catalysts show that coke deposits increased in the order of AI2 < AS3 < H-Y < H-Mord, as shown in Figure 3 (right) and Table 4. Relative to AI2, the higher degree of coking on AS3 is consistent with its higher cracking activity. These results indicate that hydrocracking activity of Al-MCM-41 samples or their coking tendency may not always parallel with the acidity values measured from n-BA TPD. Finally, a word of caution, one can not rank the hydrocracking activity of Al-MCM-41 samples based on the order of their acidity derived from n-BA TPD. For example, n-BA TPD indicated higher acidity values of AI2 or SA2 compared to AS3, but the former gave lower TIPB conversion.

SUMMARY

Four series of Al-MCM-41 samples were synthesized using different Al sources, and characterized by several techniques including XRD, porosity, ^{27}Al NMR, n-BA TPD and hydrocracking of TTBB and TIPB. The results clearly indicated that the type of Al source has a significant influence in determining Al incorporation and the acidic characteristics of Al-MCM-41. Al-MCM-41 prepared with sodium aluminate and Al isopropoxide under proper conditions can show improved acidic characteristics than those with pseudo boehmite (alumina) and Al sulfate.

At SiO_2/Al_2O_3 feed ratio of ~50, both aluminum isopropoxide and sodium aluminate are good sources for incorporating Al in the MCM-41 framework to produce acidic materials in their proton-form. The acidity of such materials measured by n-BA TPD between 240-500°C is comparable to that of microporous sieves such as ultra-stabilized H-Y and H-Mord with similar SiO_2/Al_2O_3 ratios. The protonated form of mesoporous Al-MCM-41 materials synthesized with proper aluminum source are much more active than H-Y or H-Mord for hydrocracking of large-sized molecules such as 1,3,5-TTBB. On the other hand, the protonated form of Al-MCM-41 materials from sodium aluminate and aluminum isopropoxide gave similar conversions of 1,3,5-TIPB compared to that with H-Y. Relative to an ultra-stabilized H-Y, the Al-MCM-41 synthesized using sodium aluminate gave higher conversion in 1,3,5-TTBB hydrocracking with lower amount of coke, in the tests using either fresh new catalysts or used (partially coked) catalysts. These results suggest that properly synthesized mesoporous Al-MCM-41 molecular sieves can have acidity comparable to ultra-stabilized H-Y, and could have real catalytic applications if properly engineered.

ACKNOWLEDGMENT

We are grateful to Dr. H. H. Schobert for his encouragement and support. Financial support was provided by the US Department of Energy, the Pittsburgh Energy Technology Center, and the US Air Force, Wright Laboratories.

References

1. C. T. Kresge, M. E. Leonowicz, W. J. Roth, J. C. Vartuli, and J. S. Beck, Nature, 359 (1992) 710.
2. J. S. Beck, J. C. Vartuli, W. J. Roth, M. E. Leonowicz, C. T. Kresge, K. D. Schmitt, C. T. W. Chu, D. H. Olson, E. W. Sheppard, S. B. McCullen, J. B. Higgins, and J. C. Schlenker, J. Am. Chem. Soc., 114 (1992) 10834.
3. C.-Y. Chen, S. L. Burkett, H.-X Li, and M. Davis, Microporous Mater., 2 (1993) 17; C.-Y. Chen, H.-X Li, and M. Davis, Microporous Mater, 2 (1993) 27.
4. J. C. Vartuli, K. D. Schmitt, C. T. Kresge, W. J. Roth, M. E. Leonowicz, S. B. McCullen, S. D. Hellring, J. S. Beck, J. L. Schlenker, D. H. Olson, and E. W. Sheppard, Chem. Mater., 6 (1994) 2317.
5. S. A. Bagshaw, E. Prouzet, T. J. Pinnavaia, Science 269 (1995) 1242.
6. A. Tuel and S. Gontier, Chem. Mater., 8 (1996) 114.
7. A. Sayari, C. Danumah, and I. L. Moudrakovski, Chem. Mater., 5 (1995) 813.
8. C-.F. Cheng, H. He, W. Zhou, J. Klinowski, J. A. S. Goncalves, and L. F. Gladden, J. Phys. Chem., 100 (1996) 390.
9. T. Blasco, A. Corma, T. Navarro, and J. Perez-Pariente, J. Catal. 156 (1995) 65.
10. K. M. Reddy, I. Moudrakovski, and A. Sayari, J. C. S., Chem. Commun., (1994) 1059.
11. T. K. Das, K. Chaudhari, A. J. Chandwadkar, and S. Sivasanker, J. C. S., Chem. Commun., (1995) 2495.
12. J-.B. Kim and T. Inui, Catal. Lett., 36 (1996) 255.
13. A. Corma, A. Martinez, V. M-. Soria, and J. B. Monton, J. Catal., 153 (1995) 25.
14. N. L. Quang, N. J. C. Hill, U.S. Patent 5 191 134 (1993).
15. J. S. Beck, R. F. Socha, D. S. Shihabi, J. C. Vartuli, U. S. Patent 5 143 707 (1992).
16. K. M. Reddy and C. Song, Catal. Lett., 36 (1996) 103.
17. K. M. Reddy and C. Song, Catal. Today, 31 (1996) 137.
18. K. M. Reddy and C. Song, Mat. Res. Soc. Symp. Proc., 454 (1997) 125.
19. K. M. Reddy and C. Song, Catal. Today, (1998) in press.
20. C. Song, W-.C. Lai, A. D. Schmitz, and K. M. Reddy, Am. Chem. Soc. Div. Fuel Chem. Prepr., 41 (1996) 71.
21. A. Ghosh, and G. Curthoys, J. Phys. Chem., 88 (1984) 1130.
22. C. Song, T. Nihonmatsu and M. Nomura, Ind. Eng. Chem. Res., 30 (1991) 1726.

MESOPOROUS MOLECULAR SIEVES 1998
Studies in Surface Science and Catalysis, Vol. 117
L. Bonneviot, F. Béland, C. Danumah, S. Giasson and S. Kaliaguine (Editors)

NMR Relaxation and Self-diffusion Measurements on Small Molecules in MCM-41 with Different Pore Size

F. P. Matthae, W. D. Basler and H. Lechert

Institute of Physical Chemistry, University of Hamburg,
Bundesstr. 45, D-20146 Hamburg, Germany

The diffusion of water and small organic molecules in the pores of Na-MCM-41 with diameters between 2.8 and 6.5 nm has been studied by H-NMR relaxation times T_1 and T_2 and the pulsed field gradient technique. The translational diffusion of all organic molecules (benzene, cyclohexane, 1,4-dioxane, n-pentane, neopentane and n-dodecane) is comparable to the bulk liquids. But no freezing has been observed down to 135 K. Only benzene in samples hydrolized at pH = 12 and 373 K showed freezing, as confirmed by DSC, with a freezing point depression of 60–70 K.

The water diffusion in Na-MCM-41, only obtained from T_1 and T_2, is 1000 times slower than in bulk water and comparable with water sorbed in zeolites. No freezing has been found, too.

1. INTRODUCTION

In micro- and mesoporous catalysts the molecular transport is one of the most important factors which determines the rate of chemical reactions. But its influence on the kinetics is often difficult to obtain from rate studies. However, the molecular motion may be measured or at least estimated by various NMR methods. The molecular translation and rotation can be obtained from the temperature dependence of the nuclear relaxation times T_1 and T_2. The self-diffusion coefficient may be measured by the pulsed field gradient technique in favourable cases.

A critical survey of these NNR methods applied to molecules sorbed in zeolites has been given by Pfeifer [1].

Whereas in zeolites the diameter of the molecules and the pores are comparable and a strong reduction of molecular diffusion as compared with the bulk liquids is expected and always observed, a more bulk-like behaviour is foreseen in the larger channels of the mesoporous alumosilicates of the M41S family.

We studied the motion of the small water molecules and some organic molecules of different shape and polarity (benzene, cyclohexane, 1,4-dioxane, n-pentane, neopentane and n-dodecane) in MCM-41 of pore diameters between 2.8 and 6.5 nm.

2. EXPERIMENTAL

MCM-41 samples with a constant Si/Al-ratio of 30 and different pore sizes were prepared following the synthesis procedure with optimized parameters for a small pore size distribution [2]. The silica source was Cab-osil M-5 (Fluka). Pore diameters of 4.0 nm were obtained with surfactant octadecyltrimethylammoniumchloride and template tetramethylammoniumhydroxide. Pore diameters of 2.8 nm were obtained with surfactant octyltrimethylammoniumbromide and template tetrapropylammoniumhydroxide, all in basic solution of ammoniumhydroxide. Pore sizes up to 6.5 nm were achieved by adding the expander molecule mesitylene which dissolves in the hydrophobic region of the micelles of octadecyltrimethylammoniumchloride, thus increasing their size. The synthesis gels were stirred for half an hour and then heated to 363 K for 72 hours.

The as-synthesized products were washed with twice distilled cold water and calcined at 853 K for 10 hours. The hexagonal phase was verified by XRD ($2\,\Theta = 0.5$–$10°$). The surface area and pore size distribution were monitored by nitrogen sorption and calculated according to the BET-equation or the Kelvin-equation, respectively.

H-MCM-41 was prepared by first ion-exchanging with an excess of 1 M ammonium nitrate solution at room temperature and then heating the dried NH_4-sample to 853 K.

Na-MCM-41 samples were prepared by repeated ion-exchange in 10^{-4} M NaOH solution at room temperature until pH > 9. Then they were rinsed with twice distilled water and dried in air at ambient temperature.

The iron content of the final MCM-41 samples was 3–5 ppm by wet chemical analysis. All samples were degassed under shallow bed conditions at 623 K and under a pressure of less than 1 mPa for at least 10 hours. Adsorption was performed using vapors of carefully degassed liquid.

The H-NMR relaxation times T_1 and T_2 were measured at 60 MHz for T = 135–450 K by conventional methods with a BRUKER BKR 322 spectrometer. The self-diffusion coefficients have been determined by the pulsed field gradient technique (PMFG) as the crystal size was between 10 and 20 μm.

3. RESULTS AND DISCUSSION

3.1. Water in Na-MCM-41(24), Na-MCM-41(34) and Na-MCM-41(70)

In Figure 1 the temperature dependence of the H-NMR relaxation times T_1 and T_2 of water in Na-MCM-41, water acivity $a_w = 0.75$ giving about 400 mg H_2O / g Na-MCM-41, is shown as lg T_i versus 1/T. Only a shallow T_1-minimum is observed in the range of 250 to 300 K. The increase of T_2, i.e. the motional narrowing of the rigid lattice spectrum, starts above 200 K. The H-NMR signal intensity was constant between 135 K and 450 K. No phase transition has been observed down to 135 K. Thus even in the pores of 7 nm diameter no freezing of water to ice takes place. This is in contrast to results of Hansen [3, 4], who observed abrupt changes of H-NMR intensity and relaxation times near 0°C and -29°C and hysteresis using MCM-41 samples "saturated with water". The H-NMR relaxation times are only slightly affected by the pore size. The values are shifted to lower temperatures by increasing pore diameter showing a somewhat faster diffusion in the 7 nm pores.

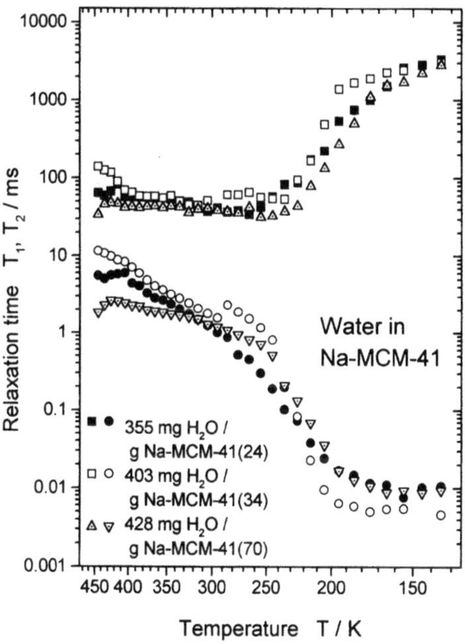

Figure 1. Temperature dependence of the H-NMR relaxation times T_1 and T_2 for water in Na-MCM-41 with different pore diameters given in round brackets.

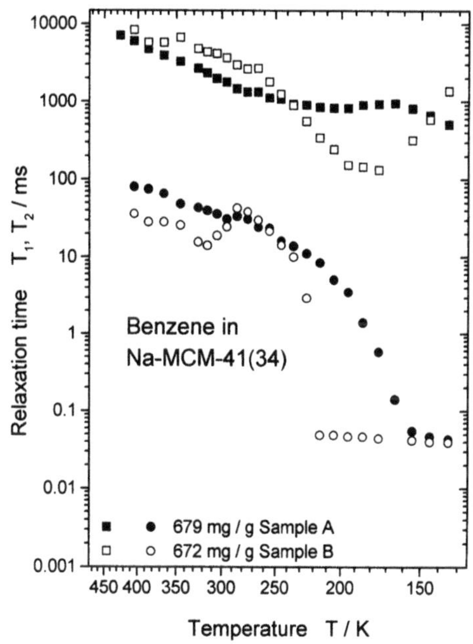

Figure 2. Temperature dependence of the H-NMR relaxation times T_1 and T_2 for benzene in Na-MCM-41(34).

The temperatures of T_1-minima and T_2-narrowing are quite similar to those found in other zeolites. Thus the molecular diffusion is not significantly faster in MCM-41 than in e.g. faujasites despite the considerably larger pore diameter. From the temperatures of T_1-minimum and T_2-narrowing an activation energy of about 30 kJ/mol is calculated. Due to the short T_2 values the diffusion coefficient could not be determined by the PMFG technique. From the T_1-minimum a diffusion coefficient of $D = 2 \times 10^{-12}$ m^2/s is estimated which is 1000 times smaller than in bulk water. The temperature dependence of T_1 and T_2 typical for isotropic reorientation, as found in the 3-dimensional connected pores of faujasites, is not observed in the straight channels of MCM-41.

Concluding, water in 2–7 nm pores of Na-MCM-41 is a sorbate and not a liquid.

3.2. Benzene in Na-MCM-41(24) and Na-MCM-41(34)

In Figure 2 the H-NMR relaxation times T_1 and T_2 of benzene in Na-MCM-41(34), activity 0.7 giving about 670 mg benzene / g Na-MCM-41, are shown for two samples with different pretreatment: Sample A was washed with 0.1 mM NaOH at ambient temperature for 0.5 hour, sample B with 10 mM NaOH at 373 K for 1 hour.

The H-NMR signal intensity was constant between 135 K and 435 K. Sample A showed T_1 to be nearly independent of the temperature and a smooth T_2-narrowing above 160 K. Sample B gave a T_1-minimum near 180 K and an abrupt change of T_2, dropping to the rigid lattice value of rotating benzene. This stepwise change of T_2 is caused by a first order phase transition, e.g. freezing of benzene, as was verified by differential scanning calorimetry showing a strong sharp peak at 220 K, too. The very large freezing point depression of 60 K shows that benzene is sorbed in probably very large pores of sample B, but not at the outer surface of the crystals.

Below the freezing point of 220 K the 100% rigid lattice signal of sample B did not show any additional component corresponding to sample A. Therefore, the treatment by NaOH must have transformed the total pore system of the original MCM-41 sample. By nitrogen adsorption no pores of 3.4 nm were found anymore in sample B and the BET surface was reduced from 813 to 280 m^2/g.

The self-diffusion coefficient of benzene in sample A could be measured by the PMFG technique and is given in Table 1. The value of 1.7×10^{-9} m^2/s at 300 K is 75% of the bulk value. Thus the diffusion of benzene in Na-MCM-41(34) is nearly liquid-like, whereas the diffusion of benzene in Na-X faujasite is 1000 times slower.

The activation energy of the diffusion by PMFG is 23 kJ/mol in the range 300 K to 390 K compared with 8.4 kJ/mol of the bulk liquid. Practically the same results were

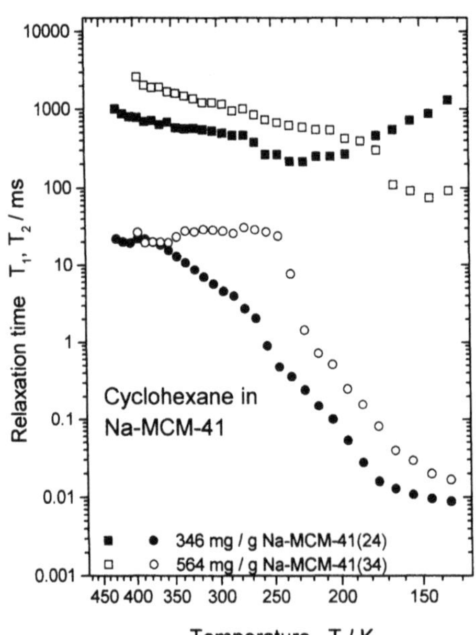

Figure 3. Temperature dependence of the H-NMR relaxation times T_1 and T_2 for cyclohexane in Na-MCM-41 with different pore diameters.

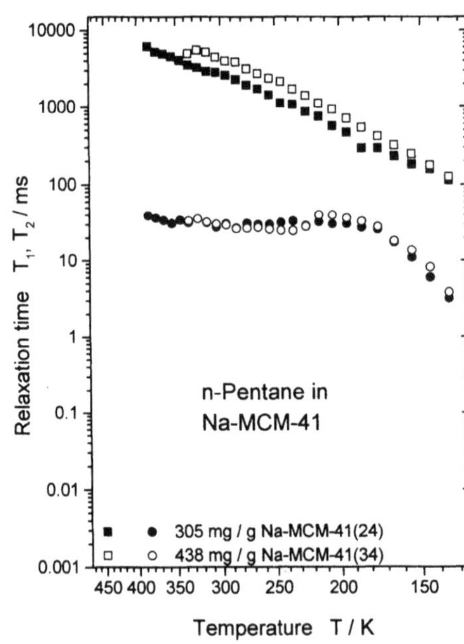

Figure 4. Temperature dependence of the H-NMR relaxation times T_1 and T_2 for n-pentane in Na-MCM-41 with different pore diameters.

found for benzene in the smaller pores of Na-MCM-41(24). Only the temperature of the phase transition was shifted by 10 K down to 210 K in sample B probably caused by slightly smaller pores. The PMFG diffusion coefficient is reduced to 1.1×10^{-9} m²/s in the 2.4 nm pores, which is 50% of the bulk value. Its activation energy is 17 kJ/mol.

3.3 Cyclohexane and 1,4-dioxane in Na-MCM-41(24) and Na-MCM-41(34)

The remarkably different diffusion of benzene (slow in Na-X faujasites, but like bulk liquid in Na-MCM-41) may be caused by the smaller pores of faujasite as compared with MCM-41 or/and by the specific interaction of the π-electrons of benzene with the numerous Na^+-ions of Na-X which is nearly missing in Na-MCM-41 with Si/Al = 30, i.e. far more benzene molecules than available Na^+-ion sorption sites. Therefore we studied cyclohexane which has molecular dimensions comparable with benzene but no π-electrons and the polar 1,4-dioxane. The coverage was adjusted to 346 mg cyclohexane / g Na-MCM-41(24) and 564 mg / g Na-MCM-41(34) for an equal number of cyclohexane molecules as compared with benzene.

The H-NMR relaxation times T_1 and T_2 of cyclohexane are presented in Figure 3. The PMFG diffusion coefficient is 1.1×10^{-9} m²/s in Na-MCM-41(34) at 300 K which is

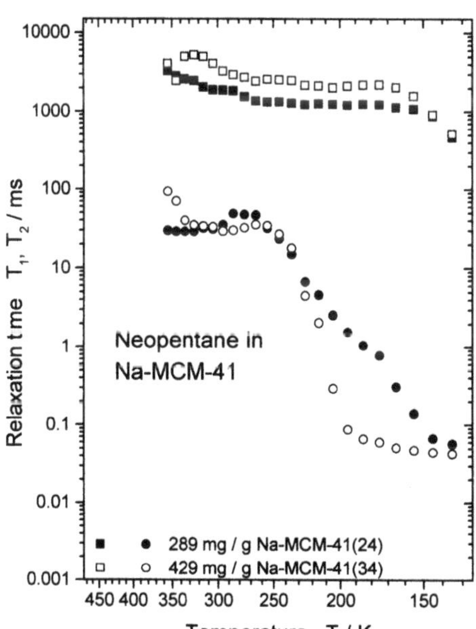

Figure 5. Temperature dependence of the H-NMR relaxation times T_1 and T_2 for neopentane in Na-MCM-41 with different pore diameters.

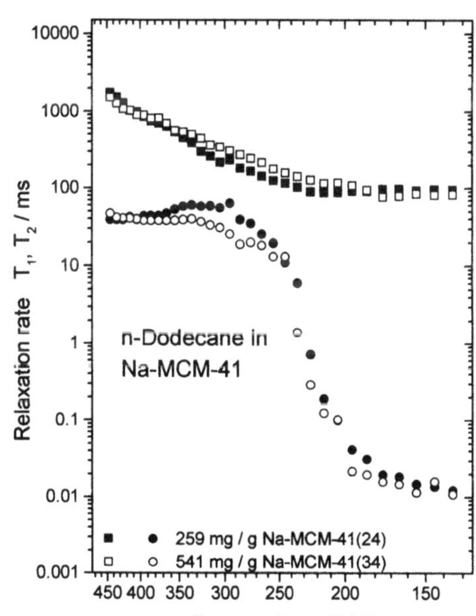

Figure 6. Temperature dependence of the H-NMR relaxation times T_1 and T_2 for n-dodecane in Na-MCM-41 with different pore diameters.

73% of the bulk value. Therefore the diffusion of both benzene and cyclohexane is slowed down only by 25% in the 3.4 nm pores of Na-MCM-41 relative to the bulk liquids. This slight reduction is certainly a steric effect, no specific π-electron interaction in case of benzene is present. The diffusion of cyclohexane is even more reduced to $0.4x10^{-9}$ m^2/s in the 2.4 nm pores which is 25% of the bulk value.

This comparably rapid diffusion of both benzene and cyclohexane is in agreement with the fact that the narrowing of T_2 starts in both cases at 160 K in the 3.4 nm pores.

Even the polar 1,4-dioxane molecule diffuses liquid-like in the MCM-41 pores: The PMFG diffusion coefficient at 300 K is $1.1x10^{-9}$ m^2/s in Na-MCM-41(34), which is 98% of the bulk value, and $0.9x10^{-9}$ m^2/s in Na-MCM-41(24) e.g. 81% of the bulk value. The activation energies are 23 kJ/mol (3.4 nm), 27 kJ/mol (2.4 nm) compared with 22 kJ/mol of the bulk liquid.

The temperature dependence of the H-NMR relaxation times T_1 and T_2 of 1,4-dioxane is very similar to that of cyclohexane and therefore not shown.

3.4. n-Pentane, Neopentane and n-Dodecane in Na-MCM-41(24) and Na-MCM-41(34)

Finally the three unpolar hydrocarbons n-pentane, neopentane and n-dodecane have been investigated. The H-NMR relaxation times T_1 and T_2 of 305 mg n-pentane / g Na-MCM-41(24) resp. 438 mg / g Na-MCM-41(34) are given in Figure 4. T_2 is still about 10 ms, e.g. far longer than the rigid lattice value even at 150 K. This is in agreement with the PMFG diffusion coefficient of $1.9x10^{-9}$ m^2/s at 300 K (33% of bulk) and its low activation energy of 15 kJ/mol which extrapolates to a diffusion coefficient of $5x10^{-12}$ m^2/s at 150 K, giving the observed 100-fold T_2-narrowing. Even T_1 decreases monotonically with temperature and the T_1-minimum must be located below 135 K. No significant difference is found between the 2.4 and 3.4 nm pores.

Table 1
PMFG self-diffusion coefficients of liquids in Na-MCM-41 samples with different pore size in comparison with the bulk liquids (T = 300 K).

	Self-diffusion coefficient [10^{-9} m^2/s]		
Liquid	bulk liquid	Na-MCM-41(34)	Na-MCM-41(24)
Benzene	2.26	1.7	1.1
Cyclohexane	1.51	1.1	0.4
1,4-Dioxane	1.08	1.1	0.9
n-Pentane	5.82	—	1.9
Neopentane	5.01	2.6	1.7
n-Dodecane	0.88	0.2	0.09

Table 2
Activation energy of the PMFG self-diffusion of liquids in Na-MCM-41 samples with different pore size in comparison with the bulk liquids (T = 300–390 K).

	Activation energy of the diffusion [kJ/mol]		
Liquid	bulk liquid	Na-MCM-41(34)	Na-MCM-41(24)
Benzene	8.37	23	17
Cyclohexane	19.1	44	23
1,4-Dioxane	21.7	23	27
n-Pentane	6.45	—	16
Neopentane	11.1	—	19
n-Dodecane	14.9	37	29

Changing from the linear n-pentane to branched neopentane, loading 289 mg / g Na-MCM-41(24) and 429 mg / g Na-MCM-41(34), the rigid lattice T_2 is now reached at 200 K (3.4 nm) resp. 145 K (2.4 nm), see Figure 5. All T_1 and T_2 values are shifted by +100 K compared with n-pentane.

The PMFG diffusion coefficient of neopentane is 1.7×10^{-9} m²/s (2.4 nm, 34% of bulk) and 2.6×10^{-9} m²/s (3.4 nm, 52% of bulk). Taking the PMFG diffusion at 300 K and an estimated diffusion coefficient of 10^{-14} m²/s at the T_2 narrowing at 145 K, an activation energy of about 30 kJ/mol is calculated for 2.4 nm pores, which disagrees with the experimental PMFG activation energy of 19 kJ/mol. Probably the H-NMR of sorbed neopentane has been disturbed by larger amounts of intercrystalline gas phase neopentane.

Finally the long-chain nonpolar n-alkane n-dodecane has been investigated. The H-NMR T_1 and T_2 values are shown in Figure 6. No significant difference between 2.4 nm and 3.4 pores is found. But the PMFG diffusion coefficient changes from 0.2×10^{-9} m²/s in the 3.4 nm pores (23% of bulk) to only 0.09×10^{-9} m²/s (10% of bulk) in the 12.4 nm pores. Whereas T_1 and T_2 are determined by reorientation of the alkane segments, which are independent of the pore diameter, the PMFG diffusion is given by the translational displacement of the whole molecule, which is apparently more hindered in the 2.4 nm pores. The relative high temperature of the T_2-narrowing (about 200 K) and the steep increase of T_2 with temperature indicate a high activation energy of n-dodecane motion. This agrees with the activation energy of the PMFG diffusion of 37 kJ/mol in the 3.4 nm pores and 29 kJ/mol in the 2.4 nm pores, but only 15 kJ/mol in bulk n-dodecane.

A slightly indicated shallow T_1-minimum near 250 K shows again the slower diffusion of n-dodecane as compared with n-pentane where both T_2 narrowing and T_1-minimum are below 135 K.

4. CONCLUSIONS

The diffusion of small hydrocabons (benzene,cyclohexane, n-pentane, neopentane, 1,4-dioxane) in the mesopores of Na-MCM-41 is nearly as fast as in the bulk liquids, no significant influence of polarity or π-electrons has been found in Na-MCM-41 with Si/Al = 30. Only the diffusion of n-dodecane in 2.4 nm pores is reduced to 10% of the bulk liquid.

In all cases no thermodynamic freezing is observed in properly treated samples. Only the crystallization of benzene with a large freezing point depression of 60 K has been found after a sample treatment with 10 mM NaOH at 373 K.

However, the diffusion of water in the mesopores of Na-MCM-41 is 1000 times slower than in bulk water and equals the water diffusion in the micropores of zeolites.

5. ACKNOWLEDGEMENTS

We thank the "Deutsche Forschungsgemeinschaft" for financial support of our work.

REFERENCES

1. H. Pfeifer, "NMR, Basic Principles and Progress" (Edited by P. Diehl, E. Fluck and R. Kosfeld), Vol 7, p. 55, Springer, Berlin, Heidelberg, New York (1972)
2. F. P. Matthae, D. Genske, Ch. Minchev and H. Lechert, Poster: "Synthesis and Characterization of MCM-41with Different Pore Size and Si/Al-ratio", ISMMS, Baltimore 1998
3. R. Schmidt, M. Stöcker, E. Hansen, D. Akporiaye and O. H. Ellestedt, Microporous Mater., 3 (1995) 443–448
4. D. Akporiaye, E. W. Hansen, R. Schmidt and M. Stöcker, J. Phys. Chem., 98 (1994) 1926–1928

MESOPOROUS MOLECULAR SIEVES 1998
Studies in Surface Science and Catalysis, Vol. 117
L. Bonneviot, F. Béland, C. Danumah, S. Giasson and S. Kaliaguine (Editors)

Structural properties of simple sorbed gases (N_2 , Kr , D_2O) confined in MCM-41 sample (Ø = 40 Å) .

J.P. Coulomb[a] , C. Martin[a] , Y. Grillet[b] , P.L. Llewellyn[b] and G. André[c] .

[a] C.R.M.C[2] - CNRS, Campus de Luminy, Case 901, 13288 Marseille Cedex 9 - France.
[b] C.T.M.- CNRS, 26 rue du 141° R.I.A., 13003 Marseille - France.
[c] Laboratoire Léon Brillouin, C.E.N. - Saclay, 91191 Gif-sur Yvette - France.

We have studied by neutron diffraction the structural properties of simple gases (N_2, Kr , D_2O) sorbed on the model mesoporous material MCM-41 (Ø = 40 Å). Previously, the thermodynamic properties of the considered sorbates have been investigated. In the low and medium loading regime, the sorbed phases are characterized by a short range order ($L_{coher.}$ ≃ 20 Å). In the high loading regime, different behaviours have been observed depending on the sorbed species and temperature ranges. For example, Kr confined phase in MCM-41 sample (Ø = 40 Å) presents a strong increase of the coherence length $L_{coher.}$ around T = 85 K . This result points out the large displacement of the freezing point temperature for confined liquid in small pores. On the other hand, N_2 sorbed phase doesn't present any solidification signature in all the temperature range 30 - 86 K .

1. INTRODUCTION

MCM-41 synthesized in 1992, by the Mobil Research Laboratory (Princeton - USA) is the first model mesoporous material in the pore range 20 - 100 Å [1,2]. Such an aluminosilicate material is characterized by a porosity composed by a hexagonal structure of cylindrical pores. The pore diameter Ø depends greatly of the template micelles, it can be monitored with accuracy in a large size range . In addition, pure silicic MCM-41 samples can be prepared. Nevertheless, MCM-41 samples are not crystalline porous materials, their walls are composed of amorphous silica . As a consequence the MCM-41 inner surface is not as energetically homogeneous as the zeolite inner surfaces are .

Due to the model shape (cylindrical) and the diameter size (20 ≤ Ø ≤ 100 Å) of its mesopores, MCM-41 is a model sorbant in the analysis of the Kelvin law domain of validity. At the present time, it is the sorption properties of MCM-41 material which are the most widely investigated. Sorption isotherms have been measured for a great variety of sorbate species, as for instance, van der Waals gases (Ar, Kr, CH_4, Xe), polar molecules (N_2, CO, CO_2) or molecules characterized by hydrogen bonds (H_2O, CH_3OH) [3 - 8]. We have represented on figure 1, methane sorption isotherms for a series of MCM-41 samples of different pore diameters (Ø = 40 Å, Ø = 25 Å and Ø ≃ 15 Å). It is an illustrative example of the pore size influence on the capillary condensation phenomenon. When the MCM-41 pore diameter decreases, the sorption isotherm step is strongly displaced towards the low relative pressure values (sorption capacities have been normalized to MCM-41 (25 Å) one's and only the sorption branches are represented). The agreement with the Kelvin law is only qualitative (around 60 %). In addition to the capillary phase, an adsorbed film is also observed in the low relative pressure range. The filling mechanism of MCM-41 mesopores is a two stages process (adsorbed film + capillary phase). Among the simple gases, currently used in sorption isotherm measurements, N_2 is certainly one of the more suited to determine the MCM-41 sorption characteristics (specific surface, pore size distribution, pore filling mechanism) [3 - 9, 13 - 16]. At the present time, only few studies have been devoted to the structural analysis of sorbed phases on MCM-41 samples [17 - 20]. In this paper we present a structural property investigation concerning N_2, Kr and D_2O confined phase in a MCM-41 (Ø = 40 Å) host sample.

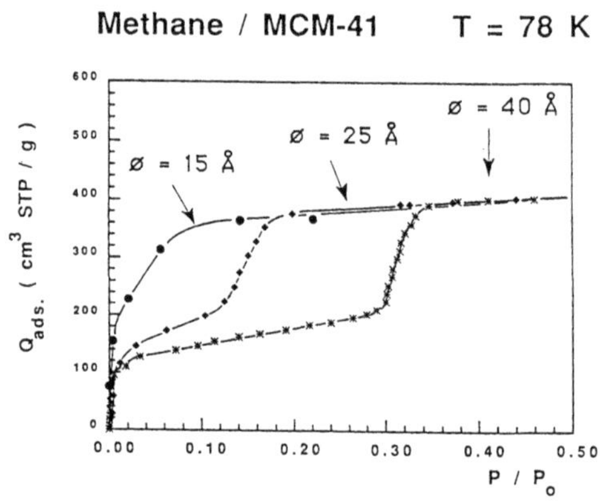

Figure 1. Methane sorption isotherms measured at T = 78 K for a series of MCM-41 samples (40 Å, 25 Å, 15 Å).

2. EXPERIMENTAL SECTION

The silicic MCM-41 (Ø = 40 Å) sample used in our neutron diffraction measurement (m = 0.405 g) was synthesized at the laboratory of Prof. K.K. Unger (Maintz - RFA). Sorption isotherms have been measured during the diffraction experiment to define the sorbate loadings with accuracy and to control that the thermodynamic equilibrium is reached before starting the diffractogram recording . The neutron diffraction experiments were performed at the Léon Brillouin Laboratory (C.E.A - Saclay). Prior to sorption isotherm measurements, the sample has been outgased at around 200 °C under vacuum ($P \leq 10^{-6}$ Torr) during 12 hours.

3. RESULTS AND DISCUSSION

We have measured the neutron diffractograms of nitrogen sorbed phase on MCM-41 (Ø = 40 Å) for different N_2 sorbate loadings in a large temperature range 30 - 85 K. The different N_2 loadings corresponding to the measured diffractograms (the sorption was done at T = 77.5 K) are represented on figure 2 . During the sorption, we have observed that the characteristic pattern of the " empty " MCM-41 (40 Å) sample is strongly modified as shown on figure 3 . In the low loading regime ($Q_{ads.}$ = 38 %) during the film adsorption on the mesopore inner surface, the (100) diffraction peak intensity I_{100} increases. On the other hand, those of the (110) and (200) peaks decrease .

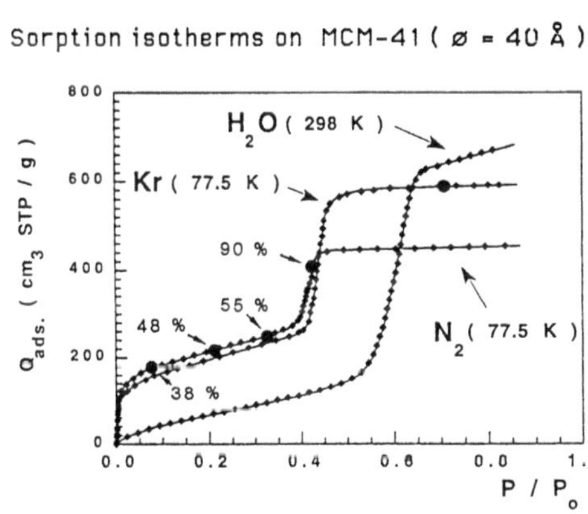

Figure 2. Sorption isotherms of N_2 , Kr and H_2O on MCM-41 (40 Å) measured at 77.5 K 77.5 K and 298 K. The loadings where diffractograms have been recorded are represented by dots.

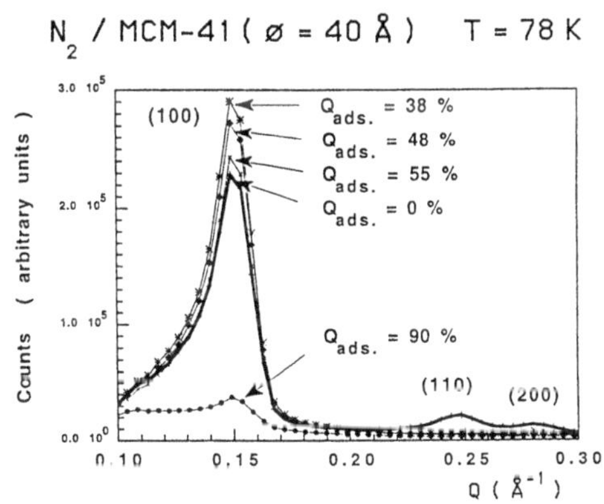

Figure 3. Diffractograms of N_2 sorbed phase on MCM-41 (40 Å) at 77.5 K for different loadings $Q_{ads.}$: 38 % , 48 % , 55 % and 90 % .

Such a behaviour has already been observed during the sorption of CD_4 on MCM-41 (Ø = 25 Å) [18] . In our opinion, the adsorbed film at the early stage of its growth presents a uniform thickness. The adsorbed film phase both decreases the MCM-41 pore diameter and thickens the pore walls. The scattered signals from the composite system, adsorbed film + MCM-41, generate constructive interferences which intensify I_{100} and destructive one's which reduce I_{110} and I_{200}. In the medium loading regime ($48 \leq Q_{ads.} \leq 55$ %), I_{100} begins to decrease weakly. We think that as a consequence of the inner surface heterogeneity, the film doesn't grow with a uniform thickness (according to a layer by layer mode) but in a more irregular way leading to a rough film. Finally, in the high loading regime during the N_2 capillary condensation ($Q_{ads.} = 90$ %) I_{100} strongly decreases . The interference effects between the N_2 sorbed phase and the MCM-41 host material are almost fully destructive. The diffracted signals from the confined nitrogen phase in MCM-41 (Ø = 40 Å) at T = 77.5 K for different loadings are represented on figure 4. In the diffractograms, such signals appear as large bumps located in the scattering wave vector range $1.78 \leq Q \leq 1.86$ A^{-1}. The

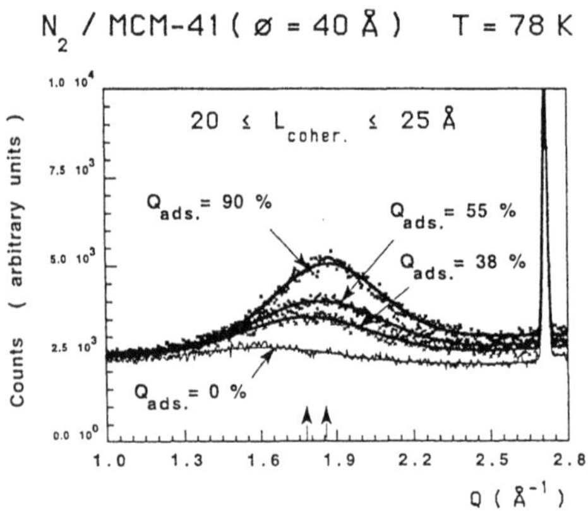

Figure 4. Diffratograms of N_2 sorbed phase on MCM-41 (40 Å) at 77.5 K for different loadings $Q_{ads.}$: 0 %, 38 %, 55 % and 90 %.

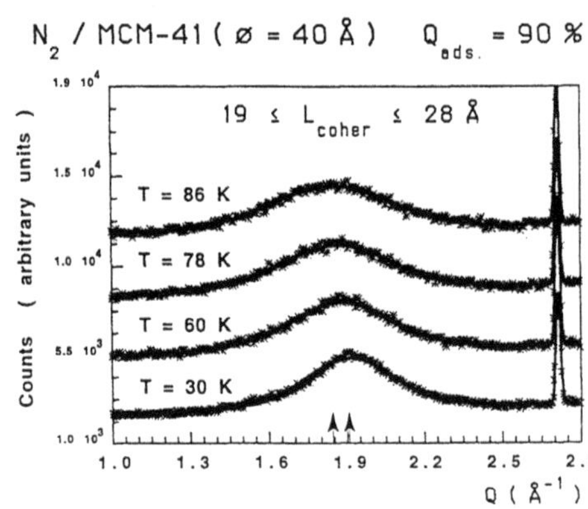

Figure 5. Diffratograms of N_2 sorbed phase on MCM-41 (40 Å) for different temperatures : 86 , 78 , 60 and 30 K.

bump intensity increases when increasing the N_2 loading (in that Q range we note that the MCM-41 sorbant signal is very small) . The coherence lengths deduced from the diffracted bump widths are rather small $20 \leq L_{coher.} \leq 25$ Å . Such $L_{coher.}$ values indicate that the molecular organization in the sorbed N_2 phase is characterized by a short range order (as for liquid phase or amorphous solid phase). For the nitrogen loading equal to 90 % (when the capillary phase is almost completely developed) we have measured the diffractograms in the temperature range 30 - 86 K . The results are shown on figure 5. Surprisingly, we don't observed notable difference in the diffraction patterns. Even at T = 30 K, the coherence length is rather small $L_{coher.} = 28$ Å. Far below the bulk triple point temperature (T_{3t} = 63 K), the confined nitrogen phase in cylindrical mesopores of 40 Å diameter doesn't present any crystallisation phenomenon. The same result has been obtained for CD_4 confined phase but the mesopore diameter is smaller Ø = 25 Å [18]. We have noticed that the intensity of the MCM-41 (100) diffraction peak strongly increases when the temperature decreases in the range 60 - 86 K, as represented on figure 6. Such a variation is too large to be interpreted as a consequence of the Debye - Waller factor. We think that such an observation can be correlated to the crossing of the capillary critical temperature T_{cc}. For the considered system N_2 / MCM-41 (Ø = 40 Å) T_{cc} is located around 73 K [16]. Neutron diffractograms have been also measured for Kr and D_2O sorbed on MCM-41 (40 Å) at T = 77.5 K and T = 208 K respectively, as represented on figure 7. On the contrary to the nitrogen sorbed phase which is characterized by a short range order, both krypton and heavy water confined phases in MCM-41 (40 Å) are characterized by a medium range order ($L_{coher.}$ ≃ 60 Å). We recall that, during the sorption of the Kr capillary phase at T = 77.5 K on MCM-41 (Ø = 40 Å) an exothermic heat peak has been observed in the calorimetric signal [12]. Such a result is interpreted for krypton as a capillary " solidification " phenomenon. The detailed study, by neutron diffraction, of the structural property of sorbed krypton phase versus temperature in the range 30 - 95 K, points out that the Kr confined phase in MCM-41 (40 Å) undergoes a phase transition around T = 85 K. The Kr coherence length $L_{coher.}$ dependence on temperature is summarized on figure 8 (detailed results are published elsewhere [21]).

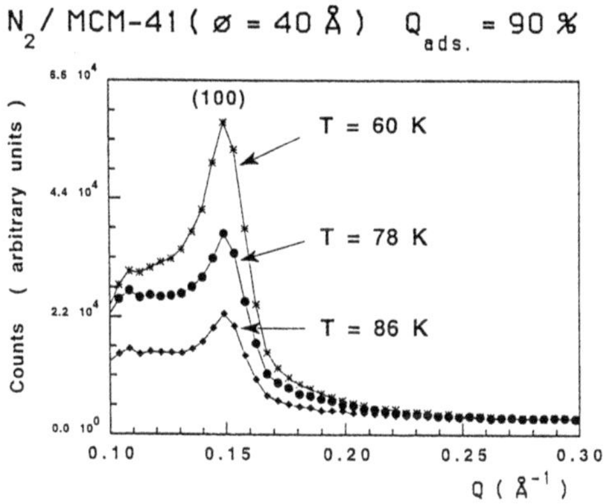

Figure 6. Diffratograms of N_2 sorbed phase on MCM-41 (40 Å) for different temperatures : 60 K, 78 K and 86 K.

The Kr / MCM-41 (Ø = 40 Å) system is particularly interesting. It is the first example of capillary " solidification " which can be observed in suited temperature and pressure ranges (T = 77.5 K , P ≃ 0.8 Torr ; as indicated on figure 2). The melting temperature of the confined Kr solid phase T_m = 85 K in cylindrical pore of 40 Å diameter are strongly displaced towards the low temperature in comparison of the Kr bulk triple point T_{3t} = 116 K. The melting temperature T_m = 85 K can be considered as a " capillary " triple point temperature T_{ct} for the Krypton confined phase in MCM-41 (40 Å). Three confined phases coexist at such a temperature T_{ct} : a capillary gas ⇔ a capillary solid ⇔ a capillary liquid . At the present time, we don't have measured the isotherm desorption branch of the Kr / MCM-41 (40 Å) system at 77.5 K. It will be interesting to know if the pore filling by the capillary solid phase gives rise to a hysteresis loap phenomenon. The growth mode of such a capillary solid phase will be also interesting to analyse . Solidification of the confined water in MCM-41 (Ø = 40 Å) has been also observed below T = 233 K [20,22]. We have to note that for water sorbed species, only the capillary phase develops (no adsorbed film is observed in the low pressure range as shown on figure 2). For confined water the pore diameter is effectively equal to 40 Å . On the other hand, for krypton the pore diameter is reduced by the physisorbed film thickness (for a Kr monolayer Ø ≃ 32 Å). For smaller MCM-41 mesopore size (Ø = 25 Å) we have observed the confined water solidification around T = 215 K [22]. On the other hand, we don't observe any solidification for hydrogen and methane sorbed phases on MCM-41 (Ø = 25 Å) [18] . It is interesting to understand why certain sorbed species undergo a phase transition and some others don't . Such a difference is well illustrated by the two systems : Kr / MCM-41 (40 Å) and N_2 / MCM-41 (40 Å). On figure 8, we have represented the coherence length dependence on temperature in the range 30 - 95 K for

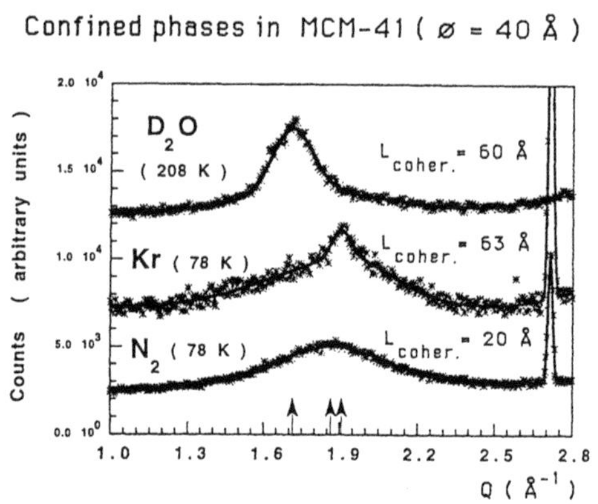

Figure 7. Diffractograms of N_2, Kr and D_20 sorbed phases on MCM-41 (40 Å) at 78 K, 78 K and 208 K in the high loading regime.

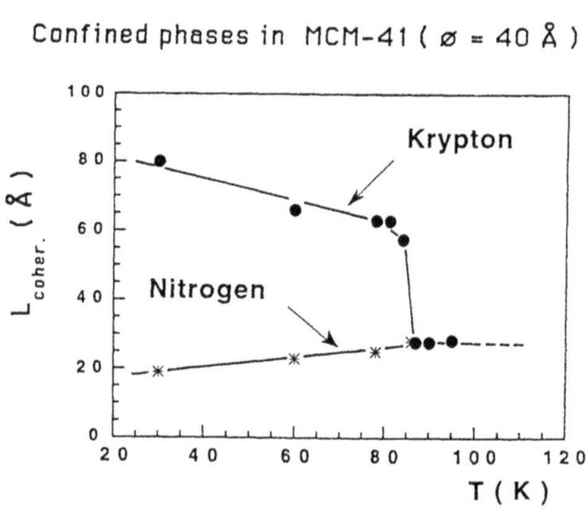

Figure 8. Coherence length $L_{coher.}$ versus temperature of Kr and N_2 confined phases on MCM-41 (Ø = 40 Å).

the two considered sorbates. The polar character or the morphology of the N_2 molecules might be the reason of the behaviour difference between the two confined phases. The phase transition existence for confined Kr and D_2O phases in MCM-41 (Ø = 40 Å) is well established now. But an open question remains concerning the confined solid phase nature. Is it a crystalline solid or an amorphous one ? In our opinion , it is difficult to answer due to the fact that there is no correlation between the MCM-41 mesopores (as a consequence of the amorphous character of the walls). The diffraction peaks might be broadened in any cases .

Now, we plan to investigate the dynamics properties of the two sorbed phases (confined liquid and confined solid) in the case of the water / MCM-41 system, for a series of diameter values (Ø = 40 Å , Ø = 25 Å and Ø ≃ 15 Å).

ACKNOWLEDGEMENTS

The authors would like to thank kindly Prof. K.K. unger (Maintz - RFA) for MCM-41 ́sample supplies .

REFERENCES

1. C.T.Kresge, M.E. Leonowicz, W.J. Roth, J.C. Vartuli and J.S. Beck, Nature, Vol. 359 (1992) 710 .
2. J.S. Beck, J.C. Vartuli, W.J. Roth, M.E. Leonowicz, C.T.Kresge, K.D. Schmitt, C.T-W. Chu, D.H. Olson, E.W. Sheppard, S.B. McCullen, J.B. Higgins and J.L. SchlenKer, J. Am. Chem. Soc., 114 (1992) 10834 .
3. O. Franke, G. Schulz-Ekloff, J. Rathousky, J. Starek and A. Zukal, J. Chem. Soc., Chem. Commun., 9 (1993) 724.
4. P.J. Branton, P.G. Hall and K.S.W. Sing, J. Chem.Soc. Chem. Commun. (1993) 1257.
5. P.L. Llewellyn, Y.Grillet, F. Schüth, H. Reichert and K.K. Unger, Microporous Material, 3 (1994) 345.
6. P.J. Branton, P. G. Hall and K.S.W. Sing, H. Reichert, F. Schüth and K.K. Unger, J. Chem. Soc. Faraday Trans., 90 (1994) 2965.
7. J. Rathousky, A. Zukal, O. Franke and G. Schulz-Ekloff, J. Chem. Soc. Faraday Trans., 90 (1994) 2821.
8. P.J. Branton, P. G. Hall and K.S.W. Sing, Adsorption, 1 (1995) 77.

9. R. Schmidt, M. Stöcker, E. Hansen, D. Akporiaye and O. H. Ellestad, Microporous Materials, 3 (1995) 443.
10. P.L. Llewellyn, F. Schüth, Y. Grillet, F. Rouquerol, J. Rouquerol and K. K. Unger, Langmuir, 11 (1995) 574.
11. R. Schmidt, E.W. Hansen, M. Stöker, D. Akporiaye and O.H. Ellestad, J. Am. Chem. Soc., 117 (1995) 4049.
12. P.L. Llewellyn, Y. Grillet, J. Rouquerol, C. Martin and J.P. Coulomb, Surface Science, 352 - 354 (1996) 468.
13. A.V. Neimark, P.I. Ravikovitch, S.C. Domhnaill, F. Schüth and K.K. Unger, in Fundamentals of Adsorption, Edt. M.D. LeVan, Kluwer Academic Publishers (1996) p. 667.
14. M. Kruk, M. Jaroniec and A. Sayari, J. Phys. Chem., B 101 (1997) 583.
15. P.I. Ravikovitch, D. Wei, W.T. Chueh, G.L. Haller and A.V. Neimark, J. Phys. Chem., B 101 (1997) 3671.
16. K. Morishige, H. Fujii, M. Uga and D. Kinukawa, Langmuir, 13 (1997) 3494.
17. B.S. Schirato, M.P. Fang, P.E. Sokol and S. Komarneni, Science, 267 (1995) 369.
18. J.P. Coulomb, C. Martin, Y. Grillet, P. Llewellyn and G. André, Studies in Surface Sciences and Catalysis, Vol. 105 (1997) 1827.
19. J.M. Baker, J.C. Dore and P. Behrens, J. Phys. Chem., B 101 (1997) 6226.
20. K. Morishige and K. Nobuoka, J. Chem. Phys., 107 (1997) 6965.
21. J.P. Coulomb, Y. Grillet, P. Llewellyn, C. Martin and G. André (Submitted to Fundamentals of Adsorption 6th Conference, May - 1998, Giens - France).
22. J.P. Coulomb, C. Martin, N. Floquet, Y. Grillet, P. Llewellyn and G. André (Submitted to the 12th Intern. Zeolite Conference, July - 1998, Baltimore - USA).

MESOPOROUS MOLECULAR SIEVES 1998
Studies in Surface Science and Catalysis, Vol. 117
L. Bonneviot, F. Béland, C. Danumah, S. Giasson and S. Kaliaguine (Editors)

FT-IR spectroscopic studies of the acidic properties of mesoporous molecular sieves by adsorbed carbon monoxide

L. Pasqua, F. Testa and R. Aiello

Dipartimento di Ingegneria Chimica e dei Materiali, Università degli Studi della Calabria, 87030 Rende (CS) ITALY

Purely siliceous, gallium, iron and aluminum containing silicates mesoporous materials were synthesized in fluoride media at a pH near neutrality and characterized by FT-IR spectroscopy using CO as probe molecule. The sample containing gallium degassed at 500°C shows a peak at 3662 cm^{-1} attributed to OH stretching of non-bridged Brønsted acid sites. The peak is extending on the high frequencies side, with the prominent peak centered at 3680 cm^{-1} when the sample is degassed at 700°C. This is probably due to migration of gallium atoms into extra-framework positions with the formation of free Ga–OH groups on the surface of gallia whose stretching mode is centered at 3680 cm^{-1}. The migration of gallium atoms is also confirmed from Lewis acidity presents only in the spectra of the high temperature degassed sample. In the aluminum containing sample degassed at 700°C occurrence of Brønsted acid site type is showed only indirectly by adsorption of CO. Total migration of iron into aggregates of Fe_2O_3 is recognized from spectra of iron-containing sample degassed at 500°C.

1. INTRODUCTION

The synthesis of mesoporous materials by Mobil researchers [1] opened new perspectives in the field of the catalysis thanks to the possibility to operate with molecular sieves having homogeneous porosity and, at the same time, dimensions of channels greater than zeolites.

The properties of catalysts depend on the chemical composition and it is well known that acidity of zeolites decrease when aluminum, gallium, iron and boron substitute isomorphically silicon in the framework [2]. FT-IR spectroscopy of probe molecules is an important tool for the comprehension of the surface properties of catalysts. In the last years the acid characteristics of H-ZSM-5 and related gallium and iron substitute silicalites have been the subject of a relevant number of studies [3-6]. Recently, the physico-chemical properties of Al, Ga and Fe substituted mesoporous materials synthesized in basic medium were reported [7]. The trivalent elements have been found in tetrahedral coordination in the framework although the activation of the products promotes the formation of octahedrally coordinated extra-framework species. The strength of Brønsted acidity measured by the IR spectra of adsorbed pyridine decreases in the series Al>Ga>>Fe. In the hexagonal mesoporous materials tetrahedra do not form a crystalline structure and this influence the properties of these materials. MCM-41 aluminosilicates have been characterized by FT-IR and MAS-NMR [8] spectroscopy evidencing that the flexibility of the amorphous framework is responsible for the differences with zeolitic aluminosilicates.

Mesoporous silica precursors have been synthesized in fluoride media [9] using ionic- and non-ionic surfactants. F^- anions promote the formation into solution in the form of SiF_6^{2-} anions, favour the policondensation of tetrahedra and act as a countercation when the negative charge density of the silica walls is too low to compensate the charge of cetyltrimethylammonium cations.

In this paper, we report the synthesis conditions and the characterization by FT-IR spectroscopy of CO adsorbed on pure siliceous, gallo-, iron- and alumino-silicates mesoporous samples synthesized in the presence of ammonium fluoride using cetyltrimethylammonium cations as surfactant.

2. EXPERIMENTAL

The reactants were: tetraethoxysilane (TEOS, purum, 98%, Aldrich), NH_4F (purum, >99%, Carlo Erba), cetyltrimethylammonium bromide (CTMABr, Aldrich), $Al(NO_3)_3 9H_2O$ (RPE, >99%, Carlo Erba), $Fe(NO_3)_3 9H_2O$ (purum, >98%, Aldrich), $Ga(NO_3)_3$ (purum,99.9%, Sigma) and distilled water. As an example, the procedure for the synthesis of mesoporous gallosilicate precursor was the following: 191.4 g of distilled water were heated at the temperature of 50°C, then 5.48 g of CTMABr, 10.99 g of NH_4F and 0.94 g of $Ga(NO_3)_3$ were carefully added in this order waiting for a complete dissolution of each reactant. Finally, 15.09 g of TEOS were added. The resulting mixture was kept at 50°C for 4 hours. The solid formed was filtered, washed first with distilled water, then with ethanol and dried at 100°C for 2 hours. The molar ratios of the initial mixture were: $CTMABr/SiO_2=0.21$; $NH_4F/SiO_2=4.1$; $H_2O/SiO_2=146.5$ and $Ga/SiO_2=0.05$.

The samples were activated at 550°C for 8 hours with a heating rate of 1°C/min in a air flow of 15 cc/min. The X-ray diffraction powder patterns were acquired on a Philips PW1730 diffractometer using CuKα radiation. The samples were scanned from 1 to 15 2θ in steps of 0.005° with a counting time of 1 s per step. For IR studies, thin self-supporting wafers of each sample were prepared and outgassed under vacuum, in a home-made cell, at temperatures of 773 and 973 K for two hours.The IR spectra were taken on a FT-IR Bruker IFS-28 with a resolution of 2 cm^{-1}. The N_2 adsorption-desorption isotherms at 77 K were measured on a Micromeritics Asap 2010. Samples were pre-treated in vacuum for 6 hours at 200°C.

3. RESULTS AND DISCUSSION

3.1. Structural Characterization

In Table 1 are reported the chemical and physical properties of the synthesized samples. The d_{100} values obtained for all the samples are in good agreement with those showed by Voegtlin et al [9]. In fact, products obtained from batches having a pH near neutrality do not show a significant decrease of the d_{100} values when the organic fraction is removed. The rates of hydrolysis and polycondensation are similar and the cells around the surfactant are well formed. On the other hand, at these pH values, the mechanism of aggregation $S^+(I^\circ X^-)$ due to the weak interaction between the cationic surfactant (S^+) and the weakly charged anionic framework (I°) mediated by the counterion Br^-, F^-, NO_3^- (X^-) does not allow a high degree of organization. It was previously demonstrated that similar materials still have hexagonal order [10].

Table 1
Chemical and physical properties of synthesized mesoporous samples

T/SiO_2 initial	Synthesis Temp. (°C)	d_{100} as made (Å)	d_{100} calcined (Å)	S_{BET} (m^2/g)	T (moles %)
Pure Silica	50	46.1	48.1	415	-
Ga/SiO_2=0.05	50	46.1	46.1	481	0.05
Fe/SiO_2=0.05	50	48.1	46.1	465	0.06
Al/SiO_2=0.04	50	44.2	44.2	395	0.05

3.2. Pure Siliceous Mesoporous Sample

The background spectrum of the purely siliceous mesoporous sample degassed at 500°C for 2 hours shows in the OH stretching region an important absorption at 3742 cm^{-1} attributed to isolated internal silanols, as a product of nest of [SiO_4] building units [11]. Upon CO treatment, on the high CO pressure range (1-100 mbar), a slight decrease in intensity of the peak at 3742 cm^{-1} can be noted, while a band centered at 3660 cm^{-1} assigned to the stretching of silanols hydrogen-bonded to the CO molecules originates on the low frequency side of 3742 cm^{-1} peak. In the CO stretching region the peak at 2160 cm^{-1} is attributed to the stretching of the CO interacting with silanols while the peak at 2138 cm^{-1} (which is present only in the higher CO pressure spectrum) is due to liquid-like CO.

3.3. Gallosilicate Mesoporous Sample

FT-IR of Gallosilicate degassed at 500°C. The background spectrum (Fig.1(a), lower spectrum) of gallosilicate mesoporous sample outgassed at 500°C for 2 hours shows a peak located at 3749 cm^{-1} attributed to isolated silanols and a peak in the Brønsted acid sites region at 3662 cm^{-1} and attributed to the stretching modes of OH groups on partially extra-framework gallium atoms generated by hydrolysis or incomplete formation of Si–O–Ga bridged Brønsted acid sites [3]. The upper spectra in Figure 1(a) show that upon CO interaction, the formation of a new band centered at 3465 cm^{-1} and with a large shoulder at 3520 cm^{-1} gradually occurs. This new band, that disappears below 0.5 mbar of CO pressure, is assigned to the interaction of non-bridged Brønsted acid groups through hydrogen bonding with CO molecules that shifts the OH stretching at lower wavenumbers ($\delta\nu \approx -200$ cm^{-1}).

In Figure 1(b) the IR spectra in the CO stretching region are reported:

i) the peak at 2172 cm^{-1} corresponds to the stretching vibration of the CO molecules interacting with the Brønsted acid centers;

ii) the peak at 2138 cm^{-1} is present only in the spectra acquired at high equilibrium CO pressure and is assigned to the stretching modes of CO in a liquid-like state.

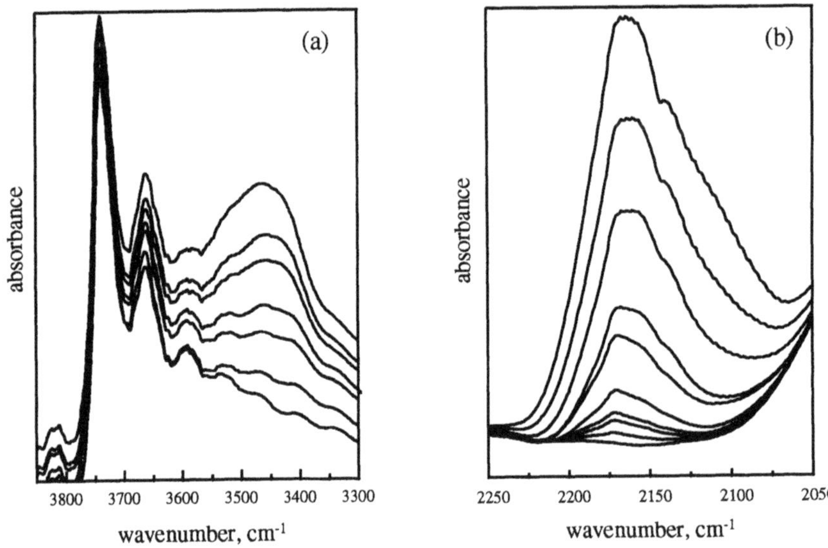

Figure 1. FT-IR spectra of CO adsorbed on gallosilicate mesoporous sample degassed at 500°C: (a) OH stretching region; (b) CO stretching region

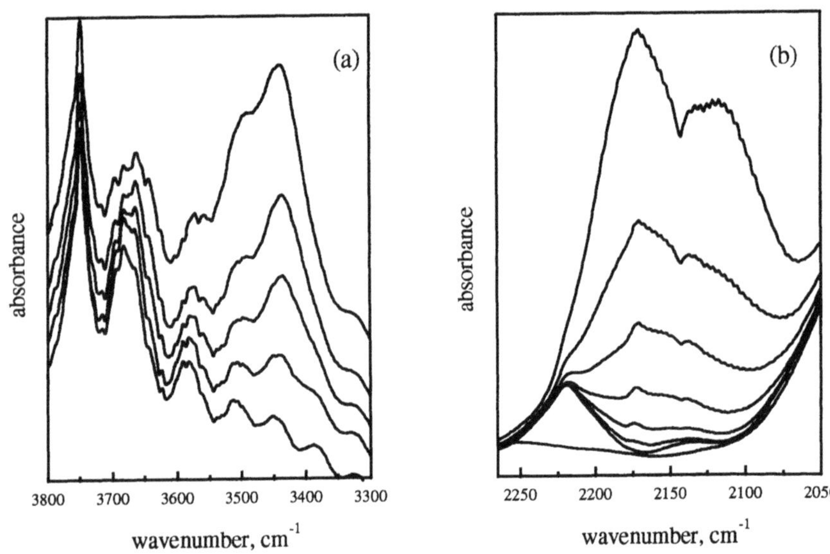

Figure 2. FT-IR spectra of CO adsorbed on gallosilicate mesoporous sample degassed at 700°C: (a) OH stretching region; (b) CO stretching region

FT-IR of Gallosilicate degassed at 700°C. In order to follow the migration of gallium atom into extraframework position, the gallosilicate mesoporous sample was degassed at 700°C for 2 hours. The background spectrum (Fig. 2(a), lower spectrum) in the hydroxyl stretching region shows that the peak of Brønsted acid sites has enlarged and transforms into a band with a maximum absorption centered at 3680 cm^{-1} and a low-frequency shoulder centered at 3662 cm^{-1}.

Interaction with CO (Fig. 2(a), upper spectra) causes a marked erosion of the band particularly for the high frequency component. The decrease of the 3662 cm^{-1} component is not easily evaluated because, unlike zeolites, the massive interaction of Brønsted acid sites region needs CO pressure values at which erosion of the silanols bands causes the formation of a new peak at ≈3660 cm^{-1} ($\delta\nu$= -90cm^{-1}). In addition to the Brønsted acid sites band erosion, the formation of a complex absorption composed of a peak centered at 3440 cm^{-1} and a large shoulder at 3496 cm^{-1} occurs and is attributed to the stretching of Brønsted acid sites interacting with CO. At lower CO pressures, the comparison between the spectra of the sample degassed at 500°C and 700°C shows that, for the last sample, both the 3465 cm^{-1} and the 3520 cm^{-1} components (still present, see above for the assignment) have enlarged on the low frequency side with the appearance of a band respectively at 3440 cm^{-1} and 3496 cm^{-1}. In the CO stretching region, reported in Fig. 2(b), a relevant effect of the increased temperature of activation is revealed by the presence, upon CO treatment, of a peak at 2220 cm^{-1} disappearing for CO pressure below 0.05 mbar and attributed to the stretch of CO interacting with strong Lewis acid sites. The appearance of Lewis acidity is indicative of the formation of Ga_2O_3 microcrystals. The peak at 3680 cm^{-1} is assigned to stretching modes of OH groups on the surface of gallia generated from more severe activation, while the peak at 3662 cm^{-1} is assigned to non-bridged Brønsted T–OH sites as in the sample degassed at 500°C. For both experiments, the particular structure of the band generated from stretching of OH groups interacting with CO is justified on the basis of the results of Otero Arean *et al* [3]. They demonstrated the formation of bifurcated hydrogen bonds between bridged OH groups and oxygen atoms at defective sites of Ga-ZSM-5. We suppose the formation of a similar interaction between acidic hydroxyl groups, CO and oxygen atoms less intense than the corresponding one of CO with free acid sites in agreement with the high frequency shoulder.

3.4. Ironsilicate Mesoporous Sample

The ironsilicate mesoporous sample was treated in vacuum for 2 hours at 500°C. The resulting background spectrum (Fig.3(a), lower spectrum) shows a peak at 3742 cm^{-1} assigned to isolated internal hydroxyl while there is no evidence of Brønsted acid sites type. Interaction of CO with internal hydroxyl on the high pressure range shifts the OH stretching at 3660 cm^{-1} (Fig.3(a), upper spectra). In the CO stretching region, reported in Fig.3(b), are recognizable the 2174 cm^{-1} absorption assigned to the stretching of CO interacting with the Lewis acid groups of extra-framework Fe_2O_3 [4], hidding the stretching of CO interacting with terminal silanols that would be located at 2160 cm^{-1}, and the 2138 cm^{-1} absorption assigned to the stretch of the CO liquid-like adsorbed. The peak at 2174 cm^{-1}, shown in Fig. 3(c), is also originated from interaction of CO at room temperature. The predominance of the 2174 cm^{-1} component on the expected 2180 cm^{-1} band attributed to interaction of CO with more coordinatively insatured Fe^{3+} located at edges and corners of nanoclusters of Fe_2O_3, suggests a considerable aggregation of extra-framework iron.

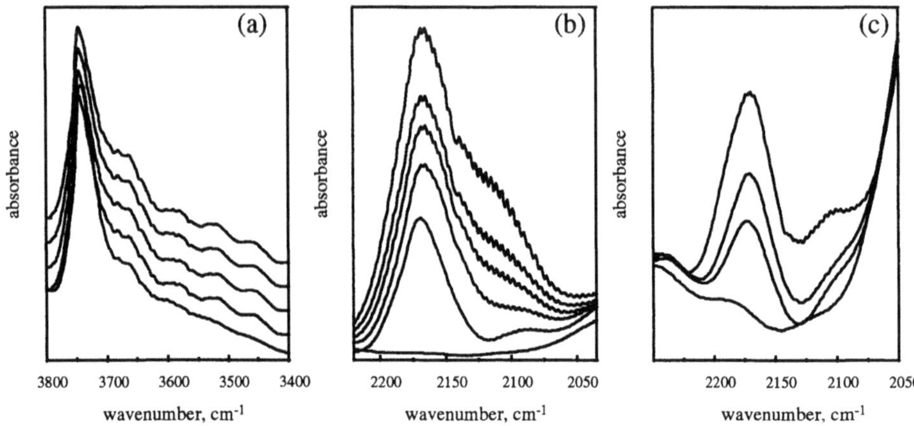

Figure 3. FT-IR spectra of CO adsorbed on ironsilicate mesoporous sample degassed at 500°C: (a) OH stretching region at a nominal temperature of 77K; (b) CO stretching region at a nominal temperature of 77K; (c) CO stretching region at room temperature

3.5. Aluminosilicate Mesoporous Sample

FT-IR Aluminosilicate degassed at 500°C. The aluminosilicate mesoporous sample was degassed at 500°C in vacuum for two hours. The resulting spectrum did not present relevant differences from that of a pure silica sample treated in the same conditions. In fact, a low-frequency tail of the silanol absorption peak extending down to 3400 cm^{-1} is found, attributed to stretching of internal hydroxyls interacting through hydrogen bonding. The IR study of the interaction with CO does not give any valuable difference from the pure siliceous sample.

FT-IR Aluminosilicate degassed at 700°C. Aluminosilicate sample was degassed at 700°C for two hours to better appreciate every change in the FT-IR spectra due to CO interaction. The background spectrum (Fig. 4(a), lower spectrum) shows two adjacent neat peaks at 3742 cm^{-1} and 3749 cm^{-1} assigned, respectively, to the stretching of internal isolated silanols and external silanols while there is no evidence of absorption corresponding to Brønsted acid sites. When CO interacts with the sample at room temperature (Fig. 4(b)) two new absorptions occur at 2173 cm^{-1} and 2230 cm^{-1} and are assigned to stretching of CO interacting respectively with weak and strong Lewis acid sites [8]. The effects of the interaction with decreasing pressure of CO at the nominal temperature of 77 K on IR spectra correspond to those already reported by Viale et al. [8]. At high CO pressures (Fig. 4(a), upper spectra) the peak assigned to silanols slightly decreases in intensity while the formation of a peak centered at 3660 cm^{-1} assigned to the stretching of silanols interacting with CO molecules occurs. The formation of a new large band in the range 3360-3520 cm^{-1}, on the high CO pressure range, is assigned to stretching of "hidden" Brønsted acid sites on partial extra-framework aluminum atoms interacting with CO. In the CO stretching region (Fig. 4(c)) the peak at 2173 cm^{-1} corresponds to the stretching of CO interacting with "hidden" Brønsted acid sites and weak Lewis acid sites, and the peak centered at 2138 cm^{-1} corresponds to stretching of liquid -like CO. Upper spectra in Fig. 4(a) show that, for intermediate CO pressures, the band assigned to the stretching of Brønsted acid sites assumes a better defined profile and we can detect two main absorptions at 3432 cm^{-1} and 3491 cm^{-1} disappearing for P_{CO}<0.5 mbar. The 3432 cm^{-1} peak is assigned to the OH stretching of Brønsted acid group interacting with CO. The assignment of the shoulder at 3491 cm^{-1} follows the arguments given above for the gallosilicate.

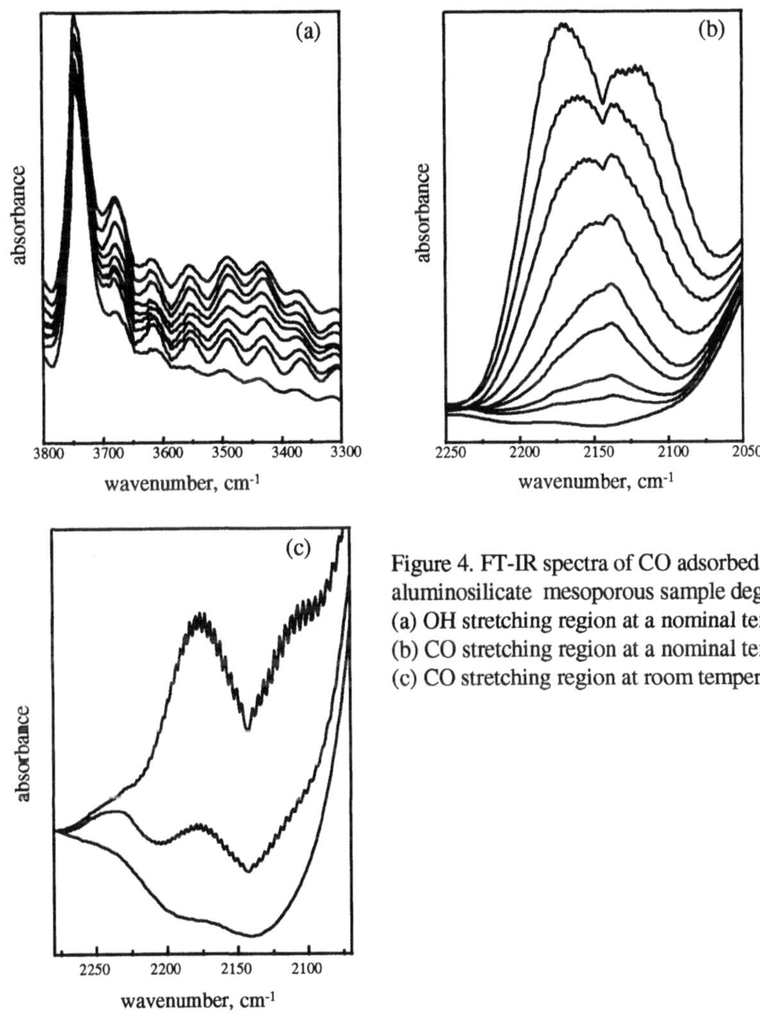

Figure 4. FT-IR spectra of CO adsorbed on aluminosilicate mesoporous sample degassed at 700°C:
(a) OH stretching region at a nominal temperature of 77K;
(b) CO stretching region at a nominal temperature of 77K;
(c) CO stretching region at room temperature

4. CONCLUSIONS

The interpretation of the complex IR absorption attributed to the OH stretching of Brønsted acid sites interacting with CO in gallo- and alumino-silicates is strictly connected to analogous FT-IR studies of CO adsorbed on H-ZSM-5 and Ga-ZSM-5. We suppose that the occurrence of the shoulder at higher frequency is a consequence of the peculiarity of CO molecule that is capable of promoting bifurcated interaction both with zeolites and with mesoporous materials. Also, we observe that, the frequency gap between the shoulder and the main peak is directly dependent on the main shift (the difference in wavenumber between the Brønsted acid sites peak and the main peak of the complex absorption generated from the CO interaction) in both the materials. Gallium containing samples show a neat absorption in the Brønsted acid sites region so far unknown for this kind of amorphous materials. Aluminum-containing sample, upon CO interaction, shows significant presence of Brønsted acid sites partially in extra-framework position. Fluoride synthesis route appears able, for this kind of materials, to produce important differences in catalytic requirements regard to other synthesis procedures. Further spectroscopic studies will give a more defined knowledge about the properties of these fascinating materials.

5. REFERENCES

1. J.S. Beck, J.C. Vartuli, W.J. Roth, M.E. Leonowicz, C.T. Kresge, K.D. Schmitt, C.T.W. Chu, D.H. Olson, E.W. Sheppard, S.B. McCullen, J.B. Higgins and J.L. Schlenker, J. Am. Chem. Soc., 114 (1992) 10834.
2. F. Trifirò, Proc. Of the III Convegno Nazionale di Scienza e Tecnologia delle Zeoliti, R. Aiello Ed., Cetraro 28-29 September 1995, 187.
3. C. Otero Areán, G. Turnes Palomino, F. Geobaldo and A. Zecchina, J. Phys. Chem., 100 (1996) 6678.
4. Zecchina, F. Geobaldo, C. Lamberti, S. Bordiga, G. Turnes Palomino and C. Otero Areán, Catalysis Letters, 42 (1996) 25 and references therein.
5. I. Mirsojew, S. Ernst, J. Weitkamp and H. Knözinger, Catalysis Letters, 24 (1994) 235.
6. A. Zecchina, S. Bordiga, G. Spoto, D. Scarano, G. Petrini, G. Leofanti, M. Padovan and C. Otero Areán, J. Chem. Soc. Faraday Trans., 88 (1992) 2959.
7. H. Kosslick, G. Lischke, G. Walther, W. Storek, A. Martin and R. Fricke, Microporous Materials, 9 (1997) 13.
8. S. Viale, E. Garrone, F. Di Renzo, B. Chiche and F. Fajula, Stud. Surf. Sci. Catal., 105 (1997) 533.
9. A.C. Voegtlin, F. Ruch, J.L. Guth, J. Patarin and L. Huve, Microporous Materials, 9 (1997) 95.
10. P.T. Tanev and T. J. Pinnavaia, Chem. Mater., 8 (1996) 2068 and reference therein.
11. A. Zecchina, S. Bordiga, G. Spoto, L. Marchese, G. Petrini, G. Leofanti and M. Padovan, J. Phys. Chem., 96 (1992) 4991.

MESOPOROUS MOLECULAR SIEVES 1998
Studies in Surface Science and Catalysis, Vol. 117
L. Bonneviot, F. Béland, C. Danumah, S. Giasson and S. Kaliaguine (Editors)

Adsorption methods for characterization of surface and structural properties of mesoporous molecular sieves

M. Jaroniec,[a] M. Kruk [a] and A. Sayari [b]

[a] Department of Chemistry, Kent State University, Kent, OH 44242, U.S.A.

[b] Department of Chemical Engineering and CERPIC, Universite Laval, Ste-Foy, Quebec G1K 7P4, Canada

Application of nitrogen adsorption technique to study porous structures and surface properties of unmodified and chemically modified mesoporous molecular sieves (e.g., MCM-41 and MCM-48) is presented and recommendations for their proper characterization are made. It is shown that comparative plot methods (e.g., t-plot and α_s-plot) are useful to calculate not only the primary mesopore volume and external surface area, but also the micropore volume and total surface area. The pore size of MCM-41 can be accurately evaluated from the primary mesopore volume and the XRD interplanar spacing on the basis of geometrical considerations. An accurate method for calculation of pore size distributions for mesoporous molecular sieves, especially siliceous ones, is presented. Moreover, it is shown that low pressure adsorption data provide information about surface properties of ordered mesoporous materials, which is particularly useful for characterization of samples with coated/bonded organic ligands, allowing to detect functional groups and estimate their surface coverage and uniformity.

1. INTRODUCTION

Since 1992, mesoporous molecular sieves (MMS), e.g., MCM-41 and MCM-48, have attracted considerable and still growing attention due to their fascinating structures and wide perspectives of application as advanced materials [1-5]. In order to achieve further progress in the field of nanostructured materials, such as MMS, reliable characterization techniques need to be developed. Although many analytical methods, for example X-ray diffraction (XRD), transmission and scanning electron microscopy, nuclear magnetic resonance spectroscopy and gas adsorption, have been extensively applied [1-5], each of them has its limitations and allows to elucidate only certain structural aspects of MMS. For instance, the powder XRD is invaluable for determination of symmetry of porous structures and for evaluation of sizes of their unit cells. However, XRD may indicate ordering even if only a part of the material's structure is periodical. This problem is exemplified by data for a calcined large pore (5 nm) MCM-41 material [6], which was treated in pure water at 373 K to assess its hydrothermal stability [7]. The presence of sharp (100) peaks on the XRD spectra (insert in Fig. 1a) suggests the retention of quite uniform porous structures characteristic of MCM-41 after both 12 and 22 hours of heating in water. However, nitrogen adsorption data (Fig. 1a and b) showed that actually the 22 hour treatment led to the development of large pores and extensive degradation of the initial porous structure, whereas the 12 hour treatment resulted only in broadening of the pore size distribution. Therefore, XRD technique may in some cases be misleading, since it may be difficult to distinguish, whether a given sample is ordered as a whole or it only has some ordered domains. Similarly, transmission electron microscopy is useful to probe the local structure of materials, but it is difficult to effectively determine quality of the entire porous network.

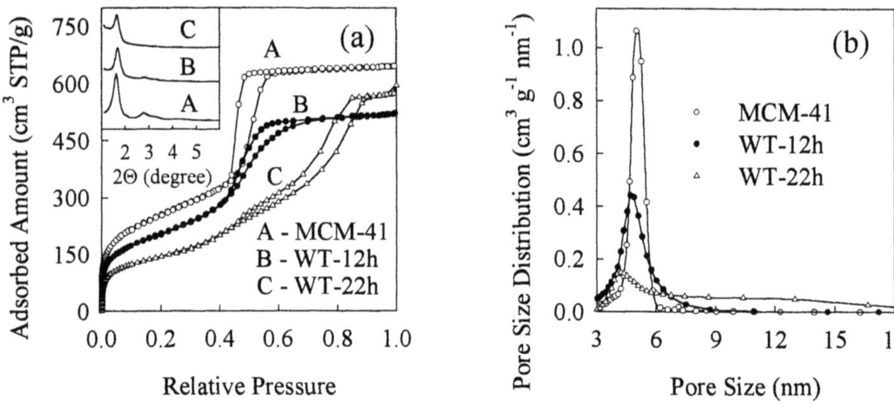

Figure 1. XRD spectra, nitrogen adsorption isotherms and pore size distributions (calculated from the BJH method with Eq. 3) of MCM-41 sample before and after treatment in water at 373 K.

In contrast, adsorption techniques [6-30] are not applicable to assess the symmetry and periodicity of materials structure [11], but provide information about global characteristics of samples and are extremely useful for evaluation of the surface area, pore size distribution and surface properties. In the current work, the application of adsorption methods to characterize mesoporous molecular sieves will be presented. The emphasis will be placed on the interpretation of nitrogen adsorption data, which provide significant amount of information about porosity and surface properties, but they can easily be misinterpreted, leading to inaccurate assessment of the pore size and pore wall thickness, misleading picture of the pore size distribution, etc. Moreover, in order to facilitate studies in the direction of synthesis and application of functionalized MMS with bonded organic phases, which recently gained much interest [1, 30-34], it will be shown how low pressure adsorption data can be used to study the surface functionality and the degree of the surface coverage of these advanced materials.

2. RESULTS AND DISCUSSION

2.1. Surface area and pore volume evaluation

The Brunauer-Emmett-Teller (BET) method is commonly applied to calculate the specific surface area on the basis of nitrogen adsorption isotherm measurements at 77 K [8-10]. Usually, data in the relative pressure range from 0.05 to 0.3 are used [9]. The BET model assumes multilayer adsorption of gas on the adsorbent's surface. Therefore, caution needs to be exercised in the data selection for the surface area analysis of MMS, in which the capillary condensation rather than multilayer adsorption may take place in the above pressure region, especially for small pores (2-3 nm in diameter) or poorly ordered materials [11, 12]. Consequently, the data in the capillary condensation pressure range should be excluded during calculations to avoid a significant overestimation of the specific surface area. To facilitate a comparison of the results reported by different laboratories, the pressure interval used in the BET calculations should be specified.

In the case when low pressure adsorption data (below relative pressure 0.01) are available, one can use comparative plot methods (t-plot, α_s-plot) [6, 8, 11, 13-15] to assess the specific surface area. Comparative plots can easily be obtained by drawing the amount adsorbed for the solid under study as a function of the amount adsorbed for an appropriate macroporous reference solid at the same pressure values. The plots are linear when the adsorption process on these two materials proceeds in the same way (i.e., by monolayer-multilayer adsorption) and exhibit deviations when adsorption mechanism on the compared solid is different (e.g., micropore filling, capillary condensation). Due to high similarity of nitrogen adsorption properties at 77 K for siliceous ordered mesoporous materials and chromatographic grade macroporous silica gels [12, 16], the latter are suitable as reference solids [6, 11, 13-14].

Initial parts of the comparative plots are usually linear for ordered mesoporous materials (see Fig. 2b, open circles) [6, 11, 13-14, 17] and their slope can be used to calculate the specific surface area [6, 11] using low pressure data. The latter are not affected by capillary condensation phenomena, what is especially important for small-pore or poorly ordered samples [11-12]. Moreover, the comparative plot analysis appears to be the best currently available method to detect and estimate microporosity in MMS, which may be present for example as a result of extended hydrothermal treatments [6-7, 14]. In these cases, the initial segment of the low pressure part of the comparative plot markedly deviates from linearity and is often followed by a linear part, before the upward swing caused by the capillary condensation in mesopores takes place (Fig. 2b, filled circles). The intercept of the line drawn through this linear part with the "Amount Adsorbed" axis of the graph provides the volume of gas adsorbed in micropores, which can easily be converted to the micropore volume by multiplying it by a proper conversion factor (0.0015468 for nitrogen adsorption at 77 K, when the volume adsorbed and the micropore volume are expressed in cm^3 STP/g and cm^3/g, respectively). Moreover, the slope of the linear part can be used to estimate the total surface area of mesopores and macropores.

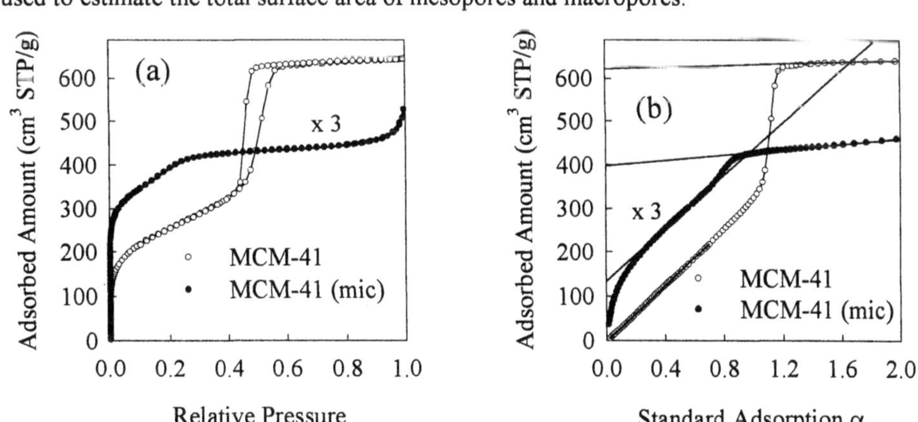

Figure 2. Nitrogen adsorption isotherms and the corresponding high resolution α_s-plots for a large pore MCM-41 material and a microporous-mesoporous MCM-41.

Linear high pressure part usually observed on comparative plots are suitable for evaluation of the primary mesopore volume (i.e., that of uniform pores characteristic of MMS) and the external surface area (i.e., that of exterior of particles of the material and possibly amorphous parts of it) [6, 11]. The secondary mesopore volume (i.e., that of mesopores and small macropores below ca. 100 nm, which do not form ordered structures characteristic of MMS) can be estimated as a difference between the total pore volume and primary mesopore volume. The former can be calculated from the maximum amount adsorbed at pressures close to the saturation vapor pressure by converting it to the volume of liquid adsorbate, analogously to the micropore volume evaluation.

2.2. Pore size analysis

Accurate assessment of the pore size distribution is crucial for the development of tailor-made mesoporous molecular sieves for advanced applications. Transmission electron microscopy allows for a direct measurement of the pore size and its uniformity for some materials, such as MCM-41, but the method probes only a small part of the structure and it is difficult to estimate the exact pore diameter and pore wall thickness. If one assumes a constant pore wall thickness (usually ca. 1 nm), the pore size of MCM-41 can be estimated on the basis of the XRD interplanar spacing. In most cases, this seems to be a reasonable approximation, but does not allow for high accuracy and in some cases may be grossly misleading due to large variations in the pore wall thickness from one sample to another [6, 14]. Nitrogen adsorption is a standard and most commonly used method to assess pore size distributions [8-9]. However, it is well-known that the calculation of pore diameters of small mesopores (which are present in MMS) on the basis of adsorption data is a difficult task due to low accuracy of classical approaches (i.e., based on the Kelvin equation), which lead to a serious underestimation of the pore size [8, 12-13, 18-19]. More recent Horvath-Kawazoe approach [21] indicates the presence of nonexisting microporosity for mesoporous and even macroporous materials [22, 31] and was developed for estimation of microporosity, and therefore is not expected to be accurate for the mesopore size estimation. The advanced DFT software [23] developed recently for porous carbons with slit-like pores provides a rather accurate assessment of the pore size, but may incorrectly indicate bimodal microporous-mesoporous pore size distributions for MMS (e.g., MCM-41) [14]. Advanced procedures intended for MMS, based on computer simulations and density functional theory, are still in the stage of development [17, 20] and are currently not available for most of the scientists in the field. Therefore, there is a need for simple and accurate approaches to the pore size analysis of ordered mesoporous materials. Such methods were recently developed and will be described below.

In the case of MCM-41 materials, which exhibit hexagonal arrays of uniform pores [1-4], one can derive a formula for the pore diameter based on purely geometrical considerations of the ratio between the pore volume and pore wall volume in a honeycomb structure [12, 35]:

$$w_d = cd\left(\frac{\rho V_p}{1+\rho V_p}\right)^{\frac{1}{2}} \tag{1}$$

where w_d is the pore size, V_p is the primary mesopore volume, ρ is the pore wall density (ca. 2.2 cm^3/g for siliceous materials [35-36]), d is the XRD (100) interplanar spacing and c is a constant dependent on the assumed pore geometry and equal to 1.213 for cylindrical [1] and 1.155 for hexagonal [37] models. In the latter, w_d is defined as a distance between centers of opposite sides of the hexagonal pore cross-section. The choice of pore geometry does not have a significant influence on the pore size determined and the apparent difference of ca. 5 % expected on the basis of c-constants for cylindrical and hexagonal geometry arises mostly from different definitions of pore size in these two cases. Eq. 1 can also be used to determine the pore wall thickness by subtracting the pore size w_d from the distance between pore centers ($2/3^{1/2}$ d). This allows for an assessment of the pore wall thickness from purely geometrical considerations and therefore is expected to be much more accurate than previously reported methods and indeed was shown to provide highly consistent results [6, 11-14]. Eq. 1 can also be modified in order to account for the presence of microporosity in pore walls of ordered mesoporous materials [14]:

$$w_d = cd\left(\frac{V_p}{1/\rho + V_p + V_{mi}}\right)^{\frac{1}{2}} \quad \textbf{(2)}$$

where V_{mi} is the micropore volume. Eq. 2 was successfully used to explain why very large interplanar spacing and relatively small primary mesopore size values were observed for bimodal microporous-mesoporous MCM-41 materials synthesized in the presence of octyltrimethylammonium bromide using the postsynthesis hydrothermal restructuring method [14].

Eqs. 1 and 2 are rather difficult to generalize for other MMS structures and cannot be used for assessment of pore size distributions. However, they are suitable for accurate assessment of pore sizes of MCM-41 materials, especially those of high structural uniformity. Therefore, one can use a series of good quality MCM-41 samples characterized by means of Eq. 1 as model materials to calibrate adsorption methods for the mesopore analysis. It is convenient to establish relations between: i) the capillary condensation or capillary evaporation pressure and the pore size, and ii) pressure and the statistical thickness of adsorbate on the surface (t-curve) [13, 25]. Once these relations are found, they can be used in various methods of the pore size analysis [8], among which the Barrett-Joyner-Halenda (BJH)[24] approach appears to be particularly suitable [13, 25].

Our recent study [13] showed that for nitrogen at 77 K, adsorption data are much more suitable for the pore size analysis than desorption branches of isotherms. The desorption data were found to be considerably affected by quality of MCM-41 materials, presumably due to pore connectivity effects [26]. Moreover, the transition from adsorption-desorption reversibility to irreversibility at relative pressures ca. 0.4-0.45 [8] appeared to cause a rapid decline of desorption branches of isotherms for materials with pore sizes of ca. 4-5.5 nm. This effect may erroneously suggest a high degree of pore size uniformity even for materials of relatively low quality and cause other problems in the pore size distribution evaluation [13].

On the basis of adsorption data for good quality MCM-41 materials, an experimental relation was found for nitrogen adsorption in cylindrical pores at 77 K between the capillary condensation pressure and the pore size in the range from ca. 2 to 6.5 nm. The analytical expression for this relation was established, which is approximately equivalent to the Kelvin equation in the limit of large pore sizes (ca. 10-100 nm) and accurately describes our experimental data, allowing for a reliable assessment of pore sizes from 2 to ca. 100 nm, which includes the entire mesopore range:

$$r(p/p_0)\ [nm] = 0.416\ [\log(p_0/p)]^{-1} + t(p/p_0) + 0.3 \quad \textbf{(3)}$$

where r is the pore radius, p and p_0 are the equilibrium and saturation vapor pressures. The statistical film thickness curve $t(p/p_0)$ derived in our study [13] was based on an adsorption isotherm for a macroporous silica gel measured in the relative pressure interval from ca. 10^{-5} to 0.995 and has an analytical expression, which is very accurate in the relative pressure range from 0.1 to 0.95:

$$t(p/p_0)\ [nm] = 0.1 \left[\frac{60.65}{0.03071 - \log(p/p_0)}\right]^{0.3968} \tag{4}$$

Eq. 3 was used in the BJH method to calculate pore size distributions for numerous MCM-41 materials [13] and was found to be in an excellent agreement, usually within ± 0.1 nm, with pore sizes assessed on the basis of geometrical considerations (Eq. 1). Moreover, the total pore volume estimated on the basis of the corrected BJH approach was very close to the total pore volume estimated from the maximum adsorbed amount (as described above), which shows the consistency of the method. In contrast, we often observed that the standard BJH method with the Kelvin equation and previously derived statistical film curves significantly overestimated the total pore volume of ordered mesoporous materials. Eq. 3 is especially suitable to calculate pore sizes of siliceous materials with approximately cylindrical pores (MCM-41, MCM-48, HMS, KIT-1 [1-5]). It also provides a good method to study framework- and/or surface-modified ordered silicas [1, 3, 30-34], conventional silica gels and nonsiliceous ordered mesoporous materials [5].

2.3. Application of low pressure nitrogen adsorption to study surface properties of MMS

It is well-known [10] that low pressure adsorption provides valuable information about surface heterogeneity of porous materials. Therefore, it was interesting to explore the influence of pore size, inclusion of heteroatoms and bonded ligands on the low pressure adsorption properties of mesoporous molecular sieves. Our numerous studies of different siliceous ordered mesoporous materials, such as MCM-41, HMS and KIT-1 showed that these properties are in all cases very similar and resemble those of chromatographic silica gels [6, 11-13, 16]. Small, but noticeable enhancement of low pressure adsorption with decrease in the pore size from ca. 3.8 to 2.0 nm was observed for siliceous MCM-41 materials synthesized using the same procedure with different surfactants [12]. A systematic increase in the low pressure adsorption was also caused by incorporation of small amounts of heteroatoms (i.e., titanium) in siliceous HMS samples, but the effect was not very pronounced [16]. In contrast, surface modifications of silica-based materials by chemical bonding of organosilanes dramatically influence the course of low pressure adsorption. Our preliminary studies showed that it is possible to attribute peaks on the adsorption energy distributions [10, 27] to certain functional groups of the silica surface and organosilane modifiers [28-30], which provides an opportunity to identify surface species and estimate the uniformity of the surface coverage [28-29] and, in some cases, qualitatively assess the surface coverage of organic groups [30]. Shown in Fig. 3a are adsorption energy distributions (AEDs) for a pure-silica MCM-41 material before and after coating with organic species (4'-cyano-4-biphenyl [4-(4-pentenyloxy)]benzoate denoted as CBPB). For comparison, the AED is provided for a reference silica-based material, whose surface was covered by large amount of chemically bonded and subsequently, physically coated CBPB. Therefore, the AED for the latter sample is expected to closely reflect that of pure CBPB. Although the loading of the organic modifier was relatively large, the AED for the modified MCM-41 is more similar to that of the unmodified MCM-41 than to that of the reference sample. This allows to draw a conclusion that the coating procedure applied was not successful and did not provide uniform coverage of CBPB. It was also possible to use the low pressure adsorption data to estimate the fraction of surface covered by CBPB molecules, which actually appeared to be much lower than that, which would have been attained if CBPB molecules were distributed uniformly on the siliceous surface [30].

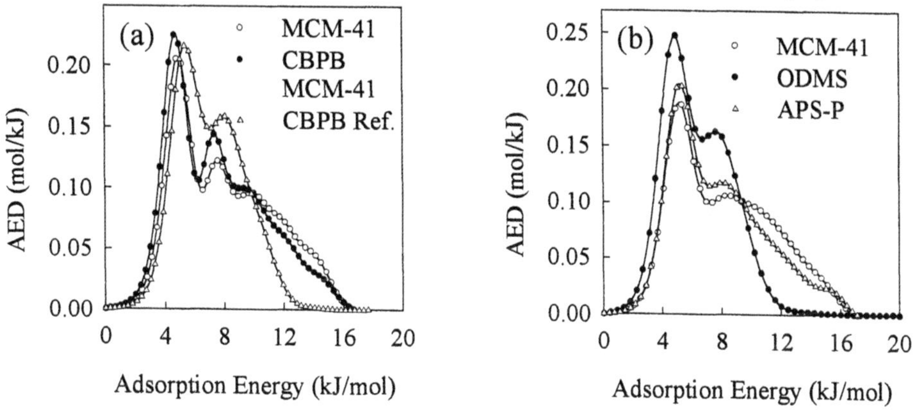

Figure 3. Adsorption energy distributions for unmodified and modified MCM-41 samples.

Shown in Fig. 3b are AEDs for large-pore MCM-41 samples with chemically bonded octyldimethylchlorosilane (ODMS) and 3-aminopropyltriethoxysilane (APS-P). One can notice that the introduction of large nonpolar groups (i.e., octyl aliphatic chains) led to a depletion and/or shielding of high energy adsorption sites at 11-17 kJ/mol, which are present on the surface of the unmodified MCM-41. The presence of surface 3-aminopropyl groups caused much smaller changes in AEDs. One can identify the highest energy adsorption sites (13-17 kJ/mol) as silanol groups (most likely non hydrogen-bonded ones) on the silica surface [28-29] and also on the APS-P ligands, which may have silanols as a result of hydrolysis of side ethoxy groups [29]. The adsorption sites of energy from ca. 10 to 13 kJ/mol can be attributed to the presence of polar groups (e.g., amine functionalities) in the structure of organosilane [28- 30].

3. CONCLUSIONS

It was shown that nitrogen adsorption technique is suitable for a thorough characterization of porous structures and surface properties of mesoporous molecular sieves. High pressure adsorption data provide information about the surface area and mesopore size distribution, whereas low pressure measurements allow for characterization of surface properties, especially for MMS modified by bonding or coating of organic ligands. However, methods of adsorption data analysis need to be carefully chosen to obtain reliable information about structures of MMS.

REFERENCES

1. J. S. Beck, J. C. Vartuli, W. J. Roth, M. E. Leonowicz, C. T. Kresge, K. D. Schmitt, C. T.-W. Chu, D. H. Olson, E. W. Sheppard, S. B. McCullen, J. B. Higgins and J. L. Schlenker, J. Am. Chem. Soc., 114 (1992) 10834.
2. G. D. Stucky, Q. Huo, A. Firouzi, B. F. Chmelka, S. Schacht, I. G. Voigt-Martin and F.

Schuth, Stud. Surf. Sci. Catal., 105 (1997) 3.
3. N. K. Raman, M. T. Anderson and C. J. Brinker, Chem. Mater., 8 (1996) 1682.
4. A. Sayari, Stud. Surf. Sci. Catal., 102 (1997) 1.
5. A. Sayari and P. Liu, Microporous Mater., 12 (1997) 149.
6. A. Sayari, P. Liu, M. Kruk and M. Jaroniec, Chem. Mater., 9 (1997) 2499.
7. M. Kruk, M. Jaroniec and A. Sayari, Microporous Mater., in press.
8. S. J. Gregg and K. S. W. Sing, Adsorption, Surface Area and Porosity, Academic Press, London, 1982.
9. K. S. W. Sing, D. H. Everett, R. A. W. Haul, L. Moscou, R. A. Pierotti, J. Rouquerol and T. Siemieniewska, Pure Appl. Chem., 57 (1985) 603.
10. M. Jaroniec and R. Madey, Physical Adsorption on Heterogeneous Solids, Elsevier, Amsterdam, 1988.
11. M. Kruk, M. Jaroniec, R. Ryoo and J. M. Kim, Microporous Mater., 12 (1997) 93.
12. M. Kruk, M. Jaroniec and A. Sayari, J. Phys. Chem. B, 101 (1997) 583.
13. M. Kruk, M. Jaroniec and A. Sayari, Langmuir, 13 (1997) 6267.
14. A. Sayari, M. Kruk and M. Jaroniec, Catal. Lett., 49 (1997) 147.
15. M. Jaroniec and K. Kaneko, Langmuir, 13 (1997) 6589.
16. M. Kruk, M. Jaroniec and A. Sayari, Microporous Mater., 9 (1997) 173.
17. P. I. Ravikovitch, D. Wei, W. T. Chueh, G. L. Haller and A. V. Neimark, J. Phys. Chem. B, 101 (1997) 3671.
18. R. F. Cracknell, K. E. Gubbins, M. Maddox and D. Nicholson, Acc. Chem. Res., 28 (1995) 281.
19. P. I. Ravikovitch, S. C. O. Domhnaill, A. V. Neimark, F. Schuth and K. K. Unger, Langmuir, 11 (1995) 4765.
20. M. W. Maddox, J. P. Olivier and K. E. Gubbins, Langmuir 13 (1997) 1737.
21. G. Horvath and K. Kawazoe, J. Chem. Eng. Jpn., 16 (1983) 470.
22. M. Kruk, M. Jaroniec and J. Choma, Adsorption, 3 (1997) 209.
23. J. P. Olivier, W. B. Conkin and M. v. Szombathely, Stud. Surf. Sci. Catal., 87 (1994) 81.
24. E. P. Barrett, L. G. Joyner and P. P. Halenda, J. Am. Chem. Soc., 73 (1951) 373.
25. H. Naono, M. Hakuman and T. Shiono, J. Colloid Interface Sci., 186 (1997) 360.
26. H. Liu, L. Zhang and N. A. Seaton, J. Colloid Interface Sci., 156 (1993) 285.
27. M. v. Szombathely, P. Brauer and M. Jaroniec, J. Comput. Chem., 13 (1992) 17.
28. M. Kruk, M. Jaroniec, R. K. Gilpin and Y. W. Zhou, Langmuir, 13 (1997) 545.
29. C. P. Jaroniec, R. K. Gilpin and M. Jaroniec, J. Phys. Chem. B, 101 (1997) 6861.
30. M. Jaroniec, M. Kruk, C. P. Jaroniec and A. Sayari, Adsorption, in press.
31. J. C. Vartuli, K. D. Schmitt, C. T. Kresge, W. J. Roth, M. E. Leonowicz, S. B. McCullen, S. D. Hellring, J. S. Beck, J. L. Schlenker, D. H. Olson and E. W. Sheppard, Chem. Mater., 6 (1994) 2317.
32. X. Feng, G. E. Fryxell, L.-Q. Wang, A. Y. Kim, J. Liu and K. M. Kemner, Science, 276 (1997) 923.
33. L. Marcier and T. J. Pinnavaia, Adv. Mater., 9 (1997) 500.
34. M. H. Lim, C. F. Blanford and A. Stein, J. Am. Chem. Soc., 119 (1997) 4090.
35. T. Dabadie, A. Ayral, C. Guizard, L. Cot and P. Lacan, J. Mater. Chem., 6 (1996) 1789.
36. R. K. Iler, The Chemistry of Silica, Wiley, New York, 1979.
37. V. Alfredsson, M. Keung, A. Monnier, G. D. Stucky, K. K. Unger and F. Schuth, J. Chem. Soc., Chem. Commun., (1994) 921.

MESOPOROUS MOLECULAR SIEVES 1998
Studies in Surface Science and Catalysis, Vol. 117
L. Bonneviot, F. Béland, C. Danumah, S. Giasson and S. Kaliaguine (Editors)

The creation of MO_x surface species on pure silica MCM-48 using gas- and liquid-phase modifications with M-acetylacetonate complexes.

P. Van Der Voort[a], M. Mathieu[a] ,M. Morey[b], G.D. Stucky[b] and E.F. Vansant[a]

[a] University of Antwerp (U.I.A.) - Dept. of Chemistry - Laboratory of Inorganic Chemistry, Universiteitsplein 1, 2610 Wilrijk, BELGIUM.

[b] University of California at Santa Barbara, Dept. of Chemistry, Santa Barbara, CA 93106, USA.

Pure silica MCM-48 is prepared by a novel synthesis method, using the $[C_{18}H_{37}N^{+}(CH_3)_2 - (CH_2)_{12} - N^{+}(CH_3)_2C_{18}H_{37}].2Br^{-}$ surfactant, abbreviated as GEMINI 18-12-18. The MCM-48, obtained after careful calcination, is a highly crystalline, mesoporous material, with the characteristics of the Ia3d cubic phase, a surface area exceeding 1000 m^2/g and a narrow mesoporous pore size distribution (r = 1.4 nm; FWHH < 0.2 nm).

This MCM support is grafted with VO_x species, using a designed dispersion of $VO(acac)_2$ in a gas deposition reactor. In a first step, the complex is anchored to the support, in a subsequent step the adsorbed complex is thermolyzed to yield chemically bonded VO_x surface species. The final material contains 1.7 mmol V/g (8.7 w% V), still has narrow pore size distribution and a surface area of 800 m^2/g. It is observed that all silanols are consumed during the adsorption of the $VO(acac)_2$ complex to the MCM support. Therefore, the maximum achievable number of surface V-species is limited by the silanol number, and not by the geometrical surface, which has a higher capacity. After calcination of the adsorbed complex, the supported VO_x species are present in a strictly tetrahedral configuration, but mainly as chains of linked tetrahedra and not as isolated species.

1. INTRODUCTION

Since the discovery of the M41S family of materials [1] many publications have appeared on the synthesis, characterization and potential uses of the hexagonal MCM-41 mesoporous substrate [2].

The cubic MCM-48, however, has received less attention, since its synthesis is considered to be more difficult and less reproducible. Recently, an alternative for the alkyltrimethylammonium route has been reported by Huo, using the so-called gemini surfactants, with general formula $[C_nH_{2n+1}N^{+}(CH_3)_2 - (CH_2)_s - N^{+}(CH_3)_2C_mH_{2m+1}].2Br^{-}$, abbreviated as gemini n-s-m [3]. Huo[3] concluded that the gemini 16-12-16 surfactant favors the formation of the cubic (Ia3d) phase, irrespective of the synthesis temperature and even without adding an organic additive.

Figure 1. The Q^{230} network and the gyroid surface representing how the surfactant intertwines with the MCM-48 walls. (From ref. 5)

The structure of MCM-48 is consistent with the Q^{230} model, proposed by Mariani for water-surfactant systems [4]. Figure 1 visualizes the Q^{230} rod network and surface , using the representation of Anderson [5]. The three dimensional pore structure of the MCM-48 substrates make them potentially very interesting and promising supports for heterogeneous catalysts, by grafting the surface of the support with catalytic active species [6].

The use of $VO(acac)_2$ (vanadyl-acetylacetonate) to graft VO_x groups on a catalytic support by means of the molecular designed dispersion method has been described in detail in some of our previous publications [7 ,8]. In principle, the complex is anchored to the hydroxyl groups of the support by either a hydrogen bonding or by a ligand exchange mechanism. The adsorbed complex is called the precursor. A treatment in air at elevated temperatures converts the adsorbed acetylacetonate complex into metal oxide species, that are chemically bonded to the surface. The process and the reagent are visualized in Figure 2.

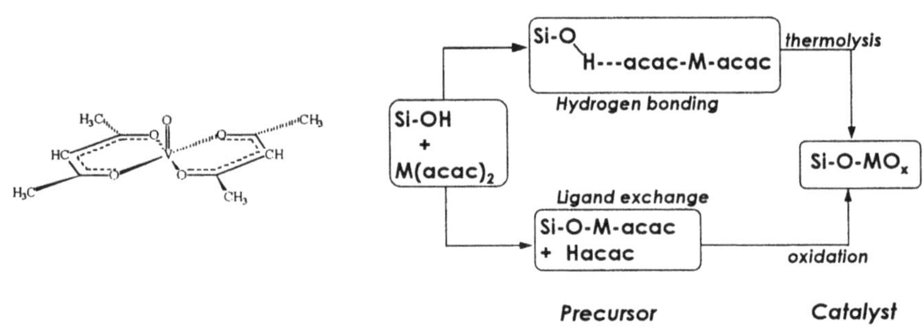

Figure 2. Left: $VO(acac)_2$; Right: general overview of the molecular designed dispersion process.

The gas phase deposition of the complexes at elevated temperatures and high vacuum conditions, has several intrinsic advantages over the liquid phase adsorption. Haukka [9] and Babitch [10] have shown that fairly inert and stable complexes, which are unreactive in the liquid phase (*e.g.* $Cr(acac)_3$) can be successfully deposited using a gas-phase reactor. Moreover, problems of solvent-effects and solvent recovery are non-existent. Finally, from a fundamental point of view, gas-phase reactions allow the capture of the reaction products and thus a detailed elucidation of the reaction mechanism.

2. EXPERIMENTAL

The MCM-48 is this study was prepared using the 18-12-18 gemini surfactant, with formula $[C_{18}H_{37}\text{-}N^{+}(CH_3)_2\text{-}C_{12}H_{25}\text{-}N^{+}(CH_3)_2\text{-}C_{18}H_{37}].2Br^{-}$. The detailed synthesis procedure can be found elsewhere [11]. The reader is also referred to ref. 3 for a more general approach of the gemini synthesis.

VO_x supported catalysts were prepared in a vacuum deposition reactor, as explained in detail in previous publications [11]. The sublimed $VO(acac)_2$ complex reacts with the heated substrate. Reaction products are collected in a cryotrap. Reaction is performed until there is saturation of the substrate, which is visible by the formation of crystals of complex on top of the colder parts of the reactor. Subsequently, the samples are purged for 2 h at reaction temperature in a dry nitrogen flow, to remove all physisorbed species. At this stage, the precursor is formed. Calcination of this precursor to form the catalyst was performed in a programmable oven (25°C to 500°C, heating rate: 5°C/min, isothermal period: 17h) in ambient air.

X-Ray diffraction patterns were collected on a Philips PW1840 powder diffractometer (45kV, 30mA), using a Ni-filtered Cu K_α radiation. Porosity and surface area studies were performed on a Quantachrome Autosorb-1-MP automated gas adsorption system. Infrared spectra were recorded at room temperature on a Nicolet 5 DBX Fourier Transform spectrometer with photo-acoustic detection. Thermogravimetric measurements were performed on a Mettler TG50 thermobalance, equipped with a M3 microbalance and connected to a TC10A processor. UV-VIS diffuse reflectance measurements were recorded on a Unicam 8700 spectrometer, equipped with a special designed diffuse reflectance accessory.

The ratio of adsorbed acac ligands to the adsorbed vanadium species is expressed as the R-value:

$$R = \frac{\text{mmol acac per g sample}}{\text{mmol V per g sample}} = \frac{n_{acac}}{n_V}$$

The analytical procedures have been described elsewhere [7].

3. RESULTS AND DISCUSSION

3.1. Characterization of the support

The nitrogen isotherm and the pore size distribution of the synthesized MCM-48 is presented in Figure 3. The two most important features in the isotherm are the very sharp increase in the adsorbed gas volume in the region (p/p_0) = 0.3-0.4 and the (almost) absence of hysteresis in the region (p/p_o) = 0.4-1.0. The sharp increase in adsorbed gas volume already suggests a narrow pore size distribution: in this region the smaller mesopores are filled by capillary condensation. Hysteresis in the higher pressure region (discrepancy between the adsorption and the desorption branch of the isotherm) is caused by particle-particle porosity or by significant larger pores. Particle-particle porosity does not create a sharp pore distribution, because of the different sizes of the particles which create the mesopores between them.

The pore size distribution in the mesoporous region was calculated using the well established method of Barret, Joyner and Halenda [12].

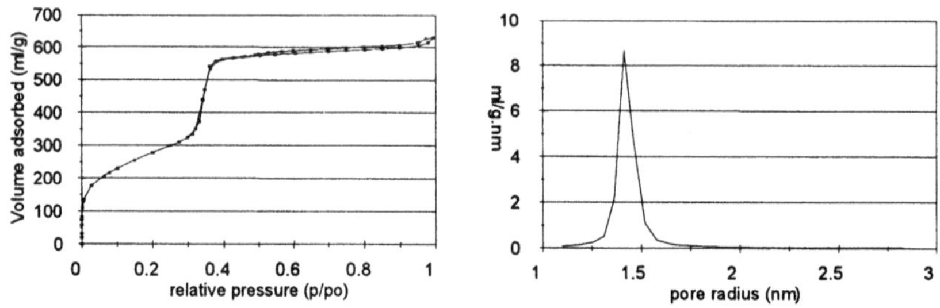

Figure 3. Nitrogen isotherm (77K) and pore size distribution of the parent MCM-48.

The pore size distribution in Figure 3 shows a narrow, mesoporous distribution, with a pore maximum of r = 1.4 nm. T-plot analysis confirmed the absence of micropores (r < 1 nm). The total surface area, as calculated by the BET method, is 1025 m^2/g and the total pore volume at p/p_o=0.98 is 0.95 ml/g.

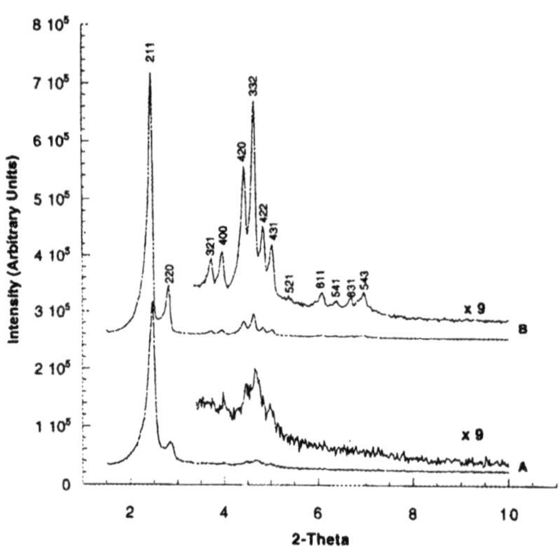

Figure 4 shows the X-Ray diffractograms of the MCM-48 sample, which had been autoclaved prior to calcination for 10 and 21 days respectively, illustrating nicely how the 'crystallinity' improves as a function of time. The cubic lattice parameter *a* ($a = d.(h^2+k^2+l^2)^{1/2}$) is 8.9 nm, which is slightly higher than the value obtained for a MCM-48, prepared by the cetyltrimethylammonium route (a = 8.1 nm) [13] .

Figure 4. XRD of the MCM-48, autoclaved for 10 days (A) and 21 days (B).

3.2. Characterization of the precursor

The precursors were prepared by subliming an excess of $VO(acac)_2$ in the vacuum deposition reactor at 150°C. Saturation of the substrate was achieved after almost 16 h of reaction. The catalyst has a V-loading of 1.7 mmol/g V or 8.7 w%V.

Figure 5 shows the FTIR-PAS of the blank MCM-48 (thermally treated at 500°C, spectrum (a)), the MCM-48 saturated with $VO(acac)_2$ (the precursor, spectrum (b)) and the same sample after calcination at 500°C (the catalyst, spectrum (c)).

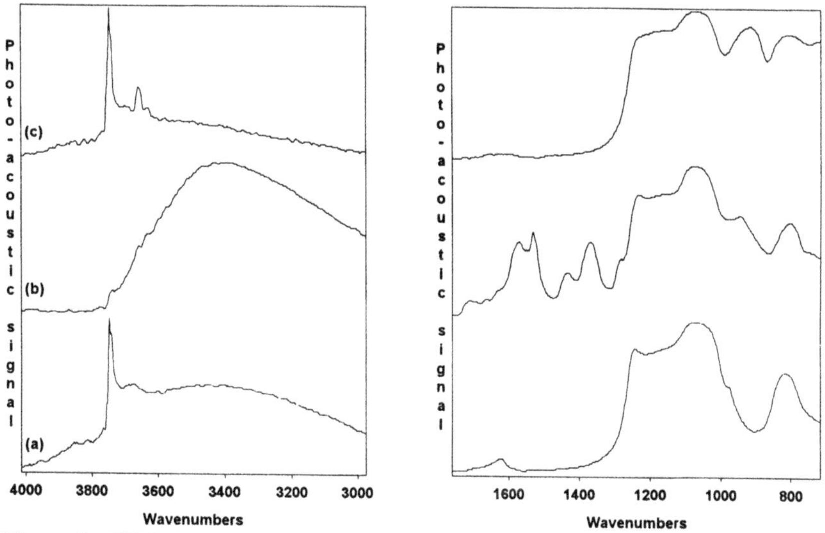

Figure 5. FTIR-PA spectra of (a) MCM-48 (500°C); (b) MCM-$VO(acac)_2$ precursor and (c) MCM-VO_x catalyst. Left side: hydroxyl region; right side: acac region.

Spectrum (a) shows a narrow band at 3745 cm^{-1}, assigned to free silanols and a broader feature at 3600-3200 cm^{-1}, attributed to the presence of some bridged silanols or possibly physisorbed water trapped in the pore system of the MCM [14]. After the gas-phase reaction, all isolated silanols have disappeared, due to either a ligand exchange mechanism or a hydrogen bond interaction (spectrum (b)). In the case of hydrogen bond interaction, the silanol vibration shifts to a lower wavenumber (app. 3400 cm^{-1}).

The complete disappearance of the isolated silanol band leads to the conclusion that the saturation of the surface is governed by the availability of surface hydroxyls and not by geometric, sterical considerations. The $VO(acac)_2$ complex has a mean cross-sectional area of 0.60 nm^2. A full monolayer would therefore correspond to a V-loading of 2.8 mmol/g V. The actual loading is only 60% of this monolayer capacity.

The silanol content of the blank MCM-48 support can be estimated, based on theoretical considerations. Alami et al.[15] have calculated that the 'head group area' for a gemini surfactant with a spacer of 12 C-atoms amounts to 2.26 nm^2. Each headgroup participates in 2 $Si\text{-}O^-\text{-}NR_4^+$ functions during the synthesis of the MCM, that are converted to Si-OH functions upon calcination. The theoretical silanol density of a MCM-48, prepared by the 18-12-18 gemini surfactant and a 100% silica source is therefore 2 OH per 2.26 nm^2 or 0.9 OH / nm^2.

The vanadium loading of the precursor sample is 1.7 mmol V / g sample or 1.0 V/nm^2. Therefore, within a margin of error of 10%, the $VO(acac)_2$ complex interacts in a 1:1 ratio with the surface silanols of the silica support.

The R-value (mmol acac per g sample / mmol V per g sample) for these precursors is 1.5 ± 0.1. This means that -on average- 1.5 acac ligands are associated with each grafted V center. In other words, 50% of the $VO(acac)_2$ is bonded by means of the ligand exchange mechanism (only 1 acac associated with a V center, the other evolved as Hacac) and 50% is bonded by the hydrogen bonding interaction mechanism (both acac ligands associated with the V-center). The infrared spectrum in Figure 5(b) indeed shows a strong contribution of hydrogen bond interacting species, that were not present in the blank MCM-48.

3.3. Characterization of the grafted VO_x surface species (the catalyst)

Figure 5(c) shows the infrared spectrum of the final catalyst after calcination of the precursor at 500°C in ambient air. All acac vibration have disappeared due to thermolysis or oxidation. A new band has appeared at 3660 cm^{-1}, assigned to VO-H surface species [8].

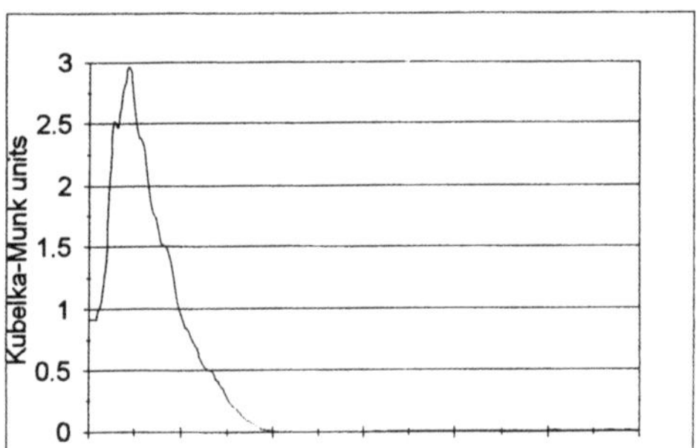

Figure 6. UV-VIS diffuse reflectance spectrum of the MCM-VO_x catalyst.

Additional structural information on the grafted VO_x species can be obtained from UV-VIS diffuse reflectance measurements of the catalyst (Figure 6). The spectrum is recorded in extremely dry conditions, to avoid hydrolysis of the V-centers, which results in a color change from white to orange [8]. It is well known that the position of the O → V charge transfer bands provides a very good indication on the coordination of the central V^{5+} center.

The tetrahedrally surrounded V^{5+} centers have their main absorption bands in the region 240-350 nm, square pyramidal configurations in the region 350-450 nm and octahedral (or pseudo-octahedral as in V_2O_5) have absorptions in the region 450-600 nm [8]. It can be inferred from Figure 6 that this particular catalyst is exclusively covered with tetrahedral species. Moreover, non-oxidized V^{IV} centers should give strong d-d transition bands, which are not observed.

Although UV-VIS has established that the surface of the MCM is exclusively covered with tetrahedrally surrounded VO_x centers, it can not differentiate between monomers, one-dimensional chains and two-dimensional chains (Scheme 1).

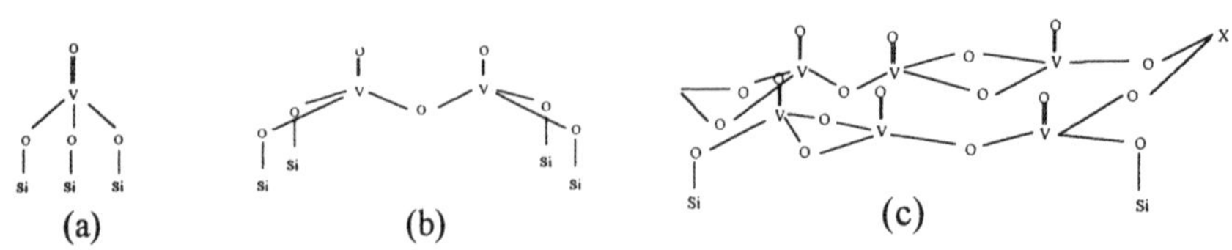

Scheme 1

It can be reasoned though, that a surface coverage with exclusively tetrahedral monomers (Scheme 1) is impossible, since such a coverage would require (1.7*3) = 5.1 mmol OH/g, whereas the actual silanol concentration is only 0.9 OH/nm^2 or 1.5 mmol OH/g. Moreover, infrared spectrum 5(c) indicates that a large fraction of the original isolated silanols (3745 cm^{-1}) are restored after calcination of the precursor at 500°C in ambient (wet) air. This observation suggests that the formation of V-O-V bonds during the calcination of the precursor is favored relative to the formation of Si-O-V bonds.

An objective criterion to evaluate the clustering of the grafted VO_x species is the *ensemble size*, defined as:

$$E.S. = \frac{\text{total number of V centers}}{\text{total number of reacted silanols}} = \frac{V\ (mmol/g)}{OH_r\ (mmol/g)}$$

When used with extreme care, the FTIR-PAS band of the isolated silanols can be integrated, normalized to the 1950-1766 cm^{-1} silica structure vibration, to assess the fraction of silanols that has disappeared due to the formation of Si-O-V bonds[16]. Applying this procedure to spectra 5(a) and 5(c) indicates that 80% of the silanols of the blank MCM-48 is restored after calcination of the precursor. Therefore, the number of silanols that have disappeared is 20% of 1.5 mmol/g = 0.3 mmol / g. The vanadium loading is 1.7 mmol/g and thus the ensemble size is 1.7 / 0.3 = 5.6. This means that -on average- 5.6 V centers are bonded to one silanol anchor.

Van Hengstum et al.[17] have calculated the cross-sectional area of a $VO_{5/2}$ group to be 0.103 nm^2. This would mean that only 20% of the geometrical MCM surface is covered with VO_x species and provides an additional geometrical argument for the formation of linked tetrahedral VO_x chains: the cross sectional area of a $VO_{5/2}$ groups is obviously too small to form bidentate or tridentate bondings with the silanols, having a density of only 0.9 OH/nm^2.

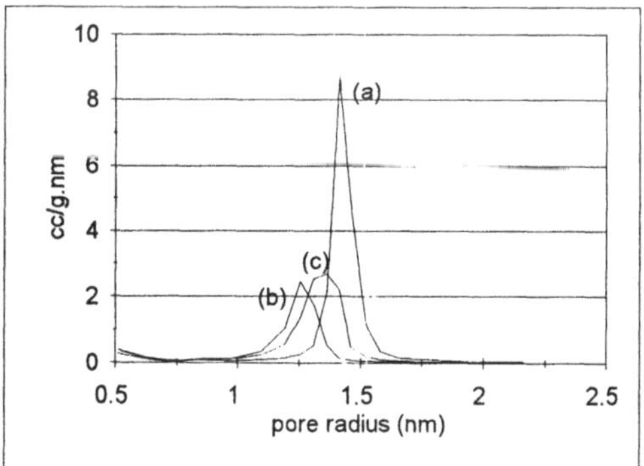

Figure 7 shows the pore size distributions of the original MCM-48, the precursor and the catalyst. The surface area and pore volume are listed in Table 1.

When the MCM is loaded with the bulky $VO(acac)_2$ complex, the porosity and surface area are drastically reduced, but after calcination and oxidation of the acac ligands, the surface area is restored to almost 80% of the original value.

Figure 7. Pore size distribution of blank MCM-48 (a); MCM-$VO(acac)_2$ precursor (b); MCM-VO_x catalyst (c).

Although the pore size distribution of the final catalyst has broadened at little, the FWHH is still smaller than 0.3 nm. The slight broadening of the PSD might be explained by the fact that this particular sample is only covered with a submonolayer of vanadium species. Therefore, uncovered patches will be altered with monomers and oligomers of tetrahedral VO_x species, inducing a broadening of the overall pore size distribution. Due to the sharp PSD of the original MCM and the pore radius of almost 1.5 nm, pore blocking effects can be neglected.

Table 1. Surface area and total pore volume of the blank MCM-48, the MCM-48 with adsorbed $VO(acac)_2$ species (the precursor) and the MCM-48 with grafted VO_x species (after calcination of the precursor).

Sample	Surface area (m^2/g)	total pore volume (ml/g)
MCM	1025	0.95
MCM-$VO(acac)_2$	530	0.45
MCM-VO_x	800	0.67

4. CONCLUSIONS

MCM-48 has been prepared by a novel synthesis route, using the $[C_{18}H_{37}N^+(CH_3)_2 - (CH_2)_{12} - N^+(CH_3)_2C_{18}H_{37}].2Br^-$ (GEMINI 18-12-18) surfactant. The resulting support has a surface area of 1025 m^2/g and a very narrow mesoporous pore distribution around radius = 1.4 nm.

$VO(acac)_2$ adsorbs to the MCM-48 support in a gas-phase deposition reaction by a combination of a ligand exchange mechanism (evolving one ligand as Hacac) and a hydrogen bond interaction between a surface silanol and the pseudo π system of the acac ring. The saturation of this reaction is not governed by the surface area, but by the availability of silanols at the surface.

After calcination of the adsorbed $VO(acac)_2$ species (called the precursor), chemically bonded VO_x species are grafted to the surface, in a concentration of 1.7 mmol V/g (8.7 w% V). The VO_x species are present on the surface of the MCM-48 support as tetrahedral coordinated oligomers (linked tetrahedra), with an average 'ensemble size' of 5.6. The original surface area of 1025 m^2/g is reduced to 800 m^2/g by this treatment.

Even at this high loading, still only 20% of the geometrical surface is covered with VO_x species. For the precursor, maximum 60% of the geometrical surface was covered with $VO(acac)_2$ species.

ACKNOWLEDGEMENTS

P. Van Der Voort acknowledges the F.W.O. Vlaanderen (Fund for Scientific Research - Flanders - Belgium) for financial support.

The authors thank Mrs. F. Quiroz for her aid in the experimental work.

REFERENCES

1. C.T. Kresge, M.E. Leonowicz, W.J. Roth, J.C. Vartuli and J.S. Beck, Nature, 359 (1992) 710; C.T. Kresge, M.E. Leonowicz, W.J. Roth and J.C. Vartuli, US Patent 5 098 684 (1992).
2. A. Firouzi, F. Atef, A.G. Oertli and G.D. Stucky, J. Am. Chem. Soc., 119 (1997) 3595; A. Firouzi, D. Kumar, L.M. Bull and T. Besier, Science, 267

(1995) 1138; C.F. Cheng, D.H. Park and J. Klinowski, J. Chem. Soc. Faraday Trans., 93 (1997) 193; A. Monnier, F. Schüth, Q. Huo, D. Kumar, D. Margolese, R.S. Maxwell, G.D. Stucky, M. Krisnamurty, P. Petroff, A. Firouzi, M. Janicke and B.F. Chmelka, Science, 261 (1993) 1299; K.J.Edler, P.A. Reynolds, J.W. White and D. Cookson, J. Chem. Soc. Faraday Trans., 93 (1997) 199; C.F. Cheng, W. Zhuo, D.H. Park, J. Klinowksi, M. Hargreaves, and L.F. Gladder, J. Chem. Soc. Faraday Trans., 93 (1997) 359; J. Rathousky, A. Zukal, O. Franke and G. Schulz-Elkoff, J. Chem. Soc. Faraday Trans., 90 (1994) 2821; C.Y. Chen, S.L. Burkett, H.X. Li and M.E. Davis, Microporous Materials, 2 (1993) 27.

3. Q. Huo, R. Leon, P.M. Petroff and G.D. Stucky, Science, 268 (1995) 1324.; Q. Huo, D.E. Margolese and G.D. Stucky, Chem. Mater., 8 (1996) 1147.
4. P. Mariani, V. Luzatti and H. Delacroix, J. Mol. Biol., 204 (1988) 135.
5. M.W. Anderson, Zeolites, 19 (1997) 220.
6. M. Morey, A. Davidson, H. Eckert and G.D. Stucky, Chem. Mater., 8 (1996) 486.
7. P. Van Der Voort, K. Possemiers and E.F. Vansant, J. Chem. Soc. Faraday Trans., 92 (1996) 843; P. Van Der Voort, I.V. Babich, P.J. Grobet, A.A. Verberckmoes and E.F. Vansant, J. Chem. Soc. Faraday Trans., 92 (1996) 3635; P. Van Der Voort, M.G. White and E.F. Vansant, Interface Science, 5 (1997) 179.
8. P. Van Der Voort, M.G. White, M.B. Mitchell, A.A. Verberckmoes and E.F. Vansant, Spectrochimica Acta A, 53 (1997) 2181.
9. S. Haukka, E.L. Lakomaa and T. Suntola, Appl. Surf. Sci., 75 (1994) 220; S. Haukka, E.L. Lakomaa and T. Suntola, Appl. Surf. Sci., 82 (1994) 548.
10. I.V. Babich, Yu. V. Plyuto, P. Van Der Voort and E.F. Vansant, J. Coll. Interf. Sci., 189 (1997) 144.
11. P. Van Der Voort, M. Morey, G.D. Stucky, M. Mathieu and E.F. Vansant, J. Phys. Chem. B, 102 (1998) 585; P. Van Der Voort, M. Mathieu, E.F. Vansant, S.N.R. Rao and M.G. White, J. Porous Materials, 1998, in press.
12. E.P. Barret, L.G. Joyner and P.P. Halenda, J. Am. Chem. Soc., 73 (1951) 373.
13. J.C. Vartuli, K.D. Schmitt, C.T. Kresge, W.J. Roth, M.E. Leonowicz, S.B. McCullen, S.D. Hellring, J.S. Beck, J.L. Schlenker, D.H. Olson and E.W. Sheppart, Zeolites and related microporous Materials, Weitkamp, J.; Karge, H.G.; Pfeifer H.; Hölderich W. eds, Studies in Surface Science and Catalysis, 1994, 84, Elsevier Science, Amsterdam.
14. E.F. Vansant, P. Van Der Voort and K.C. Vrancken, Characterization and Chemical Modification of the Silica Surface, Studies in Surface Science and Catalysis, 1995, 93, Elsevier Science, Amsterdam.
15. E. Alami, G. Beinert, P. Marie and R. Zana, Langmuir, 9 (1993) 1465.
16. P. Van Der Voort, I. Gillis-D'Hamers, K.C. Vrancken and E.F. Vansant, J. Chem. Soc. Faraday Trans., 87 (1991) 3899; P. Van Der Voort, I. Gillis-D'Hamers and E.F. Vansant, J. Chem. Soc. Faraday Trans., 86 (1990) 3751.
17. A.J. Van Hengstum, J.G. Van Ommen, H. Bosch and P.J. Gellings, Appl. Catal., 5 (1983) 207.

MESOPOROUS MOLECULAR SIEVES 1998
Studies in Surface Science and Catalysis, Vol. 117
L. Bonneviot, F. Béland, C. Danumah, S. Giasson and S. Kaliaguine (Editors)

Surface properties of Cu_2O/MCM-41 mesoporous systems

A. Zecchina[a], D. Scarano[a], G. Spoto[a], S. Bordiga[a], C. Lamberti[a] and G.Bellussi[b]

[a]Dipartimento di Chimica Inorganica, Chimica Fisica e Chimica dei Materiali
Università di Torino, via Pietro Giuria 7, 10125 Italy

[b]ENIRICERCHE laboratories of San Donato Milanese (Mi) Italy.

The surface properties of MCM-41 samples with Si/Al ratio ≅ 20 and pore diameter ≅ 4 nm, with 1%, 5% and 10% (w/w) Cu_2O/MCM-41 ratios, were studied by FTIR spectroscopy, at ≅ 100K using CO as probe molecule. At low copper content, atomically dispersed Cu(I) ions, anchored to the framework, are predominant (like in Cu(I)-ZSM5). These coordinatively unsaturated ions can coordinate up to three CO molecules. At higher copper concentrations, Cu_2O particles are progressively formed: consequently besides the spectral features due to isolated Cu(I) species, the spectroscopic manifestations associated with surface atoms of Cu_2O microparticles are observed.

1. INTRODUCTION

Mesoporous molecular sieves with hexagonal structure of the MCM-41 type, first described by Kresge et al [1], represent an interesting family of materials, which, because of their large internal surface area, high porosity, controllable and narrowly distributed pore sizes, can be promising candidates as catalysts or as supports for small active metals and metal oxide clusters [2-6], as adsorbents [7-11] and as host for host-guest interactions to develop advanced composite materials[12, 13], which are important for non linear optical properties. In this connection it was demonstrated that for nanoparticles encapsulated within micro-mesoporous materials, the optical band gap depends on size-quantization and on interactions between particles and host [14, 15].

A generally accepted model for MCM-41 consists of an hexagonal arrangement of cylindrical pores in the 2-10 nm range, embedded in a matrix of amorphous silica, the pore diameter being controlled during the synthesis procedure by suitable template molecules [2, 16]. In particular, as already observed for liquid-crystals, where the diameter of the hexagonal liquid-crystal phases prepared with anionic surfactants is dependent on the alkyl chain length of the surfactant, it was proposed that the formation of these materials takes place by means of a liquid-crystal 'templating' mechanism, in which the silicate material forms inorganic walls between ordered surfactant micelles [1]. From electron micrographs of products obtained by hydrothermal reactions of aluminosilicate gels in presence of quarternary ammonium surfactants, regular hexagonal arrays of uniform channels are evidenced, while from electron and X-ray diffraction patterns the periodicity of the structure is confirmed. More recently, to better understand the structure and in particular the relation between the pore size and the wall thickness, model structures for MCM-41 materials have been obtained, by means of molecular dynamic simulation techniques and the results were compared with the observed diffraction patterns [16]. On the basis of these methods a detailed description of the structure of the MCM-41 material has been reported, including the density of T-sites, the concentration of silanols, the distribution of the ring sizes and the calculation of the X-ray diffraction patterns [16].

As far as their catalytic activity is concerned, the best prospective for MCM-41 type materials was their use in acid-catalyzed reactions, but quite early it was realized that the acidity was not at all on the same level as for zeolites [3]. In fact TPD and IR spectroscopy experiments revealed that the concentration as well as the strength of the Brönsted acidic sites in MCM-41 is remarkably lower than for zeolites. The acidic properties of MCM-41, however, seem closer to those of amorphous aluminosilicates, but with a very high concentration of Lewis acidic sites, which are attributed to aluminum species removed from the walls. These similarities of acidity between MCM-41 and amorphous aluminosilicates supports the essentially amorphous nature of the pore walls in MCM-41 [3a, 17-20].

On the contrary the adsorption ability of MCM-41 for polar molecules (H_2O, methanol, ammonia) was, to a large extent, demonstrated and related to the density of surface silanol groups [11]. Many studies concerning the characterization of surface SiOH groups on mesoporous MCM-41 were performed by means of FTIR and gravimetric methods, combined with solid state ^{29}Si nuclear magnetic resonance experiments, using magic angle spinning and TPD, TGA of pyridine [21-23] and ammonia [24].

Although the acidic properties of MCM-41 were somewhat disappointing with respect to catalytic applications, these materials can represent catalyst supports whose thermal stability was probed up to 1273 K, as recently reported [9b, 25]. For this reason, after impregnation with different loadings of metal or metal oxides [26-28] and of ligand stabilized clusters [4, 29], these materials improve their catalytic activity as compared to more common catalysts. It was recently reported [5] that some elements like for example vanadium can be incorporated in the MCM-41 framework in form of monodispersed ions (VO^{2+} and VO^{3+}) in tetrahedral coordination and likely grafted on the surface of the tubular walls. This occurs by condensation with silanol groups of walls, which are similar to defect sites of microporous zeolites and can plausibly act as anchors for the attachment of transition metal species. In presence of water vapour the oxidation of (VO^{2+}) to form V_2O_5 clusters was promoted, instead of isolated VO^{3+} species. In this way , besides dispersing and stabilizing vanadium oxides, the structural features of mesoporous matrices affect the form of vanadium species and its catalytic behaviour.

An other interesting possibility was the incorporation of ligand stabilized clusters into the well defined pores of MCM-41 type materials, whose sizes match the cluster size and, acting as a macroligand, improve the stability of the clusters compared to free ones in many catalytic reactions [4].

In our laboratory the impregnation of MCM-41 with 1%, 5% and 10% (w/w) Cu_2O was studied, in order to understand how an increased copper loading can affect the nature and the distribution of the active species. In particular the aim of this work was to distinguish between the presence of isolated copper species, grafted on the walls of MCM-41 and/or the formation of Cu_2O clusters encapsulated into the channels system.

2. EXPERIMENTAL

MCM-41-supported Cu_2O samples were prepared as follows. MCM-41 samples with Si/Al ratio ≅ 20 and pore diameter ≅ 40, characterized by an high external surface area (≅ 1000 m^2g^{-1}) and kindly supplied by ENIRICERCHE laboratories of San Donato Milanese (Mi) Italy, have been used. MCM-41 samples were impregnated with an aqueous solution of $Cu(NO_3)_2$ (Aldrich, analytical grade) in the appropriate amount to obtain 1%, 5% and 10% (w/w) Cu_2O/MCM-41 ratios. These materials were then dessicated at 350 K and thin self-supported pellets of solid samples were prepared and activated in vacuum (residual pressure < 0.013 Pa) at 873 K for 1h inside an IR cell which allowed *in situ* high temperature activation and low temperature CO dosage to be carried out. This thermal treatment causes the self-reduction of CuO (formed by thermolysis of copper nitrate) to Cu_2O, since Cu(II) oxide is known to lose oxygen, when heated at high temperature [30]. Fourier Transmission IR spectra were recorded, at 2 cm^{-1} resolution, using a Bruker IFS 48 spectrometer equipped with a cryogenic detector and having the sample pellet permanently cooled with liquid nitrogen, the actual temperature under the IR beam being of about 100 K.

3. RESULTS AND DISCUSSION

FTIR spectra of CO adsorbed at ≅ 100K on the surface of samples with 1%, 5% and 10% Cu_2O/MCM-41 ratios are reported in Figs.1 a-c, while those corresponding to CO adsorbed on pure MCM-41 are shown in Fig.1d. In all cases the top spectrum corresponds to 2.66 kPa of CO in equilibrium with the sample pellet at nominally liquid nitrogen temperature. Successive spectra we obtained by gradually decreasing the CO pressure inside the IR cell, the bottom spectrum corresponding in all cases to residual CO pressure smaller than 0.266 Pa. Similar sequences of spectra could be obtained following the opposite procedure i.e. by gradually increasing the CO pressure (from 0 to 2.66 kPa). At lowest CO equilibrium pressure, the spectrum of CO adsorbed on samples with 1%(w/w) Cu_2O/MCM-41 ratio (Fig.1a) shows a single IR absorption band centred at ≃ 2159 cm^{-1}, which is quite resistant to ougassing. By increasing the CO coverage, the intensity of the 2159 cm^{-1} band gradually increases and new components, at 2152 cm^{-1} and at 2180 cm^{-1} and 2230 cm^{-1} start to appear. At even higher CO pressures, the components at 2180 cm^{-1} and 2152 cm^{-1}progressively decrease, while new bands at 2194 cm^{-1}, 2171 cm^{-1} and 2138 cm^{-1} are progressively formed. All these bands are quite reversible and can be easily removed by ougassing at RT.

Although the same basic features (with small differences in the intensity ratios) are in general observed in the spectra of CO adsorbed on samples with 5% and 10% Cu_2O/MCM-41 ratios (Fig.1b, c), it can be immediately noticed that, on samples having higher copper contents, a new component at 2125-2120 cm^{-1} appears. At the same time the 2230 cm^{-1} band disappears.

In general it can be noticed that the intensity and the halfwidth of the most prominent bands in Figs.1a,b,c, increase with the copper concentration. Also on the most concentrated samples, the 2230 cm^{-1} band is absent.

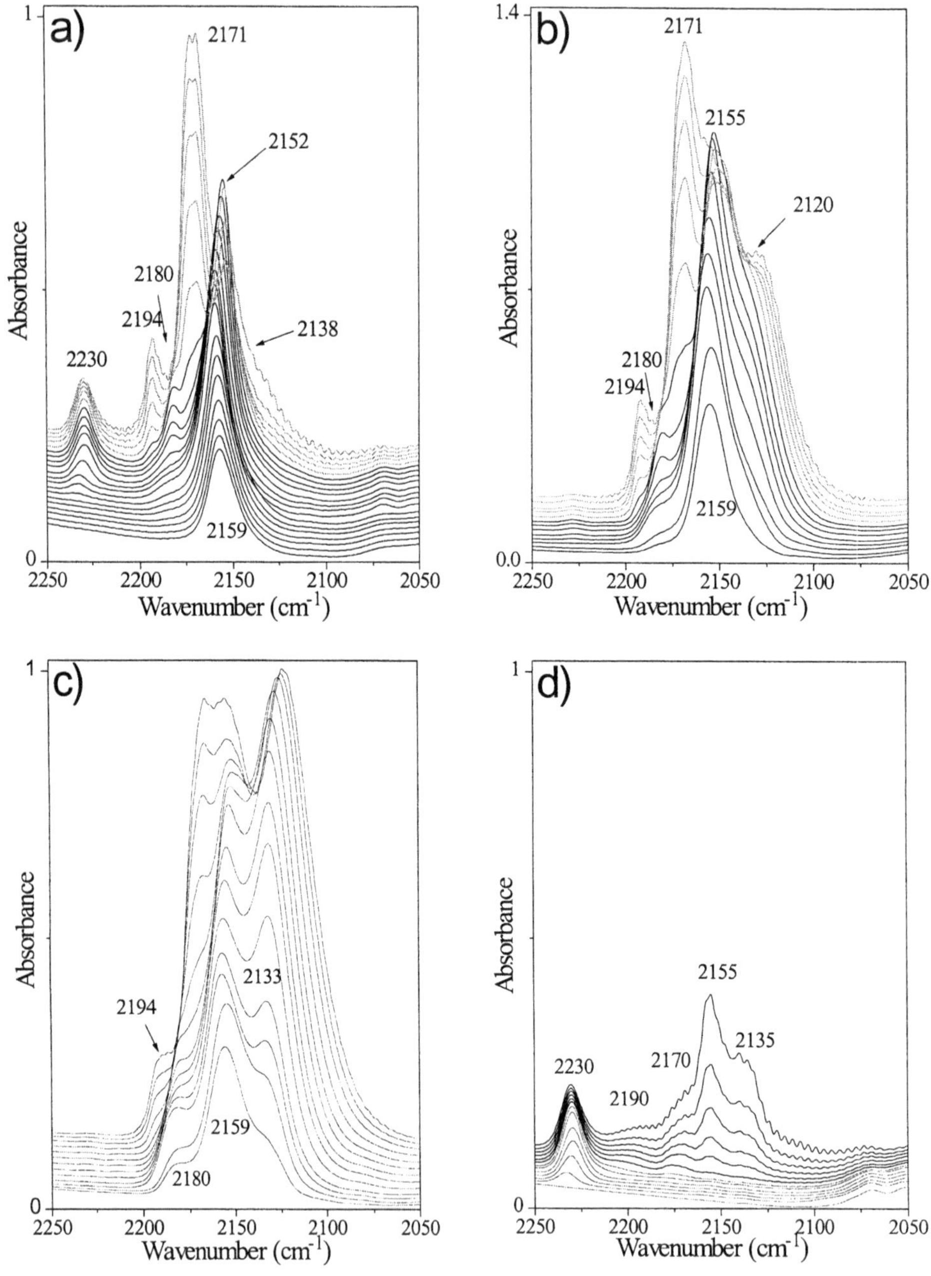

Fig. 1. FTIR spectra of CO adsorbed at ≅ 100K on Cu_2O/MCM41 for CO pressures ranging from 0.133 Pa to 2.66 kPa: a) 1% (w/w) ratio; b) 5% (w/w) ratio; c) 10% (w/w) ratio; d) pure MCM41.

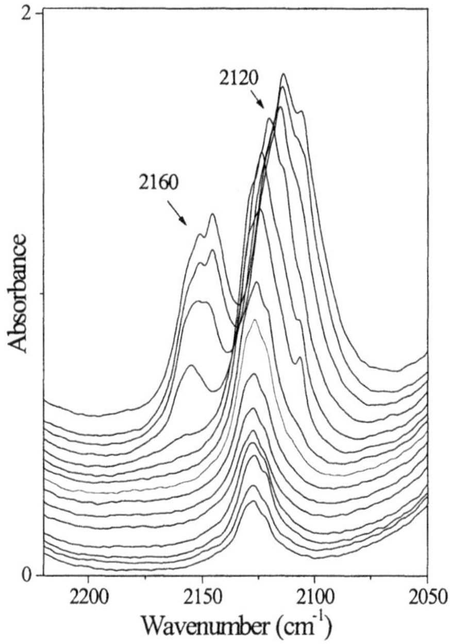

Fig 2: FTIR spectra of CO adsorbed at ≅ 100K on Cu_2O for CO pressures ranging from 0.133 Pa to 2.66 kPa.

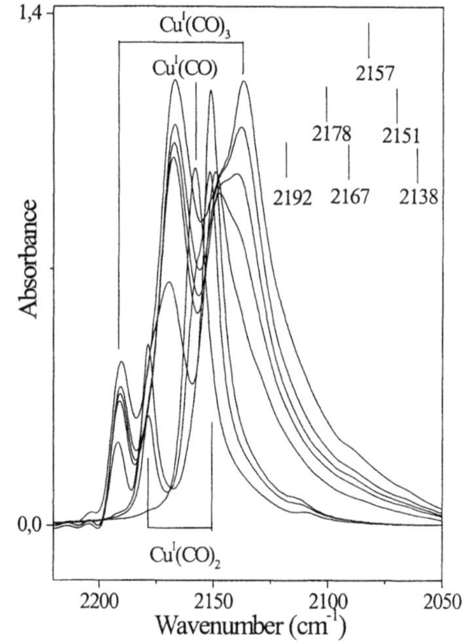

Fig. 3: FTIR spectra of CO adsorbed at ≅ 80K on Cu-ZSM-5 for CO pressures ranging from 0.133 Pa to 2.66 kPa.

In order to assign accurately the previously described spectral features, it can be useful to discuss the spectra obtained by dosing CO on pure MCM-41 samples (Fig.1d), obtained in the same experimental conditions as the copper containing samples.
Five low intensity bands at 2230 (m), 2190 (v w), 2170(sh), 2155(m) and 2135(w) cm^{-1} are observed. On the basis of ref.[31], they are assigned to CO adsorbed on Brönsted sites (2170 cm^{-1}), to hydrogen bonded to silanols (2155 cm^{-1}) [23] and to physically adsorbed species within the channels (2135 cm^{-1}) [32]. On the contrary the 2230 cm^{-1} band is attributed to CO interacting with strong Lewis sites (partially extraframework Al). This band is also observed on 1% copper containing samples. The corresponding ν(OH) modes of silanols before and after interaction with CO are not illustrated in detail for sake of brevity.

Starting from these preliminary and concise considerations, concerning the interaction of CO with pure MCM-41 sites, it is possible to interpret more accurately the spectra of CO on Cu_2O/MCM-41 samples, whose remarkable complexity requires also the comparison with the results obtained on Cu-ZSM5, where copper species are atomically dispersed [33 and refs. therein], together with those obtained on pure Cu_2O [34].

The first conclusion which can be derived by comparing Figs.1a-c with Fig.1d is that the spectral features in the 2200-2130 cm^{-1} range, observed for the CO-Cu_2O/MCM-41 systems

(Fig.1a-c) can be reasonably associated with isolated copper sites and Cu_2O microclusters, like those observed in Cu/ZSM5 systems and Cu_2O, as reported in recent contributions [33, 34]. To support these assignments, we report in Fig.3 the IR spectra of CO adsorbed on Cu/ZSM5. The extremely high similarity of the spectra obtained by dosing CO on 1%(w/w)Cu_2O/MCM-41 samples (Fig.1a) with those reported in Fig.3 indicates, without ambiguity, that, like in copper exchanged zeolites, MCM-41 samples with low copper content contain highly dispersed Cu(I) species, which are able to coordinate up to two or three ligands. This demonstrates that atomically dispersed Cu(I) species have low coordination number and high coordinative unsaturation.
In particular, as shown in the Fig. 1a, the 2159 cm^{-1} band appearing at the lowest coverage, can be assigned to Cu(I)CO complex, which by increasing the CO coverage, is progressively eroded and two strong and distinct components at 2180 cm^{-1} and 2152 $cm^{-1}$are growing in a proportional way. These correspond respectively to the symmetric and asymmetric stretching modes of $Cu(I)(CO)_2$ complexes. By further addition of CO the transformation of the $Cu(I)(CO)_2$ dicarbonyls in $Cu(I)(CO)_3$ cis-tricarbonyls complexes is occurring, as it results from the appearance of new components at 2194 cm^{-1}, 2171 cm^{-1} and 2135 cm^{-1}, accompanied by the parallel decrease of the 2180 cm^{-1} and 2152 cm^{-1} doublet due to the dicarbonyl species.
As described before, the most relevant feature, observed on spectra corresponding to samples with higher copper content (5% and 10% w/w Cu_2O) (Fig.1b,c), is the appearance of the broad 2125-20 cm^{-1} band. As this absorption is the main feature observed when CO is dosed on pure Cu_2O (Fig.2) and Cu_2O supported on silica (not reported for sake of brevity), we associate the presence of this band with the formation of Cu_2O clusters. In agreement with this hypothesis, it can be observed that the intensity of this band increases progressively with the copper content.

Up to now we are not able to distinguish whether Cu_2O particles are located within the channels and/or on the external surfaces. However in experiments, not reported here for sake of brevity, concerning samples activated at higher temperatures, it was noticed that the intensity of all the bands is decreased by the high temperature treatments: this fact can be probably explained with the partial occlusion of some channels, associated with the increase of the microparticles size and with partial collapse of the structure.

As far as a detailed discussion of the 2125-20 cm^{-1} band is concerned, it will be the subject of an incoming paper [34].
We only remember that it can be attributed to CO adsorbed on monocoordinated Cu(I) on the most predominant (111) faces of Cu_2O microcrystals. A further remarkable observation needing a comment is that on samples containing a high concentration of copper (5% and 10%), the 2230 cm^{-1} band (due to CO adsorbed on low coordinated Al^{3+}sites in partially extralattice positions) is totally missing. An explanation of this fact is not straightforward. A possible explanation is that the Cu_2O aggregates are siting on the Al^{3+}Lewis centres, inhibiting their adsorptive properties towards CO.

On the basis of the previous considerations we can conclude that the introduction of copper compounds within the MCM-41 lattice causes, at low copper content, the formation of highly isolated species, like already observed on copper exchanged zeolites. For copper content higher than 1%, the formation of Cu_2O particles is observed, whose sizes are depending on the copper loading. These results confirm the possibility to employ profitably these systems for dispersing, in a controlled way, Cu(I) and Cu_2O microcrystals.

4. CONCLUSIONS

In this work we have discussed the FTIR spectra of CO adsorbed on MCM-41 samples, with 1%, 5% and 10% (w/w) Cu_2O/MCM-41 ratios. The copper concentration plays an important role in determining the dispersion state and the spectroscopic results. In fact at low copper percentage, atomically dispersed Cu(I) ions, anchored to the framework, are predominant and the spectral features are quite similar to those of copper exchanged zeolites (Cu-ZSM5), where the coordinatively unsaturated Cu(I) coordinate up to three CO molecules. At higher copper content, Cu_2O aggregates are formed and the spectral manifestations associated with CO adsorbed on the surface sites located on the most predominant (111) faces of Cu_2O microcrystals are simultaneously observed. From these results it is inferred that the aggregation state of Cu(I) in MCM-41 can be easily controlled by varying the copper concentration and the temperature of activation.

ACKNOWLEDGEMENTS

This contribution has been supported by MURST and CNR.

REFERENCES

1. C.T. Kresge, M.E. Leonowicz, W.J. Roth, J.C. Vartuli and J.S. Beck, Nature, 359 (1992) 710.
2. J.S. Beth, J.C. Vartuli, W.J. Roth, M.E. Leonowicz, C.T. Kresge, K.D. Schmitt, C.T-W Chu, D.H. Olson, E.W. Sheppard, S.B. McCullen, J.B. Higgins and J.L. Schlenker, J.Am.Chem.Soc., 114 (1992) 10834.
3. a) F. Schüth, Ber. Bunsenges. Phys. Chem., 99 (1995) 1306; b) J. Weglarski, J. Datka, H. Hhe and J. Klinowski, J.Chem.Soc., Faraday Trans., 92(24) (1996) 5161.
4. U. Junges, F. Schüth, G. Schmid, Y. Uchida and R. Schlögl, Ber. Bunsenges. Phys. Chem., 101 No.11 (1997) 1631.
5. K.J. Chao, C.N.Wu, H. Chang, L.J. Lee and Shu-fen Hu, J.Phys. Chem. B, 101 (1997) 6341.
6. R.K. Kloetstra, H.van Bekkum, J.Chem.Res., (1995) 26.
7. a) P.J. Branton, P.G. Hall and K.S.W. Sing, J.Chem.Soc., Chem.Commun., (1993) 1257; b) P.J. Branton, P.G. Hall, M. Treguer, K.S.W. Sing, J.Chem.Soc., Faraday Trans., 91 (1995) 2041.
8. Z. Luan, H. He, W. Zhou, Chi-Feng and J. Klinowsky, J.Chem.Soc.Faraday Trans.,91(17) (1995) 2955.
9. a) A. Corma, V. Fornes, M.T. Navarro and J. Perez-Pariente, J.Catal., 148 (1994) 569; b) A. Corma, A. Martínez, V. Martínez-Soria, J.B Montón, J.Catal., 153 (1995) 25.
10. N. Coustel, F.Di Renzo and F. Fajula, J.Chem.Soc., Chem.Commun, (1994) 967.
11. X.S. Zhao, G.Q. Lu, A.K. Whittaker, G.J. Millar and H.Y. Zhu, J.Phys.Chem.B, 101 (1997) 6525.
12. C.G. Wu, T. Bein, Science, 266 (1994) 1013; C.G. Wu, T. Bein, Science, 264 (1994) 1757; C.G Wu, T. Bein, Chem.Mater., 6 (1994) 1109.
13. C. Huber, K. Moller and T. Bein, J:Chem.Soc., Chem.Commun., (1994) 2619.
14. A. Jentys and R.W. Grimes, J.Chem.Soc., Faraday Trans., 92(12)(1996) 2093.

15. M. Wark, N.I. Jaeger, W. Lutz and O.P. Tkachenko, Ber.Bunsenges.Phys.Chem., 101 (1997) 1635.
16. B.P. Feuston and J.B. Higging, J.Phys.Chem.,98 (1994) 4459.
17. M. Janicke, D. Kumar, G.D. Stucky, B.F. Chmelka, in Zeolites and Related Microporous Materials: State of Art 1994, Studies in Surface Science and Catalysis, J. Weitkamp, H.G. Karge, H. Pfeifer and W. Hölderich (eds.) 84 (1994) 243.
18. J. Man Kim, J. Hun Kwak, S.J un and R. Ryoo, J.Phys.Chem., 99 (1995) 16742.
19. M. Busio, J. Jänchen, J.H.C. van Hoff, Microporous Materials, 5 (1995) 211.
20. Bellussi, C. Perego, A. Carati, S. Peratello, E. Previte Massara, G. Perego in Zeolites and Related Microporous Materials: State of Art 1994, Studies in Surface Science and Catalysis, J. Weitkamp, H. G. Karge, H. Pfeifer and W. Hölderich (eds.) 84 (1994) 85.
21. T. Ishikawa, M. Matsuda, A. Yakusawa, K. Kandori, S. Inagaki, T. Fukushima and S. Kondo, J.Chem.Soc., Faraday Trans., 92 (1996) 1985.
22. P.C. Llewellyn, F. Schüth, Y Grillet, F. Rouquoerol, J. Rouquoerol and K.K. Unger, Langmuir, 11 (1995) 574.
23. A. Jentys, N.H. Pham, H. Vinek, J.Chem.Soc., Faraday Trans., 92 (1996) 3287.
24. A. Liepold, K. Roos, W. Reschetilowski, A.P. Esculas, J. Rocha, A. Philippou and M.W. Anderson, J.Chem.Soc., Faraday Trans., 92(22) (1996) 4623.
25. U. Junges, W. Jacobs,, I. Voigt-Martin, B. Krutzsch and F. Schüth, J.Chem.Soc. Chem.Commun., (1995) 2283.
26. A. Corma, M.T. Navarro and J. Perez-Pariente, J.Chem.Soc.Chem.Commun., 147 (1994).
27. P.T. Tanev, M. Chibwe and T.J. Pinnavaia, Nature, 368 (1994) 321.
28. K.M. Reddy, I. Moudrakowski and A. Sayari, J.Chem.Soc.Chem.Commun., 1059 (1994).
29. D. Brunel, A. Cauvel, F. Fajula and F.Di Renzo, in Zeolites:A Refined Tool for Designing Catalytic Sites, L.Bonneviot and S.Kaliaguine (eds.) (1995) 173.
30. A.M.M. Gadalla and J. White, Trans.Brit.Ceram.Soc., 63 (1964) 39, 119.
31. A. Zecchina, S. Bordiga, G. Spoto, D. Scarano, G. Petrini, G. Leofanti, M. Padovan and C. Otero Areán, J.Chem.Soc., Faraday Trans., 88 (1992) 2959.
32. F. Di Renzo, B. Chiche, F. Fajula, S. Viale and E. Garrone, in 11[h] International Congress on Catalysis (eds.Hightower), 101 (1996) 851.
33. a) C. Lamberti, S. Bordiga, M. Salvalaggio, G. Spoto, A. Zecchina, F. Geobaldo, G. Vlaic, and M. Bellatreccia, J.Phys.Chem.B, 165 (1997) 129; b) A. Zecchina, S. Bordiga, M. Salvalaggio, G. Spoto, D. Scarano and C. Lamberti, J.Catal., (1998) in press.
34. D. Scarano, S. Bordiga, C. Lamberti, G. Spoto, G. Ricchiardi and A Zecchina, submitted to Surface Science for publication.

MESOPOROUS MOLECULAR SIEVES 1998
Studies in Surface Science and Catalysis, Vol. 117
L. Bonneviot, F. Béland, C. Danumah, S. Giasson and S. Kaliaguine (Editors)

Promoter (PO_4^{3-}) assisted efficient synthesis of all silica, alumino-silicate and titanium-silicate analogues of MCM-41 type mesoporous materials

Priyabrata Mukherjee[a,+], Rajiv Kumar[a,*] and Ulf Schuchardt[b,+]

[a]Catalysis Division, National Chemical Laboratory, Pune - 411 008, India
(Fax:+91-212-334761 or -330233; email: rajiv@ems.ncl.res.in)

[b]Instituto de Quimica, Universidade Estadual de Campinas, 13083-970 Campinas-SP, Brazil

A new and efficient method for the preparation of MCM-41 type mesoporous silicas using phosphate as promoter under reflux conditions is reported. All silica (Si-MCM-41), alumino-silicate (Al-MCM-41) and titanium-silicate (Ti-MCM-41) mesoporous materials were studied. Instead of following conventionally used autoclave method at autogeneous pressure, the synthesis was carried out by reflux method under atmospheric pressure. Addition of small quantity of phosphate ions (PO_4^{3-}), used as promoters, significantly reduced the synthesis time of all these mesoporous materials. The quite high surface areas (930-1480 $m^2\ g^{-1}$) of all these MCM-41 samples were typical of MCM-41 type ordered mesoporous materials.

1. INTRODUCTION

Ordered mesoporous materials such as MCM-41 [1] have hexagonal arrangement of unidimensional pores, the diameter of which may range from 20-100 Å. This discovery has not only enlarged the domain of zeolites and molecular sieves but also advanced the materials synthesis in general. These mesoporous materials are of interest to a broad range of areas like selective separation, catalysis, fine chemicals, support, carriers, chemical sensors etc. New applications of these interesting materials are expected to grow in the near future. Hence, it is no wonder that the synthesis of these mesoporous materials is attracting increasing attention.

Recently, Stucky et al. [2] have reviewed the surfactant-directed synthesis of MCM-41 type mesoporous and other organic / inorganic composite materials. Generally, a surfactant like cetyltrimethyl ammonium cation is used as an organic additive in the hydrothermal synthesis of MCM-41 under autogeneous pressure at ca. 150 °C [1,2].

* Address for correspondence (Fax: +91-212-334761 or -330233, email: rajiv@ems.ncl.res.in).
+ P.M. gratefully acknowledges C.S.I.R., New Delhi for grant of a Senior Research Fellowship and U.S. acknowledges FAPESP, Brazil for allowing him a sabbatical leave at NCL Pune, India.

However, Carvalho et al. [3,4] have described an alternate procedure for the synthesis of mesoporous metallosilicate MCM-41 type materials using low temperature reflux method under atmospheric pressure. This method was found to be particularly useful for preparing stable and catalytically active redox metal ions incorporated MCM-41 type materials [3,4].

Recently, it has been observed [5] that the addition of small amount of certain oxyanions such as phosphate, perchlorate, arsenate, chlorate, bromate etc, used as promoters, to the synthesis precursor mixture of zeolite and related silica based molecular sieves, reduces their crystallization time considerably. This concept of promoter was found to be quite general in the synthesis of a variety of silica based microporous materials including small, medium and large pore molecular sieves. Further, it was also applicable to all-silica, aluminosilicate and metallo-silicate (like TS-1) molecular sieves.

In the present communication, we report, for the first time, the promoter assisted synthesis of mesoporous materials. The main aim of this work is to find out whether the concept of promoter-assisted synthesis of silica based microporous materials is applicable to the synthesis of mesoporous materials also or not. The synthesis of Si-MCM-41, Al-MCM-41 and Ti-MCM-41 type mesoporous molecular sieves is studied under reflux conditions in the presence and absence of phosphate ions used as promoter.

2. EXPERIMENTAL

2.1 Synthesis

Synthesis of Si-MCM-41, Al-MCM-41 and Ti-MCM-41 was carried out by following the procedure of Carvalho et al. [3,4]. For the synthesis of Si-MCM-41 in the absence of promoter (sample A, Si-MCM-41-S) a mixture of two types of silica sources namely sodium meta silicate, Na_2SiO_3, (s.d.fine chem. India) and amorphous silica (Areosil) were employed where the ratio of Na_2SiO_3 / SiO_2 was kept 0.124 : 1. At first silica sources were suspended in aq. solution of TMAOH (25 %, Aldrich). This suspension was then combined with a solution of cetyltrimethyl ammonium bromide (CTMABr, Aldrich) in water. The initial gel composition was: 1 SiO_2 : 0.22 NaOH : 0.1 NaCl : 0.8 TMAOH : 0.21 CTMABr : 0.125 H_2O. For preparing the Si-MCM-41 using promoter, (sample B, Si-MCM-41-P), a solution of NaH_2PO_4 (s.d.fine chem. India) in water was added in the final gel where same procedure was used as mentioned above. To maintain the Si / Na molar ratio in the gel of Si-MCM-41-S and Si-MCM-41-P samples the same, a required amount of NaCl was added in the former case. In the case of Si-MCM-41-P sample, the molar gel composition was: 1 SiO_2 : 0.22 NaOH : 0.1 NaH_2PO_4 : 0.8 TMAOH : 0.21 CTMABr : 0.125 H_2O.

The synthesis of Al-MCM-41 samples was also carried out in the same way as that of Si-MCM-41 samples except that the source of Al_2O_3 (Al_2SO_4 . 16 H_2O, s.d.fine chem., India) was also added. The molar ratio SiO_2/Al_2O_3 was 40. Remaining gel composition was the same as that of Si-MCM-41 samples. Here also NaH_2PO_4 was used as promoter.

For the synthesis of Ti-MCM-41, the procedure reported by Carvalho et al. [4] (method II) was adopted as it gave quite active and stable catalyst. In a typical synthesis of Ti-MCM-41-S (standard) the similar procedure and gel composition was used as given above for Si-MCM-41 samples, except that a solution of tetra n-butoxy titanate (TNBT, Aldrich) in isopropanol was added, adapting from Thangaraj procedure for TS-1 [6]. The molar gel composition was: 1 SiO_2 : 0.018 TiO_2 : 0.8 TMAOH : 0.21 CTMABr : 0.125 H_2O. For preparing Ti-MCM-41-P sample, H_3PO_4 (85 %, s.d.fine chem. India) was also added in the final gel as prepared for Ti-MCM-41-S sample. The SiO_2 : H_3PO_4 molar ratio was : 1 : 0.1. The rest of the gel composition was same as that of Ti-MCM-41-S sample.

The synthesis was carried out by refluxing the gels in a glass flask. After refluxing for the required time the solid products were collected by filtration, washed several times with distilled water, and dried at 373 K. Finally the products were calcined at 813 K for 1 h under a flow of nitrogen followed by 6 h under air flow.

2.2 Characterization

The samples were characterized by X-ray diffraction (Regaku D Max III, VC diffractometer), scanning electron microscopy (SEM, Jeol JSM 5200), FTIR (Perkin Elmer 1600, Nujol on KBr disc technique) and BET surface area (Omnisorb CX-100).

3. RESULTS AND DISCUSSION

Figure 1 depicts the synthesis time (in hours) taken by different samples such as Si-MCM-41-S (A), Si-MCM-41-P (B), Al-MCM-41-S (C), Al-MCM-41-P (D), Ti-MCM-41-S (E) and Ti-MCM-41-P (F). It is clear from Fig. 1 that there is a considerable reduction in the synthesis time required for all the samples prepared in the presence of phosphate promoter compared to those prepared in the absence of promoter. However, the time taken for the synthesis of various MCM-41 materials followed the trend: Si-MCM-41 ≅ Ti-MCM-41, < Al-MCM-41 both in the presence as well as in the absence of promoter. Based on liquid state NMR studies on the synthesis mixture of Silicalite-1 [5], it was found that the presence of promoter oxyanions enhances the condensation process of silicate-template composite species, mainly involving Q^4 silicate entities. High polarizing ability of the promoter anions probably polarizes water spheres formed around silicate and template molecules towards itself and thereby facilitating the overlap of silicate and template moieties resulting in the formation of composites species. It is reasonable to envisage a similar type of role of phosphate ions in the synthesis of MCM-41 type of mesoporous materials also.

XRD of all the samples exhibited weak 110, 200 and 210 hkl lines along with main 100 reflection indicating high degree of hexagonal ordering. Figure 2 depicts the scanning electron micrograph (SEM) of sample B (Si-MCM-41-P) exhibiting small elongated particles of submicron size (0.5-0.8 μm). The Si-MCM-41 (sample A) is reported to have worm like morphology having average particle size 2 x 10 μm [3]. Although, the particle shape of both the samples (sample A and B) is quite similar, there is a significant reduction of the particle size of sample B (prepared in the presence of promoter) compared to that of sample A (prepared in the absence of promoter). The same was found to be true for other samples. This is quite expected because of significant increase in the ordering rate in the presence of promoter even at the initial stages of the synthesis of MCM-41 type mesoporous materials.

In Table 1 the surface area of these mesoporous materials is given. It is quite clear from the high BET surface areas that all the samples have highly ordered mesoporous structure. Sample B (Si-MCM-41-P) exhibited higher surface area (1480 cm^2 g^{-1}) compared to that of sample A (1296 cm^2 g^{-1}), which may be attributed to the smaller particle size of the former. Almost similar trend is also observed for sample C and D (Al-MCM-41 materials). However, the reduced surface area of sample F compared to that of sample E (Ti-MCM-41 materials) indicates the presence of some amorphous material in sample F which is also confirmed by its scanning electron micrograph exhibiting a small contribution of amorphous material.

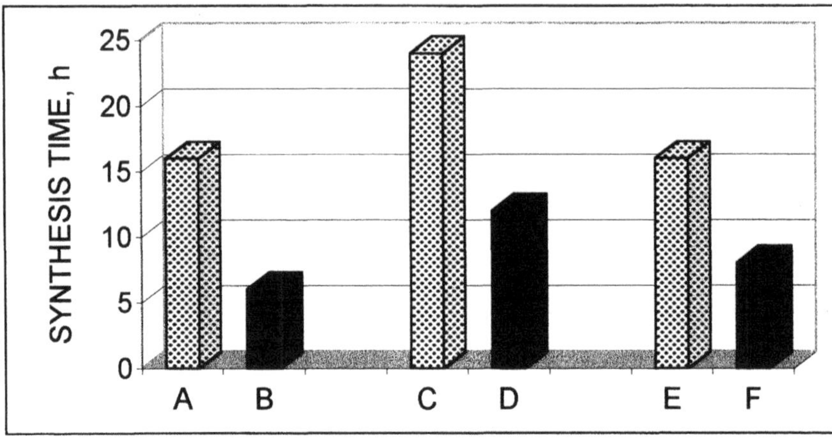

Figure 1. Synthesis time (h) of Si-MCM-41-S (A), Si-MCM-41-P (B), Al-MCM-41-S (C), Al-MCM-41-P (D), Ti-MCM-41-S (E) and Ti-MCM-41-P samples.

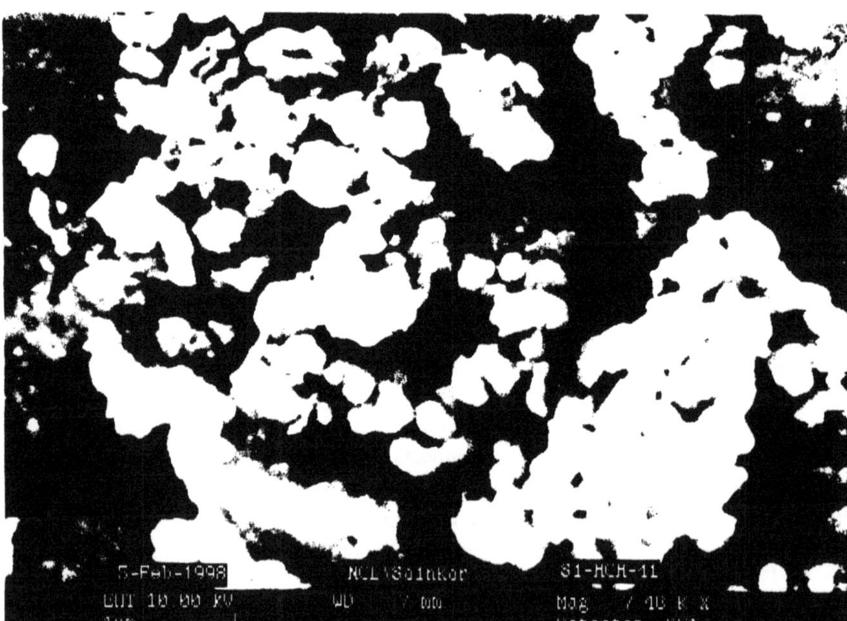

Figure 2. SEM of sample B (Si-MCM-41-P)

Table 1
Characterization of MCM-41 type mesoporous materials

Sample	Synthesis time (h)	Surface area (m^2g^{-1})	Average particle size (μm)
A (Si-MCM-41-S)	16	1296	2 x 10
B (Si-MCM-41-P)	6	1480	0.5 x 0.8
C (Al-MCM-41-S)	24	930	n.d.
D (Al-MCM-41-P)	12	1030	0.8 x 1.2
E (Ti-MCM-41-S)	16	1245	2 x 10
F (Ti-MCM-41-P)	8	1050	0.4 x 0.6

Figures 3 and 4 depict, respectively, the adsorption isotherms of nitrogen and the infrared spectra (1300 - 400 cm^{-1} range) of MCM-41 samples prepared in the presence of promoter (samples B, D and F), exhibiting characteristic patterns of ordered MCM-41 type mesoporous materials [7-9].

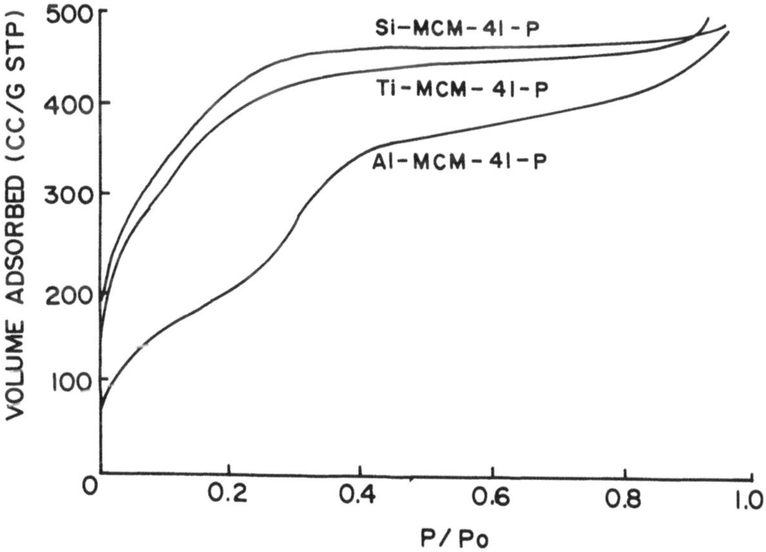

Figure 3. Adsorption isotherms of nitrogen of Si-, Al- and Ti-MCM-41-P materials (samples B, D and F) prepared in the presence of phosphate ions as promoters

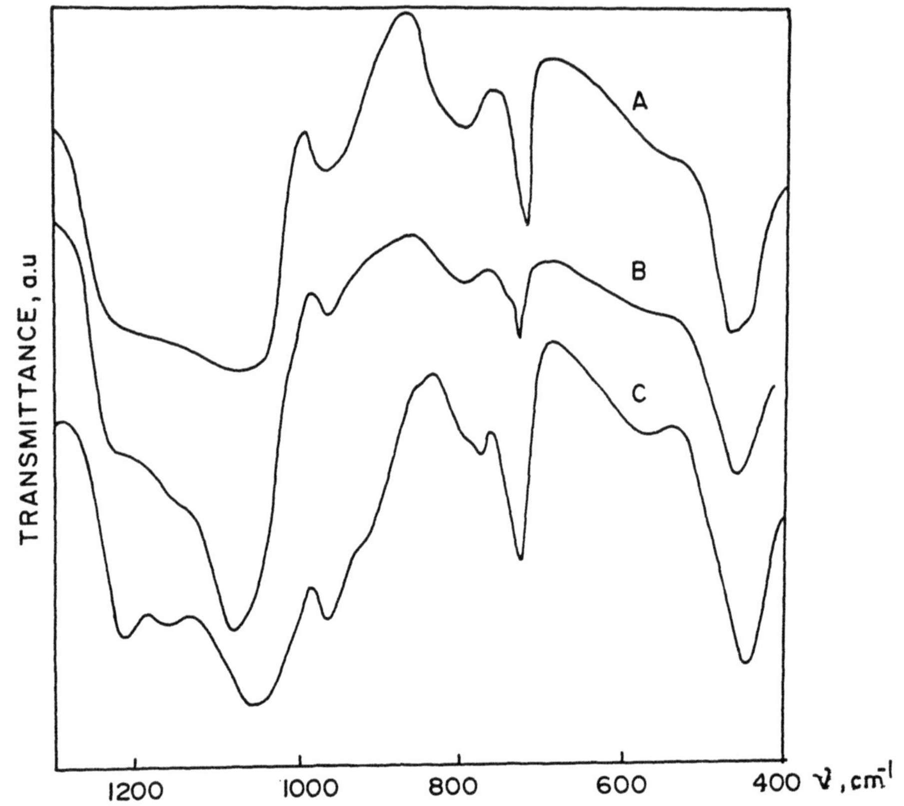

Fig. 4. Infrared spectra of MCM-41 materials prepared in the presence of phosphate ions (promoter) curves A: sample B (Si-MCM-41-P), B: sample D (Al-MCM-41-P) and sample F (Ti-MCM-41-P).

REFERENCES

1. C.T. Kresge, M.E. Leonowicz, W.J. Roth, J.C. Vertuli and J.S. Beck, Nature, 359 (1992) 710.
2. G.D. Stucky, Stud. Surf. Sci. Catal., 105A (1997) 1.
3. W.A. Carvalho, P.B. Varaldo, M. Wallau and U. Schuchardt, Zeolites, 18 (1997) 408.
4. W.A. Carvalho, M. Wallau and U. Schuchardt, Bull. Catal. Soc. India, 7 (1997) 1.
5. R. Kumar, A. Bhaumik, R.K. Ahedi and S. Ganapathy, Nature, 398 (1996) 298.
6. A. Thangaraj, R. Kumar, S.P. Mirajkar and P. Ratnasamy, J. Catal., 131 (1991) 1.
7. R. Anwander and R. Roesky, J. Chem. Soc., Dalton Trans., (1997) 137.
8. C.-Y. Chen, S-Q. Xiao and M.E. Davis, Microporous Materials, 3 (1995) 1.
9. R. Schmidt, M. Stoeker, E. Hansen, D. Akporiaye and O.H. Ellestad, Microporous Materials, 3 (1995) 443.

MESOPOROUS MOLECULAR SIEVES 1998
Studies in Surface Science and Catalysis, Vol. 117
L. Bonneviot, F. Béland, C. Danumah, S. Giasson and S. Kaliaguine (Editors)

Coordinative anchoring of zinc phthalocyanine in Ti-MCM-41

A.Ortlam[a], M.Wark[a], G.Schulz-Ekloff[a], J.Rathouský[b] and A.Zukal[b]

[a]Institute of Applied and Physical Chemistry, University Bremen, Germany

[b]J.Heyrovský Institute of Physical Chemistry, Academy of Sciences of the Czech Republic, Prague, Czech Republic

Titanium-containing MCM-41 materials were prepared by *in situ* incorporation of Ti and zinc phthalocyanine (ZnPc) during the hydrothermal synthesis (denoted as Ti-MCM-41-syn-ZnPc) and impregnation of siliceous MCM-41 either with tetrabutylorthotitanate (denoted as Ti-MCM-41-TBOT) or with titanyl acetylacetonate (denoted as Ti-MCM-41-$TiO(acac)_2$). As a reference an aluminum-containing MCM-41 was used, obtained from impregnation of Si-MCM-41 with aluminum acetylacetonate (denoted as Al-MCM-41-$Al(acac)_3$).

Perfect ordering of the hexagonal structures is revealed from XRD patterns and nitrogen adsorption isotherms. Uniform dispersion of the Ti in the pores is demonstrated by UV/VIS diffuse reflectance spectra (DSR). The degree of dispersion of ZnPc is concluded from the analysis of the Q-band structure in the DRS. In titanium-containing MCM-41, ZnPc exists predominantly in a non-aggregated, monomeric state whereas aggregates are present in Al-MCM-41 exclusively. ZnPc-impregnated Ti-MCM-41 samples do not exhibit fluorescence of the dye. The effects are interpreted with a peculiar interaction between the Ti centers and the ZnPc molecule.

1. INTRODUCTION

After discovery of the M41S family of mesoporous molecular sieves by Beck and coworkers [1,2], numerous different transition metals with known redox properties were incorporated and tested for their catalytic properties [3]. In most cases the metals were incorporated by *in situ* synthesis of the M41S materials aiming at an isomorphous substitution. By this method a significant fraction of the metal is buried in the amorphous walls and not accessible to reduction, as found e.g. for V-MCM-41 [4]. For this reason, in the following, Ti-MCM-41 materials are prepared by impregnation of siliceous MCM-41, using organic titanium compounds. An additional impregnation of the obtained Ti-MCM-41 with ZnPc is carried out with the objective to achieve charge transfer couples of the donor ZnPc and the acceptor titanium oxide. Photoionization via a single-photon process was recently reported for titania-encapsulated zinc-9,18,27,36-tetra(4-tolyl)-tetrabenzoporphin [5], whereas photophysical hole-burning appears with as-synthesized ZnPc in Ti-MCM-41, where ZnPc was added with the surfactant to the synthesis batch [6].

2. EXPERIMENTAL

2.1. Materials

The sources of silica and titania were Ludox AS-40 (DuPont, 40 wt.-% colloidal silica in water) and tetrabutylorthotitanate (TBOT, Merck, p.a.). As quaternary ammonium surfactant compound hexadecyltrimethylammonium chloride (HDTMACl, Fluka, 25 wt.-% water solution) was used. Other compounds of the synthesis mixture were tetraethylammonium hydroxide (TEAOH, Merck, 20 wt.-% solution in water) and isopropanol (Merck, p.a.). The dye was obtained from Aldrich (zinc phthalocyanine, ZnPc). All chemicals were used without further purification.

2.2. Synthesis procedures

Ti-MCM-41-syn-ZnPc with an atomic Ti/Si ratio of 1:47 was synthesized from the pre-cooled components where the reaction mixture corresponded to a molar ratio of, e.g., 1 TBOT : 47 SiO_2 : 14 TEAOH : 13 HDTMACl : 1308 H_2O. 9.63 g of Ludox AS-40 was combined with 9.26 g of a TEAOH solution under stirring (1000 rpm) in a 400 ml polypropylene beaker. $4x10^{-5}$ moles of the dye per gram of product was directly dissolved in 23.54 g of a HDTMACl solution. The ultrasonic vibration was applied for 10 min after mixing the dye solution with the surfactant. The dye/surfactant solution was added immediately to the synthesis mixture. Finally, 0.45 g of TBOT, dissolved in 0.95 g of isopropanol, was added to the mixture. The dissolution of TBOT in isopropanol prevents the early hydrolysis of the titanium component. The gel which was formed during stirring (5 min.) was transferred into a 250 ml polypropylene bottle and let crystallize without stirring at 106 °C for 3 days. The resulting solid product (a yield of ca. 14 g) was recovered by filtration and washed with 150 ml of water. A fraction of as-synthesized product was extracted with ethanol and calcined on air at 600°C for 22 hours. The calcined Ti-MCM-41-syn-ZnPc was characterized with respect to its structure properties and the dispersion of titanium.

The Si-MCM-41, used as host for the impregnation, was synthesized from Ludox AS-40, TEAOH, HDTMACl and distilled water. 18.5 g of Ludox AS-40 were combined with 18.5 g TEAOH solution and 25 g distilled water under stirring (ca. 300 rpm) in a 400 ml polypropylene beaker for 20 h at room temperature. Afterwards, 25 g HDTMACl solution was added to the reaction mixture, transferred to a 250 ml polypropylene bottle and left one day at room temperature for ageing. Finally, the synthesis gel was heated up to 100°C with a heating rate of 10°C/h and left there for crystallization at 100°C for 40 h. The resulting solid product was recovered by filtration and washed with 150 ml of distilled water. To obtain Si-MCM-41 as host for the impregnation, the as-synthesized products were extracted with ethanol and water and calcined in air at 600°C for 22 h.

2.3. Impregnation procedures

1g of Si-MCM-41 was impregnated with solutions of $TiO(acac)_2$ or TBOT in isopropanol giving Ti loadings ranging from 1.7-8.6 wt.-%. The suspension was stirred (ca. 300 rpm) at room temperature for 24 h. Afterwards, the solvent was removed by evaporation. Finally, the samples were calcined in air at 600°C for 24 h.

All Ti-MCM-41 samples, prepared by the incipient wetness procedure, were impregnated with a solution of ZnPc in dried dimethylformamide (DMF). Afterwards the solvent was removed by evaporation.

A reference sample of Al-MCM-41-Al(acac)$_3$-ZnPc was obtained by impregnation of siliceous MCM-41 with Al(acac)$_3$ in ethanol and subsequent impregnation with ZnPc in DMF.

2.4. Characterization

The adsorption isotherms of nitrogen at -196 °C were measured with an Accusorb 2100E instrument (Micromeritics). Details concerning the processing of adsorption data are given elsewhere [7].

Powder X-ray diffraction data were obtained on a Philips PW-3400 diffractometer in the Bragg-Brentano geometry arrangement using Cu-Kα_1 radiation and a germanium primary monochromator PW 3092/00.

UV/VIS spectra in the range from 200 to 800 nm were obtained on a Cary-4 spectrometer (Varian) equipped with a diffuse reflectance attachment (Praying Mantis).

The Soxhlet extraction was used in order to remove physisorbed dye from the external surface of the support and to obtain information on the stability of the bonding of dye molecules to the pore walls.

Fluorescence excitation spectra of as-synthesized samples were obtained using 150 W Xe arc-lamp radiation, dispersed by a Spex 1402 monochromator with a resolution of 3 cm^{-1}, and a CRC C31034 A photomultiplier as well as a SI single-photon counting system. Appropriate cut-off filters were used to select fluorescence light only.

3. RESULTS

3.1. Quality of the MCM-41 materials

The quality of the materials can be judged from the XRD patterns and the adsorption isotherms. From the five reflections, found in the range of $2\theta < 8°$ for the siliceous MCM-41, the d values (nm) of the Bragg reflections (in parenthesis) 4.07(100), 2.33(110), 2.02(200) and 1.33(300) and the lattice constant $a_0 = 4.6$ nm were calculated, indicating a perfect ordering of the hexagonal structure. The Ti-MCM-41 samples (1.7-2.2 wt.-% Ti) exhibit less well-resolved X-ray diffractograms and smaller lattice constants ranging from $a_0 = 4.00$-4.20 nm and indicating a lattice contraction.

The nitrogen isotherms confirm the perfect structure of the siliceous MCM-41 yielding a total surface of 1020 m^2/g, an external surface of 33 m^2/g, a pore volume of 0.88 cm^3/g and a pore diameter of 3.6 nm. The calcined Ti-MCM-41-syn-ZnPc (1.7 wt.-% Ti) exhibits smaller pores (3.26 nm) and pore volumes (0.66 cm^3/g), whereas the Ti-impregnated samples (1.7-2.2 wt.-% Ti) show corresponding values being larger but smaller than those of the siliceous MCM-41, i.e., pore diameters of 3.3-3.5 nm and pore volumes of 0.71-0.74 cm^3/g. Also for the reference Al-MCM-41-Al(acac)$_3$ smaller values, i.e., $a_0 = 4.05$ nm and $D_P = 3.25$ nm, were found.

3.2. Dispersion of titanium

Information on the titanium dispersion can be revealed from DRS (Fig. 1a and b). The onset of the red flank of the calcined sample Ti-MCM-41-syn-ZnPc (1.7 wt.-% Ti) is located around 280 nm, i.e., between the values usually found for TS-1 (245 nm) and for titanium-impregnated faujasites (320-340 nm) having comparable contents of titanium [8]. An increase of the titanium contents (4.3-5.6 wt.-% Ti) results in a red shift of this flank, as usual. The values are significantly different from that of anatase (373 nm), so that a high dispersion of the titanium oxide species, resulting after the procedures of impregnation and calcination, can be concluded from DRS.

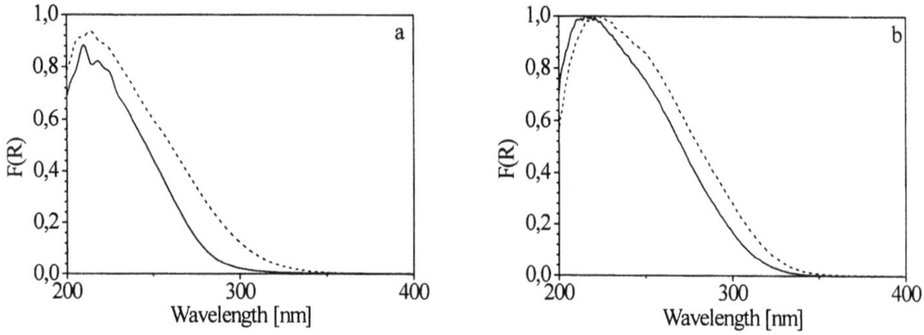

Figure 1. DR spectra of calcined Ti-MCM-41-syn-ZnPc with 1.7 wt.-% (—) and 5.6 wt.-% (- - -) (a) and Ti-MCM-41-TiO(acac)$_2$ with 2.2 wt.-% (—) and 4.3 wt.-% (- - -) (b).

3.3. Impregnation of Ti-MCM-41 with ZnPc

The impregnation of Ti-MCM-41-TBOT (3.4 wt.-%) and Ti-MCM-41-TiO(acac)$_2$ (4.3 wt.-%) by ZnPc strongly reduced the intensity of the X-ray reflections but not their positions. The initial intensity is, however, restored following a removal of the ZnPc by calcination.

The pore volumes are reduced to 30-50% of the initial values after impregnation with ZnPc. Since the incorporated amount of ZnPc would occupy only 1% of the pore volume, the found strong reduction of the pore volume has to be related to a partial plugging of the pores.

Impregnation of Al-MCM-41-Al(acac)$_3$ by ZnPc (1.2x10^{-5} mol/g) resulted in a reduction of the pore volume to 15%.

3.4. Dispersion of ZnPc

The dispersion of ZnPc on the different Ti-MCM-41 samples or the Al-MCM-41 reference following the impregnation and the removal of the solvent can be revealed from the analysis of the Q-band in the DRS (Fig. 2). Excellent resolution is found for the samples Ti-MCM-41-syn-ZnPc (Fig. 2a) and Ti-MCM-41-TiO(acac)$_2$-ZnPc (Fig. 2b), where the 0-1 excitation (ca. 605 nm) of the monomers is clearly distinguished from the dimer band (ca. 645 nm). A large fraction of dimers can be concluded from the poorer resolution for the Ti-MCM-41-TBOT-ZnPc (Fig. 2c), whereas the predominant presence of aggregates is concluded from the unresolved Q-band of ZnPc on Al-MCM-41-Al(acac)$_3$-ZnPc (Fig. 2d).

3.5. Fluorescence spectroscopy of Ti-MCM-41-syn-ZnPc

Fluorescence spectra of as-synthesized zinc phthalocyanine-loaded Ti-MCM-41-syn samples (Fig. 3) were obtained by irradiation at 400 nm (Soret band). The fluorescence maximum of the Q-band (675 nm) exhibits a negligible Stokes shift due to the small energy difference between ground and excited states. No fluorescence is obtained from the impregnated samples Ti-MCM-41-TBOT-ZnPc and Ti-MCM-41-TiO(acac)$_2$-ZnPc.

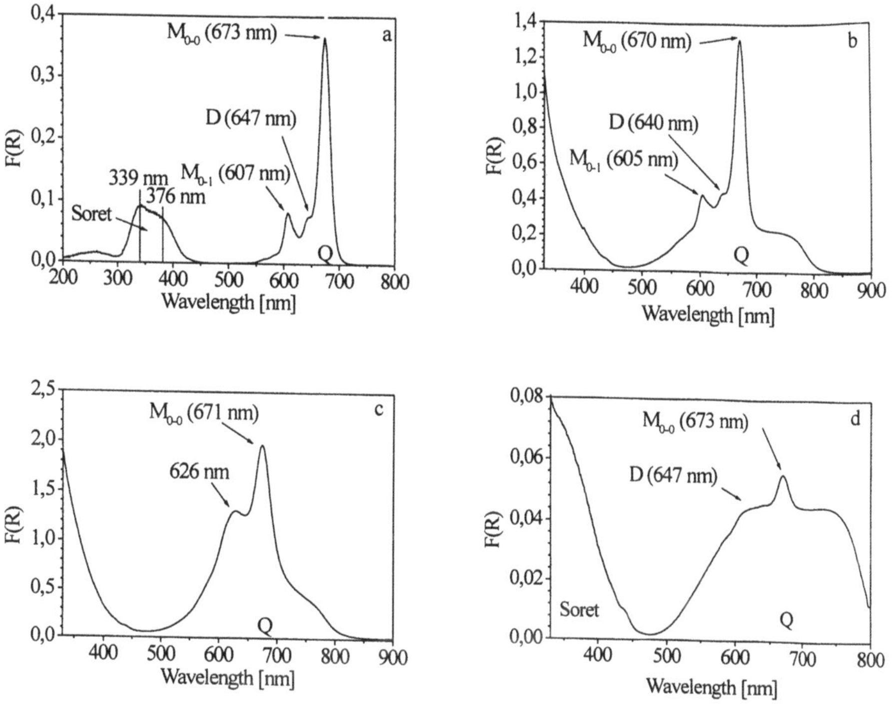

Figure 2. DR spectra of ZnPc-loaded Ti-MCM-41-samples: Ti-MCM-41-syn-ZnPc (5.6. wt.-% Ti) (a)*, Ti-MCM-41-TiO(acac)$_2$ (4.3 wt.-% Ti) (b) and Ti-MCM-41-TBOT (3.4 wt.-% Ti) (c) and Al-MCM-41-Al(acac)$_3$ (3.2 wt.-% Al) (d).*The spectrum of Ti-MCM-41-syn was substracted.

4. DISCUSSION

4.1. Titanium dispersion

The DRS band structure of the calcined sample Ti-MCN-41-syn-ZnPc (Fig. 1a), exhibiting a maximum around 210 nm and an onset, obtained from the intersect between the abscissa and the tangent through the point of inflection of the red flank, around 280 nm, points to a dominating fraction of tetrahedrally coordinated Ti, presumably, located inside the walls, and a smaller fraction of octahedrally coordinated Ti, presumably, present at the surface of the pore walls. The latter might be responsible for the red shift. The amount of titanium was kept below the limit (4 wt.-%), beyond which strong red shifts with onsets ≥ 350 nm and strong shoulders in the range 250 - 300 nm appear. This means, that the exclusive presence of mononuclear TiO_x species can be assumed [8-10]. This conclusion is also valid for the impregnated Ti-MCM-41 samples, exhibiting similar band structures for similar Ti contents.

4.2. The ZnPc dispersion

In the as-synthesized sample Ti-MCM-41-syn-ZnPc (5.6 wt.-% Ti) the dye molecule is obviously uniformly distributed as monomer (Fig. 2a). Such a uniform distribution can be referred to a complete incorporation of the hydrophobic ZnPc in the hydrophobic center of the surfactant micelle. This assumption is supported by the fluorescence activity of the ZnPc (Fig 3).

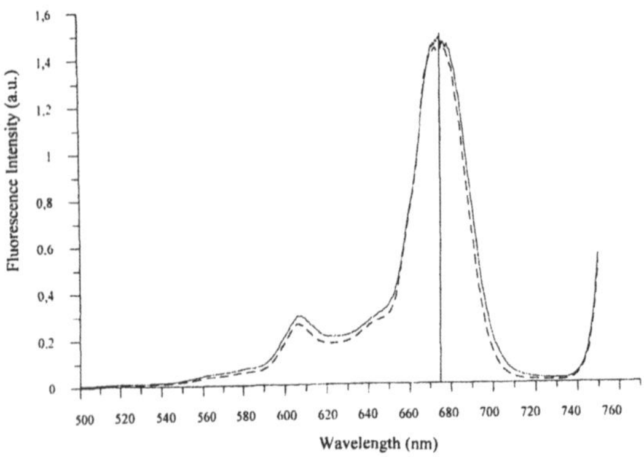

Figure 3. Fluorescence spectrum of Ti-MCM-41-syn-ZnPc (1.7 wt.-% Ti) at 2 K (- - -) and 200 K (—).

The surfactant-free ZnPc-impregnated samples, i.e., Ti-MCM-41-TBOT-ZnPc and Ti-MCM-41-$TiO(acac)_2$-ZnPc, exhibit differences with respect to the dispersion of the ZnPc. Whereas the latter exhibits predominantly monomers (Fig. 2b), the former shows a larger fraction of dimers (Fig. 2c). This high fraction of dimers points to a less uniform distribution of the Ti in the sample impregnated with TBOT. Differences in the titanium dispersion, using either TBOT or $TiO(acac)_2$, can be expected if the nature of interaction between the titanium compounds and the SiOH groups of the pore walls would be different. Such a different type or

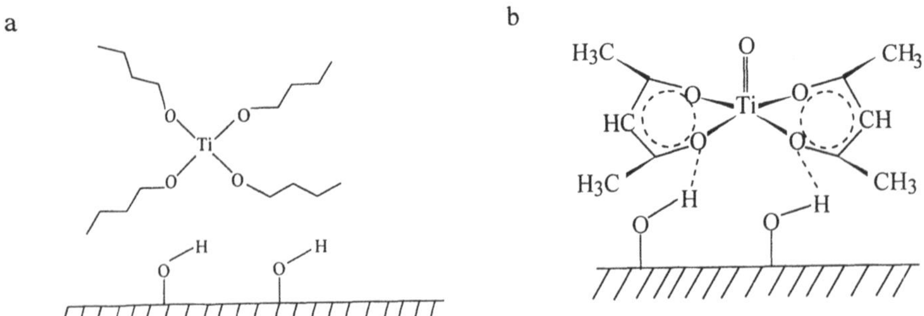

Figure 4. Model of the interaction of the titanium compounds with the pore walls of the S[illegible] MCM-41: TBOT (a) and $TiO(acac)_2$ (b).

strength of interaction can be, indeed, expected, since the hydrophobic hydrocarbon ligands of TBOT (Fig. 4a) exhibit a much weaker interaction with the silanol groups than the hydrophilic acetylacetonate ligands (Fig. 4b), resulting in a more uniform titanium dispersion in the incipient wetness procedure for the $TiO(acac)_2$.

4.3. Coordinative anchoring of ZnPc

ZnPc in Ti-MCM-41 maintains its ability to fluoresce as long as it is caped by the micelles and shielded from the titanium oxide centers of the pore walls by the head groups of the surfactant molecules. Direct contact between these titanium oxide centers and the ZnPc is, however, possible for the samples Ti-MCM-41-TBOT-ZnPc and Ti-MCM-41-$TiO(acac)_2$-ZnPc. This direct contact enables a dispersion of the ZnPc depending on the preceding Ti dispersion. Poor ZnPc dispersions are achieved for the Al-MCM-41-$Al(acac)_3$-ZnPc sample (Fig. 2d), pointing to a negligible interaction between the aluminum oxide centers and ZnPc.

ZnPc anchored at the titanium oxide centers does not exhibit fluorescence, presumably, due to charge transfer interactions.

5. CONCLUSIONS

Transition metal centers, being fully accessible for redox processes [4] or catalysis, can be generated in the pores of M41S materials using incipient wetness procedures. The dispersion of the metal oxides on the pore walls depends sensitively on the type of metal organic compound used for the impregnation. The degree of dispersion of the transition metal oxide determines the dispersion of the subsequently introduced molecules strongly interacting with the transition metal centers.

ACKNOWLEDGMENT

The authors express their gratitude to Dr. F. W. Deeg (University of Munich) for measuring the fluorescence spectra. The financial support of the Volkswagen Foundation (Grant No I/72134) is gratefully acknowledged.

...CES

...onowicz, C.T. Kresge, K.D. Schmitt, C.T.-W.
..., J.B. Higgins and J.L. Schlenker, J. Am.

...Vartuli and J.S. Beck, Nature, 359 (1992)

... M. Wark and A. Zukal, Microporous and

...l. Phys. Lett., 66 (1995) 1240.

7. J. Rathouský, A. Zukal, O. Franke and G. Schulz-Ekloff, J. Chem. Soc. Faraday Trans., 90 (1994) 2821.
8. J. Klaas, G. Schulz-Ekloff and N.I. Jaeger, J. Phys. Chem., B 101 (1997) 1305.
9. R. Hoppe, A. Ortlam, J. Rathouský, G. Schulz-Ekloff and A. Zukal, Microporous Materials, 8 (1997) 267.
10. G. Grubert, M. Wark, N.I Jaeger and G. Schulz-Ekloff, J. Phys. Chem., B 102 (1998) in press.

MESOPOROUS MOLECULAR SIEVES 1998
Studies in Surface Science and Catalysis, Vol. 117
L. Bonneviot, F. Béland, C. Danumah, S. Giasson and S. Kaliaguine (Editors)

Synthesis and characterization of metal incorporated H_I SiO_2 materials

S.G. Fiddy and J.M. Corker

Department of Chemistry, University of Southampton, Highfield, Southampton, Hampshire, UK, SO17 1BJ

The synthesis of metal incorporated mesoporous materials (H_I M/SiO_2 where M = V, Cr) and their characterization by PXRD, surface area measurements, TEM, ^{29}Si and ^{51}V MAS NMR, DRS, EPR, Fourier Transform Raman spectroscopy and EXAFS are reported. The metal incorporated derivatives (M = V, Cr) exhibit a hexagonal arrangement of pores analogous to that found in the pure H_I SiO_2 material. The increase of d_{100} observed by powder XRD on the addition of a metal to the synthesis mixture supports the incorporation of the metal into the silicate framework. EXAFS has also indicated that the chromium in H_I Cr/SiO_2 is attached to the silicate framework.

1. INTRODUCTION

In 1992, researchers from the Mobil Oil Company discovered a new family of mesoporous materials denoted as M41S [1,2] exhibiting channel diameters typically in the range 25-150Å. Since that time, the synthesis of a variety of hexagonal mesoporous silicates have been reported in the literature with the aim of developing new catalytic materials or utilising the tailored nanostructure to host a variety of species for potential applications as fuel cells, batteries or sensors. In 1995, the synthesis of a hexagonal mesoporous silicate, H_I SiO_2, prepared using a high surfactant concentration was reported by Attard et al [3]. The Attard method consists of a modified sol-gel process conducted in type I hexagonal mesophases formed by non-ionic surfactants. It was found that the Attard method provided a much greater degree of control over the nanostructure of the inorganic material since the long range order of the liquid crystalline phase was retained throughout the progress of the reaction. H_I SiO_2 materials typically exhibit particle sizes >300μm which is an important advantage over conventional hexagonal mesoporous materials, where particle size is typically <2μm. Transition metal incorporation into zeo-type materials is of great interest from a catalytic point of view and although there have been many reports on the incorporation of transition metals into mesoporous materials, there are no reports to date of the modification of H_I SiO_2. In the present study, we report on the heteroatom substitution of Si(IV) in H_I SiO_2 by vanadium(IV) and chromium(III) to extend the applications of these materials further. The majority of studies reported previously on modified mesoporous materials have lacked complete characterization data; we have thus used a wide variety of characterization techniques in order to establish the location and state of the heteroatom.

2. EXPERIMENTAL

H_I M/SiO_2 materials (where M = V, Cr) were prepared using tetramethyl orthosilicate (TMOS) (Fluka), octaethylene glycol monododecyl ether ($C_{12}EO_8$) (Fluka), and distilled water acidified to pH2 with sulphuric acid. The metal salt used was either vanadyl sulphate or chromium nitrate. Solution A was prepared by dissolving the required amount of metal salt (2% mol Si) with tetramethyl orthosilicate. Solution B was prepared by adding octaethylene glycol monododecyl ether to the distilled water acidified to pH2. Solutions A and B were then mixed and methanol was released from the hydrolysis of the TMOS. The methanol was removed using gentle vacuum after which condensation of the silica network occurs. The sample was then dried at 323K for 24hours and calcined at 723K in air for 24 hours followed by 24 hours in oxygen. Pure hexagonal mesoporous silica (H_I SiO_2) was prepared using the same procedure except no metal salt was added

Powder X-ray diffraction data was recorded on a Siemens θ-2θ D5000 diffractometer using CuKα radiation with a standard scintillation counter. TEM images of the various mesoporous materials was carried out on a JEOL FX 2000 transmission electron microscope. The TEM was operated at 200kV, using spot size 2. The samples were prepared by crushing the material in distilled water and dropping the finely ground material onto a carbon grid on a copper disc. N_2 BET surface area measurements were determined following the BET method for nitrogen adsorption at 77K, the data being collected with a Micromeritics GEMINI III 2375 surface area analyser. The materials were further characterized by Fourier transform Raman spectroscopy (Perkin Elmer System 2000 NIR FT Raman spectrometer, Nd-YAG laser with a wavelength of 1064nm), EPR spectroscopy (Bruker ECS106 X-band spectrometer operating at a microwave frequency of 9.35GHz and a temperature of 298K) and solid state diffuse reflectance spectroscopy (DRS) using a Lambda 19 UV/Vis/NIR spectrometer set in KUBELKA-MUNK mode. Solid state MAS NMR spectra were recorded at room temperature on a Bruker AM300 NMR spectrometer using either 7mm zirconia or 5mm boron nitride rotors. ^{29}Si : 59.6MHz, spinning speed: 4kHz. The lines were referenced to tetramethyl silane. ^{51}V: 79.1MHz, spinning speed: 14kHz. The spectra were recorded in power mode due to the difficulty of phasing the spectrum and chemical shifts are given relative to external vanadyl chloride ($VOCl_3$). X-ray absorption spectra were recorded on Station 7.1 of the Synchrotron Radiation Source at the Daresbury Laboratory, U.K. (operating at 2GeV with an average current of ca. 180mA, using a Si(111) monochromator). Spectra were recorded at room temperature in fluorescence mode using a single fluorescence detector (Na/Tl scintillation counter). Samples were housed in thin aluminium cells equipped with Kapton windows.

3. RESULTS AND DISCUSSION

The greater degree of control over the nanostructure of these materials allows an element of reproducibility into the synthesis procedure. This makes the synthesis of these materials much easier and quicker than the conventional mesoporous materials. The surfactant is completely removed by calcination in air and oxygen with no degradation of the structure. Alternatively, the surfactant can be removed by solvent extraction but this does lead to some degradation of the structure.

3.1 Powder X-ray Diffraction.

Powder XRD patterns for calcined H_I V/SiO_2 and H_I Cr/SiO_2 are shown in Figure 1 and are typical of the patterns seen for mesoporous materials. Analysis of the X-ray diffraction patterns show that both the as-synthesised and calcined materials exhibit a very strong peak between 33Å and 43Å. The d values (Table 1) are in good agreement with those reported by Attard et al [3] for pure H_I SiO_2. The d values for both H_I M/SiO_2 materials are larger than for pure H_I SiO_2, which indicates a degree of metal incorporation into the structure.

Table 1
Synthesis and characterization data of the reported as-synthesised and calcined materials

Sample	Metal Salt Used	d-value [100] for as-synthesised material (Å)	d-value [100] for calcined material (Å)	N_2-BET surface area (m^2/g)
H_I SiO_2	-	-	33.7	1400
H_I V/SiO_2	Vanadyl Sulphate	39.4	33.9	1208
H_I Cr/SiO_2	Chromium Nitrate	42.4	34.6	1176

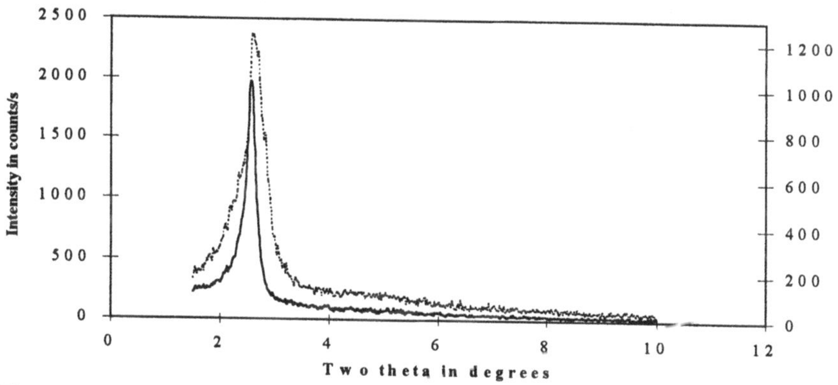

Figure 1: X-ray powder diffraction pattern of calcined H_I Cr/SiO_2 (solid line, left hand scale) and H_I V/SiO_2 (dotted line, right hand scale).

3.2 Transmission Electron Microscopy.

The transmission electron micrographs show that these materials possess a hexagonal array of pores with a typical hole to hole distance of approximately 40Å. Figure 2A shows a view of H_I Cr/SiO_2 where the beam is parallel to the length of the tubules whereas figure 2B shows a view where the beam is perpendicular to the length of the tubules. According to Chenite et al [4], the lines are consistent with a fringe system representing a hexagonal array of very long silicate tubules rather than a lamellar phase. The equidistance between the darker regions is $a\sqrt{3}/2$, where a is the hole to hole distance. Therefore, according to figure 2B, the approximate hole to hole distance for H_I Cr/SiO_2 is 38.5Å which is consistent with the PXRD analysis and the hole to hole distance from

Figure 2A. The TEM micrograph of H_I V/SiO_2 (Figure 3) also shows a hexagonal array of holes with an approximate hole to hole distance of 39Å.

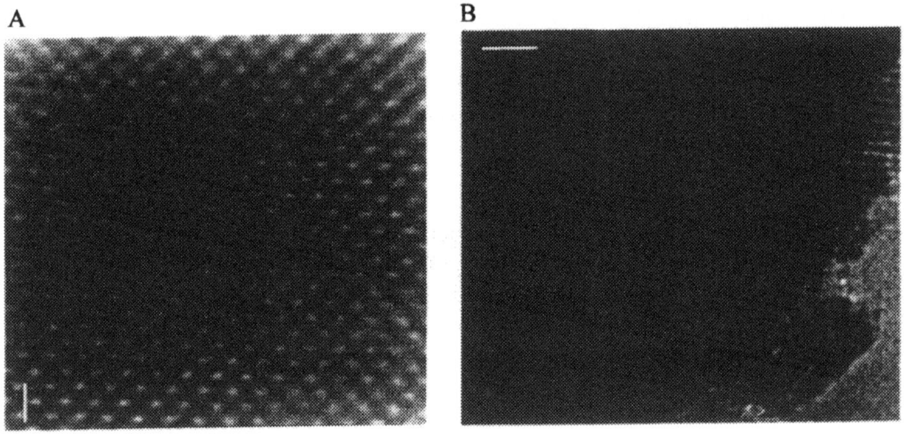

Figure 2: TEM micrographs of H_I Cr/SiO_2.

Figure 3: TEM micrograph demonstrating the hexagonal array of pores of H_I V/SiO_2.

3.3 Solid State MAS NMR Spectroscopy.

The ^{29}Si and (where appropriate) ^{51}V MAS NMR were obtained for the as-synthesised and calcined materials. For the as-synthesised H_I Cr/SiO_2, ^{29}Si MAS NMR indicates the presence of three different Si environments (Figure 4). The ratio of the Q2 [$Si(OSi)_2(OH)_2$], Q3 [$Si(OSi)_3(OH)$] and Q4 [$Si(OSi)_4$], represented by the peaks at -95, -104 and -111ppm respectively, is approximately 1 : 6 : 4. After calcination at 450°C, the ratio changes to 1:5:7 and the number of Q3 groups decreases as expected [5]. ^{29}Si MAS NMR also shows that a small percentage of Q2 groups remain after calcination suggesting the possible presence of $Si(OSi)_2(OCr)_2$ groups and therefore, implying the incorporation of chromium into the silicate framework. The ^{29}Si MAS NMR spectrum for as-synthesised H_I V/SiO_2 shows three bands at -94, -103 and -112ppm in the approximate ratio of 1: 7: 5 The spectrum for the calcined material shows an increase in the number of Q4 environments and a decrease in the number of Q3 environments with no resolvable bands that could be associated with vanadium in the framework unlike the H_I Cr/SiO_2 material. It is suggested by Steel et al [5] that a Q3/Q4 ratio of approximately 1 suggests a wall composed of two Q4 layers bordered by one Q3 layer. The ^{51}V MAS NMR of H_I V/SiO_2 indicates that vanadium pentoxide is present. However, the ^{51}V MAS NMR spectrum also exhibits two signals around –500 and –525ppm which can be attributed to V(V) in a tetrahedral coordination [6].This suggests that a percentage of the vanadium is indeed attached to the silicate framework.

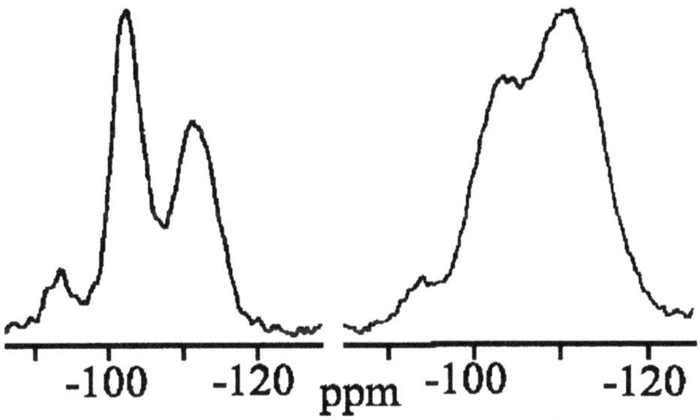

Figure 4: ^{29}Si MAS NMR spectra of as-synthesised (left) and calcined (right) H_I Cr/SiO_2

3.4 Diffuse Reflectance Spectroscopy.

Three absorption bands centred at 576, 416 and a broad shoulder at 320nm dominate the DRS spectrum in the UV-Visible region of as-synthesised H_I Cr/SiO_2. According to Weckhuysen et al [7], the three absorption bands are typical for $[Cr(H_2O)_6]^{3+}$ species which has bands at 588, 417 and a shoulder at 303nm. When H_I Cr/SiO_2 is calcined at 450°C, a colour change from green to dark yellow occurs and the DRS spectrum of the calcined material contains three main bands centred at 445(sh), 353 and 263nm. The shoulder at 445nm can be attributed to the presence of dichromate and/or polychromates species. Weckhuysen et al [7] also observed these bands for a number of different chromium containing zeolites attributing them to O →Cr(IV) in chromate and octahedral Cr(III) species. The O →Cr(IV) bands can be assigned to $1t_1 \rightarrow 2e$, $1t_1 \rightarrow 7t_2$ and $6t_2 \rightarrow 2e$ where the first band is spin forbidden and the remaining bands are all spin allowed. This suggests that there are chromate species present but it is unclear as to whether there is any attachment to the silicate framework. The DRS spectrum of as-synthesised H_I V/SiO_2 exhibits bands at 206 and 247nm. The band at 247nm can be assigned to a charge transfer band and is in the range expected for V(IV). There is also a very broad band around 700nm that can be assigned to d-d transitions. On calcination at 450°C, the band positions shift to 211 and 256nm respectively with a new band appearing ca. 292nm. The broadness of the bands indicates the presence of V(V) but also prevents accurate determination of band positions.

3.5 Electron Paramagnetic Resonance Spectroscopy.

The EPR spectrum of the as-synthesised H_I Cr/SiO_2 contains a broad isotropic signal with a width of several hundred Gauss centred at 1.98g. According to Weckhuysen and Schoonheydt [8,9], this signal is characteristic of Cr(III) in an octahedral environment which is consistent with the DRS analysis. The spectrum for the calcined material contains a broad signal typical of Cr(III) but no resolvable signal which could be associated with Cr(V). Weckhuysen and Schoonheydt [8,9] suggested that on calcination of various chromium containing zeolites, Cr(III) was partially oxidised to a mixture of Cr(V) and Cr(VI). Thus it appears that H_I Cr/SiO_2 contains predominately Cr(VI) with a small amount of Cr(III) in contrast to Weckhuysen observations. This may have an important bearing on the catalytic properties of this material. The EPR spectrum for as-synthesised H_I V/SiO_2 is characteristic of vanadyl ions in an axially symmetric crystal field and the EPR parameters are consistent with those reported by Chien et al [10] for vanadium containing zeolite beta and by Gontier and Tuel [11] for vanadium containing mesoporous silica: g(parallel) = 1.937, g(perpendicular) = 1.996, A(parallel) = 200G, A(perpendicular) = 87G. As expected on calcination, no signal is resolvable indicating the complete oxidation of V(IV) to V(V), which is EPR invisible.

3.6 Fourier Transform Raman Spectroscopy.

The Raman spectrum for calcined H_I Cr/SiO_2 shows no peaks consistent with the presence of chromium oxide or other extra framework chromium species indicating that the chromium may be attached to the silicate walls. Raman spectroscopy is a very useful technique for the analysis of vanadium containing mesoporous materials as the technique is very sensitive to impurities such as vanadium pentoxide. The Raman spectrum for calcined H_I V/SiO_2 contains three main peaks centred at 995, 700 and 281cm^{-1}, characteristic of vanadium pentoxide [12] (which is also seen by ^{51}V MAS NMR). However, no resolvable peaks could be found in the Raman spectrum that could be associated with the presence of vanadium in a framework environment.

3.7 Extended X-ray Absorption Fine Structure Spectroscopy.

EXAFS is a very useful technique for characterising metal incorporated mesoporous materials since it can give direct structural information on the environment of the metal and in some cases, can establish whether there is attachment to the silicate framework by allowing the fit of a second shell at a further distance.

Analysis of the Cr k-edge EXAFS of as-synthesised H_I Cr/SiO_2 gave a first coordination sphere of six oxygen atoms at 1.95Å (Figure 5A; Table 2). Additional shells between 3.20 - 3.40Å could be tentatively assigned to a complex mixture of silicons and chromiums. The EXAFS is therefore not consistent with the formation of chromium (III) oxide as proposed by Rey et al [13]. Calcination of the as-synthesised material resulted in a significant change in the Cr k-edge EXAFS, with analysis (Figure 5B) revealing a first coordination sphere of three oxygen atoms at 1.60Å and one oxygen atom at 1.95Å. However, the R factor was found to decrease significantly (>20%) when an additional shell of three silicons at 3.16Å was included in the fit, suggesting the attachment of the chromium to the silicate framework. The Cr-O distances of 1.60Å and 1.95Å are typical for terminal and bridging Cr-O bonds respectively. Taking the EXAFS Cr-O and Cr-Si distances to be 1.60 and 3.16Å respectively and assuming the a Si-O distance of 1.60Å, the Cr-O-Si angle can be estimated to be 160° which is larger than those observed for many chromate – silicate species which are typically between 128 and 147° although the Cr-O and Cr···Si distances are still within the ranges expected. The V k-edge EXAFS spectra of as-synthesised H_I V/SiO_2 can be best fitted to a model of one oxygen at 1.61Å, four longer oxygens at 2.01Å and one oxygen at 2.35Å. This structure is typical for vanadyl sulphate, the starting metal salt, which suggests that the vanadium species is occluded in the framework mesopores. This is consistent with the EPR characterization

Table 2.
EXAFS derived parameters of H_I Cr/SiO_2; The R-factor is defined as $\{ \int [\chi^T - \chi^E] k^3 dk\} / \{\int [\chi^E] k^3 dk\} \times 100\%$ where χ^T and χ^E are the theoretical and experimental EXAFS and k is the photoelectron wave vector; The Debye-Waller factor is given as $2\sigma^2$, where σ is the root-mean-square internuclear separation. The values given in parentheses represent the statistical errors generated in EXCURVE.

Material	Shell	Coordination number	Distance (R/Å)	R-factor (%)	$2\sigma^2$ (Å^2)
As-synthesised H_I Cr/SiO_2	O	6	1.952(4)	32.8	0.010(1)
	Si	3	3.251(4)		0.030(9)
	Cr	3	3.317(7)		0.013(2)
Calcined H_I SiO_2	O	3	1.600(2)	17.2	0.009(1)
	O	1	1.945(7)		0.010(2)
	Si	3	3.160(9)		0.025(5)

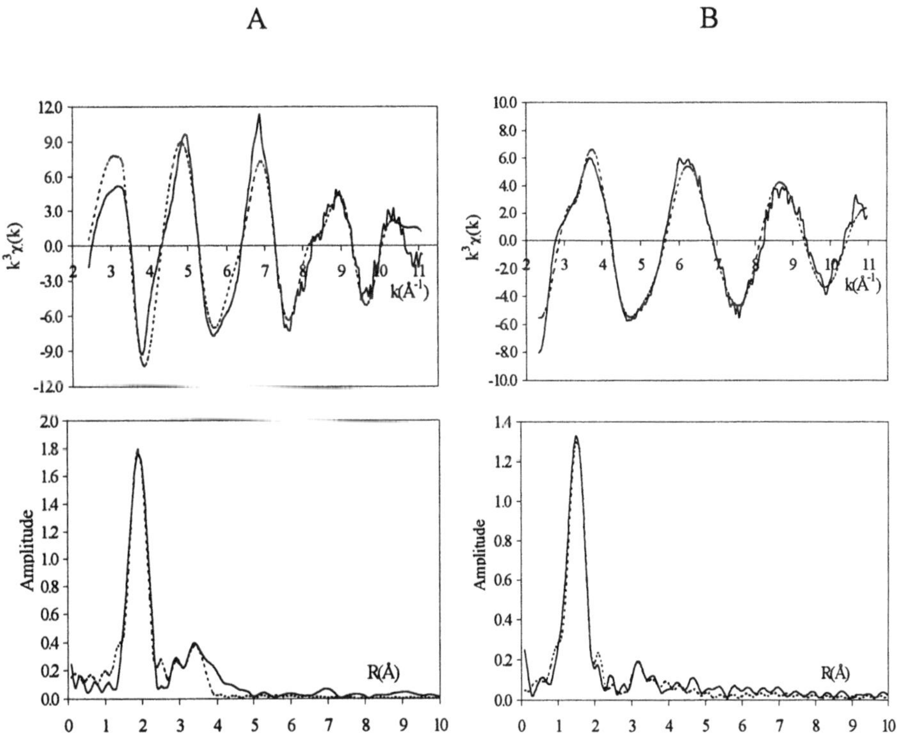

Figure 5: Cr k-edge EXAFS spectrum and Fourier transform, phase corrected for chromium, of as-synthesised (A) and calcined (B) H_I Cr/SiO_2 from experiment (solid lines) and spherical wave theory (dotted lines).

and previous EXAFS work carried out on V-MCM-41 [13]. An additional shell of silicons could not be successfully fitted to the EXAFS data suggesting that there is no attachment to the silicate framework before calcination. Analysis of the EXAFS data for calcined H_I V/SiO_2 was inconclusive, as the model for vanadium oxide could not be fitted successfully to the EXAFS data. This suggests a mixture of species, which is consistent with the ^{51}V MAS NMR data that showed the calcined material to contain vanadium pentoxide as well as tetrahedral vanadium species. Similar problems have been reported by Rey et al for other vanadium containing catalysts [13] and were thought to be due to the large variations in V-O distances.

4. CONCLUSIONS

New mesoporous chromium and vanadium containing materials were prepared using the synthesis method proposed by Attard et al [3]. These materials were found to be easier and quicker to synthesise than conventional metal incorporated mesoporous materials. Due to the larger particle size, these materials are also significantly easier to handle. H_I Cr/SiO_2 and H_I V/SiO_2 have been shown to be isostructural with the pure H_I SiO_2 material. For metal salt concentrations greater than ca. 2%mol Si, extra framework metal oxide species are formed and the importance of using a variety of techniques to identify the metal environment has been demonstrated. EXAFS, in particular, has shown the chromium in calcined H_I Cr/SiO_2 to be attached to the silicate framework suggesting a monomeric dispersed chromium species which could be an important feature with regard to the catalytic properties of this material.

ACKNOLEDGEMENTS

A research grant from the EPSRC, UK is gratefully acknowledged. The authors are indebted to Dr B. Cressey for her assistance with the TEM studies and the Daresbury Laboratory, U.K. for the provision of synchrotron radiation facilities.

REFERENCES

1. J.S. Beck, J.C. Vartuli, W.J. Roth, M.E. Leonowicz, C.T. Kresge, K.D. Schmitt, C.T.W. Chu, D.H. Olson, E.W. Sheppard, S.B. McCullen, J.B. Higgins and J.L. Schlenker, *J. Am. Chem. Soc.*, 114 (1992) 10834.
2. C.T. Kresge, M.E. Leonowicz, W.J. Roth, J.C. Vartuli and J.S. Beck, *Nature*, 359 (1992) 710.
3. G.S. Attard, J.C. Clyde and C.G. Göltner, *Nature*, 378 (1995) 366.
4. A. Chenite, Y. Le Page and A. Sayari, *Chem. Mater*, 7 (1995) 1015.
5. A. Steel, S.W. Carr and M.W. Anderson, *Chem. Mater*, 7 (1995) 1829.
6. S. Gontier and A. Tuel, *Microporous Mater.*, 5 (1995) 161.
7. B.M. Weckhuysen, H.J. Spooren and R.A. Schoonheydt, *Zeolites*, 14 (1994) 450.
8. B.M. Weckhuysen and R.A. Schoonheydt, *Stud. Surf. Sci. Catal.*, 84 (1994) 965.
9. B.M. Weckhuysen and R.A. Schoonheydt, *Zeolites*, 14 (1994) 360.
10. S-H. Chien, J-C. Ho and S-S. Mon, *Zeolites*, 18 (1997) 182.
11. S. Gontier and A. Tuel, *Stud. Surf. Sci. Catal.*, 97 (1995) 157.
12. D.C.M. Dutoit, M. Schneider, P. Fabrizioli and A. Baiker, *J. Mater. Chem.*, 7 (1997) 271.
13. F. Rey, G. Sankar, T. Maschmeyer, J.M. Thomas, R.G. Bell and G.N. Greaves, *Topics in Catal.*, 3 (1996) 121.

MESOPOROUS MOLECULAR SIEVES 1998
Studies in Surface Science and Catalysis, Vol. 117
L. Bonneviot, F. Béland, C. Danumah, S. Giasson and S. Kaliaguine (Editors)

Mesoporous molecular sieve immobilized enzymes

M. E. Gimon-Kinsel, V. L. Jimenez, L. Washmon and K. J. Balkus, Jr.*

Department of Chemistry, The University of Texas at Dallas, PO Box 830688, Richardson, Texas 75083-0688 United States

The immobilization of the globular enzyme horse heart cytochrome *c* (~30Å) in mesoporous molecular sieves was studied. Cytochrome *c* was physically adsorbed into the all silica MCM-41 (32Å), aluminosilicate MCM-41 (32Å), all silica MCM-48 (34Å) and Nb-TMS1 (33Å). The physical adsorption of the enzyme showed clear dependences on molecular sieve composition. The cytochrome *c* cannot be removed from the molecular sieves unless the pH is > 9. Silanation of the molecular sieve pore openings prevented significant leaching even at high pH values. The reduction/oxidation activities of the immobilized cytochrome *c* are also reported.

1. INTRODUCTION

The support of biologically active molecules onto mesoporous molecular sieves combines the high selectivities of enzymatic reactions with the chemical and mechanical properties of these inorganic materials. Molecular sieves offer properties such as high surface area, tailorable pore size and chemical inertness which should make them attractive materials for enzyme immobilization. The uniform pore size of molecular sieves offers a distinct advantage over more conventional sol-gel derived oxide supports. The wide range of mesopores that can be engineered allows for the adsorption of quite a variety of small biomolecules. The pores are also amenable to modification using organosilanes in order to truly entrap the enzymes. We have previously shown that the enzyme trypsin immobilized by this method in aluminosilicate MCM-41 not only maintained its activity but also showed an increase in enzyme stability [1].

The compositional and structural variances of mesoporous molecular sieves offer many interesting support properties such as insulator or semiconductor electrical behavior and one-dimensional (1-D) or three-dimensional (3-D) channel structures. The all silica and aluminosilicate MCM-41 [2] materials possess a regular hexagonal array of 1-D channels with a narrow distribution of pore sizes that can be varied between 20 to 100Å. The cubic all silica MCM-48 [2] has a well defined 3-D channel network with pore sizes typically reported to range from 19 to 36 Å. The transition metal oxide Nb-TMS1 [3] has a hexagonal mesostructure with pore sizes that can be varied between 22 to 33Å. Because transition metals may have several accessible oxidation states, mesoporous Nb-TMS1 materials could be used as an electroactive support.

The present work examines the immobilization of cytochrome *c* in a variety of mesoporous molecular sieves including all silica and aluminosilicate MCM-41, MCM-48 and

* Author to whom correspondence should be addressed

Nb-TMS1. Cytochrome *c* (specifically horse heart cytochrome *c*) is a small globular protein somewhat ellipsoidal in shape with an average molecular diameter of 30Å and a mass of 12,382 Daltons [4]. Cytochrome *c* consists of a single peptide chain containing an embedded iron heme which has an Fe(II)/Fe(III) couple that is readily characterized by cyclic voltammetry. The heme group is not centrally located, but rather it is positioned to one side of the protein at the center of a highly positively charged region which is rich in the basic amino acid lysine. Cytochrome *c* is an important biological enzyme which functions in numerous charge transport reactions [5]. In the present study the immobilized cytochrome *c* was characterized both electrochemically and spectroscopically (UV-vis).

2. EXPERIMENTAL

2.1. General Characterization Methods

Solution UV-vis spectra were obtained using Shimadzu UV-1601PC UV-vis spectrophotometer. Powder X-ray diffraction patterns were collected on a Scintag XDS 2000 diffractometer using CuKα monochromatic radiation. CaF_2 was used as an internal standard. Single point BET surface areas were calculated using a Micromeritics Flowsorb II 2000 instrument which was calibrated using a silica-alumina standard (198 m^2/g, Micromeritics lot # A-501-18).

2.2. Molecular sieve synthesis

The all silica and aluminosilicate MCM-41 materials were synthesized according to the published procedures [2, 6]. A typical preparation for the siliceous synthesis gel involved the mixing of reagents in the following molar ratios, 1 SiO_2: 0.17 $(CTMA)_2O$: 70 H_2O where CTMA is cetyltrimethylammonium ion. The source of silicon was a sodium silicate solution (27% silica, Aldrich). The resulting gel was heated under static conditions at 150°C for 24 hours. The aluminosilicate MCM-41 was prepared by mixing the reagents in the following molar ratios, 1 SiO_2: 0.1 $Al_2(SO_4)_3$: 0.26 $(TMA)_2O$: 0.27 $(CTMA)_2O$, respectively, where TMA is the trimethylammonium ion. The sources of silicon and aluminum were Cab-O-Sil M5 fumed silica (Cabot) and aluminum sulfate (Aldrich). The resulting gel was heated under static conditions at 150°C for 48 hours. The solid product was isolated by suction filtration, washed with deionized water and dried in air at room temperature. The all silica and aluminosilicate MCM-41 were calcined in air at 540°C for 24 hours to remove the organic template.

All silica MCM-48 was synthesized using a literature procedure [7]. A synthesis gel with the following molar composition, 1 TEOS: 0.25 Na_2O: 0.64 CTMACl, was placed in a polypropylene bottle and heated under static conditions at 100°C for 72 hours. The white solid product was collected by suction filtration, washed with deionized water, and dried in air at room temperature. The organic template was removed by calcination in air at 540°C for 16 hours.

The transition metal oxide mesoporous molecular sieve Nb-TMS1 was synthesized according to the published procedure [3]. A typical preparation of the synthesis gel involved the mixing of niobium (V) ethoxide (CHEMAT Technology, Inc.) and octadecylamine (Aldrich) with the molar ratio 1:X. The resulting mixture was heated at 50°C for approximately for 1 minute, then 10 ml ethanol followed by 10 ml deionized water were added which resulted in the precipitation of a white solid. The solid was aged in the supernatant for 72 hours. The white solid and supernatant were loaded into a 23 ml Teflon lined stainless steel autoclave and heated for 24 hours at 80°C, 24 hours at 100°C, followed by 7 days at 180°C. The white product was isolated by suction filtration, washed with deionized water, ethanol and diethyl ether and dried overnight at room temperature. The organic template was removed by stirring the Nb-TMS1 in a 3:1 isopropanol:water solution (acidified to pH 1.5 using nitric acid) at 80°C for 24 hours.

2.3. Adsorption and immobilization of Cytochrome *c*

Horse heart ferricytochrome *c* (>95%) was obtained as a lyophilized powder and used without further purification (Sigma). Cyctochrome *c* was adsorbed into each template-free molecular sieve by adding 100-150 mg of the molecular sieve to 5.0 ml of a 15 μM cytochrome *c* solution (pH 6 phosphate buffer) in a 12 ml centrifuge tube. The mixture was stirred at 4°C for 4 hours. The supernatant was separated from the solid materials by centrifugation and filtered through a 0.45 μm particle syringe filter (Whatman). The enzyme loaded molecular sieves were washed with deionized water and dried in air. These materials were then stirred in pH 7, 9, and 10 buffer solutions for 2 hours or overnight. The supernatant from each experiment was subjected to UV-vis spectroscopy to determine the amount of enzyme leached from the mesoporous material.

The pore openings and external surface of the all silica MCM-41 and Nb-TMS1 supported cytochrome *c* were subsequently silanated. 0.75g of the enzyme loaded molecular sieve was combined with 50.0ml of dry methylene chloride and 2.5ml of 3-aminopropyltriethoxysilane (Hüls). The mixture was stirred at 0°C for 7 hours. The silanated solid material was collected using suction filtration, washed with methylene chloride, pH 6 buffer and deionized water then air dried.

2.4. Reduction/oxidation of immobilized Cytochrome *c*

Cyclic voltammetry of the enzyme loaded molecular sieves was carried out at room temperature using a EG&G Princeton Applied Research VERSASTAT potentiostat. A standard three-electrode cell was used in the electrochemical experiments. Electrode potentials were referenced to a Ag/AgCl electrode. A platinum wire was used as the counter electrode. The working electrode was prepared by pressing a sample/graphite mixture [8] onto a platinum gauze (100 mesh, 0.25 cm^2) using a laboratory press. 4, 4'-bipyridyl (Aldrich), pH 7.0 phosphate buffer (Fisher) and sodium perchlorate were used as received.

3. RESULTS AND DISCUSSION

The adsorption of cytochrome *c* into the molecular sieve supports likely involves numerous interactions between the support surface and the exterior amino acid residues of the enzyme. These interactions can depend on the composition of the molecular sieve and the tertiary structure of the enzyme. For example, the presence of (negative) charge sites as in aluminosilicate MCM-41 can promote adsorption through electrostatic forces through (positively) charged amino acid residues on the enzyme. Hydrogen bonding to surface hydroxyl groups can also have an influence on enzyme adsorption. Both of these interactions can be altered by changes in the enzyme buffer solution pH. Therefore, cytochrome *c* adsorption studies were conducted using a variety of mesoporous molecular sieves and the enzyme-support interactions were examined as a function of pH.

Four mesoporous molecular sieves were included in the present study which have different compositions and/or structures: all silica MCM-41, aluminosilicate MCM-41, all silica MCM-48 and Nb-TMS1. Table 1 summarizes several of the physical properties of the molecular sieves. While the MCM-41 and Nb-TMS1 materials exhibited the expected powder XRD patterns (not shown) after template-removal, the measured surface areas were lower than those reported in the literature. The degree of molecular sieve crystallinity will undoubtedly affect the adsorption of cytochrome *c*. Figure 1 shows the UV-vis absorption spectra of the buffered solution before and after adsorption into the all silica MCM-41.

Table 1
Physical properties of mesoporous molecular sieves

Molecular sieve	Structure	Pore size (Å)[a]	Surface area (m^2/g)	
			Calcined	Enzyme loaded
Silica MCM-41	1-D channels	32	300 (>700[b])	290
Aluminosilicate MCM-41	1-D channels	32	400 (>700[b])	380
Silica MCM-48	3-D channels	34	1120 (>700[b])	700
Nb-TMS1	1-D channels	33	86 (>600[c])	11

[a] calculated from powder XRD patterns, [b] see reference 2a; [c] see reference 3

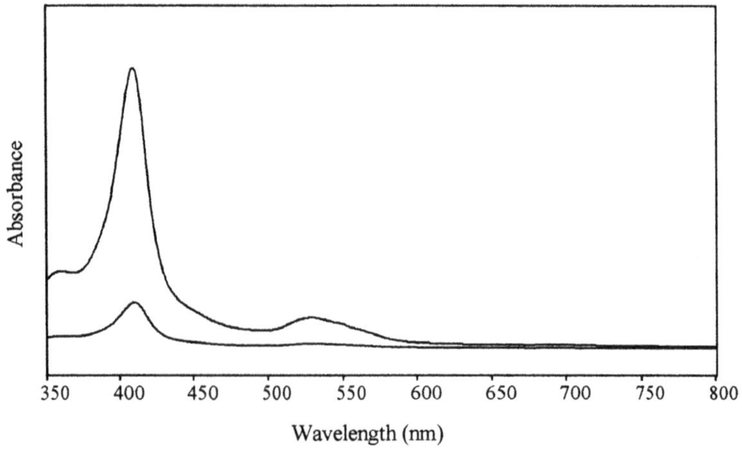

Figure 1. UV-vis absorption spectra of 15 μM cytochrome *c* buffer (pH 6.0) solution before and after adsorption into all silica MCM-41.

The amount of adsorbed enzyme was calculated from the difference in the supernatant UV-vis absorption at 409 nm (92000 M^{-1} cm^{-1}) before and after the stirring with the molecular sieve (Table 2). A decrease in surface area was measured for the enzyme loaded molecular sieves. Significantly lower surface areas were measured for the all silica MCM-48 and Nb-TMS1 while that recorded for the MCM-41 materials was slightly smaller. In the case of the 1-D channel type mesophases, one might expect lower loading and smaller changes in surface area compared with MCM-48 because of diffusional issues. The enzyme loading estimated from UV-vis spectroscopy and the molecular sieve surface area before enzyme adsorption provide an estimate of the enzyme coverage. For example, 8.7 mg of cytochrome c was loaded per 1.0 g of the all silica MCM-48. Neglecting the slight elliptical shape of cytochrome *c*, a monolayer of this enzyme would consist of approximately 2 x 10^{18} molecules. This value is very close to that calculated from the cytochrome *c* loading of 4.2 x 10^{17} molecules for MCM-48.

Table 2
Cytochrome *c* immobilization as a function of pH

Molecular Sieve	Cyt. *c* solution conc. (μM)	Cyt. *c* loading ($mg/g_{molecular\ sieve}$)	Cyt. *c* leaching (μM) pH 7	pH 9	pH 10
Silica MCM-41	15	6.2	-[a]	0.02	0.08
Aluminosilicate MCM-41	15	5.7	0.44	0.66	0.73
Silica MCM-48	15	8.7	-[a]	0.02	0.04
Nb-TMS1	15	6.2	0.03	-[a]	0.46

[a] below detection limit

The effect of pH on cytochrome *c* adsorption into the various molecular sieves was also examined and the results are also shown in Table 2. The isoelectric point of cytochrome *c* is at pH 10 [5]; therefore, the enzyme is positively charged below this value. The order of stability in terms of enzyme retention is all silica MCM-48 ~all silica MCM-41 > Nb-TMS1 > aluminosilicate MCM-41. At pH $\leq$10, no appreciable enzyme (~0.3-0.5%) is leached out of the all silica MCM-41 or MCM-48 after 2 hours. Similarly, trace amounts of enzyme (~3%) leached out of the Nb-TMS1 at pH <10. The aluminosilicate MCM-41 lost the most enzyme (~3-5%) regardless of pH. The strong adsorption of the cytochrome *c* to the all silica supports suggests that the enzyme-support interaction is dominated by hydrophobic/hydrophilic properties rather than electrostatic interactions. The reduction of the pore size through silanation prevents significant loss of the cytochrome *c* even for extended exposure to high pH buffer solutions (results not shown). As demonstrated previously [1], the organosilanes employed for pore reduction are not expected to react with (i.e., silanate) the adsorbed enzyme.

The cyclic voltammograms of the Nb-TMS1 immobilized cytochrome *c* obtained in the presence of a 4, 4'-bipyridyl mediator indicate two apparent quasi-reversible processes (Figure 2A). Two cathodic peaks, P_{C1} and P_{C2}, were found at -6mV and +87mV vs. Ag/AgCl. Two anodic peaks, P_{A1} and P_{A2}, were found at +49mV and +140mV vs. Ag/AgCl. The redox potential of the iron heme has been reported at 260mV vs. SHE [9]. These peaks are due to the reduction and oxidation of the (as received) cytochrome *c* as shown in Figure 2B; the reduction of the 4, 4'-bipyridyl occurs at much lower potentials around -0.86V vs. Ag/AgCl and do not interfere. The separation between the reduction and oxidation peaks increased when the scan rate was increased from 50mV/s to 100mV/s. This suggests that the reduction/oxidation of the molecular seive immobilized cytochrome *c* is not a diffusion controlled process. This may indicate an advantage of the mesoporous molecular sieves as a support over self-assembled monolayers which show diffusion dependence [10]. In previous studies on the electrochemistry of cytochrome *c* at solid electrodes, the 4, 4'-bipyridyl functioned as a mediator between the electrode surface and active site of the

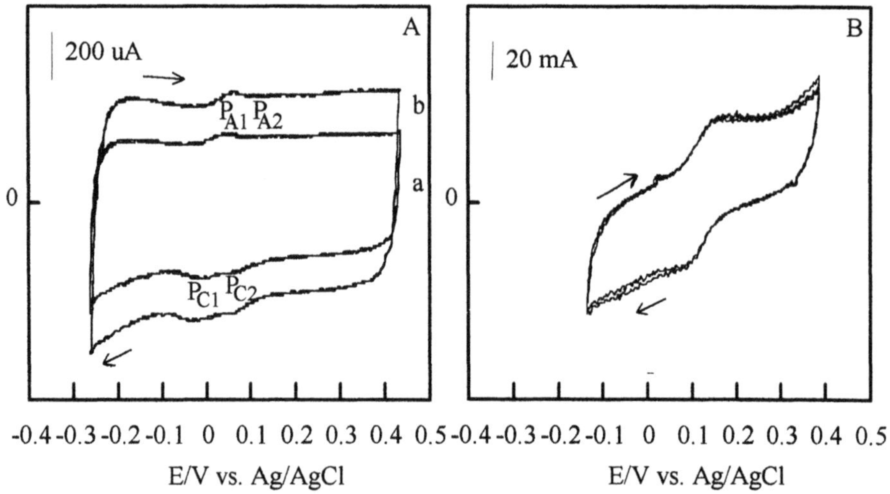

Figure 2. (A) Cyclic voltammogram of cytochrome *c* immobilized in Nb-TMS1 in $NaClO_4$ (0.1mM), pH 7.0 phosphate buffer in the presence of 4, 4'-bipyridyl (10 μM) in the potential range +0.4 to -0.3V vs. Ag/AgCl, sweep rate (a) 50 mV/s and (b) 100 mV/s. (B) Cyclic voltammogram of 0.38 mM cytochrome *c* in the presence of 15 mM 4,4'-bipyridyl and 0.1 M $NaClO_4$.

enzyme [11]. The role of the 4, 4'-bipyridyl in the present experiment may only be to interface between the carbon and the pore openings because the enzymes likely fill the channels. However, it appears that the 4,4'-bipyridyl is necessary to observe redox activity in the molecular seive immobilized enzymes. We anticipate that the origin of the multiple redox processes will be revealed upon purification of the cytochrome *c* [12]. Nevertheless, it is clear that we have immobilized cytochrome *c* in several mesoporous molecular sieves and the cytochrome *c* retains its activity.

4. CONCLUSIONS

The physical immobilization of the globular enzyme cytochrome *c* in a variety of mesoporous molecular sieves has been demonstrated. Clear differences in cytochrome *c* immobilization are observed depending on molecular sieve composition and pore size. The variability of molecular sieve crystallinity also likely plays an important role; future studies will focus on materials such as the all silica MCM-48 which can be synthesized with a high degree of order. Silanation of the molecular sieve pore openings after enzyme adsorption prevented enzyme leakage at pH>9 without significantly affecting the electrochemical behavior. The present results may encourage the application of molecular sieve immobilized enzymes in areas such as electrochemical biosensors. The immobilized enzyme is stable for at least several weeks and further long term stability studies are in progress. We are currently examining the catalytic activity of the mesoporous molecular sieve supported cytochrome *c* in other systems as well as identifying new enzymes for immobilization.

ACKNOWLEDGMENTS

We thank the National Science Foundation and the Texas Advanced Technology Program for supporting this research.

REFERENCES

1. J. F. Díaz and K. J. Balkus, Jr. J. Mol. Catal. B, Enzymatic, 2 (1996) 115.
2. (a) J. S. Beck, J. C. Vartuli, W. J. Roth, M. E. Leonowicz, C.T. Kresge, K. D. Schmitt, C. T.-W. Chu, D. H. Olsen, E. W. Sheppard, S. B. McCullen, J. B. Higgins and J. L Schlenker, J. Am. Chem. Soc., 114 (1992) 10834. (b) C. T. Kresge, M. E. Leonowicz, W. J. Roth, J. C. Vartuli and J. S. Beck, Nature, 359 (1992) 710.
3. D. M. Antonelli, A. Nakahira and J. Ying, Inorg. Chem., 35 (1996) 3126.
4. D. W. Dixon, X. Hong and S. E. Woehler, Biophys. J., 56 (1989) 339.
5. D. Durham and F. S. Millett, in R. B. King (Ed.), Encyclopedia of Inorganic Chemistry, Vil. 4, John Wiley and Sons, Chichester, 1994.
6. Z. Luan, C.-F. Cheng, W. Zhou, J. Klinowski, J. Phys. Chem., 99 (1995) 1018.
7. R.Schmidt,. Stöcker, D.Akporiaye, E.H. Torstad and A. Olsen, Micro. Mater., 5 (1995) 1.
8. F. Bedioui, E. DeBoysson, J. Devynck and K. J. Balkus, Jr., J. Chem. Soc., Faraday Trans., 87 (1991) 3831.
9. F. M. Hawkridge and T. Kuwana, Anal. Chem., 45 (1973) 1021.

10. A. E. Kasmi, J. M. Wallace, E. F. Bowden, S. M. Binet and R. J. Linderman, J. Am. Chem. Soc., 120 (1998) 225.
11. (a)P. Yeh and T. Kuwana, Chem. Letts., (1977) 1145. (b) M. J. Eddowes and H. A. O. Hill, J. Am. Chem. Soc., 101 (1978) 4461. (c) N. S. Lew and M. S. Wrighton, Science, 211 (1981) 944. (d) E. F. Bowden, F. M. Hawkridge and H. N. Blount, J. Electroanal. Chem., 161 (1984) 355. (d) L.-H. Guo and H. A. O. Hill, in Advances in Inorganic Chemistry, Vol. 36, Academic, New York, 1991,pp 341-375.
12. E. F. Bowden, F. M. Hawkridge, J. F. Chlebowski, E. R. Bancroft, C. Thorpe and H. N. Blount, J. Am. Chem. Soc., 104 (1982) 7641.

MESOPOROUS MOLECULAR SIEVES 1998
Studies in Surface Science and Catalysis, Vol. 117
L. Bonneviot, F. Béland, C. Danumah, S. Giasson and S. Kaliaguine (Editors)

Structure Modification of MCM-41, HMS and MSU-X Mesoporous Silicates by *ex-situ* Post-synthesis Hydrothermal Treatment.

Stephen A. Bagshaw

Advanced Materials Group, Industrial Research Limited, P O Box 31-310, Lower Hutt, New Zealand.

MCM-41[1,2], HMS [3] and MSU-2 [4] silicates were prepared at *ambient temperature* via electrostatic ($S^+ I^-$), anion mediated ($S^+ X^- I^+$), neutral $S^\circ I^\circ$ and non-ionic and $N^\circ X^- I^+$ assembly pathways. These materials were then restructured by *ex-situ* hydrothermal treatment in water at 373 and 423 K. Both beneficial and destructive effects are discussed.

1. INTRODUCTION

The recent development of the M41S, HMS and MSU-X mesoporous silicate materials has made available robust porous structures possessing templated pore diameters in the 2-5 nm range. Generalised syntheses to these materials utilising various inorganic precursors, templates, reaction media and crystallisation temperatures have now been developed [5]. Tailoring of the pore diameters of M41S materials was originally reported through the direct use of surfactants of increasing chain length [1]. Subsequently, pore diameters were further increased *via* the addition of supplementary organic compounds that expanded the templating micelles [1]. Using this method however, structural regularities were found to be lower than those materials prepared by direct synthesis.

Two post-synthesis treatment methods were simultaneously reported that resulted in pore diameter expansions and improved structural regularities. The first, reported by Khushalani *et al.*[6] involved increasing the synthesis temperature of the mother liquor and sample *in situ* from 343 to 423 K for 1-10 days. A model involving dissolution and redeposition of silica in areas of high curvature was proposed. Sayari *et. al.* recently described this system in some detail [7]. They described pore diameter expansion and wall thickening but discovered that the method has pore expansion limits and induces pore blocking defects. An *ex situ* method was briefly described by Huo *et al.*[8] and involved dispersing a dried, but not calcined, MCM-41 sample in water and heating over 14 days at 373 K.

In the present communication, we describe the effect of *ex situ* hydrothermal treatment on [Al]-MCM-41 materials prepared at *ambient temperature* by both S^+I^- and $S^+X^-I^+$ assembly, and on [Al]-MSU-X and [Si]-HMS materials prepared by non-ionic $N^\circ X^- I^+$ and neutral $S^\circ I^\circ$ templating pathways.

2. EXPERIMENTAL

2.1. Syntheses

[Al]-MCM-41 materials were prepared from $C_{16}(Me)_3N^+Br^-$, silicon tetra-ethoxide (TEOS), and $Al_2(SO_4)_3$, *via* the S^+I^- and $S^+X^-I^+$ assembly pathways, [Si]-HMS was synthesised from $C_{12}NH_2$ and TEOS *via* the S^oI^o assembly route at *ambient temperature* following the reaction protocols described by Zhang *et al.* [9], [Al]-MSU-2 materials were prepared using the $N^oX^-I^+$ assembly route from Triton®-N101 ($C_9Ph(PEO)_{10}$) surfactant, TEOS and $Al_2(SO_4)_3$ following the reaction pathway recently described by Bagshaw *et al.*[10].

The precipitates formed from each reaction were filtered, washed and air dried at approximately 353 K. After drying, the as-prepared materials were treated following the procedure described by Huo *et al.* [8], where a sample was loaded into a 250 mL polypropylene bottle (Nalgene®) containing distilled water (pH 7) at solid/water ratio 1:30 w/w. Each bottle was then sealed and heated in an oven at 373 K for 10 days, after which the samples were cooled, filtered, washed and dried in air. An S^+I^- [Al]-MCM41 sample was treated at 423 K to investigate the effect of the higher treatment temperature. Both untreated and hydrothermally treated samples were calcined in air at 873 K for 4 hours to removed the occluded template.

2.2. Characterization

XRD patterns were measured on a Philips PW 1700 series APD diffractometer equipped with an automatic variable divergence slit and Co K_α radiation (λ = 0.179 nm). ^{29}Si MAS NMR and Cross-polarisation (CP) MAS NMR spectra were recorded on a Varian INOVA 200 spectrometer using a DOTY Scientific multi-nuclear probe and 7mm sapphire rotors, ^{29}Si resonance frequency of 39.745 MHz. Standard Q^x notation is used for the Si speciation, where x denotes the number of Si atoms surrounding the central Si atom bridged by oxygens. ^{27}Al MAS NMR spectra were obtained with a Varian Unity 500 MHz spectrometer using a DOTY Scientific multi nuclear probe and 5 mm ZrO_2 rotors, ^{27}Al resonance frequency, 130.244 MHz. Thermal gravimetric data TGA/DTA were performed in air with a Rheometric Scientific STA 1500H analyser between 298 and 1073 K. N_2 adsorption/desorption isotherms were obtained with a Micromeritics ASAP 2010 sorptometer at 77.2 K. FTIR spectra were recorded from KBr discs between 450 and 4000 cm^{-1} using a Digilab FTS-7 FTIR spectrometer.

3. RESULTS AND DISCUSSION

3.1. Non-treated materials

3.1.1. [Al]-MCM-41

XRD patterns, ^{29}Si MAS NMR spectra and N_2 sorption isotherms of S^+I^- assembled materials were characteristic of well defined, hexagonally symmetric MCM-41 materials (Figs. 1A, 2A, 3A). Calcination produced characteristic unit cell contractions as determined from shifts of diffraction lines and increases in the intensities of X-ray diffraction lines. In the ^{29}Si MAS NMR spectra, calcination increased the intensity of the Q^4 Si resonance, while at the same time the intensity of the Q^3 signal was maintained. CP-NMR data indicated strong interactions of the template with surface Q^3 Si atoms. XRD, NMR and N_2 sorption of $S^+X^-I^+$ assembled MCM-41 also confirmed the formation of well defined hexagonal materials (Figs. 1B, 2B, 3B). ^{29}Si NMR data indicated a significantly greater contribution from fully condensed Q^4 silicons and interactions of the template with Q^2, Q^3 and Q^4 silicons. Weight losses from as-prepared samples (Table 1) were consistent with the literature, at 58 - 60 wt % for S^+I^- materials and 50 - 52 wt % for $S^+X^-I^+$ materials.

3.1.2. HMS

XRD of as-prepared HMS mesostructures (Fig. 1C) displayed strong single d_{100} peaks and broad, low intensity *hkl* shoulders characteristic of small particle, hexagonal pore structure materials. ^{29}Si MAS NMR spectra of as-prepared [Si]-HMS (Fig. 2C) showed approximately equal intensity Q^3 and Q^4 Si resonances indicating greater cross-linking of the framework than in MCM-41 materials. CP-signals were very weak, suggesting only weak interactions between surface Si atoms and neighbouring protons from the template. TG weight losses (Table 1) were between 45 and 50 wt% for the as-prepared HMS samples. N_2 sorption isotherms (Fig. 3C) were typical of HMS materials with well defined pores.

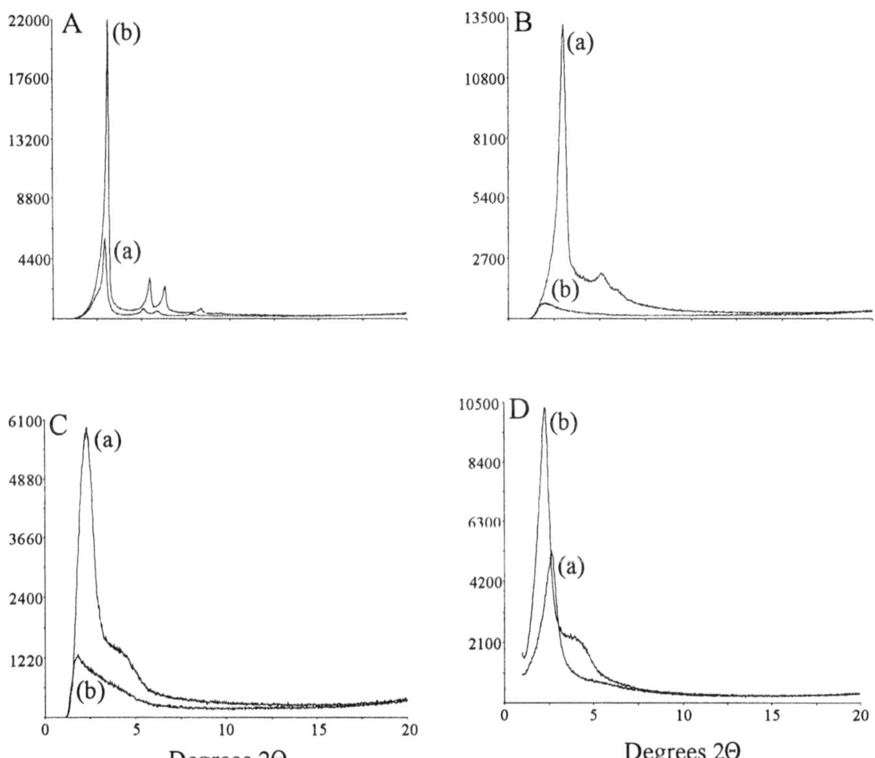

Figure 1. XRD patterns of **(a)** calcined un-treated and **(b)** calcined hydro-treated mesostructures. **A)** S^+I^- [Al]-MCM41, **B)** $S^+X^-I^+$ [Al]-MCM41, **C)** [Si]-HMS, **D)** [Al]-MSU-2.

3.1.3. MSU-X

MSU-2 samples produced single peak XRD patterns with low intensity, very broad *hkl* shoulders (Fig. 1D) indicative of uniform diameter three dimensional wormhole-like pores. ^{29}Si MAS NMR spectra (Fig. 2D) were similar to those of S^+I^- MCM-41 samples. Strong Q^2 signals however, implied that significant levels of unhydrolysed TEOS remained after synthesis. From CP spectra we inferred that, ethoxide ligands remain bound to the Si atoms in the as-prepared materials, that the [Al]-MSU-X surface was highly hydroxylated or that there were strong interactions of the PEO template with all Si species within the silica structure. N_2 sorption produced isotherms (Fig. 3D) describing materials with uniform pores in the small mesoporous region (Table 1).

3.2. Hydrothermal treatment

3.2.1. MCM-41

$S^+ I^-$ assembled MCM-41 materials exhibited shifts of X-ray diffraction lines to smaller angles, indicating an expansion of the unit cell parameter (Table 1, Fig. 1A). This expansion suggests that either the pore diameter or the wall thickness have been increased. Substantial increases in peak intensities and considerable improvements in the definitions of the *hkl* reflections were also observed. The Full-Width-at-Half-Maxima (FWHM) of all the observed reflections decreased sharply, suggesting an improved contrast between the wall structure and the pore void/template

Table 1. Physico-chemical parameters of as-prepared, hydrothermally treated and calcined mesostructures

Sample	d_{100} As- prepared (nm)	d_{100} Calcined (nm)	Weight loss (%)	BET surface area ($m^2\ g^{-1}$)	BJH ads. pore dia. (nm)	BJH ads. pore vol. ($cc\ g^{-1}$)
[a][Al]-MCM-41 [$S^+ I^-$]	4.09 (3.85)[e]	3.50 (3.15)	45 (57)	1060 (990)	2.9 (2.6)	0.89 (0.63)
[a][Al]-MCM-41 [$S^+ I^-$] (423 K)	4.45 (3.85)	3.85 (3.15)	37.5 (57)	680 (830)	2.9 (2.6)	[f]0.52 (0.63)
[b][Al]-MCM-41 [$S^+ X^- I^+$]	5.55 (3.53)	5.96 (3.55)	5.5 (51)	120 (900)	14 (2.6)	0.43 (n.o)
[c][Al]-MSU-2 [$N^oX^-I^+$]	4.65 (4.50)	4.45 (3.79)	36 (43)	910 (1100)	2.6 (2.3)	0.65 (0.21)
[d][Si]-HMS [S^oI^o]	5.56 (4.07)	6.00 (3.35)	22 (52)	280 (730)	5.3 (2.6)	0.61 (0.75)

a. 0.19 Template: 1.0 TEOS: 0.5 NaOH: 0.01 MO_2: 150 H_2O. M = Al, Ti.
b. 0.19 Template: 1.0 TEOS: 2.0 HCl: 0.05 MO_2: 150 H_2O. M = Al, Ti.
c. 1.0 Template: 8.0 TEOS: 0.04 MO_2: 400 H_2O: *x*HCl. M = Al, Ti.
d. 0.27 Template: 1.0 TEOS: 9 EtOH: 30 H_2O.
e. Values in parentheses represent data from non-hydrothermally treated samples.
f. BJH pore volume calculated from desorption isotherm owing to large hysteresis.

space. We suggest that the change in relative intensities of the d_{110} and d_{200} *hkl* lines may indicate the formation of some regions of lamellar phase during the treatment. Calcination of hydro-treated MCM-41 samples produced unit cell contractions of the same order as the un-treated materials (Table 1). TGA data showed weight losses between 40 and 50 wt% for *hydro-treated* $S^+ I^-$ assembled materials (Table 1), indicating that the template was retained within the porous structure. Some variations in response to the treatment were observed depending on the presence of substituent metals. Al substituted materials appeared to respond better than Ti substituted samples. Nevertheless, after post synthesis treatment, all S^+I^- samples showed unit cell expansions and some degree of hydrothermal restructuring as determined from the XRD patterns.

^{29}Si MAS NMR spectra (Fig. 2A), show strong Q^3 resonances along with increased intensity Q^4 signals. The indications are therefore, of continued hydrolysis and condensation of the silicate structures during treatment. CP spectra revealed that even after the hydrothermal treatment, the template remained intimately associated with the Q^3 Si atoms. These data firmly suggest that the template is strongly bound within the structure in the S^+T^- system. ^{27}Al MAS NMR spectra indicated that the Al atoms in the main part remained in Td coordination throughout the treatment.

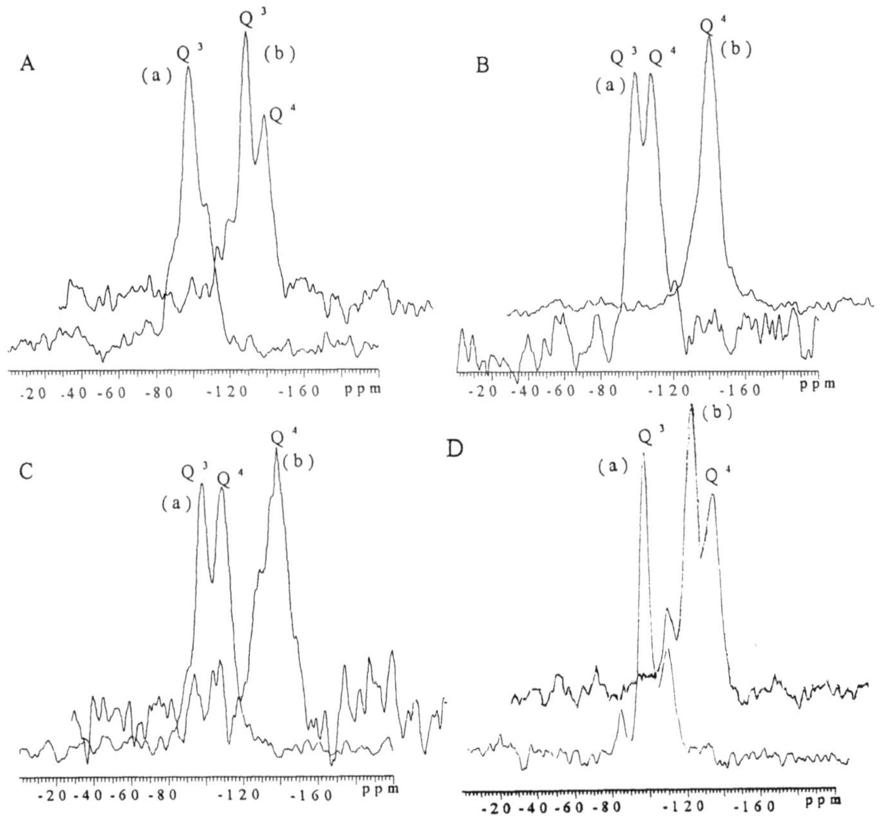

Figure 2. ^{29} MAS NMR spectra of mesostructures **(a)** before and **(b)** after, hydrothermal treatment. **A)** S^+T^- [Al]-MCM-41, **B)** S^+XI^+ [Al]-MCM-41, **C)** [Si]-HMS, **D)** [Al]-MSU-2. Spectra of hydro-treated samples are offset for clarity.

Nitrogen sorption (Fig. 3A) confirmed the formation of lamellar regions, with the appearance of well defined desorption hysteresis over wide relative pressure range. Such isotherms have been observed for super-microporous Porous Clay Heterostructure (PCH) [11] materials and are characteristic of porous systems prepared from layered phyllo-silicates.

In strong contrast to the S^+I^- system, the well defined XRD reflections of the S^+XI^+ assembled materials were completely destroyed by the hydrothermal treatment (Fig. 1B) leaving only a weak and very broad single peak. TGA gave a total weight loss of the order of only 5.5% for the hydrothermally treated samples (Table 1), indicating that the surfactant had been completely extracted from the structure. ^{29}Si MAS NMR spectra (Fig. 2B) showed virtually complete condensation of the structure to Q^4 silicon with only a very weak shoulder assignable to Q^3 silicon and almost complete loss of CP signal. Nitrogen sorption produced an isotherm (Fig. 3B) indicative of a low surface area macroporous alumino-silicate. These data therefore, strongly indicate the collapse of the ordered pore structure of S^+XI^+ MCM-41 as a result of template extraction *during the hydro-treatment*.

3.2.2. 423 K treatment

Hydrothermal treatment of S^+I^- assembled [Al]-MCM-41 at 423 K results in greater interplanar spacing expansion than at 373 K (Table 1), but both long and short-range structural degradation (Fig. 3B). TGA showed only a further 2.5 % weight loss after

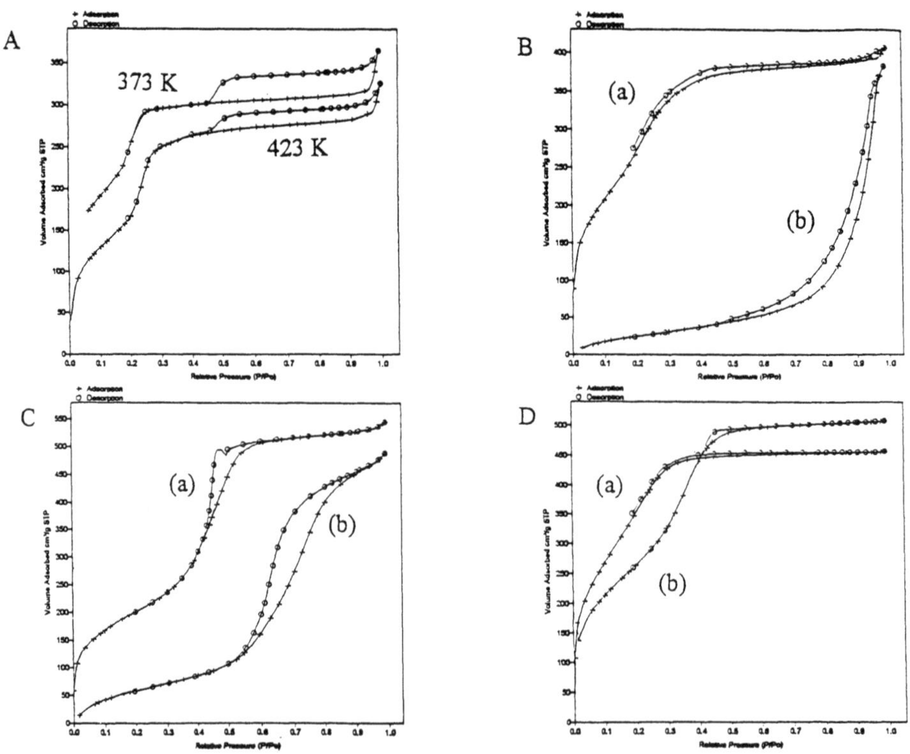

Figure 3. Nitrogen sorption isotherms of calcined **(a)** untreated and **(b)** hydrothermally treated mesostructures. **A)** S^+I^- [Al]-MCM41, **B)** S^+XI^+ [Al]-MCM41, **C)** [Si]-HMS, **D)** [Al]-MSU-2.

treatment at 423 K (Table 1). This, together with the CP MAS NMR spectra which showed only a Q^3 signal, provides further evidence that the template molecules remained bound to surface Q^3 Si atoms even during treatment at 423 K.

3.2.3. HMS and MSU-X

The response of HMS materials to the hydrothermal treatment was almost exactly the same as that of the $S^+X^-I^+$ assembled MCM-41 materials. The results will therefore not be discussed in great detail here (Figs. 1-3C). An interesting result however was obtained from nitrogen sorption. The isotherm did not show destruction of the porous structure such as that seen in the $S^+X^-I^+$ MCM-41 isotherm, and which was expected from the XRD result, but the formation of very large mesopores with a relatively narrow pore size distribution. This result, suggests that the HMS structure was able to retain some integrity inspite of the removal of template from the structure during treatment. We ascribe this to the greater hydrolytic resistance to collapse of the thicker pore walls in HMS materials [9]. MSU-X $N^oX^-I^+$ materials on the other hand, responded in a manner very much like the S^+I^- MCM-41 system. XRD patterns (Fig. 1D) of the treated materials showed, shifts of d_{100} diffractions to lower angles, intensity increases, reductions of FWHM and enhancement of the definition of higher angle *hkl* shoulders. Increased ordering of the worm-like composite pore structures and increased condensation of the silicate framework are therefore indicated. One of the more interesting observations was that the basal spacing contractions after calcination of the *hydro-treated* samples were notably reduced from those of *un-treated* samples (Table 1). This would suggest that the hydrothermal treatment favorably condenses the silicate structure to a state where thermal condensation is minimised. This in turn allows a much closer correlation to be drawn between the as-prepared d_{100} spacing and the final pore diameter.

^{29}Si MAS NMR spectra also demonstrated behaviour similar to that of (S^+I^-) MCM-41. Q^2 resonances were removed, strong Q^3 resonances were maintained and Q^4 contributions were increased (Fig 2D). CP spectra indicated well defined proton interactions with all three different Si species and in fact showed almost no change from that observed in the as-prepared state. ^{27}Al MAS NMR spectra of [Al]-MSU-2 indicated results the same as those observed for S^+I^- [Al]-MCM-41 materials wherein the Al atoms remained largely in Td coordination.

TGA of as-prepared MSU-X materials indicated weight losses of approximately 50 wt%. After hydrothermal treatment, the weight loss was reduced to approximately 38%. The 12 wt% weight loss differences compare very closely with those found for *hydro-treated* S^+I^- assembled MCM41 samples and suggest that the template is largely retained within the pore network during treatment.

N_2 isotherms show significant improvement in the size and regularity of the framework confined mesopores (Fig. 3D). Significant increases in the BET surface areas, pore volumes and importantly, pore diameters (Table 1).

3.3. Template conformational analysis

Vibrational spectroscopic analyses of the occluded MCM41 template indicated conformational changes of the alkyl-chain between the as-prepared and hydrothermally treated materials. The spectral changes in the fingerprint C-C stretching and C-H bending regions suggested the removal of *gauche* defects from the hydrocarbon tail. This results in the alkyl chain straightening out and thus forming greater space filling micelles.

4. PROPOSED MECHANISM

A model for the MCM-41 surface was recently proposed by Edler *et al.* [12] where they described the silicate wall as being constructed from two zones of varying density silicate. In this model, the interior of the pore wall is constructed from highly condensed silicate and resembles that of bulk, amorphous SiO_2. The second zone is a diffuse, possibly dendritic, region where the density of the silica decreases as it extends from the wall into the micellar region. Given this scenario, the pore surface may be thought of as being a 'corona' of low density silica which forms at a micelle/silica 'interphase' instead of a well defined interface as has been proposed to date. This zone may extend some 1.2 nm into the pore space and is not strongly differentiated by N_2 adsorption. It is however differentiated by H_2 adsorption, which appears to describe a continuous interconnected void system.

We explain our results in terms of the Edler model. We offer that the *ex-situ* hydrothermal treatment, causes the low density, uncondensed dendritic silicate at the pore/pore wall interphase to gradually redissolve under the hydrothermal conditions. At this point, only in the S^+I^- and $N^oX^-I^+$ systems, the dissolved silicate and template molecules remain intimately associated with the pore wall interphase region. As the energy of the system and the void space available increase, the template conformation changes adopting a more space filling form. Simultaneously, as the local concentration of dissolved silicate increases, it is directed by the template to reprecipitate into a more dense form at the organic/inorganic interphase. This would allow for the observed interplanar spacing expansions and also for the enhanced definition of the X-ray diffraction lines as the silicate/void interface becomes less diffuse. The mechanism of reconstruction appears to be imperfect, as both the XRD and N_2 sorption data indicate the formation of lamellar-like defects. The mechanism as described would appear to be valid for mesostructured silicates formed by all assembly pathways. As we have asserted however, for $S^+X^-I^+$ and S^oI^o assembled materials, the template is not intimately bound to the silicate surface and dissolves from the structure into the treatment medium. This has the result that the dissolved silicate cannot be re-templated and the porous structure gradually disintegrates. The thicker HMS walls resist this dissolution better than the $S^+X^-I^+$ MCM-41 walls.

We suggest therefore, that in S^+I^- MCM-41, the pore diameter and the wall thickness are simultaneously increased by hydrothermal treatment as the 'dendritic' interphase region consolidates into dense wall silicate. Furthermore, it should be possible to fine tune the pore diameter of S^+I^- assembled MCM-41 by carefully controlling the treatment temperature between 353 and perhaps 473 K and the length of the treatment. There will however, eventually be a compromise between pore expansion and structural degradation, as the intrinsic hydrolytic instability of the silicate framework would limit the real practicability of the technique.

For $N^oX^-I^+$ assembled MSU-2 materials, we believe these results raise interesting questions regarding the nature of the organic/inorganic interphase. The bonding between the template and the silica precursors has been assumed to be simply hydrogen bonding in nature at a shell-like micelle interface. This work indicates that this is clearly not the complete story. The surfaces of PEO micelles are considerably more diffuse than other surfactants, creating a corona rather than a shell as other surfactant micelles are simply described. A tentative model therefore involves the formation of a highly diffuse micelle/silicate interphase wherein the PEO segment of the template molecules is incorporated into the pore wall. As the pore surface silicate dissolves the template either remains incorporated with the walls or complexes the dissolved silicate units which allows the template to re-deposit them at the pore wall interphase in a more dense form.

REFERENCES

1. C. T. Kresge, M. E. Leonowicz, W. J. Roth, J. C. Vartuli and J. S. Beck, Nature, 359 (1992) 710.
2. Q. Huo, D. I. Margolese, U. Ciesla, P. Feng, T. E. Gier, P. Sieger, R. Leon, P. M. Petrof, F. Schüth and G. D. Stucky, Nature, 368 (1994) 317.
3. P. T. Tanev and T. J. Pinnavaia, Science, 267 (1995) 865.
4. S. A. Bagshaw, E. Prouzet and T. J. Pinnavaia, Science, 269 (1995) 1242.
5. A. Corma, Chem. Rev, 97 (1997) 2373 and references therein.
6. D. Khushalani, A. Kuperman, G. A. Ozin, K. Tanaka, J. Garces, M. M. Olken and N. Coombs, Adv. Mater., 7 (1995) 842.
7. A. Sayari, P. Liu, M. Kruk and M. Jaroniec, Chem. Mater., 9 (1997) 2499.
8. Q. Huo, D. I. Margolese and G. D. Stucky, Chem. Mater., 8 (1996) 1147.
9. W. Zhang, M. Fröba, J. Wang, P. T. Tanev, J. Wong and T. J. Pinnavaia, J. Am. Chem. Soc., 118 (1996) 9164.
10. S. A. Bagshaw, T. Kemmitt and N. Milestone, Micro. Meso. Mater. (1998) Submitted.
11. A. Galarneau, A. Barodawalla and T. J. Pinnavaia, Nature, 374 (1995) 529.
12. K. J. Edler, P. A. Reynolds, P. J. Bruton, R. F. Trouw and J. W. White, J. Chem. Soc. Faraday Trans., 93 (1997) 1667.

MESOPOROUS MOLECULAR SIEVES 1998
Studies in Surface Science and Catalysis, Vol. 117
L. Bonneviot, F. Béland, C. Danumah, S. Giasson and S. Kaliaguine (Editors)

Ordered mesoporous materials as catalyst supports

Ursula Junges[a], Stephan Disser[a], G. Schmid[b] and Ferdi Schüth[a]

[a]Institut für Anorganische Chemie; Johann Wolfgang Goethe-Universität Frankfurt; Marie Curie Straße 11; 60439 Frankfurt; Germany

[b]Institut für Anorganische Chemie; Universität Essen; Universitätsstraße 5-7; 45117 Essen; Germany

1. INTRODUCTION

The discovery of a new family of mesoporous molecular sieves, denoted M41S, has attracted the attention of many scientists working in the field of catalysis [1, 2]. The large pores of MCM-41 combined with the acidic nature of the inorganic walls were expected to catalyse the cracking of large, organic molecules. However, independent of the aluminum content, MCM-41 materials show only weak acidity, comparable to amorphous aluminosilicates. More success has been achieved when transition metal elements were introduced in the framework of MCM-41. For example, it has been found that Ti-MCM-41 shows high activity in the selective oxidation of large organic molecules.

The high surface area, the thermal stability and the presence of groups able to be functionalized makes MCM-41 very attractive as a catalysts support. Thus metals, metal oxides and organometallic compounds were supported on the high surface area MCM-41 material resulting in very high dispersions of the active phase [3-6].

This study focuses on the incorporation of nickel, cobalt, palladium, platinum and gold into the pore structure of MCM-41, partly introduced as preformed metal clusters. The catalysts were tested in the liquid phase hydrogenation of crotonaldehyde under atmospheric pressure and the CO oxidation in the gas phase.

2. EXPERIMENTAL

The catalysts were synthesized via four different pathways. In addition to conventional preparation techniques like incipient wetness impregnation (IW) and ion exchange (IE), the *in situ* incorporation (IS) of the metal during the synthesis has been studied. Moreover, catalysts with high metal dispersion were obtained by the incorporation of ligand stabilized Au_{55} and Pd_{1415} clusters. The clusters were synthesized in the size range of the well defined pores of MCM-41, so that the pore walls can act as a stabilizing macroligand.

The following preparation techniques are valid for all catalysts investigated in this study, concentrating on catalysts based on palladium for simplicity.

MCM-41 was synthesized under alkaline or acidic conditions according to Kresge, Linden and Huo et al.[7-9].

Incipient wetness impregnation and ion exchange were performed using solutions of $[Pd(NH_3)_4](NO_3)_2$ or $[Pd(NH_3)_4]Cl_2$, respectively. The impregnated samples were dried at different temperatures between room temperature and 673 K. For the ion exchange a solution of $[Pd(NH_3)_4](NO_3)_2$ was used, varying the solution-to-solid ratio, the exchange temperature and the exchange time.

The direct incorporation of palladium during the synthesis was performed as previously published with $[Pd(NH_3)_2Cl_2]$ and $Pd(acac)_2$ as a precursor [5]. All palladium catalysts were reduced in a flow of hydrogen at 673 K for a minimum of 3 hours.

The ligand stabilized cluster that matches best with the pore size of MCM-41 was the Pd7/8(phen) cluster. It was prepared according to the procedure described by Schmid et al.[10].

The cluster was dissolved in a water/pyridine mixture and loaded on MCM-41 either during the synthesis or by incipient wetness.

Generally the loadings of the catalysts were varied between 0.5 and 5 % by mass.

The catalysts were characterized by X-ray diffraction (Stoe, StadiP, CuKα radiation, λ = 0.15404 nm) in transmission mode, Nitrogen sorption experiments at 77.35 K (Micromeritics ASAP 2000) and Transmission Electron Microscopy (TEM) (Philips CM 200 FEG at 200kV). The metal content was analyzed by X-ray fluorescence (Philips, PW 1400) and by EDX in an Atomika/Amray 1920 Eco electron microscope. The thermal stability of the clusters was investigated by TG/DTA measurements in air with a heating rate of 10 K min^{-1}.

As catalytic test reactions the hydrogenation of crotonaldehyde and the oxidation of carbon monoxide have been studied.

The CO oxidation was performed with synthetic air in an open flow reactor (16 mm) at a gas hourly space velocity of 25 000 h^{-1} over pelleted and sieved catalyst particles between 500 and 1000 μm in size. Non-dispersive IR analysis was used to measure conversion.

Crotonaldehyde hydrogenation in liquid phase was carried out in a round bottom flask equipped with a condenser. At 303 K and a stirring speed of 600 rpm a hydrogen flow of 150 ml/min was allowed to bubble through a suspension of 1 g catalyst in 100 ml methanol as solvent and 10 ml crotonaldehyde. Through a septum samples were extracted and analyzed in a gas chromatograph (Siemens, Sichromat 1-4) provided with a Stabilwax column (60 m, 0.25 mm ID, Restek).

3. RESULTS AND DISCUSSION

All preparation methods result in palladium containing catalysts with the loadings dependent on the preparation technique. For incipient wetness and ion exchange aluminum containing MCM-41 with different Si:Al ratios was tested. It was found that the nominal loadings were independent of the aluminum content. Even in aluminum free MCM-41, palladium could be deposited by a process, here denoted as an equilibrium adsorption process. Ion exchange at a temperature of 333 K led to structural breakdown after 5 h. At room temperature a maximum of 5% of palladium by mass could be deposited; however, this is appreciably lower than expected for the aluminum content of MCM-41. This is due to the fact that the synthesis procedure described above only results in very little incorporation of aluminum on tetrahedral sites, as shown by pyridine adsorption followed by IR spectroscopy.

As opposed to the ion exchange method, results for the *in situ* synthesis show that 80-100% of the offered palladium was incorporated into the final catalyst.

The different influence of palladium on the pore structure of siliceous and Al containing MCM-41, and *in situ* Pd/MCM-41 respectively becomes evident from X-ray diffraction patterns. The siliceous parent material synthesized according to Linden (a in figure 1) and *in situ* Pd/MCM-41 (c in figure 1) can be indexed as (100), (110), (200), corresponding to the hexagonal unit cell. The synthesis described by Linden uses butanol as a co-surfactant and therefore results in smaller d-spacings than the traditional alkaline syntheses described by Kresge. From figure 1c, it is apparent that palladium introduced via the *in situ* route in MCM-41 provides a well-ordered material whose three peak diffraction pattern is retained as compared to a metal free sample.

The same applies to Al-MCM-41 subjected to impregnation and ion exchange. The diffraction pattern of the parent material still has three peaks, but the (100) peak shows a shoulder to lower angles. Different results were found for the siliceous MCM-41 impregnated with palladium by incipient wetness and ion exchange. Impregnation as well as equilibrium adsorption lead to materials that have just one diffraction peak located approximately at the former (100) position (b in figure 1.). This broad peak also displays a shoulder to lower angles.

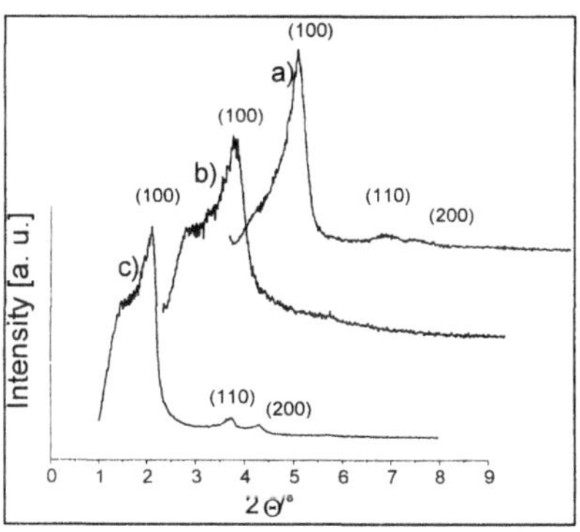

Sample	Reflection		
	(100)	(110)	(200)
a)	3.68 nm (2.40°)	2.23 nm (4.22°)	1.87 nm (4.89°)
b)	3.57 nm (2.48°)	-	-
c)	4.21 nm (2.10°)	2.37 nm (3.73°)	2.06 nm (4.30°)

Figure 1. X-ray diffraction pattern of a) siliceous MCM-41, b) IS-Pd-MCM-41 and c) IW-Pd-MCM-41.

The addition of cluster-containing solution to the synthesis or the calcined MCM-41 material do not affect the hexagonal structure of MCM-41. In all cases samples containing the hexagonal phase were obtained with at least three well resolved peaks.

The wide angle region was also investigated to gain information about the stability of clusters. The diffraction pattern does not show any distinct reflections which excludes agglomeration of the clusters.

To study the thermal stability of the clusters the samples were treated at different temperatures (293-823 K) for 1 hour. Figure 2 shows that sharp reflections can not be observed below 573 K. The broad peaks of low intensity correspond to Pd (111), (200) and (220) reflections.

At higher temperatures (823 K) reflections of PdO can be identified, indicating coalescence of the Pd7/8(phen) clusters and subsequent oxidation.

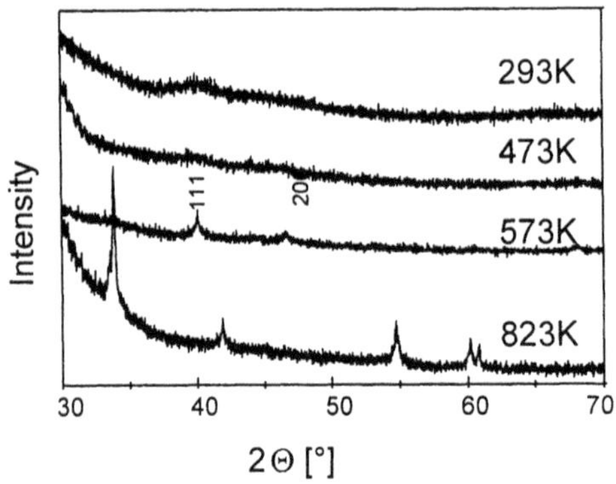

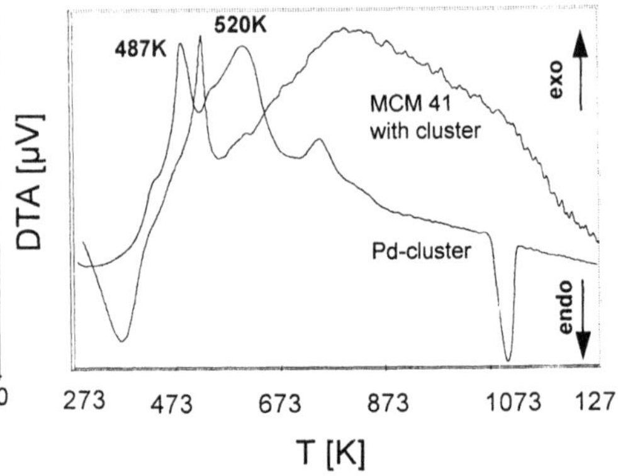

Figure 2. X-ray diffraction pattern of MCM-41 impregnated with Pd7/8(phen) clusters in dependence of temperature.

Figure 3. DTA signal of pure Pd7/8(phen)clusters in comparison with MCM-41 loaded with Pd7/8(phen) clusters.

Figure 3 shows the DTA signal of the Pd7/8(phen) clusters in comparison with the impregnated samples. The pure clusters show an exothermic peak at 487 K that probably corresponds to the loss of the stabilizing ligands. The exothermic peak of the impregnated samples is shifted to higher temperatures. This suggests strong interactions exist between the cluster and the framework of MCM-41, in effect stabilizing the clusters.

Depending on the preparation techniques, the introduced metals have different effects on the pore structure of MCM-41 as shown by nitrogen adsorption experiments.

Incorporation of nickel into MCM-41 via the *in situ* route affords materials whose nitrogen adsorption isotherms change with the amount of incorporated metal. As depicted in figure 4, all samples show a type IV isotherm according to IUPAC classification. However, depending on the nickel content of the material, a decrease of the height of the capillary condensation step in the nitrogen adsorption isotherm is observed simultaneously with a BET surface area decrease (see inset of figure 4). At 5 wt% nickel loading, the materials completely lack a distinct pore filling step. The same is observed for cobalt and palladium containing MCM-41. Obviously, the condensation process of the silicate framework is strongly influenced by the addition of metal into the synthesis solution. After exceeding a particular metal content (approximately 5% [gg^{-1}]), the pore structure is lost. Note, the still high BET surface area of 400 m^2g^{-1} is probably due to increasing textural porosity resulting from the decreasing particle size. The closure point of the type H1 hysteresis at high relative pressures (> 0.6) reflects the size of particles. This point is shifted to lower relative pressures with increasing nickel content. From this it can be concluded that the particle sizes of the material decreases with increasing metal content. This agrees with the conjecture of the condensation process being disturbed by metal complexes in these synthesis gel conditions.

If palladium is supported on MCM-41 by impregnation, the pore structure is influenced in a different way. As displayed in figure 5, the pore filling step of the nitrogen adsorption isotherm is completely removed by incorporation of the metal into the mesoporous silica.

It was shown that this effect is due to the metal and not to the catalyst preparation route: Performing the preparation with pure water in the absence of palladium had no effect on the pore structure of the parent material.

A complete pore blockage by metal particles cannot be responsible for the lack of a pore filling step in the adsorption isotherm, otherwise one would expect a dramatic decrease in BET surface area in this case. For these materials no pronounced decrease was observed. Thus it is most likely that the pores are partially blocked thus limiting the space accessible to sorptive molecules. The same effects are observed for siliceous MCM-41 materials following ion exchange.

Lastly, Al-MCM-41 shows different effects. Ion exchange and impregnation of palladium salts and palladium clusters on aluminum containing MCM-41 does not alter the sorption isotherm of the parent material.

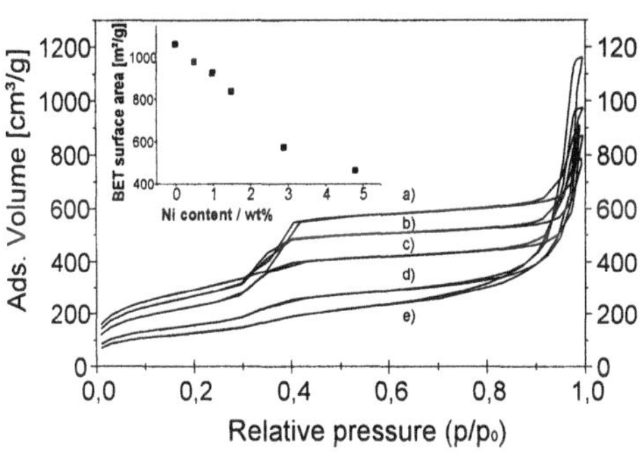

Figure 4. N_2 sorption isotherm of IS-Ni MCM-41 in dependence of the Ni content [gg^{-1}], a) 0%, b) 0.5%, c) 1.5%, d) 2.9%, e) 4.8%.

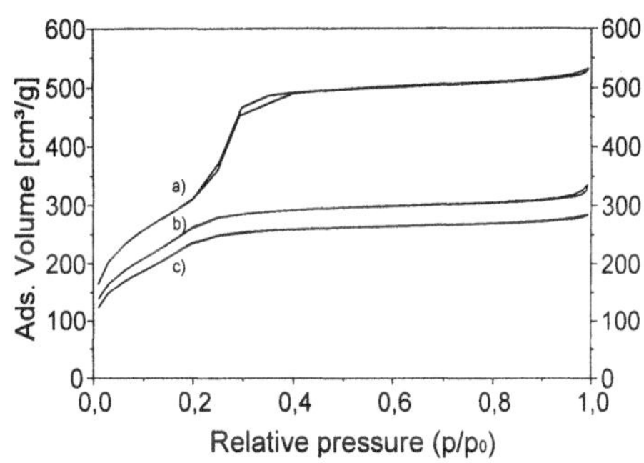

Figure 5. N_2 sorption isotherm of IW-Pd MCM-41, a) MCM-41, b) after incipient wetness, c) after catalysis.

High resolution transmission electron microscopy (HRTEM) was used to elucidate the dispersion and location of the Pd7/8(phen) cluster on MCM-41. Figure 6 shows clusters of 3-4 nm in size. No coalescence could be observed. However, the distribution of clusters is not entirely homogeneous. There are regions where the Pd7/8(phen) clusters are homogeneously distributed in the pore structure of MCM-41 as well as regions where clusters are adsorbed on the disordered amorphous surfaces, possibly due to defects introduced by the pyridine/water solution. The homogeneous distribution of the Pd clusters strongly suggests that the clusters are located in the pores and not on the external surface. Moreover, it is in these regions with the regular array of uniform channels that the clusters can interact with the mesoporous framework and thus be stabilized even after the loss of their protecting shell.

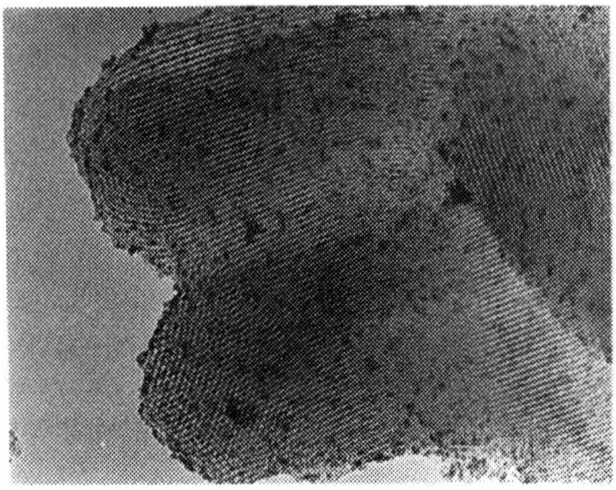

Figure 6. TEM of Pd7/8(phen) cluster distributed in the pore structure of MCM-41 [1 cm = 50 nm].

As a first catalytic test the oxidation of CO with synthetic air was analyzed. In figure 7 the performance of four catalysts, representative for each synthetic pathway, is compared. The loadings of the catalysts were approximately 3% [gg^{-1}] except for the ion exchanged sample. Here the Pd content of the tested catalysts to date has not exceeded 2% [gg^{-1}]. All catalysts show high activities in the CO-oxidation reaction. The catalysts can be divided into four different classes according to the method of preparation where the catalytic activity is different for each case, as described below.

The least active catalyst was obtained via the *in situ* incorporation of the palladium species. These catalyst reach 50% conversion at around 453 K. The palladium particle size was determined by line broadening of the Pd (111) reflection using the one line fourier analysis and/or the Scherrer equation. Results from these analysis show that very large particles of approximately 80 nm in size were present.

The catalysts prepared by the impregnation of Pd7/8(phen) cluster reach 50% conversion at 418 K. The activity of these catalysts is higher due to much smaller crystallites (4 nm).

The ion exchanged catalysts show 50% conversion at temperatures as low as 388 K. This is relatively low and can be explained by the presence of very small Pd-crystallites (< 4 nm). However, the loadings of the tested catalysts were not higher than 2%.

The best performance was achieved by incipient wetness impregnation with a solution of $[Pd(NH_3)_4](NO_3)_2$. A catalysts dried at 673 K reaches 50% conversion at 370 K. The drying temperature has no remarkable effect on the catalytic performance. The temperature deviation for 50% conversion is only around ± 5 K. The average particle size of all impregnated samples was calculated to be around 4-10 nm.

All catalysts prepared showed no agglomeration of the palladium particles after catalysis.

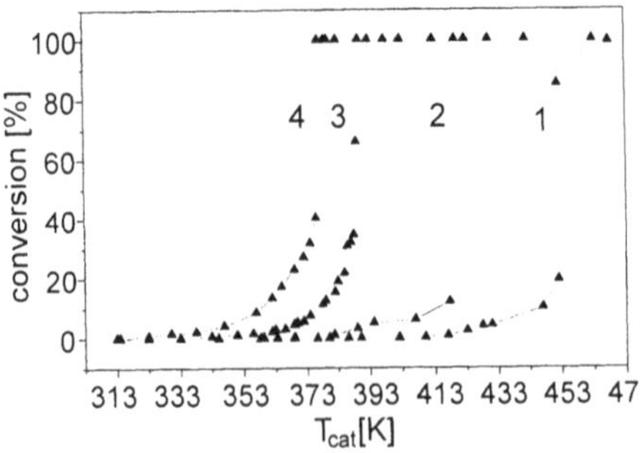

Figure 7. Activity of Pd-MCM-41, CO conversion *vs* T,
1. *in situ* ($[Pd(NH_3)_2Cl_2]$,
2. incipient wetness (Pd7/8(phen) cluster,
3. incipient wetness ($[Pd(NH_3)]_4(NO_3)_2$),
4. ion exchange ($[Pd(NH_3)]_4(NO_3)_2$).

The palladium containing catalysts were also tested in the liquid phase hydrogenation of crotonaldehyde under atmospheric pressure.

It is known that palladium has an inherent selectivity to the hydrogenation for the olefinic double bond in αβ-unsaturated aldehydes [11] which is the kinetically and thermodynamically favored reaction pathway. Thus palladium catalyzes the hydrogenation of crotonaldehyde (croHO in figure 8) to butyraldehyde (buHO) which is in equilibrium with its methylhemiacetal. BuHO can undergo alternatively a conversion to butyraldehyde dimethylacetal (buHO-acetal) or a water elimination to its methylenolether which is further hydrogenated to butylmethylether (ether). The conversion of buHO to its dimethylacetal was found to be catalyzed by acidic silanol groups of MCM-41 and the palladium surface. Ether is only formed when both hydrogen and palladium are present.

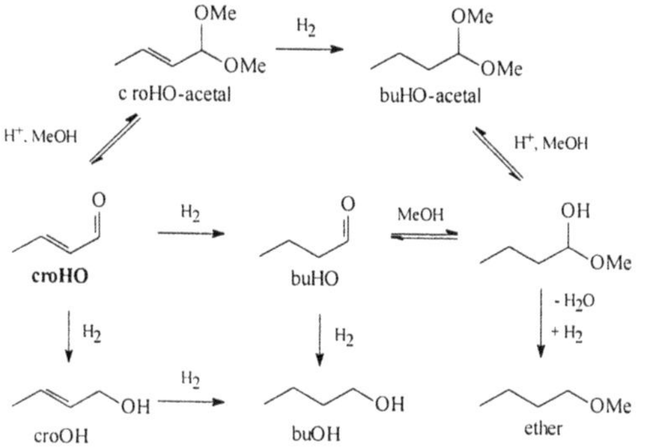

Figure 8. Reaction scheme of croHO hydrogenation in methanol.

On Pd/MCM-41 we observed the formation of buHO, its buHO-acetal, and ether. Crotyl alcohol (croOH) and butanol (buOH) are only formed in negligible amounts. The product distribution was found to be dependent on dispersion of the active component, the amount of metal loaded and the specific support used.

In order to compare the catalytic activity of Pd/SiO_2 and Pd/MCM-41 we calculated the turnover numbers of catalysts prepared by different methods for 50% of crotonaldehyde conversion. The number of active sites was derived from the particle diameter obtained by Xray broadening analysis assuming spherical particles and 50% of the metal surface being active. Figure 9 shows that the turnover frequencies increase with increasing particle size. Palladium supported on MCM-41 as well as on silica shows this behaviour. It is known that the hydrogenation of olefinic double bonds occurs preferentially on rough surfaces rich in surface steps or kinks, the overall number of which should increase with the particle diameter for the particle sizes used in this study. This explains the results given in figure 9.

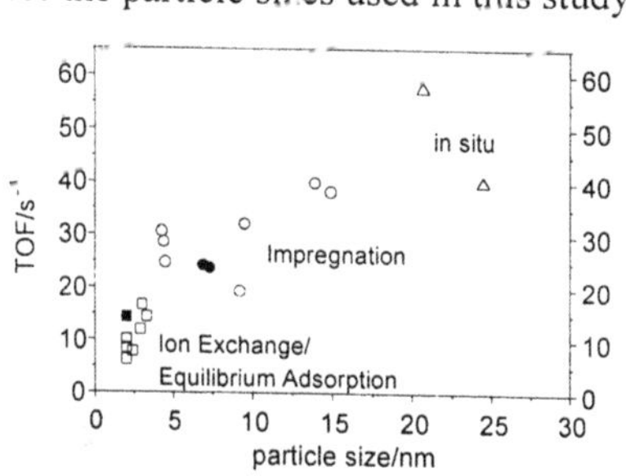

Figure 9. TOF of Pd/MCM-41 in dependence of the palladium particle size. Solid symbols refer to Pd/SiO_2.

4. CONCLUSIONS

In summary, MCM-41 and Al-MCM-41 behave differently when subjected to impregnation, ion exchange, and equilibrium adsorption of palladium, respectively. The pore structure of the MCM 41 is strongly influenced by the preparation technique while AlMCM41 is not; however, all materials retain a high BET surface area. This was confirmed by nitrogen adsorption. It was found that introducing ligand stabilized clusters into the pores of MCM-41 does not affect the pore structure and leaves the clusters intact. The palladium catalysts prepared in this study show fairly high activities in the CO-oxidation and hydrogenation of the olefinic bond of crotonaldehyde. Platinum supported on titanium containing mesoporous materials shows a higher selectivity to crotylalcohol than when deposited on pure siliceous supports, which is probably due to the SMSI effect.

It is believed that the impregnation of stabilized cluster in MCM-41 opens up new and exciting possibilities in preparing industrially important catalysts based on inorganic mesoporous supports, such as zirconia or titania [12, 13].

ACKNOWLEDGEMENTS

We would like to thank Dr. Uchida (Fritz-Haber-Institut, Berlin) for the TEM analysis, Dr. Mathiasch (Univ. of Mainz) for the X-ray fluorescence analysis and the group of Prof. Kolbesen (Univ. of Frankfurt) for the EDX measurements. Financial support by the DFG under grant number Schu 744/8-1, the BMBF and the FCI is gratefully acknowledged.

REFERENCES

1. A. Sayari, Chem. Mater., 8 (1996) 1840.
2. A. Corma, Chem. Rev., 97 (1997) 2373.
3. U. Junges, W. Jacobs, I. Voigt-Martin, B. Krutzsch and F. Schüth, J. Chem. Soc., Chem. Commun., (1995) 2283.
4. T. Maschmeyer, F. Rey, G. Sanker and J.M. Thomas, Nature, 378 (1995) 159.
5. U. Junges, F. Schüth, G. Schmid, Y. Uchida and R. Schlögl, Ber. Bunsenges. Phys. Chem., 101 (1997), 1631.
6. C.A. Koh, R. Nooney and S. Tahir, Catal, Lett., 47 (1997) 1997.
7. C.T. Kresge, M.E. Leonowicz, W.J. Roth, J.C. Vartuli and J.S. Beck, Nature, 359 (1992) 710.
8. M. Linden, S.A. Schunk and F. Schüth, Angew. Chem., in press.
9. Q. Huo, D. Margolese, U. Ciesla, P. Feng, T.E. Gier, P. Sieger, R. Leon, P. Petroff, F. Schüth and G.D. Stucky, Nature, 368 (1994) 317.
10. G. Schmid, M. Harms, J.-O. Malm, J.-O. Bovin and J. van Ruitenbeck, J. Am. Chem. Soc. 115 (1993) 2046.
11. P. Rylander, Catalytic hydrogenation in organic synthesis, Academic Press, New York, San Francisco, London, 1979, pp. 51 and 71.
12. U. Ciesla, S. Schacht, G.D. Stucky, K.K. Unger and F. Schüth, Angew. Chem., 108 (1996) 597.
13. D.M. Antonielli and J.Y. Ying, Angew. Chem., 107 (1995) 2202.

MESOPOROUS MOLECULAR SIEVES 1998
Studies in Surface Science and Catalysis, Vol. 117
L. Bonneviot, F. Béland, C. Danumah, S. Giasson and S. Kaliaguine (Editors)

Influence of alkali ion concentration on pore size distribution and vanadium content of [V]-MCM-41

B. J. Alexa Arnold, Wolfgang F. Hölderich

Department of Chemical Technology and Heterogeneous Catalysis, University of Technology RWTH Aachen, Worringerweg 1, 52074 Aachen, Germany,

The vanadium content and the d_{100} spacing of [V]-MCM-41 type materials were found to be dependent on the concentrations of $VOSO_4.5H_2O$, TEAOH (tetraethylammoniumhydroxide) and alkali metal ions Na^+ and K^+ in the synthesis gel.

1. INTRODUCTION

In 1992, Beck et al.[1] discovered a new family of mesoporous molecular sieves which was designated M41S. One member of this family of materials is MCM-41, which is built up of a hexagonal array of uniform mesopores[1]. The pore size is dependent on the chainlength of the organic template, a surfactant, thus making it possible to synthesize materials having pores in the range 15-100Å with narrow pore size distributions. In 1994, Reddy et al.[2] presented a vanadium containing analogue of MCM-41 which was an active catalyst in oxidation reactions. In this work, we describe the dependance of the planar d_{100} spacing and the V/Si ratio in [V]-MCM-41 on the cation concentration in the TEAOH-source and the TEAOH/Si ratio in the synthesis gel. Intended is a fine tuning of the pore size of [V]-MCM-41 as a catalyst in oxidation reactions.

2. EXPERIMENTAL

[V]-MCM-41 was synthesized from the gels of the following composition

$$1\ SiO_2 : x\ VOSO_4.5H_2O : 0.15\ TDTMABr : y\ TEAOH : 27\ H_2O$$

with $0.014<x<0.034$ $0.19<y<0.33$ and TDTMABr for tetradecyltrimethylammonium-bromide.

In a typical synthesis, 18% of the silica (Cab-osil M-5, Fluka) was suspended in deionised water in a polypropylene flask. TDTMABr (Fluka), an aqueous solution of TEAOH (20%, Merck; 35%, Sachem; 40%, Fluka) and $VOSO_4.5H_2O$ (Fluka) were added while stirring.

After 1 hour the remaining silica was added; stirring was continued for another 5 hours. The bottle was sealed and kept for 7 days at 373K. The product was filtered, washed with water, dried at 373K and calcined for 6 hours at 813K (heating rate $1Kmin^{-1}$).The resulting [V]-MCM-41 was analysed with XRD (Siemens D500, $Cu_{K\alpha 1,2}$) and ICP-AES (Spectroflame D), BET (ASAP 2010), and TG (Netsch STA 409).

3. RESULTS AND DISCUSSION

The XRD patterns all showed one strong and up to four weak reflections, which could be indexed as d_{100}, d_{110}, d_{200}, d_{210} and d_{300} of hexagonal phases, in agreement with the work of Beck et al.[1]. No [V]-MCM-41 could be obtained with TEAOH/Si ratios smaller than 0.21. For syntheses with sufficient amounts of TEAOH, the d_{100} of the calcined samples decreased with increasing amounts of TEAOH in the synthesis gel (Figure 1). This was possibly due to a pH difference of the synthesis gel, or to the influence the tetraethylammonium anions have on the micelle formation.

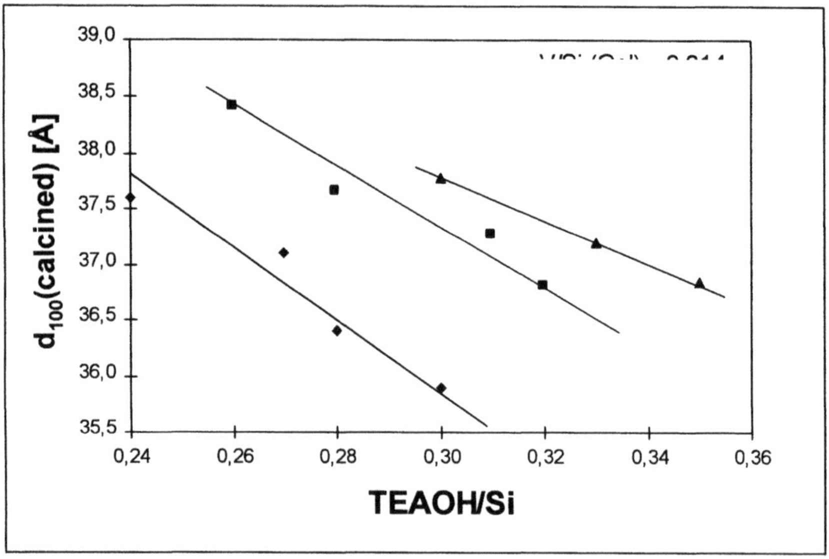

Figure 1. d_{100} of calcined [V]-MCM-41 as a function of the TEAOH-concentration at various V/Si ratios in the synthesis gel: ◆=0,014; ■=0,025; ▲=0,032.

The d_{100} spacing was also dependent on the TEAOH-source. When keeping the TEAOH/Si ratio constant a synthesis with the 40% TEAOH resulted in the smallest d_{100} spacing. The

35% and 20% TEAOH solutions gave smallest d_{100} spacings that were approximately 2Å larger.

The 40% TEAOH solution had the highest contents of sodium and potassium (390ppm and 4100 ppm, respectively); the 35% and 20% TEAOH solutions did not contain any sodium and only small amounts of potassium (3 and 8 ppm, respectively). To examine the influence of the Na and K contents in the synthesis gel on the d_{100} reflection KCl and NaCl were added to a gel containing low amounts of Na and K. Indeed the planar distances d_{100} decreased in relation to the gel containing low amounts of Na and K (Table 1.).

Table 1
Synthesis of [V]-MCM-41 using different TEAOH solutions and the influence of additional NaCl and KCl

TEAOH/Si (gel)	TEAOH 20% Merck Na <det.limit; K=8ppm			TEAOH 20% Merck (salt added) Na=84ppm; K=750ppm			TEAOH 40% Fluka Na=390ppm; K=4100ppm		
	Na/Si (gel)	K/Si (gel) $[10^{-4}]$	d_{100} [Å] calcined	Na/Si (gel) $[10^{-4}]$	K/Si (gel) $[10^{-4}]$	d_{100} [Å] calcined	Na/Si (gel) $[10^{-4}]$	K/Si (gel) $[10^{-4}]$	d_{100} [Å] calcined
0.21	0	0.3	38.5	11.4	59.0	38.0	13.9	160.4	36.2
0.22	0	0.3	37.8	11.8	61.6	37.6			
0.23	0	0.3	37.6	12.4	64.8	37.6	14.3	176.2	35.6
0.26	0	0.4	36.4	14.2	73.6	35.6			
0.31	0	0.5	35.1	17.0	88.6	35.7	19.5	241.2	34.1

det. limit = detection limit of ICP-AES

Compared to the all-silica MCM-41 of Beck et al.[1], there is an increase of respectively 6Å in d_{100} spacing and 7Å in a_0 which is also reported for the sodium containing syntheses of [V]-MCM-41 with dodecyltrimethylammoniumbromide[2] and hexadecyltrimethylammoniumbromide[3] as templates.

The vanadium uptake from the gel during crystallization of vanadium silicalites is not complete. The vanadium contents in the [V]-MCM-41 materials were found to be dependent on both the TEAOH concentration and the sources. Our Syntheses were made with three different V/Si ratios. For all these three series, the V/Si ratio in the product decreased with increasing TEAOH/Si ratio in the synthesis gel (Figure 2). At low TEAOH/Si ratios, the TEAOH-source had a strong influence on the vanadium uptake. The materials synthesized with the 40% TEAOH solution contained much more vanadium than the samples synthesised with the 35% TEAOH solution (Figure 3).

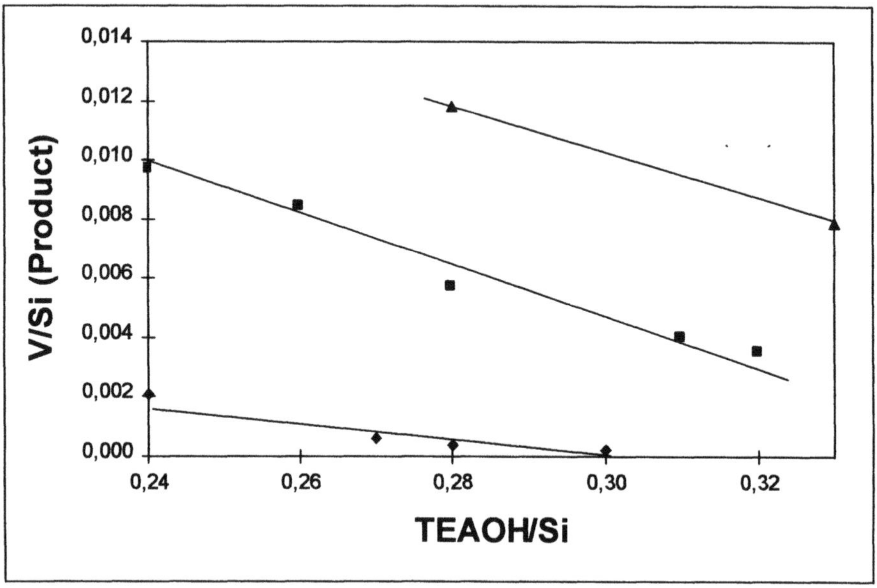

Figure 2. V/Si ratio in [V]-MCM-41 in function of the TEAOH-concentration at various V/Si ratios in the synthesis gel: ◆=0,014; ■=0,025; ▲=0,032.

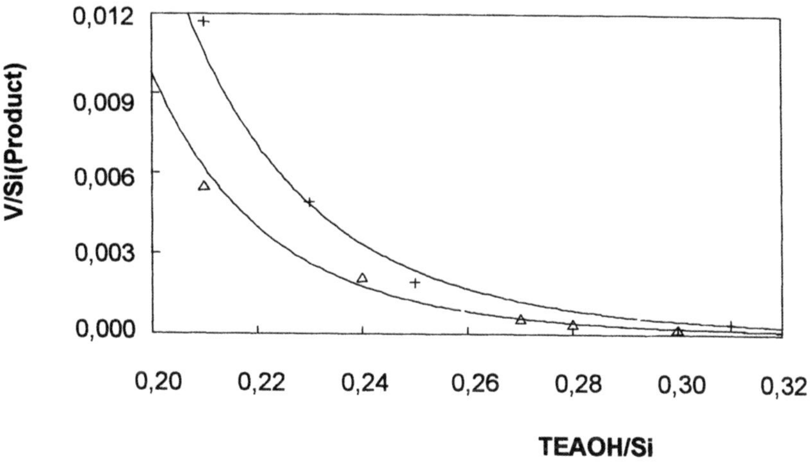

Figure 3. Influence of TEAOH content and source on the vanadium uptake (V/Si(Gel)=0,014). + TEAOH Fluka; Δ TEAOH Sachem.

4. CONCLUSIONS

An increase of the content of the alkali metal ions Na^+ and K^+ resulted in a decrease of the d_{100} planar spacing. The vanadium uptake was also found to be a function of the TEAOH source.

REFERENCES

1. J. Beck, J. Vartuli, W. Roth et al., J. Am. Chem. Soc., 114(1992) 10834.
2. K. M. Reddy, I. Moudrakovski, A. Sayari, J. Chem. Soc., Chem. Commun., (1994) 1059.
3. C. N. Wu, T. S. Tsai, C. N. Liao, K. J. Chao, Microporous Mater., 7(1996) 173.

MESOPOROUS MOLECULAR SIEVES 1998
Studies in Surface Science and Catalysis, Vol. 117
L. Bonneviot, F. Béland, C. Danumah, S. Giasson and S. Kaliaguine (Editors)

EPR and UV-visible fluorescence spectroscopic evidences for intermediate phases during the formation of micelle-templated silicas

A. Galarneau[a], D. Lerner[a], M. F. Ottaviani[b], F. Di Renzo[a] and F. Fajula[a]

[a]Laboratoire de Matériaux Catalytiques et Catalyse en Chimie Organique, UMR 5618 ENSCM-CNRS, 8 rue de l'Ecole Normale, 34296 Montpellier cedex 5, France.

[b]Dipartimento di Chimica, Università di Firenze, Via G. Capponi 9, 50121 Firenze, Italy

Electron Paramagnetic Resonance (EPR) and fluorescence techniques, combined with X-Ray Diffraction, have been used to monitor the synthesis of Micelle-Templated Structures (MTS). For the EPR study, surfactants containing nitroxide groups linked at different positions of the surfactant chain were incorporated into cetyltrimetylammonium bromide micelles. For the fluorescence study, pyrene was dissolved in micelles. Structural and mobility parameters were obtained by computer aided analysis of the EPR spectra, both before and after addition of silicate to the micellar solution over time. Analysis of the fluorescence vibronic structure provides information on microviscosity and polarity of the probe environment. These studies provide a direct insight into the modification of the micelle structure and on the dynamics of surfactant molecules during MTS synthesis.

The kinetics of MTS synthesis from silicate solutions was found to involve two main steps. In the first aggregation step, silicate species coat micelles, forming a disordered silica-micelle aggregate. The probe mobility decreases and local microviscosity increases. In the second step silica-stabilized micelles lengthen at the expense of less-stabilized ones and a hexagonal MCM-41 is formed. Cationic surfactant heads strongly interact with the condensed silica surface. In the core of micelles the anisotropic organization of the elongated micelles corresponds to a tilting of the surfactant chains.

1. INTRODUCTION

The great potential and interest of new mesoporous materials with controlled pore size in the range 15 to 100 Å and narrow pore size distribution has been widely demonstrated in literature [1]. The synthesis of these new materials used various surfactants as micellar templating agents and an inorganic source of lattice-forming elements. The interaction of inorganic units with micelles forms a hybrid organic/inorganic mesostructure, with the geometry of well-known lyotropic phases, where water is replaced by the inorganic fraction. Henceforth these mesostructures will be called Micelle-Templated Structures (MTS). The synthesis conditions are crucial to obtain MTS with different structures (hexagonal, lamellar, cubic). The present study deals with the synthesis of the hexagonal silica MTS, corresponding to the early discovered ordered MCM-41 phase [2]. The kinetic of MCM-41 formation strongly depends on the synthesis conditions and specially on the silica source and the temperature setting. Synthesis time of MCM-41 can be few minutes when the source of silica is tetraethylorthosilicate (TEOS) [3,4]. Longer synthesis times have been observed when the silica source used was a silicate solution [5]. To slow down the reaction and, therefore, to detect intermediate steps (if present) in the synthesis, we have chosen an inorganic silica source and relatively low temperature (298-333 K).

To follow modifications of micellar structure during MTS synthesis, we have used techniques widely applied in colloid chemistry, that is, Electron Paramagnetic Resonance (EPR) and fluorescence spectroscopy. EPR and fluorescent probes were dissolved in cetyltrimethylammonium bromide (CTAB) micelles. Spectral analysis allowed to measure changes in mobility, polarity and microviscosity of the probes and their environment and to correlate these changes with micelle elongation and ordering. By combining these informations with those obtained from X-Ray diffraction (XRD), a two-step mechanism was proposed, which involves the formation of a silica precipitate onto the micelle surface, which transforms into the final hexagonal MTS.

2. EXPERIMENTAL SECTION

The probes selected for the EPR study are surfactant molecules, in which the paramagnetic center is a nitroxide group. These radical-surfactants were dissolved in the cetyltrimethylammonium bromide (CTAB) micelles solution used for MTS synthesis. In order to monitor the behavior of different regions of the CTAB micelles, the probes had a nitroxide group positioned at two different sites within the surfactant molecule. The nitroxide group was in the cationic head of the surfactant in 4-cetyldimethylammonium-2,2,6,6-tetramethyl-piperidine-1-oxyl (CAT16), and near the end of the hydrophobic alkyl chain in 12-doxylstearic acid (12DXSA). The fluorescent probe was pyrene. CAT16 was dissolved in water at 313 K, whereas 12DXSA and pyrene were dissolved in chloroform. CTAB solution in water was prepared at the initial concentration of 0.11 M and mixed with the proper amount of CAT16 solution to get a molar ratio CTAB/CAT16=200. The same CTAB/12DXSA ratio was obtained by evaporating chloroform from the proper portion of 12DXSA solution and then by adding the proper amount of CTAB solution. Similarly, CTAB/pyrene=2200 (monomer probe) and CTAB/pyrene=40 (excimer probe) were prepared.

A sodium silicate solution was prepared by mixing 1.32 g SiO_2 Aerosil 200V in 20 ml NaOH 0.28M. The solution was stirred overnight at 298 K. A portion of this solution was mixed with an equal volume of 0.11M CTAB/probe solution. The addition time was taken as time zero. Samples prepared at time zero were inserted in the EPR cavity or in the fluorescence spectrometer, at the chosen temperature, and signal modifications were followed for at least 24 h. The synthesis process was repeated at different temperatures: 298, 323 and 333 K. For XRD analysis, the synthesis at each temperature was stopped at different times after CTAB-silica mixing. The resulting solid products were recovered by filtration, washed with water, dried at 353 K and analyzed by XRD.

3. RESULTS AND DISCUSSION

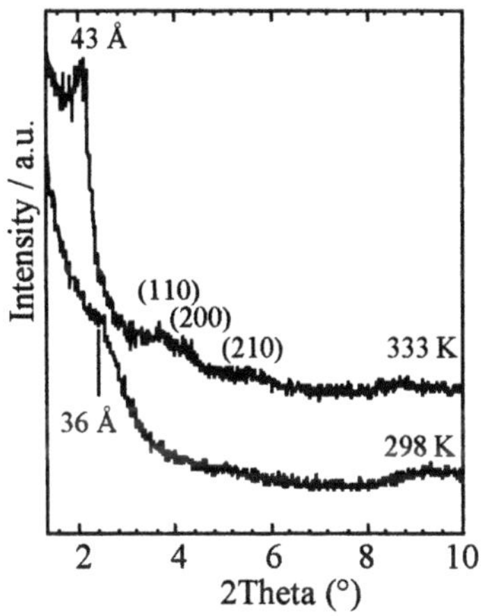

Figure 1. XRD pattern of disordered MTS synthesized at 298 K and MCM-41 type hexagonal MTS synthesized at 333 K.

3.1. XRD

XRD patterns of MTS synthesized at 298, 323, and 333 K, were recorded at different times after silicate addition. The XRD patterns of MTS synthesized at 298 K showed a broad diffraction line at ~36 Å (Fig. 1), which corresponds to a poorly ordered material. In the synthesis at 323 K, two different materials were identified over time. During the first 2 hours, a poorly ordered material was formed. It was characterized by a broad diffraction line at ~36 Å, such as for the solids synthesized at 298 K. From 2 to 3 h the d-spacing increased from 36 to 41 Å and the diffraction peaks became much sharper. At times longer than 3 h, a hexagonal structure started to form. From 5 to 24 h, the d-spacing increased slightly to 43 Å. The line width slightly decreased and the three diffraction lines due to the (110), (200) and (210) reflections of the hexagonal symmetry were clearly identified. The final material (24 h synthesis) was a long-range ordered hexagonal MTS. At 333 K, the transformation from a poorly-ordered material (peak at 36 Å) into hexagonal MTS took only 15 min. The presence of a poorly-ordered intermediate, as a first step of MTS synthesis, had already been observed by Chen *et al.* [5] for similar synthesis conditions.

3.2. Fluorescence

Pyrene is a fluorescent probe whose spectrum is sensitive to the polarity of its environment. The ratio of the intensities between the third and the first vibronic components of the pyrene emission spectrum (bands I at 373 nm and III at 385 nm) (Fig. 2), r = III/I, is a measure of the local polarity. For instance, r = 1.65 in hexane and 0.54 in acetonitrile. A value of r = 0.80 for pyrene in CTAB micelle at 298 K and 333 K was obtained, which indicated the location of pyrene in the proximity of the quaternary ammonium head groups. When the alkaline silicate solution was added, r increased to 0.87. So, it seems that the probe was slightly pushed away from the interface towards a less polar region deeper inside the micelle. This was the consequence of a shrinking of the space between trimethylammonium head groups induced by the exchange of bromide by silicate species. This tightening of the cationic layer provides the driving force for an increase of the number of surfactant molecules in each micelle and finally for the lengthening of micelles to form hexagonal arrays. The r value 0.87 remained constant during the whole MTS synthesis (1 day). Therefore, pyrene localized into the micellar core did not sensed further changes occuring during the reaction.

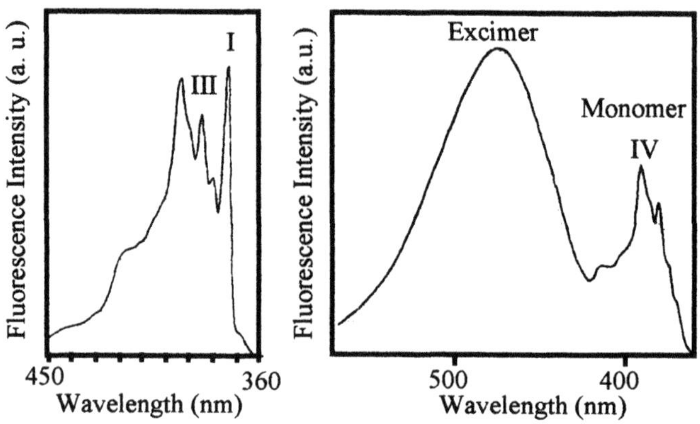

Figure 2. Typical emission spectrum of pyrene in CTAB micelles (left). The ratio between the bands I at 373 nm and III at 385 nm provides a measurement of the polarity of the environment. For higher pyrene concentration , excimers are formed and give an additional emission at 490 nm (right).

For higher pyrene concentrations, excimers were formed due to interaction of an excited pyrene with a ground state one. The ratio IV/E of emission intensities for monomer (band IV at 395 nm) and excimer (band E at 490 nm) (Fig. 2) is a direct measure of the microviscosity of the probe environment. For the synthesis at 298 K, after silicate addition to CTAB micelles, the microviscosity rapidly increased. Three quarters of the final value are reached in 3 minutes, after which the IV/E ratio slowly evolves towards an asymptotical value (Fig. 3). The marked increase in microviscosity probably resulted from the tigthening of the head groups described above, corresponding to the formation of the disordered MTS phase.

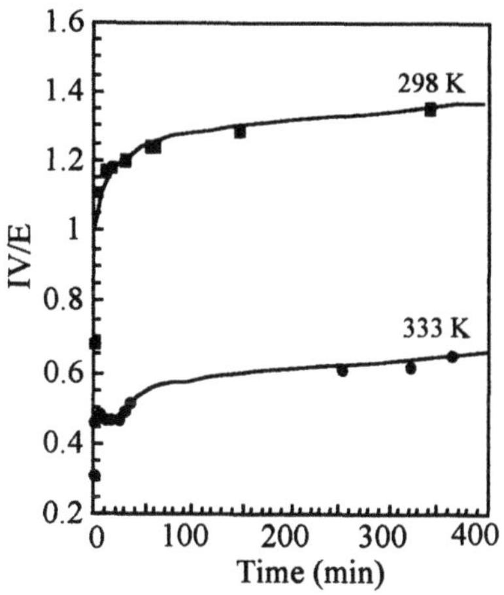

Figure 3. Microviscosity inside the micelles increases during the synthesis. Evolution of the ratio IV/E where IV is the pyrene fluorescence intensity of the monomer at 395 nm and E the fluorescence intensity of the excimer at 490 nm.

At 333 K, absolute values of microviscosity were lower due to thermal motion and, above all, the microviscosity evolution was very different. Instead of a continuous increase in IV/E, as for the synthesis at 298 K, the ratio IV/E reached a first maximum at the end of the silicate addition, after which it slightly decreased for nearly 30 minutes, and then increased again quite rapidly. In these first 30 min, the microviscosity evolution appeared to result from two opposite phenomena globally compensating each other: one decreased the microviscosity and the other increased it. By comparing this observation with XRD, we presume that the initial increase in microviscosity corresponded to the fast formation of a silica layer at the micelle surface leading to the same kind of disordered MTS formed at 298 K. The light decrease in microviscosity may be explained as destabilization of this first solid by dissolution of the silica layer at the expense of the growth of the hexagonal MTS. The subsequent microviscosity increase was likely due to the interaction with the more condensated silica of the MCM-41 type phase.

The formation of a quasi-amorphous precursor of an ordered phase is currently observed in zeolite synthesis, where an amorphous precursor material is often formed before the nucleation and crystallisation of the zeolite.

3.3. EPR

EPR results confirmed the conclusions from fluorescence results, and also provided further information on the system. The EPR spectra were recorded at 323 K, before and after addition of the silicate solution to CTAB micelles containing either CAT16 or 12DXSA. These two radical surfactants were used to monitor the evolution of two different environments: the surfactant heads interacting with silica and the region near the micellar core. The simulation of the EPR spectra allowed to calculate the correlation time for the rotational motion as a function of time (Fig. 4). To compute the 12DXSA spectra, it was necessary to introduce in the calculation a tilt of the rotational main axis which indicated an anisotropic deformation of the micelle core. Details on the calculations of EPR spectra are provided in reference [6].

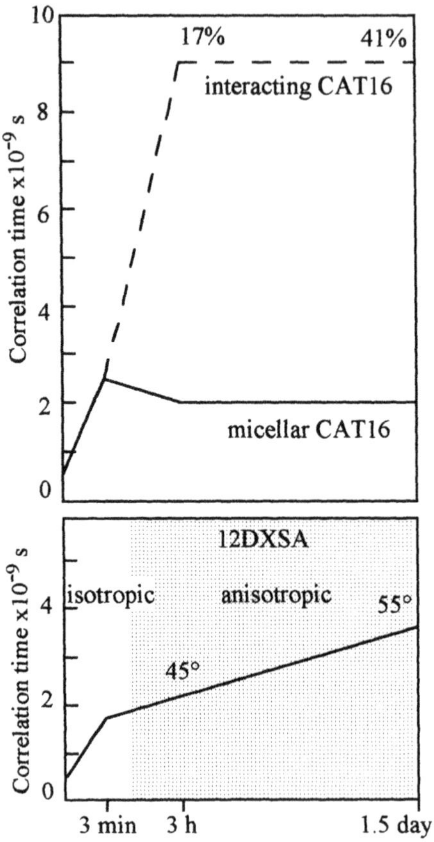

Figure 4. Evolution of correlation time of CAT16 (up) and 12DXSA (bottom). Numbers added above lines refer to the % of CAT16 interacting with the silica surface and the tilt angles of 12DXSA showing the increase in anisotropy of the micelles.

The two probes in the CTAB micelles in the absence of the silica solution showed a fast mobility (correlation time of about 0.5×10^{-9} s). In the first 3 minutes following the addition of silica solution, significant variations of the EPR line shape were found, showing a decrease in rotational mobility (correlation time of 2.5 and 1.7 $\times 10^{-9}$ s for CAT16 and 12DXSA, respectively) due to silicate-micelle interactions. This variation corresponded to the initial increase in microviscosity detected by fluorescence and to the formation of the poorly-ordered solid found by XRD. No evidences of micellar anisotropy were found in this phase, suggesting that micelles in poorly-ordered MTS are still near the isotropic shape that they present in the solution before addition of silicate. If any lengthening of micelles already took place, it was limited to a short ellipsoidal shape.

In the first 3 hours after mixing, another EPR component was identified in CAT16 spectra. This new component showed a marked decrease in rotational mobility (correlation time of 9×10^{-9} s), indicating a strong interaction between micelles and their environment. The same EPR spectrum has been observed for CAT16 directly adsorbed on calcined MCM-41. So, this component corresponded to a fraction of CAT16 directly interacting with the newly-formed silica surface. This component is designated as "interacting CAT16" in Figure 4. In the mean time, the rotational mobility of the remaining "micellar" CAT16 slightly increased (the rotational correlation time decreased to 2×10^{-9}s). This partial reversal of the trend observed for the first few minutes after the mixing of the CTAB/probe and the silicate solutions is in accordance with the results of the fluorescence experiments. Two processes occur at the same time: one increased the probe mobility - as it decreased the microviscosity from fluorescence - and the other one decreased the EPR probe mobility - as it increased the microviscosity from fluorescence. These observations confirmed the hypothesis elaborated from fluorescence, of a continuous development of the hexagonal MTS by dissolution of a poorly ordered solid. At 3 h after mixing, the hexagonal MTS was clearly detected by XRD at 323 K. At the same time, 17% of CAT16 strongly interacted with silica. The same interacting component was also recently noticed in a fast MCM-41 synthesis by Zhang *et al.* [4]. These authors used different synthetic conditions and the interacting component was immediately observed after the addition of silica precursors. This signal of CAT16 may be taken as an indicator of the progress in hexagonal MTS synthesis. At this stage of the synthesis, spectra of 12DXSA probing the micellar core had to be computed by introducing in the calculation a tilt of the rotational main axis. The tilting of the hydrocarbon chains was ascribed to the appearance of a marked anisotropy in the micelles, which is interpreted as the lengthening of the micelles from a nearly isotropic shape to a long cylindrical structure. For longer synthesis times ($t > 3h$), all parameters indicated a decrease of probe mobility: the spectral contribution of "interacting" CAT16 was growing in relative intensity at the expense of the faster contribution at $2x10^{-9}$ s (micellar CAT16). After 1.5 days, 41% of "micellar" CAT16 interacted strongly with the silica surface. This probe gave an indirect measurement of the condensation rate of the silica network.

4. CONCLUSION

EPR and UV-visible fluorescent probes are useful tools to monitor the kinetics of MTS formation. They allow to follow the structural modifications of the micelles (microviscosity and anisotropy) and the interactions between micelles and silica surface. These spectroscopic results shed some light on the precursors observed in MTS synthesis.

MTS formation rate depends on reagents and on synthesis temperature. Under our synthetic conditions, using an alkaline silicate solution as the source of silica at 298 K, only a poorly ordered structure was obtained which was characterized by a broad diffraction peak at around 36 Å. Higher temperature was necessary to synthesize hexagonal MTS. At 323-333 K XRD combined with fluorescence and EPR showed a two-step mechanism. The first step corresponds to the synthesis of a poorly ordered solid similar to the one found in the synthesis at 298 K. The anion-exchange of bromide by silicates leads to a stronger interaction of the micelles head groups with their environment, along with an increase of microviscosity. The solid formed is built by aggregation of nearly isometric micelles surrounded by silica. The second step concerns the transition from this solid precursor to the hexagonal MTS. The observed decrease in microviscosity and rotational correlation time of one micellar component corresponds to the destabilisation of the poorly ordered MTS by dissolution of the silica layer. The first phase dissolves to form the ordered hexagonal MTS, whose growth is monitored by the appearing of a new strong interaction between surfactant heads and silica, a further decrease of microviscosity and the anisotropy of the micellar core. This represents an example of the Ostwald step rule, in which a disordered solid mimicking the structure of the initial micellar solution is initially formed, and subsequently dissolves to form a more ordered and stable phase. The formation of the intermediate solid decreases the supersaturation of the system, and slows down the formation of the ordered MTS. This effect justifies the different kinetics observed when alkaline silicate solutions or TEOS are used as source of silica.

REFERENCES

1. A. Corma, Chem. Rev., 97 (1997) 2373.
2. C. T. Kresge, M. E. Leonowicz, W. J. Roth, J. C. Vartuli and J. S. Beck, Nature, 359 (1992) 710.
3. Q. Huo, D. I. Margolese, U. Ciesla, D. G. Demuth, P. Feng, T. E. Gier, P. Sieger, A. Firouzi, B. F. Chmelka, F. Schüth and G. D. Stucky, Chem. Mater. 6 (1994) 1176.
4. J. Zhang, Z. Luz, D. Goldfarb, J. Phys. Chem. B, 101 (1997) 7087.
5. C. Y. Chen, S. L. Burkett, H.-X. Li, M. E. Davis, Microporous Mater., 2 (1993) 27.
6. A. Galarneau, F. Di Renzo, F. Fajula, L. Mollo, B. Fubini, M. F. Ottaviani, J. Colloid. Int. Sci., in press.

MESOPOROUS MOLECULAR SIEVES 1998
Studies in Surface Science and Catalysis, Vol. 117
L. Bonneviot, F. Béland, C. Danumah, S. Giasson and S. Kaliaguine (Editors)

Meerwein-Ponndorf-Verley reductions mediated by lanthanide-alkoxide-functionalized mesoporous silicates

R. Anwander* and C. Palm

Institut für Technische Chemie I, Universität Stuttgart, Pfaffenwaldring 55, D-70569 Stuttgart, Germany

Lanthanide alkoxide moieties were grafted onto mesoporous silicate MCM-41 *via* a surface siloxide linkage. Two different immobilization procedures were applied, involving the treatment of dehydrated MCM-41 with rare earth alkoxide solutions and a heterogeneously performed silylamide route. The latter procedure facilitated the introduction of a wide range of alkoxide ligands *via* ligand exchange reactions of MCM-41/rare earth amide hybrid species with various alcohols. The resulting materials were characterized by elemental analysis and FTIR spectroscopy. Nitrogen physisorption measurements revealed that the effective mean pore diameter and pore volume, respectively, were markedly affected by the steric bulk of the various alkoxide ligands. The novel hybrid materials are efficient precatalysts in the Meerwein-Ponndorf-Verley reduction of 4-*tert*-butylcyclohexanone to the corresponding 4-*tert*-butylcyclohexanol isomers. The transformations were conducted in 2-propanol and conversions >95 % were accomplished within five hours even at ambient temperature, and without formation of any side products.

1. INTRODUCTION

Soluble metalorganic complexes of the rare earth elements developed into a class of highly efficient precatalyst systems in the field of both polymerization and fine chemicals [1]. However, there is a dearth of investigations on heterogenizing rare earth metal complexes [2] in order to combine the advantages of homogeneous and heterogeneous catalysis [3]. Recently, we envisaged the mesoporous silicate MCM-41 [4] as a versatile support material to examine the impact of immobilization of such catalytically relevant lanthanide species [5]. As part of this program, we present our findings of a catalytic Meerwein-Ponndorf-Verley (MPV) reduction [6], employing both molecular and grafted alkoxide species. The enhanced activity of molecular rare earth alkoxide complexes in catalytic MPV reductions compared to zirconium and aluminum alkoxides, in particular [Al(O*i*Pr)$_3$], is well documented [7] and generally related to Lewis acidity and metal size effects in combination with kinetically favored ligand exchange reactions [6]. In addition, rare earth alkoxides were reported to be quite effective in enantioselective reductions of ketones using 2-propanol [8]. Various heterogeneously catalyzed MPV reductions have been reported previously [9,10].

Given the synthetic potential of metal alkoxide precursors in sol-gel technology their utilization for the synthesis of metal-incorporated mesoporous silicas was soon envisaged [11]. Metal alkoxides were also employed for both the post-synthetic modification of the MCM-41

* Corresponding author, present address: Anorganisch-chemisches Institut, Technische Universität München, Lichtenbergstr. 4, D-85747 Garching, Germany, E-mail: anwander@arthur.anorg.chemie.tu-muenchen.de.

silicate surface, e.g., as lining agent to improve the hydrothermal stability [12], and for the synthesis of mesoporous metal oxides [13].

2. EXPERIMENTAL

2.1. General

Purely siliceous MCM-41 material was synthesized according to the literature, employing $C_{16}H_{33}N(CH_3)_3Br$ as a templating agent [4,14]. After calcination (N_2: 540 °C, 5h, heating rate 1.5 °C min^{-1}; air: 540 °C, 5h) and dehydration (280 °C, 10^{-5} torr, 4 h, heating rate 1 °C min^{-1}) the resulting material **1** was stored in a glovebox (MB Braun MB150B-G-II; <1 ppm O_2, <1 ppm H_2O) where successively all derivatization reactions and product manipulations were carried out. *n*-Hexane and THF were distilled from Na/K alloy under nitrogen. 2-Propanol (Aldrich, anhydrous) was stored for at least one week over molecular sieve (4 Å) prior to use. Hexamethyldisilazane was purchased from Aldrich and degassed (freeze-pump-thaw) before use. MCM-41,$SiMe_3$ (**2**) was prepared *via* solution impregnation with disilazane agent [5]. 4-*tert*-Butylcyclohexanone was from Aldrich and sublimed prior to use. H-Tritox, $Nd[N(SiMe_3)_2]_3$ and $Nd(tritox)_3$ were prepared according to the literature [15]. $Nd(OEt)_3$ was precipitated from a solution of $Nd[N(SiMe_3)_2]_3$ in *n*-hexane by slow diffusion of an ethanol/*n*-hexane solution.

2.2. Synthesis of the hybrid materials

Neodymium silylamide derived material **3** was synthesized according to ref. 5. Alkoxide surface species according to route *A* were prepared by reaction of a suspension of dehydrated MCM-41 **1** (0.397 g) in 15 mL of *n*-hexane with $Nd(tritox)_3$ (0.420 g, 0.57 mmol) dissolved in 2 mL of THF at ambient temperature. After stirring for 20 h, unreacted $Nd(tritox)_3$ and produced H-tritox (total 0.264 g) were removed by several washings with THF (20 mL) and *n*-hexane (50 mL). Thus obtained light-blue material **5** was dried *in vacuo* for at least 5 h at ambient temperature. Alkoxide surface species according to route *B* were obtained by addition of an alcohol (slight excess based on metal content)/*n*-hexane solution to a suspension of hybrid species **3** (0.400 g) in 20 mL of *n*-hexane. After stirring for 20 h at ambient temperature, the light-blue solids were separated by centrifugation and washed several times with *n*-hexane. The combined *n*-hexane fractions were colorless and contained no metal species. The obtained hybrid materials **4a-d** were dried *in vacuo* for at least 5 h at ambient temperature.

2.3. Characterization

The structural integrity (four-peak-pattern) of the starting MCM-41 material **1** was checked by powder X-ray diffraction on a Siemens D5000 instrument in the step/scan mode (step width: 0.01, accumulation time: 4.5 s/step, range (2Θ): 1.0 - 9.0°) using monochromatic CuKα radiation. Elemental analysis were performed on an Elementar VarioEL and an emission spectrometer plasma 400 (Perkin Elmer). FTIR spectra were recorded on a Perkin Elmer FTIR spectrometer 1760X, using Nujol mulls between CsI plates. Nitrogen physisorption measurements were performed on an ASAP 2010 volumetric adsorption apparatus (Micromeritics) at 77 K for relative pressures from 10^{-2} to 0.60 [$a_m(N_2, 77\ K) = 0.162\ nm^2$]. A Siemens routine gas chromatograph RGC202 equipped with a HP Ultra2 column (50 m; 0.2 mm; 0.33 µm film thickness) and a FID detector was employed for product identification of the catalytic MPV reductions (temperature program, 120 °C, 30 min).

2.4. General procedure of the catalytic reactions

The MPV reductions were carried out in a 100 mL round-bottomed flask equipped with a magnetic stirrer, a teflon septum and a blow-off valve. A typical experiment was the following: In a glovebox the precatalyst (1-5 mol%) and HO*i*Pr (ca. 12.5 g) were placed in the reaction vessel. Also inside the glovebox the ketone (0.78 g, 5.0 mmol) and *n*-nonane (0.1 g) as internal standard were dissolved in HO*i*Pr (ca. 12.5 g). The reaction was started outside the glovebox by injection of the substrate solution. While stirring (900 rpm) the reaction at ambient temperature, aliquots were taken and after centrifugation the course of the reaction monitored by gas chromatography. Products were identified by comparison with authentic reference compounds.

3. RESULTS AND DISCUSSION

3.1. Preparation and characterization of the MCM-41-supported rare earth alkoxide species

All hybrid materials were synthesized according to a solution impregnation technique in *n*-hexane at ambient temperature (Scheme 1). Neodymium metal complexes produce blue-colored solutions and were selected to visually follow the progress of the immobilization reactions.

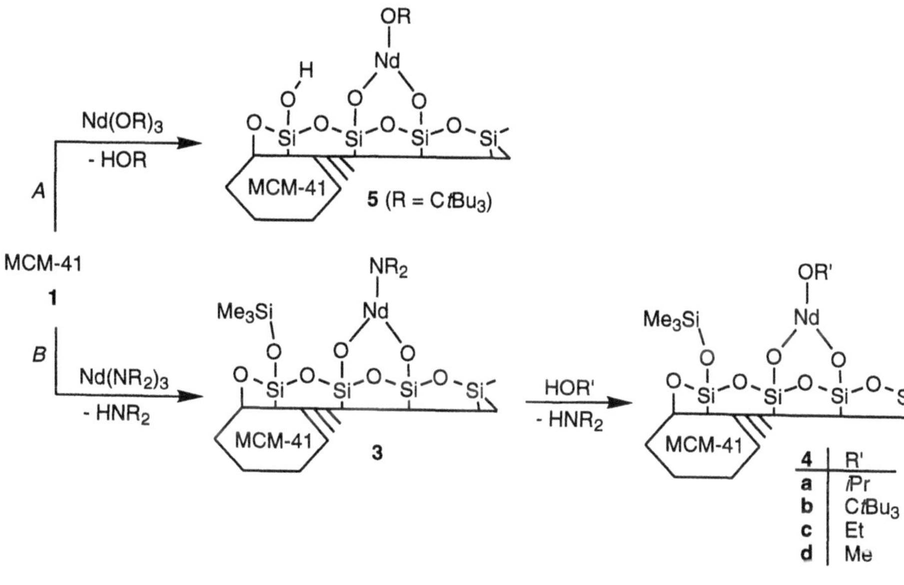

Scheme 1. Grafting of lanthanide alkoxide species onto mesoporous silicate MCM-41 according to two different procedures [$NR_2 = N(SiMe_3)_2$].

The one-step alkoxide impregnation of partially dehydroxylated MCM-41 support material **1** according to route *A* was performed with the mononuclear compound $Nd(tritox)_3$, containing the sterically very bulky ligand tri-*tert*-butylmethoxide (= tritox). The mononuclearity of this complex combined with a favorable ligand basicity (+I effect of the alkyl groups) should ease the surface reaction and suppress metal alkoxide oligomerization reactions and cluster formation on the surface. The feasibility of such alkoxide/silanol exchange reactions under the prevailing reaction conditions is supported by previous findings in homogeneous media [16]. Assuming a monofunctional surface reaction, calculations from the elemental analysis (Table 1) reveal a ligand/metal ratio of 1.4 for material **5**. Furthermore, the neodymium surface coverage can be derived as $0.4/nm^2$, considering a BET specific surface area of 1059 m^2g^{-1} for material **1**. From the amount of available isolated surface silanol groups of ca. $1.2/nm^2$ derived from a $HN(SiMe_3)_2$-silylation experiment (material **2**), it can be concluded that many SiOH surface sites are not accessible to neodymium siloxide formation. This is probably due to the steric bulk of the tritox ligand. The limited surface reaction is clearly indicated by the FTIR spectrum of material **5** which shows a sharp band at 3695 cm^{-1} attributable to a considerable amount of unmodified isolated silanol groups.

Route *B* uses a strategy reported recently by A. Choplin et al. for the system "silica/$Zr(CH_2tBu)_4$/HO*i*Pr" [9]. The initial grafting of $Nd[N(SiMe_3)_2]_3$ yielded a material **3** (Scheme 1). The bifunctional reaction of lanthanide silylamide complexes with surface silanol groups involves siloxide formation and silylation reactions which appear to be diffusion-controlled [5]. The accompanying silylation reaction implies the consumption of most of the surface silanol groups and hence significantly affects the surface hydrophobicity, producing metal centers embedded in a hydrophobic matrix of silyl groups (approximately 0.6 Nd/nm^2 and 0.6 $SiMe_3/nm^2$). The grafted silylamide species which could be identified by FTIR spectroscopy are capable of *second generation* surface reactions, i.e., subsequent exchange of the remaining amide ligands. Consequently, treatment of material **3** with various alcohols resulted in the formation of surface grafted neodymium alkoxide species **4a-d** (*B*, Scheme 1). The amide/alcohol ligand exchange reactions were monitored by gas chromatography, elemental analysis (Table 1) and FTIR spectroscopy. Examination of the combined colorless *n*-hexane washings proved the release of silylamine and the absence of any metal leaching. The disappearance of the characteristic, strong "$N(SiR_3)_2$" vibration modes (864, 846, 757, 563 cm^{-1}) which serve as excellent spectroscopic probes confirmed the exchange reaction. Powder X-ray diffraction indicated the survival of the hexagonal pore structure, however, the characteristic four-peak-pattern appeared with decreased intensity. Route *B* represents a unique method to synthesize "monometallic" alkoxide species of the lanthanide elements bearing small terminal-only ligands.

Table 1
Analytical data, pore volume, and pore diameter

Material	Components	Wt% Nd	Wt% C	$V_p{}^d$ / cm^3g^{-1}	$d_p{}^e$ / nm
1[a]	MCM-41	–	–	0.89	2.8
3[b]	MCM-41/$Nd[N(SiMe_3)_2]_3$	12.7	9.29	0.34	1.75
4a[b]	MCM-41/$Nd[N(SiMe_3)_2]_3$/HO*i*Pr	13.5	9.64	0.43	2.1
4a[b,c]	MCM-41/$Nd[N(SiMe_3)_2]_3$/HO*i*Pr	13.1	8.35	0.37	2.1
4b[b]	MCM-41/$Nd[N(SiMe_3)_2]_3$/ HOC*t*Bu_3	12.2	16.52	0.35	1.8
4c[b]	MCM-41/$Nd[N(SiMe_3)_2]_3$/HOEt	13.4	6.93	0.47	2.2
4d[b]	MCM-41/$Nd[N(SiMe_3)_2]_3$/HOMe	14.0	5.54	0.48	2.2
4d[a]	MCM-41/$Nd[N(SiMe_3)_2]_3$/HOMe	15.0	4.61	0.52	2.25
5[b]	MCM-41/Nd(OC*t*$Bu_3)_3$	9.2	15.40	0.48	1.9

[a] Pretreatment temperature 250°C, 3h, 10^{-3} torr. [b] Pretreatment temperature 100°C, 3h, 10^{-3} torr. [c] After the first catalytic run. [d] BJH desorption cumulative pore volume of pores between 1.5 and 3.5 nm diameter. [e] Pore diameter according to the maximum of the BJH pore size distribution calculated from the desorption branch.

3.2. Nitrogen physisorption

The pore size and surface texture of the parent and hybrid materials **1**, **3** and **4** was characterized by their nitrogen adsorption/desorption isotherms. As reported earlier lanthanide silylamide immobilization causes a dramatic change of the nitrogen adsorption behavior [5]. It was shown that the isotherm changes from type-IV to type-I (Fig. 1), producing a material **3** of drastically decreased pore volume and effective mean pore diameter, respectively (Table 1). Upon exchange of the remaining bulky silylamide ligands with the sterically similar bulky tritox ligand and formation of **4b** the type-I isotherm is retained. The $N(SiMe_3)_2$/HO*i*Pr ligand exchange reaction forming the sterically less encumbered isopropoxide ligand in material **4a** led to a considerable increase of the pore volume and mean pore diameter, respectively, and the type-IV isotherm reshaped (Fig. 1). Further increase of the pore volume was detected upon coordination of the smallest alkoxide ligands methoxide and ethoxide, respectively, to the metal centers. However, now the silyl moieties generated *via* route *B* (first step) seem to govern the pore texture as evidenced by effective mean pore diameters similar to that of the silylated material **2**. No significant change of pore volume and mean pore diameter occured by switching from ethoxide to methoxide ligand (Table 1).

Route *A* produces a material displaying a type-I isotherm (not shown) which is in accord with the coordination of the sterically bulky tritox ligand to the metal center. Analysis of the integral BJH desorption pore volume d*V*/d*d* of material **5** revealed one maximum, corresponding to uniformly modified mesopores. In comparison to the "silylated" Nd-tritox hybrid material **4b** which contains additional $SiMe_3$ groups, material **5** shows a significantly larger pore volume due to the lower surface coverage and absence of the silylation species.

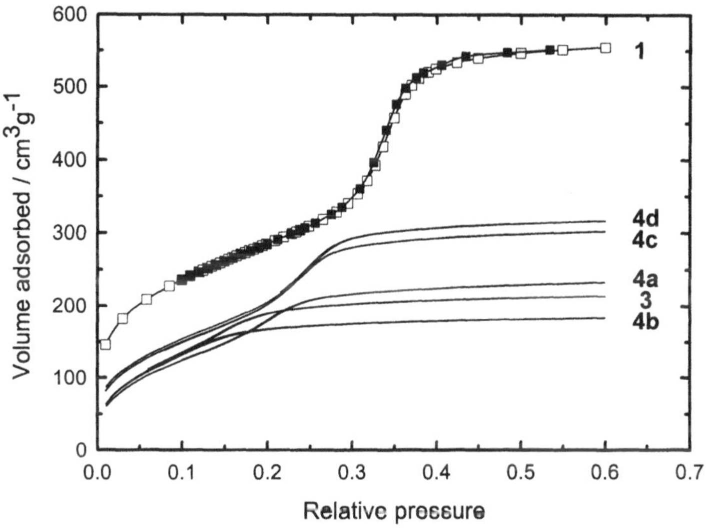

Figure 1. Nitrogen adsorption/desorption isotherms at 77.4 K (compare to Table 1; material **1**: filled symbols denote desorption)

Preliminary investigations directed towards the thermal stability of the surface grafted alkoxide species pointed to another promising option of route B. Treatment of material **4d** at 250°C/10^{-3} torr led to a material of increased pore volume and lower carbon content (Table 1). This is probably due to the separation of coordinated methanol. More detailed thermo-analytical studies in the temperature range of 25 °C to 550 °C revealed that the immobilized alkoxide ligands clearly degrade to produce materials with well-defined pore size distributions [17]. The formation of low-coordinated metal ion centers embedded in a hydrophobic matrix of silyl groups might have significant implications for catalytic applications. Further studies will show whether metal oxide cluster formation occurs under these conditions. The generation of thin metal oxide layers on thermally robust and structurally ordered mesoporous silicates (*hidden* mesoporous metal oxides) *via* alkoxide degradation [18] seems to be worthwhile, considering the efforts being made in the synthesis of *bulk* mesoporous metal oxides [13].

3.4. Catalytic MPV reduction of 4-*tert*-butylcyclohexanone

Various molecular and supported metal alkoxide species were employed in this MPV reaction and the results are summarized in Table 2 and Figure 2. The catalytic behavior of the molecular complexes is in accord with the literature data [6,7]. Under the prevailing reaction conditions $Al(OiPr)_3$ exhibited no significant catalytic activity after 5 h (entry 1). However, storing the solvent 2-propanol (= reductant) over molecular sieve and prolonged reaction times considerably increased the catalytic activity of $Al(OiPr)_3$ even at 25 °C (entry 2). Nd(tritox)$_3$ which completely dissolves in 2-propanol gave almost quantitative conversion after 50 min (entry 3). This is in accord with the high catalytic activity reported for lanthanide alkoxide complexes of type $Ln(OiPr)_3$ [7]. Surprisingly, the *insoluble* $[Nd(OEt)_3]$ proved to be also quite active in this MPV reduction (entry 4).

Table 2
Catalytic activities of various (immobilized) alkoxide species[a]

Entry	Precatalyst	Mol% M	Conversion / % (*trans* : *cis*) 5 h	24 h
1[b]	$[Al(OiPr)_3]$	5	<1	7 (4.0)
2	$[Al(OiPr)_3]$	5	16 (2.1)	59 (2.1)
3[c]	$Nd(OCtBu_3)_3$	1.4	95 (2.7)	>99 (2.6)
4	$[Nd(OEt)_3]$	1.5	96 (2.7)	>99 (2.7)
5[d]	**1**	–	–	<1
6[d]	**2**	–	–	<1
7	**4a** (OR = O*i*Pr)	1.9	90 (3.3)	>99 (3.3)
8[e]	**4a** (OR = O*i*Pr)	2.0	34 (3.9)	82 (3.9)
9	**4b** (OR = $OCtBu_3$)	1.7	96 (3.3)	>99 (3.3)
10	**4c** (OR = OEt)	2.0	29 (4.2)	68 (4.2)
11	**4d** (OR = OMe)	2.1	33 (4.1)	70 (4.2)
12	**5** (OR = $OCtBu_3$)	1.3	63 (3.3)	>99 (3.3)

[a] Conditions: 25 g HO*i*Pr, 0.1 g *n*-nonane, 0.78 g ketone, ca. 0.10 g of hybrid precatalyst, 25 °C (mol% M = 100 $n_{metal}/n_{substrate}$). [b] HO*i*Pr not predried. [c] 95 % conversion after 50 min. [d] 0.20 g mesoporous material. [e] Recovered material (no activity of the centrifugate during 48 h).

Both the parent MCM-41 material **1** and the silylated material **2** exhibited no catalytic activity under these conditions. Previous studies by van Bekkum et al. employing the same ketone revealed that even at an elevated reaction temperature (80 °C), H-MCM-41 and Na-MCM-41 (Si/Al = 15, activation temperature 450 °C) gave only 10 % conversion after 6 h.

The neodymium alkoxide supported materials **4a** and **4b** were highly efficient in this MPV reduction. Conversions higher than 90 % were observed after 5 h reaction time at ambient temperature (entry 7 and 9, Fig. 2). Apparently, the metal centers are readily accessible to the substrate molecules, maintaining the catalytic cycle given in Scheme 2. The similar catalytic activities of **4a** and **4b** suggest that the $OCtBu_3$/HO*i*Pr ligand exchange [step (i), detected by GC analysis] and adduct formation [step (ii)] are fast compared to acetone elimination [step (iii)], involving a six-membered transition state and subsequent C_α-hydrogen transfer. Hence, both systems should produce the same catalytically active species. This can be explained by the structural and electronic peculiarities of the tritox ligand, featuring enhanced Brønsted basicity, increased steric bulk, and lack of C_α-hydrogen transferability. The hybrid catalysts **4a** and **4b** can easily be separated from the reaction mixture *via* centrifugation, as shown for material **4a**. The recovered material was washed with HO*i*Pr, THF and *n*-hexane and reused in a second catalytic run (entry 8). Although the catalytic activity of the recovered material was significantly decreased (probably due to the presence of coordinated THF), a conversion of >80 % was achieved after 24 h. In contrast, the separated uncolored HO*i*Pr solution showed no further activity upon addition of new substrate solution.

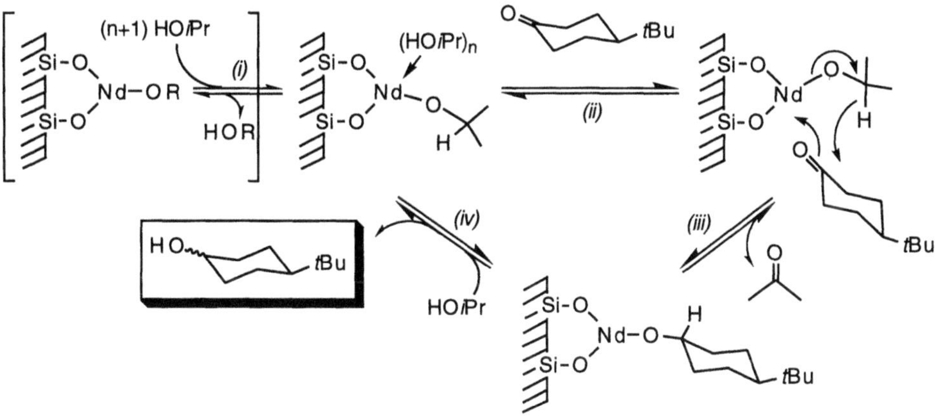

Scheme 2. Mechanistic scenario of the MPV reduction.

The decreased catalytic activity of materials **4c** and **4d** is plausible from the decreased oxidizability of the methoxide and ethoxide ligand. Although the more Brønsted acidic primary alcohols should additionally depress the alcohol exchange reactions (i) and (iv), the observed catalytic activity is quite remarkable (Table 2, Fig. 2) probably due to the pronounced ligand exchange ability of rare earth alkoxide complexes. Further studies are neccessary to explain the reduced catalytic activity of material **5**, containing Nd-OC*t*Bu_3 surface moieties comparable to material **4b**. Possible factors may be the lower concentration of active metal centers (1.3 mol%, Table 2), the changed hydrophobicity of the hybrid catalyst and metal-coordinated THF. The ratio of *cis*- and *trans*-4-*tert*-butylcyclohexanol is nearly equal to that of the thermal equilibrium state (table 2) [6], although formation of the *cis* isomer is significantly higher in the presence of a catalytically more active material (entry 3, 4, 7, 9, 12). For all reactions under investigation no side products, e.g., according to aldol type condensation occured.

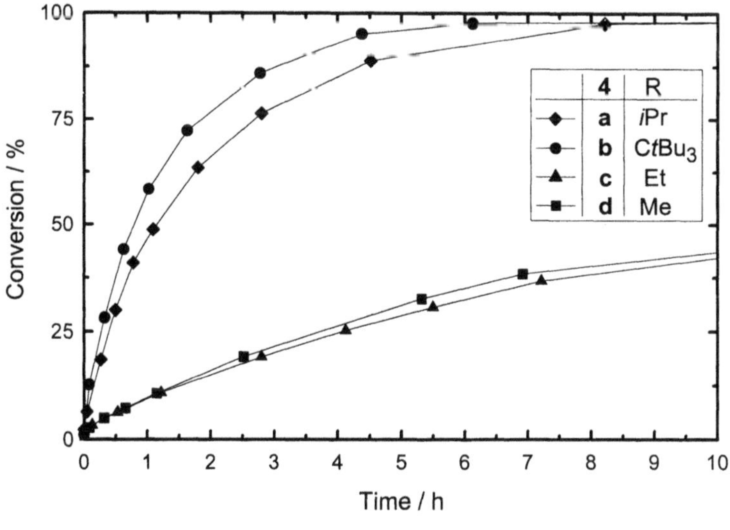

Figure 2. Effect of the alkoxide ligand on the MPV activity of rare earth hybrid catalysts **4**.

4. CONCLUSIONS

Depending on the synthesis procedure, MCM-41-supported lanthanide alkoxide species of different hydrophobicity were obtained. Nitrogen physisorption revealed that pore and surface texture can be directed by the introduction of differently-sized alkoxide ligands. The resulting hybrid materials are the most efficient solid catalysts in the MPV reduction reported so far. The catalytic activity is directed by the ligand exchange ability of the surface grafted alkoxide species. The examined hybrid catalysts display the favorable options of a heterogeneous catalyst, including ease of separation and recovery, and promising acitivity of the reused material.

ACKNOWLEDGEMENT

We thank the Deutsche Forschungsgemeinschaft for generous support.

REFERENCES

1. R. Anwander, in Applied Homogeneous Catalysis with Organometallic Compounds, B. Cornils and W.A. Herrmann (eds.), VCH Verlagsgesellschaft mbH, 1996, p. 866.
2. (a) D.E. Bergbreiter, L.-B. Chen and R. Chandran, Macromolecules, 18 (1985) 1055; (b) A. Bensalem, F. Bozon-Verduraz, M. Delamar and G. Bugli, Appl. Catal. A, 121 (1995) 81; (c) S. Kobayashi and S. Nagayama, J. Am. Chem. Soc., 118 (1996) 8977.
3. (a) A.L. Robinson, Science, 194 (1976) 1261; (b) W.M.H. Sachtler and Z. Zhang, Adv. Catal., 39 (1993) 129; (c) T.J. Marks, Acc. Chem. Res., 25 (1992) 57.
4. J.S. Beck, J.C. Vartuli, W.J. Roth, M.E. Leonowicz, C.T. Kresge, K.D. Schmitt, C. T.-W. Chu, D.H. Olson, E.W. Sheppard, S.B. McCullen, J.B. Higgins and J.L. Schlenker, J. Am. Chem. Soc., 114 (1992) 10834.
5. R. Anwander and R. Roesky, J. Chem. Soc., Dalton Trans., (1997) 137.
6. (a) A.L. Wilds, Org. React., 2 (1944) 178; (b) C.F. de Graauw, J.A. Peters, H. van Bekkum and J. Huskens, Synthesis, 10 (1994) 1007.
7. (a) J.L. Namy, J. Souppe, J. Collin and H.B. Kagan, J. Org. Chem., 49 (1984) 2045; (b) T. Okano, M. Matsuoka, H. Konishi and J. Kiji, Chem. Lett., (1987) 181.
8. D.A. Evans, S.G. Nelson, M.R. Gagne and A.R. Muci, J. Am. Chem. Soc., 115 (1993) 9800.
9. P. Leyrit, C. McGill, F. Quignard and A. Choplin, J. Mol. Catal. A, 112 (1996) 395.
10. E.J. Creyghton, S.D. Ganeshie, R.S. Downing and H. van Bekkum, J. Mol. Catal. A, 115 (1997) 457 and references therein.
11. A. Corma, Chem. Rev., 97 (1997) 2373.
12. S.B. McCullen and J.C. Vartulli, to Mobil Oil Corporation, US Pat. 5/156/829 (1992).
13. D.M. Antonelli, A. Nakahira and J. Ying, Inorg. Chem., 35 (1996) 3126.
14. T. Boger, R. Roesky, R. Gläser, S. Ernst, G. Eigenberger and J. Weitkamp, Microporous Mater., 8 (1997) 79.
15. W.A. Herrmann, R. Anwander, M. Kleine and W. Scherer, Chem. Ber., 125 (1992) 1971.
16. P.S. Gradeff, K. Yunlu, T.J. Deming, J.M. Olofson, R.J. Doedens and W.J. Evans, Inorg. Chem., 29 (1990) 420.
17. C. Palm, R. Anwander, unpublished results.
18. (a) D.C. Bradley and M.M. Factor, Trans. Faraday Soc., 55 (1959) 2117; (b) K.S. Mazdiyasni, C.T. Lynch and J.S. Smith, J. Am. Ceram. Soc., 48 (1965) 372; (c) M. Nandi, D. Rhubright and A. Sen, Inorg. Chem., 29 (1990) 3066.

MESOPOROUS MOLECULAR SIEVES 1998
Studies in Surface Science and Catalysis, Vol. 117
L. Bonneviot, F. Béland, C. Danumah, S. Giasson and S. Kaliaguine (Editors)

Diels-Alder reaction of isoprene with different dienophiles catalyzed by zeolites and MCM-41

D. Genske, K. Bornholdt and H. Lechert

Institute of Physical Chemistry of the University of Hamburg,
Bundesstr. 45, D-20146 Hamburg, Germany

Three series of mesoporous aluminosilicate molecular sieves MCM-41 with Si/Al-ratios in the range of 12 - 72 were synthesized. In the synthesis of MCM-41 three surfactants with varying lenght of the n-alkyl-chain (C_{14}TMABr C_{16}TMACl, C_{18}TMABr) were used as template to obtain different pore diameter. The samples were characterized by BET, EDAX, NH_3-TDS and ^{27}Al-MAS-NMR. These samples are used as catalysts in Diels-Alder reactions of isoprene with the dienophiles methylvinylketone, 1,4-naphthoquinone. The conversion and the regioselectivity of these reactions were compared with the same reactions using zeolite Y as catalyst. The results showed that the catalytic activity of MCM-41 is higher than that of the zeolite Y. The catalytic activity increases as the Si/Al-ratio decreases. The reactions were carried out in dichloromethane and acetonitrile as solvent. The decrease in catalytic activity in acetonitrile can be attributed to the stronger adsorption of acetonitrile at the acid sites of the catalyst compared with dichloromethane. In this context the adsorption of isoprene and the dienophiles were studied separately. The results showed that both reactants are adsorbed rather strongly at the surface of the catalyst. The adsorption of the reaction products is generally stronger than the adsorption of the methylvinylketone.

1. INTRODUCTION

Uncatalyzed pericyclic reactions sometimes require high temperature and/or high pressure. It was therefore a major breakthrough when it was found that [4+2] cycloaddition, known as the Diels-Alder reaction, could be conducted at low temperature in the presence of Lewis acid catalysts to afford often highly regioselective conversions [1,2].

The use of Lewis acids like aluminum trihalides (AlX_3), antimony pentachloride ($SbCl_5$) or trifluoride-diethyl ether (BF_3-O-$(C_2H_5)_2$) for the catalysis of Diels-Alder reactions is well accepted. Most of these reactions were carried out in homogenous phases. It was also found that the Diels-Alder reaction can be catalyzed by zeolites [3,4,5].

Because of its uniform mesopores MCM-41 might be a promising catalyst for the Diels-Alder reactions of bulkier molecules.

In this paper we have investigated the performance of the zeolite Y and the MCM-41 for the reaction of isoprene (1) with methylvinylketone (2) to 4-acetyl-1-methyl-1-cyclohexene (3) and 4-acetyl-1-methyl-2-cyclohexene (4) in comparison with the reaction of the same diene with 1,4-naphthoquinone (5) and 9,10-dioxy-2-methyl-1,4-dihydroanthracene (6) shown in scheme 1.

Results concerning the reaction (A) catalyzed by EMT, ZSM-5 and beta are described in [6].

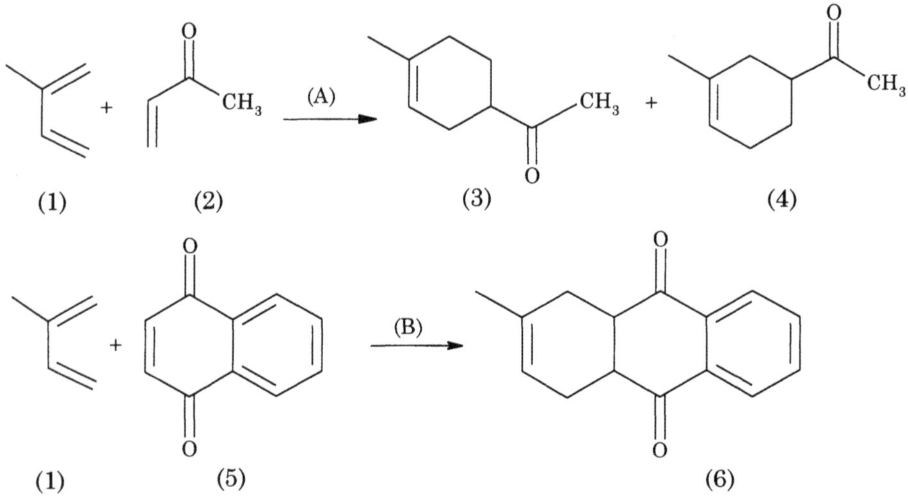

Scheme 1. Diels-Alder reactions catalyzed by zeolite Y and MCM-41

2. EXPERIMENTAL SECTION

2.1. Materials

Zeolite Y (Si/Al = 2,6) was a commercial product from Degussa. The source of silica and alumina used for the synthesis of MCM-41 were fumed silica Cab-osil M-5 (Fluka), sodium aluminate $NaAlO_2$ (Riedel-de Haën, 54 wt.-% Al_2O_3, 41 wt.-% Na_2O, 5 wt.-% H_2O). The quaternary ammonium surfactant compounds tetradecyltrimethylammoniumbromide C_{14}TMABr 99%, hexadecyl-trimethylammoniumchloride C_{16}TMACl 25 wt% in H_2O, octadecyltrimethyl-ammoniumchloride C_{18}TMACl 98% and tetramethylammoniumhydo[illegible]ide, TMAOH, 28 wt.-% aqueous NH_3 solution were obtained from Fluka.

The organic compounds used for the Diels-Alder reactions were isop[illegible] methylvinylketone, 1,4-naphthoquinone from Fluka.

2.2. Sample preparation

The MCM-41 has been synthesized following a procedure of Beck et. al.[7]. The molar composition of the resulting gels were:

$n\ SiO_2 : Al_2O_3 : 1.3\ Na_2O : m\ (TMA)_2O : p\ (C_xTMA)_2O\ q\ (NH_4)_2O : r\ HX : z\ H_2O$

n = 146 / 73 / 25, m = 22 / 11 / 4, p = 13 / 6.5 / 2, q = 7.5 /3.5 / 1.3, r = 26/13/4.5, z = 2850/1420/482

The MCM-41 samples were synthesized from gels with four different n_{Si}/n_{Al} - ratios (Si/Al = 72, 36, 12, ∞). To obtain different pore-size in the MCM-41 samples three different quaternary ammonium surfactant compounds were used as template in the synthesis. After combining the components in appropriate amounts, the reaction gels were stirred until they were homogenous, filled into a Teflon bottle and kept at 363 K for two days. The solid products were filtered and washed with deionized water. Then, they were dried at 363 K. To remove the template, the resulting products were calcined at 933 K for several hours. The MCM-41 samples and zeolite Y were triple ion-exchanged with a 1 M solution of NH_4NO_3 at 353 K. Afterwards they were washed with water and dried at 363 K. For a better handling during the further experimental procedures the samples were pressed and were sieved into particles of 0.25-0.5 mm. Prior to use all catalysts were activated at 713 K for 8 h under reduced pressure to decompose the ammonium form into the protonic form.

2.3. Sample characterization

X-Ray powder pattern were collected on a Siemens D 5000. Nitrogen adsorption-desorption isotherms to determine the BET specific surface areas were measured at 77 K using an Sorptomatic 1900 from Carlo Erba Instruments. Prior to measurements the samples were outgassed at 463 K for 4 h in vaccum (p = 0.1 Torr).

The Si/Al-ratio of the samples were determined using two EDAX and ICP-OES methods. For EDAX measurements a microscope SEM 515/ D 806 V from Phillips was used. ICP-OES measurements were carried out on Perkin-Elmer Plasma-II spectrometer. ^{1}H and ^{13}C NMR spectra were recorded on a Bruker WM-400. GC-MS analysis were performed on an Varian MAT, 70 eV type 311 A system.

2.4. Reaction procedures

The catalytic reaction was performed in a 100 mL three-necked round-bottom flask. The flask was evacuated and flushed with dry nitrogen several times and charged with different amounts of the activated catalysts to a solution containing 10 mmol diene and dienophile in dichloromethane or acetonitrile. The reactions were carried out under stirring at different temperatures (253 K, 273 K, 298 K) and stopped after 90 min. The solvent and the reactants were evaporated and the products were separated from the catalyst by extraction with pentane for 4 hours.

The extract was analyzed by withdrawing an aliquot into a Fisons Model 8160 gas chromatograph equipped with an unpolar column (length 30 m, ID 0.32 mm, 25 μm) and a FID.

The course of the reactions were monitored by high resolution GLC of samples taken from the reaction mixture at regular time intervals. The identification of the products was carried out by ^{1}H-NMR, ^{13}C-NMR and GC-MS.

3. RESULTS AND DISCUSSION

Table 1 summarizes the results of characterization of catalysts. It can be seen that for the higher Si/Al-ratios the incorporation of silicate is not complete.

Table 1
Physico-chemical characteristics of the zeolite Y and MCM-41 with different Si/Al-ratios

sample	Si/Al (EDAX)	Si/Al (ICP-OES)	BET-surface [$m^2 g^{-1}$]	acidity [mmol/g][a]
HY	2.6	-	-	1.9
MCM-41-12	12.5	16.3	632	0.89
MCM-41-30	30	27.8	1076	0.69
MCM-41-60	39	34.4	1148	0.59
MCM-41-Al-free	-	-	1132	-

[a] determined by TDS-experiments of NH_3 according to [8]

The ^{27}Al-MAS-NMR spectra (not shown) indicate that in the "as synthesized" samples only tetrahedral Al is present and in the calcined samples a very low amount of octahedral Al can be detected. After the ion exchange these samples show that the aluminum is only in tetrahedral coordination. Framework and to a less degree extraframework aluminum exists in the protonic form of MCM-41 after activation procedure at 713 K.

As can be seen in the table 2 the surface area of the MCM-41 is very high and increases with length of the surfactant n-alkyl-chain in the synthesis.

To obtain a material with larger pore diameter in the synthesis of MCM-41 using the surfactant C_{18}TMBr mesitylene was added. The adsorption-desorption measurement with nitrogen at 77 K for this sample shows a broader pore size distribution.

Table 2
Physico-chemical characteristics of MCM-41 synthesized with different templates and a constant Si/Al-ratio of 12

sample	template	Si/Al (EDAX)	BET-surface [$m^2 g^{-1}$]	pore diameter [nm]
MCM-41-C14	C_{14}TMABr	12	1065	2.8
MCM-41-C16	C_{16}TMACl	12	1132	3.0
MCM-41-C18+Mes [a]	C_{18}TMACl	12	1170	3.6 - 4.0

[a] Mes = mesitylene

The yield and selectivity of the reaction catalyzed by zeolite Y (Si/Al = 2.6) and MCM-41 with different Si/Al-ratios are shown in table 3. It is evident that the catalytic activity increases with decreasing Si/Al-ratio. Although the zeolite Y has stronger acid sites than MCM-41 the catalytic activity for Diels-Alder reactions is lower.

The characterization of Brönsted and Lewis acidity of zeolite Y and MCM-41 is well documented, particulary in the gas phase where it is possible to differentiate between the two types of acid sites [9,10,11]. Catalytic activity is only observed if aluminum is present in the catalyst where an exact differentiation of Brönsted or Lewis acid site is not possible in the present state of investigations.

To study the influence of solvent on the Diels-Alder reactions, the experiments were repeated in dichloromethane and acetonitrile. It can be seen that the conversion is decreased when the reactions were carried out in acetonitrile for which a stronger adsorption can be expected at the acid sites.

Table 3
Results obtained from the reaction of isoprene (1) with methylvinylketone (2) at 298 K in a ratio of 1:1

catalyst	amount of catalyst [g]	solvent	yield [%]	selectivity para/meta (3)/(4)[a]
H-Y	1.0	CH_2Cl_2	70	18
H-Y	1.0	CH_3CN	48	13
MCM-41-12	0.7	CH_2Cl_2	89	26
MCM-41-12	0.7	CH_3CN	81	25
MCM-41-30	1.1	CH_2Cl_2	78	24
MCM-41-30	1.1	CH_3CN	55	24
MCM-41-60	1.1	CH_2Cl_2	48	22
MCM-41-60	1.1	CH_3CN	34	22
MCM-41-Al-free Si	1.1	CH_2Cl_2	2	-

[a] determined by gas chromatography

The decrease in yields for the zeolite Y is more pronounced than for MCM-41. This suggests that the adsorption of acetonitrile is stronger on zeolite Y than on MCM-41. In this context adsorption measurements in dichloromethane with methylvinylketone and the product (3 and 4) were performed separately to determine adsorption coefficients (K_{ads}). The adsorption coefficients were calculated according to the Langmuir isotherms [12,13].

Table 4
Results obtained from adsorption measurements

catalyst	K_{ads} methylvinylketone [l/mol]	K_{ads} for product (3,4) [l/mol]
H-Y	284	664
MCM-41-C12	29	137

The adsorption coefficient for the product is higher for zeolite Y than for MCM-41. Due to the stronger adsorption of the product on zeolite Y an inhibition can be observed in the course of reaction causing the conversion to decrease.

This seems to be an indication that the product is more adsorbed on the Brönsted acid sites than on the Lewis acid sites.

To study the influence of pore diameter on conversion and selectivity of the Diels-Alder reaction some MCM-41 were used with different pore diameter and constant Si/Al-ratio.

Table 5
Results obtained from the reaction of isoprene (1) with methylvinylketone (2) at 273 K in a ratio of 1:1 using an amount of catalyst of 1 g

catalyst	pore diameter [nm]	yield [%]	selectivity para/meta (3)/(4)
MCM-41-C14	2.8	82	30
MCM-41-C16	3.0	87	32
MCM-41-C18+Mes[a]	3.6 - 4.0	85	33

[a] Mes = mesitylene

The results show that the pore diameter has no effect on the catalytic activity for the Diels-Alder reaction of isoprene with methylvinylketone. The molecular diameter of the product (3) is about 0.42 nm. Therefore the access to the active sites should not be restricted. The selectivity in this case is very high.

According to the literature the reaction of isoprene (1) with 1,4-naphthoquinone (5) was carried out at 373 K [14]. In our experiment the cycloaddition of these reactants was performed in dichloromethane at 273 K for a period of 90 minutes. Four isomers could be detected using gaschromatography. The major isomer (6) was identified by ^{1}H-NMR and GC-MS. The other isomers were not separated.

Table 6
Results obtained from the reaction of isoprene (1) with 1,4-naphthoquinone (5) at 273 K in a ratio of 1:1

catalyst	time [h]	amount of catalyst [g]	overall yield [%]
HY	24	1	70
MCM-41-12	1.5	0.9	24
MCM-41-30	1.5	0.9	30
MCM-41-60	1.5	1.1	22

At higher temperature and with the diene in excess the isomer (6) is reacting to form the products (7a,b). The molecular mass of the product (7) could be found by the GC-MS method. Further investigations are in progress. The reactions shown in scheme 2 occur in a sequence of two steps.

In the first reaction step (B) isoprene (1) and 1,4-naphthoquinone (5) react to 9,10-dioxy-2-methyl-1,4-dihydroanthracene (6) followed by the parallel reactions (P1, P2) to form the products (7a,b). All isomers are not shown in this scheme.

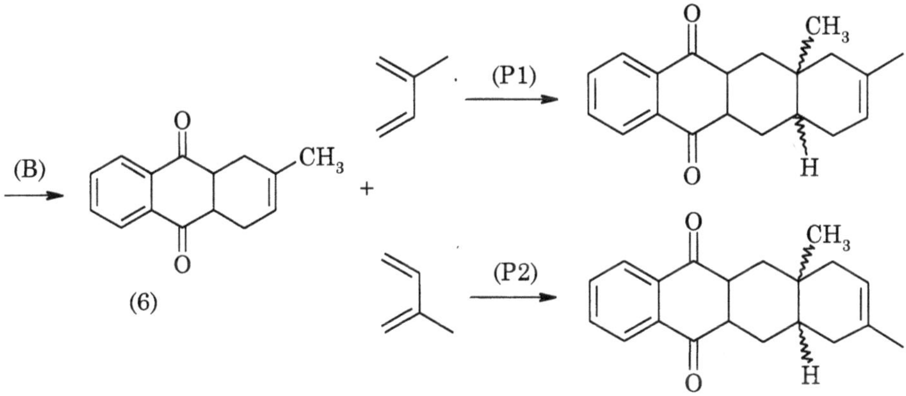

(7a)

(7b

Scheme 2. Possible reaction pathway for parallel reactions of isoprene (1) with 9,10-dioxy-2-methyl-1,4-dihydroanthracene (6) to product (7a,b)

4. CONCLUSIONS

The Diels-Alder reaction of isoprene with methylvinylketone or 1,4-naphthoquinone is catalyzed by zeolite Y and MCM-41. The MCM-41 has a dinstinctly higher activity than zeolite Y in spite of the stronger acid sites in the zeolite. Increasing aluminum content in the MCM-41 leads to higher conversion. This indicates that the number of acid sites is of greater influence than its acid strength. Adsorption measurements of the reactants show the strongest adsorption for the products. This explains that catalysts with stronger acid sites cause a more pronounced inhibition and restriction of the reaction.

ACKNOWLEDGEMENTS

The authors would like to thank Christo Minchev for the TDS-experiments of ammonia, U. Schenk, O. Groeger for the MAS-NMR-spectra. We thank the "Deutsche Forschungsgemeinschaft" for financial support.

REFERENCES

1. E.F. Lutz and G.M. Bailey, J. Am. Chem. Soc., 86 (1964) 3899.
2. T. Inukai and T. Kojima,. J. Org. Chem., 31 (1966) 1121.
3. J. Ipaktschi, Z. Naturforsch., 41b (1986) 496.
4. L. Eklund, A.K. Axelsson, A. Nordahl and R. Carlson, Acta Chem Scand., 47 (1993) 581.
5. C. Cativiela, F. Figueras, J.M. Fraile, J.I. Garcia, J.A. Mayoral, L.C. de Menorval and E. Pires, Appl. Catal., 101 (1993) 253.
6. K. Bornholdt and H. Lechert, Stud. Surf. Sci. Catal., 94 (1995) 619.
7. J.S. Beck, J.C: Vartuli, W.J. Roth, M.E. Leonowicz, C.T Kresge, K.D: Schmitt, C.T.W. Chu, D.H. Olson, E.W. Sheppard, S.B. McCullen, J.B. Higgins and J.L. Schlenker, J. Am. Chem. Soc., 114 (1992) 10834.
8. C. Minchev, Y. Neinska, V. Valtchev, V. Minkov, T. Tsoncheva, V. Penchev, H. Lechert and M. Hess, Catal. Lett., 18 (1993) 125.
9. M. Busio, J. Jaenchen and J.H.C. van Hooff, Microporous Mater., 5 (1995) 211.
10. R. Mokaya, W. Jones, Z. Luan, M.D. Alba and J. Klinowski, Catal. Lett., 37 (1996) 113.
11. A. Liepold, K. Roos and W. Reschetilowski, Chem Eng Sci., 51 (1996) 3007.
12. K. Bornholdt, Ph.D.-Thesis, University of Hamburg (1996).
13. D. Genske, diplome work, University of Hamburg (1997).
14. O. Diels and K. Alder, Chem. Ber., 62 (1929) 2357.

MESOPOROUS MOLECULAR SIEVES 1998
Studies in Surface Science and Catalysis, Vol. 117
L. Bonneviot, F. Béland, C. Danumah, S. Giasson and S. Kaliaguine (Editors)

DEHYDROGENATION REACTIONS ON MCM-41

T.Stoylkova[1] , P.Matthae[2] , C.Bezouhanova[1] and H.Lechert[2]

[1] University of Sofia, Faculty of Chemistry
1 James Bourchier Ave, Sofia 1126, Bulgaria
[2] University of Hamburg, Institute of Physical Chemistry
Bundesstrasse 45, 20146 Hamburg, Germany

The nature of the catalytically active sites in mesoporous MCM-41 molecular sieves with pore size of 40, 65, or 80 Å has been studied using cyclohexanol and benzylalcohol as probe molecules. For cyclohexanol on MCM-41 at 438 K the dehydration was accompanied by 2 % isomerization to methylcyclopentane. Benzylalcohol was converted to toluene and benzaldehyde on acid and basic sites of the catalyst, respectively. The dehydrogenation to benzaldehyde was promoted by the presence of oxygen in the carrier gas. By oxidative dehydrogenation of cyclohexane on MCM-41 cyclohexene was created. Thus the alcohol test permitted the detection of Brönsted acid sites and of basic sites in MCM-41. In the presence of oxygen oxidative dehydrogenation of hydrocarbons was achieved.

Keywords: MCM-41, cyclohexanol, benzylalcohol, cyclohexane, dehydration, dehydrogenation, cyclohexene

1. INTRODUCTION

The synthesis of mesoporous aluminosilicates of the type of MCM-41 was stimulated from the necessity of molecular sieves with pore diameter between that of microporous and macroporous materials. The diameter of the pores of MCM-41 varies between 20 and 100 Å. It was already shown that they may be used as acid catalysts for the conversion of bulky compounds [1]. Their acid properties were characterized by the adsorption of cyclohexylamine [2], acetonitrile [3], pyridine [2,4] and ammonia [5,6]. The acidity of aluminium containing MCM-41 samples was found to be weaker than that of HY zeolites [1] and related to the presence of weak and mild Lewis acid sites, that enable their activity as alkylation catalysts.
The aim of our study was to characterize the nature of catalytically active sites in mesoporous MCM-41 molecular sieves with pore size of 40, 65 and 80 Å using cyclohexanol and benzylalcohol as probe molecules. The advantage of alcohols is their ability to interact with Brönsted acid sites giving oxonium ions and the corresponding olefins via a dehydration reaction. In the presence of basic sites the alcohols can be dehydrogenated to the respective carbonyl compounds. The presence of basic sites in MCM-41 was observed by the conversion of 2-methylbut-3-yn-2-ol to acetylene and acetone at 180° C [7].

We studieded also the ability of MCM-41 molecular sieves to dehydrogenate cyclohexane.

2. EXPERIMENTAL

Catalysts

Three samples of MCM-41 were synthesized according to [8]. The Si/Al ratio and the pore dimensions of the samples are presented in Table 1.
The desired pore size was achieved by mesitylene addition during synthesis and estimated by argon physisorption.

Catalytic experiments

A fixed bed flow reactor containing 0.15 - 0.33 g of preliminary pressed, crushed and sieved to 8-14 mesh catalyst was used. All the samples were activated at 823 K under nitrogen flow for 1 hour and then under air flow during 6 hours. The experiments with cyclohexanol were performed in inert atmosphere. Benzylalcohol and cyclohexane conversion was studied under air flow of varied velocity.
The reaction products collected in a cooling water trap were analyzed by GC on a 2 m column with 15% DC 550 on silanized chromosorb W 60-80 mesh. Hewlett Packard GC-MS was used for the identification of the products.

Table 1: Characteristics of the synthesized MCM-41 samples

Sample	Pore diameter, Å	Si/Al
MCM-41 - 1	40	36.7
MCM-41 - 2	64	12.4
MCM-41 - 3	80	12.4

3. RESULTS

Cyclohexanol conversion

The conversion of cyclohexanol on the mesoporous MCM-41 catalysts proceeds at relatively low temperature of 438 K (Fig.1) and rather small contact time (18.6 h). At 573, 473 K or 438 K and longer contact time the conversion of cyclohexanol was 100 %
MCM-41-3 has considerably lower activity in these conditions. The main product is cyclohexene (Fig. 2). The activities of MCM-41-1 and MCM-41-2 are very similar in spite of the different pore dimensions and different ratio Si/Al. Probably in MCM-41-2 the lower concentration of acid sites enables their higher reactivity.

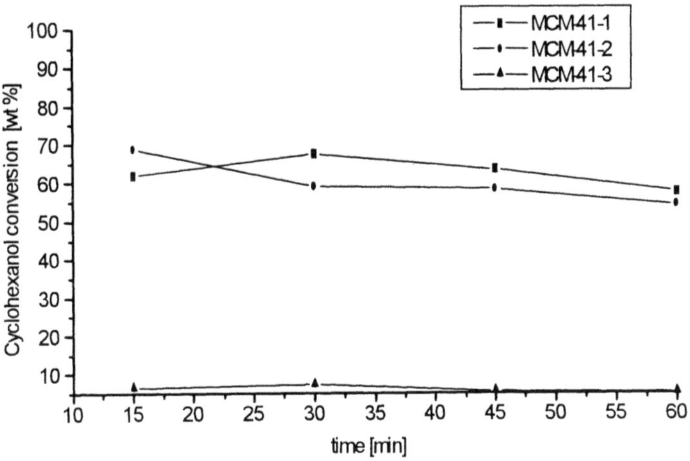

Fig.1: Cycloxexanol conversion at 438 K; WHSV 18.6 h^{-1}.

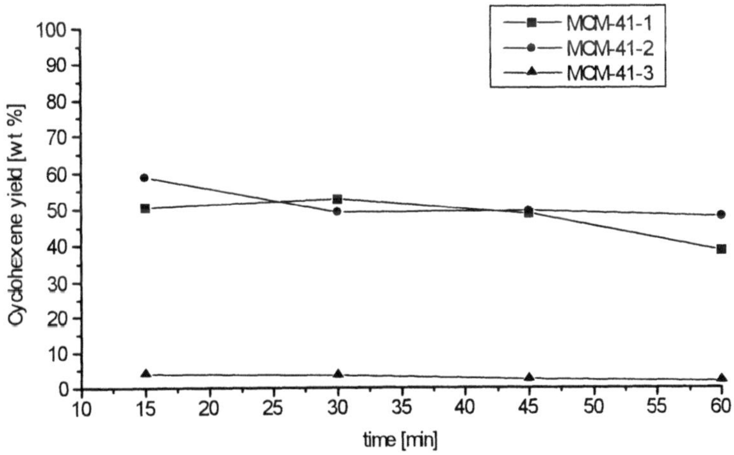

Fig.2: Cyclohexene yield at 348 K; WHSV 18.6 h^{-1}.

Benzylalcohol conversion

The conversion of benzylalcohol was performed at higher temperature compared to that for cyclohexanol - 573 K. Two reaction products were obtained - toluene and benzaldehyde. The formation of toluene (Fig. 3) may be explained by the

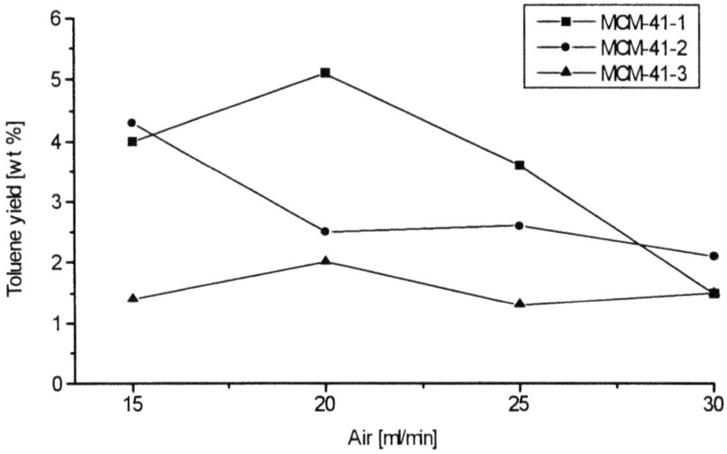

Fig.3: Toluene yield at 537 K; WHSV 2.5 h^{-1}.

stabilization of the benzylcation obtained after dehydration of the oxonium ion from protonated benzylalcohol. The quantity of toluene indicates more differences in the Brönsted acidity of the three MCM-41 samples. Again MCM-41-3 is less acidic. The production of benzaldehyde is related to the stabilization of the benzylcation via hydrogen abstraction from another alcohol molecule. The much higher quantity of aldehyde (Fig.4) in the absence of any metal component in the catalysts used suggests additional dehydrogenation, facilitated by the presence of oxygen and increasing even at lower contact time.

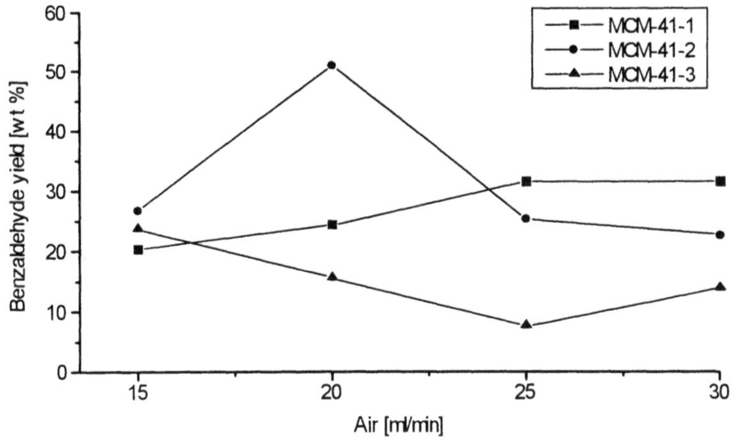

Fig.4: Benzaldehyde yield at 573 K; WHSV 2.5 h^{-1}.

Cyclohexane conversion

The conversion of cyclohexane in the same temperature range and air flow confirmed the dehydrogenation activity of the mesoporous aluminosilicates MCM-41. An essential increase of cyclohexane conversion (Fig. 5) was observed by increasing the rate of the air flow, that actually diminished the contact time. Cyclohexene was the main reaction product (Fig. 6).

The difference in the activity of the MCM-41 samples with different pore size was rather small in this reaction.

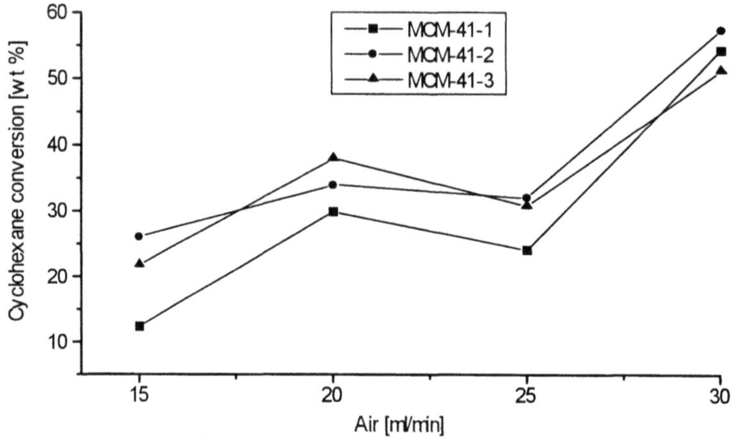

Fig. 5: Cyclohexane conversion at 623 K; WHSV 2.5 h^{-1}.

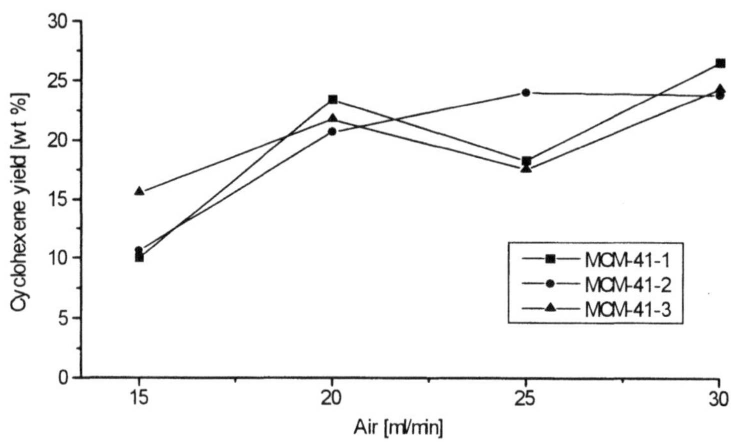

Fig.6: Cyclohexene yield at 623 K; WHSV 1.8 h^{-1}.

4. DISCUSSION

The presence of Brönsted acid sites in MCM-41 was revealed by the contact with cyclohexanol and benzylalcohol. Cyclohexanol dehydration was achieved at relatively low temperature giving evidence of strong Brönsted acidity.

The conversion of benzylalcohol with MCM-41 may be explained by the joint action of acid-base pairs of active sites. After protonation of benzylalcohol the oxonium ion is dehydrated to benzylcation, which is stabilized to toluene after hydride ion abstraction from another alcohol molecule, deprotonized in a next step on a basic site. The yield of benzaldehyde was increased in the presence of additional oxygen.

In the oxydative dehydrogenation reaction of cyclohexanethe quantity of cyclohexene increased with the air flow velocity.

A direct dehydrogenation on the oxygen atoms from the catalyst acting as basic sites may be supposed.

Experiments with catalysts of increased basicity due to the presence of alkali cations showed much lower activity. The dehydrogenation of cyclohexane may proceed via protonation, hydrogen abstraction and deprotonation of the carbenium ion, that involves again a co-operative action of acid-base pair of active sites.

In the presence of oxygen in the reaction mixture cyclohexane was partially oxidized to cyclohexanone and cyclohexene-2-one. The mechanism of oxygen activation on MCM-41 is not clear. One suggestion is the existence of oxygen vacancies in the catalyst structure.

5. CONCLUSIONS

The interaction of cyclohexanol and benzylalcohol with mesoporous MCM-41 aluminosilicates revealed the presence of Brönsted acid sites in their channels. The catalytic activity is influenced by the alcohol structure and in the case of benzylalcohol the dehydration is accompanied by dehydrogenation to benzaldehyde. The dehydrogenation is promoted by the oxygen in the reaction media. In the presence of oxygen the mesoporous aluminosilicate MCM-41 activate the dehydrogenation of cyclohexane to cyclohexene in the absence of any transition metal. In these conditions cyclohexane may be partially oxidized to cyclohexanone. The cyclohexene formed is also partially oxidized to cyclohexene-2-one. The presense of Brönsted acid sites is necessary for the oxydative dehydrogenation.

REFERENCES

1 E.Armengol, M.L.Cano, A.Corma, H.Garcia, M.T.Navarro, J. Chem. Soc., Chem.Commun., (1995), 519.

2 R.Mokaya, W.Jones, L.Zhaohua, M.D.Alba, J.Klinowski, Catal. Lett., 37, (1996), 113.
3 M.Busio, J.Janchen, J.C.H. van Hoof, Microporous Mater., 5, (1995), 211.
4 A.Corma, V.Fornes, M.T.Navarro, J.Perez-Pariente, J.Catal., 148, (1994), 569.
5 C.Y.Chen, H.X.Li, M.E.Davis, Microporous Mater., 2, 1993, 17.
6 H.Kosslick, H. Landmesser, R.Fricke, J.Chem.Soc., Far. Trans., 93, (1997), 1849.
7 A. Galarneau, A.Barodawalla, T.J.Pinnavaia, J.Chem.Soc., Chem. Commun., (1997), 1661.
8 J.S.Beck, J.C.Vartuli, W.J.Roth, M.E.Leonowicz, C.T.Kresge, K.D.Schmitt, C.T-W Chu, D.H.Olson, E.W.Sheppard, S.B.McCullen, J.B.Higgins, J.I.Schlenker, J.Am.Chem.Soc., 114, (1992),10834.
9 C.P.Bezoukhanova, Y.A.Kalvachev, Catal.Rev. - Sci.Eng., 36, (1994), 125.

MESOPOROUS MOLECULAR SIEVES 1998
Studies in Surface Science and Catalysis, Vol. 117
L. Bonneviot, F. Béland, C. Danumah, S. Giasson and S. Kaliaguine (Editors)

Catalytic properties of MCM-41 for the feedstock recycling of plastic and lubricating oil wastes

D.P. Serrano, J. Aguado, J.L. Sotelo, R. Van Grieken, J.M. Escola and J.M. Menéndez

Chemical Engineering Department, Faculty of Chemistry,
Complutense University of Madrid, 28040 Madrid, Spain

The potential application of Al-containing MCM-41 for the catalytic degradation of polyolefinic plastics and used lubricating oils has been investigated and compared with the behavior of ZSM-5 zeolite and commercial amorphous SiO_2-Al_2O_3. For all the studied raw plastics, MCM-41 presents significant higher activity than the amorphous catalyst, and even superior to that of ZSM-5 for the degradation of pure polypropylene, a mixture of three polyolefins (high density polyethylene + low density polyethylene + polypropylene) and a lube distillate. On the contrary, the ZSM-5 zeolite leads to higher conversions in the degradation of pure polyethylenic plastics and shows a slower deactivation during the lube oil cracking. Regarding the product distribution, while ZSM-5 leads mainly to light hydrocarbons (C_2-C_5), MCM-41 cracks these wastes into liquid fractions (gasoline and middle distillates), which suggests the cracking pathway is not the same with these two materials, being governed by their pore size and/or their acid strength.

1. INTRODUCTION

The development of suitable methods for the treatment and recycling of polymeric and oil wastes is one of the challenges that the technology must face in the next years. Among the different alternatives to deal with this kind of wastes, chemical recycling present a high potential since it may allow to transform these residues in different products that can be used as raw materials for the preparation of fuels and chemicals. Currently, a number of processes are being investigated for the chemical recycling of plastic wastes : alcoholysis, hydrolysis, gasification, hydrogenation, pyrolysis and catalytic cracking.

In the past, different types of acid catalysts have been applied for the catalytic cracking of polyolefinic plastics or for the conversion of heavy oils obtained by thermal cracking of the former. Among those catalysts, amorphous silica-alumina and different zeolites (X, Y and ZSM-5) have been typically used [1-3]. However, whereas zeolitic materials are advantageous in terms of acid sites distribution and strength, their application for the conversion of heavy wastes is limited by their pore size. In recent works [4,5], we have shown that these problems can be overcome with the use of MCM-41 materials as catalysts for the conversion of pure polyolefins, since they present a balanced combination of medium acid strength with high surface area and uniform pore sizes in the mesopore range (1.5 – 10 nm).

In the present work, the study of the potential use of MCM-41 materials for the catalytic conversion of polymeric and oil wastes into chemicals and hydrocarbon mixtures useful as feedstocks has been continued by using a mixture of polyolefins and a lube distillate as raw materials. In the last case, the evolution of the activity along the time on stream has provided information about the stability and the deactivation of MCM-41.

2. EXPERIMENTAL

The MCM-41 sample used in this work has been synthesized at room temperature using tetratethylorthosilicate and aluminum isopropoxide as Si and Al sources, respectively. The alkoxides are hydrolyzed in a first step with an HCl solution also containing the surfactant, cetyltrimethylammonium chloride. Subsequently the formation of the solid mesoporous phase is accelerated through the addition of an ammonia solution. The solid product so obtained is separated by filtration, dried at 110°C and calcined in air at 550°C, yielding the MCM-41 sample free of surfactant and directly in acid form, hence further ion exchange treatments are not necessary.

The ZSM-5 sample has been synthesized from ethanol-containing gels according to a published procedure [6]. A commercial amorphous SiO_2-Al_2O_3 sample was also used as reference (Südchemie, KA-3).

The three catalysts have been characterized by conventional techniques: X-ray fluorescence (XRF), X-ray diffraction (XRD), N_2 adsorption at 77 K, NH_3 temperature programmed desorption (TPD), ^{27}Al MAS NMR, etc. Thus, the XRD and N_2 adsorption measurements confirm that the MCM-41 sample has a regular mesoporosity with pore size around 2.9 nm, whereas ^{27}Al MAS-NMR measurements show that all the Al atoms present tetrahedral coordination in the as-synthesized sample. The main properties of these materials have been summarized in Table 1.

Table 1. Physicochemical properties of the catalysts

Catalyst	Si/Al	Acidity (NH_3 TPD) (mmol/g)	Acidity (NH_3 TPD) T_{max} (°C)	D_P (nm)[a]	S_{BET} (m^2/g)
MCM-41	45.6	0.22	332	2.9	1177
ZSM-5	34.1	0.52	467	0.55	467
SiO_2-Al_2O_3	2.0	0.14	304	2 -12	169

[a] pore diameter

The catalytic experiments of polymer degradation have been carried out in a batch reactor at 400°C with continuous N_2 flow. The reaction temperature was reached after 15 min heating from room temperature. The products leaving the reactor during 30 min were separated and accumulated as gaseous and liquid fractions for subsequent GC analysis. The following polyolefins were used as raw materials: high-density polyethylene (HDPE), low-density polyethylene (LDPE), polypropylene (PP, isotacticity index = 93%) and a mixture of them having a composition that resembles the distribution of these polymers existing in municipal plastic wastes (LDPE: 46.5 wt%, HDPE: 25 wt% and PP: 28.5 wt%).

The catalytic conversion of a lube distillate, formed by paraffinic hydrocarbons with boiling points in the range 450 –550°C (C_{25} – C_{50}) , has been studied in a down-flow fixed bed reactor at 400°C under atmospheric pressure. The lube distillate was fed by means of a syringe pump and mixed with a N_2 stream previously to be introduced into the reactor. The products in the effluent stream were separated in a condenser at 0°C into gaseous and liquid phases, which were further quantified and analyzed by GC.

3. RESULTS AND DISCUSSION

3.1. Catalytic degradation of pure polyolefins

MCM-41 was first tested in the catalytic conversion of three pure polyolefins: HDPE, LDPE and PP, the results obtained being compared with those corresponding to a ZSM-5 zeolite and a commercial amorphous silica-alumina. Different plastic/catalyst ratios were used for each polymer in order to obtain adequate polymer conversions.

The activities are shown and compared in Figure 1 for the different catalysts and polymers. With the three polyolefins, the less active sample is the commercial amorphous silica-alumina, which leads to conversions just slightly higher than those observed by thermal cracking. This poor activity is probably originated by the weak acidity and the low surface area present in this material (see Table 1).

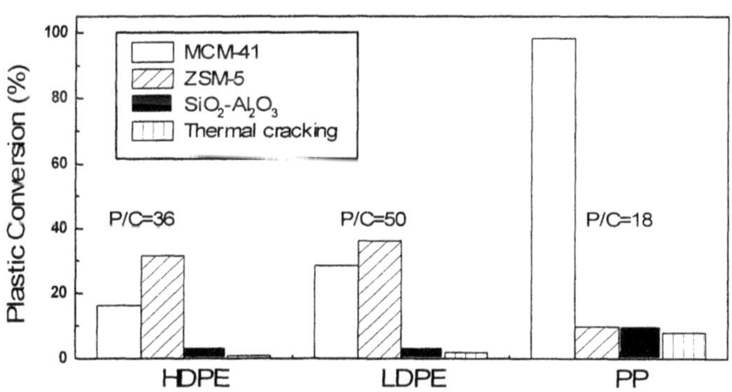

Figure 1. Activity in the cracking of pure polyolefins
(400°C, 30 min., P/C = plastic/catalyst mass ratio)

The comparison between the catalytic behavior of MCM-41 and ZSM-5 zeolite is more complex. The latter is clearly a superior catalyst for the degradation of highly linear polymers such as HDPE probably due to its stronger acidity, but it exhibits very low activity in the PP conversion, polyolefin with a high proportion of side methyl groups. In this case, the conversion obtained over MCM-41 is almost total, whereas the activity of ZSM-5 is very close to that of the thermal cracking. An intermediate result is observed in the degradation of LDPE, linear polymer having a certain degree of branching, since both materials present similar activities. The order of activity exhibited by MCM-41, both in terms of overall conversion or conversion per unit mass of catalyst, follows the expected trend according to the proportion of highly reactive tertiary carbons present in the polymeric chains: PP > LDPE > HDPE. In this way, the lack of activity observed over ZSM-5 for the PP degradation is remarkable, which confirms previous results obtained at a higher plastic/catalyst ratio [5], indicating that in this case the presence of the side methyl groups in the polymeric chain hinders its access to the narrow zeolite pores. Thus, the low PP conversion observed over ZSM-5 must take place on the external acid sites that, given the crystal size of this sample (5μm), accounts just for a very small proportion of the total acidity. These steric and diffusional limitations are not present in the polymer degradation catalyzed by MCM-41 due the larger pore size of this material.

Concerning the product distribution, Figure 2 compares the selectivity towards different hydrocarbon groups obtained with the two most active catalysts, MCM-41 and ZSM-5, in the conversion of the three polymers. In all cases, MCM-41 leads mainly to liquid fractions having boiling points in the range of gasoline and middle distillates, with overall selectivities between 70 and 85%. In contrast, the main products obtained over ZSM-5 zeolite are gaseous hydrocarbons with a high proportion of olefins.

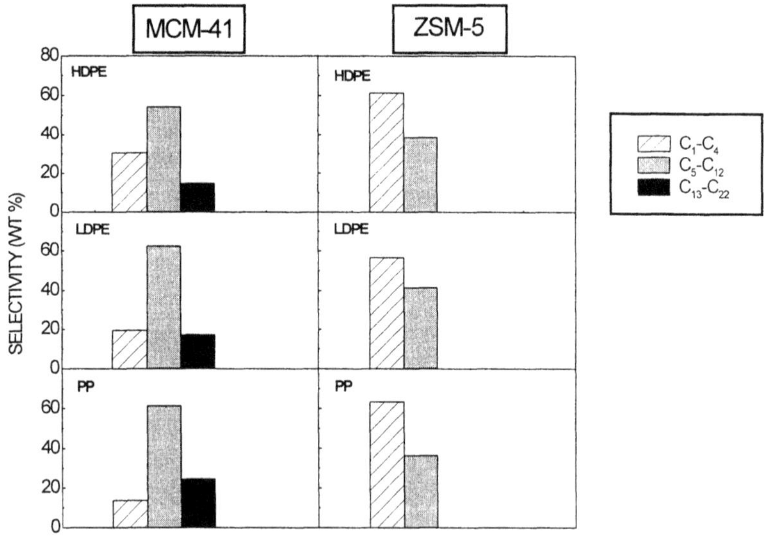

Figure 2. Product distribution in the catalytic cracking of pure polyolefins (reaction conditions as in Figure 1)

Similar conclusions can be obtained from the corresponding product distribution per carbon atom number. Thus, Figure 3 compares the distribution obtained in the LDPE cracking over MCM-41 and ZSM-5. For the latter, a strong maximum is observed between C_3 and C_6, with around 35 wt% of the products corresponding to C_4 hydrocarbons. On the contrary, the distribution obtained over MCM-41 is much wider since, although a maximum is present also at C_4, significant proportions of products are observed up to C_{20}. From these results, two different cracking pathways can be envisaged: end-chain cracking leading to gaseous hydrocarbons and random cracking at any point in the chain, which yields a wide distribution of heavy hydrocarbons. The first mechanism is predominant in the polymer degradation over ZSM-5, while both pathways take place over MCM-41. These differences are probably originated by the narrow pores of the zeolite which promote an intensive cracking of the molecules as they enter and diffuse along the channels, although the higher strength of the zeolite acid sites may also favour the end-chain cracking.

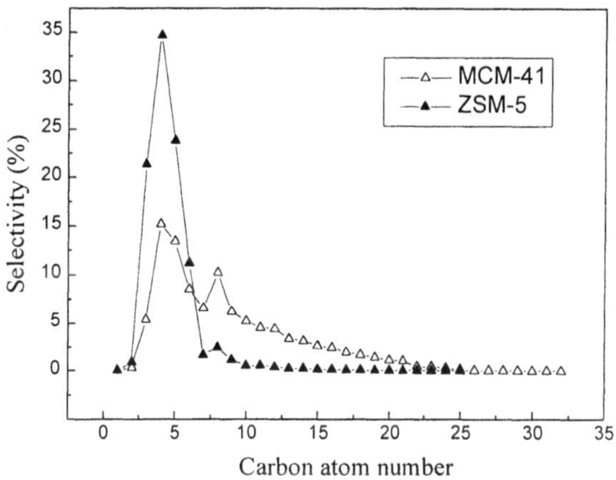

Figure 3. Product distribution per carbon atom number in the LDPE catalytic cracking (reaction conditions as in Figure 1)

Likewise, it must be pointed out that the primary products of the cracking reactions may undergo subsequent transformations leading to a higher variety of hydrocarbons. In this way, in the two distributions shown in Figure 3, a second maximum is observed at C_8 for both ZSM-5 and MCM-41, originated probably by the dimerization of C_4 olefins.

3.2. Catalytic degradation of mixed polyolefins

The three catalysts have been also checked for the degradation of a physical mixture of the three polyolefins since this may be a more realistic case when dealing with plastic wastes. The composition of the polymer mixture has been selected to be within the range of polyolefin distribution usually present in municipal plastic wastes. The results obtained in the catalytic degradation of this mixture have been summarized in Table 2.

The amorphous silica-alumina has a very low activity, as expected from the results obtained with the pure plastics. MCM-41 leads to a conversion close to 50%, remarkably higher than that obtained with the ZSM-5 sample. It is interesting to note that the ZSM-5 activity is appreciably lower than the one could be predicted based on the results with the pure polyolefins. Although ZSM-5 has shown not to be active for the PP degradation, the proportion of this polymer in the mixture (28.5 wt%) does not support the low overall plastic conversion observed with this catalyst. It seems that polypropylene has a deactivating effect on the ZSM-5 zeolite. A possible explanation to this result is the existence of a preferential contact of the ZSM-5 crystals with the PP particles, hindering the conversion of the other polyolefins present in the mixture.

Table 2. Catalytic cracking of mixed polyolefins (LDPE=46.5,HDPE=25,PP=28.5 wt%) T=400°C, t=30 min., plastic/catalyst= 50 (mass ratio)

	MCM-41	ZSM-5	SiO_2-Al_2O_3
Conversion (%)	49.0	6.8	3.8
C_1-C_4 paraffins (%)	3.9	17.3	17.9
C_2-C_4 olefins (%)	10.1	33.2	28.9
C_5-C_{12}	54.2	49.5	53.2
C_{13}-C_{22}	29.4	---	---
C_{23}-C_{40}	2.4	---	---

The product distribution obtained in the mixed polyolefin degradation over MCM-41 and ZSM-5 agrees well with that observed in the pure polymer cracking. In this case around 86% of the products obtained with MCM-41 are liquid hydrocarbons in the range of both gasoline and middle distillates.

3.3. Catalytic degradation of lube distillate

The catalytic properties of MCM-41 for the chemical recycling of used lubricating oils have been explored using as raw material a lube distillate formed by paraffinic hydrocarbons in the range C_{25}-C_{50}, the results being compared with those obtained over the ZSM-5 sample. The experiments have been carried out in a continous flow system, which has allowed us to obtain information about the catalyst deactivation.

The composition of the reactor effluent observed in the cracking of the lube oil at 400°C after 2 hours of time on stream is illustrated in Figure 4, being also included as reference the composition of the raw material. The conversion obtained in each case has been calculated from the decrease observed in the curve corresponding to the raw lube oil. Under these conditions, MCM-41 is quite more active than the ZSM-5 sample in spite of the stronger acidity of the latter, suggesting that at short times on stream the activity is determined by the pore size of the catalyst, the superior conversion over MCM-41 being related to the absence of diffusional limitations.

In regards to the product distribution, the results obtained are very similar to those commented above in the polyolefin degradation. Thus, whereas the product from the ZSM-5 cracking is formed by 63 wt% gases, mainly C_3 and C_4 olefins, the lube oil conversion over MCM-41 leads to 83 wt% of the products being in the range C_5-C_{28}. Therefore, it can be also concluded that ZSM-5 catalyzed cracking of the lube oil takes place preferently at the end of the chains, while both end-chain and random cracking are observed with MCM-41.

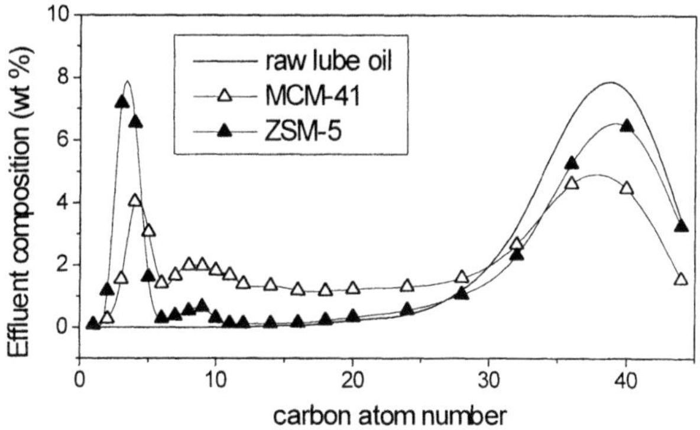

Figure 4. Effluent composition in the catalytic cracking of lube oil (400°C, WSHV=0.21 h^{-1}, t= 2 h)

The evolution of the catalytic activity along the time on stream for the lube oil cracking over MCM-41 and ZSM-5 is shown in Figure 5. Both catalysts undergo a loss of activity, but it is more accentuated for the MCM-41 sample. Thus, after 6 hours of time on stream the conversion observed with MCM-41 is lower than that of the zeolite. Although these results indicate that MCM-41 suffers an important deactivation, it is remarkable that this material can stand several hours of reaction exhibiting significant lube oil conversions. Moreover, it must be taken into account that ZSM-5 zeolite is widely known by its high resistance to deactivation by coking since the absence of large cavities in its pore structure avoid bulky polyaromatic molecules to be formed. However, this is not the case for MCM-41 due to its larger pore size.

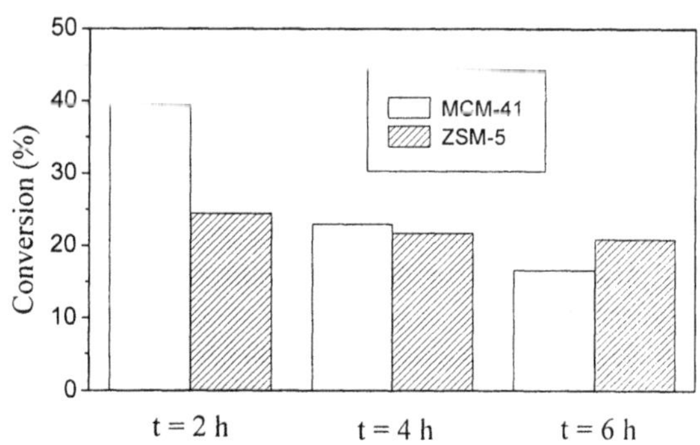

Figure 5. Catalyst deactivation during the lube oil cracking (reaction conditions as in Figure 4)

Changes in the product distribution are also observed along the time on stream. In the case of MCM-41, the relative proportion of gaseous hydrocarbons decreases as the catalyst is deactivated, whereas the formation of products higher than C_{10} is less affected. These results show that the catalyst deactivation influences in a higher extension the end-chain cracking reactions. Since coke deposition should also take place preferably on the stronger acid sites of the catalysts, it seems possible to establish a link between the acid strength and the cracking pathway. According to this, it can be proposed that the end-chain cracking reactions are favored and promoted by an stronger acidity. This conclusion would also explain the high amount of gaseous products observed in the polyolefins and lube oil degradations over ZSM-5 as a consequence of both the narrow pore size and the high acid strength of this material.

4. CONCLUSIONS

MCM-41 presents interesting catalytic properties for the degradation and feedstock recycling of both plastic wastes and used lubricating oils. For all the raw materials investigated in this work, MCM-41 leads to activities quite superior than those observed with an amorphous silica-alumina sample.

Compared to ZSM-5, the zeolite is a better catalyst than MCM-41 for the conversion of linear polymers, such as HDPE, due to its higher acid strength. However, the opposite is observed in the conversion of PP, a mixture of HDPE + LDPE + PP, and a lube oil, which can be assigned to the absence in MCM-41 of steric and diffusional hindrances. Regarding the catalyst deactivation along the time on stream, it has been observed during the lube oil cracking that MCM-41 is deactivated faster than ZSM-5 since the formation of bulky coke precursors within the mesopores is not restricted as in the zeolite. Nevertheless, MCM-41 leads to significant lube oil conversions during several hours of time on stream.

Two different cracking pathways have been observed: end-chain cracking leading to gaseous hydrocarbons and random scission at any point in the chain, which yields a wider distribution of heavier hydrocarbons. Both mechanism take place in the reactions catalyzed by MCM-41, whereas mainly end-chain cracking is observed over ZSM-5. This difference has been related to the narrow pores and the high acid strength of the zeolite.

REFERENCES

1. Y. Uemichi, Y. Kashiwaya, M. Tsukidate, A. Ayame and H. Kanoh, Bull. Chem. Soc. Jpn., 56 (1983) 2768.
2. A. R. Songip, T. Masuda, H. Kuwuhara and K. Hashimoto, Appl. Catal. B: Environ., 2 (1993) 153.
3. R. Lin and R.L. White, J. Appl. Polym. Sci., 58 (1995) 1151.
4. J. Aguado, D.P. Serrano, M.D. Romero and J.M. Escola, Chem. Commun., (1996) 725.
5. J. Aguado, J.L. Sotelo, D.P. Serrano, J.A. Calles and J.M. Escola, Energy Fuels, 11 (1997) 1225.
6. M.A. Uguina, A. de Lucas, F. Ruiz and D.P. Serrano, Ind. Eng. Chem. Res., 34 (1995) 451.

MESOPOROUS MOLECULAR SIEVES 1998
Studies in Surface Science and Catalysis, Vol. 117
L. Bonneviot, F. Béland, C. Danumah, S. Giasson and S. Kaliaguine (Editors)

Guanidine linked to micelle-templated mesoporous silicates as base catalyst for transesterification.

A. Derrien, G. Renard and D. Brunel*

Laboratoire des Matériaux Catalytiques et Catalyse en Chimie Organique. UMR 5618-CNRS-ENSCM, 8 rue de l'Ecole Normale, 34296-Montpellier cedex 5 (France)
Fax number: +33 (0)4 67 14 43 49

Micelle-Templated Silicas (MTSs) are raw materials for the design of new hybrid porous solids through covalent grafting of organic moieties onto the mineral surface. Anchorage of organic nitrogen-containing molecules having high base strength such as 1,5,7-triazabicyclo[4.4.0]dec-5-ene (TBD) is reported. It was performed through nucleophilic displacement of chlorine by TBD on chloropropylsilane chain previously grafted onto the surface. The end-capping of the chloropropylsilyl-MTS surface was also carried out using hexamethyldisilazane (HMDZ) in vapor phase before the substitution of chlorine in order to prepare homologous series of TBD-grafted MTS possessing hydrophobic behaviour. Material characterization comprised FTIR spectroscopy, thermogravimetric and elemental analyses. Textural properties of the modified MTS were determined by X-ray diffraction and by analyses of nitrogen adsorption and desorption isotherms after each treatment. The regular mesoporous structure of the mineral support was preserved during the various chemical modifications. HMDZ treatment favored the nucleophilic displacement of chlorine by TBD. Activity of the TBD-anchored MTS silicas was evaluated in the base-catalyzed transesterification reaction of ethylpropionate in *n*-butanol. The improved catalytic activity of TBD-anchored silicas after HMDZ treatment is explained by the lowering of interactions between the basic site of the guanidine and the surface.

1. INTRODUCTION

Micelle-Templated Silicas (MTSs) discovered by Mobil's researchers [1] provide excellent opportunities for the design of new hybrid nanostructured solids through covalent grafting of organic moieties onto this supporting mineral surface [2,3]. In this respect, we recently reported immobilization of organic functions such as amine groups [4,5], transition metal ligands [6] and β-aminoalcohols [7] on MTS surface in order to take benefit fromadvantages of both potential catalytic properties of these organic moieties and the ordered monodisperse mesopore system. Thus, primary and tertiary amine functions linked to

MTS surface showed base catalytic activity in Knoevenagel condensation reaction [5]. Moreover, these supported catalytic sites were designed to be both very active in addition of fatty acid to glycerol, leading to the highly selective synthesis of α-monoglyceride, and stable to leaching during reactions performed in liquid phase using organic solvents [6]. Nevertheless, the major drawback of such catalysts is their limited applications to non-demanding reactions due to the low base strength of amine-immobilized functions. In order to overcome this limitation, we have planned to anchor guanidine molecules which are stronger base than amine functions. They have previously been either used as homogeneous catalyst [8] or heterogeneized on polymer support for demanding base catalyzed reactions [9]. In the course of this work, two heterogeneization modes of guanidines have been reported in the literature. They were carried out either by encapsulation in hydrophobic zeolite Y [10] or by grafting on silica surface [11]. The different applications of these composite materials described in these works emphasize the great scope of these new basic and mesoporous solids as catalysts.

The present study deals with the anchorage of 1,5,7-triazabicyclo[4.4.0]dec-5-ene (TBD) by covalent linkage on MTS surface. We might expect that the polarity of the uncovered surface will affect the base strength of the grafted guanidine molecule through H-bonding between residual silanol groups and basic sites and therefore will influence the catalytic properties of these materials. We have studied in previous works the beneficial effect of the end-capping of grafted MTS on their catalytic efficiency during monoglyceride synthesis. Hence the aim of this work is also to determine the effect of such a treatment on the catalytic activity of the grafted guanidine molecules in the transesterification reaction of ethylpropionate in *n*-butanol and on the stability of these immobilized catalysts.

2. EXPERIMENTAL SECTION

2.1. Materials.

- MTS silica was prepared according to a procedure derived from the Mobil's method [1] consisting in addition of silica (Ludox HS-40 Aldrich) (100 g) to a stirred solution of cetylammonium bromide (25 g, 0.93 mol) and sodium hydroxide (7 g, 0.175 mol) in deionized water (350 mL) at 70°C for 16 h. After filtration, and successive washings with water and ethanol, the solid was dried at 50°C in air, then calcined at 550°C for 7 h to eliminate the organic template.
- 3-chloropropylsilyl-MTS (Cl-MTS) was prepared by addition of 3-chloropropylmethoxy-silane (3 g, 18 mmol) to a stirred suspension of MTS silica, previously activated at 180 °C under vacuum, in refluxing dry toluene (50 mL). Fractions of toluene containing traces of methanol and dimethyl ether were distilled from the reaction mixture during the functionalization procedure. The modified solid was recovered by filtration, washed in a Soxhlet apparatus using a 1:1 diethyl ether - dichloromethane mixture.

- End-capped 3-chloropropylsilyl-MTS (AZA-Cl-MTS) was obtained by treatment of evacuated (10^{-1} Torr) Cl-MTS heated at 175 °C with hexamethyldisilazane (HMDZ) in gas phase under dynamic vacuum (1 Torr) for 3 h. The solid was then reevacuated at 10^{-1} Torr for 1 h. Absence of physisorbed HMDZ or NH_3 on the AZA-Cl-MTS was controlled by elemental analyses.
- 3-7,N(1,5,7-triazabicyclo[4.4.0]dec-5-ene)propylsilyl-MTS (G-Cl-MTS and G-AZA-Cl-MTS) were obtained by reaction of stirred suspensions of activated Cl-MTS (1 g) and AZA-Cl-MTS (1 g) respectively, with a refluxing toluene solution (100 mL) of 1,5,7-triazabicyclo[4.4.0]dec-5-ene (2 g, 13 mmol) for 7 h. After filtration, the solids were successively washed with methanol, THF and acetone, then thoroughly washed in a Soxhlet apparatus with methanol.

2.2. Material characterization

Textural features of the materials were determined by X-ray diffraction and from nitrogen adsorption -desorption isotherms. Compositions of the organic-inorganic hybrid solids were determined by elemental analyses from Si/C/Cl and Si/C/Cl/N ratios for both Cl-MTS and Cl-AZA-MTS and both G-Cl-MTS and G-AZA-Cl-MTS, respectively. These results have been correlated with those obtained by thermogravimetry using organic over mineral weight ratio. Chemical structures of the organics were monitored by FTIR and ^{13}C MAS-NMR.

2.3. Transesterification reactions

Reactions of ethyl propionate in n-butanol catalyzed by G-Cl-MTS and G-AZA-MTS catalysts were carried out in a stirred (rotation speed : 700 rpm) batch reactor under an N_2 atmosphere at 100°C. Typical conditions were as follows : ethylpropionate (1.15 mL, 10 mmol) and n-butanol (8.85 mL, 124 mmol) were added to the catalyst (300 mg) previously evacuated at 130°C overnight. Reactions were followed and quantified by gas chromatography using decane as external standard.

3. RESULTS AND DISCUSSION

3.1. Preparation of 1,5,7-triazabicyclo[4.4.0]dec-5-ene (TBD) anchored onto MTS surface

The 3-chloropropylsilane moiety was previously grafted on MTS surface in order to subsequently anchor TBD molecule through nucleophilic displacement of the chloro atom by the 7-NH group (Scheme 1). This grafting reaction was performed on well-dehydrated MTS silica and using an anhydrous and apolar solvent such as toluene in order to avoid sol-gel polymerisation of the silanating reagent leading to an island-type coverage. These conditions should favor a better distribution of the grafted chains over the mesoporous surface [12].

The coupling reaction with TBD was carried out at relatively low temperature (120°C) and drawn up by elimination of evolved HCl using an large excess of TBD

Scheme 1

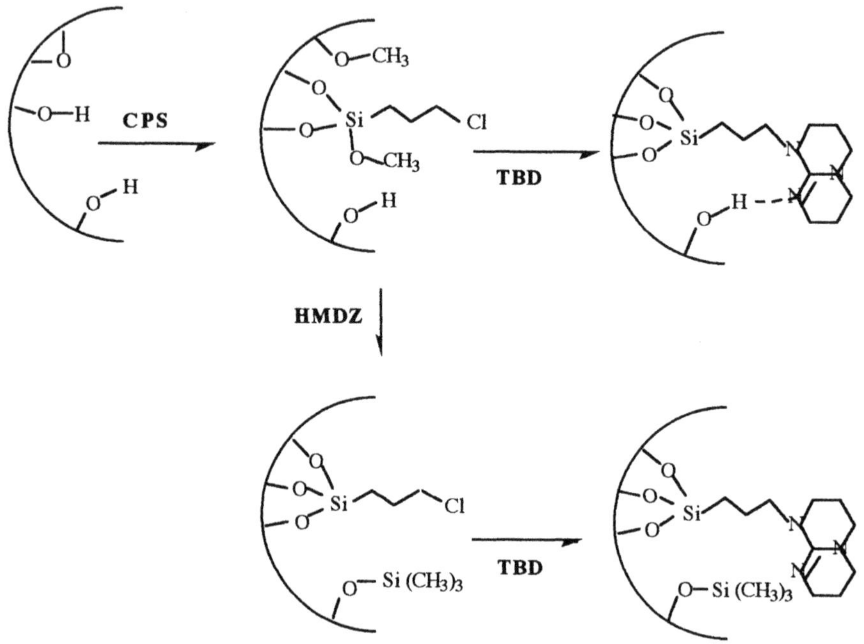

The high base strength of guanidine molecules constitutes one of the most important pitfalls during the anchorage procedure on silica surface due to the possible alteration of the regular ordered mesoporous structure. In order to avoid possible hydrolysis of the silicic walls through hydroxide formation, anhydrous experimental conditions *i.e* evacuated solids, anhydrous solvent and dry nitrogen atmosphere, were used. On the other hand end-capping of the residual surface silanols or silane silanols of Cl-MTS might improve the hydrolysis-resistance of the uncovered walls. In this respect, another Cl-MTS sample was treated by hexamethydisilazane (HMDZ) vapor under dynamic vacuum at 175 °C [13], prior to guanidine coupling reaction. This relevant modification was also carried out to reduce possible strong H-interactions between residual silanol groups and imine-type nitrogen atoms, possessing the highest electron density [14] as shown in Scheme 1, and consequently to preserve the base strengh of anchored guanidine molecules.

To ensure elimination of guanidinium chlorine and possible physisorbed TBD and MTBD on the MTS surface, modified solids resulting from the coupling reaction were thoroughly washed with methanol in a Soxhlet apparatus after washings on glass-frit in succession with methanol, THF and acetone.

3.2. Characterization of materials

X-ray diffraction spectra of modified MTSs compared to the ones of the parent MTS showed the preservation of the ordered hexagonal mesoporous texture during grafting experiments. Lattice parameters **a** of the hexagonal array determined from d_{100} are given in Table 1. These results were confirmed by nitrogen sorption-desorption studies showing a sharp, reversible step at around 0.3-0.4, typical of the filling of regular mesoporous systems. The surface area S resulting from analyses of these isotherms by BET equation in the p/p_0 range 0.05 to 0.2 , are given in Table 1 as well as the mesoporous volume Vm, measured at the top of the pore-filling step.

Table 1
Textural properties and organic contents of modified MTS

Materials	**a** (Å)	S ($m^2.g^{-1}$)	Vm ($mL.g^{-1}$)	organic chain content ($mmol.g^{-1}$) -$(CH_2)_3$-Cl	- $(CH_3)_3$	-$(CH_2)_3$-TBD
MTS	46.3	1020	0.83	-	-	-
Cl-MTS	46.3	900	0.59	1.0	-	-
G-Cl-MTS	46.3	783	0.55	0.1		0.40
AZA-Cl-MTS[a)]	47.6	617	0.4	0.45	2	-
G-AZA-Cl-MTS[a)]	46.3	757	0.54	0.1	1	0.35

a) synthesized from another parent Cl-MTS sample prepared with a chloropropyl chain loading of 0.5 $mmol.g^{-1}$ and subsequently washed with 2% HCl solution in order to hydrolyze methoxysilane groups before end-capping treatment.

Identification of grafted moieties was achieved by combining ^{13}C MAS-NMR and FTIR spectroscopic studies performed on the modified MTSs and comparison with the corresponding groups in solution.

The infrared spectrum of Cl-MTS showed stretching vibration bands around 2930-2890 cm^{-1} assigned to the -CH_2-of the propyl chain and a band at 2965 cm^{-1} attributed to asymmetric CH_3 vibrations. Moreover, bands of isolated silanols at 3747 and 3742 cm^{-1} are much less intense than the ones of the parent MTS despite the fact that bands of terminal and H-bonded silanols at 3715 and 3536 cm^{-1}, respectively, are less affected during the grafting as previously reported [12]. On the other hand, the spectrum of AZA-Cl-MTS exhibited an intense band at 2960 cm^{-1} assigned to the CH_3-Si groups with a strong reduction of isolated terminal silanol bands. In addition to characteristic vibration bands of the parent Cl-MTS and AZA-Cl-MTS, spectra of the G-Cl-MTS and G-AZA-Cl-MTS showed a band at 1643 cm^{-1} assigned to C=N stretching vibrations. On the other hand, the spectra exhibited another band at 1600 cm^{-1} which is very less intense for G-AZA-Cl-MTS than for G-Cl-MTS. These results would be consistent with a higher interaction of C=N groups with residual surface silanols, which is also featured by a broader band of interacting silanols in the range 3530-3700 cm^{-1}. Nevertheless one cannot totally evidence such an assumption due to the presence of incompletely desorbed water whose bending vibrations appeared at 1630 cm^{-1}.

Contents and structures of grafted organic chains were determined by elemental analysis combined with thermogravimetry analysis. Results are reported in the last columns of Table 1. It should be noted that coupling reactions with guanidine were no complete. Moreover a noticeable decrease in the functional organic chain balance (Cl- + G-) is observed during the coupling reaction and washing, carried out on the first serie. This phenomenom could result from subsequent quaternization at the 5-N atom by neighbour 3-chloropropyl chain and $-CH_2-N^+\equiv$ bond cleavage as shown in Scheme 2.

Scheme 2

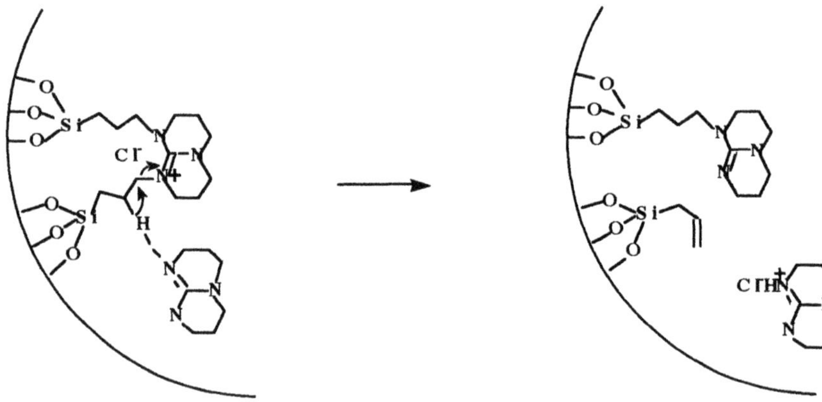

Indeed, this quaternization reaction could be favored by higher density of chloropropyl chains. Thus increases in surface area and in mesoporous volume would be in good accordance with such an elimination of chorine atoms. On the other hand, partial elimination of linked organic chain can occur by hydrolysis of linkage siloxane bonds during the methanol washing of G-MTS samples in Soxhlet apparatus which was performed on modified solids exposed to moisture. Nevertheless the latter hypothesis cannot satisfactorily explain the increase in the surface area and mesoporous volume observed during these last modifications. No significant loss of the ordered mesoporous structure has been observed as shown by both X-ray diffraction and sharp filling step on nitrogen sorption isotherms even though elimination of some organic chains would be a local reaction due to the proximity of the basic site.

3.3. Catalytic activity of linked TBD onto MTS in transesterification reaction

Transesterification is one of the most important reactions in the development of clean process technology, in particular for synthesis of methyl- and ethyl-transesterified vegetable oils. These reactions have been performed by either acid or base media, but reaction rates of alcoholysis are faster using base catalysts, such as alkaline metal hydroxide rather than acid ones [15]. Nevertheless, in non-anhydrous media, saponification reaction of fatty acids resulting from hydrolysis of fatty esters, have lowered the transesterification yield. Hence, the use of strongly basic and porous solids would constitute alternative catalysts to perform such a reaction in heterogeneous and anhydrous conditions.

In this work , transesterification of ethyl propionate in *n*-butanol was studied as a model reaction :

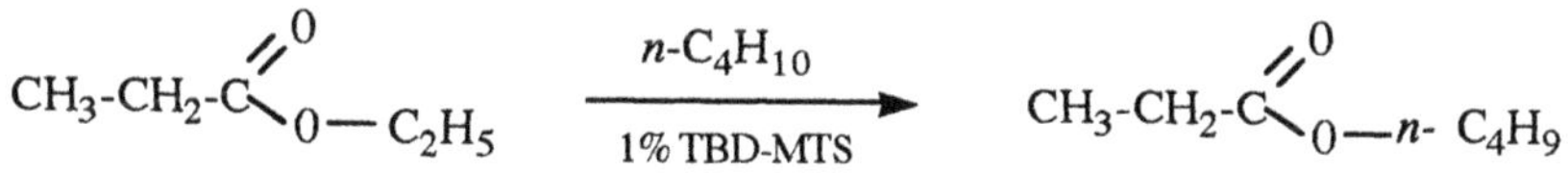

Conversions of ethyl propionate in *n*-butyl propionate and turn-over numbers using G-Cl-MTS (TBD loading: 0.4 mmol.g^{-1}) and G-AZA-Cl-MTS (TBD loading: 0.35 mmol.g^{-1}) as catalysts are reported in Table 2. Data obtained with 7-methyl-1,5,7-triazabicyclo[4.4.0]dec-5-ene (MTBD) in homogeneous catalysis are also mentioned for comparison.

Table 2
Catalytic data for transesterification of ethyl propionate in *n*-butanol using G-Cl-MTS and G-AZA-Cl-MTS as catalysts.

Catalyst	**G-Cl-MTS[a]**		**G-AZA-Cl-MTS[a]**		**MTBD[b]**	
Time (h)	% conversion	TON	% conversion	TON	% conversion	TON
1	-	-	2	2	6	6
6	10	8	13.5	13	26.5	26.5
24	13	11	43	41	56	56
48	-	-	66	63	73	73

Reaction conditions: a) catalyst: 300 mg, ethylpropionate : 1.15 mL (10 mmol); *n*-butanol : 8.85 mL (125 mmol) ; Temperature :100°C ; b) same conditions except homogeneous catalyst : MTBD: 16 mg (0.1 mmol)

It is worth noting that the immobilized TBD on end-capped Cl-AZA-MTS showed higher catalytic activity than the homologous modified MTS without end-capping treatment although they possess comparable TBD loadings. This result would be explained by an improvement of the basic strength of the guanidine molecule due to the reduced interaction of grafted basic sites with polar residual silanol groups. Nevertheless, the heterogeneized guanidine is slightly less effective than the homogeneous counterpart. Such a result would be consistent with non selective anchorage of the TBD on the grafted propyl chain through 7-N linkage, with probably some little contribution of quaternization reaction on 5-N atom. The latter reaction would be more important during the preparation of G-Cl-MTS due to higher chloropropyl chain density.

On the other hand, recycling of G-AZA-Cl-MTS after filtration, washing with methanol and reactivation led to a strong decrease of the catalytic activity in a second run. This activity loss was in good accordance with leaching of guanidine molecules observed during the recycling procedure.

4. CONCLUSION

Because of their very high surface areas and ordered mesoporous structure, MTS materials constitute excellent supports for the grafting of homogeneous catalytic sites. Nevertheless, anchorage of strong organic base onto MTS surfaces is not easily performed because of interaction with the supporting surface and leaching during catalytic runs in liquid phase. Thus, although such heterogeneized strong bases are promizing catalysts, this work reveals that in-depth studies are required to circumvent the problems met in the repeated reaction-regeneration cycles.

ACKNOWLEDGEMENTS

The authors thank their colleagues Drs Monique Laspéras and Francesco Di Renzo and Dr Jacques Tellier (Elf-GRL) for fruitful discussions and are indebted to Elf-Aquitaine for financial support.

REFERENCES

1 J.S. Beck, J.C. Vartuli, W.J. Roth, M.E. Leonowicz, C.T. Kresge, K.D. Schmitt, C.T.-W. Chu, D.H. Olson, E.W. Sheppard, SB. McCullen, J.B. Higgins and J.L. Schenker, J. Am. Chem. Soc., 114 (1992) 10834.
2 D. Brunel, A. Cauvel, F. Fajula and F. Di Renzo, Sud. Surf. Sci. Catal., 97 (1995) 173.
3 A. Cauvel, D. Brunel, F. Di Renzo and F. Fajula, Amer. Inst. Phys., 354 (1996) 477.
4 A. Cauvel, G. Renard and D. Brunel, J. Org. Chem., 62 (1997) 749.
5 M. Laspéras, T. Llorett, L. Chaves, I. Rodriguez, A. Cauvel and D. Brunel, Stud. Surf. Sci. Catal., 108 (1997) 75.
6 P. Sutra and D. Brunel, J. Chem. Soc., Chem. Comm., (1996) 2485.
7 N. Bellocq, D. Brunel, M. Laspéras and P. Moreau, Stud. Surf. Sci. Catal., 108 (1997) 485.
8 U. Schuchardt, R. M. Vargas and G. Gelbard, J. Mol. Catal., A: Chemicals, 99 (1995) 65.
9 U. Schuchardt, R. M. Vargas and G. Gelbard, J. Mol. Catal., A: Chemicals, 109 (1996) 37.
10 R. Sercheli, A.L.B. Ferreira, M.C. Guerreiro, R. M. Vargas, R.A. Sheldon and U. Schuchardt, Tetrahedron Lett., 38 (1997) 1325.
11 Y.V. Rao, D.E. De Vos and P.A. Jacobs, 3rd Inter. Symp. "Supported Reagents and Catalysis in Chemistry" Limerick 8-11 july 1997. Royal Soc. Chem. Ed. in press.
12 D. Brunel, A. Cauvel, F. Di Renzo, E. Garrone and B. Fubini, in preparation
13 S. Haukka and A. Root, J. Phys. Chem., 98 (1994) 1695.
14 M. Abiuso, PhD, n° 310-95 Claude Bernard-Lyon I University, december, 1995, 12th.
15 J. Graille, P. Lozano, D. Pioch, P. Geneste, A. Finiels and C. Moreau, Rev. Francaise Corps Gras, 32 (1985) 311.

MESOPOROUS MOLECULAR SIEVES 1998
Studies in Surface Science and Catalysis, Vol. 117
L. Bonneviot, F. Béland, C. Danumah, S. Giasson and S. Kaliaguine (Editors)

Probing the active sites of aluminated mesoporous molecular sieve MCM–41 by secondary synthesis in the conversion of cyclohexanol

Salasiah Endud, Hadi Nur and Halimaton Hamdan

Department of Chemistry, Faculty of Science,Universiti Teknologi Malaysia
KB 791, 80990 Johor Bahru, Malaysia

The active sites of H-Al-MCM-41 aluminated by secondary synthesis have been probed by the conversion of cyclohexanol and compared with those of H-Al-MCM-41 prepared by direct synthesis, purely siliceous MCM-41, $AlPO_4$-5 and H-ZSM-5. Conversion of cyclohexanol produces cyclohexene, cyclohexanone and 3-methylpentane in the presence of Brönsted, basic and Lewis acid sites respectively. Whereas cyclohexanol is converted to polyaromatic compounds in the presence of very strong acid sites as those found in H-ZSM-5. The formation of cyclohexene as the only product of conversion over H-Al-MCM-41 prepared by direct and secondary synthesis indicates the presence of Brönsted acid sites in both systems, which is not observed in the purely siliceous MCM-41 and $AlPO_4$-5 samples. The larger amount of cyclohexene formed over H-Al-MCM-41 by secondary synthesis suggests that there is a higher degree of Brönsted acidity in this system than that of H-Al-MCM-41 by direct synthesis. However, the strength of acidity in H-Al-MCM-41 by secondary synthesis is weaker than that of H-ZSM-5. In addition, the results of this reaction, supported by IR spectroscopy study, do not indicate the presence of Lewis acidity in H-Al-MCM-41 prepared by secondary synthesis.

1. INTRODUCTION

Mesoporous molecular sieve MCM-41 attracts much attention because of its unique properties [1,2] and potential as catalysts in reactions involving large organic molecules [3-6]. Purely siliceous MCM-41 does not show significant catalytic activity as it lacks the ion exchange capability due to its electrically neutral framework. However, the isomorphous substitution of the framework Si by aluminium or other trivalent cations offers a mechanism for generating acidity and for modifying the chemical properties of the purely siliceous MCM-41. The incorporation of aluminium into the framework of MCM-41 is particularly important as it gives rise to Brönsted acid sites. It has been reported that calcined Al-MCM-41 has acid properties similar to those of amorphous silica [7]. However, it has not been firmly established whether the acidity is due to the Brönsted acidic protons which are ion-exchanged in order to balance the necessary number of framework negative charges introduced by the incorporation of aluminium, or attributed to the formation of extraframework aluminium which are associated with the presence of Lewis acidity.

In a recent communication [8], we have succeeded in preparing mesoporous aluminosilicate MCM-41 materials by isomorphous substitution via secondary synthesis. The success of preparing Al-MCM-41 by such a technique offers great advantages in tailoring Al-MCM-41 as catalyst with variable acidic properties for specific reactions. Therefore, it is of interest to identify and understand the nature of active sites present in these materials; whether it is acidic, basic or cationic. In previous acidity studies [9,10], pyridine or ammonia adsorption is most often used to determine the concentration and strength of the Brönsted and Lewis acid sites in zeolitic materials. In our study presented here, the cyclohexanol conversion has been chosen as the probe reaction.

In the conversion of cyclohexanol, two reactions can occur: the dehydration into cyclohexene and water and the dehydrogenation leading to cyclohexanone. The dehydration of cyclohexanol occurs due to the presence of Brönsted acid sites, whereas the dehydrogenation of cyclohexanol takes place in the presence of surface basic sites (Figure 1). Therefore besides identifying the nature of active sites, it is possible to estimate the relative acidity of H-Al-MCM-41 prepared by both direct and secondary synthesis based on this diagnostic test reaction. In addition, the catalytic behaviour of these materials in cyclohexanol conversion is compared with other catalysts: $AlPO_4$-5 that possesses some weak basic sites and H-ZSM-5, which is strongly acidic and has been shown to be an excellent catalyst for the conversion of alcohol to gasoline [11,12].

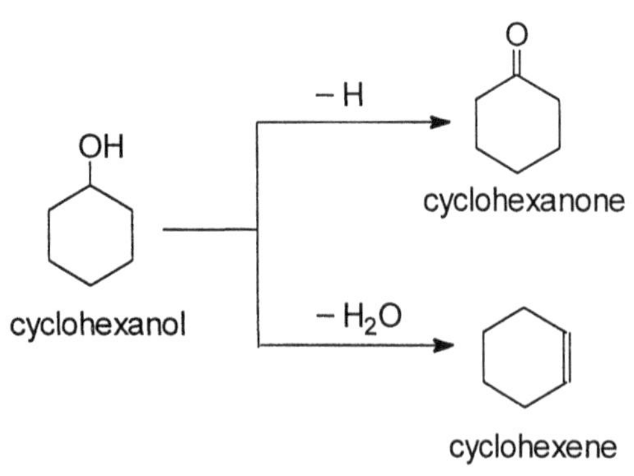

Figure 1. Schematic representation of the dehydrogenation of cyclohexanol to cyclohexene and the dehydration of cyclohexanol to cyclohexene.

2. EXPERIMENTAL

Sample of H-ZSM-5 was supplied by N. E. Chemcat, Japan and $AlPO_4$-5 molecular sieve was synthesized as described elsewhere [13]. The synthesis, treatment and characterization of purely siliceous MCM-41 and aluminosilicate MCM-41 (Si/Al = 2.8, 3.6 and 3.8) is made according to the procedure described previously [8]. Calcined Al-MCM-41 sample (Si/Al = 3.6), is used for the protonic ion exchange procedure performed with 1 M NH_4NO_3 solution at 90 °C (with liquid-to-solid ratio of 50 ml/g). H-Al-MCM-41 was obtained by deammoniating the ammonium form of Al-MCM-41 at 550 °C in air. Sample of Al-MCM-41 by direct synthesis was prepared according to a literature method [2]. The cyclohexanol conversion reaction was performed over 0.5 g of H-Al-MCM-41, H-ZSM-5 (Si/Al ratio ≈ 38), $AlPO_4$-5 and purely siliceous MCM-41 using a continuous flow, fixed bed microreactor at 400 °C and atmospheric pressure, with a feed rate of 0.05 ml/min. Cyclohexanol was a BDH product with 99% purity and used as received. Product distribution was determined by a Shimadzu GC-14B gas chromatograph equipped with an FID detector and a 50 m fused silica capillary column with helium as carrier gas. The products were identified on the basis of their mass spectra using Hewlett Packard 5890 Series II gas chromatograph and Hewlett Packard 5982A mass spectrometer. For IR studies on MCM-41 materials, thin self supporting wafers of about 10 mg/cm^2 were prepared and activated in situ in the IR cell fitted with CaF_2 windows at 400 °C in vacuum for 16 h. The infrared spectra were recorded in the range of 4000–1300 cm^{-1} using a Perkin Elmer 1600 spectrometer with a resolution of 2 cm^{-1}.

3. RESULTS AND DISCUSSION

The infrared spectra of hydroxyl groups of the purely siliceous MCM-41 and secondary aluminated MCM-41 are shown in Figure 2(i) in the region of stretching vibrations of O-H bands. A very intense band at 3744 cm^{-1} was observed in the IR spectrum of the parent MCM-41 corresponding to terminal silanol groups. These groups have been previously observed on amorphous silica and microporous zeolites and reported to reveal a very weak acid character [14]. The IR results clearly show that the MCM-41 samples have very large amount of silanol groups, the largest population being in the purely siliceous MCM-41.

It is seen that incorporation of aluminium in the purely siliceous MCM-41 samples leads to a decrease in the intensity of the IR band assigned to terminal Si-OH hydroxyls (3744 cm^{-1}) which is shifted to 3742 cm^{-1}. This is presumably connected with structural redistributions resulting from the insertion of Al species into the mesoporous structure during alumination. After adsorption of pyridine at room temperature, the OH stretch band at 3742 cm^{-1} shows a loss in intensity and broadens as seen in Figure 2(ii), due to the interaction with the pyridine molecules, indicating the presence of weak acid sites.

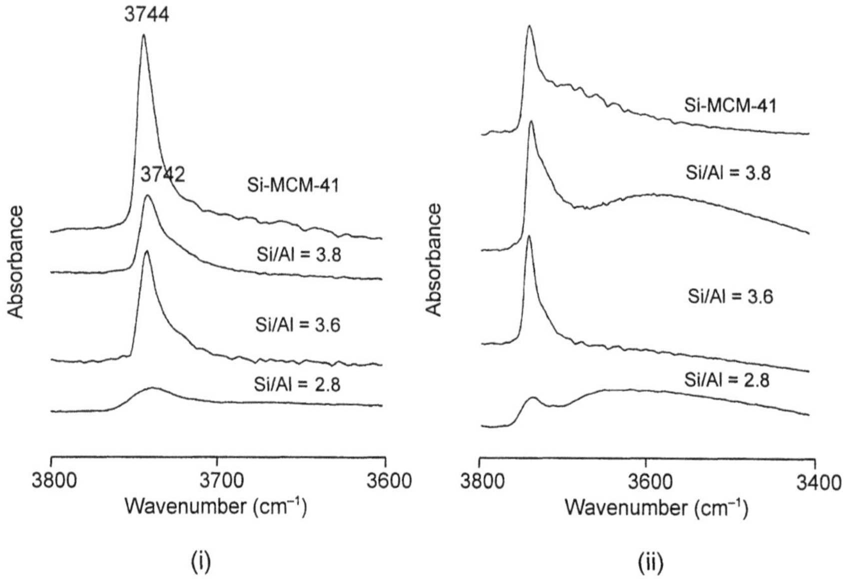

Figure 2. FTIR spectra of the hydroxy groups of (i) activated MCM-41 samples and (ii) after pyridine adsorption.

Figure 3(i) depicts the spectrum of pyridine chemisorbed on the purely siliceous MCM-41, showing no Brönsted acidity and a weak Lewis acidity. The same IR bands due to pyridinium ions and to pyridine bonded to Lewis sites but of lower intensities were observed with the desorption of pyridine at higher temperatures. Secondary aluminated MCM-41 of various concentrations exhibit two IR bands associated with Brönsted acidity (B) at 1540 cm^{-1} and Lewis acidity (L) at 1450 cm^{-1} respectively as shown in Figures 3(ii)–(iv). It is observed that the amount of Brönsted acidity increases as the amount of aluminium in the framework of secondary Al-MCM-41 increases. The ratio of observed intensity of the Brönsted band, I_B, to the Lewis band, I_L, for Al-MCM-41 with the Si/Al = 2.8 is I_B/I_L = 2.3. The formation of a small amount of strong Lewis acid sites, *i.e.* octahedrally coordinated Al species in the aluminosilicate sample may be affiliated with the presence of octahedral Al in the MCM-41 framework [15]. We note that these acid sites are rather strong in Al-MCM-41 as pyridine is still present on the sample after outgassing at 400 °C. Moreover, when the experiment was performed with purely siliceous MCM-41, the peaks hardly persist upon evacuation of the sample at 250 °C. That was an additional evidence that strong Lewis acid sites were generated upon substitution of framework Si^{4+} by Al^{3+}.

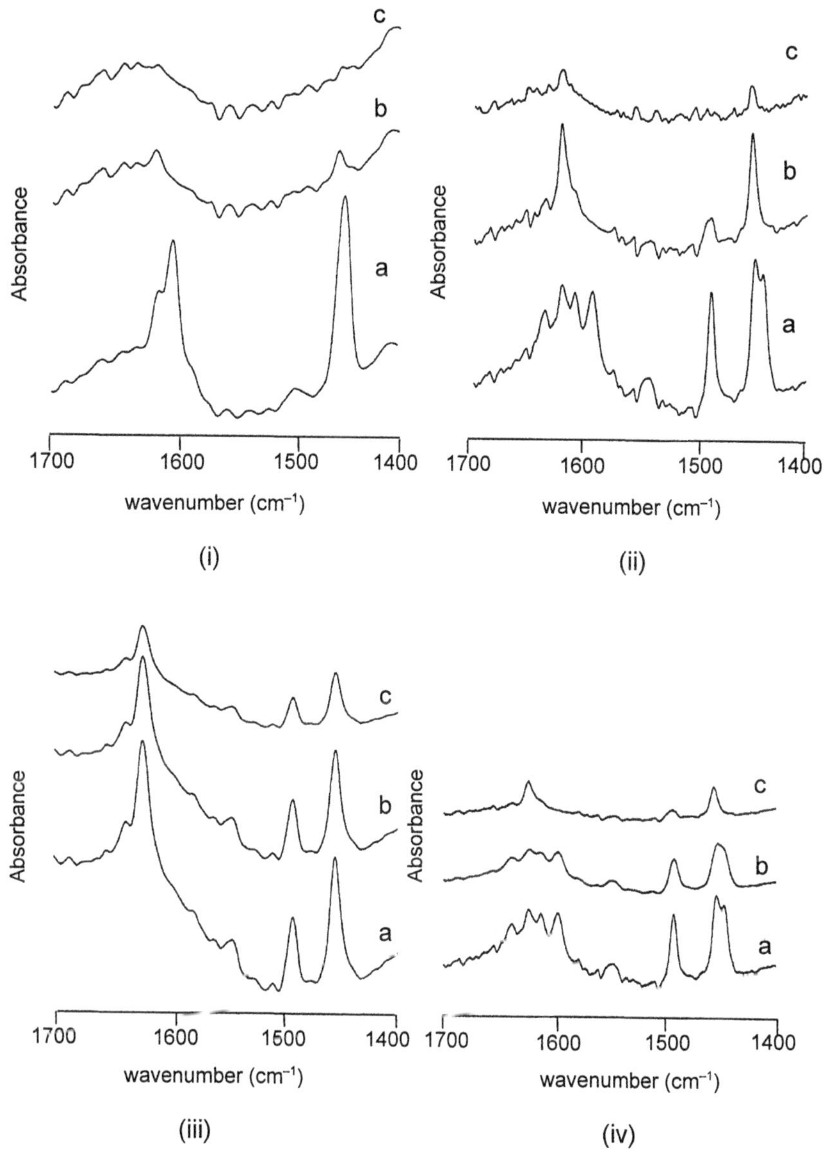

Figure 3. FT-IR spectra of adsorbed pyridine on (i) purely siliceous MCM-41 and Al-MCM-41 samples with (ii) Si/Al = 2.8, (iii) Si/Al = 3.6 and (iv) Si/Al = 3.8 evacuated at a = 25 °C, b = 250 °C and c = 400 °C

Table 1 shows the values of total conversion of cyclohexanol to cyclohexene at 400°C over H-Al-MCM-41, Si-MCM-41, $AlPO_4$-5 and H-ZSM-5. The results show that at 400 °C cyclohexanol conversion is the highest over secondary aluminated H-Al-MCM-41 which produces cyclohexene as the only product at nearly 100 % conversion. This confirms that secondary aluminated H-Al-MCM-41 consists only of Brönsted acid sites. H-Al-MCM-41 prepared by direct synthesis yielded a lower rate of dehydration of cyclohexanol, suggesting a lower concentration of Brönsted acid sites in this system, confirming the poor incorporation of aluminium in the framework. As expected, the purely siliceous MCM-41 did not show any catalytic activity towards the conversion of cyclohexanol to cyclohexene due to the absence of Brönsted acidity. This material has been characterized to have mainly Lewis acidity by means of IR spectroscopy using pyridine adsorption. The only product detected from the test reaction is 3-methylpentane. In contrast, cyclohexanol was converted to cyclohexene and cyclohexanone in the presence of $AlPO_4$-5. The formation of cyclohexanone in this reaction indicates that some basic sites are also present in $AlPO_4$-5, as already reported by other workers [16,17]. In the case of H-ZSM-5, cyclohexanol is mostly converted to cyclohexene with a small amount of polyaromatic compounds as minor products confirming its well-known acidic nature. In this context, H-Al-MCM-41 prepared by secondary synthesis is less acidic as compared to H-ZSM-5, which inhibits consecutive disproportionation and isomerization reactions of cyclohexene.

4. CONCLUSIONS

Secondary aluminated Al-MCM-41 contains weak Brönsted acid sites that are not present in the purely siliceous sample. The concentration of the Brönsted acid sites increases with the amount of aluminium inserted into the

Table 1
Total conversion of cyclohexanol to cyclohexene

Catalyst	Conversion wt. [%]		other products
	cyclohexene	cyclohexanone	
H-Al-MCM-41 (secondary)	100	none	none
H-Al-MCM-41 (direct)	83	none	none
Si-MCM-41	none	none	3-methylpentane
H-ZSM-5	98	none	polyaromatics
$AlPO_4$-5	78	22	None

MCM-41 framework. Catalytic conversion of cyclohexanol to cyclohexene result demonstrates that H-Al-MCM-41 prepared by secondary synthesis contains only Brönsted acid sites with a higher concentration than the one present in the directly synthesized H-Al-MCM-41. The Al-MCM-41 materials in this study is hydrothermally stable and is a potential catalyst in acid-catalyzed conversion of alcohols at moderate temperatures.

REFERENCES

1. C. T. Kresge, M. E. Leonowicz, W. J. Roth, J. C. Vartuli and J. S. Beck, Nature, 359 (1992) 710.
2. J. S. Beck, J. C. Vartuli, W. J. Roth, M. E. Leonowicz, C. T. Kresge, K. D. Schmitt, C. T–W. Chu, D. H. Olson, E. W. Sheppard, S. B. McCullen, J. B. Higgins and J. L. Schlenker, J. Am. Chem. Soc., 114 (1992) 10834.
3. A. Corma, A. Martínez, V. Martínez-Soria and J. B. Montón, J. Catal. 153 (1995) 25.
4. K. Roos, A. Liepold, W. Reschetilowski, R. Schmidt, A. Karlsson and M. Stöcker, Stud. Surf. Sci. Catal., 94 (1995) 389.
5. K. M. Reddy and C. Song, Catal. Today, 31 (1996) 137.
6. J. Aguado, D. P. Serrano, M. D. Romero and J. M. Escola, J. Chem. Soc., Chem. Commun., (1996) 725.
7. C–Y. Chen, H–X. Li and M. Davis, Microporous Mater., 2 (1993) 17.
8. H. Hamdan, S. Endud, H. He, M. N. M. Muhid and J. Klinowski, J. Chem. Soc., Faraday Trans., 92 (1996) 2311.
9. J. Weglarski, J. Datka, H. He and J. Klinowski, J. Chem. Soc., Faraday Trans., 92 (1996) 5161.
10. J–H. Kim, M. Tanabe and M. Niwa, Microporous Mater., 10 (1997) 85
11. C. D. Chang, Stud. Surf. Sci. Catal., 36 (1988) 127.
12. S. A. Tabak and S. Yurchak, Catal. Today, 6 (1990) 307.
13. S. T. Wilson and E. M. Flanigen, European Patent Application. 0043562, 1982.
14. A. Corma, V. Fornés, M.T. Navarro and J. Pérez-Pariente, J. Catal., 148 (1994) 569.
15. E. P. Parry, J. Catal., 2 (1963) 371.
16. J. B. Moffat, R. Vetrival and B. Viswanathan, J. Mol. Catal., 30 (1985) 171.
17. S. P. Elangovan and V. Murugesan, J. Mol. Catal.,118 (1997) 301.

MESOPOROUS MOLECULAR SIEVES 1998
Studies in Surface Science and Catalysis, Vol. 117
L. Bonneviot, F. Béland, C. Danumah, S. Giasson and S. Kaliaguine (Editors)

Fe-containing Mesoporous Molecular Sieves as Benzylation Catalysts

Jieming Cao, Nongyue He, Cun Li, Jialu Dong, and Qinhua Xu*

Department of Chemistry, Nanjing University, Nanjing 210093, P. R. China
State Key Laboratory of Coordination Chemistry, Nanjing University, Nanjing 210093, P. R. China

Fe-containing mesoporous molecular sieves (MCM-41, HMS) possessing high activity and selectivity (100%) for benzylation of benzene with benzyl chloride are reported. It is very interesting that only Fe-containing molecular sieves have high activity and that no good connection between activity and acidity of different heteroatom-containing catalysts has been found. The activities of samples with different pore diameters suggest that pore size plays an important role in this reaction. The relationship between the catalytic properties and the states of Fe species in mesoporous molecular sieves has been well discussed. We consider the high activity of iron oxide highly dispersed on wall surface, because of the nanosize pore and extreme large surface area of these mesoporous materials.

1. INTRODUCTION

Strongly acidic reagents are normally employed as catalyst in Friedel-Crafts alkylation reactions. However, not only polyalkylation and rearrangement are generally difficult to avoid [1], but also these homogeneous catalysts (e.g. HF, H_2SO_4, $AlCl_3$) themselves are not environmentally benign. One important consequence for chemistry of new environmental legislation and the drive towards "Clean Technology" will be the use of "Environmentally Friendly Catalysts", typically involving the use of solid catalysts [2]. Their use should lead to minimal pollution and waste.

In 1970's, Arata et al. investigated the Friedel-Crafts benzylation and benzoylation of toluene and benzene with alkyl chloride [3,4]. They found that calcined iron sulphate and iron oxide were excellent catalysts for the above reactions.

In 1989, Clark et al. [5] reported montmorillonite supported zinc (clayzic) and nickel chloride were highly active and selective catalyst for Friedel-Crafts alkylations. Rhodes et al. [6] pointed out that the highest catalytic activities are associated with samples with significant pore volume of 10-12 nm in diameter, which is generated by acid treatment. However, although clayzic is a good example of an environmentally friendly catalyst, its thermal stability can be expected to be low and will lead to regeneration problems.

The recent discovery of mesoporous inorganic materials, especially MCM-41, by the scientists of Mobil Oil Corporation [7,8] has attracted considerable interest concerning the potential application for catalysis, adsorption, and materials design. Mesoporous molecular sieves exhibit excellent catalytic ability for macro molecular reactions because of their large surface area and tunable nanosize pore. They also offer new opportunities for transition metal

* To whom correspondence should be addressed.

incorporation into silica frameworks [9,10]. We have reported the synthesis and characterization of Fe-containing MCM-41 [11] and HMS [12]. We also studied the Fe states in mesoporous materials [13].

Here we report in detail that Fe-containing mesoporous molecular sieve materials (MCM-41, HMS, etc.) are high activity and selectivity (100%) catalysts for benzylation of benzene with benzyl chloride.

2. EXPERIMENTAL

The synthesis and characterization of Fe and other heteroatom-containing mesoporous silica materials (MCM-41, Fe-MCM-41 and HMS, Fe-HMS etc.) have been described elsewhere [11,12]. We also prepared template-free samples through solvent extraction by following the procedure in ref. 13 and ref. 14. $Fe(NO_3)_3$ was respectively impregnated and ion-exchanged on calcined MCM-41 and HMS. They were calcined at 600℃ for 4 h to prepare Fe(III)-loaded samples.

The synthesis procedure of Zn, Cr, Ni, Co- containing mesoporous silica materials was similar to Fe-samples as described above except $ZnSO_4 \cdot 7H_2O$, $Cr(NO_3)_3 \cdot 9H_2O$, $NiCl_2 \cdot 6H_2O$ and $CoCl_2 \cdot 6H_2O$ were substituted for $Fe(NO_3)_3 \cdot 9H_2O$, respectively.

FeY [15], FeKL [16], FeUSY [17] were all synthesized and characterized in our laboratory.

The as-synthesized, calcinated, extracted, impregnated and ion-exchanged samples are denoted as ori-, cal-, ext-, imp-, exc-Me-MCM-41(or HMS)(n) respectively, where Me=Fe, Zn etc. and n is the Si/Me ratio in the precursor gel.

The benzylation reaction was carried out by rapid stirring of 60 mg of catalyst which was activated at 200℃ for 1 h with a mixture of 65 mmol of benzene and 3.25 mmol of benzyl chloride to produce diphenylmethane. The activity and selectivity of the samples were measured by gas chromatography.

The samples were characterized by low-angle X-ray powder diffraction (Rigaku, D/max-rA) with Cu-Kα radiation, FT-IR (Nicolet, 5DX), EPR (Bruker EP 200-D-SRC), by Mössbauer spectroscopy (recorded on a constant acceleration Mössbauer spectrometer with a source of ^{57}Co in Pd matrix), SEM (JSM 6300 scanning microscope) and by nitrogen adsorption at 77 K (Micromeritics ASAP2400 instrument).

3. RESULTS AND DISCUSSION

The benzylation catalytic properties of several calcined mesoporous molecular sieves incorporated with no or different heteroatoms are listed in Table 1. The siliceous mesoporous molecular sieves sample MCM-41 or HMS does not catalyze the reaction of benzene with benzyl chloride to form diphenylmethane. The HUSY only shows very low activity. However, although α-Fe_2O_3 shows no activity, the benzylation activities of Fe-containing mesoporous molecular sieves are very high. For example, rapid stirring of a mixture of Fe-SiMCM-41 and benzyl chloride in benzene at 60℃ for 1 h gave 61% conversion of benzyl chloride and a selectivity of diphenylmethane of 100% .

It is very interesting that only Fe-containing catalysts have high activity (*e.g.* 100% conversion for cal-Fe-HMS-25) and selectivity (100%) at 60℃ in 4 h. By contrast, no other

Table 1
Compositions of samples and conversion of benzyl chloride*

Sample	Si/Me	Fe_2O_3 %(wt.)	Pore size (A)	Conversion(%)
MCM-41	----	----	32 [a]	0
HMS	----	----	25 [a]	0
Ori-Fe-HMS(50)	50	2.66	-----	0
Cal-Fe-HMS(100)	100	1.33	25 [a]	71.5
Cal-Fe-HMS(50)	50	2.66	25 [a]	73.7
Cal-Fe-HMS(25)	25	5.32	25 [a]	100
Imp-Fe-HMS(100)	100	1.33	25 [a]	51.4
Imp-Fe-HMS(50)	50	2.66	25 [a]	76.0
Ext-Fe-HMS(50)	50	2.66	25 [a]	3.0
Ext-Fe-HMS(25)	25	5.32	25 [a]	19.4
Al-HMS(10)	10	----	25 [a]	0
Zn-HMS(100)	100	----	25 [a]	0
Ni-HMS(100)	100	----	25 [a]	0
Co-HMS(100)	100	----	25 [a]	0
Cr-HMS(100)	100	----	25 [a]	0
Fe-SiMCM-41(62)	62	3.0	33 [a]	80
Fe-SiMCM-41(31)	31	5.9	33 [a]	96
Fe-SiMCM-41(16)	16	11.0	33 [a]	100
Na-Fe-MCM-41(31)	31	5.9	33 [a]	57
Imp-Fe-SiMCM-41	----	6.2	32 [a]	80
Exc-Fe-SiMCM-41	----	1.5	33 [a]	66
FeKL	7.0	9.3	~7.1	0
Imp-Fe-KL	----	6.5	~7.1	0
HY	----	----	~7.4	0
FeY	----	3.0	~7.4	8.0
HUSY	----	----	~26.4 [b]	6.6
FeUSY	----	13	~26.4 [b]	95
Imp-Fe-USY	----	13	~26.4 [b]	99
α-Fe_2O_3	----	>99	----	0

* 60℃ in 4 h; Selectivity is 100%; Benzene:benzyl=20:1(molar) ; Catalyst: 1%(wt.)

[a] Determined from low temperature N_2 adsorption with BJH method.

[b] Diameters of secondary pores measured by ^{129}Xe NMR method[18].

heteroatom (Zn, Co, Cr, Al, Ni, etc)-containing mesoporous materials have significant activity under this reaction condition. Only Zn-HMS (100) has 38.0% conversion after reaction at 60 ℃ overnight. So we considered that there is a relationship between this excellent benzylation activity of Fe-containing mesoporous catalysts and the Fe species in these samples.

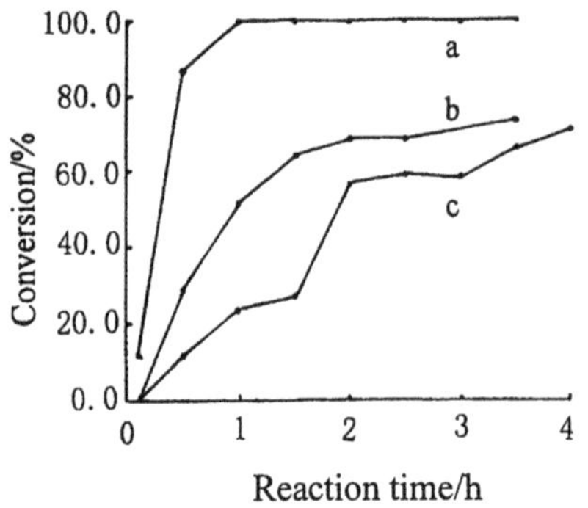

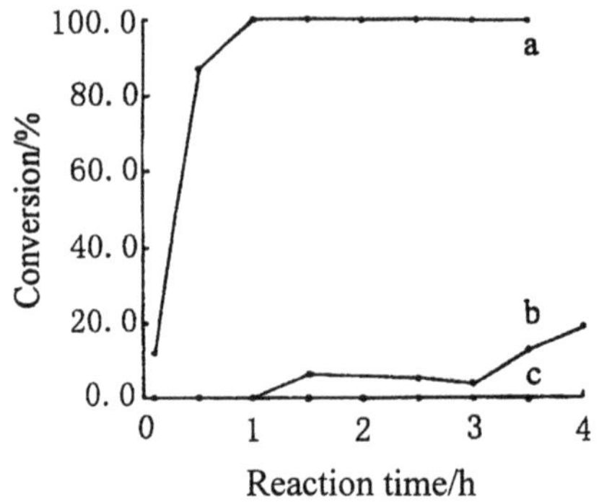

Figure 1. Catalytic activities of cal-Fe-HMS(n) with different Fe contents.
a) n=25; b) n=50; c) n=100.

Figure 2. Catalytic activities of cal-Fe-HMS(50) at different temperatures.
a) 60℃ ; b) 50℃ ; c) 40℃.

From Table 1, it is also easy to see that the catalytic activity of the Fe-containing materials increased with the increasing Fe content of catalysts. For example, the conversion of cal-Fe-HMS (25) reaches 100% within 1 h while that of cal-Fe-HMS (100) only 71.5% after 4 h (Figure 1).

We have investigated the acidic properties of the samples by means of adsorbed pyridine IR spectra, NH_3-TPD and NH_3-microcalorimeteric methods. Some samples with stronger acidity showed very low or no reaction activity. For instance, Al-MCM-41 and Al-HMS did not show any conversion at 60℃ in 4 h, although the amount and strength of their acidity are all higher than that on the same Si/Me ratio Fe-containing samples. No good correlation between the acidic properties and the activities for the different heteroatom-containing samples was found. However, it is suggested that the Brönsted acid sites favor the above reaction for the Fe-containing samples because Fe-MCM-41 gave a much higher activity than Na-Fe-MCM-41 (see Table 1). We noticed that the influence of the temperature is very important (Figure 2). The conversion of cal-Fe-HMS(50) reached 100% at 60℃ in 1 h, while there is still no activity at 40℃ after 4 h.

From Table 1 and Figure 3, we found that the larger pore is more beneficial to this reaction. While the Fe-containing microporous zeolites (Y, KL) have a very low activity, the high activity of Fe-impregnated USY which possesses 39% of secondary mesopores in volume measured by low-temperature nitrogen adsorption is also demonstrated. It is suggested that the pore size plays an important role in this reaction. We considered that zeolite micropores could not provide sufficient space for the reaction of these macromolecules. This effect is also proved by the longer time taken on FeUSY to reach high conversion than that on FeMCM-41 (see Figure 3). The activity of MCM-41 samples, which have larger pores, is also larger than that of HMS impregnated with the same Fe contents. For benzylation reaction , the catalytic activities of MCM-41 and HMS samples are as follows: cal-Fe-MCM-41≈imp-Fe-MCM-41>exc-Fe-MCM-41> ori-Fe-MCM-41; cal-Fe-HMS≈imp-Fe-HMS > ext-Fe-HMS > ori-Fe-HMS respectively, see Figure 4.

In our previous work [11], EPR and Mössbauer spectra showed that Fe was mainly incorporated in the channel wall of as-synthesized Fe-MCM-41 sample. However, during

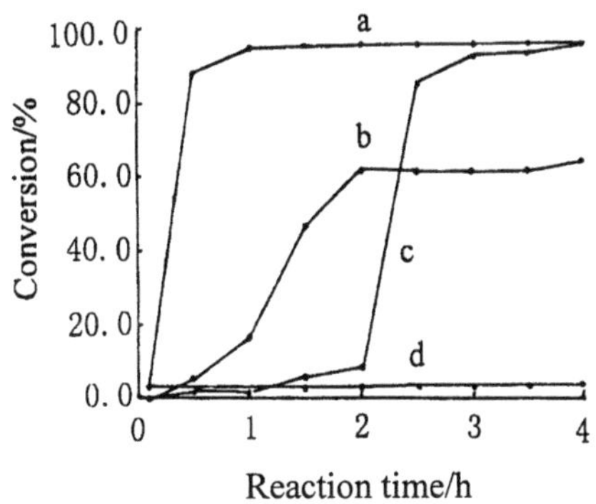

Figure 3. Benzylation reaction over different Fe-containing catalysts.
a) cal-Fe-MCM-41(31); b) cal-Fe-HMS(33); c) FeUSY; d) FeY.

calcination for template removing, the tetrahedrally coordinated Fe (III) species gradually converted to octahedrally coordinated Fe (III) species, which were mainly highly dispersed on the channel walls. As for HMS-type sample which was obtained by using non-ionic surfactant (primary amines) as template through sol-gel procedure at room temperature, it was found that not only Fe was mainly incorporated in the framework of ori-Fe-HMS sample, but also, unlike the Fe-MCM-41, the Fe (III) ions still partially existed in a tetrahedral coordination environment after removal of the template by calcination. The state of Fe ions in the as-synthesized Fe-HMS sample is more thermally stable because of its thicker wall and different synthesis mechanism. Only a part of the Fe transferred to the wall surface [12]. Moreover in the case using acidified alcohol to extract the surfactant, it is more effective to retain Fe in the tetrahedral coordinated framework. Fe species on wall surfaces are highly dispersed when removing the surfactant at high temperature [13,14]. It has also been found that the Fe ions of impregnated samples are mainly octahedrally coordinated Fe (III) species, which are highly dispersed on the channel walls.

Regarding the state of Fe ions in some mesoporous molecular sieves and the benzylation catalytic activities of these samples, we suggested that the Fe species are highly dispersed on wall surfaces due to nanosize pores and extreme large BET surface area of MCM-41 and HMS materials seemed to be responsible for their high benzylation catalytic activities.

The very high activity of calcined synthesized Fe-MCM-41 can be well explained because it was found that Fe species are highly dispersed on wall surfaces of these samples [11]. The high activities of the imp-Fe-MCM-41 samples are similar to that of the cal-Fe-MCM-41 samples, indicating that the framework Fe is probably not the main active specie. We noticed that the Fe content of these excellent catalysts is quite low (Si/Fe>10).

Compared to MCM-41, HMS sample has thicker channel walls, while its BET surface area is relatively smaller (~1000cm^3/g *vs.*~1300cm^3/g for MCM-41). Consequently the catalytic conversion of Fe-containing HMS sample is smaller. The template-removed sample by extraction, which is more effective for retaining Fe in tetrahedral coordinated framework, has a lower activity than that of the calcinated Fe-HMS sample.

Figure 5 shows the activity of some imp-Fe-HMS samples prepared with different iron sources. It is observed that the anion also remarkably influences the activity.

We also found that none of Fe-containing mesoporous materials have a XRD pattern

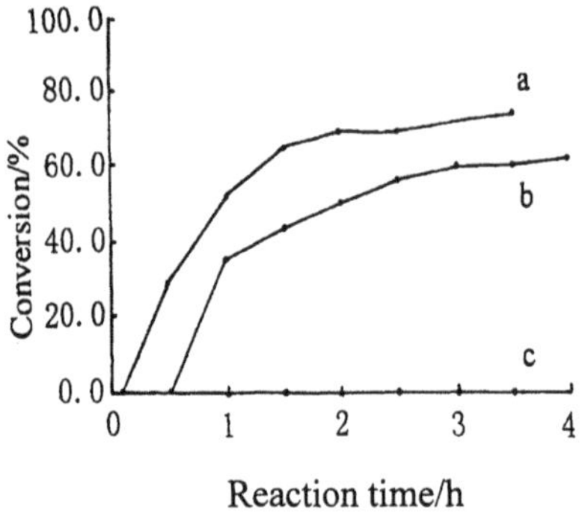

Figure 4. Catalytic activities of Fe-HMS (50) prepared by different methods.
a) cal; b) imp; c) ext.

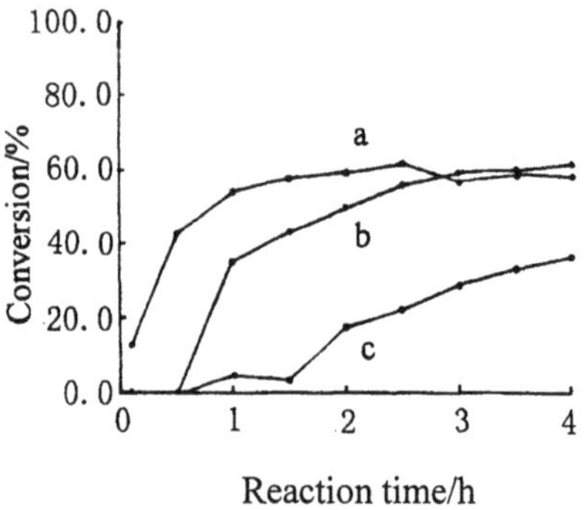

Figure 5. Catalytic activities of imp-Fe-HMS(50) prepared by different iron resources.
a) $FeCl_3$; b) $Fe(NO_3)_3$; c) $Fe_2(SO_4)_3$.

characteristic of α-Fe_2O_3 except mix-Fe/HMS which was obtained by mechanical mixing of Si-HMS with α-Fe_2O_3 and showed no activity. Then we considered that the high activity might be attributed to amorphous iron oxide species, which formed with highly dispersed Fe species and other species on the channel surface.

The excellent catalytic ability and the convenient preparation method of these mesoporous molecular sieves with such low Fe content indicated that these catalytic materials have a remarkable potential for industrial application.

4. CONCLUSION

We report here that Fe-containing materials (MCM-41, HMS) are highly active and selective (100%) catalysts for benzylation of benzene with benzyl chloride. Only Fe-containing catalysts have a high activity while the mesoporous materials incorporated with other(Zn, Co, Cr, Al, Ni, etc) or without heteroatoms have no significant activity at 60℃ for 4 h. No good connection between activity and acidity of different heteroatom-containing samples has been found. The influence of temperature is important. The high activity of Fe-impregnated USY which possesses secondary mesopores is demonstrated in contrast with very low activity for Fe-containing microporous zeolites (Y, KL). The catalytic activities of Fe-MCM-41 and Fe-HMS samples are as follow: cal-Fe-MCM-41≈imp-Fe-MCM-41>exc-Fe-MCM-41; cal-Fe-HMS≈imp-Fe-HMS>ext-Fe-HMS respectively. We considered that the amorphous iron oxide species highly dispersed on wall surfaces due to nanosize pores and extremely large BET surface area of MCM-41 and HMS materials seemed to be responsible for their high benzylation catalytic activities.

REFERENCES

1. G.A. Olah, "Friedel-Crafts Chemistry", 1973, Wiley, New York.

2. J.H. Clark, S.R. Cullen, S.J. Barlow, and T.W. Bastock, J. Chem. Soc. Perkin Trans., (1994) 1117.
3. K. Arata, I. Oyoshima, Chem. Lett., (1974) 929.
4. K. Arata, K. Yabe, I. Oyoshima, J. Cata., 44 (1976) 385.
5. J.H. Clark, A.P. Kybett, D.J. Macquarrie, S.J. Barlow and P. Landon, J. Chem. Soc. Chem. Commun., (1989) 1353.
6. C.N. Rhodes and D.R. Brown, J. Chem. Soc. Faraday Trans., 89 (1993) 1387.
7. C.T. Kresge, M.E. Leonowicz, W.J. Roth, J.C. Vartuli and J.S. Beck, Nature, 359 (1992) 710.
8. J.S. Beck, J.C. Vartuli, W.J. Roth, M.E. Leonowicz, C.T. Kresge, K.D.Schmitt, C.T-U. Chu, D. H. Olson, E.W. Sheppard, S.B. McCullen, J.B. Higgins and J.L. Schlenker, J. Am. Chem. Soc., 114 (1992) 10834.
9. A. Sayari, Chem. Mater., 8 (1996) 1840.
10. A. Sayari, Stud. Surf. Sci. Catal., 102 (1996) 1.
11. N. He, B. Bao and Q. Xu , Stud. Surf. Sci. Catal., 105 (1997) 85.
12. N. He, J. Cao, B. Bao and Q. Xu , Materials Letters, 33 (1997) 133.
13. J. Cao, N. He, C. Li, J. Dong and Q. Xu , to be submitted.
14. S. Gontier and A.Tuel , Stud. Surf. Sci. Catal., 105 (1997) 29.
15. A. Zhang, Z. Hu, S. Bao and Q. Xu, Chinese Science Bulletin , Vol.41, 17 (1996) 1436.
16. N. He, S.Bao, and Q. Xu, Chinese J. Chem. , Vol.15, 1 (1997) 42.
17. D. Liu, S. Bao and Q. Xu, Appl. Catal. A: General, 158 (1997) 93.
18. D. Liu, S. Bao, Q. Xu and C. Hu, Chinese Science Bulletin , Vol.41, 6 (1996) 487.

MESOPOROUS MOLECULAR SIEVES 1998
Studies in Surface Science and Catalysis, Vol. 117
L. Bonneviot, F. Béland, C. Danumah, S. Giasson and S. Kaliaguine (Editors)

Oxidation of cholesterol over mesoporous MCM-41 materials

K. A. Vercruysse, D. M. Klingeleers, T. Colling and P. A. Jacobs

Centre for Surface Chemistry and Catalysis, Faculty of Agricultural and Applied Biological Sciences, Katholieke Universiteit Leuven, Kardinaal Mercierlaan 92, B-3001 Leuven, Belgium Telephone ++32-16.32.16.48
Telefax: ++32-16.32.19.98; E-mail: karen.vercruysse@agr.kuleuven.ac.be

The incorporation of transition metals in the mesoporous sieve MCM-41 makes this material very attractive for heterogeneous catalysis, especially with bulky substrates. We report on the use of mesoporous MCM-41 in the oxidation of cholesterol with *t*-BHP. The selectivity of the reaction depends on the metal incorporated in the pore walls of the sieve. When Zr-MCM-41 is used, 3-ketocholesterol is the main product, whereas the incorporation of Ti leads to the formation of 5,6-epoxycholesterol.

1. INTRODUCTION

The discovery in 1992 by Mobil researchers to use supramolecular assemblies of surfactant molecules as templates in the synthesis of molecular sieves resulted in a new family of materials called M41S [1]. Characteristics as an uniform pore diameter which can be tuned between 2 and 10 nm, an ordered hexagonal (MCM-41) or cubic (MCM-48) structure, high thermal and hydrothermal stability and the possibility to introduce hetero-elements as Al, Ti and Zr make these mesoporous sieves attractive for heterogeneous catalysis especially in applications where the use of bulky substrates is demanded [2]. In this report we describe the use of Zr containing MCM-41 in the oxidation of cholesterol with *tertiair*-butylhydroperoxide (*t*-BHP).
In literature, $Zr(iPrO)_4$ is mentioned as a homogeneous catalyst for the oxidation of primary and secondary alcohols while double and triple bounds do not react [3]. The same authors also tested the heterogenized form of $Zr(iPrO)_4$ on silica. The property to oxidize alcohols was retained after heterogenization. Based on these results we treated Si-MCM-41 with $Zr(iPrO)_4$ in the same way. Next to this procedure, Cp_2ZrCl_2 was used to graft Zr onto the walls of MCM-41 [4]. Tuel *et al.* reported that Zr-HMS was a good catalyst in the oxidation of alkenes whereas the oxidation of alcohols was not mentioned [5]. Therefore, this catalyst was also synthesized and tested in the oxidation of cholesterol.
As Ti-MCM-41 is also known as an oxidation catalyst, its activity and selectivity was tested and compared with Zr-MCM-41 [6].

2. EXPERIMENTAL

2.1. Synthesis of materials

The hydrothermal synthesis of Si-MCM-41 was carried out by mixing Ludox AS-40 (40 wt% SiO_2, DuPont), tetraethylammonium hydroxide (TEAOH, 20 wt% in water, Fluka) and hexadecyltrimethylammonium chloride (HDTMACl, 25 wt% in water, Aldrich). The molar composition of the gel was SiO_2: $(TEA)_2O$: $(HDTMA)_2O$: H_2O = 1: 0.1: 0.15: 28. The crystallization was carried out in stainless steel autoclaves at 383 K under dynamic conditions for 24 h. After cooling down, the product was filtered, washed with distilled water, dried at 333 K and calcined at 813 K for 10 h.

Zr-MCM-41 was obtained in two different ways. The first method is analogous to the impregnation of Zr on SiO_2 with $Zr(iPrO)_4$ [7]. 1 g of Si-MCM-41 and 0.6 g of $Zr(iPrO)_4$.iPrOH (Acros) were refluxed in 50 ml iPrOH (Acros) for 5 h. After cooling down, the product was filtered and dried under vacuum for 2 h. The second method is based on the work of Maschmeyer *et al.* [4]. 1 g of Si-MCM-41 was successively contacted with 0.48 g Cp_2ZrCl_2 (Acros) in 35 ml $CHCl_3$ (Biosolve), 50 ml triethylamine (Acros) and 200 ml $CHCl_3$. The resulting solid was dried and calcined at 820 K for 10 h.

Zr-HMS was synthesised at room temperature using hexadecylamine as surfactant molecule following a procedure described by Tuel *et al.* [5]. A solution of tetraethylortosilicate (TEOS, Acros) and $Zr(iPrO)_4$ was mixed with hexadecylamine (HDA, Acros) solved in ethanol (Biosolve) and water. The molar composition was as follows: SiO_2: HDA: EtOH: ZrO_2: H_2O = 1: 0.3: 6.5: 0.1: 30. The gel was aged for 12 h under static conditions. The product was separated by centrifugation and dried at 333 K. The template was removed by solvent extraction in ethanol followed by calcination at 820 K.

Ti-MCM-41 was prepared in the same way as Zr-MCM-41, using 0.415 g of Cp_2TiCl_2 (Acros) for 1 g of Si-MCM-41.

2.2. Characterization

Powder X-ray diffraction patterns were obtained on a Siemens D5000 diffractometer using CuKα radiation.

Diffuse Reflectance Spectra were measured on a Varian Cary 05 spectrometer with a $BaSO_4$ integration sphere. Samples were not outgassed before measurement.

IR spectra were recorded on a Nicolet FT-IR 730 spectrometer.

^{29}Si CP NMR measurements were carried out on a Bruker AMX 300. The magnetic field and the resonance frequency were 7 T and 59.628 MHz respectively.

2.3. Catalytic Set-up

Reactions were carried out in a round bottom flask of 250 ml. 1.90 mmol cholesterol (Sigma), 1.26 mmol *t*-BHP 5-6 M in n-decane (Aldrich) were mixed in 50 g of toluene (Biosolve). After the reaction temperature was reached, 0.075 g of catalyst was added.

Analysis of the reaction mixture was done by HPLC. As column a Licrosphere 100 CN 5 Micron with a length of 250 mm and internal diameter of 4.6 mm was used. The mobile phase was a 95:5 mixture of n-hexane and isopropanol [8]. Signals were detected by a refractive index detector. Calibration curves were determined for cholesterol (I), 5α,6α-epoxycholesterol (II), 3-ketocholesterol (III), 7-hydroxycholesterol (IV) and 7-ketocholesterol (V). The structure formulas are shown in scheme 1. The pure products were all obtained from Sigma.

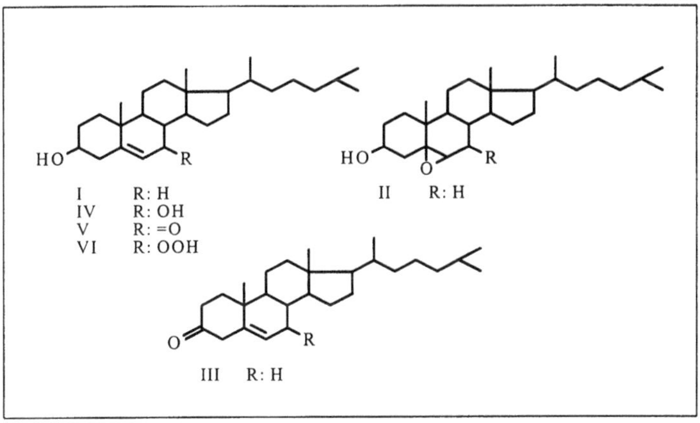

Scheme 1. Structure formula of cholesterol and some oxygenated products.

3. RESULTS AND DISCUSSION

3.1. Physico-chemical characterization

The XRD patterns of Zr-MCM-41 and Zr-HMS are shown in figure 1. It can be seen that the incorporation of Zr did not disturb the MCM-41 structure, as the typical hexagonal pattern is retained. Zr-HMS gives only one peak in XRD, which is in agreement with the fact that HMS has a less ordered hexagonal arrangement [9].

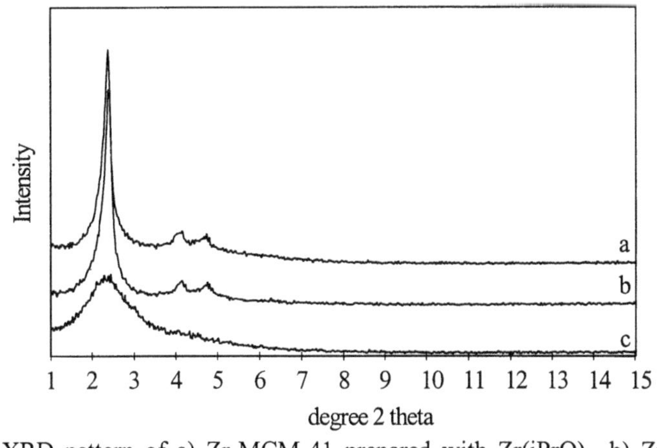

Figure 1. XRD pattern of a) Zr-MCM-41 prepared with $Zr(iPrO)_4$, b) Zr-MCM-41 prepared with Cp_2ZrCl_2 and c) Zr-HMS.

The DRS spectra of Zr-HMS and Zr-MCM-41 show a band around 210 nm which points to a good dispersion of the Zr species.

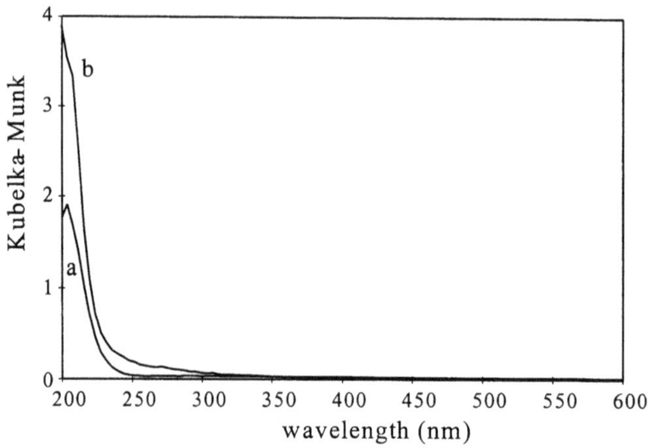

Figure 2. DRS spectra of a) Zr-HMS and b) Zr-MCM-41 prepared with $Zr(iPrO)_4$.

The presence of the alkoxy groups of the adsorbed Zr-precursor ($Zr(iPrO)_4$) can be seen from the C-H stretch vibration appearing in the IR spectrum of Zr-MCM-41 around 3000 cm^{-1}.

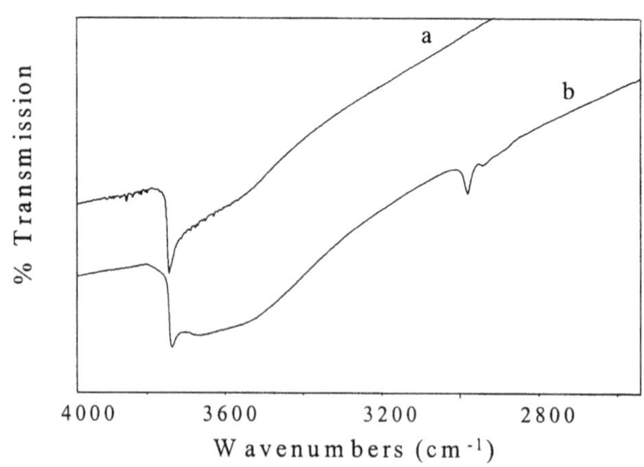

Figure 3. IR spectra of a) Si-MCM-41 and b) Zr-MCM-41 prepared with $Zr(iPrO)_4$; prior to analysis both samples were heated up to 573 K to remove sorbed water, while avoiding oxidation of the organic molecules present.

The anchoring of the Zr species on the Si-OH groups of MCM-41 can be deduced from ^{29}Si-CP NMR. In figure 4 the ^{29}Si-CP NMR spectra of MCM-41 before and after treatment with $Zr(iPrO)_4$ are shown. In both spectra 3 bands can be seen. The peak at -90 ppm is attributed to Q_2-atoms (2 T-atom neighbours), Q_3-atoms (3 T-atom neighbours) resonate around -100 ppm and Si atoms with 4 T-atom neighbours give a peak at -110 ppm (Q_4).
As can be seen, the $(Q_2+Q_3)/Q_4$ ratio decreases after complexation which means that there has been an interaction between Si-OH groups on the wall and the isopropoxy groups of the Zr-precursor.

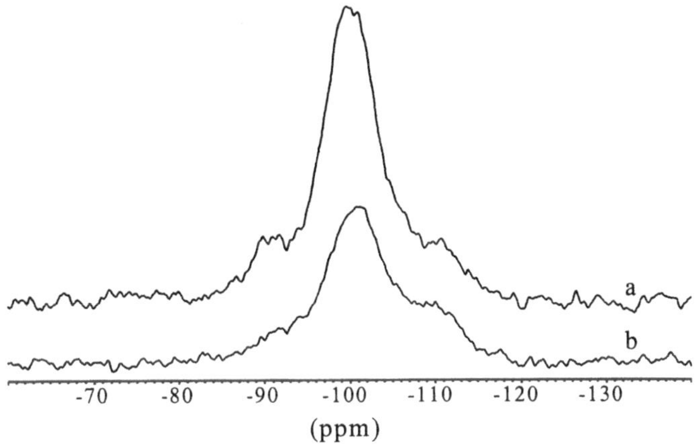

Figure 4. ^{29}Si CP NMR spectra of a) Si-MCM-41 and b) Zr-MCM-41 prepared with $Zr(iPrO)_4$.

3.2. Catalytic results

The most important products formed in the oxidation reaction of cholesterol with *t*-BHP are shown in scheme 1. The catalytic results are summarized in table 1.

Krohn *et al.* [3] proposed an Oppenauer mechanism for the oxidation of alcohols with $Zr(iPrO)_4$ and *t*-BHP. In this mechanism an important step is a ligand exchange between an isopropoxy group and *t*-BHP on the metal center. Therefore, the presence of isopropoxy ligands on Zr seems to be necessary for activity [3]. This mechanism is not in agreement with our results. Indeed, homogeneous $Zr(iPrO)_4$ oxidizes cholesterol with 3-ketocholesterol as main product (entry 6). Heterogenization of $Zr(iPrO)_4$ leads to a higher conversion but the selectivity for the ketone decreases (entry 2). Epoxidation of the double bond and allylic oxidation become important side reactions. When Zr is grafted on Si-MCM-41 following the procedure of Maschmeyer *et al.* [4] no organic ligands are left on the metal center. However this catalyst gives a conversion comparable with the homogeneous catalyst and a high selectivity for 3-ketocholesterol. No epoxide is formed (entry 3). The fact that organic ligands are not required for oxidation activity was confirmed by using Zr-

MCM-41 prepared with $Zr(iPrO)_4$ and calcined afterwards. Hence this catalyst also lacks organic ligands (which was confirmed by IR and thermogravimetric analysis). The conversion reached with the calcined catalyst is lower than with the catalyst which had no extra calcination but the selectivity for the ketone is twice as high and no epoxidation was detected (entry 4).
Conclusively, the mechanism underlying the oxidation reaction in homogeneous conditions can not be directly transferred to a heterogeneous system.
Tuel *et al.* described Zr-HMS as an oxidation catalyst for alkenes [5]. In the oxidation of cholesterol, however, no epoxidation of the double bond was found and 3-ketocholesterol was the main product (entry 5). To ascertain this result all Zr-catalysts were tested in the oxidation of cyclooctene and of 2-octanol with *t*-BHP (using the same conditions as with cholesterol). None of them were able to convert cylooctene whereas 2-octanol was converted in 2-octanone. These results point out that Zr is a suitable transition metal for the oxidation of alcohols but not for alkene epoxidation.
As Ti-catalysts are known as epoxidation catalysts, Ti was grafted on Si-MCM-41 and compared with Zr-MCM-41 (entry 7). A high conversion is reached and the 5,6-epoxycholesterol is the main product. No oxidation of the alcohol function was measured.
This result was confirmed by using cyclooctene and 2-octanol as substrate. With Ti-MCM-41 no conversion of the alcohol could be found whereas cyclooctene was readily converted in epoxycyclooctane. These results show that the transition metal incorporated plays a crucial role in the selectivity of the oxidation reactions.

Table 1
Conversion and product selectivities in the oxidation of cholesterol with *t*-BHP at 90°C

n°	Catalyst	Conversion (%)[d]	Product selectivity (%)			
			II	III	IV+V+VI	others
1	none	0.3	-	-	-	-
2	Zr-MCM-41[a]	26	20	21	36	23
3	Zr-MCM-41[b]	13	0	54	29	17
4	Zr-MCM-41[c]	18	0	40	42	18
5	Zr-HMS	9	0	63	4.5	32.5
6	$Zr(iPrO)_4$	14.3	0	50.3	27.3	22.4
7	Ti-MCM-41	48	53	0	10	37

[a] prepared with $Zr(iPrO)_4$
[b] prepared with Cp_2ZrCl_2
[c] prepared with $Zr(iPrO)_4$ and calcined afterwards to remove the non reacted isopropoxy ligands
[d] % of theoretical maximum conversion of cholesterol, after 24 h of reaction

4. CONCLUSIONS

The mesoporous characteristics of MCM-41 are exploited in the oxidation reaction of a bulky substrate as cholesterol. Both Zr-MCM-41 and Ti-MCM-41 are able to convert cholesterol to its oxygenated products. The key factor in the product selectivity is the transition metal used. Incorporation of Zr leads mainly to the oxidation of the alcohol function whereas with Ti-MCM-41 the epoxidation of the double bound is dominant.
Further research of the functional selectivity of transition metal containing MCM-41 is needed as the selective formation of oxygenated products of cholesterol may be important for the pharmaceutical industry [10].

ACKNOWLEDGMENTS

The authors acknowledge sponsoring from the Belgian government in the frame of a IUAP-PAI program on supramolecular chemistry and supramolecular catalysis.

REFERENCES

1. J.S. Beck, J.C. Vartuli, W.J. Roth, M.E. Leonowicz, C.T. Kresge, K.D. Schmitt, C.T-W. Chu, D.H. Olson, E.W. Sheppard, S.B. Mc Mullen, J.B. Higgins and J.L. Schlenker, J. Am. Chem. Soc., 114 (1992) 10834.
2. J.S. Beck, I.D. Johnson, C.T. Kresge, M.E. Leonowicz, W.J. Roth and J.C. Clarke, WO91/11390 (1991).
3. K. Krohn, I. Vinke and H. Adam, J. Org. Chem., 61 (1996) 14967.
4. T. Maschmeyer, F. Rey, G. Sankar and M. Thomas, Nature, 378 (1995) 159.
5. A. Tuel, S. Gontier and R. Teissier, J. Chem. Soc. Chem. Commun., (1996) 651.
6. A. Corma, M.T. Navarro and J. Pérez-Pariente, J. Chem. Soc. Chem. Commun., (1994) 147.
7. K. Inada, K. Shibagaki, Y. Nakariski and H. Matsushita, J. Chem. Soc. Japan, (1993) 1795.
8. B.H. Chen and Y.C. Chen, J. Chrom., A 661 (1994) 127.
9. P.T. Tanev, M. Chibwe and T.J. Pinnavaia, Nature, 368 (1994) 321.
10. E.J. Parish, V.B.B. Nanduri, H.H. Kohl and F.R. Taylor, Lipids, 21(1) (1986) 27.

MESOPOROUS MOLECULAR SIEVES 1998
Studies in Surface Science and Catalysis, Vol. 117
L. Bonneviot, F. Béland, C. Danumah, S. Giasson and S. Kaliaguine (Editors)

Alumoxo-MCM-41 as Host in the Co-oligomerization of Ethene and Propene with $Et(Ind)_2Zr(CH_3)_2$

Lieven K. Van Looveren[a], Daneel F. Geysen, Karen A. Vercruysse[a], Bart H. Wouters[a], Piet J. Grobet[a] and Pierre A. Jacobs[a]

[a] Center for Surface Chemistry and Catalysis, Faculty of Agricultural and Applied Biological Sciences, K.U. Leuven, Kardinaal Mercierlaan 92, B-3001 Heverlee (Belgium)
Tel.: ++32-16.32.14.63
Fax : ++32-16.32.19.98
E-mail: lieven.vanlooveren@agr.kuleuven.ac.be

The uniform array of tubular mesopores in addition to the high concentration of silanol groups and surface area makes the mesoporous molecular sieve MCM-41 very attractive in heterogeneous catalysis. We report on the anchoring of MAO generated by *in situ* hydrolysis of trimethylaluminium as a highly active and shape selective host in the co-oligomerization of ethene and propene with *ansa*-metallocenes. Induction of acidity in the pore walls of the sieve changes the docking characteristics of MAO, whereas the extension of the alumoxane layer alters the catalytic behavior of the metallocene encapsulated guest.

1. INTRODUCTION

Mesoporous molecular sieves of the M41S family are characterized by a regular array of uniform pores with a diameter which can be tailored by altering the alkyl chain length of the surfactant molecules or by adding organic molecules into the synthesis medium [1,2]. Therefore, the pore diameter can be tuned between 2 and 10 nm and hetero-atoms such as Al or Ti can be incorporated in the host to develop materials with potential catalytic applications [3].

In the polymerization of olefins with single site metallocene catalysts, the reduction of the excessive amounts of methylalumoxane (MAO), required to activate the metallocene compound, is one of the ongoing challenges. In general, MAO is composed out of three-dimensional cage or lattice structures [4] and acts as a weakly coordinating anion [5] for the metallocene catalyst. Impregnation of this alumoxane compound on amorphous silica supports suppresses both the binuclear deactivation mechanisms of the metallocene and the internal cluster coordination of the alumoxane structure [6,7]. However, an important disadvantage of the alumoxane impregnated silica is the rather low attainable surface area for anchoring the alumoxane compounds and the need to add supplemental MAO to obtain polymerization activity [8]. We report now on the use of a MCM-41 type mesoporous host for the *in situ* synthesis of MAO.

2. EXPERIMENTAL

2.1. Synthesis of mesoporous materials (MCM-41)

The synthesis of Si-MCM-41 was carried out by mixing tetraethylorthosilicate (TEOS, 98%, Acros), sodium hydroxide (1.7 wt% in water) and hexadecyltrimethylammonium chloride (C_{16}TMACl, 25 wt% in water, Aldrich). The molar composition of the gel was SiO_2 : C_{16}TMACl : NaOH : H_2O = 1.00 : 0.11 : 0.49 : 54.09. The mixture was vigorously stirred at room temperature for 24 h, filtered and washed with distilled water and ethanol. The sample obtained was dried overnight at 333 K and calcined at 813 K for 10 h with a heating rate not exceeding 1 K/min.

Al-MCM-41 materials were synthesized in two different Si/Al-ratios (20 and 40) by adding variable amounts of sodium aluminate to the gel mixture. Cab-O-Sil-M5 (97+%, Fluka), aqueous ammonia, tetramethylammonium hydroxide (TMAOH.5H2O, 97%, Aldrich) and C_{16}TMACl were mixed in a molar gel ratio of SiO_2 : $NaAlO_2$: NH_3 : TMAOH : C_{16}TMACl : H_2O = 1.00 : 0.05 : 0.66 : 0.24 : 0.36 : 30.57. The gel mixture was transferred in a rotated stainless steel autoclave and crystallized at 383 K for 24 h. The calcined samples were exchanged with a 1.0 M NH_4Cl solution and subsequently calcined at 773 K for 2 h with a heating rate not exceeding 1 K/min.

Ti-MCM-41 was synthesized by mixing Ludox AS-40 (40 wt% SiO_2, DuPont), tetrabutylorthotitanate (TBOT, 98+%, Merck-Schuchardt), tetraethylammonium hydroxide (TEAOH, 20 wt% in water, Fluka) and C_{16}TMACl in a typical molar ratio of SiO_2 : TBOT : TEAOH : C_{16}TMACl : H_2O : iPrOH = 1.00 : 0.02 : 0.20 : 0.29 : 14.37 : 0.11. During the synthesis, the gel mixture was cooled and isopropanol was added to the titanium source to overcome anatase formation.

2.2. Synthesis of MAO-MCM-41

The calcined MCM-41 samples were equilibrated against a relative humidity of 79.3 % at room temperature corresponding to 22 - 31 wt% of sorbed water. The water concentration on Si-MCM-41 samples was reduced by partial dehydration at 313 K. Such samples were suspended in 270 ml toluene in a stainless steel reactor, which was purged with nitrogen and cooled to 273 K under continuous stirring for 1 h. Trimethylaluminium (TMA, 2.0 M solution in toluene, Aldrich) was diluted in 20 ml toluene and added slowly with a syringe to the suspension, resulting in a water/Al ratio of 1. Methane formation and a slight exothermicity was noticed during the first 30 min of the hydrolysis. The MAO-MCM-41 was filtered, washed with toluene and dried under inert atmosphere.

2.3. Characterization

Powder X-ray diffraction measurements were performed on a Siemens D5000 diffractometer using a CuKα source with a wavelength λ = 0.1542 nm.

Diffuse Reflectance Spectroscopic measurements (DRS) were recorded on a Varian Cary 05 UV-VIS-NIR with a $BaSO_4$ integration sphere.

Electron Probe Microanalysis (EPMA) was obtained on a JEOL JXA 733 scanning electron microscope using SiO_2 and Al_2O_3 as standard materials.

^{29}Si-MAS-NMR was performed on a Bruker AMX 300 spectrometer operating at 59.62 MHz with excitation pulses of 3.5 μs and a spinning frequency of 4 kHz. ^{29}Si-CP-MAS-NMR was measured with a contact time of 1 ms.

The number average molecular weight (M_n) and the molecular weight distribution (MWD) of the co-oligomer (10 wt% solution in THF) were determined on a GPC-column of Chrompack (Microgel 3 mix) with a refractive index detector and correlated to polystyrene standards

2.4. Oligomerization set-up

The final catalyst was obtained by adding 0.036 mmol of ethylenebis(1-indenyl)zirconium dimethyl [$Et(Ind)_2Zr(CH_3)_2$] to the suspension of *in situ* synthesized MAO-MCM-41 using various Al/Zr-ratios under inert atmosphere. $Et(Ind)_2Zr(CH_3)_2$ was obtained from the metallocene dichloride via alkylation with TMA in a typical Al/Zr-ratio of 100.

The catalytic co-oligomerization reactions were performed in a 600 ml water-cooled batch reactor (Parr Instruments) for 75 min. The reactor was continuously fed with a flow of methane (491 ml/min), ethene (700 ml/min), propene (1,400 ml/min), nitrogen (40 ml/min) and hydrogen (500 ml/min) at an overall pressure of 0.7 Mpa. The solvent and the gases, used in the catalytic experiments, were carefully dried over a molecular sieve (5A, Merck). The gas-outlet was monitored by GC to calculate the conversion with methane as internal standard. After 75 min, the reaction medium was neutralized by the addition of 100 ml of a 1 N HCl solution under continuous stirring.

3. RESULTS AND DISCUSSION

3.1. Physico-chemical characterization

The X-ray diffraction patterns of the synthesized MCM-41 materials are shown in Figure 1. Si-MCM-41 and Ti-MCM-41 show the typical pattern of an hexagonal arrangement. The Al-MCM-41 samples show a less intensive XRD pattern which should be ascribed to the formation of smaller crystallites rather than to the synthesis of a less ordered material. Alumoxane anchoring on Si-MCM-41 reduces the d_{100}-spacing from 3.48 to 3.30 nm.

The synthesis of alumoxane inside the mesopores of the siliceous host is verified with DRS-analysis. During TMA treatment, the O-H overtone at 6,900 cm^{-1} (1,450 nm) as well as the combination band at 5,208 cm^{-1} (1,920 nm) of the water containing MCM-41 disappear, while a C-H overtone and a combination band at 5,952 cm^{-1} (1,680 nm) and 4,386 cm^{-1} (2,280 nm) appear [9]. The broad signal around 300 nm can be ascribed to a charge transfer between Al and O of the alumoxane anchored structure [10]. Subsequent interaction of the MAO-Si-MCM-41 host with the dimethyl metallocene compound generates a new charge transfer around 400 nm between anchored alumoxane and the electrophilic metallocenium species.

The synthesis and the specific docking of MAO on the pore walls of the mesoporous host can be analyzed by NMR spectroscopy. ^{13}C-CP-MAS-NMR of MAO-Si-MCM-41 reveals the presence of a broad band at -7.4 ppm which can be assigned to the methyl group of MAO (Figure 3). No sharp signal of coordinated or associated TMA is superimposed on this broad band [11] pointing to the complete conversion of TMA on the surface of the pore walls. Methoxy groups are noticed at 51.1 ppm as a result of oxidation or hydrolysis during sample pre-treatment and traces of toluene are visible as well at 128.7, 125.4 and 20.9 ppm.

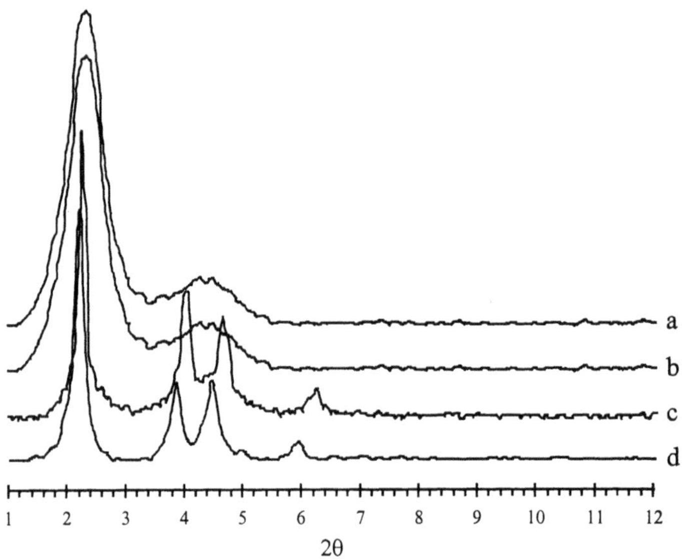

Figure 1. XRD patterns of Al-MCM-41 with a Si/Al of (a) 20 and (b) 40, (c) Si-MCM-41 and (d) Ti-MCM-41.

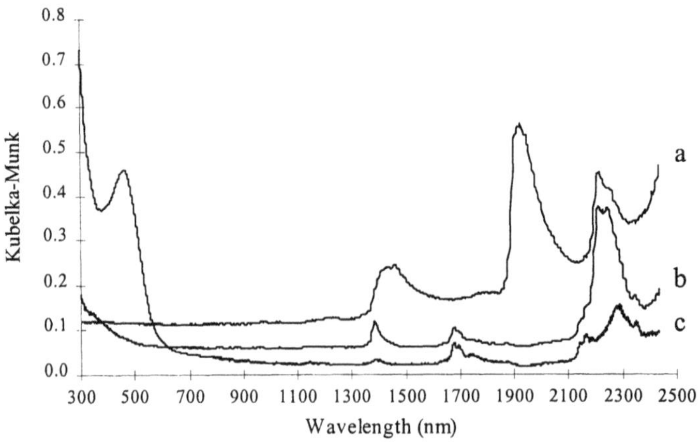

Figure 2. DRS spectra of (a) Si-MCM-41, (b) MAO-Si-MCM-41 and (c) $Et(Ind)_2Zr(CH_3)_2$/MAO-Si-MCM-41.

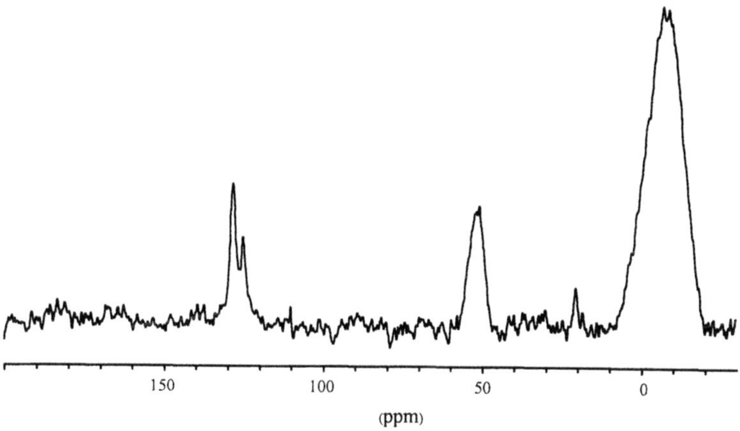

Figure 3. ^{13}C-CP-MAS-NMR spectrum of MAO-Si-MCM-41.

The anchoring of alumoxane on the silanol groups of Si-MCM-41 can be deduced from ^{29}Si-CP-MAS-NMR. After TMA treatment, the Q_2 $[Si(OSi)_2(OH)_2]$ and Q_3 $[Si(OSi)_3(OH)]$ resonances at -91.6 ppm and -100.6 ppm disappear which clearly demonstrates the covalent bonding of the alumoxane to the silanol groups of the mesoporous host (Figure 4). Si atoms with four neighboring T-atoms (Q_4) are therefore silent in the CP mode.

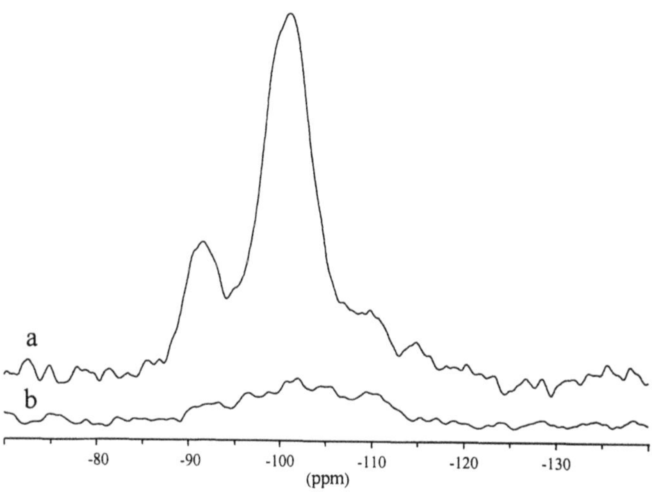

Figure 4. ^{29}Si-CP-MAS-NMR spectrum of (a) Si-MCM-41 and (b) MAO-Si-MCM-41.

EPMA-measurements of the Al-MCM-41 samples revealed a Si/Al ratio of 19.0 ± 0.3 and 43.9 ± 3.5 approximating the gel compositions. EPMA-analysis of MAO-Si-MCM-41 reveals a Si/Al ratio of 0.88 with a small deviation of 0.11. Therefore, 72 mol % of the added amount of TMA is converted into MAO and a more or less homogeneous distribution of alumoxane on the host was accomplished.

3.2. Oligomerization reactions

Table 1 shows the catalytic potential of the *ansa*-metallocene $Et(Ind)_2Zr(CH3)_2$ supported on different MAO-MCM-41 hosts in the co-oligomerization of ethene and propene. The MWD of the oligomers produced is always around 2 which points to the formation of only one active site on the metallocene catalyst encapsulated in the mesoporous host.

Table 1
Co-oligomerization of ethene and propene with the *ansa*-metallocene $Et(Ind)_2Zr(CH_3)_2$ on the alumoxane anchored mesoporous host.

Exp. [a]	MCM-41 Support [b]		Al/Zr [c]	Activity [d]	M_n [e]	M_w/M_n [e]	mol % P [f]
1	Si-MCM-41	(TEOS, 22.0 wt%)	360	4.45	790	1.87	56.4
2	Si-MCM-41	(TEOS, 22.0 wt%)	180	4.34	830	2.08	55.9
3	Si-MCM-41	(TEOS, 22.0 wt%)	90	4.30	940	2.13	55.8
4	Si-MCM-41	(TEOS, 29.0 wt%)	40	2.20	3,590	1.71	44.3
5	Si-MCM-41	(TEOS, 22.0 wt%)	40	3.88	1,280	1.94	55.3
6	Si-MCM-41	(TEOS, 17.0 wt%)	40	2.29	2,350	1.75	51.9
7	Si-MCM-41	(TEOS, 8.5 wt%)	40	0.34	5,320	1.74	69.2
8	Al-MCM-41	(Cab-O-Sil, Si/Al = 20)	90	2.63	1,910	2.11	50.9
9	Al-MCM-41	(Cab-O-Sil, Si/Al = 40)	90	3.20	1,800	2.00	53.2
10	Ti-MCM-41	(Ludox, Si/Ti = 50)	90	3.78	1,410	1.97	53.7

[a] The oligomerization experiment was performed at 343 K in toluene
[b] The mesoporous supports were synthesized with their corresponding silicium sources (TEOS, Cab-O-Sil or Ludox).
[c] The Al/Z-ratio was determined on the Al content of the alumoxane compound
[d] Co-oligomerization activity in 10^6 g mol $(Zr)^{-1}$ h^{-1} of the zirconocene after 75 min reaction time.
[e] M_n in g mol^{-1} and M_w/M_n of the co-oligomer were determined based on GPC-analysis
[f] Propene in mol % incorporated in the co-oligomer chain based on GC-conversion after 75 min.

At lower Al/Zr-ratios, the oligomerization activity decreases and the MW increases due to a lower incorporation of propene in the co-oligomer chain (entries 1, 2, 3 and 5). Hereby the physical properties of the product are changed from liquid (pour-point 250 K) to wax (pour-point 269 K) which is an important issue in the development of synthetic lubricants.

The *in situ* synthesis of alumoxane on the mesoporous host is largely dependent on the concentration of sorbed water on the support. Low water concentrations give rise to the formation of alumoxane islands with a high docking degree, whereas high concentrations produce an alumoxane layer with a low docking degree but with more internal coordination. An optimal oligomerization activity was noticed at a water concentration of 22 wt% for the siliceous mesoporous host. An increase (entry 4) or a decrease (entry 6 and 7) of the concentration slowed down the oligomerization activity and increased the MW. The constitution of the alumoxane layer also influences the incorporation of the metallocene molecules, which is translated in an effect on the MW.

In the anchoring process of alumoxane on the pore walls of a siliceous mesoporous host, a smooth balance exists on the alumoxane structure between the charge delocalization and the cluster formation by docking. By altering the synthesis procedure of the siliceous mesoporous host, the amount of silanol groups can be varied which has its effect on the docking behavior of alumoxane on the host and on the oligomerization activity of the metallocene [12]. Moreover, the idea was postulated that a softening of the docking points of the host could enhance the charge delocalization on the alumoxane structure. Acidity in the host was hereby induced by the incorporation of Al or Ti into the framework of MCM-41. So, the oligomerization activity of Al-MCM-41 has decreased with respect to the Si based structure and a drop in activity was observed for a lower Si/Al ratio (entries 3, 8 and 9). Incorporation of Ti in the mesoporous host (entry 10) leads to a more active system and decreases the MW compared to the Al-MCM-41 sample (entry 9) but still exhibits a lower oligomerization activity than the Si based host (entry 3). Therefore, it seems more appropriate to soften the docking points by the introduction of Ti in the framework hereby generating Lewis acidity in the host which softens neighboring anchoring points or induces coordinative sites for the O atoms of the alumoxane structure. The lower activity of the Ti-MCM-41 host with respect to the Si-MCM-41 sample can probably be ascribed to the different Si source employed which changes the amount of docking points on the host.

4. CONCLUSIONS

The beneficial properties of MCM-41 as a host for *ansa*-metallocenes have been demonstrated in the co-oligomerization of ethene and propene. The extension of the alumoxane layer as well as the docking characteristics of the mesoporous host play an important role in the oligomerization activity and selectivity of the encapsulated metallocene catalyst. Fine tuning of the mesoporous host changes the physico-chemical properties of the oligomerization product which opens perspectives in the development of new synthetic lubricants.

ACKNOWLEDGMENTS

The authors acknowledge sponsoring by the Belgian government in the frame of IUAP-PAI network on Supramolecular Chemistry and Catalysis. P.J.G. and L.K.V.L. acknowledge FWO for a research position and IWT for a fellowship, respectively.

REFERENCES

1. C.T. Kresge, M.E. Leonowicz, W.J. Roth, J.C. Vartuli and J.S. Beck, Nature, 359 (1992) 710.

2. J.S. Beck, J.C. Vartuli, W.J. Roth, M.E. Leonowicz, C.T. Kresge, K.D. Schmitt, C.T.-W. Chu, D.H. Olson, E.W. Sheppard, S.B. McCullen, J.B. Higgins and J.L. Schlenker, J. Am. Chem. Soc., 114 (1992) 10834.
3. A. Corma, Chem Rev., 97 (1997) 2373.
4. M.R. Mason, J.M. Smith, S.G. Bott and A.R. Barron, J. Am. Chem. Soc., 115 (1993) 4971.
5. S.H. Strauss, Chem. Rev., 93 (1993) 927.
6. J.C.W. Chien and D. He, J. Polym. Sci.(A), 29 (1991) 1603.
7. Y.-X. Chen, M.D. Rausch and J.C.W. Chien, J. Polym. Sci.(A), 33 (1995) 2093.
8. W. Kaminsky, Macromol. Chem. Phys., 197 (1996) 3907.
9. X.S. Zhao, G.Q. Lu, A.K. Whittaker, G.J. Millar and H.Y. Zhu, J. Phys. Chem. (B), 101 (1997) 6525.
10. E. Giannetti, G.M. Nicoletti and R. Mazzocchi, J. Polym. Sci., 23 (1985) 2117.
11. D.-H. Lee and S.-Y. Shin, Macromol. Symp., 97 (1995) 195.
12. L.K. Van Looveren, D.F. Geysen, K.A. Vercruysse, B.H Wouters, P.J. Grobet and P.A. Jacobs, Angew. Chem., 110 (1998).

MESOPOROUS MOLECULAR SIEVES 1998
Studies in Surface Science and Catalysis, Vol. 117
L. Bonneviot, F. Béland, C. Danumah, S. Giasson and S. Kaliaguine (Editors)

Hydrodemetallization of nickel tetraphenylporphyrin over Mo catalysts supported on mesoporous molecular sieves

M. Cheng[a], F. Kumata[a], T. Saito[a], T. Komatsu[b] and T. Yashima[b*]

[a]Advanced Catalysts Research Laboratory, Advanced Technology & Research Institute, Petroleum Energy Center. 3-2-1, Sakato, Takatsu-ku, Kawasaki, Kanagawa 213, Japan

[b]Department of Chemistry, Tokyo Institute of Technology, 2-12-1,Ookayama, Meguro-ku, Tokyo 152, Japan

The influence of Mo modification on the structure of Al-MCM-41 and the conversion of nickel tetraphenylporphyrin over Mo loaded Al-MCM-41 catalysts were investigated. The loading of Mo caused the loss of mesoporous structure of high silica Al-MCM-41 catalysts. Mo^{6+} existed as octahedral species which may be polymolybdena clusters and crystallites. The conversion of nickel tetraphenylporphyrin on Mo/Al-MCM-41 catalysts increased with aluminum content and pore diameter, but did not change significantly with Mo loading.

1. INTRODUCTION

Nickel tetraphenylporphyrin (Ni-TPP) is one of the model metal compounds used for the study of hydrodemetalization (HDM) of petroleum oil. HDM of Ni-TPP has been reported to involve a sequence of reaction steps including hydrogenation, dehydrogenation and hydrogenolysis [1,2]. These reactions are mainly catalyzed by Mo active sites on traditional $Mo/\gamma\text{-}Al_2O_3$ or $Co(Ni)Mo/\gamma\text{-}Al_2O_3$ catalysts. Acid sites may catalyze the hydrogenolysis of reaction intermediates, which may occur on the weak acid sites of H_2S adsorbed on catalyst surface. This suggests that the utilization of acidic support may accelerate the hydrogenolysis of reaction intermediates.

As a big molecule with a maximum diameter of 1.9 nm, Ni-TPP has also been used to evaluate the influence of intraparticle diffusion which may exist in some hydrotreating processes. On large catalyst particles, the conversion rate of Ni-TPP was greatly suppressed by intraparticle diffusion [3] and HDM of Ni-TPP resulted in a higher nickel deposition on the outer shell than on the inner shell of catalyst after the reaction as the case in commercial HDM process [4].

MCM-41 invented by Mobil scientists possesses hexagonal array of uniform mesopores whose pore diameter can be varied from 1.6 nm to 10 nm with different tactics [5]. It is also possible to introduce aluminum into the framework of MCM-41 during hydrothermal synthesis or by post-synthetic alumination to generate some acid sites. This novel material has been reported to have potential usage in hydrotreating such as hydrocracking [6-8], hydrodesulfurization [9] and HDM [10]. The objective of present study is to evaluate the catalytic properties of Mo loaded Al-MCM-41 with various Al contents for the HDM of petroleum using Ni-TPP as a model compound.

2. EXPERIMETAL

2.1. Catalyst Preparation

Al-MCM-41 molecular sieves with different SiO_2/Al_2O_3 ratio and a pore size of 3.4 nm were synthesized using cetyltrimethylammonium bromide as template with repeated pH adjustment procedures according to the literature [11]. Al-MCM-41(SiO_2/Al_2O_3 = 27) samples with different pore size were synthesized with dodecyltrimethylammonium bromide, cetyltrimetylammonium bromide, and cetyltrimethylammonium bromide with trimethyl benzene as templates, respectively. The as-synthesized samples were washed 5-6 times with hot water and once with ethanol, then dried at 80 ° C overnight. The dried samples were then calcined in air at 550 ° C for 10 h. NH_4Al-MCM-41 was obtained by an ion-exchange with 0.1N NH_4NO_3 solution at room temperature. Mo/Al-MCM-41 catalysts were prepared by impregnation using ammonium heptamolybdate solution and NH_4Al-MCM-41. All the catalysts were dried at 120 ° C overnight and then calcined in air at 500 ° C for 5 h. A commercial 3wt%Mo/γ-Al_2O_3 catalyst with a average pore size of 18 nm was used as a reference catalyst.

2.2. Catalyst Characterization

Powder X-ray diffraction patterns of Mo/Al-MCM-41 catalysts were collected on Rigaku RINT- 2400. Surface area, pore volume and N_2 adsorption-desorption isotherms were measured on COULTER SA3100, and the pore diameters of all the catalysts were calculated following the simple relation between surface area (S) and the pore volume (V) for cylindrical pores: D = 4V/S. UV-visible diffuse reflectance spectra were recorded on Shimadzu UV-2400.

2.3. HDM Reaction of Ni-TPP

The catalytic reactions were performed in a down-flow fixed-bed reactor. Mo/Al-MCM-41 or Mo/γ-Al_2O_3 catalyst diluted with fine quartz powder (≥230 mesh, diluent/catalyst ratio of 30 by weight) was packed into catalyst bed. All the catalysts were sulfided in situ with flowing 5%H_2S/H_2 at 400 ° C for 1 h under atmospheric pressure before each reaction. A model oil system consisting of 11±1 ppm nickel of Ni-TPP and 300 ppm dimethyl disulfide dissolved in liquid paraffin was used as a liquid feed. The reaction was carried out at 310 ° C under 3.5 MPa of hydrogen with a space velocity of 0.006 g-cat·h/ml-oil and an H_2/liquid volume ratio of 300. The activities reported were obtained at 9 h on stream. The concentration of Ni-TPP was measured from the absorbance at 526 nm on a Shimadzu UV-1200 spectrophotometer. Conversions were determined from the disappearance of Ni-TPP.

3. RESULTS AND DISCUSSION

3.1. Characterization of Catalysts

3.1.1. 6%Mo/Al-MCM-41 Catalysts with Different SiO_2/Al_2O_3 Ratio

The specific surface area, pore volume and pore diameter of Al-MCM-41 with different SiO_2/Al_2O_3 ratios before and after loading 6%Mo and the N_2 adsorption-desorption isotherms are shown in Table 1 and Fig. 1 respectively. The pore diameters of Al-MCM-41 samples were all about 3.4 nm and their N_2 isotherms showed a sharp inflection at P/P_0 of 0.3, which indicates a uniform mesopore system. In the case of Mo/Al-MCM-41(81) and Mo/Al-MCM-41(57), the low surface area and pore volume, the decrement in pore diameter and the disappearance of the sharp inflection in N_2 isotherms indicate the loss of the uniform mesopore system. Mo/Al-MCM-41(39) catalyst also lost part of mesoporous structure. But the mesoporous structure of Mo/Al-MCM-41(27) was hardly influenced by Mo impregnation. The destruction of pore structure depicts the low thermal stability of high silica Al-MCM-41 in the presence of molybdate anions.

The nature of Mo species supported on Al-MCM-41 was characterized with UV-visible spectra and XRD patterns. The UV-visible spectra of Mo/Al-MCM-41 catalysts are shown in Fig. 2 in comparison with the spectra of bulk MoO_3 and Mo/γ-Al_2O_3. The absorption in the 300-320 nm region obtained with MoO_3 is attributed to octahedral Mo^{6+} and Mo/γ - Al2O3, which contains mainly tetrahedral Mo6+, gave absorption at 230-250 nm and 260-280 nm. All Mo/Al-MCM-41 catalysts showed almost the same spectra, which suggests the same Mo species on these catalysts. The high relative absorbance of Mo/Al-MCM-41 above 300 nm implies that Mo^{6+} is mainly present as octahedrally coordinated species such as polymolybdena clusters or MoO_3 crystallites, but no diffraction peaks of MoO_3 was found in

XRD patterns.

Table 1
Specific surface area (S), pore volume (V) and pore diameter (D) of Al-MCM-41 and 6%Mo/Al-MCM-41 catalysts.

	Al-MCM-41				Mo/Al-MCM-41 catalysts			
SiO_2/Al_2O_3 ratio	81	57	39	27	81	57	39	27
S ($m^2\ g^{-1}$)	1021	991	937	826	446	442	663	814
V ($cm^3\ g^{-1}$)	0.876	0.844	0.802	0.741	0.253	0.255	0.496	0.656
D (nm)	3.43	3.41	3.42	3.59	2.27	2.31	2.99	3.22

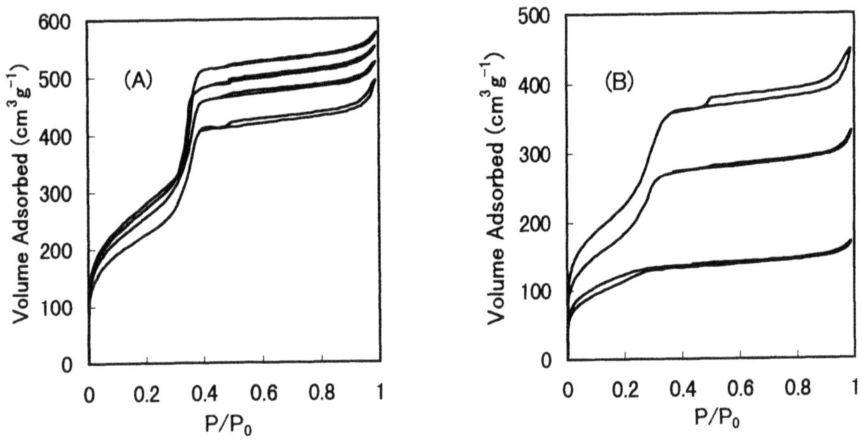

Figure 1. N_2 adsorption-desorption isotherms on (A) Al-MCM-41 with SiO_2/Al_2O_3 ratio of 81, 57, 39 and 27 from top; and (B) 6%Mo/Al-MCM-41 catalysts with SiO_2/Al_2O_3 ratio of 27, 39, 57 and 81 from top.

3.1.2. Mo/Al-MCM-41(27) Catalysts with Different Mo Loading

Mo loading up to 10% did not cause a significant loss of the uniform mesopore system of Al-MCM-41(27) as depicted by the pore diameter in Table 2, but led to an increment of the relative absorbance above 300 nm (in Fig. 3). The appearance of MoO_3 phase on high Mo loading catalysts is shown in Fig. 4.

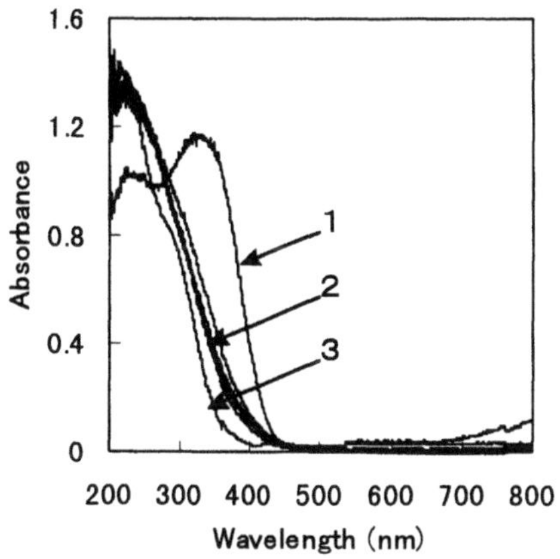

Figure 2. UV-visible spectra of 1, MoO_3; 2, 6%Mo /Al-MCM-41 catalysts including four samples after calcination in air and 3, Mo/γ- Al_2O_3.

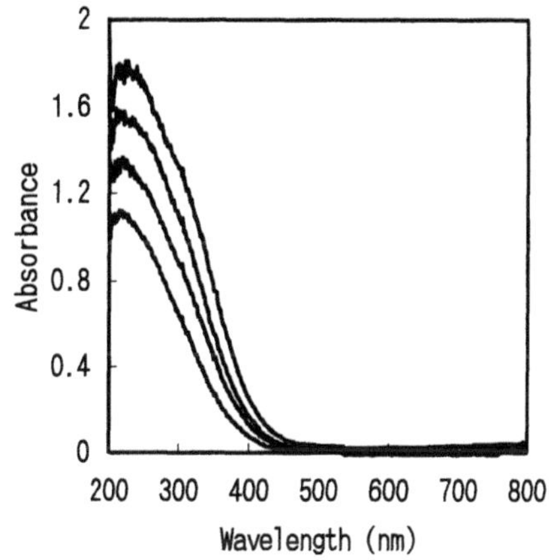

Figure 3. UV-visible spectra of Mo/Al-MCM-41 catalysts with Mo loading of 3, 6, 10 and 20% from bottom after calcination in air.

Table 2
Specific surface area, pore volume and pore diameter of Mo/Al-MCM-41(27) catalysts.

Mo (%)	3	6	10	20
S ($m^2 g^{-1}$)	842	813	740	562
V ($cm^3 g^{-1}$)	0.715	0.656	0.623	0.388
D (nm)	3.39	3.23	3.36	2.76

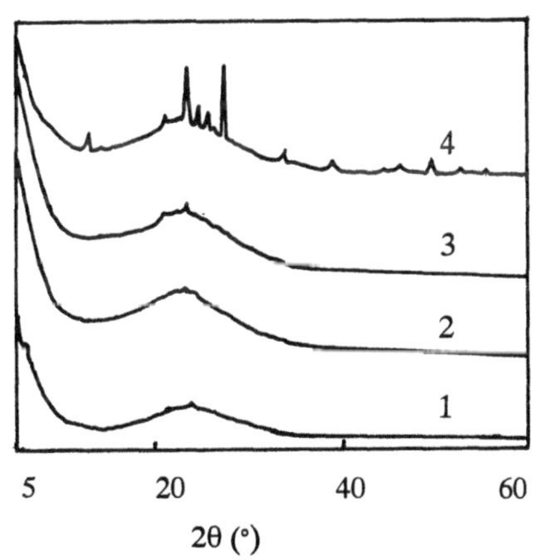

Figure 4. XRD patterns of Mo/Al-MCM-41(27) with different Mo loading.1, 3%Mo; 2, 6%Mo; 3, 10%Mo and 4, 20%Mo.

3.2. Conversion of Ni-TPP

3.2.1. Conversion of Ni-TPP on 3%Mo/γ-Al_2O_3

The activity of Mo/γ-Al_2O_3 was obtained for 75-90 μm particles on which the influence of intraparticle diffusion is very little [1,4]. The conversion of Ni-TPP was 90.5%. The high activity would be attributed to the active MoS_2 species prepared from highly dispersed Mo oxide precursors on γ-Al_2O_3 .

3.2.2. Effect of Aluminum Content in 6%Mo/Al-MCM-41 Catalysts

The effect of aluminum content in 6%Mo/Al-MCM-41 catalysts is shown in Fig. 5. The conversion of Ni-TPP increased with the aluminum content though two low aluminum catalysts, Mo/Al-MCM-41(81) and Mo/Al-MCM-41(57), showed almost the same activities. The increment of aluminum in Al-MCM-41 may result in an increment of acid sites, uniform mesopores after Mo loading and well-dispersed MoO_3 precursors. All these factors can lead to an increment in activity. Because the UV-visible spectra of Mo precursors were almost the same on all the 6%Mo/Al-MCM-41 catalysts, the increment of activity would not be caused by the difference in Mo active species. Because Mo/Al-MCM-41 has a lower hydrogenating activity than Mo/γ-Al_2O_3, the role of acid sites in hydrogenolysis would be limited. The increment of conversion of Ni-TPP would be largely attributed to the change of pore structure. Due to the low thermal stability of low aluminum Al-MCM-41, the collapse of pore structure must block some active sites. In fact, both Mo/Al-MCM-41(81) and Mo/Al-MCM-41(57) catalysts, which had the same specific surface area and pore volume, showed the same conversion of Ni-TPP.

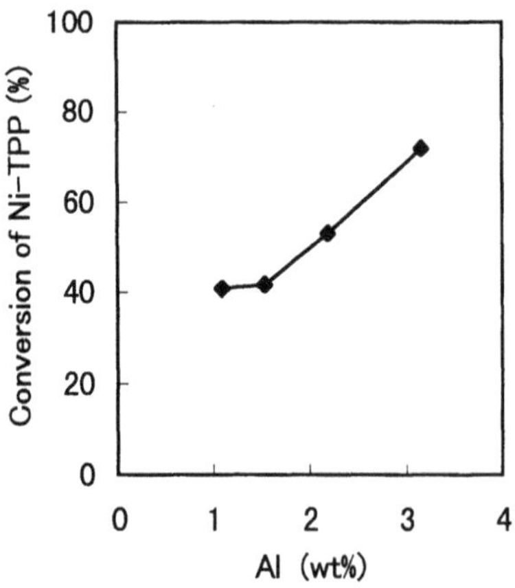

Figure 5. Effect of aluminum content on the conversion of Ni-TPP on 6%Mo/Al-MCM-41 catalysts.

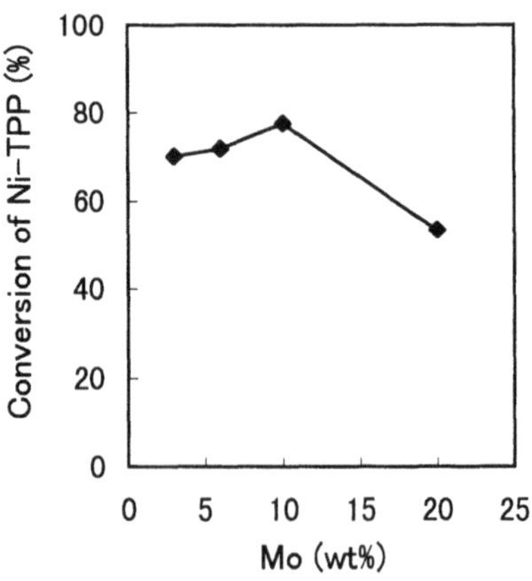

Figure 6. Effect of Mo loading on the conversion of Ni-TPP on Mo/Al-MCM-41(27) catalysts.

3.2.3. Effect of Mo Loading on Mo/Al-MCM-41(27) Catalysts

The effect of Mo loading was investigated on Mo/Al-MCM-41(27) catalysts whose pore structure was hardly affected by the Mo loading. The conversion of Ni-TPP in Fig. 6 showed a maximum at 10 wt% Mo. Though Mo is the main active species for the reaction, the effect of Mo loading was not significant as expected. This may be due to the low dispersion of MoO_3 precursors on Al-MCM-41, that is, crystallites of MoO_3 exist on high Mo loading catalysts. In addition, the reaction rate may also be suppressed by intrapore diffusion of Ni-TPP.

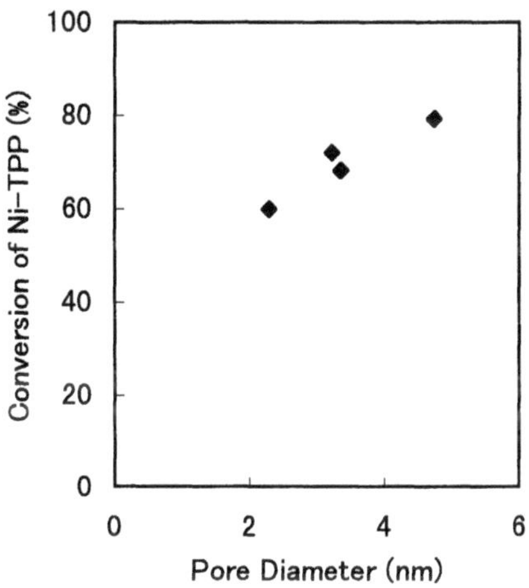

Figure 7. Effect of pore diameter on the conversion of Ni-TPP.

3.2.4. Effect of Pore Size on Mo/Al-MCM-41(27) Catalysts

In order to elucidate the influence of intrapore diffusion, Al-MCM-41 samples with different pore sizes were synthesized with the same SiO_2/Al_2O_3 ratio of 27, and the same content of Mo was loaded. Fig. 7 shows that the conversion of Ni-TPP increased with pore diameter. Because the pore diameter of all the catalysts is larger than the molecular dimension of Ni-TPP, theoretically speaking, Ni-TPP can enter into the mesoporous channels of all these catalysts. But due to the restricted effect of pore wall and the blocking effect of MoS_2 crystallites, the intrapore diffusivities of Ni-TPP may be significantly different on these catalysts. The lower conversion of Ni-TPP on small pore catalysts compared to that on large pore catalysts was attributed to the suppression of intrapore diffusion of Ni-TPP.

4. CONCLUSIONS

The impregnation of ammonium heptamolybdate solution and subsequent calcination in air caused the loss of mesoporous system of high silica Al-MCM-41. Al-MCM-41 with a SiO_2/Al_2O_3 ratio of 27 showed a good thermal stability in the presence of molybdate anions. On Mo/Al-MCM-41 catalysts, Mo^{6+} mainly existed as octahedrally coordinated species with low dispersion. The conversion of Ni-TPP on Mo/Al-MCM-41 catalysts increased with aluminum content and pore diameter, but did not change significantly with Mo loading.

ACKNOWLEGMENT

This work has been carried out as a research project of The Japan Petroleum Institute commissioned by the Petroleum Energy Center with the subsidy of the Ministry of International Trade and Industry.

REFERENCES

1. R.A. Ware and J. Wei, J. Catal., 93 (1985) 100.
2. J.P. Janssens, G. Elst, E.G. Schrikkema, A.D. van Langeveld, S.T. Sie and J.A. Moulijn, Recl. Trav. Chim. Pays-Bas, 115 (1996) 465.
3. S.Y. Lee, J.D. Seader, F.E. Massoth, S.Y. Lee and J.D. Seader, Ind. Eng. Chem. Res., 30 (1991) 29.
4. R. Agrawal and J. Wei, Ind. Eng. Chem. Process Des. Dev., 23 (1984) 515.
5. C.T. Kresge, M.E. Leonowicz, W.J. Roth, J.C. Vartuli and J.S. Beck, Nature, 359 (1992) 710.
6. A. Corma, A. Martínez, V. Martínez-Soria and J.B. Montón, J. Catal., 153 (1995) 25.
7. M.R. Apelian, T.F. Jr. Degnan, D.O. Marler and D.N. Mazzone, US Patent No. 5 264 116, (1993).
8. T.F. Jr. Degnan, K.M. Keville, M.E. Landis, D.O. Marler and D.N. Mazzone, US Patent No. 5 290 744 (1994).
9. K. M. Reddy, B. Wei and C. Song, Am. Chem. Soc. Div. Petro. Chem. Prepr., (1997) 396.
10. C.T. Kresge, M.E. Leonowicz, W.J. Roth, J.C. Vartuli, K.M. Keville, S.S. Shih, T.F. Jr. Degnan, F.G. Dwyer and M.E. Landis, US Patent No. 5 183 561 (1993).
11. R. Ryoo and J. M. Kim, J. Chem. Soc., Chem. Commun., (1995) 711.

MESOPOROUS MOLECULAR SIEVES 1998
Studies in Surface Science and Catalysis, Vol. 117
L. Bonneviot, F. Béland, C. Danumah, S. Giasson and S. Kaliaguine (Editors)

New Mo and NiMo catalysts supported on MCM-41/Alumina for thiophene hydrodesulfurization

T. Klimova[a], J. Ramírez[a], M. Calderón[a] and J.M. Domínguez[b]

[a]Departamento de Ingeniería Química, Facultad de Química, Universidad Nacional Autónoma de México, Cd. Universitaria, C.P. 04510, México D.F., México

[b]Instituto Mexicano de Petróleo, Eje Central Lázaro Cárdenas, 152, C.P. 17730, México D.F., México

A series of Mo and NiMo catalysts supported on MCM-41-Al_2O_3 with different amounts of MCM-41 was prepared and tested in the thiophene hydrodesulfurization reaction. In the Mo unpromoted catalysts, an increase in the content of MCM-41 caused a decrease in the thiophene conversion. In contrast, for the NiMo catalysts, where Ni was impregnated first, an increase in the thiophene conversion was observed with the MCM-41 loading. These different trends in HDS catalytic activity of unpromoted and Ni promoted Mo catalysts were related, according to the characterization results, to changes in the dispersion of the Mo phase induced by the presence of the MCM-41 catalyst component, which also change when Ni is impregnated prior to Mo. It was also found for the reaction studied here, that the hydrodesulfurization/hydrogenation ratio increases with MCM-41 content in the NiMo catalysts.

1. INTRODUCTION

In recent years, with the aim of improving the hydrodesulfurization catalysts, new materials have been tried as catalytic supports. Among these materials, the molecular sieve MCM-41, which contains hexagonal pores of about 30-35 Å in diameter and a surface area of about 700-1000 m^2/g, has attracted some attention as catalytic support for hydrotreating catalysts [1-3]. Recently, some applications of this material to hydrotreating catalysts have been reported [4-6]. However, to the present day there has been no report on the behavior of promoted and unpromoted NiMo and Mo hydrodesulfurization catalysts, neither on the effects caused in the hydrodesulfurization activity of Mo and NiMo catalysts, by the incorporation of variable amounts of MCM-41 into the alumina support (matrix). This last effect is of importance since the MCM-41 by itself does not have the appropriate mechanical properties for industrial application. It is the object of the present work to study the effect that the incorporation of variable amounts of MCM-41 into the alumina support has on the thiophene hydrodesulfurization activity of Mo and NiMo catalysts.

2. EXPERIMENTAL

The mesoporous MCM-41 material (SiO_2/Al_2O_3 molar ratio = 6.7) was synthesized according to the procedure reported earlier [7]. The catalytic supports were prepared by the incorporation of different amounts of MCM-41 (0.0, 10.0 and 30.0 wt. %) into a γ-alumina matrix, using pseudo-boehmite as binder. The impregnation of the Mo and NiMo (Ni first) catalysts was made by the pore volume method using aqueous solutions of ammonium heptamolybdate (pH=5.5) and nickel nitrate (pH=4.5). The nominal composition of the unpromoted catalysts was 12 wt % MoO_3 and for the promoted ones 12 wt % MoO_3 and 3 wt % NiO. After each impregnation the catalysts were dried (373 K, 24 h) and calcined (673 K, 2h). Additionally, to study the effect of Mo loading in the NiMo catalysts, several catalysts samples prepared with the support containing 30 wt % MCM-41 were prepared with different amounts of Mo (0.47, 0.98 and 1.86 Mo atoms per square nanometer of support surface), equivalent to 3, 6 and 12 wt % MoO_3. In all cases the atomic ratio Ni/(Ni+Mo) was kept at 0.32. Hereafter the catalysts will be denoted Ni(x)Mo(y)/MCM(z), where x, y and z represent the weight percent of NiO, MoO_3 and MCM-41 respectively.

The surface area, pore diameter and pore volume of the catalysts were obtained by nitrogen physisorption in a Micromeritics ASAP-200 apparatus. The UV-vis diffuse reflectance spectra were obtained in a Varian Cary 5E-NIR spectrophotometer, and the high resolution electronic microscopy observations were made in a JEOL-2010 microscope with a point to point resolution of 1.9 Å. For HREM observations the catalyst sulfided samples were ground in an agate mortar, ultrasonically dispersed in n-heptane, a drop of the supernatant liquid was placed on a carbon covered copper grid. The thiophene (TH) hydrodesulfurization catalytic activity tests were conducted in a flow microreactor operating at atmospheric pressure in the 503-563 K temperature range. Before the activity tests, the catalysts were sulfided in situ (673 K, 4 h), in a stream of H_2S (15 vol. %)-H_2. The reaction products were analyzed by on-line gas chromatography using a 30 m capillary HP-1 column coupled to a flame ionization detector.

3. RESULTS AND DISCUSSION

The results from textural characterization show that, as expected, the incorporation of MCM-41 in the alumina matrix increase the support surface area (see Table 1).

Table 1
Surface area and porosity of catalytic supports

Sample	Surface area (m^2/g)	Total pore volume (cm^3/g)
MCM-41	716	0.885
Al_2O_3	187	0.422
MCM(10)-Al_2O_3	210	0.434
MCM(30)-Al_2O_3	277	0.428

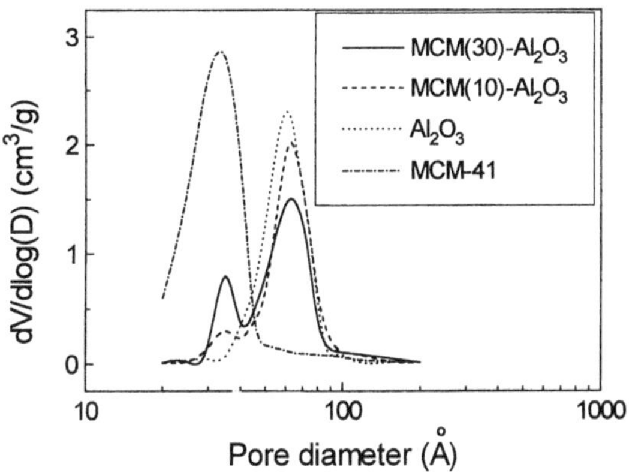

Figure 1. Pore volume distributions in the supports.

The results of the pore size distribution for the pure Al_2O_3 and MCM-41, and of the MCM(z)-Al_2O_3 supports (with z = 0.0, 10 and 30), shown in Figure 1, indicate that the incorporation of MCM-41 to the alumina matrix was adequately achieved without substantial blockage of the MCM-41 pores since the pure alumina and MCM-41 components give monomodal pore size distributions while the alumina supports with MCM-41 incorporated give bimodal pore size distributions with the maxima corresponding well with the single component systems.

When the active phases are impregnated on the supports, a decrease in surface area and pore volume is observed (see Table 2). This effect is more pronounced in the catalysts with greater MCM-41 content.

Table 2
Textural properties and HDS activity of Mo and NiMo catalysts

Catalyst	Surface area (m^2/g)	Total pore volume (cm^3/g)	Thiophene conversion* (%)
Mo(12) / MCM(0)	180	0.367	3.4
Ni(3)Mo(12) / MCM(0)	175	0.348	10.0
Mo(12) / MCM(10)	174	0.350	2.0
Ni(3)Mo(12) / MCM(10)	180	0.347	12.5
Mo(12) / MCM(30)	238	0.376	1.5
Ni(0.75)Mo(3) / MCM(30)	244	0.386	8.0
Ni(1.5)Mo(6) / MCM(30)	220	0.350	10.6
Ni(3)Mo(12) / MCM(30)	197	0.320	14.5

* TH conversion determined at 563 K, amount of catalyst used was 0.1 g for Mo catalysts and 0.03 g for NiMo catalysts.

Regarding the hydrodesulfurization activity of the catalysts, the results, also included in Table 2 as percent thiophene conversion, indicate that, for the unpromoted Mo catalysts, the incorporation of MCM-41 into the alumina support leads to a decrease in HDS activity as the MCM-41 content is increased. In contrast, the NiMo catalysts show an increase in HDS activity as the MCM-41 content in the support is increased (Figure 2). This increased trend in HDS catalytic activity is preserved even if the results are expressed as moles of thiophene converted per second per Mo atom, (intrinsic activity), in order to make a correction for the different Mo surface concentrations in the catalysts due to their different surface areas.

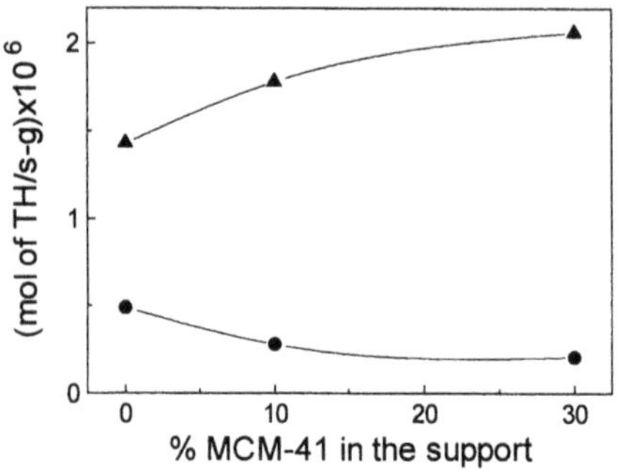

Figure 2. Catalytic activity of Mo (●) and NiMo (▲) catalysts in thiophene hydrodesulfurization (563 K, atmospheric pressure).

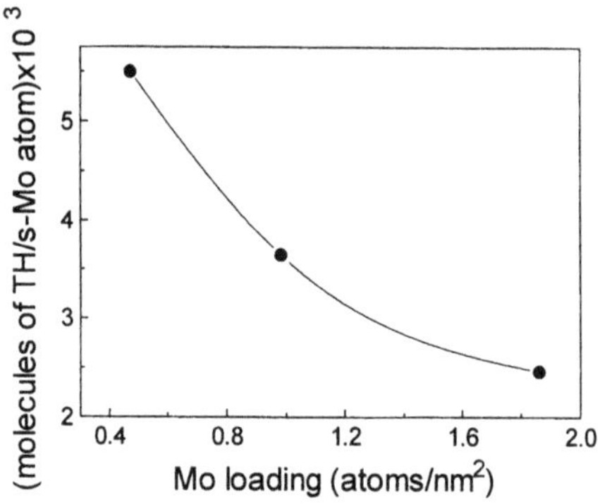

Figure 3. Catalytic activity of Ni(x)Mo(y)/MCM(30) catalysts in thiophene hydrodesulfurization (563 K, atmospheric pressure). Mo loading effect.

The activity results from NiMo catalysts with different Mo loading indicate, for the catalyst with 30 wt % MCM-41, that an increase in the thiophene conversion from 8 to 14.5 % is obtained when the Mo loading is varied from 3 to 12 % (Table 2). Clearly, the increase in activity is not proportional to the MCM-41 loading. Moreover, the catalytic activity results of catalysts with different Mo loading show that as the amount of Mo in the catalyst is increased, their intrinsic activity decreases (see Figure 3). These results clearly indicate that the Mo dispersion in the catalyst decreases with MCM-41 loading, but that, at the same time, the promotional effect of Ni is great enough to produce the result shown in Figure 2 where the HDS activity is increased as the MCM-41 content in the support is increased. In line with this, it is observed that the promotional effect of Ni, given by the ratio of the promoted to unpromoted HDS reaction rates of the catalysts, changes from 2.93 in the catalyst supported on pure alumina to 6.36 in the catalyst with 10 % MCM-41 and to 9.86 in the catalyst with 30 % MCM-41, for the different reaction rates evaluated at 563 K.

Another interesting effect observed as the MCM-41 content in the catalyst was increased, was a decrease in the production of hydrogenated product (butane), in the thiophene HDS reaction. The results indicate that, for the same total thiophene conversion (10 %), the percent of butane in the products changed from 15.8 to 13.5 and to 10.5 % when the MCM-41 content in the catalyst was varied from 0.0 to 10 and to 30 % respectively (see Figure 4). Clearly this observation is interesting since it foresees the possibility of achieving high HDS rates with low hydrogen consumption. This effect could be related to the changes in the morphology of the MoS_2 phase induced by the incorporation of MCM-41 in the catalyst support, leading to a change in the proportion of active sites responsible for hydrodesulfurization and hydrogenation.

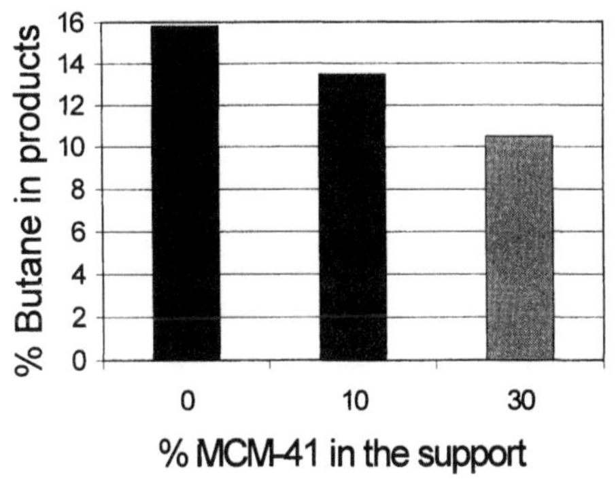

Figure 4. Percent of butane in the products of thiophene HDS as a function of MCM content in the support. (NiMo catalysts, total TH conversion equal to 10%.)

These suggested MoS_2 morphology changes were corroborated by high resolution electron microscopy observations of the sulfided catalysts. The micrographs of the Mo and NiMo catalysts supported on pure alumina and on the support containing 30 % MCM-41 (see Figures 5 and 6), showed a decrease in the dispersion of the MoS_2 structures when the MCM-41 was incorporated to the catalyst. In the MCM-41-containing catalysts, growing of the lamellar MoS_2 structures in length and stacking was observed specially on the edges of the support particles. This observation was related to the migration of the sulfided Mo phases supported on the MCM-41 to the pore mouths with the consequent decrease in MoS_2 dispersion. The observed activity trends for Mo and NiMo catalysts could then be the result of two effects, the first is the poor dispersion of Mo achieved on the MCM-41, which is in line with the activity trend observed in unpromoted catalysts, and the second, a better promotion of the Mo phase by Ni in the NiMo catalysts possibly due to the fact that Ni was impregnated first and was well dispersed in the MCM-41, although some detrimental effect is expected when Ni is impregnated first since the possibility of the entering into the alumina matrix is enhanced.

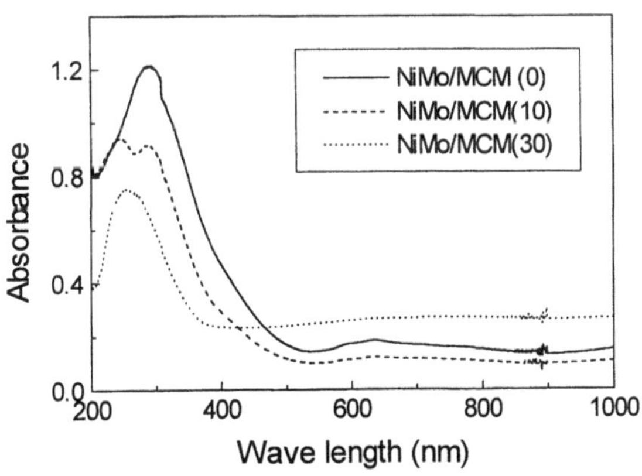

Figure 7. UV-Visible spectra of Ni(3)Mo(12)/MCM(z) catalysts.

In contrast to the above, the diffuse reflectance spectra of the oxidic precursors (Figure 7) indicate that as the MCM-41 content is increased in both, the promoted and unpromoted catalysts, less Mo in octahedral coordination is observed. Also, the intensity of the Mo band decreases with MCM-41 content. This result indicates that in the oxidic precursors there is a good dispersion of Mo since otherwise the band corresponding to Mo in octahedral coordination would increase due to the formation of agglomerated MoO_3. Regarding the nickel promoter, the intensity of the tetrahedral and octahedral Ni bands decrease with MCM-41 content. The decrease in the intensity of the Ni bands, and also those of Mo, could be related to the difficulty of the technique to detect the metals inside the

(a)

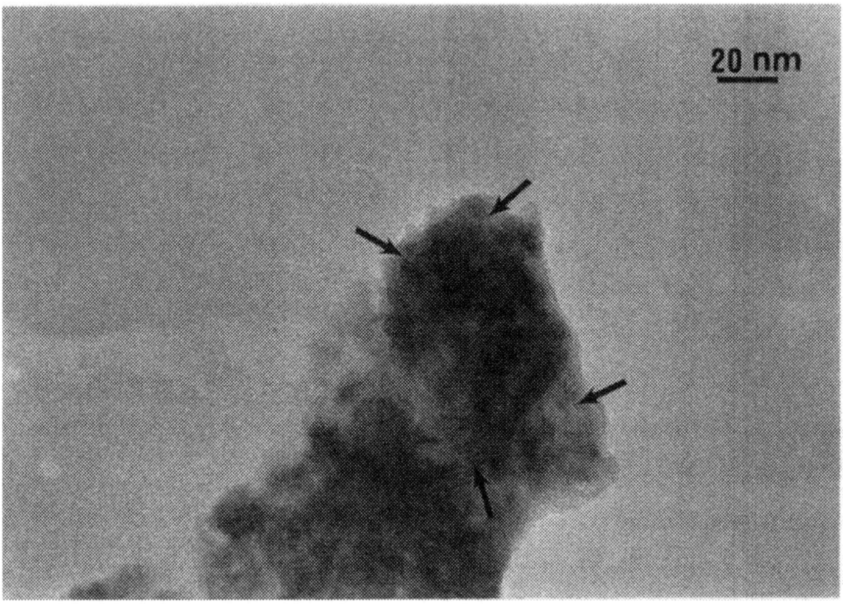

(b)

Figure 5. HREM micrographs of Mo catalysts supported on the pure Al_2O_3 (a) and MCM(30)-Al_2O_3(b). The Mo loading corresponds to 12 wt. % MoO_3.

(a)

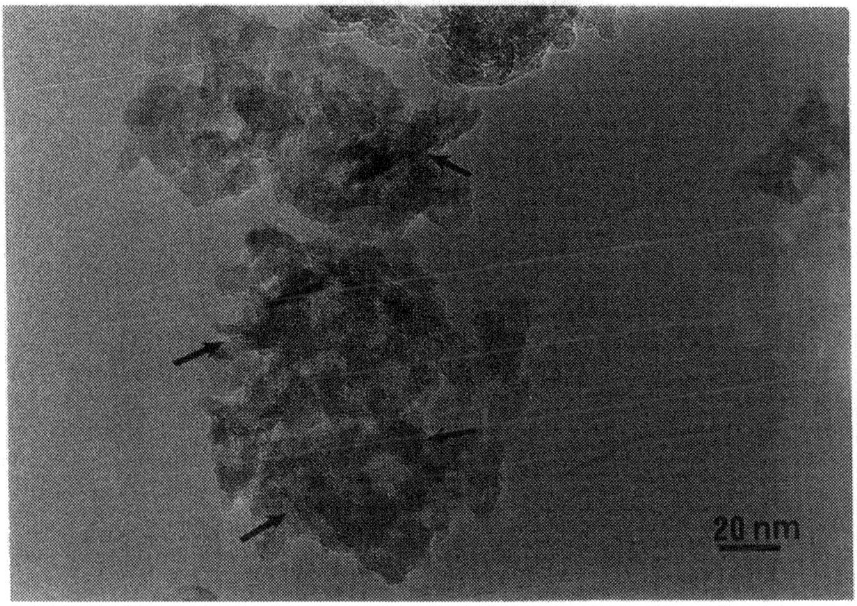

(b)

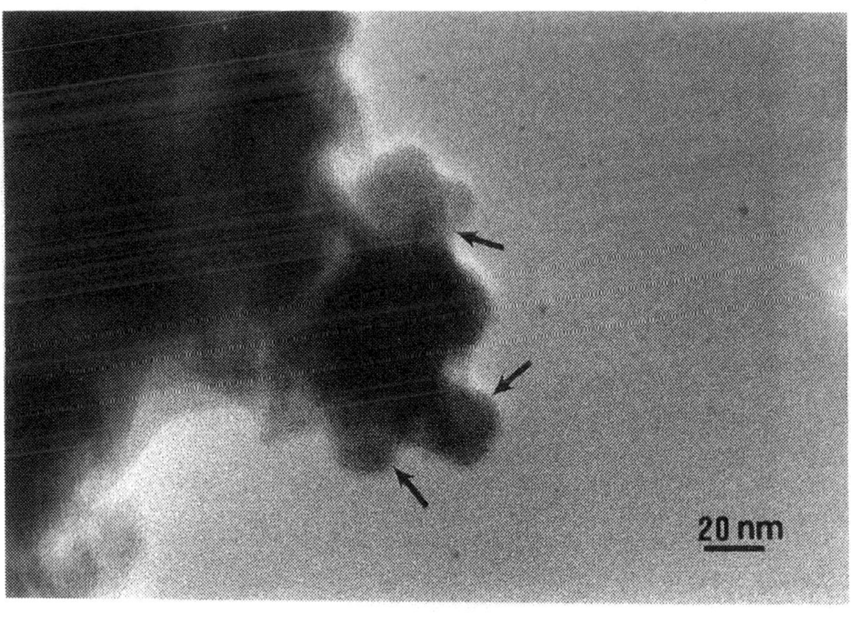

Figure 6. HREM micrographs of NiMo catalysts supported on the pure Al_2O_3 (a) and MCM(30)-Al_2O_3(b). The Ni and Mo loadings correspond to 3 wt. % NiO and 12 wt. % MoO_3.

pores of the MCM-41. It appears then that in the oxided state, the metals (Mo and Ni) are well dispersed on the catalysts containing MCM-41 but that during the sulfidation procedure this dispersion is partially lost, giving rise to some agglomeration of the Mo sulfided phase, as the HREM results indicate. This suggestion is in line with the higher intrinsic activity (activity per Mo atom) of the catalyst containing low Mo loading compared (3 wt. % MoO_3), with those with high Mo content (12 wt. % MoO_3), as shown in Figure 3.

4. CONCLUSIONS

From the above results, the following may be concluded:

1. The incorporation of MCM-41 to the alumina support in NiMo hydrotreating catalysts leads to increased HDS activity compared with the catalysts supported on pure alumina.
2. The intrinsic activity of NiMo catalysts supported on Al_2O_3-MCM-41 decreases with Mo loading indicating the loss of active Mo by a drop in dispersion, as the HREM micrographs show, or by the reaction of Mo species with the MCM-41 matrix leading to inactive Mo species. This is in line with the results of unpromoted Mo catalysts in which the HDS activity decreases with MCM-41 content.
3. For the reaction studied here, the HDS/HYD ratio in the NiMo catalysts increases with MCM-41 content.
4. According to the DRS and HREM results, it appears that although in the oxidic state a good dispersion of the Mo and Ni phases is achieved, during sulfidation there is a decrease in the dispersion, at least of Mo species.

ACKNOWLEDGMENTS

Financial support for this work by DGAPA-UNAM, CONACyT (Mexico) and IMP (FIES program) is gratefully acknowledged. We would like to thank Ivan Puente-Lee for obtaining the HREM micrographs.

REFERENCES

1. J. S. Beck and J. C. Vartuli, Current Opinion in Solid State and Materials Science, 1 (1996) 76.
2. J. S. Beck, J. C. Vartuli, W. J. Roth, M. E. Leonowicz, C. T. Kresge, K. D. Schmitt, C. T-W. Chu, H. D. Olson, E. W. Sheppard, S. B. McCullen, J. B. Higgins and J. L. Schlenker, J. Am. Chem. Soc., 114 (1992) 10834.
3. O. Franke, J. Rathousky, G. Schulz-Ekloff and A. Zukal, in Preparation of Catalysts VI, Elsevier Science, Amsterdam, p.309 (1995).
4. A. Corma, A. Martínez, V. Martínez-Soria and J.B. Montón, J. Catal., 153 (1995) 25.
5. A. Corma, M.S. Grande, V. Gonzalez-Alfaro and A.V. Orchilles, J. Catal., 159 (1996) 375.
6. A. Corma, Chem. Rev., 97 (1997) 2373.
7. E. Terrés, S.S. Ramírez, J.M. Domínguez, A. Montoya, J. Navarete and M. Gómez-Cicneros, Mat. Res. Soc. Symp. Proc., 431 (1996) 111.

MESOPOROUS MOLECULAR SIEVES 1998
Studies in Surface Science and Catalysis, Vol. 117
L. Bonneviot, F. Béland, C. Danumah, S. Giasson and S. Kaliaguine (Editors)

Diastereoselective Hydrogenation of some Prostaglandins Intermediates and Compounds over MCM-41 Supported Ru

S.Coman[a], F.Cocu[a], J.F.Roux[b], V.I.Pârvulescu[a] and S.Kaliaguine[b]

a- University of Bucharest, Faculty of Chemistry, Department of Chemical Technology and Catalysis, B-dul Republicii 13, Bucharest 70346, Romania

b- Universite Laval, Departement de Genie Chimique, Ste-Foy, QC, Canada G1K 7P4

Ru-MCM-41 catalysts prepared via deposition of Ru from $[Ru(NH_3)_6]Cl_3$ were investigated. Samples containing 3, 5 and 10 wt.% respectively, were characterised using adsorption-desorption measurements of N_2 at 77 K, H_2-TPR, TPO, H_2-chemisorption, XRD and XPS. It was found that the metal is homogeneously distributed inside of the support but the reduced Ru exists as different species. After deposition and activation the support keeps the structural features of the MCM materials. Hydrogenation of a prostaglandin intermediate and of a cyclic ketoester over these catalysts indicate that the reactions occur inside the mesoporous tubes. Under these conditions a diastereoselection of the hydrogenation pathways was found even in the absence of the chiral coadsorbates.

1. INTRODUCTION

The recent literature reflects a keen interest for enantio- and diastereoselective catalytic hydrogenation in the presence of heterogeneous catalysts [1-2]. Because of its practical applications, the enantioselective hydrogenation of β-ketoesters and α-ketoesters has received a special attention. In this respect two catalytic systems have mainly been investigated: i) nickel catalysts, usually a Raney Ni modified by α-hydroxy or α-amino-acids [3,4] and ii) platinum catalysts modified with cinchona alkaloid derivates [5-6]. Recent studies of Baiker et al. [7-8] and Sheldon et al. [9] showed that not only cinchona alkaloids but also structurally simple chiral amines and amino-alcohols or vinca-type alkaloids are efficient modifiers of Pt in stereoselective hydrogenation. All these studies refer to the use of modifying coadsorbates in heterogeneous catalysis which act in a way analogous to the ligand effects in homogeneous catalysis [10].

Only scarce informations on the heterogeneous stereoselective hydrogenation in the presence of other noble metals like Ru or Rh were given in the literature, even though these metals are extensively used in homogeneous chiral catalysis for such reactions. Our previous studies on the hydrogenation of prostaglandin intermediates showed that Ru deposition on different molecular sieves leads to effective catalysts, yielding prostaglandin structures with natural-like configuration [11].

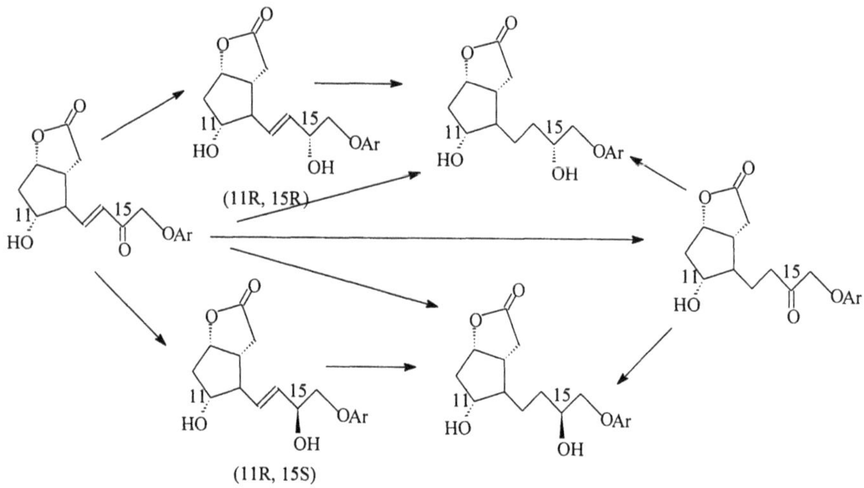

Scheme 1. Reaction scheme for diastereoselective hydrogenation of substrate #A.

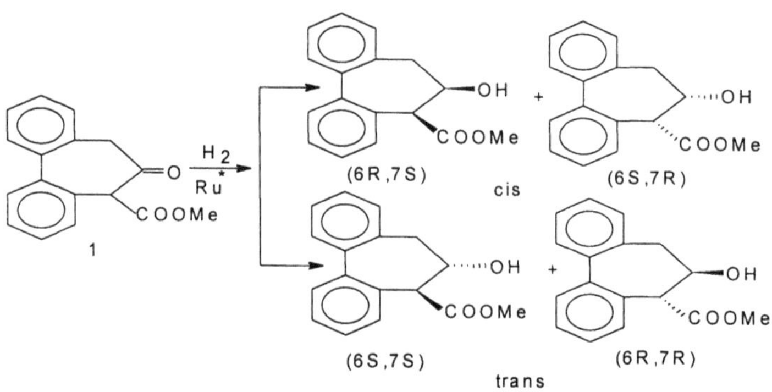

Scheme 2. Reaction scheme for diastereoselective hydrogenation of substrate #B.

Most of the studies using Pt as active species, considered Al_2O_3 as support [12]. But the highest enantiomeric excess reported to date in hydrogenation of a β-ketoester was achieved using polystyrene [13]. Good results were also obtained using SiO_2 [14].

The aim of the present study was to check if the implantation of the active species inside the mesoporous materials, i.e. on the walls of the mesoporous tubes, could lead, as an effect of shape selectivity, to the increase of the selectivity in the hydrogenation of large molecules. It was a particular interest to examine if the configuration of the active sites inside the mesoporous structure could replace, at least to a small extent, the effect of coadsorbates in diastereoselection of the hydrogenation pathway in the reactions of a prostaglandin intermediate (substrate #A) and of 7-carboxymethyl-5H-6,7-dihydrobenzo[α, γ]-cycloheptan-6-one (substrate #B) (Schemes 1 and 2). In this respect Ru-MCM-41 catalysts have been prepared. The effect of chiral modifiers such as tartaric as well as pivalic acids as has also been considered.

2. EXPERIMENTAL

Ru-MCM-41 catalysts were prepared by Ru deposition from a 0.4 M solution of $[Ru(NH_3)_6]Cl_3$ (Fluka purity). The MCM-41 silicate support was prepared in acidic medium using tetraethylorthosilicate as the source of silicon and CTAC as the template. The support was calcined in air at 500 °C. Before its exposure to the Ru solution the support was evacuated under vacuum at room temperature. After Ru deposition, the catalysts were washed with distilled water, dried at room temperature in a vacuum oven over night and then reduced in flowing hydrogen (30 ml min^{-1}) for 5 hours at 500 °C. The Ru loading was determined using atomic emission spectroscopy with inductively coupled plasma (ICP-AES). Following the mentioned procedure, samples with ruthenium content of 3, 5 and 10 wt.% were prepared. These are denoted by Ru-3, Ru-5, and Ru-10 respectively. The above catalysts were characterised using several techniques. Adsorption and desorption curves of N_2 at 77 K were obtained with a Micromeritics ASAP 2000 apparatus. H_2-Chemisorption was carried out using a Micromeritics ASAP 2010C. A Micromeritics PulseChemisorb 2705 apparatus was used in TPO experiments in the presence of 50 ml min^{-1} O_2 (5%) - He flow. Reduced ruthenium was determined from the oxygen balance assuming that at 500 °C Ru^0 is converted to RuO_2.This assumption was confirmed also by XRD and XPS. TPO results were used to determine the metal dispersion.Temperature programmed reduction (TPR) was carried out in the same installation than TPO. XPS spectra were recorded using a SSI X probe FISONS spectrometer (SSX -100/206) with monochromatic Al Kα radiation. The bands assigned to Ru_{3p}, Si_{2p} and O_{1s} were recorded. XRD measurements of the samples were made using a SIEMENS D-5000 diffractometer.The diffractograms were recorded in the range 2θ: (0-80°) using CuKα radiation (l = 1.5418 Å).

Standard experiments used 10 mg substrate dissolved in 10 ml ethanol or THF (in the case of the modified catalyst with L(+)-TA). Hydrogenation of substrate #A was carried out in a stainless steel stirred autoclave under 2 - 10 atm hydrogen pressure at 20 °C using a solution of prostaglandin intermediate in anhydrous methanol or THF. Analysis of the reaction products was carried out by HPLC using a Nucleosil 5C18 column as well as by ^{13}C and ^{1}H-NMR. The diastereoselective excess (d.e.) was defined using equation 1:

$$d.e. = [(15R) - (15S)/ (15R) + (15S)] \times 100 \quad (1)$$

We defined by 15R and 15S the allylic chiral products from the reaction scheme 1. Hydrogenation of substrate #B was carried out in an autoclave under 10 atm and at a temperature range of 60-80 °C. Conversions and selectivities were determined from GC MS data. Reaction products were analyzed using a Carlo-Erba HRGC 5300 equipped with two capillary columns, one with permethyl-β-cyclo-dextrin and the other with dimethyl-penthyl-β-cyclo-dextrin. The diastereoisomeric excess (d.e.) of the *cis*-(6S,7R) and (6R,7S) products were calculated as:

$$d.e.\ \% = \left| [\,6S,7R\,] - [\,6R,7S\,] \right| / [\,6S,7R\,] + [6R,7S\,] \times 100 \quad (2)$$

or:

$$d.e.\ \% = \left| [\,6R,7R\,] - [\,6S,7S\,] \right| / [\,6R,7R\,] + [6S,7S\,] \times 100 \quad (3)$$

for the *trans*-(6R,7R) and (6S,7S) products. In the case where pivalic acid was added, prior to analysis, the reaction product was neutralized with K_2CO_3, extracted in 100 ml anhydrous ethylic ether and dried on Na_2SO_4.

3. RESULTS

3.1. Catalysts characterisation

The main characteristics of the investigated used catalysts are given in Table 1.

Table 1.
Characteristics of the used Ru-MCM-41 catalysts

Property	Catalyst		
	Ru-3	Ru-5	Ru-10
BET Surface area, $m^2\ g^{-1}$	807	739	579
Average pore diameter, Å	21	19	18
Ru dispersion, %	9.51	7.89	6.31
Average particle diameter, Å	14.1	17.0	21.2
XPS Ru/Si ratio	0.09	0.11	0.22
ICP-AES Ru/Si ratio	0.10	0.13	0.23

XRD patterns of the catalysts as well as the shape of the adsorption-desorption curves indicated that no damage of the MCM-41 occured during the deposition of Ru or during activation of the catalysts. The excellent concordance between the Ru-to-Si atomic ratio determined from XPS and from ICP-AES allows us to suggest that Ru is distributed inside the mesoporous tubes and that there are not Ru local concentrations at least in the limits of XPS depth of analysis. The fact that t-plot micropore surface area was zero for all three catalysts also corroborates this conclusion. One should note that the average particle diameter determined from H_2-chemisorption data shows the existence of rather small Ru particles for such high metal concentrations.

3.2. Hydrogenation of prostaglandin intermediate

Figure 1 shows the catalytic performances of the chiral non-modified Ru-MCM-41 catalyts in the hydrogenation of the prostaglandin intermediate. The activity of the non-modified Ru-MCM-41 catalysts was found to increase with the Ru content. In the same time it was found that the reaction occurs with the formation of a d.e. in the epi form (15R) even though the catalyst was not chiraly modified. Previous studies in the presence of other supports [14], when the reaction occured on the external surface of the catalysts, showed that on the non-chiral modified catalysts only the racemic form had been obtained. The e.d. increases from the catalyst containing 3 wt.% Ru to the catalyst containing 10 wt.% Ru from 35.9% to 100%. For the same catalysts increasing the hydrogen pressure results in a significant enhancement of the reaction rate. Thus, for the catalyst containing 3 wt.% Ru the reaction rate increases 5 times when the pressure increases from 2 to 8 atm, for 5 wt.% Ru of 3.8 times and for 10 wt.% Ru 2.9 times. This enhancement in the reaction rate is accompanied by an enhancement in selectivity. Under these conditions, for 3 wt.% Ru catalyst the selectivity to allylic alcohol increases from 5.84 % (2 atm) to 27.3 % (8 atm). The change in stereoselectivity under these conditions is even more significant. Thus for 3 wt.% Ru catalyst this increase in pressure determines an increase of d.e. from 35.9 to 67.5% while for 5 wt.% Ru from 59.3 to 81.2%. In the case of 10 wt.% Ru d.e. remains 100 %.

Modification of the catalyst with tartaric acid leads to an important decrease of the reaction rate. Thus in the case of 3 wt.% Ru the decrease of the reaction rate is of about 16 %, 38 % for 5 wt.% Ru and 68 % for 10 wt.% Ru (Figure 2). Modifying with tartaric acid leads however to an increase of both the selectivity to alcohol and d.e. in the epi (15R) form. Again this last increase is more evident at lower Ru content. The enhancement in selectivity to allylic alcohol was 22 % and of d.e. of 14.4 % for this catalyst. Introduction in the reaction mixture of the pivalic acid has as an effect a supplementary reduction of the reaction rate. But the main effect of the pivalic acid is that its presence prevents the hydrogenation of the CO group and in such conditions only the hydrogenation of the alkenic bond occurs. One should note that in the experiments carried out using Ru supported on other molecular sieves like L, ZSM-5 or APO and also on Al_2O_3 and C, the introduction of pivalic acid enhanced the asymmetric hydrogenation [11].

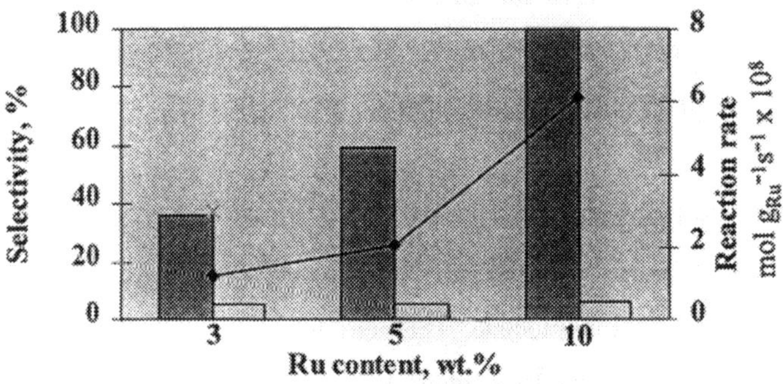

Figure 1. Variation of the reaction rate (◆), selectivity to allylic alcohol (□) and diastereoselective excess (▩) for non-modified Ru-MCM-41 catalysts. 20 °C, 2 atm, 4h.

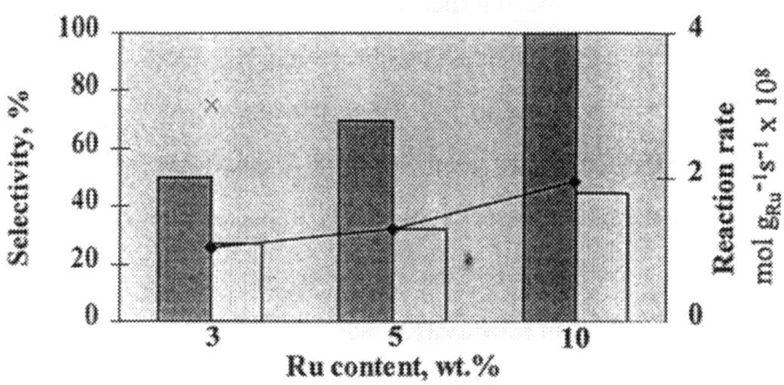

Figure 2. Variation of the reaction rate (◆), selectivity to allylic alcohol (□) and diastereoselective excess (▩) for tartaric acid modified Ru-MCM-41 catalysts. 20 °C, 2 atm, 4h.

3.3. Hydrogenation of 7-carboxymethyl-5H-6,7-dihydrobenzo[α, γ]-cycloheptan-6-one

Figure 3 shows the evolution of the reaction rate and of the selectivity to the hydroxyester on the investigated catalysts, in the absence of the chiral modifier and of ethanol as solvent. Data given in Figure 3 shows that the increase of the Ru content leads to a strong diminution of the reaction rate. Hydrogenation of substrate #B could proceed at different sites and this behaviour could be expressed by: i) chemoselectivity when we can distinguish among hydrogenation of ketonic C=O bond, keto-esteric C=O bond, and of the aromatic ring, ii) regioselectivity, when we can distinguish between *cis*- and *trans*-isomers, and iii) diastereoselectivity, when we can discern between the R and S configurations. In such conditions the selectivity was also strongly influenced by Ru content. In the case of Ru-3 the hydrogenation to *cis*-isomer occurs with a 10% d.e. in the (6R,7S)-configuration and to *trans*-isomer with a 16% d.e. in the (6R,7R)-configuration. For the oher two catalysts the d.e. was smaller than 4%. Modification of the catalysts with tartaric acid has as an effect a very strong diminution of the reaction rate, even for Ru-3.

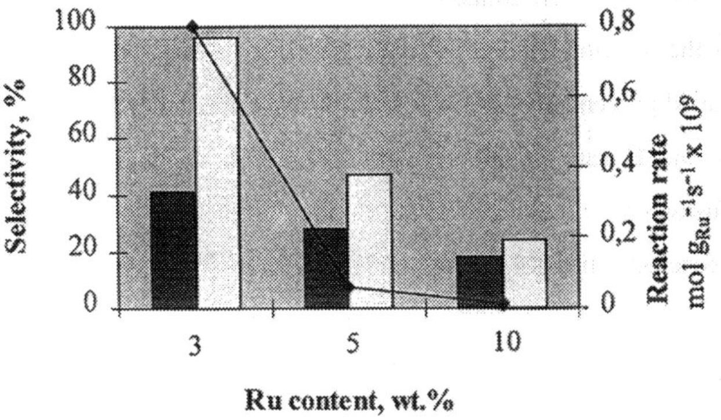

Figure 3. Evolution of the reaction rate (◆), of the chemoselectivity to hydroxyester (■) and of the regioselectivity to *cis*-hydroxyester (□). 80 ^{0}C,10 atm., 4 h.

4. DISCUSSION

Textural and structural characterisation of the Ru-MCM-41 catalysts show that ruthenium is docked inside of the mesoporous tubes without damages of the mesoporous material. The increase of the metal content leads to an increased number of metal particles without blocking of the tubes entrance.

Hydrogenation of substrate #A occurs with a selectivity to allylic alcohol which increases with the Ru content. More important is the fact that in the absence of the any modifier coadsorbate the reaction occurs with a d.e. in the epi form which could reach 100%. To our knowledge it is the first reported case in which diastereoselective hydrogenation in heterogeneous conditions occurs with so high a selectivity in the absence of chiral modifiers. The behavior of the catalysts during the pressure increase leads us to suppose that indeed the reaction takes place inside the mesoporous tubes. Increasing the pressure from 2 to 8 atm results in an increase of both the reaction rates and the selectivity to allylic alcohol. More important is the fact that these are accomplished by a d.e. increase. Modification of the catalysts with tartaric acid and furthermore the adding of pivalic acid modify the catalytic performances in a sense which indicates that reaction occurs indeed inside the mesoporous tubes. Thus the modification with tartaric acid leads to an important decrease of the reaction rate but to an increase of the selectivity to alcohol and of d.e. The introduction of pivalic acid yields a supplementary reduction of the reaction rate and, ist presence prevents the hydrogenation of the CO bond. In such conditions only the hydrogenation of the allylic bond occured.

Hydrogenation of substrate #B brings additional arguments in the favour of the reaction inside the mesoporous tubes. Ru-5 and Ru-10 were found to be almost inactive in this case but on Ru-3, even with low reaction rates the reaction occured with a high d.e. in absence of the chiral modifiers. The presence of tartaric acid prevents the catalytic reaction. It is interesting to note that the reaction rates of substrates A and B vary in opposite directions with the increase in Ru loading. This suggests that the diffusion of the bulkier substrate B is rate limiting in Ru-5 and Ru-10. With both reactions high stereoselectivities for the cis isomers are observed when internal diffsuion is not rate limiting.

In conclusion the investigation of the stereocontrolled hydrogenation of some large molecules using Ru-MCM-41 catalysts showed the concerted influence of the metal loading and of the mesoporous structure.

REFERENCES

1. H.U.Blaser, M.Garland and H.P.Jallet, J.Catal., 144 (1993) 569.
2. G.Proess and L.Hevesi, J.Mol.Catal., 80 (1993) 393.
3. T. Osawa, T. Harada, A. Tai, J. Mol. Catal., 87 (1994) 333.
4. H. U. Blaser, Tetrahedron:Asymmetry, 2 (1991) 843.
5. B. Minder, M. Schurch, T. Mallat and A. Baiker, Cat. Lett., 31 (1995) 143.
6. R. L. Augustine and S. K. Tanielyan, J. Molec. Cat. A:Chemical, 118 (1997) 79.
7. G.Wang, T.Heinz, A.Pfaltz, B.Minder, T.Mallat and A.Baiker, Chem. Commun. (1994) 2047.
8. T.Heinz, G.Wang, A.Pfaltz, B.Minder, T.Mallat and A.Baiker, Chem.Commun. (1995) 1421.
9. A.Tungler, T.Mathe, K.Fodor, R.A.Sheldon and P.Gallezot, J.Mol.Catal. A, 108 (1996) 145.
10. J.U.Köhler and J.S.Bradley, Catal.Lett., 45 (1997) 203.
11. F. Cocu, S. Coman, D. Macovei, V. I. Pârvulescu, Stud. Surf. Sci. Catal., 108 (1997) 207.
12. T.Mallat, M.Bodmer and A.Baiker, Catal.Lett., 44 (1997) 95.
13. S.Bhaduri, U.S.Darshane and D.Mukesh, Chem.Commun. (1992) 1738.
14. W.A.H. Vermeer, P.Johnston and P.B.Wells, Chem.Commun. (1993) 1053.

MESOPOROUS MOLECULAR SIEVES 1998
Studies in Surface Science and Catalysis, Vol. 117
L. Bonneviot, F. Béland, C. Danumah, S. Giasson and S. Kaliaguine (Editors)

Synthesis and transformation of thiols and organic sulfides on MCM-41 mesoporous molecular sieves

M. Ziolek, I. Nowak, P. Decyk and J. Kujawa

A. Mickiewicz University, Faculty of Chemistry, Grunwaldzka 6, 60-780 Poznań, Poland

Hydrosulfurisation of methanol and ethanol and the decomposition of ethanethiol and diethyl sulfide on MCM-41, AlMCM-41, NbMCM-41 and their hydrogen and Ni-impregnated forms were studied by means of GC analysis. The results were compared with those on the zeolites. The mesoporous materials were characterized by XRD, N_2 adsorption-desorption isotherms, and their acidity - by the isopropanol conversion. The catalytic studies of the CH_3OH + H_2S reaction have shown a higher selectivity in the synthesis of methanethiol on HNbMCM-41 than that on HAl - material which was related to various acidity of the samples. Nickel impregnated mesoporous sieves were active in the transformation of ethanethiol and diethyl sulfide and their activity and selectivity were influenced by the H_2S (the reaction product) chemisorption.

1. INTRODUCTION

Industrial synthesis of light thiols and organic sulfides is carried out in the hydrosulfurisation process, i.e., the catalytic reaction between alcohol and hydrogen sulfide. The Brønsted and Lewis acidic centres of the catalysts as well as basic sites can be employed in this reaction [1,2]. However, the same kind of the active sites can be involved in the competitive reactions, namely: i) the decomposition of alcohol to olefins and oligomers, ii) the transformation of the products, i.e., thiols and organic sulfides. Therefore, the optimisation of the catalysts needs a wide study of the synthesis and the transformation of sulfur organic compounds.

The aim of our study was to check whether mesoporous molecular sieves of MCM-41 type can be applied in the hydrosulfurisation of methanol and ethanol as well as to measure their activity in the decomposition of ethanethiol and diethyl sulfide. The latter is important also in the study of the removal of toxic sulfur compounds from the exhaust gases.

The earlier studies indicated that low or medium strength acidic centres together with basic sites are involved in the formation of CH_3SH from CH_3OH and H_2S [1,2]. Therefore, the study on the use of the mesoporous molecular sieves were concentrated on the Nb- containing MCM-41 materials, because their hydrogen forms are less acidic than Al- containing one [3].

NiY zeolite exhibited the highest activity (among others Y zeolites modified with various cations) in the ethanethiol dehydrosulfurisation [4]. Moreover, NiMCM-41 material was active in the hydrodesulfurisation of thiophene [5]. These observations provoked us to start our study from the transformation of ethanethiol and diethyl sulfide on mespoporous sieves modified with nickel.

2. EXPERIMENTAL

Silica (Ventron GMBH), NaY (Katalistiks) with Si/Al = 2.56 and mesoporous molecular sieves of MCM-41 type were used as parent materials for preparation of the modified catalysts. Mesoporous silica (MCM-41), alumino-silica (AlMCM-41 with Si/Al = 16) and niobio-silica (NbMCM-41 (16), NbMCM-41 (32) and NbMCM-41 (64) - numbers in brackets indicate the Si/Nb ratio) were prepared according to the procedure described in our earlier paper [6] and based on that presented by Beck et al. [7].

Hydrogen forms of all these materials were obtained *via* ion exchange with 0.25 M solution of NH_4Cl followed by heating *in situ* at 673 K in a flow of pure, dried helium. Nickel nitrate was used for impregnation of NaY, SiO_2, MCM-41 and AlMCM-41 samples. 5 % wt of Ni was incorporated to the samples at room temperature. The samples were dried at 393 K for 12 h and calcined at 673 K for 2 h.

The characterization of the prepared materials has been accomplished using XRD, sorption measurements and a test reaction. X-ray powder diffraction was carried out on a TUR 42 diffractometer using CuK_{α} radiation. Nitrogen adsorption-desorption isotherms were determined at 77 K with Micrometrics 2010 using a conventional volumetric technique.

The isopropanol dehydration and dehydrogenation (a test reaction) was performed using a pulse microreactor with a helium flow rate of 24 x 10^{-4} m^3 h^{-1}. The catalyst bed (0.05 g) was first activated for 2 h at 673 K in helium flow. 5 μl pulses of alcohol were introduced at 523 K. Isopropanol and the reaction products were analysed on a CHROM 5 gas chromatograph on line with the micro reactor.

A pulse technique was also applied in the study of the ethanethiol and diethyl sulfide transformation. The microreactor was filled with 0.02 g of the dehydrated form of the catalyst. Before the reaction the samples were activated for 2 h at 673 K in a helium flow. The reactions were carried out at 623 K. Pulses of 1 μl of thiol or sulfide were introduced. Products were analysed using an on line gas chromatograph equipped with a FID detector.

The hydrosulfurisation of methanol and ethanol was carried out in the flow system at 623 K. Portions of 0.1 g of dehydrated catalysts were used. For each catalytic experiment, the materials were pelleted without binder, ground, sieved to 0.5 - 1.0 mm diameter range, and activated *in situ* for 2h under a pure and dried helium flow at 673 K. Mixtures containing hydrogen sulfide (3 vol. %), alcohol (1.5 vol. %) and helium as a carrier gas were passed through the catalyst bed into a gas chromatograph. The total pressure was 1 atm and the flow rate 24 x 10^{-4} m^3 h^{-1}. The reagents and products were analysed on line using a gas chromatograph, model SRI with a FID and a sulfur FPD detectors. The catalytic activity is presented as percentage of methanol conversion.

3. RESULTS AND DISCUSSION

3.1. Characterization of materials

Powder X-ray diffraction patterns of niobium containing mesoporous materials were shown in our earlier papers [6,8]. They confirmed the presence of the hexagonal MCM-41 phase showing the decrease in the (100) peak intensity from Si/Nb = 64 to Si/Nb = 16. That feature was also registered in the nitrogen adsorption / desorption isotherms presented in Figure 1.

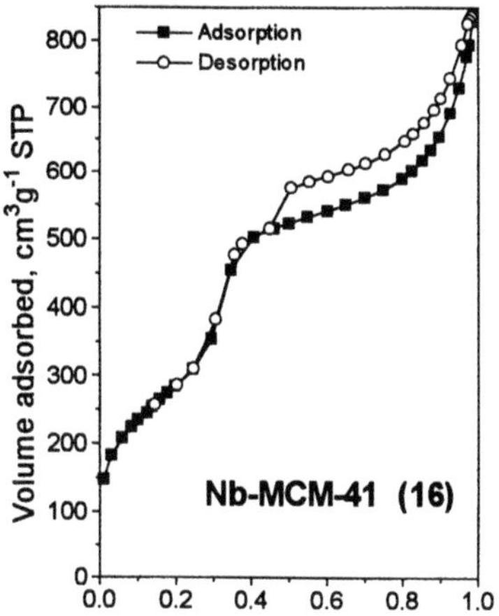

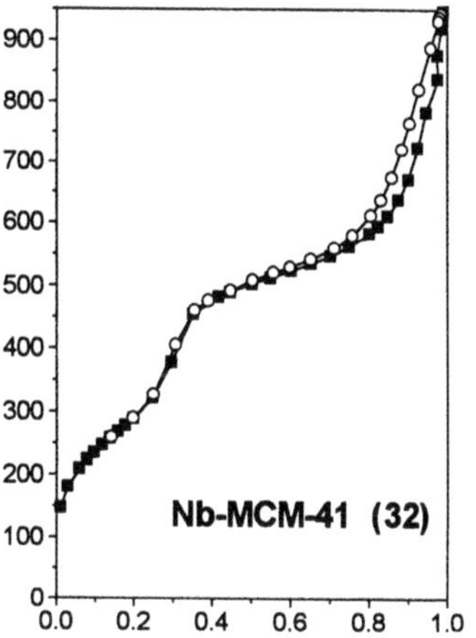

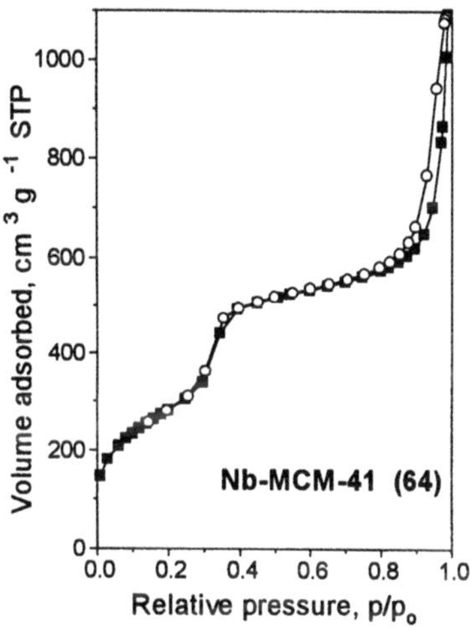

Figure 1. Nitrogen adsorption isotherms for niobium containing MCM-41 materials with various Si/Nb ratio (16, 32, 64).

The hysteresis loop caused by a capillary condensation in the secondary inter particle mesopores was higher when the concentration of Nb was higher (compare the isotherms for Si/Nb = 16 and Si/Nb = 64). That indicates the less ordering of the materials consisting the higher amount of niobium (Si/Nb = 16).

The impregnation of mesoporous sieves with $Ni(NO_3)_2x6H_2O$ did not destroyed the hexagonal phase of MCM-41 materials which was evidenced by XRD patterns.

The acidity of MCM-41, AlMCM-41, NbMCM-41 and their hydrogen forms was presented in a previous paper [3]. It was shown that the number and strength of the Brønsted acid sites were much higher on HAlMCM-41 than that on HNbMCM-41. Moreover, the latter exhibited a higher concentration of Lewis acid sites than HAl- related materials. In this paper, the isopropanol conversion and the selectivity of this reaction characterized the acidic properties of the materials. Table 1 shows the results. Some samples were also tested after sulfidation carried out in the flow of hydrogen sulfide and helium at 623 K for 1 h. Hydrogen sulfide is a reagent or product in the reactions studied and its chemisorption can change the surface properties of the catalysts as was often observed [for example: 10-12].

Table 1
Isopropanol conversion at 523 K

Catalyst	i-propanol conv., %	
	pure sample	after H_2S ads.
HNaY	100	
HAlMCM-41	88	
HNbMCM-41 (16)	37	
Ni/NaY	95	
Ni/AlMCM-41	29*	70
Ni/MCM-41	3.5	4*
Ni/SiO_2	3	16*
MCM-41	2.5	
SiO_2	0.5	

* - acetone in the reaction products

The main reaction product, on all investigated samples, was propene formed either on the Brønsted acid sites or on the pairs of Lewis acid and basic centres [9]. Moreover, diisopropyl ether, produced with the participation of medium or strong Lewis acid base pairs, was registered on HNbMCM-41, Ni/MCM-41 and Ni/SiO_2 samples. Acetone was observed only in few experiments marked by a star. The results presented in Table 1 indicate that Ni impregnated materials exhibited rather low acidity which significantly increased after the sulfidation with H_2S (especially for the Ni/AlMCM-41 material). The increase of the Brønsted acidity of NiNaY after the sulfidation with H_2S was observed by Koranyi at al. [13]. The same can be proposed for Ni supported on mesoporous materials and silica taking into account an increase of the isopropanol conversion after the H_2S pretreatment. The fact that sulfided Ni/MCM-41 and Ni/SiO_2 showed a selectivity of about 10 % towards acetone suggests that the sulfidation change also the basicity of these samples.

The acidity of the materials studied, concluded on the basis of the isopropanol conversion, was much higher for Y zeolites than MCM-41 materials. Among the mesoporous materials, HAlMCM-41 exhibited the acidity comparable with that of the HNaY and Ni/NaY zeolites.

3.2. Hydrosulfurisation of alcohols

Figure 2 exhibits the results of the reaction between methanol and hydrogen sulfide on hydrogen forms of niobium- and aluminum - containing MCM-41 materials. The highest methanol conversion on HAlMCM-41 was accompanied by the formation of various products: methanethiol, dimethyl sulfide, dimethyl ether and hydrocarbons (not plotted in the figure). The selectivity to sulfur organic products was about 50 %. That means that the reactions competitive to hydrosulfurisation (i.e. the transformation of methanol in ether and hydrocarbons) took place on this catalyst. Moreover, the yields of thiol and sulfide were not very different so, one cannot obtain a high selectivity to one of these products. The same behaviour was observed earlier on HNaY zeolites [2,14]. Contrary, HNbMCM-41 materials showed a lower activity than HAlMCM-41 sample and a very high selectivity towards methanethiol (38 % yield for ~40 % of the methanol conversion). The activity of HNbMCM-41 materials and the yield of thiol decreased with the growth of Si/Nb ratio. In the same order the number of Lewis acid sites [3] and the selectivity to thiol increased.

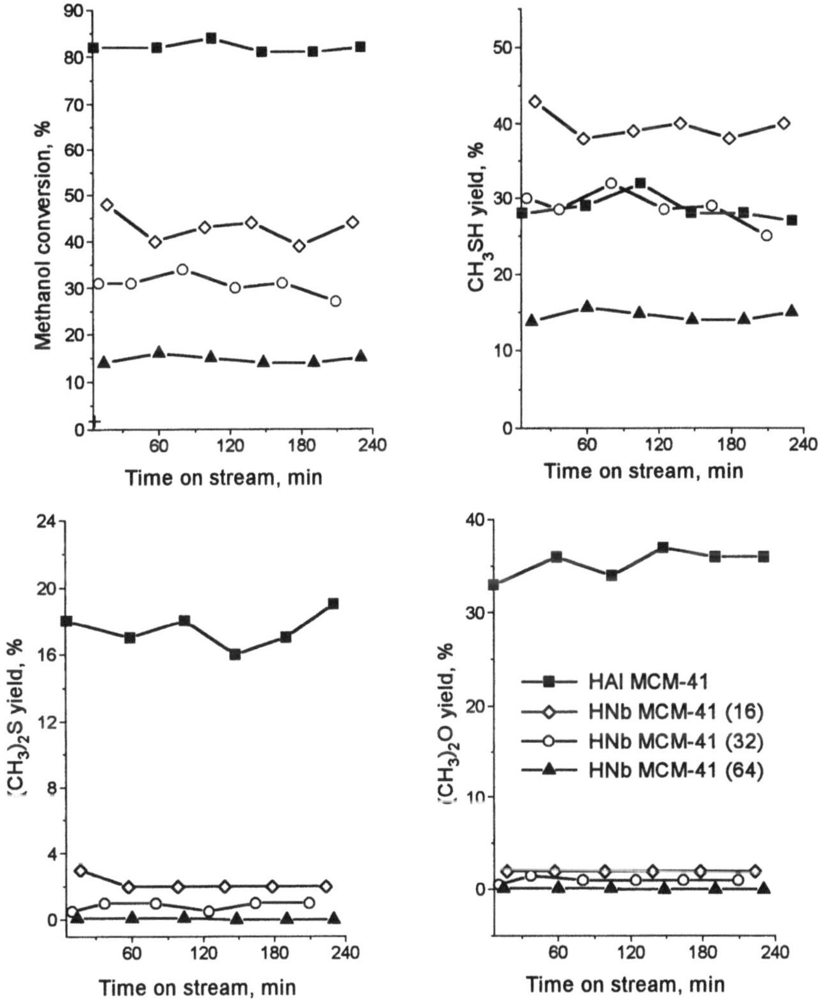

Figure 2. Activity and selectivity of hydrogen forms of mesoporous molecular sieves in the reaction between methanol and hydrogen sulfide at 623 K.

The participation of Lewis acid sites in the methanethiol formation was stressed many times in the literature [for example: 1, 2, 15-17]. It is important to note that NbMCM-41 materials showed very low activity in this reaction which can be correlated with the low number of the Lewis acid sites [3] (147 and 208 x 10^{17} per 1 g for Si/Nb = 16 and 32, respectively) to compare with HNbMCM-41 (493 and 522 x 10^{17} per 1 g for Si/Nb = 16 and 32, respectively).

The reaction between ethanol and hydrogen sulfide hardly occurred at 623 K. Table 2 shows that the conversion of ethanol on hydrogen forms of Y zeolite and niobium containing MCM-41 material was very high but the yield of sulfur organic compounds was very low. That behaviour can be due to the high dehydration activity of these samples. Nickel impregnated MCM-41 materials presented very low activity in the hydrosulfurisation of ethanol.

Table 2
The activity and the yield of products in the reaction between ethanol and hydrogen sulfide at 623 K (the results at the stationary state)

Catalyst	Ethanol	Yield, %			
	conv., %	Ethene	Etahnethiol	Diethyl sulfide	Other HC*
HNaY	100	97	0.3	0.1	2.6
HNbMCM-41(16)	97	83	8	-	6
Ni/MCM-41	5	0.3	3	-	1.7

* HC = hydrocarbons

3.3. Transformation of ethanethiol and diethyl sulfide

Hydrogen forms of mesoporous molecular sieves which were active in the dehydration of ethanol (in the $C_2H_5OH + H_2S$ reaction), did not exhibit a high activity in the transformation of ethanethiol (Table 3). It can be due to the low concentration of Brønsted acid sites in the mesoporous materials [3] (much lower than in HNaY) and the absence of extra lattice cations strong enough for coordination of sulfur organic compounds. An impregnation of these materials with Ni nitrate caused a great increase in the transformation of both ethanethiol and diethyl sulfide (Table 3). All Ni impregnated mesoporous materials and silica showed a higher activity in the transformation of ethanethiol than that for diethyl sulfide. The main reaction products were hydrocarbons and hydrogen sulfide. As the results collected in Table 3 show the formation of other sulfur products was rather casual.

Table 3
Transformation of sulfur organic compounds at 623 K - the results for the first pulse

Catalyst	C_2H_5SH	Yield, %		$(C_2H_5)_2S$	Yield, %	
	conv., %	$(C_2H_5)_2S$	C_4H_4S	conv., %	C_2H_5SH	C_4H_4S
NaY	26	3	0.4	24	5	0.3
MCM-41	5	0.2	traces	7	4	0.1
HAlMCM-41	11	2	traces	13	5	traces
HNbMCM-41(16)	8	1.5	traces	8	2	0.3
Ni/NaY	100	traces	4	94	traces	6
Ni/MCM-41	97	9	1	25	0.2	2
Ni/AlMCM-41	100	5	2.5	48	0.5	1.6
Ni/SiO_2	98	3	0.5	33	0.8	0.3

However, it is important to note that all Ni impregnated materials were not stable in the transformation of ethanethiol. Their activity decreased with the number of pulses. As nickel play the role of active sites [4] the decrease of the activity should be due to the poisoning effect of the active centres. The toxicity effect of hydrogen sulfide and thiophene was described in the literature [13] for the benzene hydrogenation on Ni/SiO_2. As the main reaction products in our studies were hydrocarbons and hydrogen sulfide, and the decrease of the activity was registered for the transformation of ethanethiol, we supposed that hydrogen sulfide could be responsible for the deactivation of Ni- containing catalysts. Therefore, the experiments with the hydrogen sulfide pretreated catalysts (in a helium flow at 623 K) were performed. The results were similar for all samples studied and are shown in Figure 3. The deactivation of Ni/NaY zeolite is not as fast as that of mesoporous materials. The chemisorption of hydrogen sulfide led to a significant decrease of the activity in spite of the increase in the catalyst acidity. This shows the dominant role of nickel in the active site. The hydrogen sulfide preadsorption causes the decrease in diethyl sulfide conversion too. In this reaction, the H_2S chemisorption influences the selectivity. Figure 4 shows the example plotted for the Ni/AlMCM-41 mesoporous sieve. First, one should stress that the activity of this material in the conversion of diethyl sulfide was stable on both pure and sulfided sample. Moreover, the transformation of diethyl sulfide into ethanethiol increased with the number of pulses on the pure material contrary to that on the sulfided catalyst on which the selectivity to ethanethiol was high in the first pulse and decreased with the number of pulses. That behaviour can be correlated with the generation of the Brønsted acid sites after chemisorption of hydrogen sulfide [13]. The Brønsted acid centres are involved in the transformation of diethyl sulfide to ethanethiol. One cannot exclude the participation of the chemisorbed hydrogen sulfide as a reagent in the reaction with diethyl sulfide:

$(CH_3)_2\ S + H_2S \rightarrow 2\ CH_3SH.$

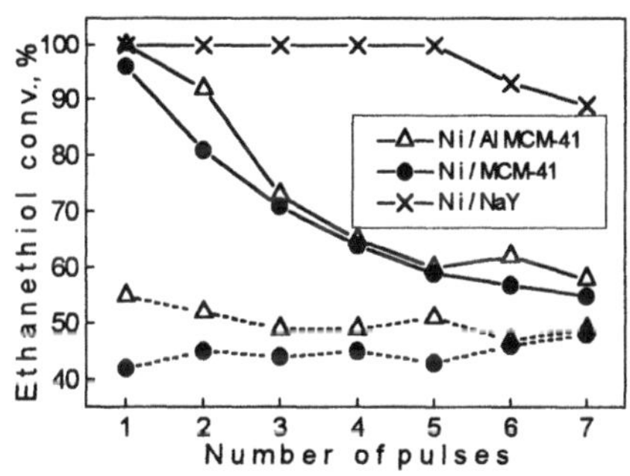

Figure 3. Ethanethiol conversion at 623 K on Ni-impregnated materials (solid line - pure sample; dashed line - after H_2S ads.)

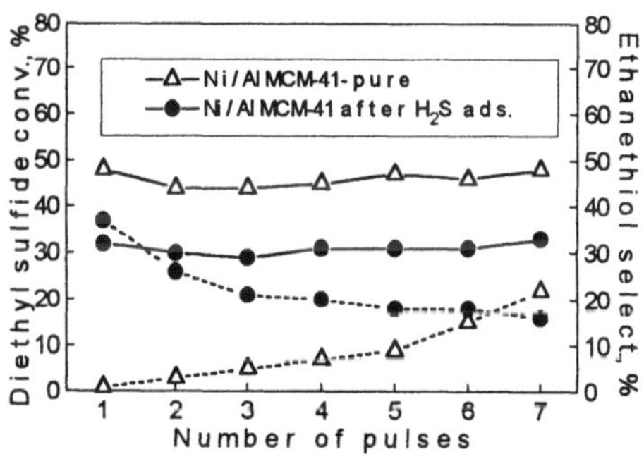

Figure 4. Activity and selectivity of Ni-impregnated AlMCM-41 in the diethyl sulfide transformation at 623 K (solid line - conversion; dashed line - selectivity).

4. CONCLUSIONS

Among the mesoporous sieves of MCM-41 type, the HAlMCM-41 sample showed the highest conversion of methanol in the reaction between CH_3OH and H_2S. However, only about half of alcohol reacted with H_2S towards both sulfur products: thiol and sulfide. The less acidic samples, i.e., HNbMCM-41 mesoporous materials exhibited a lower methanol conversion but a very high selectivity to methanethiol. The activity of HNb- containing MCM-41 materials decreased with Si/Nb ratio showing the role of Nb content in this reaction.

The hydrogen forms of the mesoporous sieves exhibited a low activity in the decomposition of ethanethiol and diethyl sulfide but an impregnation of the materials with nickel increased significantly the activity. The influence of H_2S (the reaction product) on the reaction pathway was observed. The adsorption of H_2S decreased the catalytic activity of the Ni- samples in spite of the acidity growth. The presence of hydrogen sulfide on the catalyst surface caused an increase of the ethanethiol formation in the reaction of diethyl sulfide.

The presented results did not indicate that the mesoporous molecular sieves of MCM-41 type exhibit a higher activity in the reactions studied than the traditional zeolites. However, the facts that they are active and in the case of HNbMCM-41 - very selective in the methanethiol formation, are important for the studies of the synthesis and transformation of larger sulfur organic molecules which can hardly occur in the channels of the zeolites.

ACKNOWLEDGEMENTS

This work was supported by the fund from A. Mickiewicz University, Faculty of Chemistry. The assistance of M. Kubiak, H. Nowicka and A. Rauhut in the experimental work is acknowledge.

REFERENCES

1. A.V. Mashkina, Russian Chemical Reviews, 64 (1995) 1131.
2. M. Ziolek, J. Czyzniewska, J. Kujawa, A. Travert, F. Mauge and J.C. Lavalley, Zeolites, in press.
3. M. Ziolek, I. Nowak and J.C. Lavalley, Catal. Lett., 45 (1997) 259.
4. M. Sugioka and K. Aomura, Prep. Am. Chem. Soc.,Div. Petrol. Chem., 25 (1980) 245.
5. J. Cui, Y.H. Yue, Y. Sun, W.Y. Dong and Z. Gao, Stud. Surf. Sci. Catal., 105 (1997) 687.
6. M. Ziolek and I. Nowak, Zeolites, 18 (1997) 356.
7. J.S. Beck, J.C. Vartuli, W.J. Roth, M.E. Leonowicz, C.T. Kresge, K.D. Schmitt, C.T.W. Olson, E.W. Sheppard, S.B. McCullen, J.B. Higgins and J.L. Schlenker, J. Am. Chem. Soc., 114 (1992) 10834.
8. I. Nowak and M. Ziolek, Proc. 3rd Polish German Zeolite Colloqium, (M. Rozwadowski, Ed.) NCU Press, Torun 1998, p.161.
9. A. Gervasini, J. Fenyvesi and A. Auroux, Catal. Lett., 43 (1997) 219.
10. M. Ziolek, J. Kujawa, O. Saur and J.C. Lavalley, J. Molec. Catal. A: Chemical, 97 (1995) 49.
11. M. Sugioka, React. Kinet. Catal. Lett., 41 (1990) 345.
12. A. Diaz, L.M. Gandia, J.A. Odriozola and M. Montes, J. Catal., 162 (1996) 349.
13. T.I. Koranyi, F. Moreau, V.V. Rozanov and E.A. Rozanova, J. Mol. Struc., 410-411 (1997) 103.
14. M. Ziolek and I. Bresinska, Zeolites, 5 (1985) 245.
15. A.V. Mashkina and V.N. Yakovleva, Kinetika i Kataliz, 32 (1991) 636.
16. A.V. Mashkina, V.M. Mastikhin, V.Yu. Mashkin, A.V. Nosov and V.M. Kudenkov, Kinet. Katal., 34 (1993) 880.
17. M. Ziolek, J. Kujawa, O. Saur and J.C. Lavalley, J. Phys. Chem., 97 (1993) 9761.

MESOPOROUS MOLECULAR SIEVES 1998
Studies in Surface Science and Catalysis, Vol. 117
L. Bonneviot, F. Béland, C. Danumah, S. Giasson and S. Kaliaguine (Editors)

Synthesis strategies leading to pure alumina mesophases in aqueous solution

S. Valange, J.-L. Guth and Z. Gabelica*

ENSCMulhouse, Université de Haute Alsace, Laboratoire de Matériaux Minéraux, UPESA 7016, 3 rue Alfred Werner, F-68093 Mulhouse Cedex, France

While the synthesis and characterization of the now well known mesoporous silicas or aluminosilicates of the MCM-41, MCM-48 and MCM-50 families is now being widely documented, very few studies on the existence and stabilization of the all-alumina analogs have been reported so far. In aqueous solution, only lamellar, thermally unstable mesophases could be obtained from Al^{3+} and anionic surfactants like alkylsulphonates or phosphates [1] or sodium dodecyl sulphate (SDS) [2, 3]. In non aqueous media, either lamellar [4] or pseudo hexagonal [5] alumina mesophases were prepared by reacting Al alkoxides with non ionic polyglycols and various carboxylic acids respectively. In the latter case, the calcined solids involve randomly ordered pores and a specific surface area between 400 and 700 m^2/g. Stein and Holland [6] reported the aqueous synthesis of mesoporous aluminas in the presence of Al polycations and SDS but the surface area of the calcined materials was only 18 m^2/g.

We have explored a series of novel synthesis pathways leading to (pseudo) hexagonal and thermally stable mesoporous alumina phases. The originality of our approach consisted in achieving syntheses in aqueous solutions at ambient temperature through a careful control of pH that strictly governs the final charge of the Al species for their further efficient interaction with anionic, neutral or cationic surfactant molecules.

Using Al nitrate in the pH range 1.5--6.5 with, as surfactants, SDS or long chain carboxylic acid salts (C_x-COONa) possibly admixed with C_{16}-TMABr, yielded many different, not absolutely readily reproducible mesophases. In contrast, pure and quasi hexagonal mesostructured solids involving thick alumina walls were obtained in the 7-8 pH range.

The rather versatile thermal stability of some of these solids oriented our research towards the use of prestructured Al cationic species such as the Keggin $[Al_{13}O_4(OH)_{24}(H_2O)_{12}]^{7+}$ polycations that were "neutralized" by various anionic surfactants. With SDS, thermally stable mesoporous phases were obtained within the critical pH range of 5.5-6. Similar results were obtained with C_x-COONa, the as synthesized phases showing the XRD d spacing regular increase from 26 Å (C_8-COONa) to 48 Å (C_{16}-COONa). The use of non ionic surfactants (e.g. Triton X-45) did not result in a structuration of alumina because of too weak Van der Waals type interactions. In contrast, admixing appropriate ratios of 2 surfactant molecules involving different charges, better fits the charge density and hence leads to a better stabilization of the micellar system. In such cases, highly crystalline mesoporous aluminas could be prepared. Their thermal stability was excellent, yielding, both after a slow heating under dry nitrogen flow or after a "flash" heating in dry air at 450°C, mesostructured aluminas that involved specific surface areas between 500 and 820 m^2/g.

The different synthesis routes finally resulted in the stabilization of a wide series of Al mesophases involving pores from 8 to more than 60 Å in diameter. This diameter strongly depends on the pH and the Al source and concentration. For a given surfactant, the pore diameter increases with increasing pH, with dilution and when monomeric $[Al(H_2O)_6]^{3+}$ cations are used instead of the Keggin polycations. The smaller the pore diameter, the narrower the pore size distribution.

^{27}Al-MAS NMR showed the presence of both tetrahedral and octahedral Al species in the phases synthesized in the presence of Keggin polycations, suggesting that this latter retains its structure after synthesis. In contrast, only octahedrally coordinated Al species were detected when Al monomeric cations were used as precursors. Upon calcination, in both cases, severe Al coordination changes were observed, confirming that the calcination conditions must be considered as a critical variable to control for the eventual stabilization of the final mesoporous alumina phase.

REFERENCES

1. Q. Huo, D.I. Margolese, U. Ciesla, D.G. Demuth, P. Feng, T.E. Gier, P. Sieger, A. Firouzi, B.F. Chmelka, F. Schüth and G.D. Stucky, Chem. Mater., 6 (1994) 1176.
2. M. Yada, M. Machida and T. Kijima, Chem. Commun., (1996) 769
3. M. Yada, H. Hiyoshi, K. Ohe, M. Machida and T. Kijima, Inorg. Chem., 36 (1997) 5565.
4. S.A. Bagshaw and T.J. Pinnavaia, Angew. Chem. Int. Ed. Engl., 35 (1996) 1102.
5. F. Vaudry, S. Khodabandeh and M.E. Davis, Chem. Mater., 8 (1996) 1451.
6. A. Stein and B. Holland, J. Porous Mater., 3 (1996) 83.

MESOPOROUS MOLECULAR SIEVES 1998
Studies in Surface Science and Catalysis, Vol. 117
L. Bonneviot, F. Béland, C. Danumah, S. Giasson and S. Kaliaguine (Editors)

Influence of pH adjustment on structure, stability, and catalytic activity of Al-MCM-41 mesoporous molecular sieves

S. Hitz, A. Kogelbauer, B. Lindlar and R. Prins

Laboratorium für Technische Chemie, ETH-Zürich, CH-8092 Zürich, Switzerland

1. INTRODUCTION

In view of their applicability as catalysts, the improvement of the quality of Al-MCM-41, i.e. structure, stability, and catalytic activity, is of great interest. One way of approaching this aim is the extraction of the structure directing template using acidic conditions prior to calcination [1]. The variation of the synthesis parameters represents another possibility since the stability and characteristics of the mesoporous molecular sieve Al-MCM-41 are known to be strongly affected by the synthesis conditions [2]. Along the same line, Ryoo and Kim have recently described a method to obtain pure silica MCM-41 with enhanced long-range order and greater stability by the intermediate adjustment of the pH of the synthesis gel [3]. They also demonstrated that the synthesis of Al-MCM-41 was possible using this method giving MCM of higher thermal and hydrothermal stability. In order to assess the suitability of this method for preparing better Al-MCM-41, we undertook a detailed investigation of the effects of the pH adjustment during the synthesis of Al-MCM-41 with respect to changes in structural features but also regarding catalytic activity. The results of the initial study are reported here.

2. EXPERIMENTAL

Al-MCM-41 was synthesised from a reaction mixture of sodium silicate solution, Cab-osil M-5, sodium aluminate, hexadecyltrimethylammonium chloride solution, ammonium hydroxide solution and demineralised water of the following molar composition: 1 SiO_2 / 0.013 Al_2O_3 / 0.287 Na_2O / 0.130 CTMA / 0.019 NH_3 / 34.8 H_2O. The synthesis gel was prepared according to the report by Ryoo and Kim [3], homogenised by stirring for one hour at ambient temperature and subsequently heated at 100°C for the synthesis for a total of 4 days under static conditions in polypropylene bottles. For the investigation of the pH adjustment a set of parallel syntheses was carried out which is schematically depicted in Figure 1. The separate batches were intermediately cooled to ambient temperature after 24 h of heating, stirred vigorously while the pH was adjusted in some batches to 11 ± 0.05 using acetic acid (100%) and subsequently heated again. This pH adjustment procedure was carried out up to three times for the different batches yielding Al-MCM-41 products that had undergone between 0 to 3 adjustments of their pH in 24 h intervals. It was confirmed in an independent synthesis without intermediate cooling steps that the cooling did not influence product quality. The obtained products were washed with demineralised water and ethanol.

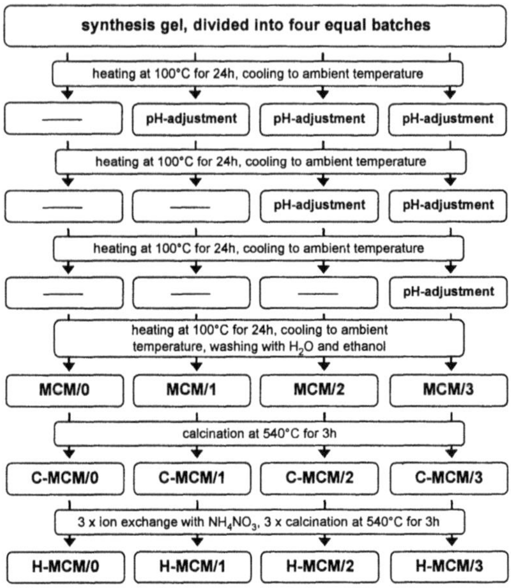

Figure 1. Schematics of the synthesis of Al-MCM-41 using various pH adjustment steps

The resulting Al-MCM-41 was dried overnight at 50°C under reduced pressure and calcined in air at 540°C for three hours. The calcined Al-MCM-41 samples were three times ion-exchanged with an aqueous ammonium nitrate solution at 60°C and finally calcined at 540°C for two hours. The resulting Al-MCM-41 materials were characterised using powder XRD, nitrogen adsorption, thermogravimetric analysis, ^{1}H and ^{27}Al MAS-NMR, and elemental analysis by AAS. The catalytic activity of the different Al-MCM-41 samples for the Friedel-Crafts acylation of 2-methoxy-naphthalene with acetic anhydride was determined following a procedure reported earlier [4].

3. RESULTS

Powder XRD patterns of the Al-MCM-41 samples obtained with increasing number of pH adjustments are depicted in Figure 2. The samples that underwent pH adjustment exhibit the typical 4 diffraction peaks characteristic for the well-ordered hexagonal MCM-41. The high-order diffraction peaks are clearly missing in MCM/0 indicating a less ordered structure. Also, the main diffraction peak of MCM/0 was found at substantially higher 2Θ values indicating a smaller d-spacing in the sample without pH adjustment. Figure 3 shows the N_2 adsorption isotherms and pore size distribution for the calcined Al-MCM-41 samples. Again, a distinct difference was observed for those samples that had undergone pH adjustment. The isotherms did not exhibit a hysteresis and showed a sharp increase in the N_2 uptake. The pore size distribution of the pH-adjusted samples is much narrower and the maximum in the average pore size was found at higher values in accordance with XRD results.

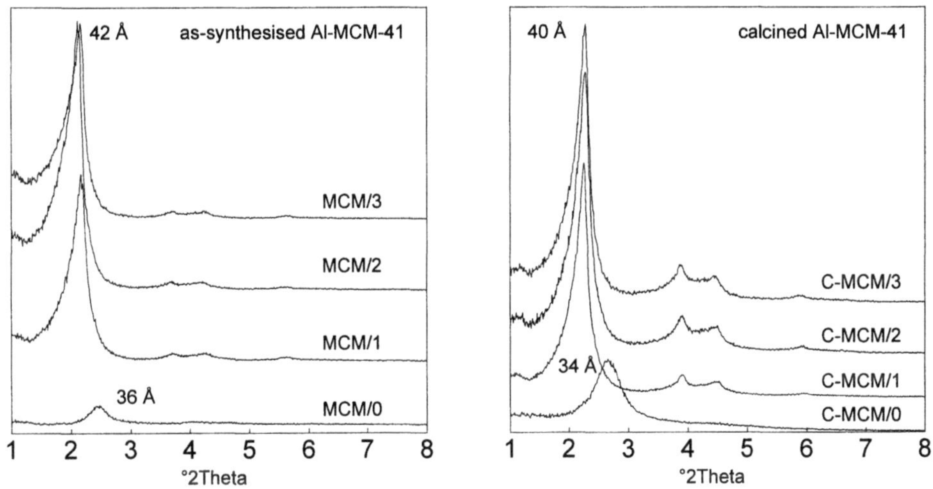

Figure 2. XRD patterns of as-synthesised and calcined Al-MCM-41

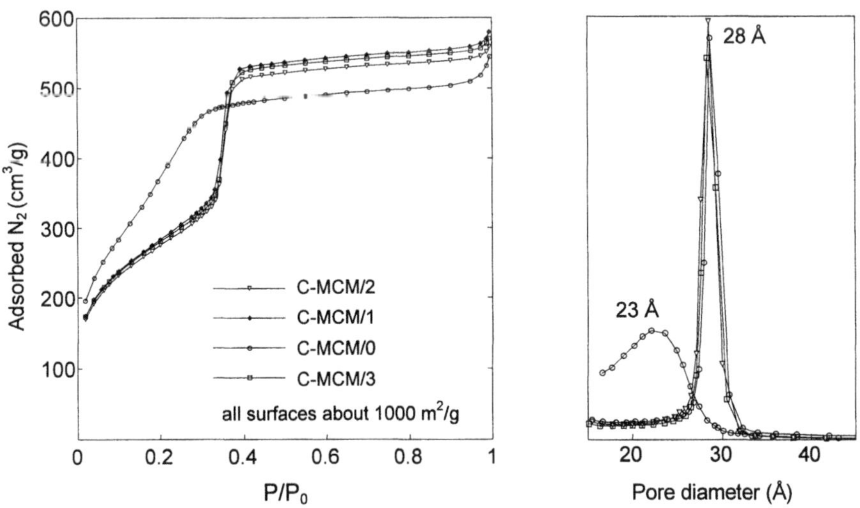

Figure 3. N_2 adsorption isotherms and pore size distribution of calcined Al-MCM-41

The impact of the pH adjustment on the chemical composition of Al-MCM-41 is compiled in Table 1. With increasing number of pH adjustments the Si/Al ratio of the samples was doubled and the Na/Al ratio increased by about an order of magnitude. In parallel, the amount of template as well as the ratio of all cationic species per silicon atom ((CTMA+Na)/Si) decreased. The yield of recovered Al-MCM-41 material based on the total number of T atoms in the synthesis gel increased markedly after the first pH adjustment reaching a final value of 82%.

Table 1
Chemical composition and yields of as-synthesised Al-MCM-41

	Si/Al	Na/Al	Template(%wt)	(CTMA+Na)/Si	Yield (%)
MCM/0	20.1	0.027	47	0.19	49
MCM/1	26.7	0.133	42	0.16	71
MCM/2	35.4	0.207	38	0.13	76
MCM/3	41.6	0.229	35	0.12	82

Despite their lower aluminium concentrations, the samples synthesised with pH adjustment exhibited higher activity for the acylation reaction than the samples without (compare Figure 4). A further increase in activity was observed for ammonium-exchanged

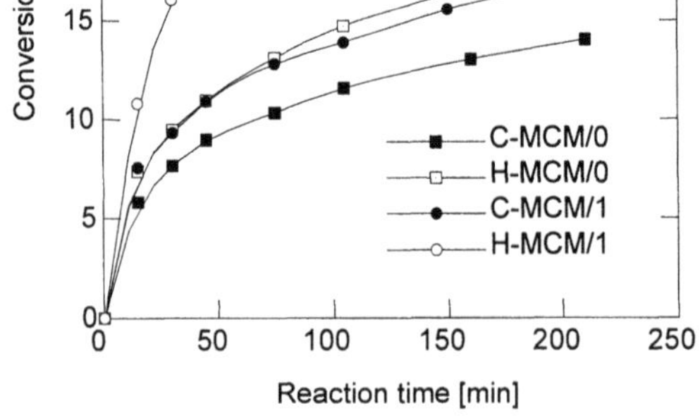

Figure 4. Activity of Al-MCM-41 (0.5g) for the acylation of 2-methoxy naphthalene (50 mmol) with acetic anhydride (25 mmol) in sulfolane (25 cm^3) at 100°C

Al-MCM-41 samples compared to the samples that had been only calcined after synthesis. This was related to the removal of residual Na ions because it was demonstrated in separate experiments that only about 3 percent conversion were obtained in the time range of the experiments without catalyst or with Al-MCM-41 that had been completely Na-exchanged. As expected for mesoporous materials, enhanced selectivity for 6-methoxy acetophenone due to shape-selectivity as was the case with zeolite beta [4] was not observed.

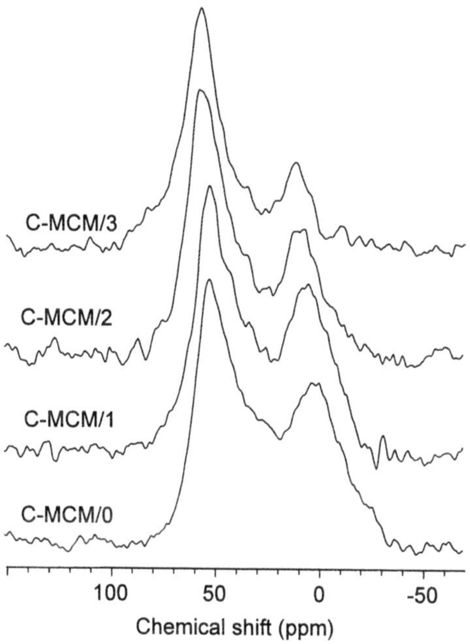

Figure 5. ^{27}Al MAS-NMR spectra of Al-MCM-41 after calcination

The ^{27}Al MAS-NMR spectra of the calcined samples are depicted in Figure 5. With increasing number of pH adjustments the relative concentration of octahedrally coordinated Al (peak at 0 ppm) decreased. A similar decrease may be inferred for penta-coordinated Al species from the disappearance of the shoulder at around 30 ppm.

4. DISCUSSION

The adjustment of the pH during the synthesis of Al-MCM-41 had a clearly observable influence on the quality of the obtained material. The higher order reflections observed in the XRD patterns indicate an increased long-range order of the MCM structure. Similar effects were reported for pure silica MCM-41 [3,5] and for an Al-MCM-41 that had undergone 2 pH

adjustments [3]. The increased ordering of the samples is also evident from the much narrower pore-size distribution determined from N_2 adsorption. All these structure improving effects were already apparent after the first adjustment and did not vary substantially with additional pH adjustments, which is in contrast to the results of Ryoo and Kim [3] for silica MCM but concurs with those of Edler and White [5].
A significant change in the chemical composition, however, was observed with increasing number of pH adjustments. Only after the third adjustment of the pH, the Si/Al ratio of MCM reached the value of the synthesis gel (Si/Al = 39) and was always lower before. This indicates preferential incorporation of aluminium during the early stages of the synthesis. Judging from the yield of MCM based on the total number of T-atoms in the synthesis mixture and the Si/Al ratio of the MCM sample without pH adjustment one may deduce that practically all the aluminium present in the synthesis mixture was incorporated prior to the first pH adjustment. The observed increase in MCM yield when the synthesis was continued with pH adjustment, which was also reported by Ryoo and Kim [3], is then related to the further incorporation of silicon. The concomitant changes in the Al environment, i.e., the increased concentration of tetrahedral Al and the overall changes in the MCM structure suggest that redissolution and recrystallisation occur leading to a more homogeneous distribution of Al in the MCM matrix. In this context, Kloetstra et al. [6] reported the formation of a dense aluminium-rich phase besides the MCM phase during the synthesis of Al-MCM-41 using high concentrations of aluminium.

These interpretations are in line with the enhanced activity observed for the samples synthesised with pH adjustment despite their lower Al concentration per unit weight. The higher activity could be related to a greater number of active sites, to a higher intrinsic activity of the sites, or to a combination of both effects. The distinct increase in the concentration of tetrahedral Al suggests that the first effect is dominating. It is therefore concluded that aluminium, probably in a more isolated environment, represents a centre that is catalytically active for the acylation reaction and that the concentration of such aluminium species is higher in Al-MCM-41 samples synthesised with pH adjustment. Ryoo and Kim [3] also proposed a more homogeneous distribution of aluminium in their sample based on ^{129}Xe NMR measurements.

It was concluded based on ^{29}Si NMR results that the stabilisation of the MCM-41 structure and the increased long-range order are due to an increased degree of condensation of silanol groups [3]. Although the presence of aluminium in our samples prevented a straightforward analysis of the ^{29}Si NMR spectra in this respect we still found evidence for an increased degree of condensation of the silanol groups. The integrated intensity of the silanol peak in ^{1}H NMR spectra of the calcined samples decreased by 15% between C-MCM/0 and C-MCM/3. Furthermore, the decrease of the total number of cationic species normalised to the number of Si atoms in the as-synthesised samples (compare Table 1) also points towards fewer silanolate groups and a more condensed framework after synthesis.

5. CONCLUSIONS

The adjustment of the pH is a simple and effective method to stabilise aluminium in the MCM-41 matrix thereby leading to catalysts with improved catalytic performance. Repetition of the pH adjustment influenced the composition and the yield of recovered Al-MCM-41 but did not lead to any further stabilisation of the matrix. The increased catalytic performance was concluded to be due to a more homogeneous distribution of aluminium in the MCM matrix brought about by recrystallisation as a consequence of the pH adjustment.

REFERENCES

1. S. Hitz and R. Prins, J. Catal., 168 (1997) 194.
2. C.-Y. Chen, H.-X. Li and M.E. Davis, Microporous Mater., 2 (1993) 17.
3. R. Ryoo and J.M. Kim, J. Chem. Soc., Chem. Commun., (1995) 711.
4. G. Harvey, G. Binder and R. Prins, Stud. Surf. Sci. Catal., 94 (1995) 397.
5. K.J. Edler and J.W. White, Chem. Mater., 9 (1997) 1226.
6. K.R. Kloetstra, H.W. Zandbergen and H. van Bekkum, Catal. Lett., 33 (1995) 157.

Improvement of the Thermal Stability of Mesostructured Metal Oxides with Zirconia as the Example

Ulrike Ciesla[a,b], Galen Stucky[b] and Ferdi Schüth[a]

[a]Institut für Anorganische Chemie, Johann Wolfgang Goethe-Universität Frankfurt, Germany
[b]Chemistry Department, University of California, Santa Barbara, USA

The preparation of high quality mesostructures made from zirconium sulfate and -phosphate and the resulting new porous structures "zirconium oxide-sulfate" and "zirconium oxo phosphate" is presented. The presence of phosphate and sulfate ions causes a crystallization delay of the amorphous zirconia phase. Therefore, by stabilizing the amorphous phase, the mesostructures are thermally stable up to 500 °C without loosing their ordered pore structure and their high surface area.

1. INTRODUCTION

The surfactant controlled synthesis, which was originally developed for the preparation of mesoporous silica and alumosilicates [1], can be also used for transition metal oxides. Mesoporous high surface area transition metal oxides are especially interesting for the use in catalysis. Subsequently, several mesostructured transition metal oxides have been successfully prepared showing hexagonal, cubic, and also lamellar mesostructures, which are also known for the silica system [2]. However, a major problem is the thermal stability of the mesostructured transition metal oxides and the surfactant removal by calcination often leads to complete structure collapse. Two reasons were proposed to be responsible for the thermal instability of mesostructured transition metal oxides: the existence of several relatively stable oxidation states resulting in oxidation and/or reduction during the calcination [2b], as well as the incomplete condensation of the species inside the pore walls [3]. The first mesoporous transition metal oxide, which could successfully be prepared, is based on TiO_2 and is thermally stable upto 350 °C [4]. Moreover, in the meantime mesoporous Nb- and Ta- oxides showing BET surface areas upto 500 m^2/g have been synthesized, however, the template removal was carried out by extraction [5].

Here, we present two porous structures based on zirconia "zirconium oxide-sulfate" and "zirconium oxo phosphate" [6] and the new mesostructure of zirconium phosphate. Zirconia is a very important catalyst with unique properties possessing basic as well as acidic sites in combination with high thermal and mechanical stability [7]. Moreover, sulfated zirconia is highly acidic and, therefore, industrially important as a cracking catalyst [8].

2. EXPERIMENTAL

2.1. Synthesis

2.1.1. Zirconium oxide-sulfate

Surfactant (alkyltrimethylammonium bromide with alkyl chain length between C_{16} and C_{22}; 6.87 mmol) and zirconium propoxide (70 wt% solution in propanol, 0.0128 mol) were dissolved in water and HCl. After the hydrolysis products had dissolved, $(NH_4)_2SO_4$ was added to the mixture. The amount of $(NH_4)_2SO_4$ was dependent on the molar ratio of SO_4 to Zr: Calcinable samples were obtained by using molar SO_4/Zr ratios of 1.2 to 0.65. The initially clear solution was then heated at 100 °C for 2 days. The colorless precipitate was filtered, dried, and then calcined at 500 °C for 5h.

2.1.2. Zirconium oxo phosphate

Surfactant (alkytrimethlyammonium bromide with alkyl chain length between C_{16} and C_{22}; 6.87 mmol) was dissolved in water, and $Zr(SO_4)_2$*$4H_2O$ (0.0128 mol), dissolved in water, was added. The colorless precipitate was stirred at room temperature for 2 h and then heated at 100 °C overnight. The precipitate was filtered and without drying added to the 0.5 M phosphoric acid solution. After 5 h the colorless product was filtered and calcined at 500 °C for 5 h.

2.1.3. Zirconium phosphate

CTAB (6.87 mmol) is dissolved in water (80 ml) and zirconium oxychloride ($ZrOCl_2$*$8H_2O$) (10 mmol), HF (48% solution, 4 ml), and phosphoric acid (85% solution, 0.672 mol) are added. The clear solution is stirred in the fume hood at room temperature. After 2 days the colorless product is formed. The solid is filtered and dried at room temperature.

2.2. Characterization techniques used

The powder X-ray diffraction (XRD) patterns were obtained on a STOE Stadi P X-ray diffractometer in Debye Scherrer geometry with a position sensitive detector covering an angle range of 7.5° (2θ) using copper Kα radiation (0.154 nm). TEM analysis of the samples was done using a Philips EM 400 T transmission electron microscope operating at 100 kV. Samples were prepared by sonication in ethanol and suspension on carbon-coated grids. The TEM measurements were carried out at the Fritz Haber-Institut, Berlin by Dr. Y. Ushida. Elemental composition of the samples was determined by X-ray fluorescence on pressed PVA-pellets using a Philips PW 1400. Sorptive analysis was carried out using a Micromeritics ASAP 2010 unit with nitrogen at 77.4 K.

3. RESULTS AND DISCUSSION

3.1. Mesostructured zirconia based composites

3.1.1. Mesostructured zirconium sulfate surfactant composites

Zirconium salts show a pronounced polyoxo ion chemistry. Livage and coworkers [9] showed that the sol-gel chemistry depends strongly on the counter ions and their complexing ability. By dissolution of zirconium sulfate in water, precursors of $[Zr(OH)_2(SO_4)_x(OH_2)_y]^{-n(2x-2)}$ are formed since sulfate ions possess a quite high complexing ability (α = 32%). Moreover, SO_4^{2-} anions behave as a network former, bridging different chains together. Therefore, charge interactions to positively charged surfactants are possible and mesostructured zirconium sulfate surfactant composites are formed. Using alkytrimethylammonium bromide with alkyl chain lengths between C_{12} and C_{22} as the surfactant, the obtained composites possess the hexagonal phase analogous to MCM-41. The X-ray diffraction pattern shows three reflections at low angles which can be indexed as (10), (11), and (20) assuming a hexagonal unit cell. Increasing the surfactant chain length leads to an increase of the lattice parameter (a=b), which can be calculated as $a=2d_{100}/\sqrt{3}$, of 0.25 to 0.4 nm with each additional C_2H_5-unit. Similar results were obtained by Beck et al. [1] in the case of MCM-41 by varying the surfactant chain length. By changing the surfactant and using benzyl-trimethylstearyl-ammonium bromide even the cubic phase, Ia3d, which is analogous to MCM-48, has been prepared.

3.1.2. Mesostructured zirconium phosphate surfactant composites

Phosphate ions show a higher complexing ability (α = 50%) than sulfate ions and zirconium phosphate precipitates even from very strong acidic solutions [9]. In the zirconium phosphate precursors, which are formed in aqueous solution, the phosphate groups possess a negative partial charge of δ = -0.5. Therefore, interactions of the zirconium phosphate precursors with positively charged surfactants should be possible. However, combining a zirconium ions containing solution with a solution of surfactant (CTAB) and phosphate ions results in a very fast precipitation of dense zirconium phosphate. Therefore, we tried to slow down the kinetic of the zirconium phosphate precipitation. This could be done by adding hydrofluoric acid to the reaction mixture of aqueous CTAB solution containing zirconium ions and phosphate ions. Fluoride ions are able to complex the zirconium ions in solution by forming the following complexes: $[ZrOF_4]^{2-}$ and/or $[ZrF_6]^{2-}$. By controlled evaporation of the HF the concentration of zirconium phosphate is always very low and before precipitation occurs, the precursor molecules are able to interact with the surfactant.

Figure 1 shows the XRD pattern of the obtained product: reflections at angles higher than 10° (2θ) could be indexed as α-$ZrH(PO_4)_2$, which is the phase normally obtained in a fluoride containing zirconium phosphate solution [10].

Additionally, a large reflection appears at very low angles showing a d-spacing of 4.15 nm which is a typical value for the (10) reflection of a mesostructured hexagonal phase using CTAB as the surfactant.

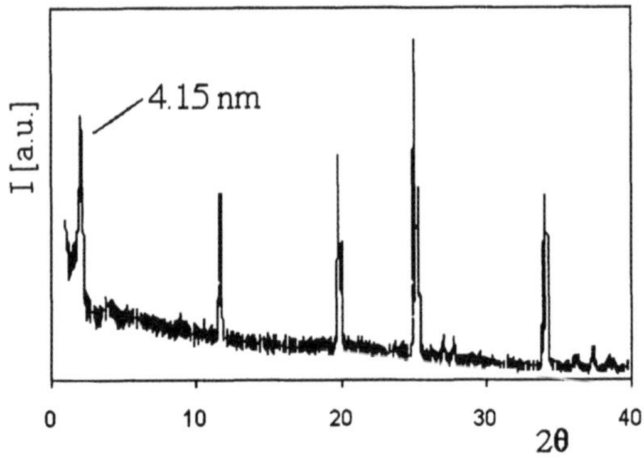

Figure 1. X-ray diffraction pattern of as-made mesostructured zirconium phosphate synthesized using $C_{16}TAB$ as surfactant.

However, from the X-ray diffraction pattern it is not clear if two coexisting phases, the mesostructure and the crystalline α-$ZrH(PO_4)_2$, are present or if the mesostructure contains crystalline walls. By using TEM analysis we were able to clarify that the product consist of two coexisting phases. Large area of mesostructured material was found, which does not show any reflections in electron diffraction except for the reflections at low angles and consists, therefore, of amorphous walls as it is also known for all other MCM-41 analogous materials (fig. 2). The dense crystalline phase, α-$ZrH(PO_4)_2$, is not stable in the electron beam and could be easily distinguished from the mesostructure. From TG/DTA analysis 17 wt% surfactant can be detected. In consideration of the results from TEM and TG/DTA (MCM-41 contains around 40 wt% surfactant) it can be estimated that the shown sample contains between 33 and 43 % mesostructured zirconium phosphate. So far, it was not possible to obtain the mesostructured phase as a pure phase, but by optimizing the synthesis conditions it should be possible to prepare the pure mesostructured zirconium phosphate.

The surfactant removal by calcination in the case of the zirconium phosphate surfactant composites was not successful and the collapse of the porous structure was observed.

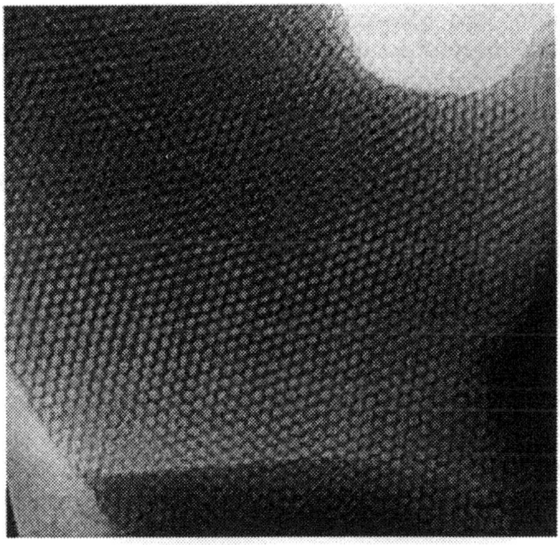

Figure 2. TEM image of as-made mesostructured zirconium phosphate synthesized using $C_{16}TAB$ as surfactant.

3.2. Porous zirconia based structures

The calcination of the zirconium sulfate surfactant composites leads to (partial) structure collapse. It was assumed that the high amount of sulfate groups was responsible for the thermal instability. Therefore, we developed two methods to decrease the sulfate amount: 1) reduction of the sulfate groups in the synthesis mixture and 2) a post-synthesis treatment of the zirconium sulfate surfactant composites with phosphoric acid solution. The reduction of the sulfate groups was carried out by decreasing the SO_4/Zr ratio from 2 to 0.5 in the synthesis mixture. It was found that with decreasing the amount of sulfate groups, the quality of the material was reduced due to less charge interactions with the surfactant. The best result was obtained by using SO_4/Zr ratios of 1.2: the as-made surfactant composites still show three reflections in the XRD pattern, but are completely calcinable without loosing structural features. The surfactant-free calcined materials are so-called zirconium oxide-sulfates due to the small amount of around 2 wt% sulfur after calcination. Due to strong shrinkage of the pores during the calcination the zirconium oxide-sulfates only show pores in the micropore region analyzed by nitrogen adsorption measurements. Additionally, TEM measurements show clearly the regular hexagonal pore arrangement, but also the less ordered regions and finger print-like regions were observed. Similar structural features could also be found in MCM-41 samples [11].

Porous zirconium oxo phosphates were prepared by developing a post synthesis procedure: mesostructured zirconium sulfate surfactant composites, made from zirconium sulfate precursors, were treated with 0.5 M phosphoric acid solution for several hours. The phosphate groups are able to replace water molecules and sulfate groups in the structure because of their high complexing ability. After calcination at 500 °C the X-ray diffraction pattern shows one distinct reflection at very low angles with a d-spacing depending on the surfactant chain length. Moreover, the (11) and (20) reflections are also visible. Nitrogen adsorption analysis shows in the case of zirconium oxo phosphate synthesized with C_{16}-surfactant only the existence of micropores. However, depending on the surfactant chain length also much larger pores can be achieved: C_{18}: 2.2 nm; C_{20}: 2.5 nm; C_{22}: 2.9 nm*. Moreover, a small pore size distribution is found which is typical for porous materials made from surfactant controlled syntheses. Taking the data of the X-ray diffraction into consideration it can be assumed that the pore walls of zirconium oxo phosphate of around 1.8 to 2.0 nm are much thicker than in the case of MCM-41. This can be confirmed by TEM analysis (fig 3).

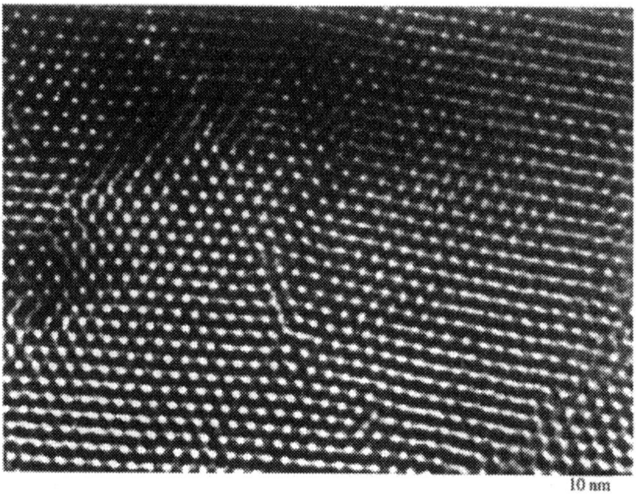

Figure 3. TEM image of calcined porous zirconium oxo phosphate synthesized using $C_{20}TAB$ as surfactant.

However, due to the focus problem it is not possible to determine exact values for the pores and wall thickness from TEM images without additional simulations. Moreover, large areas of hexagonal pore arrangement of uniformly sized pores can be seen. From X-ray analysis a composition of 4 $ZrO_2*P_2O_5$ (+SO_4)

* The pore size calculations were carried out by using the DFT method.

can be calculated. Similar results are found in the case of phosphated zirconias made from amorphous $Zr(OH)_4$ and following phosphoric acid treatment [12].

3.3. Interpretation of the results with respect to thermal stability

Zirconium sulfate and zirconium phosphate mesostructures are thermally unstable and therefore the surfactant removal by calcination is not possible. However, materials which are, in their chemical composition, similar to sulfated zirconia and phosphated zirconia, are calcinable up to temperatures of 550 °C. Also after the thermal treatment these porous structures consist of amorphous pore walls. It is known that PO_4^{3-} and SO_4^{2-} ions are responsible for a crystallization delay of amorphous zirconias [12,13]. Amorphous pore walls seem to be required for mesostructures. By heating the zirconium oxide-sulfate and the zirconium oxo phosphate to temperatures higher than 600 °C it can be observed, that the materials crystallize and, simultaneously, the pore structures collapse. We assume that the necessity of amorphous pore walls is also the reason for the structure collapse in the case of zirconium sulfate and zirconium phosphate mesostructures. The crystallization occurs below temperatures of 400 °C and there are no examples of amorphous zirconium sulfate or zirconium phosphate phases or any kind of stabilization of the amorphous phases at high temperatures. Taking these results into consideration, it becomes clear why all, so far, known mesoporous materials possess amorphous pore walls and why the preparation of mesoporous transition metal oxides is so difficult. In the case of TiO_2, for example, Raggai et al. [14] studied the crystallization of amorphous Ti-hydroxide gels. Depending on the preparation conditions the crystallization occurs between 300 and 400 °C. Therefore, TiO_2-mesostructures are only stable up to these temperatures.

4. CONCLUSION

Here, we reported the preparation of hexagonal mesostructured phases of zirconium based surfactant composites. This is highlighted by the formation of the first mesostructured zirconium phosphate surfactant composite as determined by XRD and TEM. Moreover, we succeeded in the preparation of porous zirconium oxide-sulfate and zirconium oxo phosphate. The high thermal stability of these zirconia based structures is due to a crystallization delay, because SO_4^{2-} and PO_4^{3-} ions are able to stabilize the amorphous zirconia phases. Both structures are thermally stable up to 500 °C and show ordered hexagonal pore arrangements which are analogous to MCM-41.

ACKNOWLEDGEMENT

This work was supported by the Deutsche Forschungsgemeinschaft (SCHU 744/8-1). We like to thank Dr. Y. Ushida and Prof. Dr. R. Schlögl from the Fritz Haber-Institut, Berlin for recording the TEM micrographs.

REFERENCES

1. a) C.T. Kresge, M.E. Leonowicz, W.J. Roth, J.C. Vartuli, and J.S. Beck, Nature, 359, (1992) 710. b) J.S. Beck, J.C. Vartuli, W.J. Roth, M.E. Leonowicz, C.T. Kresge, K.D. Schmitt, C.T.-W. Chu, D.H. Olson, E.W. Sheppard, S.B. McCullen, J.B. Higgins, and J.L. Schlenker, J. Am. Chem. Soc., 114, (1992) 10834.
2. a) Q. Huo, D. Margolese, U. Ciesla, P. Feng, T.E. Gier, P. Sieger, R. Leon, P. Petroff, F. Schüth, and G.D. Stucky, Nature, 368, (1994) 317. b) U. Ciesla, D. Demuth, R. Leon, P. Petroff, G. Stucky, and F. Schüth, J. Chem. Soc., Chem. Commun., (1994) 1387. c) Q. Huo, D.I. Margolese, U. Ciesla, D.G. Demuth, P. Feng, T.E. Gier, P. Sieger, A. Firouzi, B.F. Chmelka, F. Schüth, and G.D. Stucky, Chem. Mater., 6 (1994) 1176.
3. A. Stein, M. Fendorf, T.P. Jarvie, K.T. Mueller, A.J. Bebesi, and T.E. Mallouk, Chem. Mater., 8 (1995) 874.
4. D.M. Antonelli and J.Y. Ying, Angew. Chem. Int. Ed. Engl., 34 (1995) 2014.
5. a) D.M. Antonelli and J.Y. Ying, Angew. Chem. Int. Ed. Engl., 35 (1996) 426. b) D.M. Antonelli and J.Y. Ying, Chem. Mater., 8 (1996) 874.
6. U. Ciesla, S. Schacht, G.D. Stucky, K.K. Unger, and F. Schüth, Angew. Chem. Int. Ed. Engl., 35 (1996) 541.
7. H. Hattori, Chem. Rev., 95 (1995) 537.
8. A. Corma, Chem. Rev., 95 (1995) 559.
9. J. Livage, M. Henry, and C. Sanchez, Prog. Solid St. Chem., 18 (1988) 259.
10. G. Alberti and E. Torracca, J. Inorg. Nucl. Chem., 30 (1968) 317.
11. V. Alfredsson, M. Keung, A. Monnier, G.D. Stucky, K.K. Unger, and F. Schüth, J. Chem. Soc., Chem. Commun., (1994) 921.
12. K.M. Parida and P.K. Pattnayak, J. Colloid Interface Sci., 182 (1996) 381.
13. D.A. Ward and E.I. Ko, J. Catal., 150 (1994) 18.
14. J. Raggai and W. Lofti, Colloids and Surfaces, 61 (1991) 97.

MESOPOROUS MOLECULAR SIEVES 1998
Studies in Surface Science and Catalysis, Vol. 117
L. Bonneviot, F. Béland, C. Danumah, S. Giasson and S. Kaliaguine (Editors)

EPR Studies of the Formation Mechanism of the Mesoporous Materials MCM-41

Jingyan Zhang[a], H. Zimmermann[b], Zeev Luz[a] and Daniella Goldfarb[a]

[a] Department of Chemical Physics, Weizmann Institute of Science, Rehovot 76100, ISRAEL

[b] Max-Planck-Institute fur Medizinische Forschung Heidelberg, GERMANY

The formation mechanism of MCM-41 was investigated by *in situ* EPR spectroscopy using spin-probes and the location of the spin-label in the organic-inorganic interfaces was established using electron spin echo envelope modulations. It is found that the EPR line-shapes is highly sensitive to the progress of the reaction, and the mobility of the spin-probe is significantly restricted as the silicate ions polymerize at the interface. The long range ordering is established rather rapidly, while the condensation process is significantly slower.

1. INTRODUCTION

The formation mechanism of the mesoporous M41S materials[1-2] is intriguing and its details are not fully understood. The strong resemblance of the final structures to the liquid crystalline phases observed in solutions of the neat surfactant led to the formulation of a liquid crystal templating mechanism where the silicate polymerizes at the surface of the surfactant aggregates in a preformed liquid crystalline phase. This, however, could not explain the fact that the synthesis of the mesostructure is usually carried out with concentrations of 0.5- 5wt% surfactant, which is below that required for the formation of liquid crystalline phases[3,4]. Therefore, an alternative mechanism was suggested whereby the interaction between the silicate and the surfactant serves as a driving force for the generation of the structures which subsequently determine the nature of the final mesoporous materials. Three phenomena, essential for the formation of the surfactant-silicate mesostructure, were identified by Monnier *et al*[5]: (i) multidentate binding of silicate oligomers, (ii) preferred polymerization of silicates at the surfactant-silicate interface and (iii) charge density matching across the interface. A mechanism referred to as the "puckering layer model" was suggested by Steel *et al*[6]. The silicate species in the aqueous solution initially form a layered structure, further ordering results in puckering of the silicate layers forming the hexagonal channels which constitute the MCM-41 material. Another model proposed by Chen *et al*[7] is the "silicate rod assembly model". Randomly disordered rod-like micelles interact with silicate species to yield several monolayers of silica encapsulating the external surfaces of the micelles, followed by spontaneous assembling to produce the hexagonal MCM-41 structure.

Several studies were carried out in order to determine the intermediate structures generated during the synthesis of the M41S materials. ^{14}N NMR measurements on the reaction gel led to the suggestion that an hexagonal mesophase is a precursor for both the hexagonal and lamellar mesostructures[6]. Evidence for the liquid crystalline nature of a silicate-surfactant mixture was obtained also by ^{2}H NMR of deuterated surfactant[8]. Such a mesophase could be aligned in a magnetic field and MCM-41 was obtained after polymerization was induced by further treatment[9]. *In situ* FTIR measurements, monitoring the high temperature formation of MCM-41, showed that the head group ordering increased upon heating[10]. Intermediates found in the cryo-transmission electron microscopy showed that the primary building units are clusters of elongate micelles which appear already 3 minutes after mixing the gel components[11]. Although, in general these results and other[12] support the model of Monnier *et al*[5], certain details, related for example to the growth mechanism and the transformation between phases, remain unclear.

Hexagonal, lamellar and cubic silica mesostructures can also be obtained with non-ionic surfactants[13]. In these cases the required surfactant concentration is high so that a liquid crystalline phase exists prior to the addition of the silica source. Consequently, it was concluded that a liquid crystalline phase directs the formation of the mesoporous material and it was suggested that at high surfactant concentrations charge density matching and multidentate binding of charges at the interface are not essential for the process.

Considering the overall results obtained so far, it is not clear whether one universal mechanism applies for the formation of all the mesostructural materials or that under different conditions, different mechanisms are in effect. In this work we investigate the formation mechanism of MCM-41 by monitoring the micro-viscosity of the surfactant using EPR of spin-probes. While X-ray diffraction provides information regarding the setting-in of long range ordering of the mesoporous lattice, the EPR spectrum of the spin-probe is most sensitive to mobility of the organic phase[14]. The high sensitivity of the EPR method allows the addition of very small amount of spin-probes without perturbing the system. Moreover, the measurements can be performed *in situ*, thus providing direct information on the propagation of the reaction.

2. EXPERIMENTAL

Synthesis: The reagents used for the synthesis were tetraethyl-orthosilicate (TEOS, Aldrich 98%), cetyltrimethylammonium chloride (CTAC, 25wt% aqueous solution, Aldrich), cetyltrimethylammonium bromide (CTAB, Aldrich), NaOH (2N solution). The spin-probes 4-(N,N-dimethyl-N-hexadecyl)ammonium-2,2,6,6,-tetramethyl piperidine-oxyl, iodide (CAT16) and 5-doxyl-stearic acid were purchased from Molecular Probe Europe BV Company. α-d_2-CTAB was synthesized as described in the literature.[15] The synthesis procedure of MCM-41 was similar to that described in the literature,[1,16] except for the addition of the spin-probe, which was added to the surfactant solution and the mixture was stirred for several minutes until it was dissolved. The molar composition of the synthesis gel was TEOS: 0.12CTAC(B) : 0.52 NaOH : 114.87 H_2O : 1.12×10^{-3}spin-probe.

In situ EPR experiments were performed as follows: the reaction mixture without TEOS was first prepared and then TEOS was added with vigorous stirring for about 2 minutes. Part of the mixture was then quickly transferred into an EPR flat cell which was kept in the measuring cavity until the end of the measurement. The time dependent XRD measurements were carried out on a series of samples prepared by dividing the starting mixture into a number of flasks and the solid formed in each of the flasks was filtered and washed at different times after the initiation of the addition of TEOS.

Sample Characterization: Powder X-ray diffraction patterns of the final products were recorded on a Rigaku D/Max-B diffractometer, using Cu Kα radiation. The samples obtained after different reaction times were measured on a home built small-angle X-ray (SAX) diffractometer[17]. The EPR spectra were recorded at X-band using either a Varian E-12 or a Bruker ER200D-SRC spectrometer. Liquid mixtures were measured in flat cells and solid samples in 3 mm o.d. quartz tubes. Electron-spin echo envelope modulation (ESEEM) experiments were carried out on a home built spectrometer[18], using the three-pulse sequence ($\pi/2$-τ-$\pi/2$-T-$\pi/2$-τ-echo) where the echo intensity is recorded as a function of T.

3. RESULTS

Our prime interest in this study is the interface of the organic and inorganic phases and therefore, the nitroxyl radical should be located at this region. Two spin-probes were employed (Fig. 1(top)). The first is very similar to the surfactant molecules with tertiary amine as a polar head, whereas the second one has a carboxylic head group. Figure 1(left) shows the

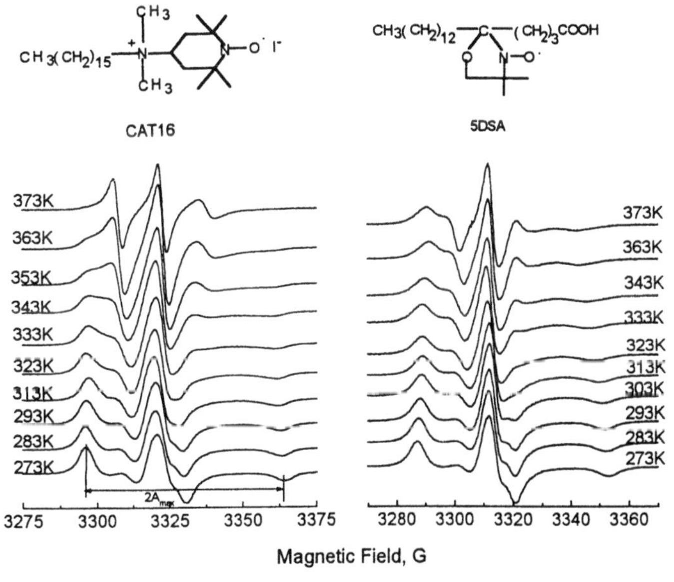

Figure. 1 Top: molecular structures of CAT16 and 5DSA. Bottom: X-band EPR spectra of the as-synthesized MCM-41(CAT16) and MCM-41(5DSA) as a function of temperature.

temperature dependence of the EPR spectra of as-synthesized MCM-41 containing CAT16, henceforth referred to as MCM-41(CAT16). At room temperature and below a rigid limit spectrum is observed, indicating no motion on the EPR time scale, while at high temperatures an isotropic-like spectrum appears. A different temperature dependence is exhibited by MCM-41(5DSA), as shown in Fig. 1(right). While its low temperature spectrum is similar to that of MCM-41(CAT16), as the temperature increases, the anisotropy is retained and it evolves into a spectrum of a partially averaged powder[14]. The rigid limit spectra of the two spin-probes at room temperature, and their temperature sensitivity make them suitable for probing the organic phase during the synthesis.

The structural similarity of CAT16 and the surfactant molecules makes it a good probe reporting the surfactant behavior. The different polar head of 5DSA, however, requires experimental evidence showing that it is located within the organic phase in a manner similar to the surfactant molecules. For this purpose MCM-41(CAT16) and MCM-41(5DSA) were synthesized with α-d_2-CTAB and the proximity of the nitroxyl group to the deuterons of the surfactant molecule was examined using ESEEM experiments which are very effective for detecting weak hyperfine interactions in solids[19]. Figure 2A shows the three-pulse ESEEM traces of MCM-41(CAT16) and MCM-41(5DSA), where modulations from ^1H and ^2H are clear in both traces. Figure 2B presents the corresponding Fourier transformed traces obtained after proper normalization allowing the comparison of the absolute peak intensities. Both spectra show two peaks, at the ^1H and ^2H Larmor frequency respectively. The intensity of the peaks is a measure of the modulation depth, which is determined by the distance between the unpaired electron and the nucleus and the number of nuclei around it[19]. The ^2H peak intensity

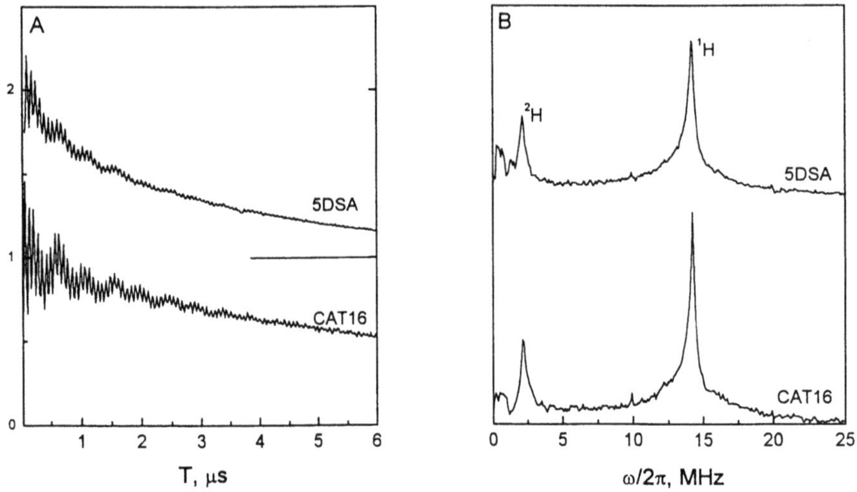

Figure 2. Three-pulse ESEEM spectra (A) MCM-41(CAT16) and MCM-41(5DSA) recorded at 3325G and τ=230ns. (B) the corresponding FT-ESEEM magnitude spectra.

is similar for the two spin-probes, indicating that the label proximities to the head-group of neighboring surfactant molecules are similar. The more intense proton peak in CAT16 is due to intra-molecular hyperfine interactions with protons of the polar head methyl groups, not present in 5DSA. The ESEEM results show that 5DSA is an appropriate spin-probe for investigating the organic-inorganic interface region during the reaction.

The spectrum of CAT16 in a water solutions consists of a sharp triplet and shows that the spin-probe is homogeneously distributed in the solution. The addition of CTAC causes a significant broadening of all three peaks, and the high field component split into two. The CTAC concentration in this solution is $4.24X10^{-2}$M, which is above the critical micellar concentration[20]. Accordingly, the splitting is interpreted in terms of a superposition of two sub-spectra due to spin-probe molecules in the micelles and in the aqueous environments, respectively[21].

A detailed description of the line-shape evolution during the formation of MCM-41(CAT16) is presented in Fig. 3A, where the top spectrum was recorded 3 min after the addition of TEOS. The temporal evolution of the spectrum shows that while the part corresponding to the aqueous solution remains practically invariant throughout the reaction, the micellar component gradually broadens until it converts completely to the rigid limit line-shape. The conversion process lasts about 2.5 h after which the spectrum remains invariant. The XRD pattern of the final product confirmed the formation of MCM-41. Subtraction of the spectrum corresponding to the aqueous phase yield spectra that reflect the changes in the micro-viscosity of the organic phase alone[22]. The temporal evolution of the micellar component is similar to that of CAT16 in a glycerol solution, recorded over the temperature range 223-353K, indicating that the spin-label is experiencing an isotropic slow down of the motion[22].

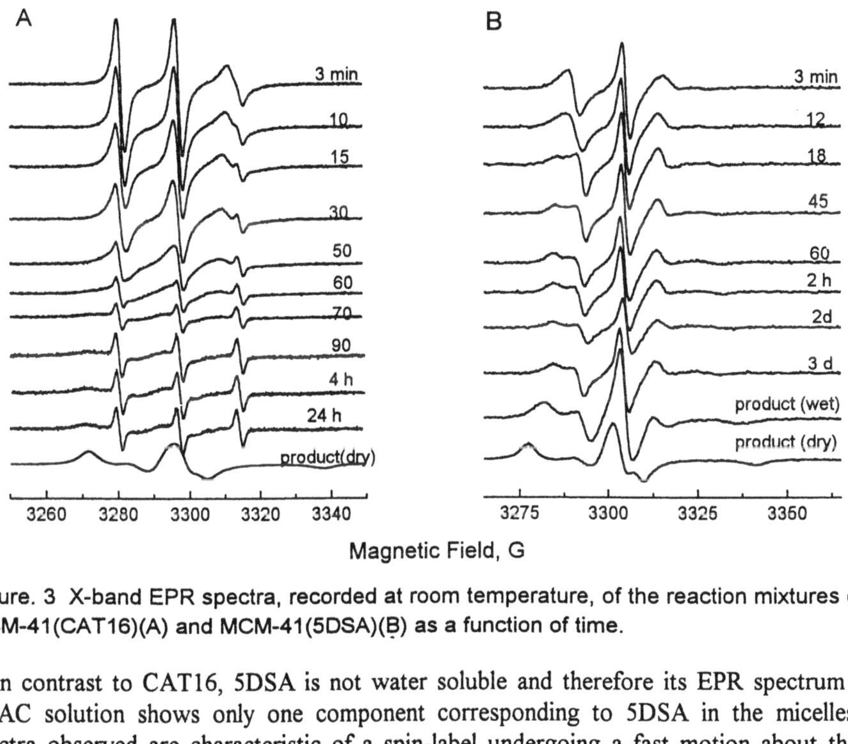

Figure. 3 X-band EPR spectra, recorded at room temperature, of the reaction mixtures of MCM-41(CAT16)(A) and MCM-41(5DSA)(B) as a function of time.

In contrast to CAT16, 5DSA is not water soluble and therefore its EPR spectrum in the CTAC solution shows only one component corresponding to 5DSA in the micelles. The spectra observed are characteristic of a spin-label undergoing a fast motion about the long molecular axis, leading to a partially averaged powder pattern[14]. The time evolution of the EPR spectrum of 5DSA during the synthesis of MCM-41 is shown in Fig. 3B. As in MCM-41(CAT16) the major changes in the line-shape occur between 12 to 18 min after the addition of TEOS. Unlike MCM-41(CAT16), the spectral evolution of MCM-41(5DSA) with time has an anisotropic character. Moreover, the final spectrum, prior to drying, indicates that the spin-label is still mobile. In contrast to MCM-41(CAT16) the width of the averaged powder pattern observed was found to be sensitive to the water content.

In parallel to the EPR experiments small angle X-ray (SAX) diffraction measurements were carried out. These, however, were not *in situ* measurements. A reaction mixture, having the same composition as that used for the EPR experiments, was divided into 10 parts, each of which was filtered and dried at different times after the addition of the TEOS. The SAX patterns so obtained are summarized in Fig. 4. Within the first few minutes a diffraction pattern, typical of an hexagonal structure is observed, with the d_{100} spacings changing from 45.7Å at 3 min to 41.8Å at 5 min. During the next 4 min it decreased further to 40.6Å and then it remained constant until the end of the process. These solid samples were also measured by EPR. These solids filtered after 3, 10 and 20 min of reaction exhibited essentially identical spectra of a rigid limit powder. This shows that the solid formed during the initial stages of the reaction is different from the one obtained after filtering and drying.

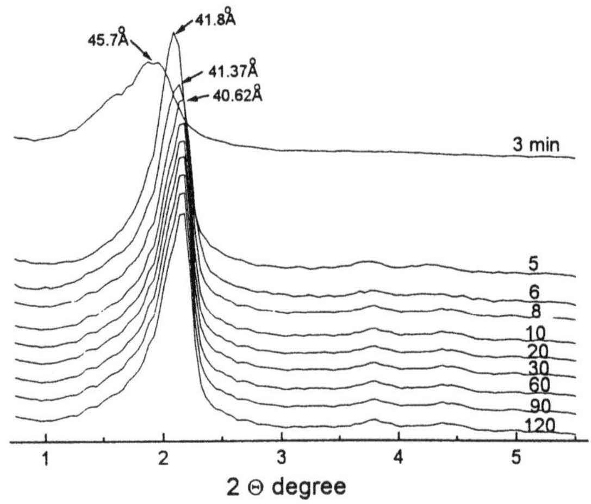

Figure 4. X-ray diffraction patterns of the solid obtained during the formation of MCM-41 at room temperature,as a function of time after the addition of TEOS.

4. DISCUSSION

Application of the EPR spin-probe methodology allows the *in situ* investigation of the room temperature formation of MCM-41. Both spin-probes used are incorporated into the micellar structure and the continuous changes in the environment of the surfactant molecules during the course of the reaction are well reflected in the EPR line-shape of the spin-probes. The spectral changes of CAT16 involving a continuous transformation from a relatively sharp triplet, characteristic of fast motion, into a rigid limit spectrum provides direct experimental evidence that micelles are indeed the precursors of the M41S final products. This is supported by the line-shape changes of 5DSA as well.

The EPR line-shapes observed during the synthesis of MCM-41(CAT16) correspond to a motional slow down similar to that observed in an isotropic environment. The reorientation motion of the micelles is too slow to average out the EPR anisotropic parameters (i.e. the g- and ^{14}N hyperfine anisotropies) and only the local motion of the spin-probe needs to be considered[23]. It is somewhat surprising that the motion leads to spectra characteristic of an isotropic motion, since the local environment of the spin-probe in rod-like micelles or in a liquid crystalline phase, presumed to appear in the course of the synthesis, is clearly not isotropic. Most likely, the geometry of the head-group of CAT16 is such that rapid reorientation about its long axis leads to a more or less complete averaging of the anisotropic magnetic interactions and hence to an isotropic spectrum. This is somewhat unfortunate because it renders CAT16 insensitive to the local ordering under fast reorientation conditions. Nevertheless, it is still sensitive to the local micro-viscosity in the range where the rate of the reorientation is of the order of the anisotropic magnetic interactions. The rigid limit EPR spectrum exhibited by MCM-41(CAT16) at room temperature indicates that the reorientation of the surfactant molecules about their long axes is strongly hindered by the silica wall, suggesting a strong interaction between the two phases.

The orientation of the nitroxyl group in 5DSA is such that a fast rotation about the long molecular axis does not fully average the anisotropies and therefore the spectrum is characteristic of orientational order. Unfortunately, also with this probe we could not detect a transition from spherical micelles to rod like micelles due to the slow lateral motion of the spin-probe. Nonetheless, the higher mobility of the 5DSA label indicates that while there is a strong interaction between the polar-head and the inorganic wall, as the organic core is approached the mobility increases and it is highly sensitive to the water content of the sample.

While the EPR spectra of the spin-probes reached their final line-shape after ~90 minutes, the X-ray diffraction pattern already exhibits its asymptotic shape of the final product after ~8 min (see Fig. 3). This indicates that the long range ordering is achieved at rather early stages of the reaction, whereas the silicate condensation, essential for the formation of the final mesostructures are significantly slower. The later causes the motion restriction of the head-group. The EPR results also indicate that the solid which appeared 3 min after initiation of the reaction is "soft" and allows fast motion of the surfactant. As the polymerization proceeds, the motion of the spin-probe slows down, leading to a solid like rigid limit spectrum. The comparison of the X-ray results, on filtered solids with those of the *in situ* EPR measurements is not straightforward though. The X-ray data were collected on dry samples while the *in situ* EPR measurements were made on the original aqueous mixtures. The EPR spectra of dry samples prepared as for the XRD measurements were quite different and exhibited rigid limit patterns even after short reaction periods. This shows that the solid observed during the initial stages of the reaction is different from that obtained after drying. In the *in situ* samples the local motion is significantly less restricted.

The EPR/XRD results are more consistent with the mechanism where the addition of the silicate generates domains with long range hexagonal order. Due to the low CTAC concentration (1.6 wt%), this mixture is not a uniform lyotropic liquid crystalline phase as obtained at high CTAC concentrations. Rather, the system can be viewed as a dispersion of ordered domains in an aqueous medium, as manifested by the persistence of an aqueous CAT16 EPR signal throughout the reaction. Only after the hexagonal domain structure is formed, or simultaneously with its formation, the silicate ions start to polymerize. In this scheme the polymerization of the silica is the rate determining step.

5. CONCLUSIONS

It was demonstrated that suitable spin-probes can be used to monitor the *in situ* formation of M41S materials by EPR. These *in situ* measurements provide direct experimental evidence that micelles serve as precursors for the mesoporous materials. The MCM-41 appears to form in two stages. The first stage, which starts immediately after the mixing of the silica source with the surfactant solution, involves the formation of domains with hexagonal order. These domains consist of micellar rods encapsulated with silicate ions oligomers. The second, much slower stage, involves polymerization of the silicate ion at the interface. It results in hardening of the inorganic phase and at the same time in restricting the motion of the surfactant molecules at the interfaces.

ACKNOWLEDGMENTS

This work was supported by a grant from the Israeli Ministry of Science and Technology.

REFERENCES

1. J.S. Beck, J.C. Vartuli, W.J. Roth, M.E. Leonowicz, C.T. Kresge, K.D. Schmitt, C.T.W. Chu, D.H. Olson, E.W. Shepard, S.B. McCullen, J.B. Higgins and J.L. Schlenker, J. Am. Chem. Soc., 114 (1992) 10834.
2. J.S. Beck, J.C. Vartuli, G.J. Kennedy, C.T. Kresge, W.J. Roth and S.E. Schranm, Chem. Mater., 6 (1994) 1861.
3. X. Auray, C. Petipas, R. Anthore, L. Rico A.J. Lattes, J. Phys. Chem., 93 (1989) 7458.
4. E.S. Blackmore, and G.J.T. Tiddy, Liq. Crys., 8 (1990) 131.
5. A. Monnier, F. Schuth, Q. Huo, D. Kumar, D. Margolese, R.S. Maxwell, G.D. Stucky, M. Krishamurty, P. Petroff, A. Firouzi, M. Janicke and B.F. Chmelka, Science, 261 (1993) 1299.
6. A. Steel, S.W. Carr, and M.W. Anderson, J. Chem. Soc., Chem. Commun., 157, 1994.
7. C.Y. Chen, S.L. Burkett, H.X. Li and M.E. Davis, Microporous Mater., 2 (1993) 27.
8. A. Firouzi, D. Kumar, L. Bull, M. T. Besier, P. Sieger, Q. Huo, S.A. Walker, J.A. Zasadizinski, V. Glinka, J. Nicol, D. Margolese, G.D. Stucky and B.F. Chmelka, Science, 267 (1995) 1138.
9. S.H. Tolbert, A. Frriouzi, C.D. Stucky, and B.F. Chmelka, Science, 278 (1997) 264.
10. D. Calabro, E.W. Valyoscik, F.X. Ryan, Microporous Mater., 7 (1996) 243.
11. O. Regev, Langmuir, 12 (1996) 4940.
12. V. Alfredsson, M. Keung, A. Monnier, G.D. Stucky, K.K. Unger, F.J. Schuth, Chem. Soc., Chem. Commun., 921, 1994.
13. G.S. Attard, J.C. Glyde, C.G. Goltner, Nature, 378 (1995) 366.
14. L.H. Freed, Spin-labelling (Theory and Applications), Academic Press, New York 1976.
15. A. Firouzi, F. Atef, A.G. Oertli, G.D. Stucky, B.F. Chmelka, J. Am. Chem. Soc., 119 (1997) 3596.
16. D. Zhao, D. Goldfarb, J. Chem. Soc., Chem. Commun., 875, 1995
17. J. J. Cheetham, E. Wachtel, D. Bach, R.M. Epand, Biochemistry, 28 (1989) 8928.
18. D. Goldfarb, J. M. Fauth, Y. Tor, A. Shanzer, J. Am. Chem. Soc., 113 (1991) 1941.
19. L. Kevan, in Modern Pulsed and Continuous Wave Electron Spin Resonance, L. Kevan, Bowman, M.K. Eds., Wiley, New York, 1990.
20. C.L. Kwan, S. Atik, L.A. Singer, J. Am. Chem. Soc., 100 (1978) 4783.
21. O.H. Griffith, L. J. Libertini and G. B. Birrell, J. Phys. Chem., 75 (1971) 3417.
22. J. Zhang, Z. Luz, and D. Goldfarb, J. Phys. Chem., 101 (1997) 7087.
23. D.D. Lasic, H. Hauser, J. Phys. Chem., 89 (1985) 2648.

MESOPOROUS MOLECULAR SIEVES 1998
Studies in Surface Science and Catalysis, Vol. 117
L. Bonneviot, F. Béland, C. Danumah, S. Giasson and S. Kaliaguine (Editors)

Variable temperature ^{129}Xe NMR studies of xenon adsorbed on mesoporous MCM-41 molecular sieves

Sung-Jeng Jong[a], Jin-Fu Wu[a], Ajit R. Pradhan[a], Hong-Ping Lin[b], Chung-Yuan Mou[b] and Shang-Bin Liu[a,*]

[a]Institute of Atomic and Molecular Sciences, Academia Sinica, P. O. Box 23-166, Taipei, Taiwan 106, R. O. C.

[b]Department of Chemistry, National Taiwan University, Taipei, Taiwan 106, R. O. C.

A comprehensive variable temperature ^{129}Xe NMR study of xenon adsorbed on mesoporous MCM-41 molecular sieves with varied pore sizes, morphology and aluminum concentration has been made. The observed ^{129}Xe NMR chemical shifts (δ) of the adsorbed Xe exhibit a strong dependent on Xe loading (ρ) and temperature (T). At low Xe loading, the ^{129}Xe chemical shifts show a concave upward dependence on Xe density which vary with temperature and pore size of the adsorbent. Whereas at high loading, a nearly linear relation between the observed chemical shifts and effective Xe density was found whose slopes of δ vs ρ_{eff} is temperature dependent. Regressional fitting based on a two-site exchange model enable us to fit the observed $\delta(T, \rho)$ data in a least squares sense to second-order polynomial with ρ. The temperature variations of the Xe-wall interaction at zero density as well as the Xe-Xe interactions were derived explicitly and compared with the results available for bulk Xe, Xe adsorbed on microporous materials and on polymer surfaces.

1. INTRODUCTION

The structure and dynamic properties of molecules adsorbed in microporous/mesoporous materials have drawn much research attention in recent decades. A number of papers have recently appeared on the application of using probe molecules to characterize the pore structure of the mesoporous adsorbent such as MCM-41 [1, 2]. In this paper, we report results of variable temperature ^{129}Xe NMR of Xe adsorbed in MCM-41 molecular sieves with varied pore size, morphology and Al concentrations. ^{129}Xe NMR [3, 4] is known to be a very sensitive probe of local environment inside porous materials. The subject has been reviewed by several authors [5-8].

There are several reasons for doing variable temperature ^{129}Xe NMR of Xe adsorbed on MCM-41. First, the straight, unconnected channels of controllable pore size make MCM-41 an ideal model adsorbent for both theoretical and experimental investigation of critical phenomena of molecules in a confined space. Second, ^{129}Xe NMR provides an excellent probe to investigate the morphology and topology of MCM-41 materials which are not well known at the present. Third, at the limit of extreme dilution, the interactions between a Xe atom (adsorbate) and the wall of mesoporous materials (adsorbent) such as MCM-41 are not known and thus demand further investigation. Furthermore, it would be interesting to compare the temperature and density dependence of ^{129}Xe chemical shifts in the present system with those obtained from bulk fluid or solid Xe [9-11] and that from Xe adsorbed on microporous materials [4, 12-15]. Such observations should provide valuable information on Xe interactions in both macroscopic and microscopic levels.

2. EXPERIMENTAL SECTION

2.1. Preparation and characterization of mesoporous materials

Powdered MCM-41 molecular sieves (Si/Al = 37) with varied pore diameters (1.80, 2.18 and 2.54 nm, denoted by Ma, Mb and Mc, respectively) were synthesized following the conventional procedure using sodium silicate, sodium aluminate and compounds such as cetyltrimethylammonium bromide (C_{16}TMAB), myristyltrimethylammonium bromide (C_{14}TMAB) and dodecyltrimethylammonium bromide (C_{12}TMAB) as the source materials for Si, Al and quaternary ammonium surfactants, respectively [16]. A pure-silica form MCM-41 sample (denoted as SM, pore diameter 2.44 nm) was synthesized with the same procedure but in the absence of sodium aluminate source material. Moreover, an ordered mesoporous MCM-41 with a hollow tubules-within-a-tubule hierarchical structure (Fig. 1) of C_{16}MTAB system (sample denoted as TM, Si/Al = 37, pore diameter 2.64 nm) was also synthesized under highly alkaline condition by careful control of the surfactant-water content and lower silica condensation rate [17]. The structure and morphology of all the synthetic MCM-41 molecular sieves were confirmed by powder X-ray diffraction and transmission and scanning electron microscopy. Their pore sizes, surface areas and pore volume were deduced from the N_2 adsorption-desorption isotherm data obtained at 77 K (Table 1).

2.2. Sample pretreatment and xenon adsorption

Sample cells were made with 7 mm i.d. pyrex tubing with a pre-pulled neck at about 23 mm from the bottom. The sample cell was first attached to a vacuum stopcock and an adapter so that it can be conveniently set up for dehydration of the sample on a vacuum apparatus and for the succeeded adsorption of Xe. Sample (ca. 0.1 g) was first placed into the necked pyrex sample cell and then subjected to dehydration. Dehydration was performed on a vacuum manifold by gradual heating (1 ^{o}C min^{-1}) to 350 ^{o}C in vacuum (< 10^{-5} Torr) and maintained at the temperature for over 40 h. Sample weight was carefully monitored during each procedure.

After sample dehydration, a known amount of Xe (natural abundant or 70% ^{129}Xe-enriched) in a calibrated volume was transferred into MCM-41 by chilling the bottom of the sample cell with liquid N_2 then the sample cell was sealed off with a mini-torch. Special cares were taken in minimizing the dead volume of each sample cell during glass sealing. For practical purposes, the amount of Xe adsorbed, ρ, is expressed as the number of Xe atoms per effective free volume of anhydrous MCM-41 at room temperature. As a result, samples with varied Xe loading ranging form 2-245 amagat (the gas density at 0 ^{o}C and 1 atm) were prepared. One should bear in mind that the Xe loading so determined represents only the *apparent* Xe density (Xe inside the sample cell). This density represents the *effective* density

Table 1
Characteristics of the MCM-41 samples

Sample	Si/Al	Pore diameter (nm)[a]	Pore volume (ml/g)[a]	$\delta_s(\rho = 0)$ (ppm)[b]	σ_1 (ppm/amagat)[b]
TM	37	2.64	1.05	56.8	0.485
Ma	37	2.54	1.15	52.4	0.122
Mb	37	2.18	0.96	58.0	0.189
Mc	37	1.80	0.96	73.0	0.271
SM	∞	2.44	0.82	55.8	0.054

[a]Derived from N_2 adsorption isotherms. [b]Obtained from Fig. 2.

(ρ_{eff}, represents Xe inside the pore of MCM-41) only when at low enough temperature and below saturation loading at which the amount of gaseous Xe (outside of the pore) is negligibly small. The *effective* Xe density, which depends on both the amount of adsorbent and the temperature, was calibrated by xenon adsorption isotherms done at specific temperatures. We note that it was calibrated *effective* density that were actually used during the regressional fitting which were performed on a personal computer using Microcal Origin developed by Microcal Software, Inc. None of our samples reached saturation loading and for temperature below ca. 230 K, nearly all of the Xe were adsorbed within the pore of MCM-41.

2.3. Variable temperature ^{129}Xe NMR experiments

All ^{129}Xe NMR measurements were performed at 83.012 MHz on a Bruker MSL-300P spectrometer. The automatic temperature regulator is accurate to at least 1 °C, with a precision of about ± 0.1 °C. The accuracy and linearity of the regulator reading with temperature were calibrated by proton chemical shift of either methanol (for T ≤ 300 K) or ethylene glycol (for T ≥ 300 K). We estimate that all temperatures were known to ± 2 °C. For room temperature (298 K) measurements, a designed sample tube was used [18]. The free-induction-decay signals were accumulated typically with a relaxation delay of 0.3 s. The ^{129}Xe chemical shift was referred to gaseous Xe at zero density [11].

3. RESULTS AND DISCUSSION

3.1. Room temperature ^{129}Xe NMR studies

3.1.1. Characterization of tubular MCM-41

Room temperature ^{129}Xe NMR spectra of the tubular (TM) and non-tubular (Ma) MCM-41 sample are shown in Fig. 1. Unlike sample Ma (Fig. 1a) whose uniform channel system exhibits only a single resonance, the spectrum of sample TM show two distinct resonance peaks (Fig. 1b). The peak appears at ca. 15 ppm, upfield from the main resonance is assigned to Xe adsorbed in macropores of the tubular MCM-41. The intensity of the macropore peak

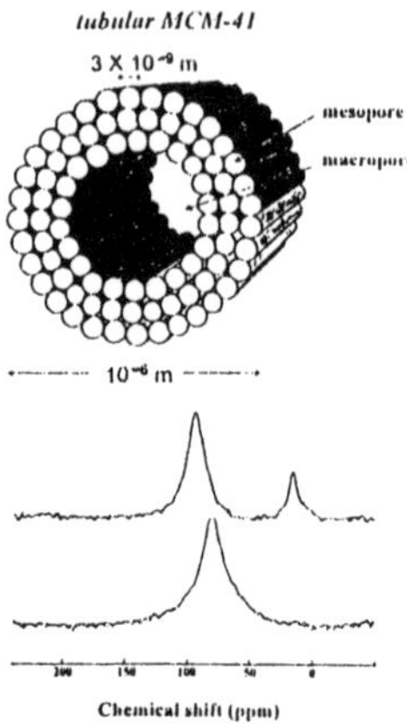

Figure 1. Schematic of the tubular MCM-41 (top) and ^{129}Xe NMR spectra of (a) MCM-41 & (b) tubular MCM-41.

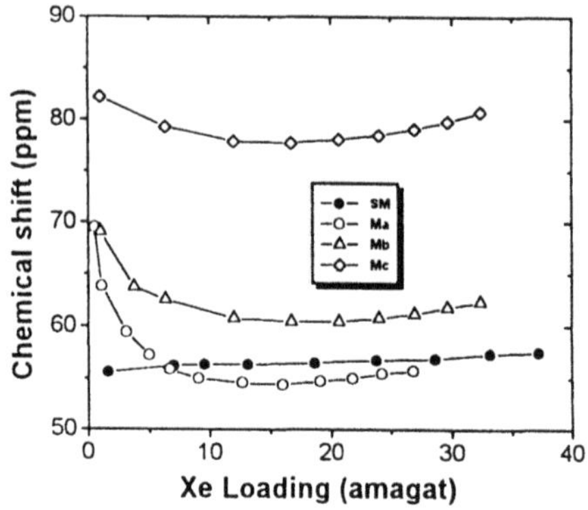

Figure 2. ^{129}Xe chemical shift vs Xe loading (298 K) for various samples.

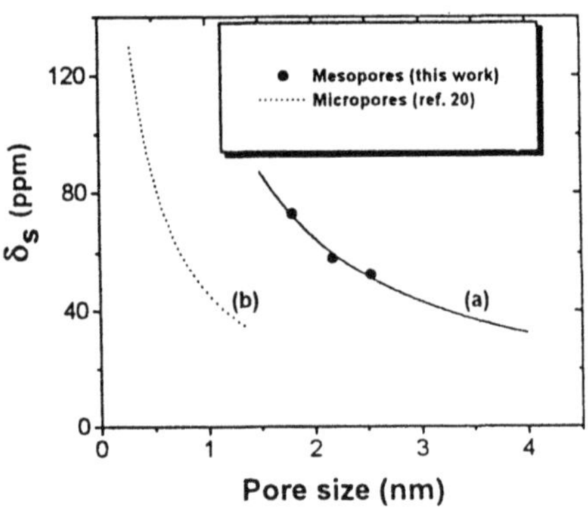

Figure 3. Correlation of δ_s with pore size for (a) MCM-41 and (b) zeolites. The solid lines indicate the fitting results.

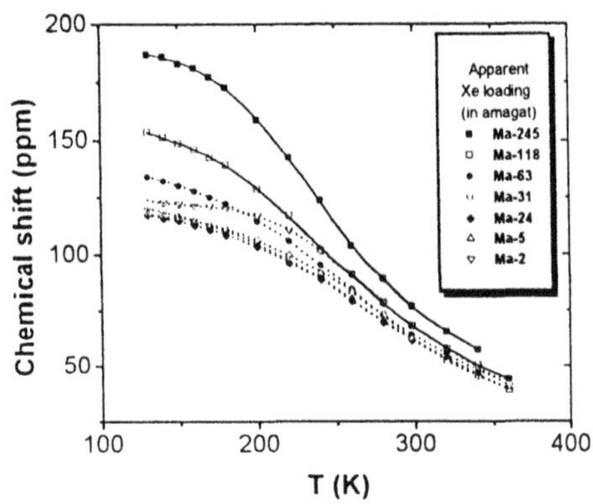

Figure 4. Variations of ^{129}Xe chemical shift with temperature for sample Ma with various apparent Xe density (ρ).

Table 2

Temperature coefficients of $\sigma_{1a}(T)$ and $\sigma_{2a}(T)$ obtained by regressional polynomial fitting[a]

Sample	A_o	A_1	A_2	B_o	B_1	B_2
Ma	0.4766	6.1×10^{-4}	-5.92×10^{-6}	- 21.4922	0.1806	-4.5×10^{-4}
TM	0.8235	-29.8×10^{-4}	3.96×10^{-6}	- 12.9020	0.1074	-2.9×10^{-4}

[a]Fitted to second-order, that is, $\sigma_{1a}(T) \approx A_o + A_1T + A_2T^2$ and $\sigma_{2a}(T) \approx B_o + B_1T + B_2T^2$.

decreases with decreasing temperature and become invisible for temperature below ca. 220 K indicating the migration of Xe from the macropore to the mesopore while lowering the temperature.

3.1.2. Effects of pore size and Al concentration

The variations of the observed ^{129}Xe chemical shift with pore size are shown in Fig. 2. At low loading, the chemical shift show a parabolic-like curvature. However, no such behavior was found for pure-silica sample (SM). Although the reasons for the parabolic-like chemical shift dependence is not fully known, it is at least partly due to the presence of Al. MCM-41 is known, thus far, to possess partially crystalline structure closely related to amorphous silica [19]. Cheung [12] ascribed the parabolic-like chemical shift behavior observed in amorphous alumina and silica-alumina at low Xe loading to the presence of a broad distribution of micropores and the inverse dependent with pore size. Ito and Fraissard [3] expressed the room temperature ^{129}Xe NMR chemical shift of Xe adsorbed in porous adsorbent as $\delta = \delta_o + \delta_s(\rho = 0) + \sigma_1\rho$, where $\delta_o = 0$ is the chemical shift reference, $\delta_s(\rho = 0)$ represents the interaction between a Xe atom and the wall of the adsorbent. The last term represents the contribution arises from binary Xe-Xe interactions which, at moderate loading, is linear with ρ. The observed values of δ_s and σ_1 are depicted in Table 1.

In Fig. 3, the values of δ_s is plotted against pore size for the non-tubular MCM-41 samples. The pore size dependence of δ_s can be described by the empirical relation: $\delta_s = 129/(d - 0.0203)$, where d is the pore diameter in nm. The empirical relation, $\delta_s = 49.9/(d - 0.2346)$, derived by Demarquay and Fraissard [20] for Xe adsorbed in microporous zeolites is also shown in Fig. 3 for comparison. It is clear that a sizable discrepancy (ca. 60 ppm in the vertical axis) exists between the two empirical relations arising form Xe adsorbed in mesoporous and microporous adsorbents. The reason for such discrepancy is not fully understood at present. However, a discrepancy of over 30 ppm was observed for a known structure of $AlPO_4$-5 [21]. The discrepancy may be due to the difference in the local environment experienced by the adsorbed Xe. Recent molecular simulation study [22] showed that the partially crystalline structure of MCM-41 [19] possesses a high skeletal density of ca. 2.7 g cm^{-3}, which is equivalent to ca. 27 T-sites per nm^3 for solid alone and 11-19 T-sites when the pore spaces are also included. The skeletal density of MCM-41 is very close to quartz (2.66 g cm^{-3}) and is indeed substantially higher than amorphous silica (2.2 g cm^{-3}) and zeolites (typically less than 2 g cm^{-3}). Thus, comparing with microporous zeolites, the observed downfield chemical shift in mesoporous MCM-41 is probably not unreasonable.

3.2. Variable temperature ^{129}Xe NMR studies

Figure 4 shows the temperature variation of ^{129}Xe chemical shifts of Xe adsorbed in non-tubular MCM-41 sample (Ma) with the *apparent* Xe loading. Similar results were also found for the tubular MCM-41 sample (TM). In Fig. 5, the observed chemical shift is plotted against the calibrated Xe loading (ρ_{eff}). At a given temperature, the observed chemical shifts at low loading also exhibit a parabolic-like curvature with a broad minimum. Whereas at high loading, a nearly linear dependent chemical shift with Xe loading is observed.

The observed ^{129}Xe chemical shift can be expressed as a weighted average between two sites in rapid exchange, namely the adsorbed and non-adsorbed (gaseous) xenon. For a

Table 3
Assorted parameters obtained from the data fitting

Sample	ΔE_{ads}(exp) (kcal/mol)	ΔE_{ads}(fit) (kcal/mol)	δ_{oa} (ppm)	δ_s(300 K) (ppm)	σ_{1a}(300 K) (ppm/amagat)
Ma	2.1 ± 0.5	2.43 ± 0.08	106.6 ± 0.6	59.0 ± 0.3	0.170 ± 0.010
TM	1.9 ± 0.5	2.39 ± 0.15	105.6 ± 1.0	60.0 ± 0.5	0.285 ± 0.005

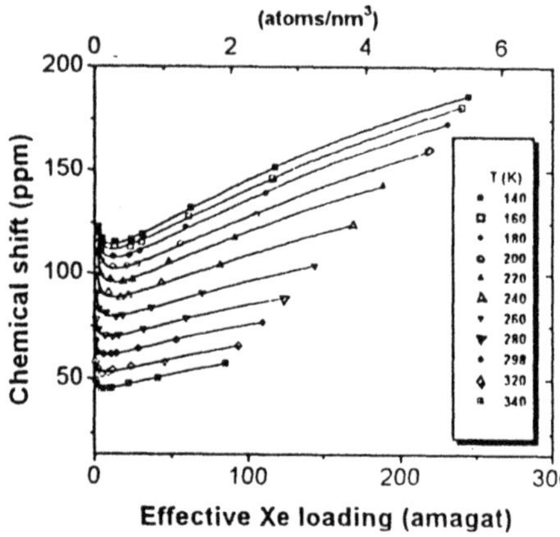

Figure 5. Variations of ^{129}Xe chemical shift with calibrated effective Xe density (ρ_{eff}) at different temperatures.

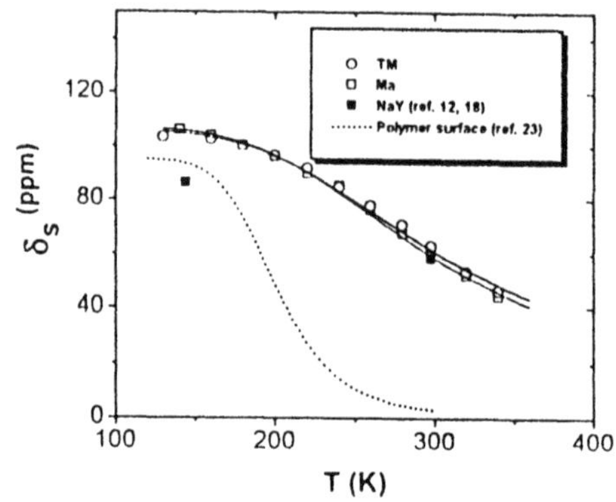

Figure 6. Plot of δ_s vs T for samples TM and Ma compared with that of Xe on NaY zeolite and polymer surface. The solid lines indicate the fitting results.

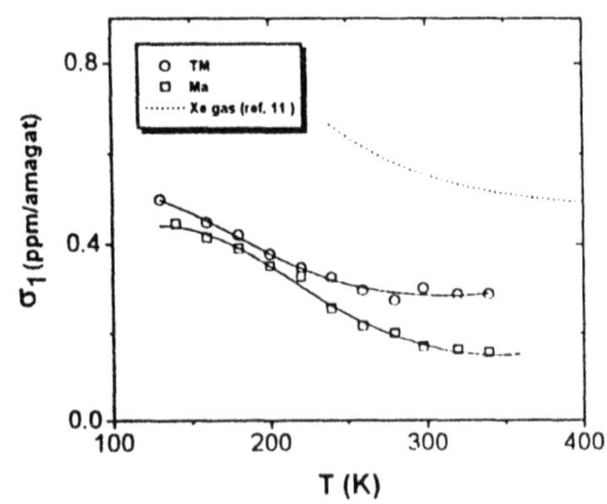

Figure 7. Plot of σ_1(T) vs T compared with results obtained from gaseous Xe. Solid lines indicate the fitting results.

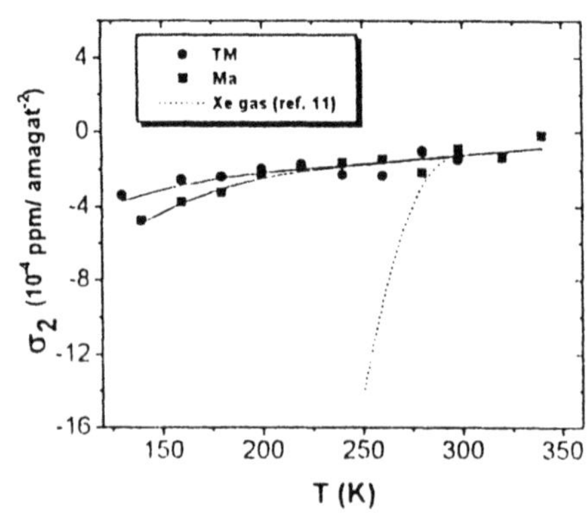

Figure 8. Plot of σ_2(T) vs T compared with results obtained from gaseous Xe. Solid line indicates the fitting results.

quantitative analysis of the experimental results, we fit the data with a polynomial function to second-order, such that

$$\delta(\rho, T) = P_a(T)\delta_a(\rho_a, T) + P_g(T)\delta_g(\rho_g, T) = \delta_s(T) + \sigma_1(T)\rho + \sigma_2(T)\rho^2 + \dots \quad (1)$$

where P_a and P_g are the probabilities of finding the adsorbed and gaseous xenon, respectively; $P_a + P_g = 1$. ρ_a and ρ_g are, respectively, the density of adsorbed and gas density; $\rho = \rho_a + \rho_g$. The probability of finding a Xe atom at the wall is given by [23]: $P_a = \rho_a/\rho = \tau_a/(\tau_a + \tau_g)$; $P_g = \rho_g/\rho = 1 - P_a$, where τ_a is the average time xenon spent on the wall, and τ_g us the reciprocal of the xenon collision rate with the surface. The average xenon sticking time on the wall is given by $\tau_a = \tau_o \exp(\Delta E_{ads}/RT)$, where τ_o is the preexponential factor and ΔE_{ads} is the energy of adsorption (Table 3), and R is the gas constant. For practical purposes, we express the adsorbed and gaseous chemical shift contributions as second-order polynomial:

$$\delta_a(\rho_a, T) \approx \delta_{oa} + \sigma_{1a}(T)\rho_a + \sigma_{2a}(T)\rho_a^2; \quad \delta_g(\rho_g, T) \approx \delta_{og} + \sigma_{1g}(T)\rho_g + \sigma_{2g}(T)\rho_g^2 \quad (2)$$

The term $\delta_g(\rho_g, T)$ can be expressed explicitly according to the relation given by Jameson and co-workers [11], where $\delta_{og} = 0$ is the chemical shift reference. Thus, we have

$$\delta_s(T) = P_a(T)\delta_{oa} \quad (3)$$
$$\sigma_1(T) = P_a^2(T)\sigma_{1a}(T) + P_g^2(T)\sigma_{1g}(T) \quad (4)$$
$$\sigma_2(T) = P_a^3(T)\sigma_{2a}(T) + P_g^3(T)\sigma_{2g}(T) \quad (5)$$

The results obtained from the data fitting are shown in Figs. 6-8 and the related parameters are summarized in Tables 2 and 3. We find that, at low temperature, $\delta_s(T < 150\ K)$ seems to reach a limiting value of $\delta_{oa} = 106$ ppm, much higher than the values found for Xe adsorbed on NaY zeolite (86 and 58 ppm at 144 and 298 K, respectively) [12, 18] or on the surface of poly(acrylic acid) polymer (95 ppm) [22]. As shown in Fig. 7, the second virial coefficient of the adsorbed Xe, $\sigma_1(T)$, is seeming smaller than $\sigma_{1g}(T)$ of the gaseous Xe and seems to vary with the morphology of the adsorbent. As shown in Table 3, $\sigma_{1a}(300\ K) =$ 0.170 and 0.285 ppm/amagat for Ma and TM, respectively, small compare to $\sigma_{1g}(300\ K) =$ 0.548 ppm/amagat [11]. Similarly, the fitting results obtained for the third virial coefficient, $\sigma_2(T)$, can also be compared with $\sigma_{2g}(T)$ of the gaseous Xe [11] as shown in Fig. 8. Finally, the fact that the fitted ΔE_{ads} values are in good agreement with the experimental values (Table 3), indicating that our model gives reasonable and meaningful quantitative description of the system.

4. CONCLUSIONS

In this work, we show that room temperature ^{129}Xe NMR of the adsorbed Xe is capable of providing structural characteristics of the mesoporous MCM-41 molecular sieves. Regressional fitting of the variable temperature ^{129}Xe chemical shift data, on the other hand, is capable of providing detailed Xe interactions in the mesoporous adsorbents. Our results indicate that the Xe interactions in mesoporous solids deviate from not only bulk Xe, but also from Xe adsorbed on microporous adsorbents or polymer surfaces. The detailed dynamics of the adsorbed Xe could also be deduced from the temperature dependence of τ_g and τ_a; this issue shall be discussed elsewhere. More experimental and theoretical studies are under way to investigate the detailed molecular phenomena when they are confined in restricted volume.

REFERENCES

1. E.W. Hansen, E. Tangstad, E. Myrvold and T. Myrstad, J. Phys. Chem., 101 (1997) 10709, and references therein.
2. K Morishige and K. Nobuoka, J. Chem. Phys., 107 (1997) 6965, and references therein.
3. T. Ito and J. Fraissard, in: L.V.C. Rees (ed.), Proc. of the 5th Inter. Conf. On Zeolites, Naples, Italy, Heyden, London, 1980, p. 510; J. Chem. Phys., 76 (1982) 5225.
4. J.A. Ripmeester, ISMAR-Ampere Inter. Conf. on Magnetic Resonance, Delft, The Netherlands, 1980 and ibid, J. Am. Chem. Soc., 104 (1082) 289.
5. J. Fraissard, Zeolites, 8 (1988) 350.
6. C. Dybowski, N, Bansal and T.M. Duncan, Ann. Rev. Phys. Chem., 42 (1991) 433.
7. P.J. Barrie and J. Klinowski, Progr. NMR Spectrosc., 24 (1992) 91.
8. D. Raftery and B.F. Chmelka, in P. Diel et al. (eds.), NMR Basic Principles and Progress, Vol. 30, Springer-Verlag, Berlin, Heidelberg, 1994, p. 111.
9. E. Kanegsberg, B. Pass and H.Y. Carr, Phys. Rev. Lett., 23 (1969) 572, and references therein.
10. W.M. Yen and R.E. Norberg, Phys. Rev., 131 (1963) 269.
11. C.J. Jameson, A.K. Jameson and S.M. Cohen, J. Chem. Phys., 59 (1973) 4540; J. Chem. Phys., 62 (1975) 4224.
12. T.T.P. Cheung, C.M. Fu and S. Wharry, J. Phys. Chem., 92 (1988) 5170; T.T.P. Cheung, J. Phys. Chem., 93 (1989) 7549; T.T.P. Cheung and C.M. Fu, J. Phys. Chem., 93 (1989) 3740.
13. Q.J. Chen and J. Fraissard, J. Phys. Chem., 96 (1992) 1809.
14. D.W. Johnson and L. Griffiths, Zeolites, 7 (1987) 484.
15. V. Pasquier, P. Levitz and A. Delville, J. Phys. Chem., 100 (1996) 10249.
16. H.P. Lin, S. Cheng and C.Y. Mou, Microporous Mater., 10 (1997) 111.
17. H.P. Lin and C.Y. Mou, Science, 273 (1996) 765.
18. S.B. Liu, L.J. Ma, M.W. Lin, J.F. Wu and T.L. Chen, J. Phys. Chem., 96 (1992) 8120; J. Phys. Chem., 98 (1994) 4393.
19. J.S. Beck, J.C. Vartuli, W.J. Roth, M.E. Lenowicz, C.T. Kresge, K.D. Schmitt, C.T.W. Chu, D.H. Olson, E.W. Sheppard, S.B. McCullen, J.B. Higgins, and J.L. Schlenker, J. Am. Chem. Soc., 114 (1992) 10834.
20. J. DeDemarquay and J.P. Fraissard, Chem. Phys. Lett., 136 (1987) 314.
21. Q.J. Chen, M.A. Spinguel-Huet and J. Fraissard, Chem. Phys. Lett., 159 (1989) 117.
22. M.W. Maddox, J.P. Oliver and K.E. Gubbins, Langmuir, 13 (1997) 1737.
23. D. Raftery, L. Reven, H. Long, A. Pines, P. Tang and J.A. Reimer, J. Phys. Chem., 97 (1993) 1649.

MESOPOROUS MOLECULAR SIEVES 1998
Studies in Surface Science and Catalysis, Vol. 117
L. Bonneviot, F. Béland, C. Danumah, S. Giasson and S. Kaliaguine (Editors)

Photochemical properties of Rhodamine-B dye molecules included within mesoporous molecular sieves

H. Yamashita[a], A. Tanaka[a], M. Nishimura[a], K. Koyano[b], T. Tatsumi[b] and M. Anpo[a]*

[a]Department of Applied Chemistry, College of Engineering, Osaka Prefecture University, Gakuen-cho 1-1, Sakai, Osaka 599-8531, Japan

[b]Engineering Research Institute, School of Engineering, The University of Tokyo, Yayoi 2-11-16, Tokyo 113-0033, Japan

To clarify the effects of the micro-environment of host-adsorbents on the electronic excited states of guest-molecules included in restricted void spaces, functional dye molecules, Rhodamine B (RhB), were adsorbed on various solid surfaces (SiO_2, ZSM-5, FSM-16, MCM-41, Ti-MCM-41) and the differences in the fluorescence properties were investigated. Analysis of the decay curves of the fluorescence of RhB indicates that RhB adsorbs on the outer surface of the microporous ZSM-5 zeolite as monomer and dimer species and its dispersion is low as well as when adsorbed on silica. On the other hand, the mesoporous molecular sieves (FSM-16, MCM-41, Ti-MCM-41) can include RhB within their mesopores where RhB exists only as a monomer species. Especially with Ti-MCM-41, a long lifetime of the fluorescence of the RhB monomer species can be observed indicating that the isolated tetrahedrally coordinated titanium oxide moieties in the mesopores are effective in anchoring RhB molecules in a highly dispersed state. The present results indicate that the large cavities of the mesoporous molecular sieves are useful in realizing the high dispersion state of dye molecules and effective in controlling the micro-environment of adsorbed dye molecules.

1. INTRODUCTION

Investigations on the photophysics and photochemistry of molecules adsorbed on solid surfaces such as SiO_2 and zeolites have been given a great deal of attention in the field of heterogeneous photochemistry and/or surface photochemistry [1-6]. One of the most interesting aims is what the influences are to the photophysical and photochemical properties of adsorbed molecules when the micro-environment of the adsorption sites is changed in their physical and/or chemical nature. It is well known that the photophysical and photochemical behaviors of adsorbed molecules are quite different from those observed in the liquid phase.

The photophysics and photochemistry of molecules included in the restricted cavities of zeolites and mesoporous molecular sieves have been the focus of many studies related to the effects of the micro-environment of the host-guest interaction on the electronic excited states and reactivities of the guest molecule adsorbed on the inner-surfaces of these porous materials [1-6]. However, the photophysical and photochemical properties of molecules adsorbed on solid surfaces and/or included in the restricted micro-cavities of porous materials are still one of the most unexplored fields in photochemistry.

The fluorescence spectra of luminescent dye molecules adsorbed on solid surfaces provide useful information not only on the electronically excited states but also on the chemical properties of solid surfaces. Rhodamine B (RhB) molecules are often used as probe molecules due to various unique properties such as the concentration dependence in the fluorescence yield [7] and the ability to change its molecular structure with the micro-environment [8]. Furthermore, RhB is one of the most readily available and useful reagents for use as an organic laser dye. It is known that the fluorescence yield of RhB is drastically decreased by their dimerization [7], therefore, the dimerization of RhB molecules should be avoided for use as a dye laser medium [9]. Avnir et al. have shown that the photostability of dye molecules embedded in silica-glass is much higher than that of dyes in water solution and has suggested that dye molecules impregnated into silica can be used as a solid-matrix laser dye [10]. Furthermore, Anpo et al. have reported that the thermal stability of dye molecules is also dramatically enhanced by their doping into silica and Ti-Si binary oxide matrices prepared by the sol-gel method [11,12].

Although the porous materials, zeolite and mesoporous molecular sieves, are considered to be good canditates as supports for dye molecules in realizing their high dispersion state, investigations of the photophysical and photochemical properties of dye molecules adsorbed on these porous materials have not yet been fully carried out. In this study, the properties of RhB dye molecules adsorbed on zeolite and mesoporous molecular sieves have been investigated by analysis of the fluorescence spectra.

2. EXPERIMENTAL

HZSM-5 (Si/Al=12, 250 m^2g^{-1}) and silica gel (Surface area: 300 m^2g^{-1}) supplied from Tosoh Ltd. and FSM-16 (950 m^2g^{-1}), MCM-41 (1000 m^2g^{-1}) and Ti-MCM-41 (Si/Ti=80, 980 m^2g^{-1}) mesoporous molecular sieves synthesized according to procedures reported previously [13,14] were used. The XAFS spectra (XANES and EXAFS) of Ti-MCM-41 were measured at the BL-7C facility of the Photon Factory at the National Laboratory for High-Energy Physics, Tsukuba. The Ti K-edge absorption spectra were recorded in the fluorescence mode at 295 K. The normalized spectra were obtained by a procedure described in previous literature [15] and Fourier transformation was performed on k^3-weighted EXAFS oscillations in the range of 3-10 Å^{-1}.

The adsorption of RhB onto these oxide matrices was carried out from the ethanol solution of RhB and the samples were dried at 325 K for 24 h. The fluorescence spectra of RhB adsorbed onto these oxide matrices were recorded with a Shimadzu RF-5000 spectrofluorophotometer at 295 K. Observation of the fluorescence decay curves was carried out using a Hamamatsu C-4334-02 streak scope with a N_2 laser.

3. RESULTS AND DISCUSSION

RhB has three distinct molecular forms, as shown in Fig. 1 [12]. The zwitterion form of RhB (Z) dominates the equilibrium in protic solvents with the lactone form (L). The cation form (C) occurs in acidic solutions. C and Z exhibit similar absorption and fluorescence bands, except for a small shift in the fluorescence spectrum of C toward longer wavelengths. L has only very weak absorption or fluorescence bands and is colorless. Figure 2 shows the changes in the fluorescence spectra of RhB dissolved in ethanol depending on the RhB concentration. The peak of the fluorescence spectra of RhB in the diluted solution at around 565 nm is attributed to the presence of the zwitterion form of RhB (Z). The intensity of the fluorescence spectra of RhB increases with an increase in the RhB concentration and a spectral shift to longer wavelengths can also be observed. As reported previously, in the diluted solution, the Z form exists in an equilibrium between the non-

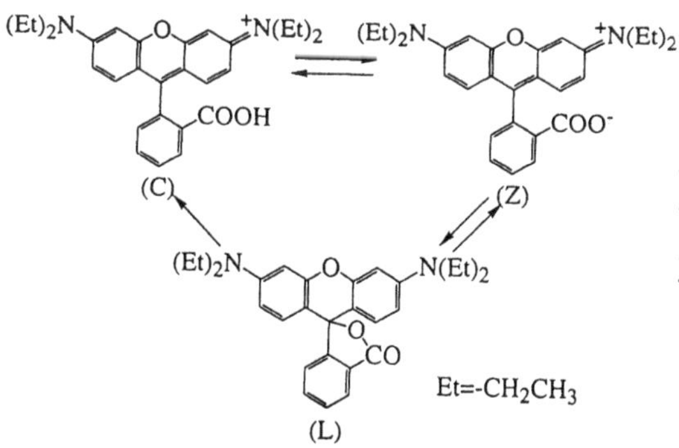

Figure 1. Equilibrating forms of RhB.
(C) cationic form
(L) lactone form
(Z) zwitterionic form

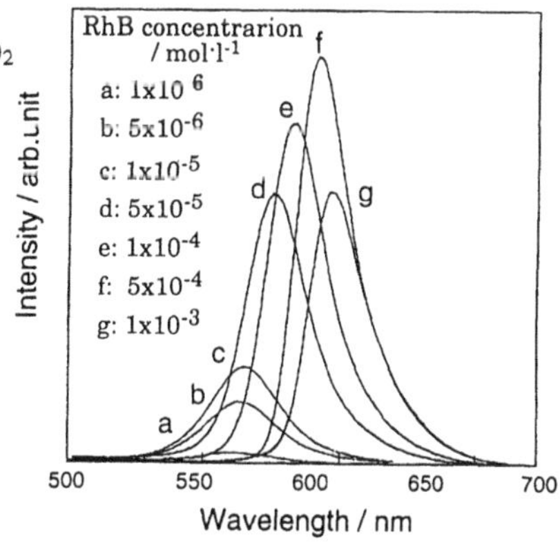

Figure 2. The effect of the concentration of RhB on the fluorescence spectra of RhB dissolved in ethanol at 295 K. Excitation wavelength: 400 nm.

fluorescent lactone form of RhB (L) [12]. With an increase in the RhB concentration, this Z=L equilibrium changes to the equilibrium between Z and the cation form of RhB (C) while the interaction between RhB molecules becomes pronounced. These changes bring about the spectral shift to the

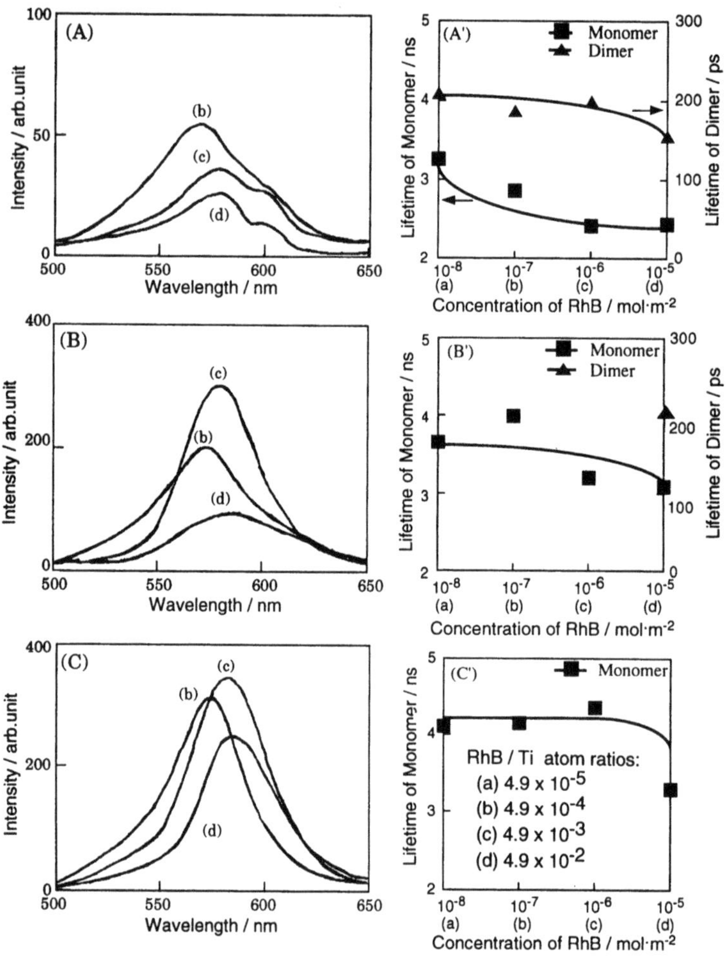

Figure 3. Fluorescence spectra of RhB adsorbed on various supports at 295 K (A-C) and the dependence of their lifetimes on the RhB concentrations (A'-C'). Systems: (A,A') RhB/ZSM-5, (B,B') RhB/MCM-41, (C,C') RhB/Ti-MCM-41. Excitation wavelength: 400 nm. Concentration of RhB: (a) 1 x10^{-8}, (b) 1 x10^{-7}, (c) 1 x10^{-6}, (d) 1 x10^{-5} mol·m^{-2}. Surface coverage with RhB estimated by assuming that the effective area of adsorbed RhB molecule is 3.48 x 10^{-19} m^2: (a) 2.1 x 10^{-3}, (b) 2.1 x 10^{-2}, (c) 2.1 x 10^{-1}, (d) 2.1.

longer wavelengths and the decrease in the intensity of fluorescence at high concentrations of 1×10^{-3} mol·l^{-1}. These characteristic optical properties can be used to monitor changes in the molecular micro-environments of RhB adsorbed on various supports.

Figure 3-A shows the fluorescence spectra of RhB adsorbed on the HZSM-5 zeolite (RhB/ZSM-5) with various RhB concentrations. An increase in the RhB concentration causes a spectral shift to longer wavelengths and decrease in the intensity of the fluorescence can be observed accompanied by the formation of a new band at around 600 nm. Although these spectral changes with the RhB/ZSM-5 system having an RhB concentration were similar to those observed with the RhB adsorbed on silica (RhB/SiO_2), the intensity was much smaller indicating that the aggregation of RhB was more dramatic on the ZSM-5 zeolite. Figure 4-A shows the fluorescence decay curve of the RhB/ZSM-5 system and the lifetimes obtained are shown in Fig. 3-A' and Table 1. The decay curves for the RhB/ZSM-5 system were of a double exponential form with lifetimes of T_1=209~155 ps and T_2=3.26~2.44 ns.

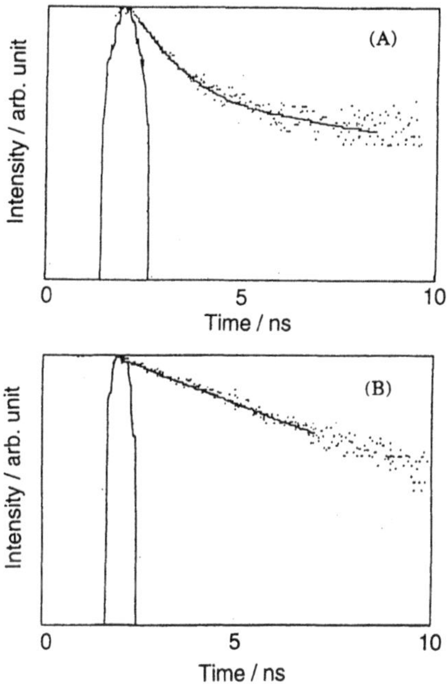

Figure 4. Fluorescence decay curve observed with the RhB/ZSM-5 (A) and RhB/MCM-41 (B) systems with RhB concentration of 1×10^{-6} mol·m^{-2}.

Table 1
Lifetimes of RhB adsorbed on various solid surfaces

RhB	Silica gel		ZSM-5		FSM-16		MCM-41		Ti-MCM-41	
(mol·m^{-2})	T_1(ps)	T_2(ns)	T_1(ps)	T_2(ns)	T_1(ps)	T_2(ns)	T_1(ps)	T_2(ns)	T_1(ps)	T_2(ns)
1×10^{-8} a)	-	3.30	209	3.26	-	3.25	-	3.65	-	4.11
1×10^{-7} b)	140	3.21	187	2.86	-	3.42	-	3.99	-	4.15
1×10^{-6} c)	138	3.16	199	2.42	-	3.21	-	3.20	-	4.35
1×10^{-5} d)	180	2.96	155	2.44	177	2.85	204	3.09	-	3.29

a)-d) Surface coverage with RhB estimated by assuming that the effective area of adsorbed RhB molecule is 3.48×10^{-19} m^2: (a) 2.1×10^{-3}, (b) 2.1×10^{-2}, (c) 2.1×10^{-1}, (d) 2.1.

d) RhB / Ti atom ratios: (a) 4.9×10^{-5}, (b) 4.9×10^{-4}, (c) 4.9×10^{-3}, (d) 4.9×10^{-2}.

The shorter lifetime species can be assigned to the RhB dimer, and the species with the longer lifetime can be assigned to the RhB monomer species in the equilibrium between the Z and C forms [11,12]. With an increase in the RhB concentration, the lifetime of the monomer species drastically decreases, suggesting that the aggregation of RhB is promoted. These results indicate that RhB having a large molecular size adsorbs on the outer surface of the microporous ZSM-5 zeolite as a mixture of the monomer and dimer species and its dispersion is very low also on silica (Scheme 1).

Figure 3-B shows the fluorescence spectra of RhB adsorbed on the MCM-41 mesoporous molecular sieve (RhB/MCM-41) with various RhB concentrations. With an increase in the RhB concentration, the fluorescence band shifts to the longer wavelengths and the intensity of the fluorescence increases at first but decreases at higher RhB concentrations. These fluorescence intensities for the RhB/MCM-41 system are much higher than those observed with the RhB/ZSM-5 and RhB/SiO_2 systems. Figure 4-B shows the fluorescence decay curve for the RhB/MCM-41 system and the lifetimes obtained by analysis of the decay curves are shown in Fig. 3-B' and Table 1. The decay curves for the RhB/MCM-41 system exist in a single exponential form with a lifetime of T_2=3.65~3.20 ns at low RhB concentration, indicating that RhB having a large molecular size can adsorb on the inside surface of the mesoporous molecular sieve as a monomer and its dispersion is much higher than those observed with the RhB/ZSM-5 and RhB/SiO_2 systems (Scheme 1). The only system at a high concentration of $1x10^{-5}$ mol·m^{-2} has a decay curve of double exponential form with lifetimes of T_1=204 ps and T_2=3.09 ns. The similar spectral changes depending on RhB concentration were also observed with the RhB adsorbed on FSM-16 mesoporous molecular sieves, as shown in Table 1.

Figure 3-C shows the fluorescence spectra of RhB adsorbed on the Ti-MCM-41 molecular sieve (RhB/Ti-MCM-41) with various RhB concentrations and the lifetimes are shown in Fig. 3-B' and Table 1. Although similar spectral changes were observed with the RhB/Ti-MCM-41 system depending on the

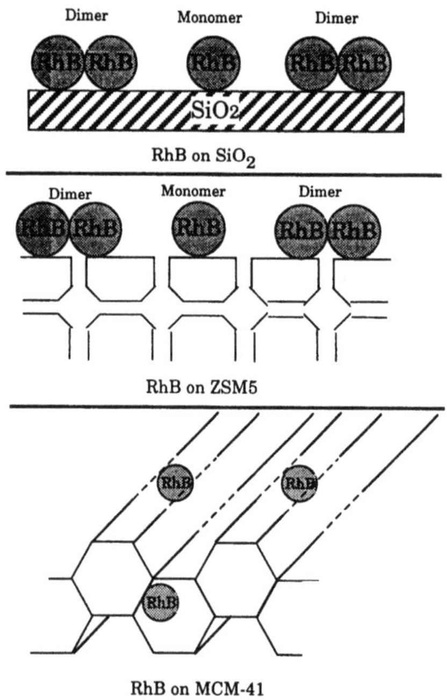

Scheme 1. Schematic diagrams of the RhB adsorbed on the various solid surfaces.

RhB concentrations as observed with the RhB/MCM-41 system, these fluorescence intensities for the RhB/Ti-MCM-41 system are higher and its lifetimes are longer than those observed with the RhB/MCM-41 system. The decay curve of a single exponential form can be observed even at a high concentration of 1x10^{-5} mol·m^{-2}. These results indicate higher dispersion can be realized with the RhB/Ti-MCM-41 system than with the RhB/MCM-41 system.

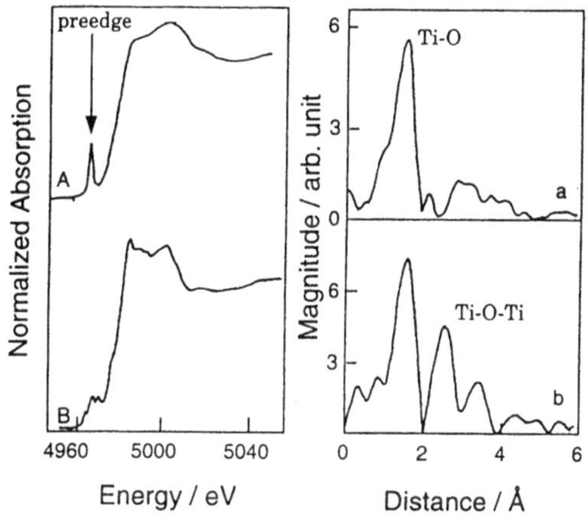

Figure 5. Ti K-edge XANES (A, B) and FT-EXAFS (a, b) spectra of Ti-MCM-41 (A, a) and TiO_2 powder (B, b).

Figure 5 shows the XAFS spectra of Ti-MCM-41. In the XANES spectra, Ti-MCM-41 shows very intense preedge peaks in contrast to bulk TiO_2. These characteristic preedge peaks indicate the presence of tetrahedrally coordinated titanium oxide moieties in the samples [15]. In the FT-EXAFS spectra, a strong peak due to the neighboring O atoms appears at about 1.5 Å (without phase correction), however, the peak due to the Ti-O-Ti moieties can not be observed with Ti-MCM-41. These results indicate that the dispersion level of the titanium oxide species is high in Ti-MCM-41 and the isolated tetrahedrally coordinated titanium oxide moieties in the mesopores can have a strong interaction with RhB molecules to be effective in anchoring RhB molecules in a highly dispersed state.

4. CONCLUSIONS

The fluorescence properties of RhB adsorbed on the zeolite and mesoporous molecular sieves were investigated to clarify the effects of the micro-environment of host-adsorbents on the electronic excited states of guest-molecules included in a restricted void spaces. It has been found that a high dispersion of dye molecules can be realized using mesoporous molecular sieves as supports. The isolated tetrahedrally coordinated titanium oxide moieties in the mesopores of the Ti-MCM-41 molecular sieves are especially useful in achieving dye molecules in a highly dispersed state. These results demonstrate that the large cavities of the mesoporous molecular sieves play an important role in realizing the high dispersion state of dye molecules and is also effective in controlling the micro-environment of adsorbed dye molecules.

ACKNOWLEDGMENT: The present work has been supported in part by the Grant-in-Aid on Priority-Area-Research on "Catalytic Chemistry of Unique Reaction Fields" (09218250) and "Electrochemistry of Ordered Interfaces" (09237258) of the Ministry of Education, Science, Sports, and Culture of Japan.

REFERENCES

1. M. Anpo, H. Yamashita, and S. Zhang, Current Opinion in Solid State & Materials Science, 1 (1996) 219.
2. M. Anpo, H. Nishiguchi, and T. Fujii, Res. Chem. Intermed., 13 (1990) 73.
3. M. Anpo (ed.), Surface Photochemistry, John Wiley & Sons, London, 1996.
4. H. Nishiguchi, K. Yukawa, H. Yamashita, and M. Anpo, Res. Chem. Intermed., 21 (1995) 885.
5. H. Nishiguchi, K. Yukawa, H. Yamashita, and M. Anpo, J. Photochem. Photobiol. A: Chem., 99 (1995) 1.
6. H. Yamashita, N. Sato, M. Anpo, T. Nakajima, M. Hada, and H. Nakatsuji, Stud. Surf. Sci. Catal., 105 (1996) 1141.
7. D. A. Hinckley and D. G. Seybold, Spectrochimica Acta, 44 (1989) 1053.
8. T. L. Chang and H. C. Cheung, J. Phys. Chem., 96 (1992) 4874.
9. F. Wakai, Y. Kodama, and S. Sakaguchi, J. Am. Ceram. Soc., 73 (1990) 457.
10. D. Avnir, Acc. Chem. Res., 28 (1995) 328.
11. N. Negishi, H. Yamashita, and M. Anpo, Chem. Lett., (1994) 1029.
12. N. Negishi, M. Fujino, H. Yamashita, M. A. Fox, and M. Anpo, Langmuir, 10 (1994) 1772.
13. K. A. Koyano and T. Tatsumi, Microporous Materials, 10 (1997) 259.
14. S. Inagaki, Y. Fukushima, and K. Kuroda, J. Chem. Soc., Chem. Commun., (1993) 680.
15. S. G. Zhang, Y. Fujii, H. Yamashita, K. Koyano, T. Tatsumi, and M. Anpo, Chem. Lett., (1997) 659.

MESOPOROUS MOLECULAR SIEVES 1998
Studies in Surface Science and Catalysis, Vol. 117
L. Bonneviot, F. Béland, C. Danumah, S. Giasson and S. Kaliaguine (Editors)

Fe and Co modifications of siliceous MCM-41 and 48 using direct or post-synthesis methods

Bousselham Echchahed, Ali-Reza Badiei, François Béland and Laurent Bonneviot

Department of Chemistry, CERPIC, Laval University, G1K 7P 4, Qc, Canada

Iron and cobalt modification of siliceous M41S type of mesoporous molecular sieves have been prepared by direct and post-synthesis routes. The solids were characterized using BET, XRD, FTIR, UV-visible and XANES spectroscopies. The iron modification leads to a phase dependent metal loading synthesis crossing the different lamellar, cubic and hexagonal mesophases of the M41S familly. Up to 6.6 % Fe/(Fe+Si) mole ratio, the iron(III) ions occupy tetrahedral sites in the as-synthesized form; however, the calcination is a critical step for iron distribution. Co(III) ions which are not stable in tetrahedral symmetry are incorporated using a new grafting approach selective to the surface silanolato groups of the as-synthesized form. Novel information about the silanolato group and template counterion concentrations is provided.

1. INTRODUCTION

The silicate and aluminosilicate materials of M41S types prepared in presence of surfactants (quaternary ammonium cations with at least one hydrophobic organic long chain) are mesophases of either lamellar; MCM-50, cubic; MCM-48, or hexagonal; MCM-41, symmetry [1]. In the as-synthesized form, the surfactants counterbalance the silanolato surface groups. It is therefore necessary to remove the template to obtain, from the cubic or the hexagonal forms, novel materials which exhibit very high surface area (~ 1000 m^2/g) with a narrow pore size distribution that can be tuned in the nanoscale range (2 - 8 nm) allowing high diffusivity of large molecules [2]. Their functionalization with transition metal ions such as the environment friendly iron (III) is particularly interesting in generating catalytic properties as for aromatic alkylation, alkene isomerization [3]. However, little is known about i) the effective metal location and symmetry in these systems, ii) the capacity of these structures to accommodate heteroatoms, particularly transition metal ions (TMIs), and iii) their effect on the mesophase stability and nature.

Not withstanding, much have been done with titanium that confers these structures with very selective oxidative properties in presence of the environment friendly oxidant, H_2O_2 [4]. The activity is related to tetrahedral Ti sites localized within the pore walls. To circumvent this drawback of the direct synthesis approach, one may force the TMIs to lie on the surface using a post-modification strategy such as the grafting of reactive complexes that react with surface silanol groups. The grafting of the biscyclopentadienyldichlorotitanium(IV) indeed leads to a more active Ti-MCM-41 since after calcination most of the titanium stay *at* the surface [5, 6]. In addition, the incorporation limit for direct synthesis which might be estimated to be ca. 3% according to various reports on Ti-MCM-41 and Ti-HMS materials [4, 7], is much higher for grafting, up to ca. 15% [5]. This is related to the high silanol concentration in these systems (ca. 3 OH/nm^2) [8]. However according to the capacity of amorphous porous silicas to host tetrahedral titanium [9], the incorporation level using direct synthesis could be as high as 15% as well.

Iron (III) ions which are known to easily adopt tetrahedral symmetry such as in silicalite-1 [10] was chosen as a probe for direct synthesis incorporation. By contrast, cobalt (III) more reluctant for such symmetry, generates very well defined inert octahedral complexes such as the *trans*-$[Co(en)_2Cl_2]^+$ which are perfect probes for surface reactant groups. These two approaches are monitored here using the techniques cited in the abstract.

More information on the structure of the sites of iron (III) and cobalt (III) in MCM-41 and MCM-48 is available in other papers [11, 6].

2. EXPERIMENTAL

3.1 Sample preparations

Various iron modified siliceous M41S were prepared according to a direct synthesis method based on a sol-gel route using a solution of cetyltrimethylammonium ($CTMA^+$, 75%Cl^- and 25% OH^-), tetraethylortosilicate (TEOS), tetramethylammonium silicate and iron nitrate, $Fe(NO_3)_3.9H_2O$. The gel composition is $1SiO_2$, 0.48 $(CTMA)_2O$, 0.083 $(TMA)_2O$, xFe_2O_3, 65 H_2O with x = 0.05 - 0.10. This recipe leads to lamellar mesophases in the absence of iron; for blank experiments, the pure silica analogs were prepared according to the procedure of Beck et al. [2]. After a hydrothermal treatment at 100°C for 48 hours, the surfactant was removed using a refluxing solution of NH_4Cl in ethanol (monitored by FT-IR , [11]). for cobalt ion exchange experiments, the pure MCM-41 was prepared according to Reddy and Song [12] using a gel composition of $1SiO_2$, 0.324 CTMABr, 0.0864 Na_2O, 0.0438 TMA_2O and 63.3 H_2O. The green *trans*-$[Co(en)_2Cl_2]^+$ was prepared according to the literature [13]. 0.5 g of the dried as-synthesized MCM-41 was treated in 50 ml of the latter complex in ethanol (95%) solution at 47°C for 20 minutes. The green solution turns to purple and, a pink solid was filtered off. Chemical analysis were performed on solutions of the digested solid and the washing solutions.

2.2 Characterization techniques

The BET measurements were performed using a Omnisorb 100. The XRD was obtained from a Phillips spectrometer. The FT-IR were recorded on a Bomem in a transmission mode using samples pressed into self supported pellets. The UV-visible reflectance spectra were recorded on a Perkin-Elmer Lambda 5 equipped with a Harrick reflectance attachment. The spectra were recorded on sample diluted in MgO and reported using the Kubelka-Munk Function $F(R_\infty)$. The XANES data were collected in transmission mode at LURE as described before for EXAFS [6] with a smaller energy step , i.e, 0.2 eV from 7100 to 7200 eV. The edge normalization was calculated using a standard software [14].

3. RESULTS AND DISCUSSION

4.1 Fe loading effect on the M41S mesophase

In a previous report, we have shown that at low Fe loadings (1.7 - 2.6%), the stable phase is MCM-48. In these samples the as-synthesized forms contain tetrahedral sites according to XANES, consistent with UV-visible spectroscopy data. In the following, the loading range has been widen from 0 up to 6.6 %. The chemical analysis and XRD data are gathered in Table 1. The chemical analysis reveals that the Fe/(Fe+Si) iron fraction is systematically higher in the solid than in the gel for the first six samples. There is indeed negligible iron left in the supernatant liquid which still

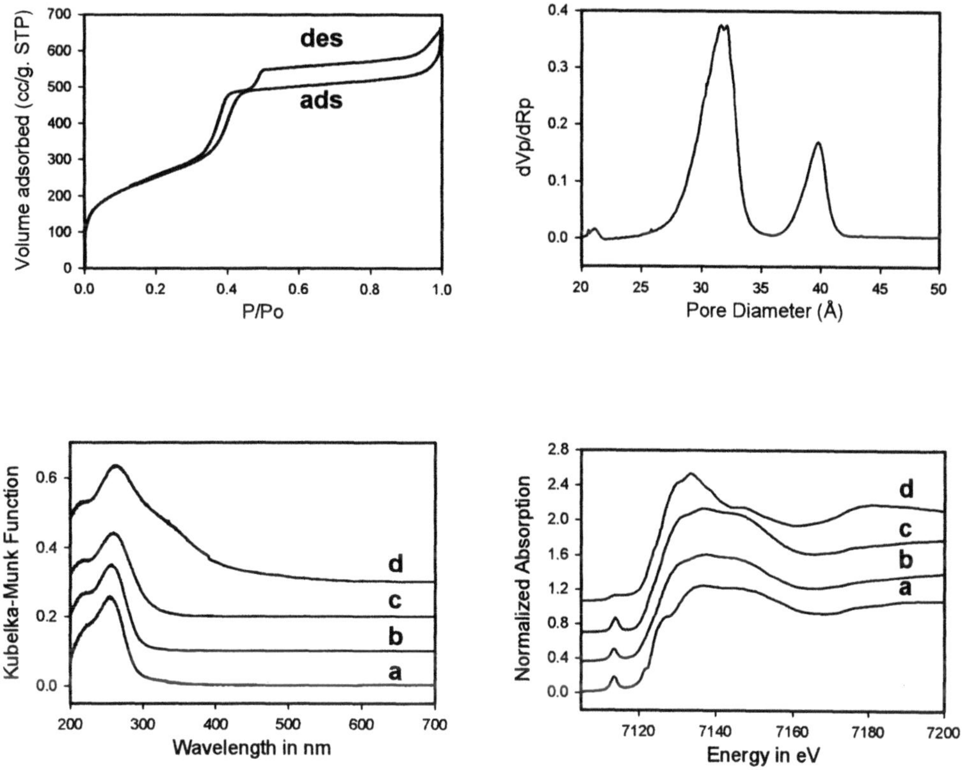

Figure 1. (top left) N_2 adsorption-desorption isotherms for Fe-MCM-41 with Si/Fe = 26:1 ion exchanged and calcined; (top right) B.J.H pore size distributions for Fe-MCM-41 with Si/Fe = 26:1 treated with NH_4Cl and calcined; (bottom left) Diffuse reflectance UV-vis spectra of Fe-MCM-41 with Si/Fe = 26:1 diluted (20%) in MgO using pure MgO as a reference: a) as-synthesized, b) treated with NH_4Cl, c) as in b) and calcined, d) calcined without template removal; (bottom right) XANES spectra of a) $FePO_4$, b) as-synthesized Fe-MCM-48 with Si/Fe = 61:1, c) as-synthesized Fe-MCM-41 with Si/Fe = 18.5:1, d) Fe_2O_3.

contains silicon for these syntheses. The higher loaded sample shows the reverse trend; this is an indication that the affinity of the siliceous mesophase for iron has a limit. More syntheses are necessary to reach this limit. All these materials exhibit a narrow distribution of pores centered at 30 ± 1 Å. However, it should be noticed that, on the desorption curves, an apparent bimodal site distribution arising at ca. 40 ± 1 Å develops when the metal loading is increased. By contrast, the adsorption curve does exhibit a single narrow pore size distribution. This apparent discrepancy is related to some distortion of the cylindrical shape of the pore in presence of iron or nitrate ions within the structure during synthesis. It is worth noting that the pure MCM-48 exhibits also an apparent pore distribution at 40 ± 1 Å, though the intensity of the second peak is rather small (<1/10 of the first peak intensity). TEM pictures not shown here indeed reveals the presence of some distorted channels among very regular ones.

The state of iron was monitored using XANES in the as-synthesized forms. It was found that the edge was very similar in all the range of metal loading investigated here and, whatever the type of mesophase produced. The pre-edge height and the edge step are consistent with a tetrahedral symmetry. The EXAFS data are consistent with a tetrahedral site occupancy inside the siliceous pore walls of the MCM-41 [6]. UV-visible spectra of the Fe-MCM-41 series are very similar to those of the Fe-MCM-48 reported earlier [11]. Nonetheless, the transition were more intense since the Fe loading was higher. The charge transfer bands is shifted toward high energy for the as-synthesized form consistent with a tetrahedral symmetry for the Fe(III) ions. After washing with NH_4Cl in ethanol, a more pronounced bathochrome shift is observed for the high loaded MCM-41 in comparison to the low loaded MCM-48 [11]. In addition, the calcined form exhibits a strong shoulder at even lower energy (higher wavelength, Figure 1). In the former case, this is consistent with the formation of octahedral Fe species extracted from their pore wall sites into the channel. The latter giving rise to small iron oxide clusters whose UV response arises at much higher energy than expected for bulk Fe_2O_3 [15]. The blue shift of the band is related to a quantum size effect. Accordingly, the more pronounced bathochrome shift for higher Fe loading is consistent with the formation of larger iron oxide clusters more likely entrapped within the channels of the MCM mesophase.

Table 1
Analytical and structure characteristics of the Fe-M41S series in function of Fe loading

gel Fe composition in %	Solid Fe composition % [a]	XRD type of structure [b]
0.5	0.86	MCM-48
1.0	1.7	MCM-48
1.5	2.6	MCM-48
2.0	2.9	MCM-41
2.5	3.8	MCM-41
5.0	5.4	MCM-41
10.0	6.6	MCM-41

a) 5% of relative accuracy from atomic absorption measurements on solution of digested solids
b) d_{211} of 38 ± 1 Å in the 48 series and d_{100} of 40 ± 1 Å in the 41 series, both in their calcined form.

4.1 Cobalt Ion exchanged template in as-synthesized MCM-41

The removal of the template is usually performed before post-modification to liberate the surface silanol groups. However, it is known that the surfactant can be removed in mild conditions using NH_4NO_3 [17], NH_4Cl [11] and eventually NaCl [16] ethanol solutions. It is clear from IR spectroscopy that NH_4^+ ions are retained in the channel during such washing treatment [11]. Sodium ions should have been also retained when NaCl is used. These characteristics are definitively related to the ion exchange properties of the surface silanolato groups of the pore walls (reaction 1). Accordingly, a positively charged complex such as $[Co(en)_2Cl_2]^+$ should also be capable to ion exchanged the positively charged surfactant (reaction 2). In addition, due to the reactivity of the Co-Cl bonds toward hydrolysis one might wonder about the possibility that a substitution of Cl^- with $\equiv SiO^-$, a surface silanolato may occur (reaction 3) and eventually followed by further evolution (for instance reaction 4). The reactions under discussion can be written as follows:

$$\{\equiv SiO^-, CTMA^+\} + NH_4^+ \longrightarrow \{\equiv SiO^-, NH_4^+\} + CTMA^+ \quad (1)$$

$$\{\equiv SiO^-, CTMA^+\} + [Co(en)_2Cl_2]^+ \longrightarrow \{\equiv SiO^-, [Co(en)_2Cl_2]^+\} + CTMA^+ \quad (2)$$

$$\{\equiv SiO^-, [Co(en)_2Cl_2]^+\} \longrightarrow \{[Co(en)_2(\equiv SiO)Cl]^+, Cl^-]\} \quad (3)$$

$$\{[Co(en)_2(\equiv SiO)Cl]^+, Cl^-]\} + H_2O \longrightarrow \{[Co(en)_2(\equiv SiO)(H_2O)]^{2+}, 2\,Cl^-]\} \quad (4)$$

The reactions performed in water leads to a drastic decrease of the XRD peaks of the materials. The subsequent experiments were performed in EtOH solution to avoid such a drawback. The BET surface area were 1050, 990 and 760 m^2/g for the calcined MCM-41 (500°C for 3 hours), the NH_4NO_3 treated and calcined MCM-41 and, the MCM-41 exchanged using 5×10^{-3}M $[Co(en)_2Cl_2]Cl$ ethanol solution. The XRD of an ion exchanged MCM-41 was quite similar to that of a directly calcined material (Figure 2). The decrease of BET surface area may be due to the presence of the cobalt complexes entrapped inside the pore channels; this point still has to be clarified. The chemical analysis data (Table 2) shows that a significant amount of cobalt is adsorbed in the MCM-41 while most the template has been removed. Consistently, the IR C-H bending and stretching bands attributed to $CTMA^+$ decrease drastically. The exchange yield of cobalt is quite high. A second treatment with the same concentration of complex solution leads to an increase of only 10% of the amount of adsorbed cobalt.

It is worth noting that much more surfactant molecules are removed than cobalt ions introduced in the MCM-41, contrary to the expectation with regard to equation 1 and 2. In fact, there is about three times more template than expected. The analysis of the exchanged solution reveals the presence of bromide in a quantity that within the measurement accuracy balances the missing charges; the full report is published elsewhere [18]. In fact, for one exchanged $CTMA^+$ ion, about two CTMABr are removed. From the number of surfactant cations and bromide ions removed simultaneously and, independently, from the number of adsorbed cobalt complexes, the surface density of the $\equiv SiO^-$ groups is found to be ca. 0.55 ± 0.05 SiO^-/nm^2 (for a surface area of 760 m^2/g,

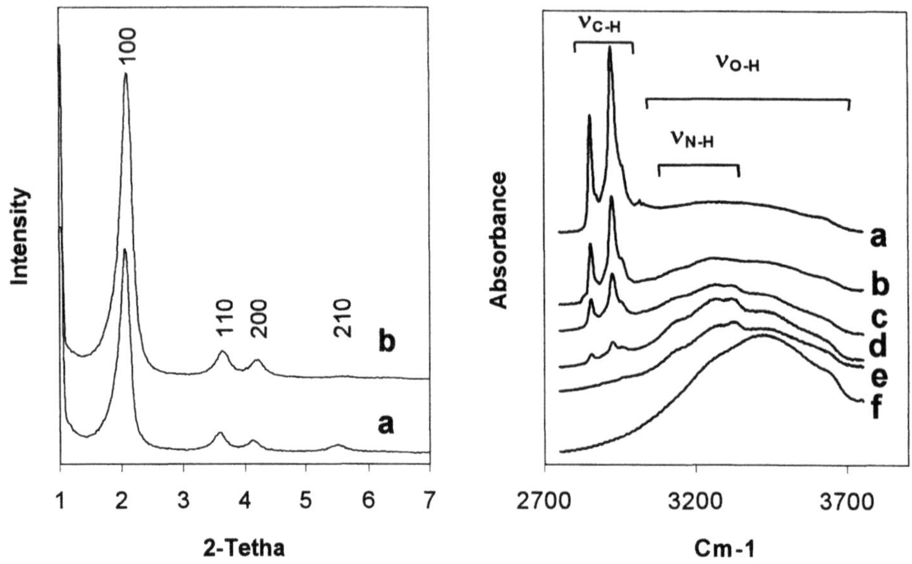

Figure 2. (left) XRD patterns of MCM41 a) as-synthesized and b) ion exchanged with *trans*-$[Co(en)_2Cl_2]^+$; (right) FT-IR spectra of MCM-41 a) as-synthesized b) 1.96 x 10^{-3}M, c) 4 x 10^{-3}M, d) 5 x 10^{-3}M, e) 5 x 10^{-3} M (treated twice) and, f) treated with a NH_4NO_3 solution followed by calcination; ν stands for stretching vibrational modes of the $-CH_2-$, $-CH_3$, $-NH_2$, -OH groups and H_2O molecule.

see above). This density matches approximately the number of isolated silanol groups (ca.0.7 SiO^- /nm^2) that react with trimethylchlorosilane [8]. These figures show that only a small fraction (ca. 1/6) of the silanol groups are effectively involved in the electrical balance of the positive charges of the electrical double layer. Consequently, this double layer between the surfactant (S^+) and the inorganic (I^-) components viewed merely as (S^+,I^-) should be now considered as a ($3S^+$, $2X^-$, I^-) association for a MCM-41 material.

When the Co ion exchanged MCM-41 is washed with a NH_4NO_3 solution, 90% of the complex is retained in the solid while IR spectroscopy clearly shows the presence of a single peak at 1380 cm^{-1} characteristic of a non-coordinated nitrate. This is consistent with grafted complexes from which outersphere Cl^- anions are displaced by NO_3^- anions. According to EXAFS data reported elsewhere [6], there is no chloride ion left in the coordination sphere of the adsorbed cobalt and only one Co-O-Si formed. Therefore, the reaction 4 is likely to represent the last step of the adsorption process of the *trans*-$[Co(en)_2Cl_2]^+$. Consequently, the nitrate anion exchange can be written as follows:

$$\{[Co(en)_2(\equiv SiO)(H_2O)]^{2+}, 2\,Cl^-]\} + 2\,NO_3^- \rightarrow \{[Co(en)_2(\equiv SiO)(H_2O)]^{2+}, 2\,NO_3^-]\} + 2\,Cl^- \quad (5)$$

The removal of at least one chlorine could have been effectively anticipated by the coloration change occurring during the exchange process. The green *trans*-$[Co(en)_2Cl_2]^+$ ethanol solution gives a pink Co exchanged MCM-41 solid. UV-visible spectroscopy shows effectively that the bands of the original complex have shifted downward in energy, consistent with the substitution of Cl^- by OH_2, OH^- or SiO^- ligands. The reaction performed at neutral pH readily favors the substitution of chlorine by basic SiO^- groups whose concentration is much higher than the OH^- concentration of similar basicity. In addition, the spectra reveals that a *trans* to *cis* isomerization has occurred [18].

Reaction 5 shows that the genuine as-synthesized MCM-41 which is a cation exchanger is converted to an anion exchanger after cobalt grafting; both mechanisms operate in neutral ethanol solution.

Table 2
Analytical characteristics of the MCM-41 treated with either cobalt complex or NH_4Cl solutions.

Co complex or NH_4^+ conc. x 10^{-3} M	Template removed mol % per Si [d]	Incorporated Co mol % per Si [e]	Template removal yield % [f]	Cobalt exchange yield % [g]
5.0 [a]	11.0	4.0	94	68
5.0 [a, b]	13.9	4.4	98	37 (7)[h]
50 [c]	14.3	0.0	99	-

a) trans-$[Co(en)_2Cl_2]^+$ complex in ethanol,
b) MCM-41 treated twice with a 5 x 10^{-3} M of a trans-$[Co(en)_2Cl_2]^+$ solution,
c) MCM-41 treated twice with a NH_4Cl ethanol solution (see text),
d) from colorimetry (see ref. 20, relative accuracy of ca. 2%), figures consistent with literature data (see refs. 1, 27, 28),
e) from atomic absorption (absolute error of ca. 0.2%),
f) from the FT-IR spectra (see text, relative accuracy of 4%),
g) Co exchanged over total Co calculated from columns 1 and 3 of the table,
h) yield for both the first and the second exchange treatment the latter in parenthesis (relative accuracy 3%).

5. CONCLUSION

A series of Fe-M41S have been prepared from the same direct synthesis route by merely changing the Fe loading. In addition, a cobalt modification of the MCM-41 siliceous form has been obtained using a ion exchange process with neutral ethanol solution of the *trans*-$[Co(en)_2Cl_2]^+$ complex. The quality of mesoporous molecular sieves material has been monitored using BET and

XRD measurments while the metal state has been investigated using FT-IR, UV-visible and XANES spectroscopies. XANES studies at the Fe K-edge show that most of the iron ions occupies tetrahedral sites in the as-synthesized form. In addition, for a given surfactant/silicon ratio, the mesophase evolves from the lamellar to the cubic and then to the hexagonal forms when the metal loading is varied in the range 0 to 6.6 Fe(Fe+Si) mol%. After calcination, a fair fraction of iron is extracted from the siliceous walls and aggregates into nanoscale oxide particles located inside the channel of the porous structure. The removal of the template using NH_4Cl ethanol solution before calcination improves the retention of iron in the pore walls. By contrast to iron, no direct incorporation of cobalt has been obtained so far but post-synthesis using direct template exchange by cobalt complexes can be achieved at 47°C without destruction of the MCM-41 channel structure. The *trans*-$[Co(en)_2Cl_2]^+$complex undergo a loss of two chloride ligands that are displaced by a surface silanolato group and more likely a water molecule. The so formed $\{[Co(en)_2(\equiv SiO)(H_2O)]^{2+}, 2\ Cl^-]\}$ are accompanied by non-coordinated anions that can be anion exchanged. Chemical analysis demonstrates that the as-synthesized MCM-41 contains a high concentration of bromide ions that has not been noticed in the literature since the discovery of these materials.

REFERENCES

1. C. T. Kresge, M. E. Leonowicz, J. C. Vartuli, W. J. Roth, J. S. Beck, Nature, 359, (1992), 710; A. Monnier, F. Schüth, Q. Kumar, D. Margolese, R. Maxwell, G. D. Stucky, M. Krishnamurty, P. Petroff, A. Firouzi, M. Janicke, B. F. Chemelka, Science, 261 (1993) 1299.
2. J. S. Beck, J. C. Vartuli, W. J. Roth, M. E. Leonwicz, C. T. Kresge, K. D. Schmitt, C. T.-W. Chu, D. H. Olson, E. W. Sheppard, S. B. McCullen, J. B. Higgins, J. L. Schlenker, J. Am. Chem. Soc., 114 (1992) 10834.
3. J.-B. Kim and T. Inui, Catal. Lett., 36 (1996) 255.
4. A. Corma, Chem. Rev., 97 (1997) 2373; D. Trong On, M. P. Kapoor, P. N. Joshi, S, Kaliaguine and L. Bonneviot, Catal. Letters, 44 (1997) 171.
5. T. Maschmeyer, F. Rey, G. Sankar, J. M. Thomas, Nature, 378 (1995) 159.
6. F. Béland, B. Echchahed, A.-R. Badiei and L. Bonneviot, in this book.
7. P. T. Tanev, M. Chibwe, T. J. Pinnavaia, Nature, 368 (1994) 321.; R. Hutter, D. C. M. Dutoit, T. Mallat, M. Schneider, A. Baiker, J. Chem. Soc., Chem. Commun., (1995) 163.
8. X. S. Zhao, G. Q. Lu, A. K. Whittaker, G. J. Millar, H. Y. Zhu, J. Phys. Chem. B, 101 (1997) 6525.
9. Z. Liu and R. J. Davis, J. Phys. Chem., 98 (1994) 1253.
10. S. Bordiga, R. Buzzoni, F. Geobaldo, C. Lamberti, E. Giamello, A. Zecchina, G. Leofani, G. Petrini, G. Tozzola, G. Vlaic, J. Catal. 148 (1996) 486.
11. B. Echchahed, A. Moen., D. Nicholson, L. Bonneviot, Chem. Mater., 9 (1997) 1716.
12. K. M. Reddy, C. Song, Catal. Lett., 36 (1996) 103.
13. J. C. Bailar, C. W. Rollinson, Inorg. Syn., 2 (1960) 183.
14. A. Michalowicz, EXAFS pour le MAC, in " Logiciels pour la Chimie", Société Française de Chimie, Paris, 1991, 101.
15. T. Abe, Y. Tachibana, T. Uematsu, M. Iwamoto, J. Chem. Soc., Chem. Commun., (1995) 1617.
16. A. Tuel, S. Gontier, Chem. Mater., 8 (1996), 114.
17. D. D. Whitehurst, U.S. Patent, 5, 143, 879, 1992.
18. A. -R. Badiei and L. Bonneviot, Inorg. Chem., submitted

MESOPOROUS MOLECULAR SIEVES 1998
Studies in Surface Science and Catalysis, Vol. 117
L. Bonneviot, F. Béland, C. Danumah, S. Giasson and S. Kaliaguine (Editors)

EXAFS multiple scattering characterization of metal modified mesoporous sieves

François Béland, Bousselham Echchahed, Ali-Reza Badiei and Laurent Bonneviot

Department of Chemistry, CERPIC, Laval University, G1K 7P4, Qc, Canada

Three examples of cobalt (III), iron (III) and titanium (IV) ions incorporated in siliceous MCM-41 mesoporous molecular sieves prepared by direct or post-synthesis modification routes are documented using multiple scattering (MS) analysis of Extended X-ray Absorption Fine Structure (EXAFS) at the respective metal K-edges. Coordination numbers, bond lengths and bond angles are characterized and related to the metal location in the bulk or at the surface of the pore walls; an emphasis is put on the number and the angle value of the M-O-Si bridges.

1. INTRODUCTION

Owing to their high area (~ 1000 m^2/g) and their very well defined pore size distribution tunable from 20 to 80 Å [1], siliceous mesoporous molecular sieves are very attractive supports for transition metal ions (TMIs) to generate novel types of metal based catalysts [2]. The pore walls are made of cross-linked tetrahedral $[SiO_4]^{4-}$ units such as in zeolites, but unlike the latter, their wall structure is amorphous. Consequently, their capacity to accommodate heteroatoms such as TMIs is expected to be higher than zeolites and their siliceous forms, i. e. silicalites. For instance, Ti^{4+} ions may substitute for Si^{4+} ions for a maximum of 1.8 % in the silicalite-1 (S-1) of ZSM-5 type of structure [3] while the incorporation level in tetrahedral sites rises up to 12% in amorphous silica [4] (here and in the following % account for M/[M+Si] mole fraction). There have been indeed several reports of HMS or MCM-41 containing up to 3% Ti in tetrahedral symmetry. The upper limit using direct synthesis has not yet been established but in the case of iron, it is clear that an as-synthesized siliceous MCM-41 mesophase may accommodate up to ~7% Fe (III) in tetrahedral sites according to XANES [5,6]. Surprisingly, there have been very few thorough investigations of the actual state of iron in these materials [5]. Not withstanding it has been shown that the Ti tetrahedral sites are not equivalent in a silicalite-1 and in an amorphous porous silica (APS) according to UV-visible and EXAFS spectroscopies [7]. From MS-EXAFS analysis at the Ti K-edge [3], it was shown that the difference lies in the angle of the Si-O-Ti bridges. In that connection, it is legitimate to investigate how MMS pore walls may host titanium and other TMIs in comparison with S-1 and APS host matrices.

To address this question, MS-EXAFS using the FEFF6 code has been applied to Co, Fe and Ti K-edges of MCM-41 obtained by post or direct synthesis; comparisons are made with Ti silicalite-1. More detailed information on the materials studied here is available in other articles [3, 5, 6].

2. BACKGROUND

2.1 Principle

During an X-ray absorption experiment, the photon energy is set below the absorption edge of an atom absorber (here, Ti Fe or Co), and scanned over ca. 1000 eV. At the K-edges, a 1s electron is either promoted into empty localized states close to the edge or ejected in the continuum above the edge. In the latter case, the incident X-ray photon energy, E, has to be larger than the energy threshold, E_o [8]. Then, an outgoing photoelectron wave is produced and back-scattered by the atoms located in the neighborhood of the absorber atom. An interference pattern is generated resulting in oscillations of the overall X-ray absorption process called EXAFS. The latter contains information on distances and bond angles around the absorber. A single scattering event takes into account only single back and forth photoelectron wave pathway. For instance, the M→O→M SS event occuring in a first coordination shell containing oxygen atoms is described by the following equation:

$$\chi_{2,O} = N_{2,O}\, S_0^2\, \frac{|f_{eff}(\pi,k,R_{MO})|}{kR_{ABj}^{\;2}} \exp(-2\sigma_{MO}^2 k^2) \exp(-2R_{MO}/\lambda) \sin(2kR_{MO}+2\delta(k)] + \phi_O[\pi,k]) \quad (1)$$

where $k = (8\pi^2 m[E-E_o]/h^2)^{1/2}$ is the photoelectron wave vector; R_{MO} is the distance between M absorber and O scatterers; $N_{2,O}$ the number of equivalent O scatterers; $f_{eff}(\pi, k, R_{MO})$ is the effective curved-wave back-scattering amplitude of the M-O pairs; $\delta(k)$ and $\phi_O(\pi, k)$ are the effective curved-wave phase shifts [9] for M and O, respectively; S_o is the many-body amplitude reduction factor; σ_{MO}, is the Debye-Waller factor (DWF) assuming the harmonic oscillator approximation; and λ is the mean free path of the electron [8]. The same formalism applies to other atoms differenciated by their nature (ex: N or Cl) or by their distance only (ex. longer R_{MO}). Each contribution is added to the other leading to a summation of superimposed oscillatory phenomena. When a second or further coordination shells of atoms are at stake, multiple scattering events are to be taken into account in the summation. Their presence is due to intervening atoms between the absorber and the scatterer atoms acting as secondary emitter-receivers.

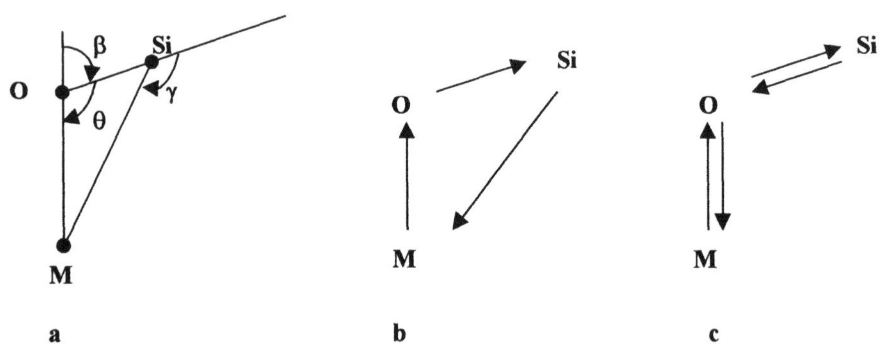

Sheme 1. a) angle definition in a M, O, Si triad of atoms and photoelectron route for dominant b) 3 and c) 4 leg pathways

Intra first shell MS events such as M→O→M→O→M indeed occur, however they are negligible. For M-O-Si type of bridges, the three and four leg pathways such as M→O→Si→M and M→O→Si→O→M respectively (**scheme 1b,c**) are the most intense and, may easily dominate the signal for M-O-Si angles, θ, larger than 120° [10]. The multiple scattering formalism includes bond angles and each interatomic distance implied in the path scheme and is derived from the single scattering formalism. For instance, the 3-leg pathway cited above is written in terms of the R_{MO}, R_{OSi}, R_{SiM} interatomic distances of each atom pairs:

$$\chi_3(k) = -N_3 S_0^2 \frac{2\cos(\beta+\gamma)\,|f_O(\beta,k,R_{MO})|\,|f_{Si}(\gamma,k,R_{MSi})|}{kR_{MO}R_{MSi}R_{OSi}} \times$$

$$\exp(-2\sigma_3^2k^2)\exp(-2R_3/\lambda)\sin(2kR_3+2\delta[k]+\phi_O[\beta,k]+\phi_{Si}[\gamma,k]) \qquad (2)$$

where, N_3 refers to the number of equivalent M-O-Si bridges, β and γ are the scattering angles defined in **scheme 1a**; σ_3 is the effective DWF factor in the harmonic oscillator approximation and $R_3 = (R_{MO} + R_{OSi} + R_{SiM})/2$. Ignoring specific normal modes and correlated atom motions, one expects a cumulative effect on the DWF so that the effective value is given by [8]:

$$\sigma_3^2 = \sigma_{MO}^2 + \sigma_{OSi}^2 + \sigma_{MSi}^2 \qquad (3)$$

For the 4-leg pathways, the formalism is again very similar with a different effective DWF, σ_4 and a scattering angle, γ, equal to π. Noticeably, for a given M-O-Si bridge, the geometrical constraint are the same as for the corresponding 3 leg pathway (scheme 1). In practice, the FEFF6 is an *ab initio* code that calculates the most significant pathways and by summation generates the EXAFS signal from the 3D coordinates of a given atom cluster. The work consist in matching the *ab initio* EXAFS with the experimental one by changing the cluster structure and adjusting the DWF associated to each path until a meaningful solution is found. In fact it is a model dependent approach based on the chemical and physical appropriateness of the solution.

3. EXPERIMENTAL

3.1 Sample preparations

Titanium silicalites were prepared from a standard sol-gel process using TPAOH as template, TEOS and TEOT as silica and titanium sources. The gel was treated in a Teflon lined autoclave at 175°C [3]. A very similar recipe was followed for Fe-S-1 except that the iron source was $Fe(NO_3)_3.9H_2O$ and the gel composition was 1: 0.02: 0.33TPAOH: 10 for SiO_2: Fe_2O_3: TPAOH: H_2O [11]. The siliceous MCM-41 was synthesized according to the literature using cetylammonium bromide (CTMABr) as templating surfactant, [1]. The Fe-MCM-41 sample was prepared according to a direct synthesis method reported earlier for the

1.5%-Fe-MCM-48. The only change was a higher Fe loading, actually Fe/[Fe + Si] was 5 % Fe/[Fe + Si] [5,6]. A 3.6 % Ti-MCM-41 was prepared from the biscyclopentadienyldichlorotitane(IV) grafted to a calcined form of a siliceous MCM-41 according to literature [12].

3.2 X-Ray absorption spectroscopy data collection

The measurements were recorded in transmission mode at the EXAFS IV station at LURE (Orsay-France). The synchrotron white radiation emitted by the DCI storage ring was monochromatized with a Si(111) double-crystal monochromator. The incident and transmitted flux were measured using low air pressure ionization chambers. The data were scanned stepwise (2eV steps) over a range of 1000 eV starting from 4800, 7000 and 7600 eV for Ti , Fe and Co K-edges , respectively. The solid samples were diluted in a fine boron nitride powder and pressed into self supported pellet such as the absorption step at the edge, $\Delta\mu$, was in the range 0.5-1. The measurements were performed at room temperature. All the samples were stable in these conditions of measurements for several hours under the beam.

3.3 EXAFS extraction, FT calculations and MS EXAFS simulations

All the data treatments were performed using a standard set of software as previously reported [3, 5, 13]. The EXAFS spectra were extracted by fitting the absorption coefficient with a polynomial curve of 5th order and normalized to the edge step. E_0 was taken at the inflexion point of the edge step. The Fourier transform was performed using a parametrized Kaiser window (37% Gaussian and 63% square functions) between k= 3.0 and 14.0 $Å^{-1}$ (otherwise stated). The back-Fourier transforms were calculated using the reciprocal window function and typically calculated in the r ranges 0.8-2.5, 2.5-3.6 and 0.8-3.6 Å where the first, second and both shells arise respectively. The theoretical amplitude and phase functions were calculated using FEFF6 applied to the cartesian coordinates extracted from crystal structure when available (model compounds) or from computer generated atom clusters, such as $[M(OSiO_3)_4]$ for a tetrahedral metal site in a siliceous matrix. The fits were performed by a standard routine using the mean square deviation minimization in k space [13]. The single scattering (SS) theory was used for the first coordination shell. The fit including the second and farther shells were performed in k space on the back Fourier transform of a filter comprising the first, the second and eventually the third shells altogether. The FEFF6 code was used to calculated the EXAFS of each SS or MS pathways individually and their summation to obtain the resulting EXAFS signal. The refinement of DWFs were performed using a standard automated minimization program where each pathways including MS pathways are taken into account individually and eventually by groups of pathways. For *a priori* unknown structure, a manual trial and error approach of the structure was necessary. The approximation and accuracy on the fit parameters are estimated using criteria based on the standard χ^2 parameter.

4. RESULTS AND DISCUSSION

4.1 Co(III) ion exchanged MCM-41

The surfactants, $CTMA^+$, of an as-synthesized form of MCM-41 counterbalance the electric charges of the pore wall silanolato groups, $\equiv$Si-O^-. They can be ion exchanged by *trans*-$[Co(en)_2Cl_2]^+$ complexes in neutral water-ethanol solutions [6] and grafted according to the UV-visible characteristic of the ion exchanged complex and its retention on the MCM-41 even after thorough washing treatment. The following equation is proposed:

$$\{\equiv SiO^-,CTMA^+\}+[Co(en)_2Cl_2]^+ \xrightarrow[-\ CTMA^+]{+\ H_2O} [Co(en)_2(\equiv SiO)(H_2O)_xCl_{1-x}]^{(1-x)+} + xCl^- \quad (4)$$

where x = 1 or 2. To test the validity of this equation, EXAFS data of the ion exchanged and the model *cis*-$[Co(en)_2(Cl)(H_2O)]^{2+}$ complexes are compared. Both EXAFS Fourier transforms exhibit a strong first FT peak at the apparent distance of ~1.5 Å that corresponds to nitrogen or oxygen atoms and two additional peaks at higher distances (Figure 1a,c,d). For the model complex, the second peak (~ 2.0 Å) is much less intense and not well resolved; it is assigned to chlorine while the following peak (~ 2.4 Å) is due to the carbon atoms of the ethylenediamine ligands. For the ion exchanged one, the FT peak of carbon is still there but the FT peak assigned to chlorine has apparently disappeared. However, a third peak arises at around 3.2 Å for the latter. The EXAFS simulations were performed to extract the real distances, to check the above assignments and to provide coordination numbers. The results indicate that there is indeed no need to use chlorine for the simulation of the ion exchanged complex while the third peak can be assigned with good confidence to a single silicon atom at 3.21 Å with an average Co-O-Si angle of 134°. The other distances (Co-O, Co-N, Co-C) are quite similar to those of the model complex and consistent with the literature [14]. It results that the ion exchange process is indeed followed by a grafting via a single Co-O-Si bridge to the pore wall. There is no chlorine left in the cobalt coordination and no indication for the presence of two Co-O-Si bridges. Furthermore, the absence of cobalt neighbors indicates that there is no clustering of cobalt complexes.

4.2 Coordination of Fe(III) in as-synthesized MCM-41 and S-1

The preparation of iron containing S-1, MCM-41 and APS matrices exhibits very similar UV-visible and XANES spectra [5,6,9]. However, there is some divergence when the samples are calcined. Only the as-synthesized samples that are consistent with tetrahedral iron sites are investigated here. Accordingly, the FT EXAFS data are almost identical for Fe (III) in each matrix. They are characterized by two well resolved peaks (Figure1e,f). The first one appears at 1.4 Å and can be calculated using four oxygen atoms at a distance of 1.85 Å consistent with the tetrahedral symmetry. The second peak is slightly more difficult to simulate. Despite the fact that a single fit already accounts for a distribution of angles in the range of ± 15°, it appeared reasonable to use two different Fe-O-Si angles. The trials were chosen according to a stepwise variation of 1° for both angles in a full range of 10° each. Each of the 121 generated simulations provide an optimum angle distribution. The best pair of angle, 135° (40%) and 154° (60%), was selected for its lower χ^2. These two angles distributions are either coming from a mixture of two populations of sites or from a single population of distorted iron sites. Since there is no loading nor matrix dependence of the angle distribution according to XANES and preliminary EXAFS studies on MCM-41, S-1 and APS matrices, the second hypothesis seems more reasonable than the first one [6, 15].

4.3 Matrix effect on Ti sites

For calcined and dehydrated Ti in silicalites, the tetrahedral Ti sites are also characterized by two Ti-O-Si angles [3]. They are somewhat larger than for the Fe-O-Si bond angles (Table 1). Values of 137° (30%) and 163° (70%) for 1.7% Ti are too large to be compatible with a matrix where the average angle of the Si-O-Si is 154° and the radius of Ti^{4+} is quite larger than Si^{4+} ions. Open sites of $[Ti(\equiv SiO)_3(OH)]$ type were assumed to account for the release of the local structural stress by breaking one of the Ti-O-Si bridges. In an APS matrix

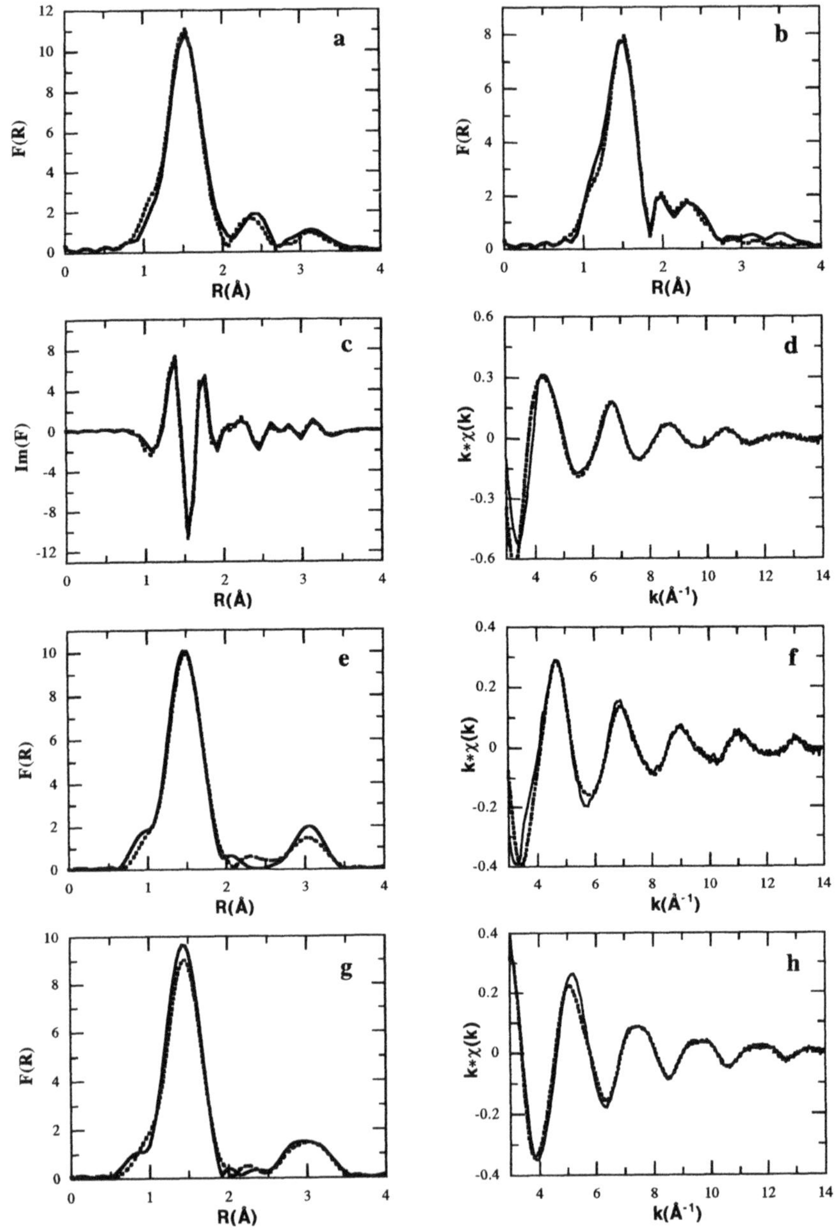

Figure 1. Experimental (—) and simulated (---) EXAFS data for 4.2% Co-MCM-41 (a,c,d), *cis*-$[Co(en)_2(H_2O)Cl]^{2+}$ (b), 5.4% Fe-MCM-41 (e,f) and, 3.6% Ti-MCM-41 (g,h) at Co, Fe and Ti K-edges respectively; k weighted oscillations (d,f,h), FT modules (a,b,e,g), and FT imaginary part (c).

containing 3% Ti, a much smaller average Ti-O-Si angle of 137° is attributed to closed sites of [Ti(≡SiO)$_4$] type [15]. By comparison, a dehydrated and calcined 3.6% Ti-MCM-41 obtained by grafting exhibits also a dual distribution of angles which are centered at 140° (30%) and 158° (70%) (Figure 1g,h). These characteristics are those of open sites of [Ti(≡SiO)$_3$(OH)] type. According to the amorphous nature of the wall and its ability to accomodate closed sites with small angles, these open sites are expected to lie on the pore wall surface. However, the presence of small angle values are an indication that some of the titanium ions have migrated into the wall during calcination and occupy [Ti(≡SiO)$_4$] closed sites.

Table 1
EXAFS simulations data

	d_{M-O} (Å)	σ^2_{M-O} (Å²)	d_{M-N} (Å)	σ^2_{M-N} (Å²)	d_{M-Si} (Å)	σ^2_{M-N} (Å²)	θ M-O-Si
Co-MCM-41[a]	1.87	0.0020	1.94	0.0020	3.21	0.0045	134
$[Co(en)_2(H_2O)Cl]^{2+}$ [b]	1.87	0.0022	1.96	0.0020	----	0.0060	----
Fe-MCM-41[c]	1.85	0.0015	----	----	3.21	0.0045	135 (40%)
					3.28		154 (60%)
Ti-MCM-41[d]	1.81	0.0021	----	----	3.22	0.0045	140 (30%)
					3.37		158 (70%)

a) as-synthesized MCM-41 ion-exchanged with $[Co(en)_2Cl_2]^+$ complex ($\chi^2 = 0.80$)
b) crystals $d_{M-Cl} = 2.28$ Å ($\chi^2 = 0.44$)
c) as-synthesized 5.4% Fe-MCM-41 ($\chi^2 = 1.05$)
d) 3.6% Ti-MCM-41 obtained by grafting on the calcined form, calcined again and dehydrated ($\chi^2 = 1.27$); precisions are ± 0.02Å and ± 3° on distances and M-O-Si angles respectively.

5. CONCLUSION

MS-EXAFS allows one to monitor the first two coordination spheres of transition metal ions incorporated in various siliceous matrices. It is shown that metal grafting by a single M-O-Si can be obtained when the CTMA$^+$ surfactant entrapped within the pore of the as-synthesized MCM-41 is ion exchanged by a transition metal complex. Moreover, this technique allows one to characterize bond lengths and average angles in M-O-Si bridges of species located within the pore wall structure or grafted at their surface despite their amorphous nature. Framework tetrahedral Fe(III) and grafted tetrahedral Ti(IV) ions have been characterized in those terms.

REFERENCES

1. J. S. Beck, J. C. Vartuli, W. J. Roth, M. E. Leonwicz, C. T. Kresge, K. D. Schmitt, C. T.-W. Chu, D. H. Olson, E. W. Sheppard, S. B. McCullen, J. B. Higgins, J. L. Schlenker, J. Am. Chem. Soc., 114 (1992) 10834.
2. A. Corma, Chem. Rev., 97 (1997) 2373.
3. L. Le Noc, D. Trong On, S. Solomykina, B. Echchahed, F. Béland, C. Cartier and L. Bonneviot, Stud. Surf. Sci. Catal., 101 (1996) 611.
4. Z. Liu and R. J. Davis, J. Phys. Chem., 98 (1994) 1253.
5. B. Echchahed, A. Moen., D. Nicholson, L. Bonneviot, Chem. Mater., 9 (1997) 1716.
6. B. Echchahed, A.-R. Badiei, F. Béland, and L. Bonneviot, in this book.
7. L. Le Noc, D. Trong On, L. Bonneviot, J. Chem. Soc., Chem. Commun., (1996) 299.
8. B.-K. Teo,EXAFS in Basic Principals and Data Analysis, Springer-Verlag, New York, 1986.
9. J. J. Rehr, R. C. Abers and S. I. Zabinsky, Phys. Rev.Lett., 69 (1992) 3397; P. A. O`Day, J. J. Rehr, S. I. Zabinsky and G. E. Brown Jr., J. Am. Chem. Soc., 116 (1994) 2938
10. S. Solomykina, PhD Thesis, Laval University, 1997; C. Lortie, D. Trong On, H. Dexpert and L. Bonneviot, Physica B, 208 & 209 (1995) 653.
11. S. Bordiga, R. Buzzoni, F. Geobaldo, C. Lamberti, E. Giamello, A. Zecchina, G. Leofani, G. Petrini, G. Tozzola, G. Vlaic, J. Catal. 148 (1996) 486.
12. T. Maschmeyer, F. Rey, G. Sankar, J. M. Thomas, Nature, 378 (1995) 159.
13. A. Michalowicz, EXAFS pour le MAC, in " Logiciels pour la Chimie", Société Française de Chimie, Paris, 1991, 101.
14. S. S. Jurisson, J. J. Benedict, R. C. Elder, R. Whittle and E. Deutsch, Inorg. Chem., 22 (1983) 1332.
15. F. Béland and L. Bonneviot, unpublished data.

MESOPOROUS MOLECULAR SIEVES 1998
Studies in Surface Science and Catalysis, Vol. 117
L. Bonneviot, F. Béland, C. Danumah, S. Giasson and S. Kaliaguine (Editors)

Hysteresis in Physical Sorption for MCM.

W.C. Conner[b], M. A. Springuel-Huet[a], J. Fraissard[a], J. Bonardet[a], T. McMahon[b], L. Boudreau[b], and J. Masciadrelli[b]

[a] Laboratoire de Chemie des Surfaces, Université P. et M. Curie, 4 Place Jussieu, Paris, FRANCE
[b] Chemical Engineering, University of Massachusetts, Amherst, MA. 01003 USA

A variety of samples of mesoporous solids were studied as prepared by the techniques giving rise to the MCM-class of materials. These were studied by physical sorption and X-ray diffraction. We found that there are significant differences between the sorption isotherms which may or may not exhibit sorption hysteresis. A correlation between X-ray and sorption analyses was not evident. Moreover, conventional analyses of desorption is not appropriate. A combination of sorption and X-ray analyses are required to characterize MCM materials.

1. INTRODUCTION

The discovery of the MCM family of solid opened up several new dimensions in the synthesis of unique materials and the characterization of mesoporous solids [1]. First a new class of materials was discovered which were, for the most part, amorphous and, yet, commonly possessed X-ray diffraction patterns characteristic of an hexagonal array of solid spacing greater than 2.0 nm. Electron microscopy confirmed that a mesoporous parallel network of (hexagonal) pores had been created. Sorption analyses initially supported the picture that the pores were uniform in dimension and parallel, i.e., there was no hysteresis between adsorption and desorption. Further, conventional analyses showed adsorption and desorption at relative pressures (P/Po, where Po is the saturation pressure of the adsorbing gas) between 0.3 and 0.6. The pore dimensions which conventionally correspond to adsorption in this range or relative pressures, span the range from 1-6 nm in diameter; however, it is believed that conventional analyses for mesoporous solids (>5 nm) should not be extrapolated this low and adsorption in microporous solids apparently occurs by a different mechanism.

Modifications to the original recipe for MCM synthesis soon gave rise to the synthesis of a large class of related mesoporous materials. As more analyses became available [2], it became evident that the MCM materials were more complex than originally perceived. Hysteresis between adsorption and desorption of N_2 at 77K was sometimes evident, and was initially believed to be due to quadrupolar interactions between N_2 and the MCM surface [3]. Other studies confirmed that a lack in hysteresis in sorption was the rule [4], not an exception. Variations in X-ray diffraction spectra confirmed the complexity.

2. EXPERIMENTAL

The MCM-41 samples were prepared from sodium silicate which is added to a basic solution (NaOH) containing the template, tetradecyltrimethyl (denoted C_{14}) or hexadecyltrimethyl (denoted C_{16}) ammonium bromide, and heated to 40 °C. By addition of a 1M HCl solution, the pH is decreased from 11.4 to about 9. The mixture is stirred 1 hour at 40 °C, then filtered, washed with distilled water, and heated at 40 °C overnight. The calcination was performed with air at 600 °C during 24 h (heating rate 24 °C/h).

For several of the samples, mesitylene has been added to the initial solution (containing the template and the silicate) at 60 °C with different template/mesitylene ratios: 1/5, 1/4, 1/3 (see Table 1) in order to obtain larger pore sizes. The solution is stirred 2 hours at 60°C after the addition of HCl to adjust the pH to 9-9.5. The sample is then filtered and washed with distilled water and is finally heated at 60 °C overnight. The calcination was performed under the same conditions as above.

N_2 adsorption-desorption isotherms were measured at 77 K with a Micromeretics instrument (ASAP 2010) using conventional volumetric techniques. The isotherms for most of these samples are discussed and shown below in Figures 2-4.

3. RESULTS & DISCUSSION

3.1 X-ray Analyses

The samples were analyzed by X-ray powder diffraction using a Phillips X-ray defractometer. These results are shown in Figure 1.

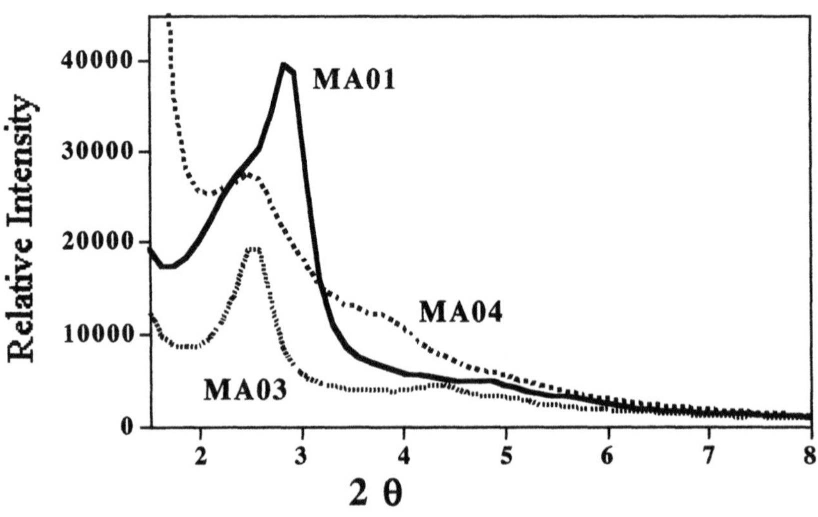

Figure 1: X-ray diffraction for samples which exhibit peaks characteristic of MCM type materials. Note that M05 and M06 showed no diffraction peaks for $2 < 2\theta < 10$.

The X-ray diffraction patterns confirmed the formation of a hexagonal structure for several of the samples as seen in Figure 1. The position of the primary peaks give lattice spacings of l = 29 - 35Å for the primary peaks. There is evidence for the secondary hexagonal peaks (at $l/\sqrt{3}$ and $l/2$) for each of the samples shown in Figure 1. There were, however, no peaks in the diffraction patterns for samples M05 and M06. As far as X-ray could detect, these were amorphous samples.

3.2. Adsorption Data

Adsorption/desorption isotherms for nitrogen adsorption at 77K are shown in Figures 2-4.

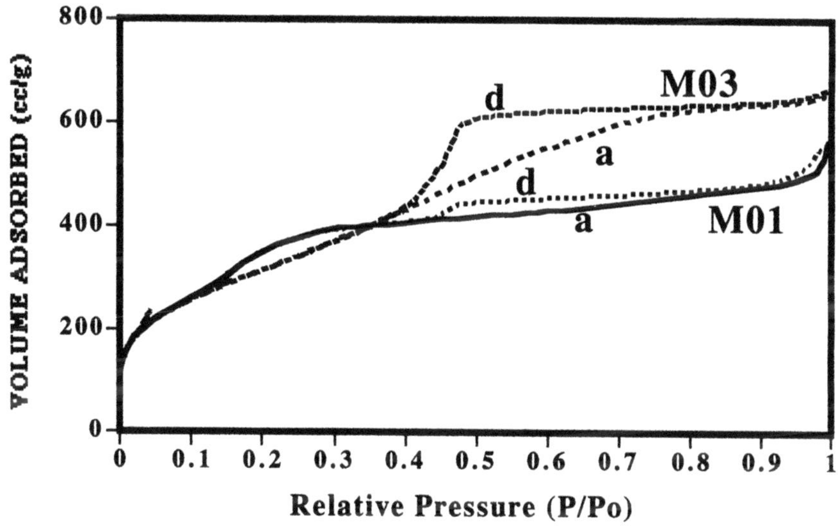

Figure 2: Sorption isotherms for nitrogen adsorption at 77K on M01 and M03. The adsorption isotherms (a) are below the desorption isotherms (d) in each case.

Comparisons between these data show that there is considerable variation in the types of isotherms found for these samples. Hysteresis in adsorption is often but not universally found. Further, the nature of the hysteresis differs significantly. As examples, the hysteresis for sample M01 occurs during desorption while the corresponding adsorption branch of the isotherm only increases slowly while the hysteresis for M03 represents a steep adsorption and only a small decrease in the desorption isotherm until P/Po goes below 0.5 whence it desorbs rapidly with a decrease in relative pressure.

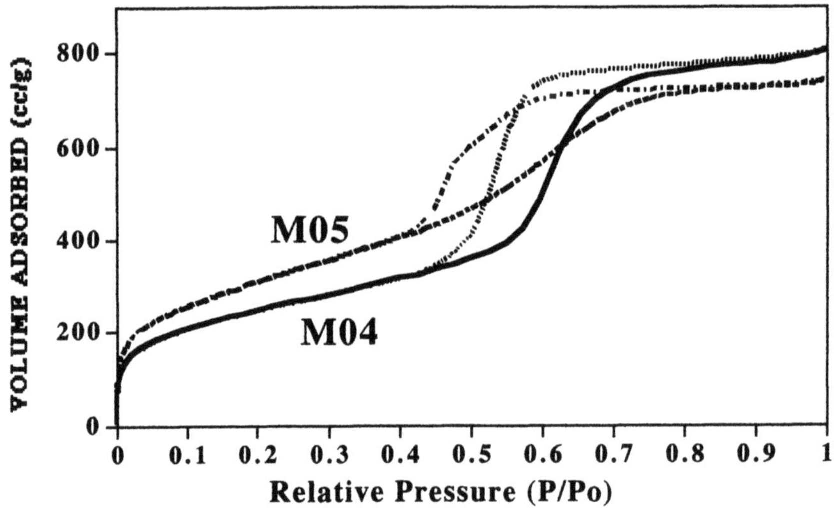

Figure 3: Sorption isotherms for nitrogen adsorption at 77K on M04 and M05. The adsorption isotherms are below the desorption isotherms in each case.

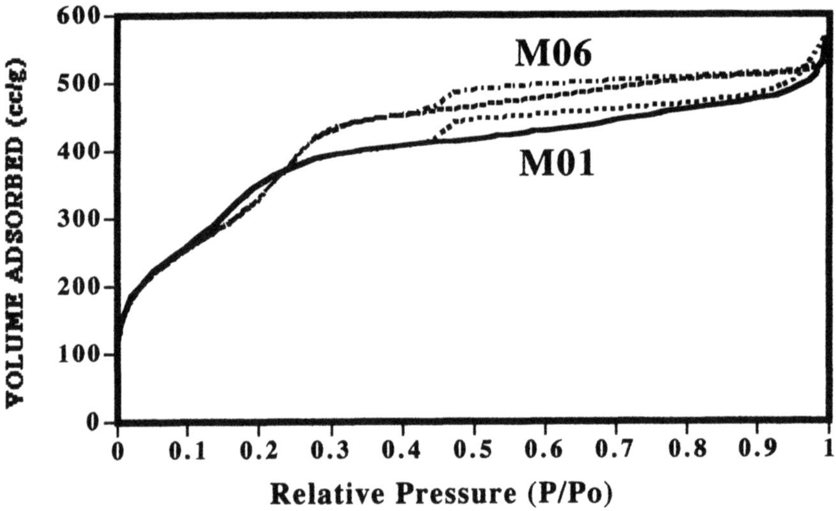

Figure 4: Sorption isotherms for nitrogen adsorption at 77K on M01 and M06. The adsorption isotherms are below the desorption isotherms in each case.

Figure 3 compares two samples with seemingly similar isotherms, M04 and M05. There is evidence for pore filling for each adsorption isotherm commencing at P/Po = 0.5 for M04 and at P/Po = 0.6 for M05. Both desorption branches desorb only slightly until P/Po < 0.6 whence considerable desorption is evident with small reduction in pressure. Whereas M04 is apparently crystalline, M05 is amorphous. Similarly, Figure 4 compares the sorption isotherms for samples M01 and M06. Without the X-ray analyses, one would conclude that the pore structures of these two samples are essentially identical. Minor hysteresis is evident at P/Po = 0.4-0.45 during desorption for each sample and there is also evidence for pore filling and emptying centered at P/Po ~ 0.17 for M01 and at P/Po ~ 0.25 for M06. These pore fillings and emptyings take place without any evidence for sorption hysteresis. Again, although the isotherms of M01 and M06 are apparently quite similar, only one sample (M01) exhibits the X-ray diffraction patterns for a "*crystalline* " MCM-type void structure. The sorption analyses is apparently more sensitive to the void structures than are the X-ray analyses.

It should be noted that hysteresis at P/Po ~ 0.4-0.45 is often evident for these samples. In some cases (M01 and M06), the hysteresis is minor while in other samples (M03, M04 and M05) this would appear to reflect the dominant void structure. In each case, automated analyses of *desorption* will show a peak in the pore size distribution curve reflecting pores 18-20Å. However, analyses of the adsorption will not always show a corresponding pore dimension (shifted up in size when hysteresis is evident). Only samples M04 and M05 exhibit peaks in the adsorption analyses which correspond to the hysteresis in desorption at P/Po ~ 0.4-0.45. We are drawn to conclude that peaks in desorption analyses which have no corresponding peak at the same (without hysteresis) or larger pore dimension (with hysteresis) may be due to the common tensile strength artifact [5].

3.3 Analyses

Pore structure analyses can be done on both adsorption and desorption isotherms. FHH analyses [5] could be performed employing the conventional values of the constants in the commonly assumed "Halsey equation" for t (the thickness of the adsorbed layer) and P/Po (the relative pressure):

$$t = 6.04\ [P/Po]^{1/3} \tag{1}$$

However, the steepness of the adsorption isotherms below the region of pore filling suggests that this "conventional" equation will over estimate the smaller pores [6]. A more appropriate relationship between the relative pressure and the thickness of the adsorbed layer was estimated by fitting the data in this region of data, below pore filling. The following relationship reflects Halsey's original concepts [7] and was fit to the data:

$$t = C\ [P/Po]^{1/HC} \tag{2}$$

The values of the relationship between the thickening of the adsorbed layer and pressure in the regions where pores are neither filled or emptied does vary for these materials. The values of the "Halsey constant" (HC in equation 2) was found to vary from ca. 1.4 to 2.3 as shown in Table 1. This is a *significant* variation and is also *substantially different* from the value of 3 used in the "conventional" [3, 8] analyses employed in all automated sorption analyses systems. It should be noted that if a larger value of this exponent is assumed in the analyses of pore structure, smaller pores will "found" in the analyses that *do not exist*. This is a common artifact in conventional analyses of MCM-type solids. We have never found that the appropriate exponent between the relative pressure and the thickness of the adsorbed layer is near 3 for MCM solids.

The results from these analyses are summarized in Table 1 which specifies the methods of preparation as detailed in the experimental section, above. The BET surface areas and the mean diameters determined using density functional theory (with cylinder geometry and Harkins & Jura equation) are summarized in columns 4 and 6. Column 5 shows the value of the C constant from the BET analyses. The characteristic dimension from X-ray analyses is shown in the last column.

Table 1
Measured properties of MCM Samples.

Sample	Temp./Add. m-mesitylene	Halsey Exp.	Surface Area M^2/g	C Constant from BET	Dp (desorp.) Å	X-ray Å
M01	C_{14}	1.44 (0.1->0.4)	1354	17	18	29
M03	C_{16}	1.80 (0.1->0.4)	1159	63	21	34
M04	C_{16}&1/4m	2.27 (0.1->0.4)	990	44	51	35
M05	C_{16}&1/3m	2.02 (0.1->0.4)	1073	68	32	amorp.*
M06	C_{16}&1/5m	1.4 (0.1-0.2)	1358	29	46	amorp.*

The template employed (C14 or C16) with or without mesitylene are listed in column 2.
The Halsey constants (fit over the range of relative pressures) are shown in column 3.
The BET surface areas and resulting C constants are shown in column 4.
The maximum in the pore diameters from desorption analyses is shown in column 5.
The lattice dimension of the largest peak in the X-ray diffraction is shown in column 6.
*The last two samples showed no evidence of X-ray diffraction, appearing amorphous.

4. ANALYSES & CONCLUSIONS

Several conclusions can be drawn from these data.

First, there is no common isotherm for adsorption and desorption for MCM-based materials.

Hysteresis in pore filling and emptying is sometimes evident but most often it is not.

Second, it is necessary to properly analyze both the adsorption and the desorption branches of the isotherms.

Conventional sorption analyses employing a "standard" isotherm (or Halsey relationship) will result in estimates of pores which do not exist.

Finally, both X-ray and sorption analyses are required to characterize the void morphology of MCM-type materials.

Samples which exhibit similar isotherms can have different X-ray diffraction patterns and samples with similar diffraction patterns can have significantly different adsorption isotherms.

The new class of mesoporous materials requires that the analyses of their morphology are modified. Neither X-ray nor sorption analyses alone give a complete definitive analysis, both must be employed. Further, both adsorption and desorption analyses are required and conventional approached to determine pore dimensions should not be used without modification. As for any new class of materials, care must be exercised to characterize these unique mesoporous solids.

5. REFERENCES

[1] J. Beck, J. Vartuli, W. Roth, M. Leonowicz, C. Kresge, K. Schmitt, C. Chu, D. Olson, E. Sheppard, S. McCullen, J. Higgins and J. Schlenker , *J. Am. Chem. Soc.* vol 114, 1992, 10834-10843

[2] Characterization of Porous Solids IV, The Royal Society of Chemistry, pbs., 1997.

[3]. P. Branton, P. Hall, and K. Sing, ***J. Chem. Soc. Chem. Commun***. 1993, 1257-1258

[4]. A. Corma, Chem. Reviews, vol 97 n° 6, 1997, 2373-2419

[5]. S. J. Gregg and K. S. W. Sing, "Adsorption, Surface Area and Porosity." Academic Press, London, 1982.

[6]. W. C. Conner, *Journal of Porous Materials*, 3-2, 1996, 1

[7]. G. Halsey, *J. Chem. Physics*, 16, 931, 1948

[8] E. Barrett, L. Joyner, and P. Halenda, ***J. Amer. Chem. Soc.*, 73**, 1951, 373

MESOPOROUS MOLECULAR SIEVES 1998
Studies in Surface Science and Catalysis, Vol. 117
L. Bonneviot, F. Béland, C. Danumah, S. Giasson and S. Kaliaguine (Editors)

Determination of the MCM-41 (25 Å and 40 Å) wall mean densities from sorption capacity and neutron diffraction measurements.

N. Floquet[a], J.P. Coulomb[a], S. Giorgio[a], Y. Grillet[b] and P.L. Llewellyn[b]

[a]C.R.M.C2 - CNRS, Campus de Luminy, Case 901, 13288 Marseille Cedex 9 - France.
[b]C.T.M. - C.N.R.S., 26 rue du 141° R.I.A., 13003 Marseille - France.

Despite the lack of periodicity along the mesopore axis, we can represent the MCM-41 sorption capacity by the number of sorbed molecules within a mesopore section of unit length (l = 1 nm). Both from sorption capacity and neutron diffraction measurements, which give in a direct way the number of sorbed molecules $N_{ads.}$ and the nearest neighbor distance l_{nn} between them respectively, we have deduced the wall mean densities of MCM-41 (Ø = 25 Å and Ø = 40 Å) samples. Surprisingly, such densities $0.8 \leq d_{MCM-41} \leq 1.6$ are much smaller than the density of the amorphous silica (d_0 = 2.2) which is currently accepted for MCM-41 material.

1. INTRODUCTION

In 1992, for the first time, the Mobil Research Laboratory (Princeton - USA) synthesized a model porous material in the mesoporous range 20 - 100 Å, the so-called MCM-41 [1,2]. Such a material can be regarded as composed of an hexagonal structure of cylindrical mesopores. But, contrarily to the zeolites which are pure crystalline materials of well known density, the MCM-41 walls are constituted of amorphous silica (SiO_2). MCM-41 samples are characterized both by a narrow distribution in pore shapes and pore sizes. Such a model material fills the gap which existed before between the zeolitic microporous materials (Ø ≤ 20 Å) and the controlled pore glasses CPG (100 Å ≤ Ø). As a consequence, MCM-41 material has given rise to an exponential number of experimental studies. A special attention has been devoted to its sorption properties which are closely related to the capillary condensation phenomenon [3-10]. Surprisingly, the hysteresis loop which is the usual signature of the capillary condensation, in certain case (as for nitrogen sorption at T = 77.5 K), disappears when decreasing the pore diameter of MCM-41 samples [3,4]. In addition, the observation or not of the hysteresis loop depends on the temperature range [9] and on the chemical nature of the cation of compensation for MCM-41 sample of low Si/Al ratio [10]. It was also evident that the Kelvi law application is questionable in the small mesopore size range 20-50 Å [11]. I appears that the confined condensed matter in the MCM-41 pores i characterized by properties which are quite different than the bulk one's. Bot the critical temperatures and the melting point of the sorbed phases ar displaced towards the low temperatures [12-14].

The sorption of different simple gases on MCM-41 has pointed out that befor the formation of the capillary phase (in the pressure range $0.1 \leq P_r \leq 0.7$), a adsorbed film grows on the inner mesopore surface in the low pressure range (F << 0.1) [15]. P_r is the relative pressure with regard to the saturated vapour one F ($P_r = P/P_0$). Depending on the considered sorbate species the film phases ar preponderant or negligible in comparison with the capillary phases, as shown o figure 1. Consequently, we have to note that the capillary phase coexists most o the time with a film phase. A schematic illustration of the two stages process i the pore filling is represented on figure 2. In special cases as for water, whic doesn't wet the inner surface of MCM-41 silicic samples , only the capillary phas

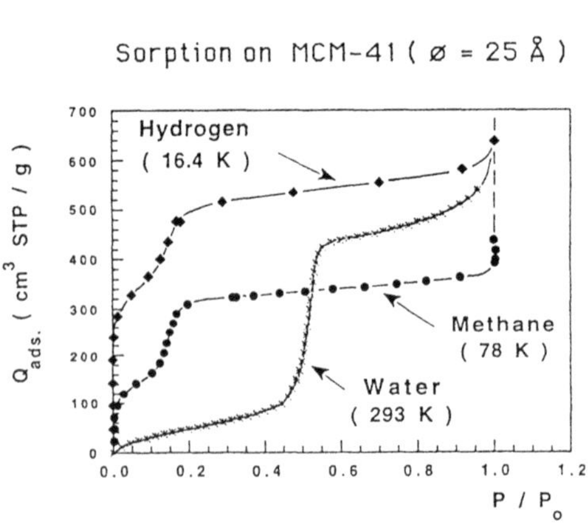

Figure 1. Sorption isotherms of hydrogen (T = 16.4 K), methane (T = 78 K) and water (T = 293 K) on MCM-41 (Ø = 25 Å).

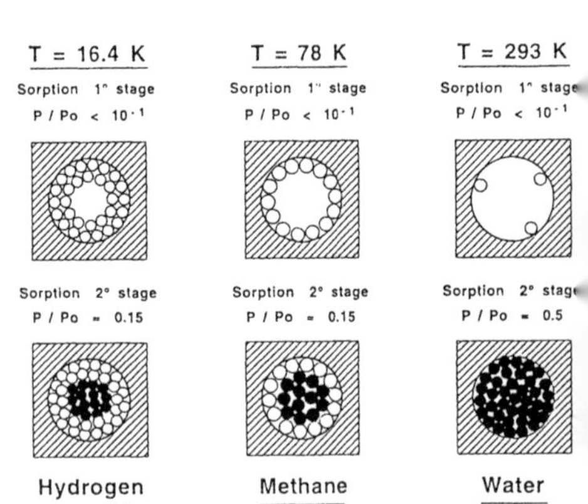

Figure 2. Schematic representatio of the two stages filling of the MCM 41 (Ø = 25 Å) pores by H_2, CH_4 an H_2O respectively.

develops during the sorption phenomenon. We have to keep in mind that th MCM-41 inner surface must be energetically heterogeneous due to its amorphou structure, characteristic which probably has a great influence on the adsorbe film properties. Another characteristic of the MCM-41 walls is that thei thickness l_w increases when the diameter of the mesopores decreases. In the por diameter range 40 - 50 Å , the wall thickness l_w is around 7 - 8 Å, but in th range 20 - 30 Å it is twice as large ($l_w \approx 16$ Å) [2,14]. The key point in the MCM 41 samples characterization is the accurate determination of the mesopor diameter value $Ø_p$ (value which can be monitored owing to the size of the template micelles). Several experimental techniques (thermoporometry, sorption isotherm measurements, high resolution electron microscopy) and theoretical models (Kelvin equation, BJH method, non-local density functional theory) have been used in the $Ø_p$ value determination [11,13]. It is usual to express the sorption capacity of zeolitic materials by the number of sorbed molecules $N_{ads.}$ within the unit cell. In such a representation, both the sorbant structure and density are needed. Despite the fact that MCM-41 doesn't present any periodicity along that mesopore axis, $N_{ads.}$ can be determined from the sorption isotherm measurement for a mesopore section of unit length (l = 1 nm). In addition of the pore diameter $Ø_p$, the hexagonal unit cell parameter a (which is the distance between the pores) must be known. Neutron or X-ray diffractogram measurement is necessary to determine the parameter a. The density value of the MCM-41 wall, d_{MCM-41} is also needed in the N_{ads} calculation. Up to now, the accepted value for d_{MCM-41} was 2.2 (density value of the amorphous silica SiO_2) [16]. As we demonstrate in our paper, such a value leads to a sorbed molecule number N_{ads} which is not acceptable for molecular overcrowding reasons.

2.EXPERIMENTAL SECTION

The two MCM-41 (Ø = 25 Å and Ø = 40 Å) samples used in our sorption isotherms and neutron diffraction measurements were synthesized at the laboratory of Prof. K. K. Unger (Mainz - R.F.A.) Sorption isotherms have been measured with a homemade apparatus already described [4]. The neutron diffraction experiments were realized at the Léon Brillouin Laboratory (C.E.A - Saclay). Prior to the different measurements, the samples have been outgased at around 200 °C under vacuum ($P \leq 10^{-6}$ Torr) during 12 hours. In addition of the two considered MCM-41 samples, we have investigated the properties of a third one synthesized from a new route at the laboratory of Prof. H. Kessler (Mulhouse – France). Such a MCM-41 sample is characterized by a porosity composed of secondary micropores (Ø ≈ 15 Å). Neutron diffractograms of the three outgased MCM-41 samples are represented on figure 3. We remark that the diffraction peaks (110) and (200) are clearly visible only for the MCM-41 (40 Å).

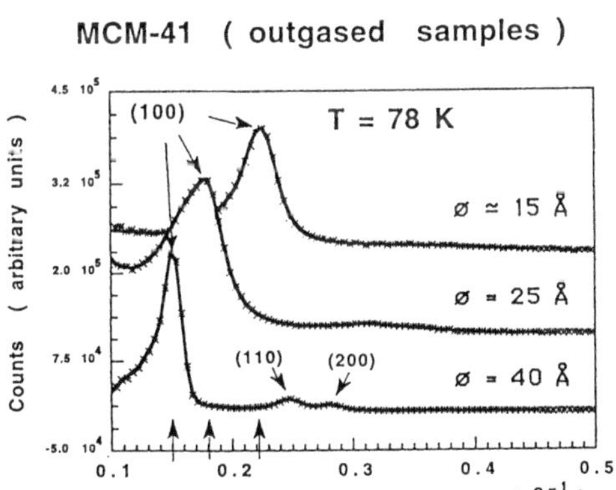

Figure 3. Neutron diffractograms of the outgased MCM-41 samples.

3. RESULTS AND DISCUSSION

From the position of the (100) diffraction peak we deduce the parameter a of the two-dimensional hexagonal unit cell of the three considered MCM-41 samples. The results are summarized in table 1. The pore diameter $Ø_p$ of the MCM-41 (25 Å and 40 Å) samples has been determined previously [4,8]. At the present time, $Ø_p$ of the "Mulhouse" sample is estimated from the micelles size. Methane sorption isotherms measured at the temperature T = 78 K for the three considered MCM-41 samples are shown on figure 4. Only the sorption branches are represented for the three MCM-41 samples. First of all, we note that the sorption capacity Q_{max} decreases strongly when $Ø_p$ decreases. At P/P_0 = 0.8, $Q_{max.}$ is equal to 562, 356 and 178 cm^3 STP/g for $Ø_p$ equal to 40, 25 and 15 Å respectively. Secondly, when $Ø_p$ decreases the step position which is the signature of the capillary condensation is strongly displaced towards the low relative pressure values P/P_0, in qualitative agreement with the Kelvin law. We note also that the methane sorbed quantity $Q_{ads.}$ during the first stage (film adsorption) decreases when the pore diameter decreases. For the two MCM-41 samples of smaller diameters, we have measured the neutron diffractograms at T = 78 K during the methane sorption for different loadings. The figure 5 represents the diffractograms measured for the CD_4 maximum loadings on MCM-41 (15 Å and 25 Å). In the considered scattering wave vector range $1 \leq Q \leq 2.8$ $Å^{-1}$ the MCM-41 samples do not present any diffraction peaks. The large bumps located at Q = 1.85 A^{-1} point out that the methane sorbed phases confined in the pores (Ø ≈ 15 Å and Ø = 25 Å) are characterized by a short range order (characteristic order of liquid phases or amorphous solid phases). Recall that, at T = 78 K for the CD_4/MCM-41 (Ø = 25 Å) system, an hysteresis loop is observed during the capillary condensation [15]. Consequently, we can deduce that the capillary

Table 1

$Ø_p$ (Å)	a (Å)	l_w (Å)
≈ 15	32.1	17
25	40.1	16
40	47.8	8

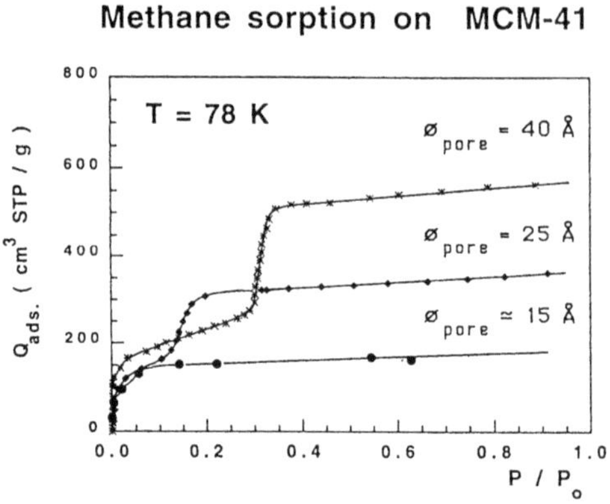

Figure 4. Sorption isotherms of methane on MCM-41 (15 Å, 25 Å and 40 Å) measured at T = 78 K.

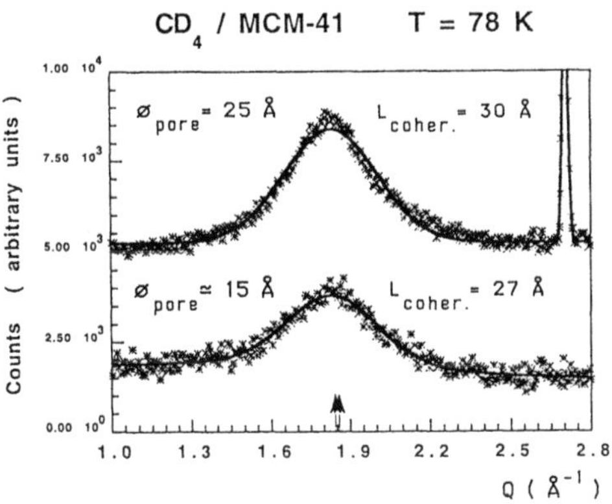

Figure 5. Neutron diffractograms of CD_4 sorbed phases on MCM-41 (Ø ≈ 15 Å) and (Ø = 25 Å) samples at T = 77.5 K.

critical temperature T_{cc} related to 25 Å pore diameter is larger than 78 K. From the bump position, we deduce the average distance between CD_4 molecules in the sorbed phases l_{nn} = 4.14 Å (distance which is in good agreement with the methane molecular size $Ø_{CH4}$ = 4.2 Å). Concerning the MCM-41 (40 Å) sample we have studied in details its sorption properties in the case of krypton, nitrogen and water sorbates species. The sorption isotherms of these three gases are shown on figure 6. We note that for Kr and N_2 the sorption phenomenon is a two stage process (film phase + capillary phase), on he other hand for water only the capillary phase grows. The sorption isotherms f Kr and N_2 present some similar features (the step position in particular) but hey differ from one another in the sorption capacities which are equal to 594 nd 453 cm^3 STP/g respectively. Krypton is an interesting sorbate species on ICM-41 (Ø = 40 Å). Indeed, for the first time, a heat peak has been observed (by alorimetric measurement) during the formation of the capillary phase at T = 7.5 K. Such a result is interpreted as a solidification of the Kr capillary phase.

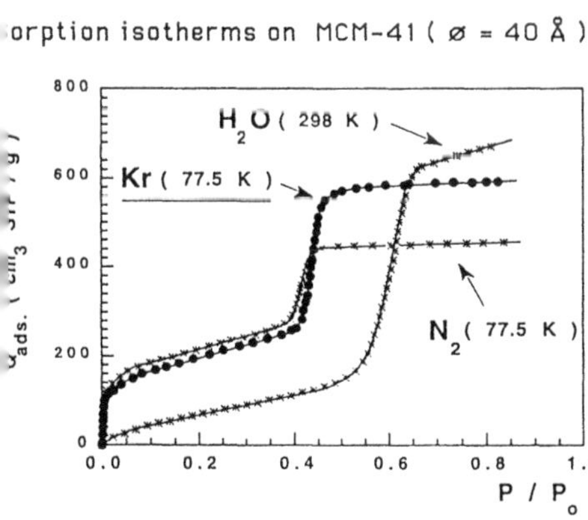

Figure 6. Sorption isotherms of Kr, N_2 and H_2O on MCM-41 (Ø = 40 Å) at 77.5 K, 77.5 K and 298 K respectively.

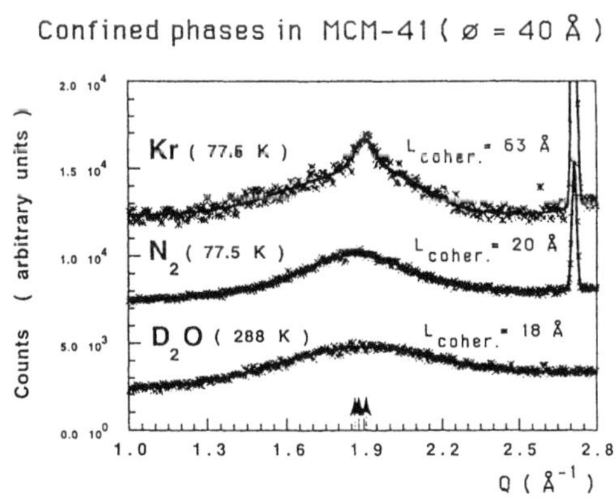

Figure 7. Neutron diffractograms of Kr, N_2 and D_2O sorbed phases on MCM-41 (Ø = 40 Å) at 77.5 K, 77.5 K and 288 K respectively.

As a consequence, due to the fact that a solid phase is denser than a liquid phase the MCM-41 (Ø = 40 Å) sorption capacity is larger for Kr than for N_2 (we recal that the molecular sizes of the two sorbates are quite close). The solid characte of the Kr sorbed phase at T = 77.5 K on MCM-41 (Ø = 40 Å) has been confirme by a neutron diffraction experiment.

We have represented on figure 7 the neutron diffractograms of Kr, N_2 and D_2(sorbed phases at 77.5 K, 77.5 K and 288 K respectively, for the maximum of thei loadings on MCM-41 (Ø = 40 Å). The diffraction signals of nitrogen and wate confined phases are very large bumps. The molecular organizations in suc phases are characterized by a short range order ($L_{coher.} \approx 20$ Å). On the othe hand, the diffracted signal of the Kr sorbed phase, which is much narrow expresses the medium range order of the atomic organization ($L_{coher.} \approx 60$ Å). W have to recall that due to the amorphous character of the MCM-41 walls, n phase correlation may exist between two nearest neighbor mesopores. As consequence, each rod shaped Kr crystals are not correlated in the different hos mesopores. In our opinion, such a feature broadens the diffracted signal of th sorbed Kr solid. We have determined the melting temperature T_m of this confine solid phase, $T_m = 85$ K [17].

From the position of the krypton diffraction peak Q = 1.91 $Å^{-1}$, we deduce th average distance between Kr atoms in the sorbed solid phase l = 4.05 Å (value i very good agreement with the van der Waals diameter of Kr atoms Ø = 3.95 Å).

In table 2, we have summarized the number of sorbed molecules within MCM-41 section of unit length (l = 1 nm) N_{ads}., deduced from the measured MCM 41 sorption capacity Q_{ads}. (cm^3 STP/g) :

Table 2.

$Ø_p$ (Å)	Q_{ads}. (cm^3STP/g)	N_{ads}. ("u.c.")
25	CD_4 : 356	190
40	Kr : 594	254

$N_{ads}. = Q_{ads}.N_A.M_{u.c.}/V_m$

$M_{u.c} = d_{MCM\text{-}41} l \, (a^2\sqrt{3}/2 - \pi Ø_p^2/4)$

N_A and V_m are the Avogadro constant and molar volume respectively

The N_{ads} values calculated with a MCM-41 wall density $d_{MCM\text{-}41}$ equal to 2.2 (th density value of the amorphous silica) lead to a molecular packing in grea disagreement with the molecular distance that we have measured by neutro diffraction. From these measured molecular distances and for a close packin molecular arrangement we have deduced the maximum number N_{max} of methan and krypton molecules confined within a pore section of unit length for diameter equal to 25 Å and 40 Å respectively. Our results are : $N_{max}(CD_4) = 63 \pm$ molecules and $N_{max}(Kr) = 168 \pm 8$ molecules. Despite the uncertainty on N_{m} values (that we have over estimated around 15%) there is a great difference i mparison with the previous N_{max} lues (difference larger than a factor o). The average density of MCM-41 lls that we deduce from the N_{max} lues are : d_{MCM-41} = 0.8 ± 0.1 (Ø = 25 Å) d d_{MCM-41} = 1.6 ± 0.1 (Ø = 40 Å). By the me method, from sorption isotherms easurements published in the erature, we have deduced the density MCM-41 samples synthesized by other laboratory [13]. The results are presented on figure 8, they are in reement with our results. Recently, tailed X-rays and neutron diffraction idies concerning MCM-41 samples em to corroborate our results [18,19].

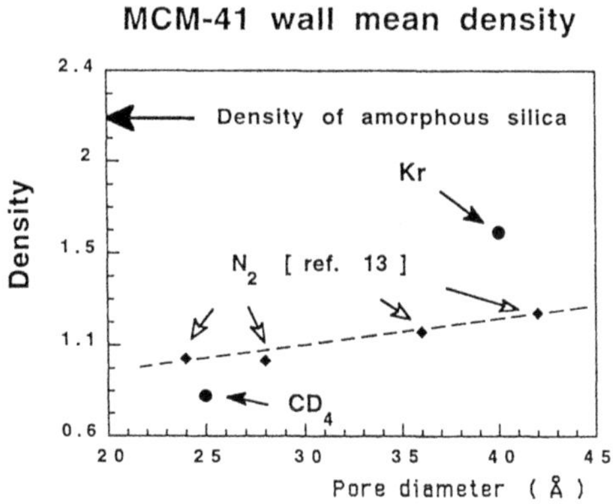

Figure 8. Mean density of the MCM-41 walls versus the mesopore diameter $Ø_p$.

CKNOWLEDGEMENTS

The authors would like to thank kindly Prof. K.K. Unger (Maintz - RFA) and . J. Patarin and Prof. H. Kessler (Mulhouse - France) for MCM-41 sample pplies and Dr. R. Denoyel (Marseille – France) for helpful discussions.

EFERENCES

C.T.Kresge, M.E. Leonowicz, W.J. Roth, J.C. Vartuli and J.S. Beck, Nature,Vol. 359 (1992) 710.

J.S. Beck, J.C. Vartuli, W.J. Roth, M.E. Leonowicz, C.T.Kresge, K.D. Schmitt, C.T-W. Chu, D.H. Olson, E.W. Sheppard, S.B. McCullen, J.B. Higgins and J.L. Schlenker, J. Am. Chem. Soc., 114 (1992) 10834.

O. Franke, G. Schulz-Ekloff, J. Rathousky, J. Starek and A. Zukal, J. Chem. Soc., Chem. Commun., 9 (1993) 724.

P.L. Llewellyn, Y.Grillet, F. Schüth, H. Reichert and K.K. Unger, Microporous Material, 3 (1994) 345.

R. Schmidt, M. Stöcker, E. Hansen, D. Akporiaye and O. H. Ellestad, Microporous Materials, 3 (1995) 443.

P.J. Branton, P. G. Hall, M. Treguer and K.S.W. Sing, J. Chem. Soc. Faraday Trans, 91 (1995) 2041.

P.L. Llewellyn, F. Schüth, Y. Grillet, F. Rouquerol, J. Rouquerol and K. K. Unger, Langmuir, 11 (1995) 574.

P.L. Llewellyn, Y. Grillet, J. Rouquerol, C. Martin and J.P. Coulomb, Surface Science, 352 - 354 (1996) 468.

A.V. Neimark, P.I. Ravikovitch, S.C. Domhnaill, F. Schüth and K.K. Unger,

in Fundamentals of Adsorption, Edt. M.D. LeVan, Kluwer Academi
Publishers, (1996) 667.
10. Z. Luan, H. He, W. Zhou, C.F. Cheng and J. Klinowski, J. Chem. So
Faraday Trans., 91 (1995) 2955.
11. U. Ciesla, M. Grün, T. Isajeva, A.A. Kurganov, A.V. Neimark, P. Ravikovitc
S. Schacht, F. Schüth and K.K. Unger, in "Access in Nanoporous Materials
Edt. T.J. Pinnavaia and M.F. Thorpe, Plenum Press, New York, (1995) 231.
12. J. Rathousky, A. Zukal, O. Franke and G. Schulz-Ekloff, J. Chem. So
Faraday Trans., 91 (1995) 937.
13. K. Morishige, H. Fujii, M. Uga and D. Kinukawa, Langmuir, 13 (1997) 3494.
14. R. Schmidt, E.W. Hansen, M. Stöker, D. Akporiaye and O.H. Ellestad, J. Am
Chem. Soc., 117 (1995) 4049.
15. J.P. Coulomb, C. Martin, Y. Grillet, P. Llewellyn and G. André, Studies i
Surface Sciences and Catalysis, Vol. 105 (1996) 1827.
16. B.P. Feuston and J.B. Higgins, J. Phys. Chem., 98 (1994) 4459.
17. J.P. Coulomb, Y. Grillet, P. Llewellyn, C. Martin and G. André (Submitted t
Fundamentals of Adsorption 6th Conference, May - 1998, Giens - France).
18. K.J. Edler, P.A. Reynolds, P.J. Branton, F.R. Trouw and J.W. White, J
Chem. Soc. Faraday Trans., 93 (1997) 1667.
19. K.J. Edler, P.A. Reynolds, J.W. White and D. Cookson, J. Chem. Soc. Farada
Trans., 93 (1997) 199.

MESOPOROUS MOLECULAR SIEVES 1998
Studies in Surface Science and Catalysis, Vol. 117
L. Bonneviot, F. Béland, C. Danumah, S. Giasson and S. Kaliaguine (Editors)

Shape selectivity of mesoporous aluminosilicates in the transformation of benzenic hydrocarbons

M. Guisnet[a], N.S. Gnep[a], S. Morin[a], J. Patarin[b], F. Loggia[c], V. Solinas[c]

[a] Laboratoire de Catalyse en Chimie Organique, UMR 6503, UFR Sciences, 40, ave. du recteur Pineau, 86022 Poitiers cedex, France
[b] Laboratoire des Matériaux Minéraux, Mulhouse, France
[c] Laboratorio Chimica Industriale, Cagliari, Italy

The activity and selectivity of three mesoporous aluminosilicates (MCM-41) with Si/Al ratios of 10, 30 and 100 and of an amorphous silica alumina (SA) were compared in toluene alkylation with methanol and in xylene transformation at 350°C. For both reactions, the order of activity is the one expected from the density and strength of the protonic sites. For toluene alkylation, the selectivity is practically independent of the catalyst : successive formation of xylenes, trimethylbenzenes, tetramethylbenzenes etc., distribution of xylenes in agreement with a Friedel-Crafts methylation of toluene, high ratio between the rates of methanol conversion into C_1-C_5 hydrocarbons and into xylenes. On the other hand, the selectivity of xylene transformation is completely different on SA and on MCM-41. With SA, the selectivity of m-xylene isomerization (p/o = 1.4) is that expected from the classical intramolecular mechanism. In agreement with this mechanism, the addition of methylcyclohexane to the reactant has no effect on the isomerization rate. With MCM-41, m-xylene isomerizes preferentially into o-xylene (p/o = 0.2-0.4) and methylcyclohexane causes an identical decrease in the rates of isomerization and disproportionation. All this can be explained by a bimolecular isomerization mechanism involving successive m-xylene disproportionation and transalkylation with trimethylbenzenes. This particular behaviour of MCM-41 samples is clearly due to the presence of regular non-interconnected long channels in which xylene molecules can undergo, before desorption, successive reactions of disproportionation and transalkylation. This new type of shape selectivity was designated as Tunnel Shape Selectivity.

1. INTRODUCTION

Zeolites are used as acid catalysts in various processes of refining and of production of petrochemicals and even of fine chemicals. One of their main advantages over the conventional acid catalysts is their shape selective properties caused by the small size of their cages and of their pore apertures [1] . However, this small size becomes a drawback in the treatment of heavy feeds or in the synthesis of large molecules. The need for acid solids with larger pores has initiated an active research with as a consequence the recent discovery of a new family of mesoporous aluminosilicates, designated as M41S, with pore diameters in the 1.5-10 nm range [2] . The potentialities of MCM-41 (one of the members of this family) as acid catalysts were explored for various reactions. Their activity in the transformation of small molecules is generally similar to that of amorphous silica alumina as it is furthermore expected from the similar acidities of MCM-41 and of silica alumina [3]. However, for the transformation of bulky molecules, e.g. cracking and hydrocracking of gas oil [3,4], much better results are obtained with MCM-41.

In this work, the catalytic properties of MCM-41 samples with Si/Al ratios of 10, 30 and 100 are compared to those of an amorphous silica alumina. The reactions which were chosen for this comparison are toluene alkylation with methanol and m-xylene transformation . It is shown that, despite the relatively small size of the reactant and product molecules, the unidirectional mesopore system of MCM-41 has a very significant effect on the selectivity of m-xylene transformation. However, this shape selectivity does not exist for toluene alkylation.

2. EXPERIMENTAL

The MCM-41 samples with Si/Al ratios of 10, 30 and 100 were synthesized following the procedure given in ref. 2 using cetyltrimethylammonium bromide (Fluka) as the template, sodium silicate (Crossfield) and sodium aluminate (Carlo Erba) as silicon and aluminium sources. In order to eliminate the templating agent, MCM-41 samples were calcined under dry air at 873K. Silica alumina (14 wt% alumina) was supplied by Ketjen. Before use, all the samples were calcined in situ under dry air flow at 500°C for 10h.

Alkylation of toluene with methanol (a) and xylene transformation (b) were carried out in a flow reactor at 350°C under the following conditions :

a) $p_{toluene}$ = 0.1 bar ; $p_{methanol}$ = 0.1 bar, p_{N2} = 0.8 bar ; WWH (weight of toluene injected per hour and per weight of catalyst) = 1.5 - 40 h^{-1}

b) p_{xylene} = 0.0625 bar ; p_{N2} = 0.9375 bar ; WWH = 1 - 11 h^{-1}.

Analysis of reaction products was performed on line by FID gas chromatography using a 30m fused silica J&W DB Wax capillary column.

3. RESULTS AND DISCUSSION

3.1. Acidity and porosity of the samples

XRD spectra of the three MCM-41 samples are typical of these materials as described by the researchers from Mobil [2]. The BET surface area is equal to 1350, 1020 and 1050 m^2 g^{-1} for MCM-41(10, (30) and (100) respectively. For all the samples, there is, in the isotherm for nitrogen adsorption at -196°C, a sharp nflection for P/P_0 = 0.35 (Figure 1), which is characteristic of MCM-41 samples. Moreover, the t-plot of data in figure 1 indicates that no micropores are present. However, large differences are found between the adsorption-desorption isotherms. With MCM-41(100), sorption curves are typical of type IV isotherms without any hysteresis, while with MCM-41(30) and especially with MCM-41(10), hysteresis are observed (Figure 1). The pore size distribution is very sharp with MCM-41(100), sharp with MCM-41(30) and relatively large with MCM-41(10). The average pore diameter is equal to 28Å for MCM-41(100), 30.5Å for MCM-41(30) and 34Å for MCM-41(10). Furthermore, TEM of the samples shows that with the latter samples, the elimination of the surfactant by oxidative treatment creates large pores owing to a partial degradation of the framework, whereas it is not the case for MCM-41(100). The number of these large pores is more significant for MCM-41(10) than for MCM-41(30). It can therefore be concluded that the differences in pore distribution of the MCM-41 samples shown by nitrogen sorption are due to differences in the extent of the framework degradation. Silica alumina has a BET surface area of 365 $m^2 g^{-1}$. Its isotherm for nitrogen sorption presents a hysteresis. The size of the pores is between 30 and 80Å, the average pore diameter being of approximatively 40Å.

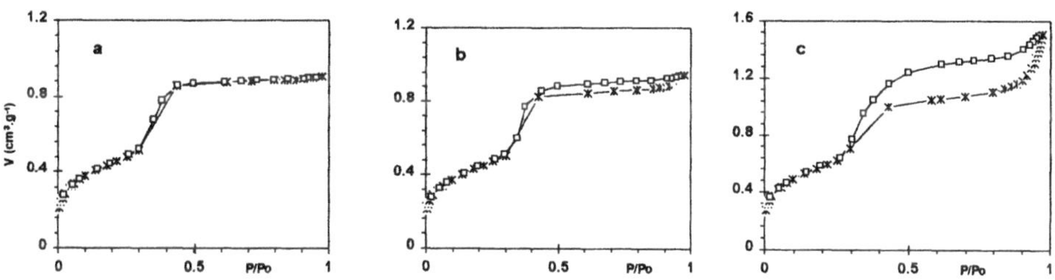

Figure 1. Nitrogen adsorption (✱) - desorption (□) isotherms of the MCM-41(100) (a), (30) (b), (10) (c) samples.

Table 1
Concentrations (mol g^{-1}) of Brönsted (C_B) and Lewis (C_L) acid sites, activities (mmol h^{-1} g^{-1}) for toluene alkylation and for m-xylene transformation of silica alumina and of the MCM-41 samples and selectivity of m-xylene transformation (D/I and p/o extrapolated at zero conversion)

Sample	Silica alumina	MCM-41(10)	MCM-41(30)	MCM-41(100)
C_B	45	53	21	19
C_L	145	210	85	70
$A_{toluene}$	0.6	0.4	0.18	0.09
$A_{m\text{-}xylene}$	4.5	1.1	0.55	0.25
D/I	0.2	0.5	0.4	0.3
p/o	1.4	0.4	0.2	0.15

The acidity of all the samples was characterized by pyridine sorption followed by IR spectroscopy. The densities of Brönsted and Lewis acid sites estimated for a desorption temperature of 150°C are given in table 1. With all the samples, the density of Lewis acid sites is higher than that of the Brönsted sites. Moreover, pyridine thermodesorption shows that the Lewis acid sites are stronger [5]. The concentrations of Brönsted acid sites of silica alumina and of MCM-41(10) are similar and approximatively 2.5 times greater than those of MCM-41(30) and MCM-41(100). However, silica alumina has stronger protonic sites than the MCM-41 samples : some pyridine remains adsorbed at 350°C over silica alumina but not over the MCM-41 samples [5]. It should be remarked that the ratio between the number of protonic sites and the total number of aluminium atoms increases with the Si/Al ratio (from 0.035 with MCM-41(10) to 0.11 with MCM-41(100)). This suggests that, as is the case with zeolites, the difficulty in the dealumination of the mesoporous aluminosilicate framework increases with their Si/Al ratio.

3.2. Alkylation of toluene with methanol

With all the samples, the reaction products are xylenes and other methylbenzenes resulting from toluene alkylation and C_1-C_5 hydrocarbons resulting from methanol transformation. For toluene alkylation, silica alumina is initially 1.5 times more active than MCM-41(10) which is 2.5 times more active than MCM-41(30) and 4 times active than MCM-41(100) (Table 1). This order in activity is the one expected from the acidity measurements : decrease in the concentration of protonic sites with the increase in the Si/Al ratio of the MCM-41 samples, quasi identical concentration of the protonic sites of MCM-41(10) and of silica alumina, but higher strength of the acid sites of silica alumina [5]. Figure 2 shows that the ratio between the amounts of methanol transformed into light hydrocarbons and into alkylation products are almost identical with silica alumina and with the MCM-41 samples. This ratio, very high (>10) at low conversion of toluene decreases with increasing this conversion.

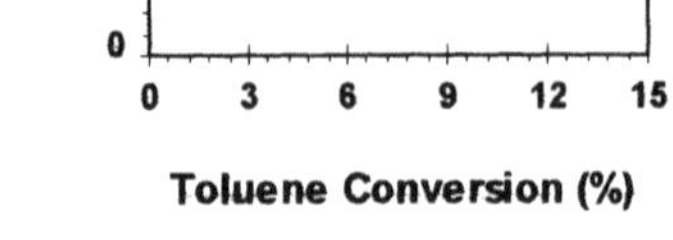

Figure 2. Toluene alkylation with methanol on MCM-41 (✻)and on silica alumina (◇). Ratio between the amount of methanol transformed into light hydrocarbons (LH) and into alkylation products (AR) as a function of toluene conversion

Similar distributions of the aromatic products and of the light products are found with silica alumina and MCM-41. The scheme of alkylation is successive, only xylenes appearing at low conversion. A very small difference in xylene distribution is observed between silica alumina and MCM-41 samples, the latter samples being slightly more selective for o-xylene as shown below by the distributions extrapolated at zero conversion :

Silica alumina : ortho 55%, meta 20%, para 25%
MCM-41 : ortho 58%, meta 19%, para 23%

These product distributions are very close to that generally obtained in Friedel-Crafts methylation of toluene : ortho : 60%, meta : 14%, para : 26% [6].

The distribution of the light products is also practically independent of the catalyst, e.g. at a toluene conversion of 4%, the distribution obtained with silica alumina and MCM-41 samples (% mol) is the following : C_1 : 2.5-4%, C_2 : 25-26%, C_3 : 34-35%, C_4 : 28-33%, C_5 : 4-7%. The only appreciable difference concerns the alkane/alkene ratio which is lower with MCM-41 than with silica alumina : e.g. 0.015 with MCM-41(30) and 0.15 with silica alumina. This indicates that hydrogen transfer is more significant with silica alumina than with MCM-41.

3.3. Xylene transformation

With all the samples, xylenes are transformed by isomerization and disproportionation, this latter reaction leading to a quasi equimolar mixture of trimethylbenzenes (in proportion close to that of thermodynamic equilibrium) and of toluene.

A detailed comparison of the catalytic properties of MCM-41 samples and of silica alumina was carried out in the case of m-xylene transformation. Initially, silica alumina was found to be 4 times more active than MCM-41(10) which is 2 times more active than MCM-41(30) and 4 times more active than MCM-41(100) (Table 1). This order in activity which is similar to the one obtained from toluene (and methanol conversions) can also be explained by the changes in acidity. However, contrary to what was obtained for the transformation of the toluene methanol mixture, large differences in selectivity can be observed between silica alumina and MCM-41 samples. The main difference concerns the (p/o) ratio which, at low conversions, is equal to 1.4 with silica alumina while it is lower than 1 with MCM-41 samples : from 0.4 with MCM-41(10) to 0.15 with MCM-41(100) (Table 1). Furthermore, with silica alumina the disproportionation/isomerization ratio (D/I) is lower than with the MCM-41 samples. With these samples, the greater the Si/Al ratio, the lower D/I (Table 1).

The large difference in isomerization selectivity suggests that the isomerization mechanisms are different over silica alumina and over MCM-41. Indeed, the p/o ratio found with silica alumina is close to that attributed to a purely intramolecular mechanism (1.2). On the other hand, the one found with MCM-41 is close to that of a bimolecular pathway (Figure 3) approximately 0.275 [7]. In agreement with this proposal, a direct isomerization of o-xylene into p-xylene (and vice versa) is observed with MCM-41(30), as could be expected from the bimolecular mechanism : e.g., m/p ratio in the products of o-xylene isomerization over MCM-41 equal to 12 at conversion close to zero (Figure 4). On the other hand, as is expected from a monomolecular mechanism, no direct interisomerization of o- and p-xylene is observed : e.g., very high initial m/p ratio over silica alumina (Figure 4).

The effect that branched alkanes such as methylcyclohexane have on the rat of xylene isomerization allowed us to discriminate between the mono and th bimolecular isomerization mechanisms. Indeed, it has been shown on FAU zeolites

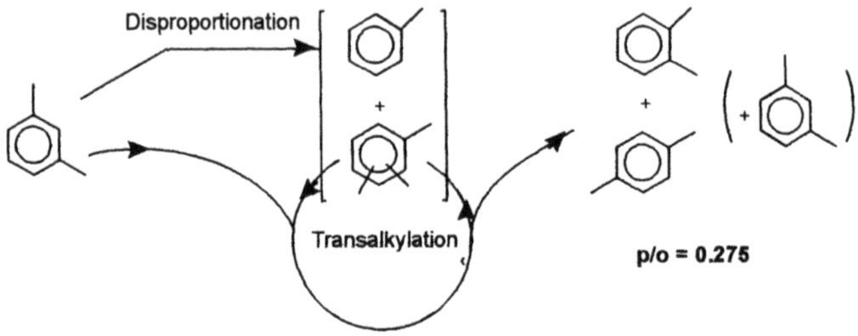

Figure 3. Bimolecular mechanism of m-xylene isomerization on MCM-41

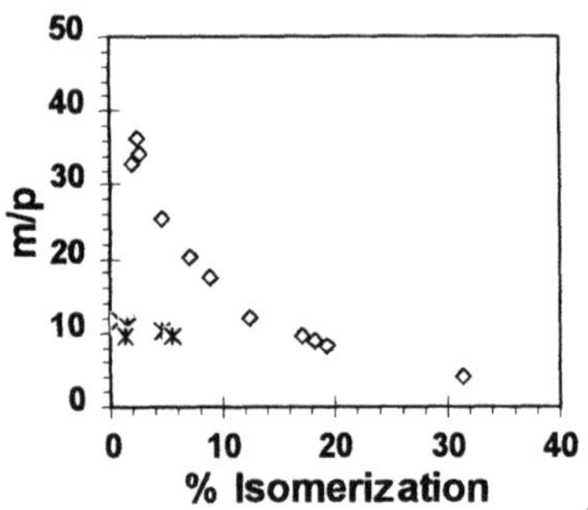

Figure 4. Isomerization (I) of o-xylen on silica alumina (◇) and on MCM-41(30) (✻). m-xylene / p-xylene (m/p) = f (%I)

that branched alkanes inhibit xylene disproportionation without affecting the monomolecular isomerization [8]. This inhibiting effect was attributed to the followin reaction :

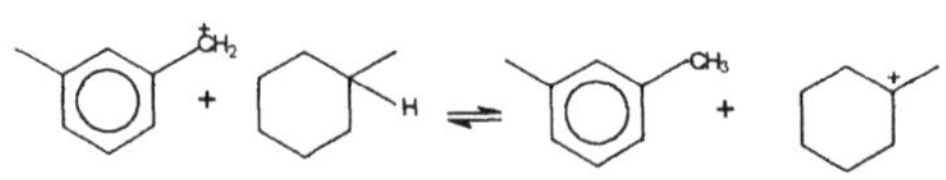

This reaction limits the concentration of benzylic carbocations which are involved a intermediates in disproportionation (and transalkylation) reactions. Figure 5 show that with silica alumina, methylcyclohexane decreases the rate of m-xylen disproportionation but has no effect on isomerization as it is expected from a unimolecular mechanism. On the other hand, methylcyclohexane decreases bo isomerization and disproportionation and exactly to the same extent. This identic effect of methylcyclohexane on the rates of these reactions is in favour of a pur bimolecular isomerization pathway.

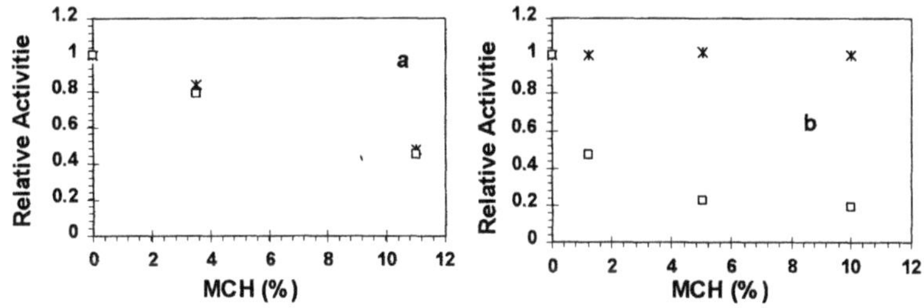

Figure 5. m-Xylene transformation on MCM-41(30) (a) and on silica alumina (b). Effect of methylcyclohexane (MCH) on the rate of isomerization (✱) and of disproportionation (□).

This very high selectivity of MCM-41 samples for bimolecular isomerization cannot be due to the characteristics of their acidity : weak protonic acidity, large concentration of strong Lewis acid sites. Indeed, no bimolecular isomerization can be observed on silica alumina which presents the same acidity characteristics. Therefore, this selectivity is certainly due to the presence of uniform non-interconnected channels with a relatively large diameter (30Å). When passing through these channels, the molecules of xylene reactant would be transformed into isomer molecules through successive reactions of diproportionation and transalkylation with trimethylbenzene molecules. We propose to call this particular selectivity of MCM-41 : **Tunnel Shape Selectivity**. Curiously, this shape selectivity cannot be observed in the narrow pores of unidirectional zeolites such as mordenites. This is probably because steric constraints strongly limit the formation of the bulky intermediates of disproportionation and transalkylation reactions. It should furthermore be emphasized that this tunnel shape selectivity was proposed in order to explain the particular distribution of the products of n-butane transformation (which involves non-bulky bimolecular intermediates) over a non-dealuminated mordenite sample [9].

It is possible through the bimolecular pathway (Figure 3) to explain the differences in D /I and p/o ratio obtained in m-xylene transformation over the MCM-41 samples (Table 1). Indeed, trimethylbenzene molecules resulting from m-xylene disproportionation can undergo during their passage through the channels, one or several successive reactions of transalkylations (known to be faster than disproportionation [7]) with production of xylene molecules. Therefore, the decrease in D/I with increasing the Si/Al ratio of the MCM-41 samples would be related to an increase in the number of transalkylation steps which lead to xylene isomers. This increase in the number of transalkylation steps could also be responsible for the decrease in the p/o ratio. Indeed, not only m-xylene but also o- and p-xylenes can undergo transalkylation reactions and the transalkylation of trimethylbenzenes with p-xylene is faster than with o-xylene [10] and moreover leads to a supplementary formation of this isomer, e.g. :

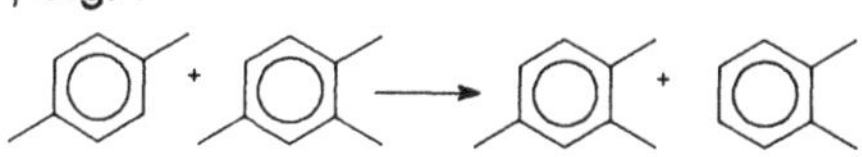

The increase in the number of transalkylation steps is most likely due to the decrease in the number of large pores created during oxidative removal of the surfactant (as was shown by TEM), hence to an increase with Si/Al of MCM-41 samples in the length of the mesoporous channels.

4. CONCLUSION

A new mode of shape selectivity has been discovered in MCM-41 mesoporous aluminosilicates. This new mode called **Tunnel Shape Selectivity** is responsible for a purely bimolecular pathway of xylene isomerization : in the non- interconnected long channels of MCM-41, reactant molecules undergo, before desorption, various successive disproportionation and transalkylation reactions with formation of xylene isomers. Curiously, no shape selectivity is observed in toluene alkylation with methanol. The selectivity of MCM-41 is identical to that of an amorphous silica alumina presenting a similar acidity : in particular there is no preferential formation of polymethylbenzenes. This could be due to the very low rate of alkylation with consequently the possibility of only one alkylation step during the passage of toluene molecules through the channels and / or to the preferential transformation of methanol into C_1-C_5 hydrocarbons.

REFERENCES

1. N.Y. Chen, W.E. Garwood, F.G. Dwyer, in Shape Selective Catalysis in Industrial Applications, Chemical Industries, H. Heinemann (ed.), vol. 36, M. Dekker, Inc., New York, 1989.
2. C.T. Kresge, M.E. Leonowicz, W.J. Roth, J.C. Vartulli and J.S. Beck, Nature, 359 (1992) 710 ; J.S. Beck, M.C. Vartulli, W.J. Roth, M.E. Leonowicz, C.T. Kresge, K.O. Schmitt, C.T.-W. Chu, D.H. Olson, E.W. Sheppard, S.B. McCullen, J.B Higgins and J.L. Schlenker, J. Amer. Chem. Soc. 114 (1992) 10834.
3. A. Corma, A. Martinez, V. Martinez-Soria and J.B. Monton, J. Catal. 153 (1995) 25.
4. A. Corma, M.S. Grande, V. Gonzalez-Alfaro and A.V. Orchilles, J. Catal. 159 (1996) 375.
5. S. Morin, P. Ayrault, S. El Mouahid, N.S. Gnep and M. Guisnet, Appl. Catal. A General 159 (1997) 317.
6. J.S. Beck and W.O. Haag, in Handbook of Heterogeneous Catalysis, G.Ertl, H. Knözinger, J. Weitkamp (eds.), V.C.H. Weinheim, vol. 5, p. 2123 ().
7. S. Morin, N.S. Gnep and M. Guisnet, J. Catal. 159 (1996) 296.
8. N.S. Gnep and M. Guisnet, React. Kin. Catal. Lett. 22 (1983) 237.
9. M.T. Tran, N.S. Gnep, G. Szabo and M. Guisnet, J. Catal. 174 (1998)
10. S. Morin, unpublished results.

AUTHOR INDEX

SUBJECT INDEX*

* Each number refers to the first page of a paper discussing the subject.

STUDIES IN SURFACE SCIENCE AND CATALYSIS

Volume 1 **Preparation of Catalysts I.** Scientific Bases for the Preparation of Heterogeneous Catalysts. Proceedings of the First International Symposium, Brussels, October 14–17, 1975
edited by **B. Delmon, P.A. Jacobs and G. Poncelet**

Volume 2 **The Control of the Reactivity of Solids.** A Critical Survey of the Factors that Influence the Reactivity of Solids, with Special Emphasis on the Control of the Chemical Processes in Relation to Practical Applications
by **V.V. Boldyrev, M. Bulens and B. Delmon**

Volume 3 **Preparation of Catalysts II.** Scientific Bases for the Preparation of Heterogeneous Catalysts. Proceedings of the Second International Symposium, Louvain-la-Neuve, September 4–7, 1978
edited by **B. Delmon, P. Grange, P. Jacobs and G. Poncelet**

Volume 4 **Growth and Properties of Metal Clusters.** Applications to Catalysis and the Photographic Process. Proceedings of the 32nd International Meeting of the Société de Chimie Physique, Villeurbanne, September 24–28, 1979
edited by **J. Bourdon**

Volume 5 **Catalysis by Zeolites.** Proceedings of an International Symposium, Ecully (Lyon), September 9–11, 1980
edited by **B. Imelik, C. Naccache, Y. Ben Taarit, J.C. Vedrine, G. Coudurier and H. Praliaud**

Volume 6 **Catalyst Deactivation.** Proceedings of an International Symposium, Antwerp, October 13–15, 1980
edited by **B. Delmon and G.F. Froment**

Volume 7 **New Horizons in Catalysis.** Proceedings of the 7th International Congress on Catalysis, Tokyo, June 30–July 4, 1980. Parts A and B
edited by **T. Seiyama and K. Tanabe**

Volume 8 **Catalysis by Supported Complexes**
by **Yu.I. Yermakov, B.N. Kuznetsov and V.A. Zakharov**

Volume 9 **Physics of Solid Surfaces.** Proceedings of a Symposium, Bechyňe, September 29–October 3, 1980
edited by **M. Láznička**

Volume 10 **Adsorption at the Gas–Solid and Liquid–Solid Interface.** Proceedings of an International Symposium, Aix-en-Provence, September 21–23, 1981
edited by **J. Rouquerol and K.S.W. Sing**

Volume 11 **Metal-Support and Metal-Additive Effects in Catalysis.** Proceedings of an International Symposium, Ecully (Lyon), September 14–16, 1982
edited by **B. Imelik, C. Naccache, G. Coudurier, H. Praliaud, P. Meriaudeau, P. Gallezot, G.A. Martin and J.C. Vedrine**

Volume 12 **Metal Microstructures in Zeolites.** Preparation - Properties – Applications. Proceedings of a Workshop, Bremen, September 22–24, 1982
edited by **P.A. Jacobs, N.I. Jaeger, P. Jirů and G. Schulz-Ekloff**

Volume 13 **Adsorption on Metal Surfaces.** An Integrated Approach
edited by **J. Bénard**

Volume 14 **Vibrations at Surfaces.** Proceedings of the Third International Conference, Asilomar, CA, September 1–4, 1982
edited by **C.R. Brundle and H. Morawitz**

Volume 15 **Heterogeneous Catalytic Reactions Involving Molecular Oxygen**
by **G.I. Golodets**

Volume 16	**Preparation of Catalysts III.** Scientific Bases for the Preparation of Heterogeneous Catalysts. Proceedings of the Third International Symposium, Louvain-la-Neuve, September 6–9, 1982 edited by **G. Poncelet, P. Grange and P.A. Jacobs**
Volume 17	**Spillover of Adsorbed Species.** Proceedings of an International Symposium, Lyon-Villeurbanne, September 12–16, 1983 edited by **G.M. Pajonk, S.J. Teichner and J.E. Germain**
Volume 18	**Structure and Reactivity of Modified Zeolites.** Proceedings of an International Conference, Prague, July 9–13, 1984 edited by **P.A. Jacobs, N.I. Jaeger, P. Jirů, V.B. Kazansky and G. Schulz-Ekloff**
Volume 19	**Catalysis on the Energy Scene.** Proceedings of the 9th Canadian Symposium on Catalysis, Quebec, P.Q., September 30–October 3, 1984 edited by **S. Kaliaguine and A. Mahay**
Volume 20	**Catalysis by Acids and Bases.** Proceedings of an International Symposium, Villeurbanne (Lyon), September 25–27, 1984 edited by **B. Imelik, C. Naccache, G. Coudurier, Y. Ben Taarit and J.C. Vedrine**
Volume 21	**Adsorption and Catalysis on Oxide Surfaces.** Proceedings of a Symposium, Uxbridge, June 28–29, 1984 edited by **M. Che and G.C. Bond**
Volume 22	**Unsteady Processes in Catalytic Reactors** by **Yu.Sh. Matros**
Volume 23	**Physics of Solid Surfaces 1984** edited by **J. Koukal**
Volume 24	**Zeolites: Synthesis, Structure, Technology and Application.** Proceedings of an International Symposium, Portorož-Portorose, September 3–8, 1984 edited by **B. Držaj, S. Hočevar and S. Pejovnik**
Volume 25	**Catalytic Polymerization of Olefins.** Proceedings of the International Symposium on Future Aspects of Olefin Polymerization, Tokyo, July 4–6, 1985 edited by **T. Keii and K. Soga**
Volume 26	**Vibrations at Surfaces 1985.** Proceedings of the Fourth International Conference, Bowness-on-Windermere, September 15–19, 1985 edited by **D.A. King, N.V. Richardson and S. Holloway**
Volume 27	**Catalytic Hydrogenation** edited by **L. Cervený**
Volume 28	**New Developments in Zeolite Science and Technology.** Proceedings of the 7th International Zeolite Conference, Tokyo, August 17–22, 1986 edited by **Y. Murakami, A. Iijima and J.W. Ward**
Volume 29	**Metal Clusters in Catalysis** edited by **B.C. Gates, L. Guczi and H. Knözinger**
Volume 30	**Catalysis and Automotive Pollution Control.** Proceedings of the First International Symposium, Brussels, September 8–11, 1986 edited by **A. Crucq and A. Frennet**
Volume 31	**Preparation of Catalysts IV.** Scientific Bases for the Preparation of Heterogeneous Catalysts. Proceedings of the Fourth International Symposium, Louvain-la-Neuve, September 1–4, 1986 edited by **B. Delmon, P. Grange, P.A. Jacobs and G. Poncelet**
Volume 32	**Thin Metal Films and Gas Chemisorption** edited by **P. Wissmann**
Volume 33	**Synthesis of High-silica Aluminosilicate Zeolites** edited by **P.A. Jacobs and J.A. Martens**
Volume 34	**Catalyst Deactivation 1987.** Proceedings of the 4th International Symposium, Antwerp, September 29–October 1, 1987 edited by **B. Delmon and G.F. Froment**
Volume 35	**Keynotes in Energy-Related Catalysis** edited by **S. Kaliaguine**

Volume 36 **Methane Conversion.** Proceedings of a Symposium on the Production of Fuels and Chemicals from Natural Gas, Auckland, April 27–30, 1987
edited by **D.M. Bibby, C.D. Chang, R.F. Howe and S. Yurchak**

Volume 37 **Innovation in Zeolite Materials Science.** Proceedings of an International Symposium, Nieuwpoort, September 13–17, 1987
edited by **P.J. Grobet, W.J. Mortier, E.F. Vansant and G. Schulz-Ekloff**

Volume 38 **Catalysis 1987.** Proceedings of the 10th North American Meeting of the Catalysis Society, San Diego, CA, May 17–22, 1987
edited by **J.W. Ward**

Volume 39 **Characterization of Porous Solids.** Proceedings of the IUPAC Symposium (COPS I), Bad Soden a. Ts., April 26–29,1987
edited by **K.K. Unger, J. Rouquerol, K.S.W. Sing and H. Kral**

Volume 40 **Physics of Solid Surfaces 1987.** Proceedings of the Fourth Symposium on Surface Physics, Bechyne Castle, September 7–11, 1987
edited by **J. Koukal**

Volume 41 **Heterogeneous Catalysis and Fine Chemicals.** Proceedings of an International Symposium, Poitiers, March 15–17, 1988
edited by **M. Guisnet, J. Barrault, C. Bouchoule, D. Duprez, C. Montassier and G. Pérot**

Volume 42 **Laboratory Studies of Heterogeneous Catalytic Processes**
by **E.G. Christoffel,** revised and edited by **Z. Paál**

Volume 43 **Catalytic Processes under Unsteady-State Conditions**
by **Yu. Sh. Matros**

Volume 44 **Successful Design of Catalysts.** Future Requirements and Development. Proceedings of the Worldwide Catalysis Seminars, July, 1988, on the Occasion of the 30th Anniversary of the Catalysis Society of Japan
edited by **T. Inui**

Volume 45 **Transition Metal Oxides.** Surface Chemistry and Catalysis
by **H.H. Kung**

Volume 46 **Zeolites as Catalysts, Sorbents and Detergent Builders.** Applications and Innovations. Proceedings of an International Symposium, Würzburg, September 4–8,1988
edited by **H.G. Karge and J. Weitkamp**

Volume 47 **Photochemistry on Solid Surfaces**
edited by **M. Anpo and T. Matsuura**

Volume 48 **Structure and Reactivity of Surfaces.** Proceedings of a European Conference, Trieste, September 13–16, 1988
edited by **C. Morterra, A. Zecchina and G. Costa**

Volume 49 **Zeolites: Facts, Figures, Future.** Proceedings of the 8th International Zeolite Conference, Amsterdam, July 10–14, 1989. Parts A and B
edited by **P.A. Jacobs and R.A. van Santen**

Volume 50 **Hydrotreating Catalysts.** Preparation, Characterization and Performance. Proceedings of the Annual International AIChE Meeting, Washington, DC, November 27–December 2, 1988
edited by **M.L. Occelli and R.G. Anthony**

Volume 51 **New Solid Acids and Bases.** Their Catalytic Properties
by **K. Tanabe, M. Misono, Y. Ono and H. Hattori**

Volume 52 **Recent Advances in Zeolite Science.** Proceedings of the 1989 Meeting of the British Zeolite Association, Cambridge, April 17–19, 1989
edited by **J. Klinowsky and P.J. Barrie**

Volume 53 **Catalyst in Petroleum Refining 1989.** Proceedings of the First International Conference on Catalysts in Petroleum Refining, Kuwait, March 5–8, 1989
edited by **D.L. Trimm, S. Akashah, M. Absi-Halabi and A. Bishara**

Volume 54 **Future Opportunities in Catalytic and Separation Technology**
edited by **M. Misono, Y. Moro-oka and S. Kimura**

Volume 55 **New Developments in Selective Oxidation.** Proceedings of an International Symposium, Rimini, Italy, September 18–22, 1989
edited by **G. Centi and F. Trifiro**
Volume 56 **Olefin Polymerization Catalysts.** Proceedings of the International Symposium on Recent Developments in Olefin Polymerization Catalysts, Tokyo, October 23–25, 1989
edited by **T. Keii and K. Soga**
Volume 57A **Spectroscopic Analysis of Heterogeneous Catalysts. Part A: Methods of Surface Analysis**
edited by **J.L.G. Fierro**
Volume 57B **Spectroscopic Analysis of Heterogeneous Catalysts. Part B: Chemisorption of Probe Molecules**
edited by **J.L.G. Fierro**
Volume 58 **Introduction to Zeolite Science and Practice**
edited by **H. van Bekkum, E.M. Flanigen and J.C. Jansen**
Volume 59 **Heterogeneous Catalysis and Fine Chemicals II.** Proceedings of the 2nd International Symposium, Poitiers, October 2–6, 1990
edited by **M. Guisnet, J. Barrault, C. Bouchoule, D. Duprez, G. Pérot, R. Maurel and C. Montassier**
Volume 60 **Chemistry of Microporous Crystals.** Proceedings of the International Symposium on Chemistry of Microporous Crystals, Tokyo, June 26–29, 1990
edited by **T. Inui, S. Namba and T. Tatsumi**
Volume 61 **Natural Gas Conversion.** Proceedings of the Symposium on Natural Gas Conversion, Oslo, August 12–17, 1990
edited by **A. Holmen, K.-J. Jens and S. Kolboe**
Volume 62 **Characterization of Porous Solids II.** Proceedings of the IUPAC Symposium (COPS II), Alicante, May 6–9, 1990
edited by **F. Rodríguez-Reinoso, J. Rouquerol, K.S.W. Sing and K.K. Unger**
Volume 63 **Preparation of Catalysts V.** Scientific Bases for the Preparation of Heterogeneous Catalysts. Proceedings of the Fifth International Symposium, Louvain-la-Neuve, September 3–6, 1990
edited by **G. Poncelet, P.A. Jacobs, P. Grange and B. Delmon**
Volume 64 **New Trends in CO Activation**
edited by **L. Guczi**
Volume 65 **Catalysis and Adsorption by Zeolites.** Proceedings of ZEOCAT 90, Leipzig, August 20–23, 1990
edited by G. **Öhlmann, H. Pfeifer and R. Fricke**
Volume 66 **Dioxygen Activation and Homogeneous Catalytic Oxidation.** Proceedings of the Fourth International Symposium on Dioxygen Activation and Homogeneous Catalytic Oxidation, Balatonfüred, September 10–14, 1990
edited by **L.I. Simándi**
Volume 67 **Structure-Activity and Selectivity Relationships in Heterogeneous Catalysis.** Proceedings of the ACS Symposium on Structure-Activity Relationships in Heterogeneous Catalysis, Boston, MA, April 22–27, 1990
edited by **R.K. Grasselli and A.W. Sleight**
Volume 68 **Catalyst Deactivation 1991.** Proceedings of the Fifth International Symposium, Evanston, IL, June 24–26, 1991
edited by **C.H. Bartholomew and J.B. Butt**
Volume 69 **Zeolite Chemistry and Catalysis.** Proceedings of an International Symposium, Prague, Czechoslovakia, September 8–13, 1991
edited by **P.A. Jacobs, N.I. Jaeger, L. Kubelková and B. Wichterlová**
Volume 70 **Poisoning and Promotion in Catalysis based on Surface Science Concepts and Experiments**
by **M. Kiskinova**

Volume 88 **Catalyst Deactivation 1994.** Proceedings of the 6th International Symposium, Ostend, Belgium, October 3–5, 1994
edited by **B. Delmon and G.F. Froment**

Volume 89 **Catalyst Design for Tailor-made Polyolefins.** Proceedings of the International Symposium on Catalyst Design for Tailor-made Polyolefins, Kanazawa, Japan, March 10–12, 1994
edited by **K. Soga and M. Terano**

Volume 90 **Acid-Base Catalysis II.** Proceedings of the International Symposium on Acid-Base Catalysis II, Sapporo, Japan, December 2–4, 1993
edited by **H. Hattori, M. Misono and Y. Ono**

Volume 91 **Preparation of Catalysts VI.** Scientific Bases for the Preparation of Heterogeneous Catalysts. Proceedings of the Sixth International Symposium, Louvain-La-Neuve, September 5–8, 1994
edited by **G. Poncelet, J. Martens, B. Delmon, P.A. Jacobs and P. Grange**

Volume 92 **Science and Technology in Catalysis 1994.** Proceedings of the Second Tokyo Conference on Advanced Catalytic Science and Technology, Tokyo, August 21–26, 1994
edited by **Y. Izumi, H. Arai and M. Iwamoto**

Volume 93 **Characterization and Chemical Modification of the Silica Surface**
by **E.F. Vansant, P. Van Der Voort and K.C. Vrancken**

Volume 94 **Catalysis by Microporous Materials.** Proceedings of ZEOCAT'95, Szombathely, Hungary, July 9–13, 1995
edited by **H.K. Beyer, H.G.Karge, I. Kiricsi and J.B. Nagy**

Volume 95 **Catalysis by Metals and Alloys**
by **V. Ponec and G.C. Bond**

Volume 96 **Catalysis and Automotive Pollution Control III.** Proceedings of the Third International Symposium (CAPoC3), Brussels, Belgium, April 20–22, 1994
edited by **A. Frennet and J.-M. Bastin**

Volume 97 **Zeolites: A Refined Tool for Designing Catalytic Sites.** Proceedings of the International Symposium, Québec, Canada, October 15–20, 1995
edited by **L. Bonneviot and S. Kaliaguine**

Volume 98 **Zeolite Science 1994: Recent Progress and Discussions.** Supplementary Materials to the 10th International Zeolite Conference, Garmisch-Partenkirchen, Germany, July 17–22, 1994
edited by **H.G. Karge and J. Weitkamp**

Volume 99 **Adsorption on New and Modified Inorganic Sorbents**
edited by **A. Dąbrowski and V.A. Tertykh**

Volume 100 **Catalysts in Petroleum Refining and Petrochemical Industries 1995.**
Proceedings of the 2nd International Conference on Catalysts in Petroleum Refining and Petrochemical Industries, Kuwait, April 22-26, 1995
edited by **M. Absi-Halabi, J. Beshara, H. Qabazard and A. Stanislaus**

Volume 101 **11th International Congress on Catalysis - 40th Anniversary.**
Proceedings of the 11th ICC, Baltimore, MD, USA, June 30-July 5, 1996
edited by **J. W. Hightower, W.N. Delgass, E. Iglesia and A.T. Bell**

Volume 102 **Recent Advances and New Horizons in Zeolite Science and Technology**
edited by **H. Chon, S.I. Woo and S. -E. Park**

Volume 103 **Semiconductor Nanoclusters - Physical, Chemical, and Catalytic Aspects**
edited by **P.V. Kamat and D. Meisel**

Volume 104 **Equilibria and Dynamics of Gas Adsorption on Heterogeneous Solid Surfaces**
edited by **W. Rudziński, W.A. Steele and G. Zgrablich**

Volume 105 **Progress in Zeolite and Microporous Materials**
Proceedings of the 11th International Zeolite Conference, Seoul, Korea, August 12-17, 1996
edited by **H. Chon, S.-K. Ihm and Y.S. Uh**

Volume 106 **Hydrotreatment and Hydrocracking of Oil Fractions**
Proceedings of the 1st International Symposium / 6th European Workshop, Oostende, Belgium, February 17-19, 1997
edited by **G.F. Froment, B. Delmon and P. Grange**

Volume 107 **Natural Gas Conversion IV**
Proceedings of the 4th International Natural Gas Conversion Symposium, Kruger Park, South Africa, November 19-23, 1995
edited by **M. de Pontes, R.L. Espinoza, C.P. Nicolaides, J.H. Scholtz and M.S. Scurrell**

Volume 108 **Heterogeneous Catalysis and Fine Chemicals IV**
Proceedings of the 4th International Symposium on Heterogeneous Catalysis and Fine Chemicals, Basel, Switzerland, September 8-12, 1996
edited by **H.U. Blaser, A. Baiker and R. Prins**

Volume 109 **Dynamics of Surfaces and Reaction Kinetics in Heterogeneous Catalysis.**
Proceedings of the International Symposium, Antwerp, Belgium, September 15-17, 1997
edited by **G.F. Froment and K.C. Waugh**

Volume 110 **Third World Congress on Oxidation Catalysis.**
Proceedings of the Third World Congress on Oxidation Catalysis, San Diego, CA, U.S.A., 21-26 September 1997
edited by **R.K. Grasselli, S.T. Oyama, A.M. Gaffney and J.E. Lyons**

Volume 111 **Catalyst Deactivation 1997.**
Proceedings of the 7th International Symposium, Cancun, Mexico, October 5-8, 1997
edited by **C.H. Bartholomew and G.A. Fuentes**

Volume 112 **Spillover and Migration of Surface Species on Catalysts.**
Proceedings of the 4th International Conference on Spillover, Dalian, China, September 15-18, 1997
edited by **Can Li and Qin Xin**

Volume 113 **Recent Advances in Basic and Applied Aspects of Industrial Catalysis.**
Proceedings of the 13th National Symposium and Silver Jubilee Symposium of Catalysis of India, Dehradun, India, April 2-4, 1997
edited by **T.S.R. Prasada Rao and G. Murali Dhar**

Volume 114 **Advances in Chemical Conversions for Mitigating Carbon Dioxide.**
Proceedings of the 4th International Conference on Carbon Dioxide Utilization, Kyoto, Japan, September 7-11, 1997
edited by **T. Inui, M. Anpo, K. Izui, S. Yanagida and T. Yamaguchi**

Volume 115 **Methods for Monitoring and Diagnosing the Efficiency of Catalytic Converters.**
A patent-oriented survey
by **M. Sideris**

Volume 116 **Catalysis and Automotive Pollution Control IV.**
Proceedings of the 4th International Symposium (CAPoC4), Brussels, Belgium, April 9-11, 1997
edited by **N. Kruse, A. Frennet and J.-M. Bastin**

Volume 117 **Mesoporous Molecular Sieves 1998.**
Proceedings of the 1st International Symposium, Baltimore, MD, U.S.A., July 10-12, 1998
edited by **L.Bonneviot, F. Béland, C. Danumah, S. Giasson and S. Kaliaguine**

www.ingramcontent.com/pod-product-compliance
Ingram Content Group UK Ltd.
Pitfield, Milton Keynes, MK11 3LW, UK
UKHW020224290926
15470UKWH00001B/364